USEFUL CONSTANTS

Speed of light in free space	c_o	2.9979×10^8	m/s	Planck's constant	h	6.6261×10^{-34}	J · s
Permittivity of free space	ϵ_o	8.8542×10^{-12}	F/m	Electron charge	e	1.6022×10^{-19}	C
Permeability of free space	μ_o	1.2566×10^{-6}	H/m	Electron mass	m_0	9.1094×10^{-31}	kg
Impedance of free space	η_o	376.73	Ω	Boltzmann's constant	k	1.3807×10^{-23}	J/°K

PREFIXES FOR UNITS

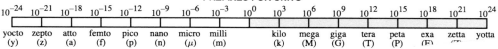

THE PHOTON

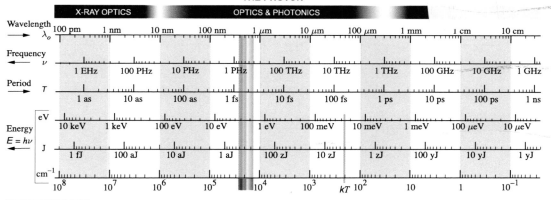

A photon of free-space wavelength $\lambda_o = 1\ \mu m$ has frequency $\nu = 300$ THz, period $T = 3.33$ fs, and energy $E = 1.24$ eV $= 199$ zJ $= 10^4$ cm^{-1}. At room temperature ($T = 300°$ K), the thermal energy $kT = 26$ meV $= 4.14$ zJ $= 209$ cm^{-1}.

OPTICAL PULSES

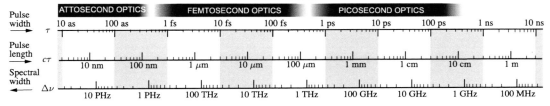

PHOTONIC STRUCTURES

FUNDAMENTALS OF PHOTONICS

THIRD EDITION

BAHAA E. A. SALEH
University of Central Florida
Boston University

MALVIN CARL TEICH
Boston University
Columbia University

This edition first published 2019
© 2019 by John Wiley & Sons, Inc.

All rights reserved. No part of this publication may be reproduced, stored in a retrieval system, or transmitted, in any form or by any means, electronic, mechanical, photocopying, recording or otherwise, except as permitted by law. Advice on how to obtain permission to reuse material from this title is available at http://www.wiley.com/go/permissions.

The rights of Bahaa E. A. Saleh and Malvin Carl Teich to be identified as the authors of the editorial material in this work have been asserted in accordance with law.

Registered Office
John Wiley & Sons, Inc., 111 River Street, Hoboken, NJ 07030, USA

Editorial Office
111 River Street, Hoboken, NJ 07030, USA

For details of our global editorial offices, customer services, and more information about Wiley products visit us at www.wiley.com.

Wiley also publishes its books in a variety of electronic formats and by print-on-demand. Some content that appears in standard print versions of this book may not be available in other formats.

Limit of Liability/Disclaimer of Warranty
In view of ongoing research, equipment modifications, changes in governmental regulations, and the constant flow of information relating to the use of experimental reagents, equipment, and devices, the reader is urged to review and evaluate the information provided in the package insert or instructions for each chemical, piece of equipment, reagent, or device for, among other things, any changes in the instructions or indication of usage and for added warnings and precautions. While the publisher and authors have used their best efforts in preparing this work, they make no representations or warranties with respect to the accuracy or completeness of the contents of this work and specifically disclaim all warranties, including without limitation any implied warranties of merchantability or fitness for a particular purpose. No warranty may be created or extended by sales representatives, written sales materials or promotional statements for this work. The fact that an organization, website, or product is referred to in this work as a citation and/or potential source of further information does not mean that the publisher and authors endorse the information or services the organization, website, or product may provide or recommendations it may make. This work is sold with the understanding that the publisher is not engaged in rendering professional services. The advice and strategies contained herein may not be suitable for your situation. You should consult with a specialist where appropriate. Further, readers should be aware that websites listed in this work may have changed or disappeared between when this work was written and when it is read. Neither the publisher nor authors shall be liable for any loss of profit or any other commercial damages, including but not limited to special, incidental, consequential, or other damages.

Library of Congress Cataloging-in-Publication Data is available.
Volume Set ISBN: 9781119506874
Volume I ISBN: 9781119506867
Volume II ISBN: 9781119506898

Cover design by Wiley
Cover image: Courtesy of B. E. A. Saleh and M. C. Teich

Printed by CPI Group (UK) Ltd, Croydon, CR0 4YY

C435200_300421

CONTENTS

PREFACE TO THE THIRD EDITION	xi
PREFACE TO THE SECOND EDITION	xx
PREFACE TO THE FIRST EDITION	xxiii

PART I: OPTICS — 1

1 RAY OPTICS — 3

1.1	Postulates of Ray Optics	5
1.2	Simple Optical Components	8
1.3	Graded-Index Optics	20
1.4	Matrix Optics	27
	Reading List	37
	Problems	38

2 WAVE OPTICS — 41

2.1	Postulates of Wave Optics	43
2.2	Monochromatic Waves	44
*2.3	Relation Between Wave Optics and Ray Optics	52
2.4	Simple Optical Components	53
2.5	Interference	61
2.6	Polychromatic and Pulsed Light	71
	Reading List	76
	Problems	77

3 BEAM OPTICS — 79

3.1	The Gaussian Beam	80
3.2	Transmission Through Optical Components	91
3.3	Hermite–Gaussian Beams	99
3.4	Laguerre–Gaussian Beams	102
3.5	Nondiffracting Beams	105
	Reading List	108
	Problems	108

4 FOURIER OPTICS — 110

4.1	Propagation of Light in Free Space	113
4.2	Optical Fourier Transform	124
4.3	Diffraction of Light	129
4.4	Image Formation	137
4.5	Holography	147
	Reading List	155
	Problems	157

5 ELECTROMAGNETIC OPTICS — 160

- 5.1 Electromagnetic Theory of Light — 162
- 5.2 Electromagnetic Waves in Dielectric Media — 166
- 5.3 Monochromatic Electromagnetic Waves — 172
- 5.4 Elementary Electromagnetic Waves — 175
- 5.5 Absorption and Dispersion — 181
- 5.6 Scattering of Electromagnetic Waves — 192
- 5.7 Pulse Propagation in Dispersive Media — 199
- Reading List — 205
- Problems — 207

6 POLARIZATION OPTICS — 209

- 6.1 Polarization of Light — 211
- 6.2 Reflection and Refraction — 221
- 6.3 Optics of Anisotropic Media — 227
- 6.4 Optical Activity and Magneto-Optics — 240
- 6.5 Optics of Liquid Crystals — 244
- 6.6 Polarization Devices — 247
- Reading List — 251
- Problems — 252

7 PHOTONIC-CRYSTAL OPTICS — 255

- 7.1 Optics of Dielectric Layered Media — 258
- 7.2 One-Dimensional Photonic Crystals — 277
- 7.3 Two- and Three-Dimensional Photonic Crystals — 291
- Reading List — 299
- Problems — 301

8 METAL AND METAMATERIAL OPTICS — 303

- 8.1 Single- and Double-Negative Media — 306
- 8.2 Metal Optics: Plasmonics — 320
- 8.3 Metamaterial Optics — 334
- *8.4 Transformation Optics — 343
- Reading List — 349
- Problems — 351

9 GUIDED-WAVE OPTICS — 353

- 9.1 Planar-Mirror Waveguides — 355
- 9.2 Planar Dielectric Waveguides — 363
- 9.3 Two-Dimensional Waveguides — 372
- 9.4 Optical Coupling in Waveguides — 376
- 9.5 Photonic-Crystal Waveguides — 385
- 9.6 Plasmonic Waveguides — 386
- Reading List — 389
- Problems — 389

10 FIBER OPTICS — 391

- 10.1 Guided Rays — 393
- 10.2 Guided Waves — 397
- 10.3 Attenuation and Dispersion — 415
- 10.4 Holey and Photonic-Crystal Fibers — 426
- 10.5 Fiber Materials — 429
- Reading List — 430
- Problems — 432

11 RESONATOR OPTICS — 433

- 11.1 Planar-Mirror Resonators — 436
- 11.2 Spherical-Mirror Resonators — 447
- 11.3 Two- and Three-Dimensional Resonators — 459
- 11.4 Microresonators and Nanoresonators — 463
- Reading List — 470
- Problems — 471

12 STATISTICAL OPTICS — 473

- 12.1 Statistical Properties of Random Light — 475
- 12.2 Interference of Partially Coherent Light — 489
- *12.3 Transmission of Partially Coherent Light — 497
- 12.4 Partial Polarization — 506
- Reading List — 510
- Problems — 512

13 PHOTON OPTICS — 514

- 13.1 The Photon — 516
- 13.2 Photon Streams — 529
- *13.3 Quantum States of Light — 541
- Reading List — 550
- Problems — 554

PART II: PHOTONICS — 559

14 LIGHT AND MATTER — 561

- 14.1 Energy Levels — 562
- 14.2 Occupation of Energy Levels — 581
- 14.3 Interactions of Photons with Atoms — 583
- 14.4 Thermal Light — 602
- 14.5 Luminescence and Scattering — 607
- Reading List — 614
- Problems — 617

15 LASER AMPLIFIERS — 619

- 15.1 Theory of Laser Amplification — 622
- 15.2 Amplifier Pumping — 626
- 15.3 Representative Laser Amplifiers — 636
- 15.4 Amplifier Nonlinearity — 645
- *15.5 Amplifier Noise — 651
- Reading List — 653
- Problems — 655

16 LASERS — 657

- 16.1 Theory of Laser Oscillation — 659
- 16.2 Characteristics of the Laser Output — 666
- 16.3 Types of Lasers — 680
- 16.4 Pulsed Lasers — 707
- Reading List — 723
- Problems — 728

17 SEMICONDUCTOR OPTICS — 731

17.1 Semiconductors — 733
17.2 Interactions of Photons with Charge Carriers — 766
Reading List — 782
Problems — 784

18 LEDS AND LASER DIODES — 787

18.1 Light-Emitting Diodes — 789
18.2 Semiconductor Optical Amplifiers — 817
18.3 Laser Diodes — 831
18.4 Quantum-Confined Lasers — 844
18.5 Microcavity Lasers — 854
18.6 Nanocavity Lasers — 862
Reading List — 864
Problems — 868

19 PHOTODETECTORS — 871

19.1 Photodetectors — 873
19.2 Photoconductors — 883
19.3 Photodiodes — 887
19.4 Avalanche Photodiodes — 895
19.5 Array Detectors — 907
19.6 Noise in Photodetectors — 909
Reading List — 935
Problems — 938

20 ACOUSTO-OPTICS — 943

20.1 Interaction of Light and Sound — 945
20.2 Acousto-Optic Devices — 958
*20.3 Acousto-Optics of Anisotropic Media — 967
Reading List — 972
Problems — 972

21 ELECTRO-OPTICS — 975

21.1 Principles of Electro-Optics — 977
*21.2 Electro-Optics of Anisotropic Media — 989
21.3 Electro-Optics of Liquid Crystals — 996
*21.4 Photorefractivity — 1005
21.5 Electroabsorption — 1010
Reading List — 1012
Problems — 1013

22 NONLINEAR OPTICS — 1015

22.1 Nonlinear Optical Media — 1017
22.2 Second-Order Nonlinear Optics — 1021
22.3 Third-Order Nonlinear Optics — 1036
*22.4 Second-Order Nonlinear Optics: Coupled Waves — 1047
*22.5 Third-Order Nonlinear Optics: Coupled Waves — 1059
*22.6 Anisotropic Nonlinear Media — 1066
*22.7 Dispersive Nonlinear Media — 1069
Reading List — 1074
Problems — 1075

23 ULTRAFAST OPTICS — 1078

- 23.1 Pulse Characteristics — 1079
- 23.2 Pulse Shaping and Compression — 1088
- 23.3 Pulse Propagation in Optical Fibers — 1102
- 23.4 Ultrafast Linear Optics — 1115
- 23.5 Ultrafast Nonlinear Optics — 1126
- 23.6 Pulse Detection — 1146
- Reading List — 1159
- Problems — 1161

24 OPTICAL INTERCONNECTS AND SWITCHES — 1163

- 24.1 Optical Interconnects — 1166
- 24.2 Passive Optical Routers — 1178
- 24.3 Photonic Switches — 1187
- 24.4 Photonic Logic Gates — 1211
- Reading List — 1220
- Problems — 1222

25 OPTICAL FIBER COMMUNICATIONS — 1224

- 25.1 Fiber-Optic Components — 1226
- 25.2 Optical Fiber Communication Systems — 1238
- 25.3 Modulation and Multiplexing — 1257
- 25.4 Coherent Optical Communications — 1266
- 25.5 Fiber-Optic Networks — 1274
- Reading List — 1281
- Problems — 1284

A FOURIER TRANSFORM — 1287

- A.1 One-Dimensional Fourier Transform — 1287
- A.2 Time Duration and Spectral Width — 1290
- A.3 Two-Dimensional Fourier Transform — 1293
- Reading List — 1295

B LINEAR SYSTEMS — 1296

- B.1 One-Dimensional Linear Systems — 1296
- B.2 Two-Dimensional Linear Systems — 1299
- Reading List — 1300

C MODES OF LINEAR SYSTEMS — 1301

- Reading List — 1305

SYMBOLS AND UNITS — 1306

AUTHORS — 1331

INDEX — 1333

PREFACE TO THE THIRD EDITION

Since the publication of the *Second Edition* in 2007, *Fundamentals of Photonics* has maintained its worldwide prominence as a self-contained, up-to-date, introductory-level textbook that features a blend of theory and applications. It has been reprinted dozens of times and been translated into German and Chinese, as well as Czech and Japanese. The Third Edition incorporates many of the scientific and technological developments in photonics that have taken place in the past decade and strives to be cutting-edge.

Optics and Photonics

Before usage of the term photonics became commonplace at the time of the *First Edition* in the early 1990s, the field was characterized by a collection of appellations that were not always clearly delineated. Terms such as quantum electronics, optoelectronics, electro-optics, and lightwave technology were widely used. Though there was a lack of agreement about the precise meanings of these terms, there was nevertheless a vague consensus regarding their usage. Most of these terms have since receded from general use, although some have retained their presence in the titles of technical journals and academic courses.

Now, more than 25 years later, the term *Optics* along with the term *Photonics*, as well as their combination *Optics & Photonics*, have prevailed. The distinction between optics and photonics remains somewhat fuzzy, however, and there is a degree of overlap between the two arenas. Hence, there is some arbitrariness in the manner in which the chapters of this book are allocated to its two volumes, *Part I: Optics* and *Part II: Photonics*. From a broad perspective, the term *Optics* is taken to signify free-space and guided-wave propagation, and to include topics such as interference, diffraction, imaging, statistical optics, and photon optics. The term *Photonics*, in contrast, is understood to include topics that rely on the interaction of light and matter, and is dedicated to the study of devices and systems. As the miniaturization of components and systems continues to progress and foster emerging technologies such as nanophotonics and biophotonics, the importance of photonics continues to advance.

Printed and Electronic Versions

The *Third Edition* appears in four versions:

1. A printed version.
2. An eBook in the form of an ePDF file that mimics the printed version.
3. An eBook in the form of a standard ePUB.
4. An eBook in the form of an enhanced ePUB with animations for selected figures.

In its *printed* form, the text consists of two volumes, each of which contains the Table of Contents and Index for both volumes along with the Appendices and List of Symbols:

- *Part I: Optics*, contains the first thirteen chapters.
- *Part II: Photonics*, contains the remaining twelve chapters.

The material in the eBook versions is identical to that in the printed version except that all 25 chapters reside in a single electronic file. The various *eBooks* enjoy the following features:

- Hyperlinked table of contents at the beginning of the text.

- Hyperlinked table of contents as an optional sidebar.
- Hyperlinked index.
- Hyperlinked section titles, equations, and figures throughout.
- Animations for selected figures in the enhanced ePUB.

Presentation

Exercises, examples, reading lists, and appendices. Each chapter of the *Third Edition* contains exercises, problem sets, and an extensive reading list. Examples are included throughout to emphasize the concepts governing applications of current interest. Appendices summarize the properties of one- and two-dimensional Fourier transforms, linear systems, and modes of linear systems. Important equations are highlighted by boxes and labels to facilitate retrieval.

Symbols, notation, units, and conventions. We make use of the symbols, notation, units, and conventions commonly used in the photonics literature. Because of the broad spectrum of topics covered, different fonts are often used to delineate the multiple meanings of various symbols; a list of symbols, units, abbreviations, and acronyms follows the appendices. We adhere to the International System of Units (SI units). This modern form of the metric system is based on the meter, kilogram, second, ampere, kelvin, candela, and mole, and is coupled with a collection of prefixes (specified on the inside back cover of the text) that indicate multiplication or division by various powers of ten. However, the reader is cautioned that photonics in the service of different areas of science can make use of different conventions and symbols. In Chapter 2, for example, we write the complex wavefunction for a monochromatic plane wave in a form commonly used in electrical engineering, which differs from that used in physics. Another example arises in Chapter 6, where the definitions we use for right (left) circularly polarized light are in accord with general usage in optics, but are opposite those generally used in engineering. These distinctions are often highlighted by *in situ* footnotes. Though the choice of a particular convention is manifested in the form assumed by various equations, it does not of course affect the results.

Color coding of illustrations. The color code used in illustrations is summarized in the chart presented below. Light beams and optical-field distributions are displayed in red (except when light beams of multiple wavelengths are involved, as is often the case in nonlinear optics). When optical fields are represented, white indicates negative values but when intensity is portrayed, white indicates zero. Acoustic beams and fields are similarly represented, but by with green rather than red. Glass and glass fibers are depicted in light blue; darker shades represent larger refractive indices. Semiconductors are cast in green, with various shades representing different doping levels. Metal and mirrors are indicated as copper. Semiconductor energy-band diagrams are portrayed in blue and gray whereas photonic bandgaps are illustrated in pink.

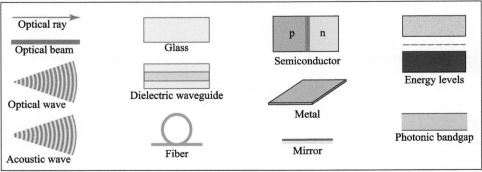

Color chart

Intended Audience

As with the previous editions, the *Third Edition* is meant to serve as:

- An introductory textbook for students of electrical engineering, applied physics, physics, or optics at the senior or first-year graduate level.
- A self-contained work for self-study.
- A textbook suitable for use in programs of continuing professional development offered by industry, universities, and professional societies.

The reader is assumed to have a background in engineering, physics, or optics, including courses in modern physics, electricity and magnetism, and wave motion. Some knowledge of linear systems and elementary quantum mechanics is helpful but not essential. The intent is to provide an introduction to optics and photonics that emphasizes the concepts that govern applications of current interest. The book should therefore not be considered as a compendium encompassing all photonic devices and systems. Indeed, some areas of photonics are not included at all, and many of the individual chapters could easily have been expanded into free-standing monographs.

Organization

The *Third Edition* comprises 25 chapters compartmentalized into six divisions, as depicted in the diagram below.

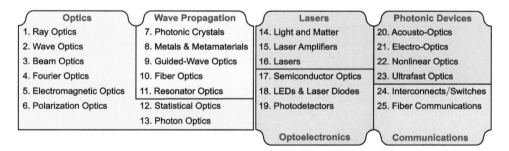

In recognition of the different levels of mathematical sophistication of the intended audience, we have endeavored to present difficult concepts in two steps: at an introductory level that provides physical insight and motivation, followed by a more advanced analysis. This approach is exemplified by the treatment in Chapter 21 (*Electro-Optics*), in which the subject is first presented using scalar notation and then treated again using tensor notation. Sections dealing with material of a more advanced nature are indicated by asterisks and may be omitted if desired. Summaries are provided at points where recapitulation is deemed useful because of the involved nature of the material.

The form of the book is modular so that it can be used by readers with different needs; this also provides instructors an opportunity to select topics for different courses. Essential material from one chapter is often briefly summarized in another to make each chapter as self-contained as possible. At the beginning of Chapter 25 (*Optical Fiber Communications*), for example, relevant material from earlier chapters describing optical fibers, light sources, optical amplifiers, photodetectors, and photonic integrated circuits is briefly reviewed. This places important information about the components of such systems at the disposal of the reader in advance of presenting system-design and performance considerations.

Contents

A principal feature of the *Third Edition* is a new chapter entitled *Metal and Metamaterial Optics*, an area that has had a substantial and increasing impact on photonics.

The new chapter comprises theory and applications for single- and double-negative media, metal optics, plasmonics, metamaterial optics, and transformation optics.

All chapters have been thoroughly vetted and updated. A chapter-by-chapter compilation of new material in the *Third Edition* is provided below.

- *Chapter 1 (Ray Optics)*. Ray-optics descriptions for optical components such as biprisms, axicons, LED light collimators, and Fresnel lenses have been added. The connection between characterizing an arbitrary paraxial optical system by its ray-transfer matrix and its cardinal points has been established. A matrix-optics analysis for imaging with an arbitrary paraxial optical system has been included.
- *Chapter 2 (Wave Optics)*. A wave-optics analysis of transmission through biprisms and axicons has been added. A treatment of the Fresnel zone plate from the perspective of interference has been introduced. An analysis of the Michelson–Fabry–Perot (LIGO) interferometer used for the detection of gravitational waves in the distant universe has been incorporated.
- *Chapter 3 (Beam Optics)*. An enhanced description of Laguerre–Gaussian beams has been provided. The basic features of several additional optical beams have been introduced: optical vortex, Ince–Gaussian, nondiffracting Bessel, Bessel–Gaussian, and Airy.
- *Chapter 4 (Fourier Optics)*. An analysis of Fresnel diffraction from a periodic aperture (Talbot effect) has been included. Nondiffracting waves and Bessel beams have been introduced from a Fourier-optics perspective. A discussion of computer-generated holography has been added.
- *Chapter 5 (Electromagnetic Optics)*. A new section on the dipole wave, the basis of near-field optics, has been incorporated. A new section on scattering that includes Rayleigh and Mie scattering, along with attenuation in a medium with scatterers, has been added.
- *Chapter 6 (Polarization Optics)*. The material dealing with the dispersion relation in anisotropic media has been reworked to simplify the presentation.
- *Chapter 7 (Photonic-Crystal Optics)*. The behavior of the dielectric-slab beamsplitter has been elucidated. A discussion relating to fabrication methods for 3D photonic crystals has been incorporated.
- *Chapter 8 (Metal and Metamaterial Optics)*. This new chapter, entitled Metal and Metamaterial Optics, provides a venue for the examination of single- and double-negative media, metal optics, plasmonics, metamaterial optics, and transformation optics. Topics considered include evanescent waves, surface plasmon polaritons, localized surface plasmons, nanoantennas, metasurfaces, subwavelength imaging, and optical cloaking.
- *Chapter 9 (Guided-Wave Optics)*. A new section on waveguide arrays that details the mutual coupling of multiple waveguides and introduces the notion of supermodes has been inserted. A new section on plasmonic waveguides that includes metal–insulator–metal and metal-slab waveguides, along with periodic metal–dielectric arrays, has been incorporated.
- *Chapter 10 (Fiber Optics)*. A discussion of multicore fibers, fiber couplers, and photonic lanterns has been added. A brief discussion of the applications of photonic-crystal fibers has been provided. A new section on multimaterial fibers, including conventional and hybrid mid-infrared fibers, specialty fibers, multimaterial fibers, and multifunctional fibers, has been introduced.
- *Chapter 11 (Resonator Optics)*. A section on plasmonic resonators has been added.
- *Chapter 12 (Statistical Optics)*. The sections on optical coherence tomography and unpolarized light have been reorganized.
- *Chapter 13 (Photon Optics)*. A brief description of single-photon imaging has been added. The discussion of quadrature-squeezed and photon-number-squeezed

light has been enhanced and examples of the generation and applications of these forms of light have been provided. A section that describes two-photon light, entangled photons, two-photon optics, and the generation and applications thereof, has been incorporated. Examples of two-photon polarization, two-photon spatial optics, and two-beam optics have been appended.

- *Chapter 14 (Light and Matter)*. The title of this chapter was changed from *Photons and Atoms* to *Light and Matter*. Brief descriptions of the Zeeman effect, Stark effect, and ionization energies have been added. The discussion of lanthanide-ion manifolds has been enhanced. Descriptions of Doppler cooling, optical molasses, optical tweezers, optical lattices, atom interferometry, and atom amplifiers have been incorporated into the section on laser cooling, laser trapping, and atom optics.

- *Chapter 15 (Laser Amplifiers)*. Descriptions of quasi-three-level and in-band pumping have been added. The sections on representative laser amplifiers, including ruby, neodymium-doped glass, erbium-doped silica fiber, and Raman fiber devices, have been enhanced.

- *Chapter 16 (Lasers)*. Descriptions of tandem pumping, transition-ion-doped zinc-chalcogenide lasers, silicon Raman lasers, and master-oscillator power-amplifiers (MOPAs) have been added. Descriptions of inner-shell photopumping and X-ray free-electron lasers have been incorporated. A new section on optical frequency combs has been provided.

- *Chapter 17 (Semiconductor Optics)*. The section on organic semiconductors has been enhanced. A discussion of group-IV photonics, including graphene and 2D materials such as transition-metal dichalcogenides, has been added. A brief discussion of quantum-dot single-photon emitters has been incorporated.

- *Chapter 18 (LEDs and Laser Diodes)*. The title of this chapter was changed from *Semiconductor Photon Sources* to *LEDs and Laser Diodes*. A new section on the essentials of LED lighting has been incorporated. Brief discussions of the following topics are now included: resonant-cavity LEDs, silicon-photonics light sources, quantum-dot semiconductor amplifiers, external-cavity wavelength-tunable laser diodes, broad-area laser diodes, and laser-diode bars and stacks. A discussion of the semiconductor-laser linewidth-enhancement factor has been added. A new section on nanolasers has been introduced.

- *Chapter 19 (Photodetectors)*. The title of this chapter was changed from *Semiconductor Photon Detectors* to *Photodetectors*. Brief discussions of the following topics have been added: organic, plasmonic, group-IV-based, and graphene-enhanced photodetectors; edge vs. normal illumination; photon-trapping microstructures; SACM and superlattice APDs; multiplied dark current; and $1/f$ detector noise. New examples include multi-junction photovoltaic solar cells; Ge-on-Si photodiodes; graphene-Si Schottky-barrier photodiodes; and SAM, SACM, and staircase APDs. A new section on single-photon and photon-number-resolving detectors details the operation of SPADs, SiPMs, and TESs.

- *Chapter 20 (Acousto-Optics)*. The identical forms of the photoelastic matrix in acousto-optics and the Kerr-effect matrix in electro-optics has been highlighted for cubic isotropic media.

- *Chapter 21 (Electro-Optics)*. New sections on passive- and active-matrix liquid-crystal displays have been introduced and their operation has been elucidated. The performance of active-matrix liquid-crystal displays (AMLCDs) has been compared with that of active-matrix organic light-emitting displays (AMOLEDs).

- *Chapter 22 (Nonlinear Optics)*. New material relating to guided-wave nonlinear optics has been introduced. Quasi-phase matching in periodically poled integrated optical waveguides, and the associated improvement in wave-mixing efficiency, is now considered. The section pertaining to Raman gain has been enhanced.

- *Chapter 23 (Ultrafast Optics)*. New examples have been incorporated that con-

sider chirped pulse amplification in a petawatt laser and the generation of high-energy solitons in a photonic-crystal rod. A new section on high-harmonic generation and attosecond optics has been added. The section on pulse detection has been reorganized.
- *Chapter 24 (Optical Interconnects and Switches)*. The role of optical interconnects at the inter-board, inter-chip, and intrachip scale of computer systems is delineated. All-optical switching now incorporates nonparametric and parametric photonic switches that operate on the basis of manifold nonlinear-optical effects. Photonic-crystal and plasmonic photonic switches are discussed. The treatment of photonic logic gates now includes an analysis of embedded bistable systems and examples of bistability in fiber-based-interferometric and microring laser systems.
- *Chapter 25 (Optical Fiber Communications)*. The material on fiber-optic components has been updated and rewritten, and the role of photonic integrated circuits is delineated. A new section on space-division multiplexing in multicore and multimode fibers has been added. The section on coherent detection has been expanded and now emphasizes digital coherent receivers with spectrally efficient coding.

Representative Courses

The different chapters of the book may be combined in various ways for use in courses of semester or quarter duration. Representative examples of such courses are presented below. Some of these courses may be offered as part of a sequence. Other selections may be made to suit the particular objectives of instructors and students.

Optics			
1. Ray Optics	8. Metals & Metamaterials	14. Light and Matter	20. Acousto-Optics
2. Wave Optics	9. Guided-Wave Optics	15. Laser Amplifiers	21. Electro-Optics
3. Beam Optics	10. Fiber Optics	16. Lasers	22. Nonlinear Optics
4. Fourier Optics	11. Resonator Optics	17. Semiconductor Optics	23. Ultrafast Optics
5. Electromagnetic Optics	12. Statistical Optics	18. LEDs & Laser Diodes	24. Interconnects/Switches
6. Polarization Optics	13. Photon Optics	19. Photodetectors	25. Fiber Communications
7. Photonic Crystals			

The first six chapters of the book are suitable for an introductory course on *Optics*. These may be supplemented by Chapter 12 (*Statistical Optics*) to introduce incoherent and partially coherent light, and by Chapter 13 (*Photon Optics*) to introduce the photon. The introductory sections of Chapters 9 and 10 (*Guided-Wave Optics* and *Fiber Optics*, respectively) may be added to cover some applications.

Guided-Wave Optics			
1. Ray Optics	8. Metals & Metamaterials	14. Light and Matter	20. Acousto-Optics
2. Wave Optics	9. Guided-Wave Optics	15. Laser Amplifiers	21. Electro-Optics
3. Beam Optics	10. Fiber Optics	16. Lasers	22. Nonlinear Optics
4. Fourier Optics	11. Resonator Optics	17. Semiconductor Optics	23. Ultrafast Optics
5. Electromagnetic Optics	12. Statistical Optics	18. LEDs & Laser Diodes	24. Interconnects/Switches
6. Polarization Optics	13. Photon Optics	19. Photodetectors	25. Fiber Communications
7. Photonic Crystals			

A course on *Guided-Wave Optics* might begin with an introduction to wave propagation in layered and periodic media in Chapter 7 (*Photonic-Crystal Optics*), and could include Chapter 8 (*Metal and Metamaterial Optics*). This would be followed by Chapters 9, 10, and 11 (*Guided-Wave Optics*, *Fiber Optics*, and *Resonator Optics*, respectively). The introductory sections of Chapters 21 and 24 (*Electro-Optics* and *Optical Interconnects and Switches*) would provide additional material.

PREFACE xvii

Lasers			
1. Ray Optics	8. Metals & Metamaterials	14. Light and Matter	20. Acousto-Optics
2. Wave Optics	9. Guided-Wave Optics	15. Laser Amplifiers	21. Electro-Optics
3. Beam Optics	10. Fiber Optics	16. Lasers	22. Nonlinear Optics
4. Fourier Optics	11. Resonator Optics	17. Semiconductor Optics	23. Ultrafast Optics
5. Electromagnetic Optics	12. Statistical Optics	18. LEDs & Laser Diodes	24. Interconnects/Switches
6. Polarization Optics	13. Photon Optics	19. Photodetectors	25. Fiber Communications
7. Photonic Crystals			

A course on *Lasers* could begin with *Beam Optics* and *Resonator Optics* (Chapters 3 and 11, respectively), followed by *Light and Matter* (Chapter 14). The initial portion of *Photon Optics* (Chapter 13) could be assigned. The heart of the course would be the material contained in *Laser Amplifiers* and *Lasers* (Chapters 15 and 16, respectively). The course might also include material drawn from *Semiconductor Optics* and *LEDs and Laser Diodes* (Chapters 17 and 18, respectively). An introduction to femtosecond lasers could be extracted from some sections of *Ultrafast Optics* (Chapter 23).

Optoelectronics			
1. Ray Optics	8. Metals & Metamaterials	14. Light and Matter	20. Acousto-Optics
2. Wave Optics	9. Guided-Wave Optics	15. Laser Amplifiers	21. Electro-Optics
3. Beam Optics	10. Fiber Optics	16. Lasers	22. Nonlinear Optics
4. Fourier Optics	11. Resonator Optics	17. Semiconductor Optics	23. Ultrafast Optics
5. Electromagnetic Optics	12. Statistical Optics	18. LEDs & Laser Diodes	24. Interconnects/Switches
6. Polarization Optics	13. Photon Optics	19. Photodetectors	25. Fiber Communications
7. Photonic Crystals			

The chapters on *Semiconductor Optics*, *LEDs and Laser Diodes*, and *Photodetectors* (Chapters 17, 18, and 19, respectively) form a suitable basis for a course on *Optoelectronics*. This material would be supplemented with optics background from earlier chapters and could include topics such as liquid-crystal devices (Secs. 6.5 and 21.3), electroabsorption modulators (Sec. 21.5), and an introduction to photonic devices used for switching and/or communications (Chapters 24 and 25, respectively).

Photonic Devices			
1. Ray Optics	8. Metals & Metamaterials	14. Light and Matter	20. Acousto-Optics
2. Wave Optics	9. Guided-Wave Optics	15. Laser Amplifiers	21. Electro-Optics
3. Beam Optics	10. Fiber Optics	16. Lasers	22. Nonlinear Optics
4. Fourier Optics	11. Resonator Optics	17. Semiconductor Optics	23. Ultrafast Optics
5. Electromagnetic Optics	12. Statistical Optics	18. LEDs & Laser Diodes	24. Interconnects/Switches
6. Polarization Optics	13. Photon Optics	19. Photodetectors	25. Fiber Communications
7. Photonic Crystals			

Photonic Devices is a course that would consider the devices used in *Acousto-Optics*, *Electro-Optics*, and *Nonlinear Optics* (Chapters 20, 21, and 22, respectively). It might also include devices used in optical routing and switching, as discussed in *Optical Interconnects and Switches* (Chapter 24).

Nonlinear & Ultrafast Optics			
1. Ray Optics	8. Metals & Metamaterials	14. Light and Matter	20. Acousto-Optics
2. Wave Optics	9. Guided-Wave Optics	15. Laser Amplifiers	21. Electro-Optics
3. Beam Optics	10. Fiber Optics	16. Lasers	22. Nonlinear Optics
4. Fourier Optics	11. Resonator Optics	17. Semiconductor Optics	23. Ultrafast Optics
5. Electromagnetic Optics	12. Statistical Optics	18. LEDs & Laser Diodes	24. Interconnects/Switches
6. Polarization Optics	13. Photon Optics	19. Photodetectors	25. Fiber Communications
7. Photonic Crystals			

The material contained in Chapters 21–23 (*Electro-Optics*, *Nonlinear Optics*, and *Ultrafast Optics*, respectively) is suitable for an in-depth course on *Nonlinear and Ultrafast Optics*. These chapters could be supplemented by the material pertaining to electro-optic and all-optical switching in Chapter 24 (*Optical Interconnects and Switches*).

Fiber-Optic Communications

1. Ray Optics	8. Metals & Metamaterials	14. Light and Matter	20. Acousto-Optics
2. Wave Optics	9. Guided-Wave Optics	15. Laser Amplifiers	21. Electro-Optics
3. Beam Optics	10. Fiber Optics	16. Lasers	22. Nonlinear Optics
4. Fourier Optics	11. Resonator Optics	17. Semiconductor Optics	23. Ultrafast Optics
5. Electromagnetic Optics	12. Statistical Optics	18. LEDs & Laser Diodes	24. Interconnects/Switches
6. Polarization Optics	13. Photon Optics	19. Photodetectors	25. Fiber Communications
7. Photonic Crystals			

The heart of a course on *Fiber-Optic Communications* would be the material contained in Chapter 25 (*Optical Fiber Communications*). Background for this course would comprise material drawn from Chapters 9, 10, 18, and 19 (*Guided-Wave Optics, Fiber Optics, LEDs and Laser Diodes*, and *Photodetectors*, respectively), along with material contained in Secs. 15.3C and 15.3D (doped-fiber and Raman fiber amplifiers, respectively). If fiber-optic networks were to be emphasized, Sec. 24.3 (photonic switches) would be a valuable adjunct.

Optical Information Processing

1. Ray Optics	8. Metals & Metamaterials	14. Light and Matter	20. Acousto-Optics
2. Wave Optics	9. Guided-Wave Optics	15. Laser Amplifiers	21. Electro-Optics
3. Beam Optics	10. Fiber Optics	16. Lasers	22. Nonlinear Optics
4. Fourier Optics	11. Resonator Optics	17. Semiconductor Optics	23. Ultrafast Optics
5. Electromagnetic Optics	12. Statistical Optics	18. LEDs & Laser Diodes	24. Interconnects/Switches
6. Polarization Optics	13. Photon Optics	19. Photodetectors	25. Fiber Communications
7. Photonic Crystals			

Background material for a course on *Optical Information Processing* would be drawn from *Wave Optics* and *Beam Optics* (Chapters 2 and 3, respectively). The course could cover coherent image formation and processing from *Fourier Optics* (Chapter 4) along with incoherent and partially coherent imaging from *Statistical Optics* (Chapter 12). The focus could then shift to devices used for analog data processing, such as those considered in *Acousto-Optics* (Chapter 20). The course could then finish with material on switches and gates used for digital data processing, such as those considered in *Optical Interconnects and Switches* (Chapter 24).

Acknowledgments

We are indebted to many colleagues for providing us with valuable suggestions regarding improvements for the *Third Edition*: Rodrigo Amezcua-Correa, Luca Argenti, Joe C. Campbell, Zenghu Chang, Demetrios Christodoulides, Peter J. Delfyett, Dirk Englund, Eric R. Fossum, Majeed M. Hayat, Pieter G. Kik, Akhlesh Lakhtakia, Guifang Li, Steven B. Lowen, M. G. "Jim" Moharam, Rüdiger Paschotta, Kosmas L. Tsakmakidis, Shin-Tson Wu, Timothy M. Yarnall, and Boris Y. Zeldovich. We are also grateful to many of our former students and postdoctoral associates who posed excellent questions that helped us hone our presentation in the *Third Edition*, including John David Giese, Barry D. Jacobson, Samik Mukherjee, Adam Palmer, and Jian Yin.

We extend our special thanks to Mark Feuer, Joseph W. Goodman, and Mohammed F. Saleh who graciously provided us with in-depth critiques of various chapters.

Amy Hendrickson provided invaluable assistance with the Latex style files and eBook formatting. We are grateful to our Editors at John Wiley & Sons, Inc., who offered valuable suggestions and support throughout the course of production: Brett Kurzmann, Sarah Keegan, Nick Prindle, and Melissa Yanuzzi.

Finally, we are most appreciative of the generous support provided by CREOL, the College of Optics & Photonics at the University of Central Florida, the Boston University Photonics Center, and the Boston University College of Engineering.

Photo Credits

Many of the images on the chapter opening pages were carried forward from the First and Second Editions. Additional credits for the chapter opening pages of the *Third Edition* include: Wikimedia Commons (Laguerre and Bessel in Chapter 3, Stokes in Chapter 6, Drude in Chapter 8, Tyndall in Chapter 9, and Born in Chapter 12); *La Science Illustrée*, Volume 4 of the French weekly published in the second period of 1889 (Colladon in Chapter 9); Courtesy of Corning Incorporated (Schultz, Keck, & Maurer in Chapter 10); Courtesy of the Faculty History Project at Bentley Historical Library, University of Michigan (Franken in Chapter 22); and Courtesy of Peg Skorpinski, photographer, and the Department of Electrical Engineering and Computer Sciences at the University of California, Berkeley (Kaminow in Chapter 25). Self-photographs were provided by Viktor Georgievich Veselago (Veselago in Chapter 8); Sir John Pendry (Pendry in Chapter 8); Paul B. Corkum (Corkum in Chapter 23); Gérard Mourou (Mourou in Chapter 23); and the authors.

BAHAA E. A. SALEH

MALVIN CARL TEICH

Orlando, Florida

Boston, Massachusetts
June 4, 2018

PREFACE TO THE SECOND EDITION

Since the publication of the *First Edition* in 1991, *Fundamentals of Photonics* has been reprinted some 20 times, translated into Czech and Japanese, and used worldwide as a textbook and reference. During this period, major developments in photonics have continued apace, and have enabled technologies such as telecommunications and applications in industry and medicine. The *Second Edition* reports some of these developments, while maintaining the size of this single-volume tome within practical limits.

In its new organization, *Fundamentals of Photonics* continues to serve as a self-contained and up-to-date introductory-level textbook, featuring a logical blend of theory and applications. Many readers of the *First Edition* have been pleased with its abundant and well-illustrated figures. This feature has been enhanced in the *Second Edition* by the introduction of full color throughout the book, offering improved clarity and readability.

While each of the 22 chapters of the *First Edition* has been thoroughly updated, the principal feature of the *Second Edition* is the addition of two new chapters: one on photonic-crystal optics and another on ultrafast optics. These deal with developments that have had a substantial and growing impact on photonics over the past decade.

The new chapter on **photonic-crystal optics** provides a foundation for understanding the optics of layered media, including Bragg gratings, with the help of a matrix approach. Propagation of light in one-dimensional periodic media is examined using Bloch modes with matrix and Fourier methods. The concept of the photonic bandgap is introduced. Light propagation in two- and three-dimensional photonic crystals, and the associated dispersion relations and bandgap structures, are developed. Sections on photonic-crystal waveguides, holey fibers, and photonic-crystal resonators have also been added at appropriate locations in other chapters.

The new chapter on **ultrafast optics** contains sections on picosecond and femtosecond optical pulses and their characterization, shaping, and compression, as well as their propagation in optical fibers, in the domain of linear optics. Sections on ultrafast nonlinear optics include pulsed parametric interactions and optical solitons. Methods for the detection of ultrafast optical pulses using available detectors, which are relatively slow, are reviewed.

In addition to these two new chapters, the chapter on **optical interconnects and switches** has been completely rewritten and supplemented with topics such as wavelength and time routing and switching, FBGs, WGRs, SOAs, TOADs, and packet switches. The chapter on **optical fiber communications** has also been significantly updated and supplemented with material on WDM networks; it now offers concise descriptions of topics such as dispersion compensation and management, optical amplifiers, and soliton optical communications.

Continuing advances in device-fabrication technology have stimulated the emergence of **nanophotonics**, which deals with optical processes that take place over subwavelength (nanometer) spatial scales. Nanophotonic devices and systems include quantum-confined structures, such as quantum dots, nanoparticles, and nanoscale periodic structures used to synthesize *metamaterials* with exotic optical properties such as negative refractive index. They also include configurations in which light (or its interaction with matter) is confined to nanometer-size (rather than micrometer-size) regions near boundaries, as in *surface plasmon* optics. Evanescent fields, such as those produced at a surface where total internal reflection occurs, also exhibit such confinement. Evanescent fields are present in the immediate vicinity of subwavelength-size apertures, such as the open tip of a tapered optical fiber. Their use allows imaging with resolution beyond the diffraction limit and forms the basis of *near-field optics*. Many of these emerging areas are described at suitable locations in the *Second Edition*.

New sections have been added in the process of updating the various chapters. New topics introduced in the early chapters include: Laguerre–Gaussian beams; near-field imaging; the Sellmeier equation; fast and slow light; optics of conductive media and plasmonics; doubly negative metamaterials; the Poincaré sphere and Stokes parameters; polarization mode dispersion; whispering-gallery modes; microresonators; optical coherence tomography; and photon orbital angular momentum.

In the chapters on laser optics, new topics include: rare-earth and Raman fiber amplifiers and lasers; EUV, X-ray, and free-electron lasers; and chemical and random lasers. In the area of optoelectronics, new topics include: gallium nitride-based structures and devices; superluminescent diodes; organic and white-light LEDs; quantum-confined lasers; quantum cascade lasers; microcavity lasers; photonic-crystal lasers; array detectors; low-noise APDs; SPADs; and QWIPs.

The chapter on nonlinear optics has been supplemented with material on parametric-interaction tuning curves; quasi-phase-matching devices; two-wave mixing and cross-phase modulation; THz generation; and other nonlinear optical phenomena associated with narrow optical pulses, including chirp pulse amplification and supercontinuum light generation. The chapter on electro-optics now includes a discussion of electroabsorption modulators.

Appendix C on **modes of linear systems** has been expanded and now offers an overview of the concept of modes as they appear in numerous locations within the book. Finally, additional exercises and problems have been provided, and these are now numbered disjointly to avoid confusion.

Acknowledgments

We are grateful to many colleagues for providing us with valuable comments about draft chapters for the *Second Edition* and for drawing our attention to errors in the *First Edition*: Mete Atatüre, Michael Bär, Robert Balahura, Silvia Carrasco, Stephen Chinn, Thomas Daly, Gianni Di Giuseppe, Adel El-Nadi, John Fourkas, Majeed Hayat, Tony Heinz, Erich Ippen, Martin Jaspan, Gerd Keiser, Jonathan Kane, Paul Kelley, Ted Moustakas, Allen Mullins, Magued Nasr, Roy Olivier, Roberto Paiella, Peter W. E. Smith, Stephen P. Smith, Kenneth Suslick, Anna Swan, Tristan Tayag, Tommaso Toffoli, and Brad Tousley.

We extend our special thanks to those colleagues who graciously provided us with in-depth critiques of various chapters: Ayman Abouraddy, Luca Dal Negro, and Paul Prucnal.

We are indebted to the legions of students and postdoctoral associates who have posed so many excellent questions that helped us hone our presentation. In particular, many improvements were initiated by suggestions from Mark Booth, Jasper Cabalu, Michael Cunha, Darryl Goode, Chris LaFratta, Rui Li, Eric Lynch, Nan Ma, Nishant Mohan, Julie Praino, Yunjie Tong, and Ranjith Zachariah. We are especially grateful to Mohammed Saleh, who diligently read much of the manuscript and provided us with excellent suggestions for improvement throughout.

Wai Yan (Eliza) Wong provided logistical support and a great deal of assistance in crafting diagrams and figures. Many at Wiley, including George Telecki, our Editor, and Rachel Witmer have been most helpful, patient, and encouraging. We appreciate the attentiveness and thoroughness that Melissa Yanuzzi brought to the production process. Don DeLand of the Integre Technical Publishing Company provided invaluable assistance in setting up the Latex style files.

We are most appreciative of the financial support provided by the National Science Foundation (NSF), in particular the Center for Subsurface Sensing and Imaging Systems (CenSSIS), an NSF-supported Engineering Research Center; the Defense Advanced Research Projects Agency (DARPA); the National Reconnaissance Office (NRO); the U.S. Army Research Office (ARO); the David & Lucile Packard Foundation; the Boston University College of Engineering; and the Boston University Photonics Center.

Photo Credits. Most of the portraits were carried forward from the First Edition with the benefit of permissions provided for all editions. Additional credits are: Godfrey Kneller 1689 portrait (Newton); Siegfried Bendixen 1828 lithograph (Gauss); Engraving in the Small Portraits Collection, History of Science Collections, University of Oklahoma Libraries (Fraunhofer); Stanford University, Courtesy AIP Emilio Segrè Visual Archives (Bloch); Eli Yablonovitch (Yablonovitch); Sajeev John (John); Charles Kuen Kao (Kao); Philip St John Russell (Russell); Ecole Polytechnique (Fabry); Observatoire des Sciences de l'Univers (Perot); AIP Emilio Segrè Visual Archives (Born); Lagrelius & Westphal 1920 portrait (Bohr); AIP Emilio Segrè Visual Archives, Weber Collection (W. L. Bragg); Linn F. Mollenauer (Mollenauer); Roger H. Stolen (Stolen); and James P. Gordon (Gordon). In Chapter 24, the Bell Symbol was reproduced with the permission of BellSouth Intellectual Property Marketing Corporation, the AT&T logo is displayed with the permission of AT&T, and Lucent Technologies permitted us use of their logo. Stephen G. Eick kindly provided the image used at the beginning of Chapter 25. The photographs of Saleh and Teich were provided courtesy of Boston University.

BAHAA E. A. SALEH

MALVIN CARL TEICH

Boston, Massachusetts
December 19, 2006

PREFACE TO THE FIRST EDITION

Optics is an old and venerable subject involving the generation, propagation, and detection of light. Three major developments, which have been achieved in the last thirty years, are responsible for the rejuvenation of optics and for its increasing importance in modern technology: the invention of the laser, the fabrication of low-loss optical fibers, and the introduction of semiconductor optical devices. As a result of these developments, new disciplines have emerged and new terms describing these disciplines have come into use: **electro-optics**, **optoelectronics**, **quantum electronics**, **quantum optics**, and **lightwave technology**. Although there is a lack of complete agreement about the precise usages of these terms, there is a general consensus regarding their meanings.

Photonics

Electro-optics is generally reserved for optical devices in which electrical effects play a role (lasers, and electro-optic modulators and switches, for example). *Optoelectronics*, on the other hand, typically refers to devices and systems that are essentially electronic in nature but involve light (examples are light-emitting diodes, liquid-crystal display devices, and array photodetectors). The term *quantum electronics* is used in connection with devices and systems that rely principally on the interaction of light with matter (lasers and nonlinear optical devices used for optical amplification and wave mixing serve as examples). Studies of the quantum and coherence properties of light lie within the realm of *quantum optics*. The term *lightwave technology* has been used to describe devices and systems that are used in optical communications and optical signal processing.

In recent years, the term **photonics** has come into use. This term, which was coined in analogy with electronics, reflects the growing tie between optics and electronics forged by the increasing role that semiconductor materials and devices play in optical systems. *Electronics* involves the control of electric-charge flow (in vacuum or in matter); *photonics* involves the control of photons (in free space or in matter). The two disciplines clearly overlap since electrons often control the flow of photons and, conversely, photons control the flow of electrons. The term *photonics* also reflects the importance of the photon nature of light in describing the operation of many optical devices.

Scope

This book provides an introduction to the fundamentals of photonics. The term *photonics* is used broadly to encompass all of the aforementioned areas, including the following:

- The *generation* of coherent light by lasers, and incoherent light by luminescence sources such as light-emitting diodes.
- The *transmission* of light in free space, through conventional optical components such as lenses, apertures, and imaging systems, and through waveguides such as optical fibers.
- The *modulation*, switching, and scanning of light by the use of electrically, acoustically, or optically controlled devices.
- The *amplification* and *frequency conversion* of light by the use of wave interactions in nonlinear materials.
- The *detection* of light.

These areas have found ever-increasing applications in optical communications, signal processing, computing, sensing, display, printing, and energy transport.

Approach and Presentation

The underpinnings of photonics are provided in a number of chapters that offer concise introductions to:

- The four theories of light (each successively more advanced than the preceding): ray optics, wave optics, electromagnetic optics, and photon optics.
- The theory of interaction of light with matter.
- The theory of semiconductor materials and their optical properties.

These chapters serve as basic building blocks that are used in other chapters to describe the *generation* of light (by lasers and light-emitting diodes); the *transmission* of light (by optical beams, diffraction, imaging, optical waveguides, and optical fibers); the *modulation* and switching of light (by the use of electro-optic, acousto-optic, and nonlinear-optic devices); and the *detection* of light (by means of photodetectors). Many applications and examples of real systems are provided so that the book is a blend theory and practice. The final chapter is devoted to the study of fiber-optic communications, which provides an especially rich example in which the generation, transmission, modulation, and detection of light are all part of a single photonic system used for the transmission of information.

The theories of light are presented at progressively increasing levels of difficulty. Thus light is described first as rays, then scalar waves, then electromagnetic waves, and finally, photons. Each of these descriptions has its domain of applicability. Our approach is to draw from the simplest theory that adequately describes the phenomenon or intended application. Ray optics is therefore used to describe imaging systems and the confinement of light in waveguides and optical resonators. Scalar wave theory provides a description of optical beams, which are essential for the understanding of lasers, and of Fourier optics, which is useful for describing coherent optical systems and holography. Electromagnetic theory provides the basis for the polarization and dispersion of light, and the optics of guided waves, fibers, and resonators. Photon optics serves to describe the interactions of light with matter, explaining such processes as light generation and detection, and light mixing in nonlinear media.

Intended Audience

Fundamentals of Photonics is meant to serve as:

- An introductory textbook for students in electrical engineering or applied physics at the senior or first-year graduate level.
- A self-contained work for self-study.
- A text for programs of continuing professional development offered by industry, universities, and professional societies.

The reader is assumed to have a background in engineering or applied physics, including courses in modern physics, electricity and magnetism, and wave motion. Some knowledge of linear systems and elementary quantum mechanics is helpful but not essential. Our intent has been to provide an introduction to photonics that emphasizes the concepts governing applications of current interest. The book should, therefore, not be considered as a compendium that encompasses all photonic devices and systems. Indeed, some areas of photonics are not included at all, and many of the individual chapters could easily have been expanded into separate monographs.

Problems, Reading Lists, and Appendices

A set of problems is provided at the end of each chapter. Problems are numbered in accordance with the chapter sections to which they apply. Quite often, problems deal with ideas or applications not mentioned in the text, analytical derivations, and numerical computations designed to illustrate the magnitudes of important quantities. Problems marked with asterisks are of a more advanced nature. A number of exercises also appear within the text of each chapter to help the reader develop a better understanding of (or to introduce an extension of) the material.

Appendices summarize the properties of one- and two-dimensional Fourier transforms, linear-systems theory, and modes of linear systems (which are important in polarization devices, optical waveguides, and resonators); these are called upon at appropriate points throughout the book. Each chapter ends with a reading list that includes a selection of important books, review articles, and a few classic papers of special significance.

Acknowledgments

We are grateful to many colleagues for reading portions of the text and providing helpful comments: Govind P. Agrawal, David H. Auston, Rasheed Azzam, Nikolai G. Basov, Franco Cerrina, Emmanuel Desurvire, Paul Diament, Eric Fossum, Robert J. Keyes, Robert H. Kingston, Rodney Loudon, Leonard Mandel, Leon McCaughan, Richard M. Osgood, Jan Peřina, Robert H. Rediker, Arthur L. Schawlow, S. R. Seshadri, Henry Stark, Ferrel G. Stremler, John A. Tataronis, Charles H. Townes, Patrick R. Trischitta, Wen I. Wang, and Edward S. Yang.

We are especially indebted to John Whinnery and Emil Wolf for providing us with many suggestions that greatly improved the presentation.

Several colleagues used portions of the notes in their classes and provided us with invaluable feedback. These include Etan Bourkoff at Johns Hopkins University (now at the University of South Carolina), Mark O. Freeman at the University of Colorado, George C. Papen at the University of Illinois, and Paul R. Prucnal at Princeton University.

Many of our students and former students contributed to this material in various ways over the years and we owe them a great debt of thanks: Gaetano L. Aiello, Mohamad Asi, Richard Campos, Buddy Christyono, Andrew H. Cordes, Andrew David, Ernesto Fontenla, Evan Goldstein, Matthew E. Hansen, Dean U. Hekel, Conor Heneghan, Adam Heyman, Bradley M. Jost, David A. Landgraf, Kanghua Lu, Ben Nathanson, Winslow L. Sargeant, Michael T. Schmidt, Raul E. Sequeira, David Small, Kraisin Songwatana, Nikola S. Subotic, Jeffrey A. Tobin, and Emily M. True. Our thanks also go to the legions of unnamed students who, through a combination of vigilance and the desire to understand the material, found countless errors.

We particularly appreciate the many contributions and help of those students who were intimately involved with the preparation of this book at its various stages of completion: Niraj Agrawal, Suzanne Keilson, Todd Larchuk, Guifang Li, and Philip Tham.

We are grateful for the assistance given to us by a number of colleagues in the course of collecting the photographs used at the beginnings of the chapters: E. Scott Barr, Nicolaas Bloembergen, Martin Carey, Marjorie Graham, Margaret Harrison, Ann Kottner, G. Thomas Holmes, John Howard, Theodore H. Maiman, Edward Palik, Martin Parker, Aleksandr M. Prokhorov, Jarus Quinn, Lesley M. Richmond, Claudia Schüler, Patrick R. Trischitta, J. Michael Vaughan, and Emil Wolf. Specific photo credits are as follows: AIP Meggers Gallery of Nobel Laureates (Gabor, Townes, Basov, Prokhorov, W. L. Bragg); AIP Niels Bohr Library (Rayleigh, Fraunhofer, Maxwell, Planck, Bohr, Einstein in Chapter 14, W. H. Bragg); Archives de l'Académie des Sciences de Paris (Fabry); The Astrophysical Journal (Perot);

AT&T Bell Laboratories (Shockley, Brattain, Bardeen); Bettmann Archives (Young, Gauss, Tyndall); Bibliothèque Nationale de Paris (Fermat, Fourier, Poisson); Burndy Library (Newton, Huygens); Deutsches Museum (Hertz); ETH Bibliothek (Einstein in Chapter 13); Bruce Fritz (Saleh); Harvard University (Bloembergen); Heidelberg University (Pockels); Kelvin Museum of the University of Glasgow (Kerr); Theodore H. Maiman (Maiman); Princeton University (von Neumann); Smithsonian Institution (Fresnel); Stanford University (Schawlow); Emil Wolf (Born, Wolf). Corning Incorporated kindly provided the photograph used at the beginning of Chapter 10. We are grateful to GE for the use of their logotype, which is a registered trademark of the General Electric Company, at the beginning of Chapter 18. The IBM logo at the beginning of Chapter 18 is being used with special permission from IBM. The right-most logotype at the beginning of Chapter 18 was supplied courtesy of Lincoln Laboratory, Massachusetts Institute of Technology. AT&T Bell Laboratories kindly permitted us use of the diagram at the beginning of Chapter 25.

We greatly appreciate the continued support provided to us by the National Science Foundation, the Center for Telecommunications Research, and the Joint Services Electronics Program through the Columbia Radiation Laboratory.

Finally, we extend our sincere thanks to our editors, George Telecki and Bea Shube, for their guidance and suggestions throughout the course of preparation of this book.

BAHAA E. A. SALEH

Madison, Wisconsin

MALVIN CARL TEICH

New York, New York
April 3, 1991

FUNDAMENTALS OF PHOTONICS

Part I: Optics
(Chapters 1–13)

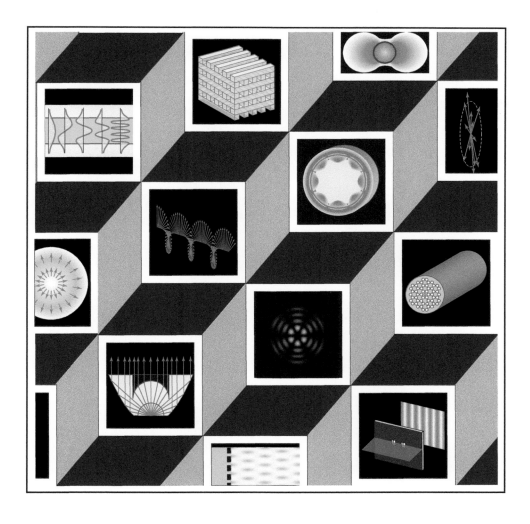

CHAPTER 1

RAY OPTICS

1.1	POSTULATES OF RAY OPTICS	5
1.2	SIMPLE OPTICAL COMPONENTS	8
	A. Mirrors	
	B. Planar Boundaries	
	C. Spherical Boundaries and Lenses	
	D. Light Guides	
1.3	GRADED-INDEX OPTICS	20
	A. The Ray Equation	
	B. Graded-Index Optical Components	
	*C. The Eikonal Equation	
1.4	MATRIX OPTICS	27
	A. The Ray-Transfer Matrix	
	B. Matrices of Simple Optical Components	
	C. Matrices of Cascaded Optical Components	
	D. Periodic Optical Systems	

Sir Isaac Newton (1642–1727) set forth a theory of optics in which light emissions consist of collections of corpuscles that propagate rectilinearly.

Pierre de Fermat (1601–1665) enunciated a rule, known as Fermat's Principle, in which light rays travels along the path of least time relative to neighboring paths.

Fundamentals of Photonics, Third Edition. Bahaa E. A. Saleh and Malvin Carl Teich.
©2019 John Wiley & Sons, Inc. Published 2019 by John Wiley & Sons, Inc.

Light can be described as an electromagnetic wave phenomenon governed by the same theoretical principles that govern all other forms of electromagnetic radiation, such as radio waves and X-rays. This conception of light is called **electromagnetic optics**. Electromagnetic radiation propagates in the form of two mutually coupled *vector* waves, an electric-field wave and a magnetic-field wave. Nevertheless, it is possible to describe many optical phenomena using a simplified *scalar* wave theory in which light is described by a single scalar wavefunction. This approximate way of treating light is called scalar wave optics, or simply **wave optics**.

When light waves propagate through and around objects whose dimensions are much greater than the wavelength of the light, the wave nature is not readily discerned and the behavior of light can be adequately described by rays obeying a set of geometrical rules. This model of light is called **ray optics**. From a mathematical perspective, ray optics is the limit of wave optics when the wavelength is infinitesimally small.

Thus, electromagnetic optics encompasses wave optics, which in turn encompasses ray optics, as illustrated in Fig. 1.0-1. Ray optics and wave optics are approximate theories that derive validity from their successes in producing results that approximate those based on the more rigorous electromagnetic theory.

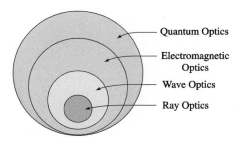

Figure 1.0-1 The theory of quantum optics provides an explanation for virtually all optical phenomena. The electromagnetic theory of light (electromagnetic optics) provides the most complete treatment of light within the confines of classical optics. Wave optics is a scalar approximation of electromagnetic optics. Ray optics is the limit of wave optics when the wavelength is very short.

Although electromagnetic optics provides the most complete treatment of light within the confines of **classical optics**, certain optical phenomena are characteristically quantum mechanical in nature and cannot be explained classically. These nonclassical phenomena are described by a quantum version of electromagnetic theory known as **quantum electrodynamics**. For optical phenomena, this theory is also referred to as **quantum optics**.

Historically, the theories of optics developed roughly in the following order: (1) ray optics → (2) wave optics → (3) electromagnetic optics → (4) quantum optics. These models are progressively more complex and sophisticated, and were developed successively to provide explanations for the outcomes of increasingly subtle and precise optical experiments. The optimal choice of a model is the simplest one that satisfactorily describes a particular phenomenon, but it is sometimes difficult to know *a priori* which model achieves this. Experience is often the best guide.

For pedagogical reasons, the initial chapters of this book follow the historical order indicated above. Each model of light begins with a set of postulates (provided without proof), from which a large body of results are generated. The postulates of each model are shown to arise as special cases of the next-higher-level model. In this chapter we begin with ray optics.

This Chapter

Ray optics is the simplest theory of light. Light is described by rays that travel in different optical media in accordance with a set of geometrical rules. Ray optics is therefore also called **geometrical optics**. Ray optics is an approximate theory. Although it adequately describes most of our daily experiences with light, there are many phenomena that ray optics cannot adequately construe (as amply attested to by the remaining chapters of this book).

Ray optics is concerned with the *locations* and *directions* of light rays. It is therefore useful in studying *image formation* — the collection of rays from each point of an object and their redirection by an optical component onto a corresponding point of an image. Ray optics permits us to determine the conditions under which light is guided within a given medium, such as a glass fiber. In isotropic media, optical rays point in the direction of the flow of *optical energy*. Ray bundles can be constructed in which the density of rays is proportional to the density of light energy. When light is generated isotropically from a point source, for example, the energy associated with the rays in a given cone is proportional to the solid angle of the cone. Rays may be traced through an optical system to determine the optical energy crossing a given area.

This chapter begins with a set of postulates from which we derive the simple rules that govern the propagation of light rays through optical media. In Sec. 1.2 these rules are applied to simple optical components, such as mirrors and planar or spherical boundaries between different optical media. Ray propagation in inhomogeneous (graded-index) optical media is examined in Sec. 1.3. Graded-index optics is the basis of a technology that has become an important part of modern optics.

Optical components are often centered about an optical axis, with respect to which the rays travel at small inclinations. Such rays are called **paraxial rays** and the assumption that the rays have this property is the basis of **paraxial optics**. The change in the position and inclination of a paraxial ray as it travels through an optical system can be efficiently described by the use of a 2×2-matrix algebra. Section 1.4 is devoted to this algebraic tool, which is known as **matrix optics**.

1.1 POSTULATES OF RAY OPTICS

Postulates of Ray Optics

- Light travels in the form of rays. The rays are emitted by light sources and can be observed when they reach an optical detector.
- An optical medium is characterized by a quantity $n \geq 1$, called the **refractive index**. The refractive index $n = c_o/c$ where c_o is the speed of light in free space and c is the speed of light in the medium. Therefore, the time taken by light to travel a distance d is $d/c = nd/c_o$. It is proportional to the product nd, which is known as the **optical pathlength**.
- In an inhomogeneous medium the refractive index $n(\mathbf{r})$ is a function of the position $\mathbf{r} = (x, y, z)$. The optical pathlength along a given path between two points A and B is therefore

$$\text{Optical pathlength} = \int_A^B n(\mathbf{r})\, ds, \quad (1.1\text{-}1)$$

where ds is the differential element of length along the path. The time taken by light to travel from A to B is proportional to the optical pathlength.

- **Fermat's Principle.** Optical rays traveling between two points, A and B, follow a path such that the time of travel (or the optical pathlength) between the two points is an extremum relative to neighboring paths. This is expressed mathematically as

$$\delta \int_A^B n(\mathbf{r})\, ds = 0, \qquad (1.1\text{-}2)$$

where the symbol δ, which is read "the variation of," signifies that the optical pathlength is either minimized or maximized, or is a point of inflection. It is, however, usually a minimum, in which case:

Light rays travel along the path of least time.

Sometimes the minimum time is shared by more than one path, which are then all followed simultaneously by the rays. An example in which the pathlength is maximized is provided in Prob. 1.1-2.

In this chapter we use the postulates of ray optics to determine the rules governing the propagation of light rays, their reflection and refraction at the boundaries between different media, and their transmission through various optical components. A wealth of results applicable to numerous optical systems are obtained without the need for any other assumptions or rules regarding the nature of light.

Propagation in a Homogeneous Medium

In a homogeneous medium the refractive index is the same everywhere, and so is the speed of light. The path of minimum time, required by Fermat's principle, is therefore also the path of minimum distance. The principle of the *path of minimum distance* is known as **Hero's principle**. The path of minimum distance between two points is a straight line so that *in a homogeneous medium, light rays travel in straight lines* (Fig. 1.1-1).

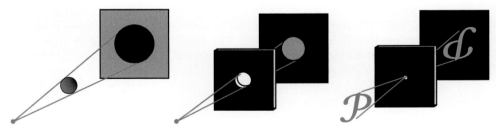

Figure 1.1-1 Light rays travel in straight lines. Shadows are perfect projections of stops.

Reflection from a Mirror

Mirrors are made of certain highly polished metallic surfaces, or metallic or dielectric films deposited on a substrate such as glass. Light reflects from mirrors in accordance with the law of reflection:

The reflected ray lies in the plane of incidence; the angle of reflection equals the angle of incidence.

The plane of incidence is the plane formed by the incident ray and the normal to the mirror at the point of incidence. The angles of incidence and reflection, θ and θ', are defined in Fig. 1.1-2(a). To prove the law of reflection we simply use Hero's principle. Examine a ray that travels from point A to point C after reflection from the planar mirror in Fig. 1.1-2(b). According to Hero's principle, for a mirror of infinitesimal thickness, the distance $\overline{AB} + \overline{BC}$ must be minimum. If C' is a mirror image of C, then $\overline{BC} = \overline{BC'}$, so that $\overline{AB} + \overline{BC'}$ must be a minimum. This occurs when $\overline{ABC'}$ is a straight line, i.e., when B coincides with B' so that $\theta = \theta'$.

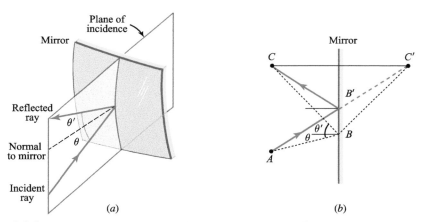

Figure 1.1-2 (a) Reflection from the surface of a curved mirror. (b) Geometrical construction to prove the law of reflection.

Reflection and Refraction at the Boundary Between Two Media

At the boundary between two media of refractive indices n_1 and n_2 an incident ray is split into two — a reflected ray and a refracted (or transmitted) ray (Fig. 1.1-3). The reflected ray obeys the law of reflection. The refracted ray obeys the law of refraction:

The refracted ray lies in the plane of incidence; the angle of refraction θ_2 is related to the angle of incidence θ_1 by Snell's law,

$$n_1 \sin \theta_1 = n_2 \sin \theta_2. \tag{1.1-3}$$
Snell's Law

The proportion in which the light is reflected and refracted is not described by ray optics.

8   CHAPTER 1 RAY OPTICS

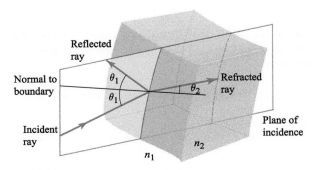

Figure 1.1-3 Reflection and refraction at the boundary between two media.

EXERCISE 1.1-1

Proof of Snell's Law. The proof of Snell's law is an exercise in the application of Fermat's principle. Referring to Fig. 1.1-4, we seek to minimize the optical pathlength $n_1 \overline{AB} + n_2 \overline{BC}$ between points A and C. We therefore have the following optimization problem: Minimize $n_1 d_1 \sec\theta_1 + n_2 d_2 \sec\theta_2$ with respect to the angles θ_1 and θ_2, subject to the condition $d_1 \tan\theta_1 + d_2 \tan\theta_2 = d$. Show that the solution of this constrained minimization problem yields Snell's law.

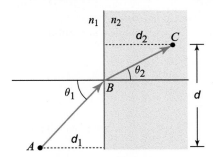

Figure 1.1-4 Construction to prove Snell's law.

The three simple rules — propagation in straight lines and the laws of reflection and refraction — are applied in Sec. 1.2 to several geometrical configurations of mirrors and transparent optical components, without further recourse to Fermat's principle.

1.2 SIMPLE OPTICAL COMPONENTS

A. Mirrors

Planar Mirrors

A planar mirror reflects the rays originating from a point P_1 such that the reflected rays appear to originate from a point P_2 behind the mirror, called the image (Fig. 1.2-1).

Paraboloidal Mirrors

The surface of a paraboloidal mirror is a reflective paraboloid of revolution. It has the useful property of focusing all incident rays parallel to its axis to a single point, called the **focus** or **focal point**. The distance $\overline{PF} \equiv f$ defined in Fig. 1.2-2 is known as the **focal length**. Paraboloidal mirrors are often used as light-collecting elements in telescopes. They are also used to render parallel the rays from a point source of light, such as a flashlight bulb or a light-emitting diode, located at the focus. When used in this manner, the device is known as a **collimator**.

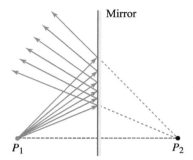

Figure 1.2-1 Reflection of light from a planar mirror.

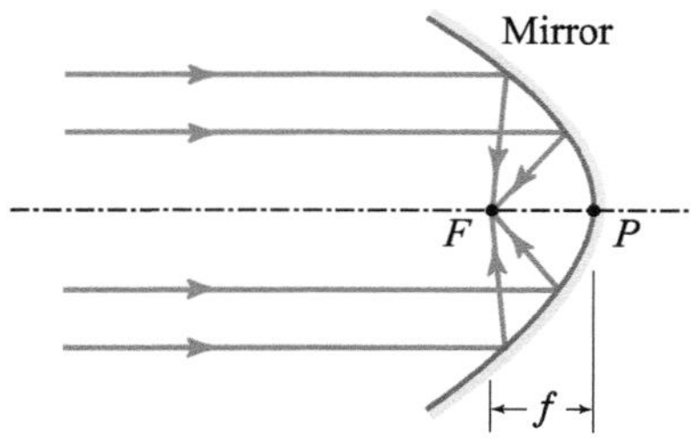

Figure 1.2-2 Focusing of light by a paraboloidal mirror.

Elliptical Mirrors

An elliptical mirror reflects all the rays emitted from one of its two foci, e.g., P_1, and images them onto the other focus, P_2 (Fig. 1.2-3). In accordance with Hero's principle, the distances traveled by the light from P_1 to P_2 along any of the paths are equal.

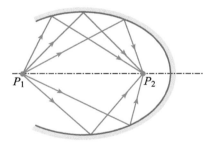

Figure 1.2-3 Reflection from an elliptical mirror.

Spherical Mirrors

A spherical mirror is easier to fabricate than a paraboloidal mirror or an elliptical mirror. However, it has neither the focusing property of the paraboloidal mirror nor the imaging property of the elliptical mirror. As illustrated in Fig. 1.2-4, parallel rays meet the axis at different points; their envelope (the dashed curve) is called the caustic curve. Nevertheless, parallel rays close to the axis are approximately focused onto a single point F at distance $(-R)/2$ from the mirror center C. By convention, the **radius of curvature** R is negative for concave mirrors and positive for convex mirrors.

Paraxial Rays Reflected from Spherical Mirrors

Rays that make small angles (such that $\sin\theta \approx \theta$) with the mirror's axis are called **paraxial rays**. In the **paraxial approximation**, where only paraxial rays are considered, a spherical mirror has a focusing property like that of the paraboloidal mirror *and* an imaging property like that of the elliptical mirror. The body of rules that results from this approximation forms **paraxial optics**, also called **first-order optics** or **Gaussian optics**.

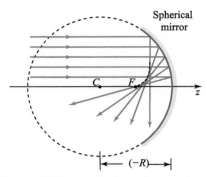
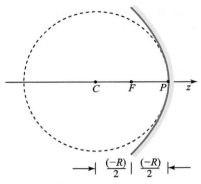

Figure 1.2-4 Reflection of parallel rays from a concave spherical mirror.

Figure 1.2-5 A spherical mirror approximates a paraboloidal mirror for paraxial rays.

A spherical mirror of radius R therefore acts like a paraboloidal mirror of focal length $f = R/2$. This is, in fact, plausible since at points near the axis, a parabola can be approximated by a circle with radius equal to the parabola's radius of curvature (Fig. 1.2-5).

All paraxial rays originating from each point on the axis of a spherical mirror are reflected and focused onto a single corresponding point on the axis. This can be seen (Fig. 1.2-6) by examining a ray emitted at an angle θ_1 from a point P_1 at a distance z_1 away from a concave mirror of radius R, and reflecting at angle $(-\theta_2)$ to meet the axis at a point P_2 that is a distance z_2 away from the mirror. The angle θ_2 is negative since the ray is traveling downward. Since the three angles of a triangle add to 180°, we have $\theta_1 = \theta_0 - \theta$ and $(-\theta_2) = \theta_0 + \theta$, so that $(-\theta_2) + \theta_1 = 2\theta_0$. If θ_0 is sufficiently small, the approximation $\tan \theta_0 \approx \theta_0$ may be used, so that $\theta_0 \approx y/(-R)$, from which

$$(-\theta_2) + \theta_1 \approx \frac{2y}{(-R)}, \tag{1.2-1}$$

where y is the height of the point at which the reflection occurs. Recall that R is negative since the mirror is concave. Similarly, if θ_1 and θ_2 are small, $\theta_1 \approx y/z_1$ and $(-\theta_2) = y/z_2$, so that (1.2-1) yields $y/z_1 + y/z_2 \approx 2y/(-R)$, whereupon

$$\frac{1}{z_1} + \frac{1}{z_2} \approx \frac{2}{(-R)}. \tag{1.2-2}$$

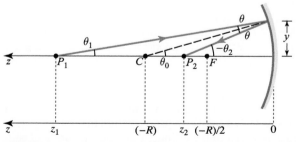

Figure 1.2-6 Reflection of paraxial rays from a concave spherical mirror of radius $R < 0$.

This relation holds regardless of y (i.e., regardless of θ_1) as long as the approximation is valid. This means that all paraxial rays originating from point P_1 arrive at P_2. The distances z_1 and z_2 are measured in a coordinate system in which the z axis points to the left. Points of negative z therefore lie to the right of the mirror.

According to (1.2-2), rays that are emitted from a point very far out on the z axis ($z_1 = \infty$) are focused to a point F at a distance $z_2 = (-R)/2$. This means that within the paraxial approximation, all rays coming from infinity (parallel to the axis of the mirror) are focused to a point at a distance f from the mirror, which is known as its focal length:

$$f = \frac{(-R)}{2},$$ (1.2-3)
Focal Length
Spherical Mirror

Equation (1.2-2) is usually written in the form

$$\frac{1}{z_1} + \frac{1}{z_2} = \frac{1}{f},$$ (1.2-4)
Imaging Equation
(Paraxial Rays)

which is known as the imaging equation. Both the incident and the reflected rays must be paraxial for this equation to hold.

EXERCISE 1.2-1

Image Formation by a Spherical Mirror. Show that, within the paraxial approximation, rays originating from a point $P_1 = (y_1, z_1)$ are reflected to a point $P_2 = (y_2, z_2)$, where z_1 and z_2 satisfy (1.2-4) and $y_2 = -y_1 z_2 / z_1$ (Fig. 1.2-7). This means that rays from each point in the plane $z = z_1$ meet at a single corresponding point in the plane $z = z_2$, so that the mirror acts as an image-formation system with magnification $-z_2/z_1$. Negative magnification means that the image is inverted.

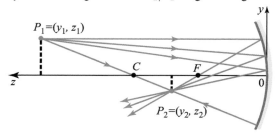

Figure 1.2-7 Image formation by a spherical mirror. Four particular rays are illustrated.

B. Planar Boundaries

The relation between the angles of refraction and incidence, θ_2 and θ_1, at a planar boundary between two media of refractive indices n_1 and n_2 is governed by Snell's law (1.1-3). This relation is plotted in Fig. 1.2-8 for two cases:

- *External Refraction* ($n_1 < n_2$). When the ray is incident from the medium of smaller refractive index, $\theta_2 < \theta_1$ and the refracted ray bends away from the boundary.

- *Internal Refraction* ($n_1 > n_2$). If the incident ray is in a medium of higher refractive index, $\theta_2 > \theta_1$ and the refracted ray bends toward the boundary.

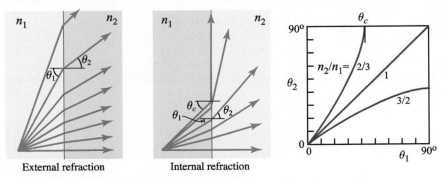

Figure 1.2-8 Relation between the angles of refraction and incidence.

The refracted rays bend in such a way as to minimize the optical pathlength, i.e., to increase the pathlength in the lower-index medium at the expense of pathlength in the higher-index medium. In both cases, when the angles are small (i.e., the rays are paraxial), the relation between θ_2 and θ_1 is approximately linear, $n_1\theta_1 \approx n_2\theta_2$, or $\theta_2 \approx (n_1/n_2)\theta_1$.

Total Internal Reflection

For internal refraction ($n_1 > n_2$), the angle of refraction is greater than the angle of incidence, $\theta_2 > \theta_1$, so that as θ_1 increases, θ_2 reaches 90° when $\theta_1 = \theta_c$, the **critical angle** (see Fig. 1.2-8). This occurs when $n_1 \sin\theta_c = n_2 \sin(\pi/2) = n_2$, so that

$$\theta_c = \sin^{-1}\frac{n_2}{n_1}. \tag{1.2-5}$$

Critical Angle

When $\theta_1 > \theta_c$, Snell's law (1.1-3) cannot be satisfied and refraction does not occur. The incident ray is then totally reflected as if the surface were a perfect mirror [Fig. 1.2-9(a)]. This phenomenon, called **total internal reflection (TIR)**, is the basis of many optical devices and systems, such as reflecting prisms [Fig. 1.2-9(b)], light-emitting diode collimators (Fig. 1.2-14), and optical fibers (Sec. 1.2D). Electromagnetic optics (Fresnel's equations in Chapter 6) reveals that all of the energy is carried by the reflected light so that the process of total internal reflection is highly efficient.

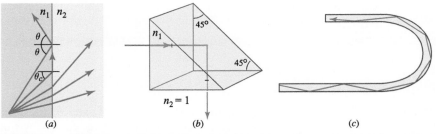

Figure 1.2-9 (a) Total internal reflection at a planar boundary. (b) The reflecting prism. If $n_1 > \sqrt{2}$ and $n_2 = 1$ (air), then $\theta_c < 45°$; since $n_2 \approx 1.5 > \sqrt{2}$ for glass, and $\theta_1 = 45°$, the ray is totally reflected. (c) Rays are guided by total internal reflection from the internal surface of an optical fiber.

Prisms

A prism of apex angle α and refractive index n (Fig. 1.2-10) deflects a ray incident at an angle θ by an angle

$$\theta_d = \theta - \alpha + \sin^{-1}\left[\sqrt{n^2 - \sin^2\theta}\,\sin\alpha - \sin\theta\cos\alpha\right]. \qquad (1.2\text{-}6)$$

This equation is arrived at by using Snell's law twice, at the two refracting surfaces of the prism. When α is very small (thin prism) and θ is also very small (paraxial approximation), (1.2-6) may be approximated by

$$\theta_d \approx (n-1)\alpha. \qquad (1.2\text{-}7)$$

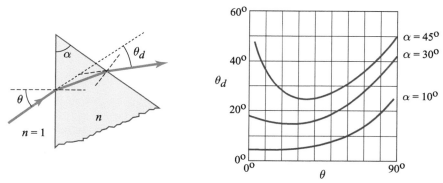

Figure 1.2-10 (*a*) Ray deflection by a prism. (*b*) Graph of (1.2-6) for the deflection angle θ_d as a function of the angle of incidence θ, for different apex angles α and $n = 1.5$. When both α and θ are small the angle of deflection $\theta_d \approx (n-1)\alpha$, which is approximately independent of θ, as is evident for the $\alpha = 10°$ curve. When $\theta = 0°$ and $\alpha = 45°$, total internal reflection occurs, as illustrated in Fig. 1.2-9(*b*).

Beamsplitters

The beamsplitter is an optical component that splits an incident ray into a reflected ray and a transmitted ray, as illustrated in Fig. 1.2-11. The relative proportions of light transmitted and reflected are established by Fresnel's equations in electromagnetic optics (Chapter 6). Beamsplitters are also frequently used to combine two light rays into one [Fig. 1.2-11(*c*)]. Beamsplitters are usually constructed by depositing a thin semitransparent metallic or dielectric film on a glass substrate. A thin bare glass plate, such as a microscope slide, can also serve as a beamsplitter although the fraction of light reflected is small. Transparent plastic materials are often used in place of glass.

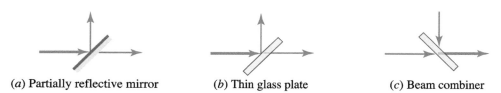

(*a*) Partially reflective mirror (*b*) Thin glass plate (*c*) Beam combiner

Figure 1.2-11 Beamsplitters and beam combiners.

Beam Directors

Simple optical components can be used to direct rays in particular directions. The devices illustrated in Fig. 1.2-12 redirect incident rays into rays tilted at fixed angles with respect to each other. The **biprism** depicted in Fig. 1.2-12(a) is the juxtaposition of a prism and an identical inverted prism. The **Fresnel biprism** portrayed in Fig. 1.2-12(b) is formed from rows of adjacently placed tiny prisms. This device is equivalent to a biprism but is thinner and lighter. The cone-shaped optic depicted in Fig. 1.2-12(c), known as an **axicon**, converts incident rays into a collection of circularly symmetric rays directed toward its central axis in the form of a cone. It has the same cross section as the biprism, namely an isosceles triangle.

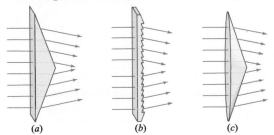

Figure 1.2-12 (a) Biprism. (b) Fresnel biprism. (c) Plano-convex axicon.

C. Spherical Boundaries and Lenses

We now examine the refraction of rays from a spherical boundary of radius R between two media of refractive indices n_1 and n_2. By convention, R is positive for a convex boundary and negative for a concave boundary. The results are obtained by applying Snell's law, which relates the angles of incidence and refraction relative to the normal to the surface, defined by the radius vector from the center C. These angles are to be distinguished from the angles θ_1 and θ_2, which are defined relative to the z axis. Considering only paraxial rays making small angles with the axis of the system so that $\sin\theta \approx \theta$ and $\tan\theta \approx \theta$, the following properties may be shown to hold:

- A ray making an angle θ_1 with the z axis and meeting the boundary at a point of height y where it makes an angle θ_0 with the radius vector [see Fig. 1.2-13(a)] changes direction at the boundary so that the refracted ray makes an angle θ_2 with the z axis and an angle θ_3 with the radius vector. With the help of Exercise 1.2-2, we obtain

$$\theta_2 \approx \frac{n_1}{n_2}\theta_1 - \frac{n_2 - n_1}{n_2}\frac{y}{R}. \qquad (1.2\text{-}8)$$

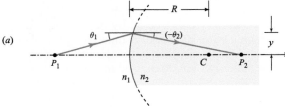

Figure 1.2-13 Refraction at a convex spherical boundary ($R > 0$).

- All paraxial rays originating from a point $P_1 = (y_1, z_1)$ in the $z = z_1$ plane meet at a point $P_2 = (y_2, z_2)$ in the $z = z_2$ plane (see Exercise 1.2-2), where

$$\frac{n_1}{z_1} + \frac{n_2}{z_2} \approx \frac{n_2 - n_1}{R} \tag{1.2-9}$$

and

$$y_2 = -\frac{n_1}{n_2}\frac{z_2}{z_1} y_1. \tag{1.2-10}$$

The $z = z_1$ and $z = z_2$ planes are said to be conjugate planes. Every point in the first plane has a corresponding point (image) in the second with magnification $-(n_1/n_2)(z_2/z_1)$. Again, negative magnification means that the image is inverted. By convention P_1 is measured in a coordinate system pointing to the left and P_2 in a coordinate system pointing to the right (e.g., if P_2 lies to the left of the boundary, then z_2 would be negative).

The similarities between these properties and those of the spherical mirror are evident. It is important to remember that the image formation properties described above are approximate. They hold only for paraxial rays. Rays of large angles do not obey these paraxial laws; the deviation results in image distortion called **aberration**.

EXERCISE 1.2-2

Image Formation. Derive (1.2-8). Prove that paraxial rays originating from P_1 pass through P_2 when (1.2-9) and (1.2-10) are satisfied.

EXERCISE 1.2-3

Aberration-Free Imaging Surface. Determine the equation of a convex aspherical (nonspherical) surface between media of refractive indices n_1 and n_2 such that all rays (not necessarily paraxial) from an axial point P_1 at a distance z_1 to the left of the surface are imaged onto an axial point P_2 at a distance z_2 to the right of the surface [Fig. 1.2-13(a)]. *Hint:* In accordance with Fermat's principle the optical pathlengths between the two points must be equal for all paths.

EXAMPLE 1.2-1. *Collimator for LED Light.* Light emitted by an LED (Sec. 18.1) is often collimated by making use of an optic whose surface takes the form of a paraboloid of revolution (Fig. 1.2-14). The LED is placed at the focus of the paraboloid by inserting its hemispherical dome (darker blue) into a recess formed in the narrow end of the optic. Rays emanating from the sides of the LED dome impinge on the paraboloidal boundary at angles of incidence greater than the critical angle and are thus reflected out of the device via total internal reflection. Rays emanating from the central portion of the LED dome are refracted out of the device at the spherical boundary. Optical systems that combine reflection and refraction are known as **catadioptric systems**.

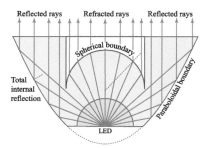

Figure 1.2-14 Cross section of a collimator for LED light. LED collimators come in many configurations but most make use of both total internal reflection and refraction to provide rays of light that are approximately parallel at the exit. Such devices are often fabricated from molded acrylic or polycarbonate plastic, which have refractive indices similar to that of glass ($n \approx 1.5$). The diameter of the narrow end of the optic illustrated is ≈ 1 cm.

Spherical Lenses

A **spherical lens** is bounded by two spherical surfaces. It is, therefore, defined completely by the radii R_1 and R_2 of its two surfaces, its thickness Δ, and the refractive index n of the material (Fig. 1.2-15). A glass lens in air can be regarded as a combination of two spherical boundaries, air-to-glass and glass-to-air.

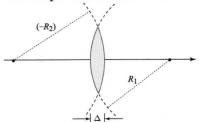

Figure 1.2-15 A biconvex spherical lens.

A ray crossing the first surface at height y and angle θ_1 with the z axis [Fig. 1.2-16(a)] is traced by applying (1.2-8) at the first surface to obtain the inclination angle θ of the refracted ray, which we extend until it meets the second surface. We then use (1.2-8) once more with θ replacing θ_1 to obtain the inclination angle θ_2 of the ray after refraction from the second surface. The results are in general complex. When the lens is thin, however, it can be assumed that the incident ray emerges from the lens at about the same height y at which it enters. Under this assumption, the following relations obtain:

- The angles of the refracted and incident rays are related by (see Exercise 1.2-4)

$$\theta_2 = \theta_1 - \frac{y}{f}, \qquad (1.2\text{-}11)$$

where f, called the **focal length**, is given by

$$\boxed{\frac{1}{f} = (n-1)\left(\frac{1}{R_1} - \frac{1}{R_2}\right).} \qquad (1.2\text{-}12)$$

Focal Length
Thin Spherical Lens

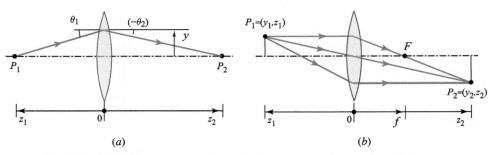

Figure 1.2-16 (a) Ray bending by a thin lens. (b) Image formation by a thin lens.

- All rays originating from a point $P_1 = (y_1, z_1)$ meet at a point $P_2 = (y_2, z_2)$ [Fig. 1.2-16(b)] (see Exercise 1.2-4), where

$$\boxed{\frac{1}{z_1} + \frac{1}{z_2} = \frac{1}{f}} \qquad (1.2\text{-}13)$$

Imaging Equation

and

$$y_2 = -\frac{z_2}{z_1} y_1.$$

(1.2-14)
Magnification

These results are identical to those for the spherical mirror [see (1.2-4) and Exercise 1.2-1].

These equations indicate that each point in the $z = z_1$ plane is imaged onto a corresponding point in the $z = z_2$ plane with the magnification factor $-z_2/z_1$. The magnification is unity when $z_1 = z_2 = 2f$. The focal length f of a lens therefore completely determines its effect on paraxial rays. As indicated earlier, P_1 and P_2 are measured in coordinate systems pointing to the left and right, respectively, and the radii of curvatures R_1 and R_2 are positive for convex surfaces and negative for concave surfaces. For the biconvex lens shown in Fig. 1.2-15, R_1 is positive and R_2 is negative, so that the two terms of (1.2-12) add and provide a positive f.

EXERCISE 1.2-4

Proof of the Thin Lens Formulas. Using (1.2-8), along with the definition of the focal length given in (1.2-12), prove (1.2-11) and (1.2-13).

It is emphasized once again that the foregoing relations hold only for paraxial rays. The presence of nonparaxial rays results in aberrations, as illustrated in Fig. 1.2-17.

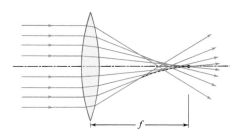

Figure 1.2-17 Nonparaxial rays do not meet at the paraxial focus. The dashed envelope of the refracted rays is known as the caustic curve.

Convex and Concave Lenses

Lenses are transparent optical devices that bend rays in a manner prescribed by the shapes of their surfaces. Most common lenses, such as the biconvex lens considered above, are spherical lenses. Lenses that consist of a single piece of material (glass and plastic are favored in the visible) are called simple lenses, while lenses that comprise multiple simple lenses, usually along a common axis, are known as compound lenses.

The surface of a lens can be convex or concave, depending on whether it projects out of, or recedes into the body of the lens, respectively, or it can be planar, indicating that it has a flat surface. A **cylindrical lens** is curved in only one direction; it thus has a focal length f for rays in the y–z plane, and no focusing power for rays in the x–z plane. A lens in which one surface is convex and the other concave is called a meniscus lens (these are often used for spectacles). A lens in which one or both surfaces have a shape that is neither spherical nor cylindrical is known as an **aspheric lens**.

Several different types of lenses are illustrated in Fig. 1.2-18. Biconvex and plano-convex lenses result in the convergence of rays and are useful for image formation, as depicted in Fig. 1.2-16. Biconcave and plano-concave lenses lead to the divergence of rays and are used in projection and focal-length expansion. A **Fresnel lens** is constructed by removing the nonrefracting portions of a conventional lens. Hence, the Fresnel-lens equivalent [Fig. 1.2-18(e)] of a plano-convex lens [Fig. 1.2-18(b)] is a flattened set of concentric surfaces with identical curvature at all locations on the surface (except at the stepwise discontinuities). The Fresnel design allows for the construction of thin, light, and inexpensive plastic lenses with sizes that range from meters to micrometers and short focal lengths. Fresnel lenses can be converging, diverging, or cylindrical.

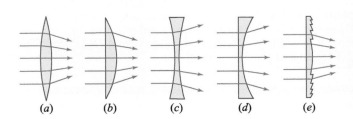

Figure 1.2-18 Lenses: (a) Biconvex; (b) Plano-convex; (c) Concave; (d) Plano-concave. (e) Fresnel-lens counterpart of the plano-convex lens displayed in (b); the curvatures are the same everywhere on the two surfaces.

D. Light Guides

Light may be guided from one location to another by use of a set of lenses or mirrors, as illustrated schematically in Fig. 1.2-19. Since refractive elements (such as lenses) are usually partially reflective and since mirrors are partially absorptive, the cumulative loss of optical power will be significant when the number of guiding elements is large. Components in which these effects are minimized can be fabricated (e.g., antireflection-coated lenses), but the system is generally cumbersome and costly.

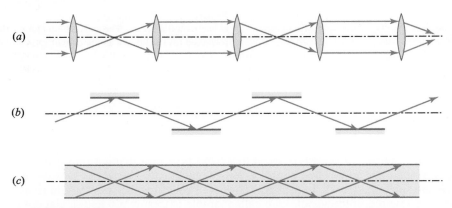

Figure 1.2-19 Guiding light: (a) lenses; (b) mirrors; (c) total internal reflection.

An ideal mechanism for guiding light is that of total internal reflection at the boundary between two media of different refractive indices. Rays are reflected repeatedly without undergoing refraction. Glass fibers of high chemical purity are used to guide light for tens of kilometers with relatively low loss of optical power.

An optical fiber is a light conduit made of two concentric glass (or plastic) cylinders (Fig. 1.2-20). The inner, called the core, has a refractive index n_1, and the outer, called the cladding, has a slightly smaller refractive index, $n_2 < n_1$. Light rays traveling in the core are totally reflected from the cladding if their angle of incidence is greater than the critical angle, $\bar{\theta} > \theta_c = \sin^{-1}(n_2/n_1)$. The rays making an angle $\theta = 90° - \bar{\theta}$ with the optical axis are therefore confined in the fiber core if $\theta < \overline{\theta_c}$, where $\overline{\theta_c} = 90° - \theta_c = \cos^{-1}(n_2/n_1)$. Optical fibers are used in optical communication systems (see Chapters 10 and 25). Some important properties of optical fibers are derived in Exercise 1.2-5.

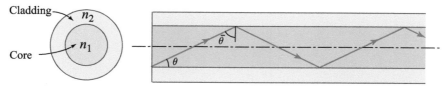

Figure 1.2-20 The optical fiber. Light rays are guided by multiple total internal reflections. Here θ represents the angle measured from the axis of the optical fiber so that its complement $\bar{\theta} = 90° - \theta$ is the angle of incidence at the dielectric interface.

EXERCISE 1.2-5

Numerical Aperture and Angle of Acceptance of an Optical Fiber. An optical fiber is illuminated by light from a source (e.g., a light-emitting diode, LED). The refractive indices of the core and cladding of the fiber are n_1 and n_2, respectively, and the refractive index of air is 1 (see Fig. 1.2-21). Show that the half-angle θ_a of the cone of rays accepted by the fiber (transmitted through the fiber without undergoing refraction at the cladding) is given by

$$\mathrm{NA} = \sin\theta_a = \sqrt{n_1^2 - n_2^2}. \qquad (1.2\text{-}15)$$

Numerical Aperture
Optical Fiber

The angle θ_a is called the **acceptance angle** and the parameter $\mathrm{NA} \equiv \sin\theta_a$ is known as the **numerical aperture** of the fiber. Calculate the numerical aperture and acceptance angle for a silica-glass fiber with $n_1 = 1.475$ and $n_2 = 1.460$. Silica glass, also known as fused silica, is amorphous silicon dioxide (SiO_2). It is widely used because of its excellent optical and mechanical properties. Moreover, its refractive index can be readily modified by doping (e.g., with GeO_2).

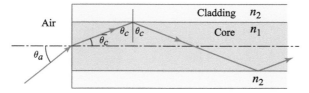

Figure 1.2-21 Acceptance angle of an optical fiber.

Trapping of Light in Media of High Refractive Index

It is often difficult for light originating inside a medium of large refractive index to be extracted into air, especially if the surfaces of the medium are parallel. This occurs since certain rays undergo multiple total internal reflections without ever refracting into air. The principle is illustrated in Exercise 1.2-6.

EXERCISE 1.2-6
Light Trapped in a Light-Emitting Diode.

(a) Assume that light is generated in all directions inside a material of refractive index n cut in the shape of a parallelepiped (Fig. 1.2-22). The material is surrounded by air with unity refractive index. This process occurs in light-emitting diodes (see Sec. 18.1B). What is the angle of the cone of light rays (inside the material) that will emerge from each face? What happens to the other rays? What is the numerical value of this angle for GaAs ($n = 3.6$)?

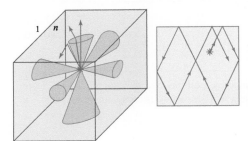

Figure 1.2-22 Trapping of light in a parallelepiped of high refractive index.

(b) Assume that when light is generated isotropically the amount of optical power associated with the rays in a given cone is proportional to the solid angle of the cone. Show that the ratio of the optical power that is extracted from the material to the total generated optical power is $3\left(1 - \sqrt{1 - 1/n^2}\right)$, provided that $n > \sqrt{2}$. What is the numerical value of this ratio for GaAs?

1.3 GRADED-INDEX OPTICS

A graded-index (GRIN) material has a refractive index that varies with position in accordance with a continuous function $n(\mathbf{r})$. These materials are often fabricated by adding impurities (dopants) of controlled concentrations. In a GRIN medium the optical rays follow curved trajectories, instead of straight lines. By appropriate choice of $n(\mathbf{r})$, a GRIN plate can have the same effect on light rays as a conventional optical component, such as a prism or lens.

A. The Ray Equation

To determine the trajectories of light rays in an inhomogeneous medium with refractive index $n(\mathbf{r})$, we use Fermat's principle,

$$\delta \int_A^B n(\mathbf{r})\, ds = 0, \qquad (1.3\text{-}1)$$

where ds is a differential length along the ray trajectory between A and B. If the trajectory is described by the function $x(s)$, $y(s)$, and $z(s)$, where s is the length of the trajectory (Fig. 1.3-1), then using the calculus of variations it can be shown that[†] $x(s)$,

---

[†] See, e.g., R. Weinstock, *Calculus of Variations: With Applications to Physics and Engineering*, 1952; Dover, 1974.

$y(s)$, and $z(s)$ must satisfy three partial differential equations,

$$\frac{d}{ds}\left(n\frac{dx}{ds}\right) = \frac{\partial n}{\partial x}, \qquad \frac{d}{ds}\left(n\frac{dy}{ds}\right) = \frac{\partial n}{\partial y}, \qquad \frac{d}{ds}\left(n\frac{dz}{ds}\right) = \frac{\partial n}{\partial z}. \qquad (1.3\text{-}2)$$

By defining the vector $\mathbf{r}(s)$, whose components are $x(s)$, $y(s)$, and $z(s)$, (1.3-2) may be written in the compact vector form

$$\boxed{\frac{d}{ds}\left(n\frac{d\mathbf{r}}{ds}\right) = \nabla n,} \qquad (1.3\text{-}3)$$
Ray Equation

where ∇n, the gradient of n, is a vector with Cartesian components $\partial n/\partial x$, $\partial n/\partial y$, and $\partial n/\partial z$. Equation (1.3-3) is known as the **ray equation**.

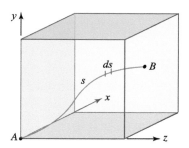

Figure 1.3-1 The ray trajectory is described parametrically by three functions $x(s)$, $y(s)$, and $z(s)$, or by two functions $x(z)$ and $y(z)$.

One approach to solving the ray equation is to describe the trajectory by two functions $x(z)$ and $y(z)$, write $ds = dz\sqrt{1 + (dx/dz)^2 + (dy/dz)^2}$, and substitute in (1.3-3) to obtain two partial differential equations for $x(z)$ and $y(z)$. The algebra is generally not trivial, but it simplifies considerably when the paraxial approximation is used.

The Paraxial Ray Equation

In the paraxial approximation, the trajectory is almost parallel to the z axis, so that $ds \approx dz$ (Fig. 1.3-2). The ray equations (1.3-2) then simplify to

$$\boxed{\frac{d}{dz}\left(n\frac{dx}{dz}\right) \approx \frac{\partial n}{\partial x}, \qquad \frac{d}{dz}\left(n\frac{dy}{dz}\right) \approx \frac{\partial n}{\partial y}.} \qquad (1.3\text{-}4)$$
Paraxial Ray Equations

Given $n = n(x, y, z)$, these two partial differential equations may be solved for the trajectory $x(z)$ and $y(z)$.

In the limiting case of a homogeneous medium for which n is independent of x, y, z, (1.3-4) gives $d^2x/dz^2 = 0$ and $d^2y/dz^2 = 0$, from which it follows that x and y are linear functions of z, so that the trajectories are straight lines. More interesting cases will be examined subsequently.

22   CHAPTER 1   RAY OPTICS

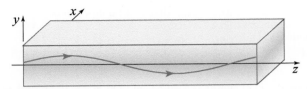

Figure 1.3-2 Trajectory of a paraxial ray in a graded-index medium.

B. Graded-Index Optical Components

Graded-Index Slab

Consider a slab of material whose refractive index $n = n(y)$ is uniform in the x and z directions but varies continuously in the y direction (Fig. 1.3-3). The trajectories of paraxial rays in the y–z plane are described by the paraxial ray equation

$$\frac{d}{dz}\left(n\frac{dy}{dz}\right) = \frac{dn}{dy}, \tag{1.3-5}$$

from which

$$\frac{d^2y}{dz^2} = \frac{1}{n(y)}\frac{dn(y)}{dy}. \tag{1.3-6}$$

Given $n(y)$ and initial conditions (y and dy/dz at $z = 0$), (1.3-6) can be solved for the function $y(z)$, which describes the ray trajectories.

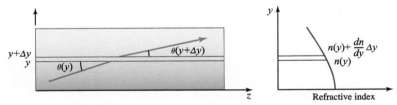

Figure 1.3-3 Refraction in a graded-index slab.

☐   **Derivation of the Paraxial Ray Equation in a Graded-Index Slab Using Snell's Law.**
Equation (1.3-6) may also be derived by the direct use of Snell's law (Fig. 1.3-3). Let $\theta(y) \approx dy/dz$ be the angle that the ray makes with the z axis at the position (y, z). After traveling through a layer of thickness Δy the ray changes its angle to $\theta(y + \Delta y)$. The two angles are related by Snell's law where θ, as defined in Fig. 1.3-3, is the complement of the angle of incidence (refraction):

$$n(y)\cos\theta(y) = n(y + \Delta y)\cos\theta(y + \Delta y)$$
$$= \left[n(y) + \frac{dn}{dy}\Delta y\right]\left[\cos\theta(y) - \frac{d\theta}{dy}\Delta y \sin\theta(y)\right], \tag{1.3-7}$$

where we have applied the expansion $f(y+\Delta y) = f(y) + (df/dy)\Delta y$ to the functions $f(y) = n(y)$ and $f(y) = \cos\theta(y)$. In the limit $\Delta y \to 0$, after eliminating the term in $(\Delta y)^2$, we obtain the differential equation

$$\frac{dn}{dy} = n\frac{d\theta}{dy}\tan\theta. \tag{1.3-8}$$

For paraxial rays θ is very small so that $\tan\theta \approx \theta$. Substituting $\theta = dy/dz$ in (1.3-8), we obtain (1.3-6).   ∎

EXAMPLE 1.3-1.   *Slab with Parabolic Index Profile.*   An important particular distribution for the graded refractive index is

$$n^2(y) = n_0^2 \left(1 - \alpha^2 y^2\right). \tag{1.3-9}$$

This is a symmetric function of y that has its maximum value at $y = 0$ (Fig. 1.3-4). A glass slab with this profile is known by the trade name SELFOC. Usually, α is chosen to be sufficiently small so that $\alpha^2 y^2 \ll 1$ for all y of interest. Under this condition, $n(y) = n_0\sqrt{1 - \alpha^2 y^2} \approx n_0(1 - \tfrac{1}{2}\alpha^2 y^2)$; i.e., $n(y)$ is a parabolic distribution. Also, because $n(y) - n_0 \ll n_0$, the fractional change of the refractive index is very small. Taking the derivative of (1.3-9), the right-hand side of (1.3-6) is $(1/n)dn/dy = -(n_0/n)^2 \alpha^2 y \approx -\alpha^2 y$, so that (1.3-6) becomes

$$\frac{d^2 y}{dz^2} \approx -\alpha^2 y. \tag{1.3-10}$$

The solutions of this equation are harmonic functions with period $2\pi/\alpha$. Assuming an initial position $y(0) = y_0$ and an initial slope $dy/dz = \theta_0$ at $z = 0$ inside the GRIN medium,

$$y(z) = y_0 \cos \alpha z + \frac{\theta_0}{\alpha} \sin \alpha z, \tag{1.3-11}$$

from which the slope of the trajectory is

$$\theta(z) = \frac{dy}{dz} = -y_0 \alpha \sin \alpha z + \theta_0 \cos \alpha z. \tag{1.3-12}$$

The ray oscillates about the center of the slab with a period (distance) $2\pi/\alpha$ known as the **pitch**, as illustrated in Fig. 1.3-4.

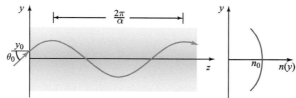

Figure 1.3-4   Trajectory of a ray in a GRIN slab of parabolic index profile (SELFOC).

The maximum excursion of the ray is $y_{\max} = \sqrt{y_0^2 + (\theta_0/\alpha)^2}$ and the maximum angle is $\theta_{\max} = \alpha y_{\max}$. The validity of this approximate analysis is ensured if $\theta_{\max} \ll 1$. If $2y_{\max}$ is smaller than the thickness of the slab, the ray remains confined and the slab serves as a light guide. Figure 1.3-5 shows the trajectories of a number of rays transmitted through a SELFOC slab. Note that all rays have the same pitch. This GRIN slab may be used as a lens, as demonstrated in Exercise 1.3-1.

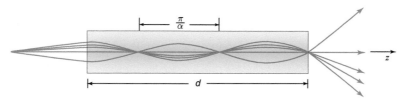

Figure 1.3-5   Trajectories of rays from an external point source in a SELFOC slab.

EXERCISE 1.3-1

The GRIN Slab as a Lens. Show that a SELFOC slab of length $d < \pi/2\alpha$ and refractive index given by (1.3-9) acts as a cylindrical lens (a lens with focusing power in the y–z plane) of focal length

$$f \approx \frac{1}{n_0 \alpha \sin(\alpha d)}. \qquad (1.3\text{-}13)$$

Show that the principal point (defined in Fig. 1.3-6) lies at a distance from the slab edge $\overline{AH} \approx (1/n_0\alpha)\tan(\alpha d/2)$. Sketch the ray trajectories in the special cases $d = \pi/\alpha$ and $\pi/2\alpha$.

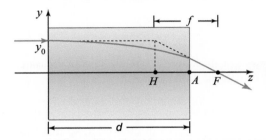

Figure 1.3-6 The SELFOC slab used as a lens; F is the focal point and H is the principal point.

Graded-Index Fibers

A graded-index fiber is a glass cylinder with a refractive index n that varies as a function of the radial distance from its axis. In the paraxial approximation, the ray trajectories are governed by the paraxial ray equations (1.3-4). Consider, for example, the distribution

$$n^2 = n_0^2 \left[1 - \alpha^2 \left(x^2 + y^2\right)\right]. \qquad (1.3\text{-}14)$$

Substituting (1.3-14) into (1.3-4) and assuming that $\alpha^2(x^2 + y^2) \ll 1$ for all x and y of interest, we obtain

$$\frac{d^2x}{dz^2} \approx -\alpha^2 x, \qquad \frac{d^2y}{dz^2} \approx -\alpha^2 y. \qquad (1.3\text{-}15)$$

Both x and y are therefore harmonic functions of z with period $2\pi/\alpha$. The initial positions (x_0, y_0) and angles $(\theta_{x0} = dx/dz$ and $\theta_{y0} = dy/dz)$ at $z = 0$ determine the amplitudes and phases of these harmonic functions. Because of the circular symmetry, there is no loss of generality in choosing $x_0 = 0$. The solution of (1.3-15) is then

$$\begin{aligned}x(z) &= \frac{\theta_{x0}}{\alpha}\sin\alpha z \\ y(z) &= \frac{\theta_{y0}}{\alpha}\sin\alpha z + y_0\cos\alpha z.\end{aligned} \qquad (1.3\text{-}16)$$

If $\theta_{x0} = 0$, i.e., the incident ray lies in a meridional plane (a plane passing through the axis of the cylinder, in this case the y–z plane), the ray continues to lie in that plane following a sinusoidal trajectory similar to that in the GRIN slab [Fig. 1.3-7(a)].

On the other hand, if $\theta_{y0} = 0$, and $\theta_{x0} = \alpha y_0$, then

$$\begin{aligned}x(z) &= y_0 \sin\alpha z \\ y(z) &= y_0 \cos\alpha z,\end{aligned} \qquad (1.3\text{-}17)$$

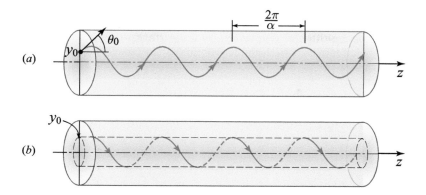

Figure 1.3-7 (a) Meridional and (b) helical rays in a graded-index fiber with parabolic index profile.

so that the ray follows a helical trajectory lying on the surface of a cylinder of radius y_0 [Fig. 1.3-7(b)]. In both cases the ray remains confined within the fiber, so that the fiber serves as a light guide. Other helical patterns are generated with different incident rays.

Graded-index fibers and their use in optical fiber communications are discussed in Chapters 10 and 25.

EXERCISE 1.3-2

Numerical Aperture of the Graded-Index Fiber. Consider a graded-index fiber with the index profile provided in (1.3-14) and radius a. A ray is incident from air into the fiber at its center, which then makes an angle θ_0 with the fiber axis in the medium (see Fig. 1.3-8). Show, in the paraxial approximation, that the numerical aperture is

$$\mathrm{NA} \equiv \sin\theta_a \approx n_0 a\alpha, \qquad (1.3\text{-}18)$$

Numerical Aperture
Graded-Index Fiber

where θ_a is the maximum acceptance angle for which the ray trajectory is confined within the fiber. Compare this to the numerical aperture of a step-index fiber such as the one discussed in Exercise 1.2-5. To make the comparison fair, take the refractive indices of the core and cladding of the step-index fiber to be $n_1 = n_0$ and $n_2 = n_0\sqrt{1 - \alpha^2 a^2} \approx n_0(1 - \frac{1}{2}\alpha^2 a^2)$, respectively.

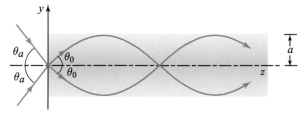

Figure 1.3-8 Acceptance angle of a graded-index optical fiber.

*C. The Eikonal Equation

The ray trajectories are often characterized by the surfaces to which they are normal. Let $S(\mathbf{r})$ be a scalar function such that its equilevel surfaces, $S(\mathbf{r}) = $ constant, are everywhere normal to the rays (Fig. 1.3-9). If $S(\mathbf{r})$ is known, the ray trajectories can readily be constructed since the normal to the equilevel surfaces at a position $\mathbf{r}$ is in the direction of the gradient vector $\nabla S(\mathbf{r})$. The function $S(\mathbf{r})$, called the **eikonal**, is akin to the potential function $V(\mathbf{r})$ in electrostatics; the role of the optical rays is played by the lines of electric field $\mathbf{E} = -\nabla V$.

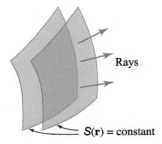

Figure 1.3-9 Ray trajectories are normal to the surfaces of constant $S(\mathbf{r})$.

To satisfy Fermat's principle (which is the main postulate of ray optics) the eikonal $S(\mathbf{r})$ must satisfy a partial differential equation known as the **eikonal equation**,

$$\left(\frac{\partial S}{\partial x}\right)^2 + \left(\frac{\partial S}{\partial y}\right)^2 + \left(\frac{\partial S}{\partial z}\right)^2 = n^2, \quad (1.3\text{-}19)$$

which is usually written in the vector form

$$\boxed{|\nabla S|^2 = n^2,} \quad (1.3\text{-}20)$$
Eikonal Equation

where $|\nabla S^2| = \nabla S \cdot \nabla S$. The proof of the eikonal equation from Fermat's principle is a mathematical exercise that lies beyond the scope of this book.[†] Conversely, Fermat's principle (and the ray equation) can be shown to follow from the eikonal equation. Therefore, either Fermat's principle or the eikonal equation may be regarded as the principal postulate of ray optics.

Integrating the eikonal equation (1.3-20) along a ray trajectory between points A and B gives

$$S(\mathbf{r}_B) - S(\mathbf{r}_A) = \int_A^B |\nabla S|\, ds = \int_A^B n\, ds = \text{optical pathlength between } A \text{ and } B.$$
$$(1.3\text{-}21)$$

This means that the difference $S(\mathbf{r}_B) - S(\mathbf{r}_A)$ represents the optical pathlength between A and B. In the electrostatics analogy, the optical pathlength plays the role of the potential difference.

To determine the ray trajectories in an inhomogeneous medium of refractive index $n(\mathbf{r})$, we can either solve the ray equation (1.3-3), as we have done earlier, or solve the eikonal equation for $S(\mathbf{r})$, from which we calculate the gradient ∇S.

[†] See, e.g., M. Born and E. Wolf, *Principles of Optics*, Cambridge University Press, 7th expanded and corrected ed. 2002.

If the medium is homogeneous, i.e., $n(\mathbf{r})$ is constant, the magnitude of ∇S is constant, so that the wavefront normals (rays) must be straight lines. The surfaces $S(\mathbf{r}) =$ constant may be parallel planes or concentric spheres, as illustrated in Fig. 1.3-10.

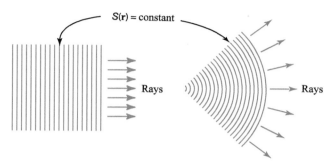

Figure 1.3-10   Rays and surfaces of constant $S(\mathbf{r})$ in a homogeneous medium.

The eikonal equation is revisited from the point-of-view of the relation between ray optics and wave optics in Sec. 2.3.

1.4 MATRIX OPTICS

Matrix optics is a technique for tracing paraxial rays. The rays are assumed to travel only within a single plane, so that the formalism is applicable to systems with planar geometry and to meridional rays in circularly symmetric systems.

A ray is described by its position and its angle with respect to the optical axis. These variables are altered as the ray travels through the system. In the paraxial approximation, the position and angle at the input and output planes of an optical system are related by two *linear* algebraic equations. As a result, the optical system is described by a 2×2 matrix called the ray-transfer matrix.

The convenience of using matrix methods lies in the fact that the ray-transfer matrix of a cascade of optical components (or systems) is a product of the ray-transfer matrices of the individual components (or systems). Matrix optics therefore provides a formal mechanism for describing complex optical systems in the paraxial approximation.

A. The Ray-Transfer Matrix

Consider a circularly symmetric optical system formed by a succession of refracting and reflecting surfaces all centered about the same axis (optical axis). The z axis lies along the optical axis and points in the general direction in which the rays travel. Consider rays in a plane containing the optical axes, say the y–z plane. We proceed to trace a ray as it travels through the system, i.e., as it crosses the transverse planes at different axial distances. A ray crossing the transverse plane at z is completely characterized by the coordinate of y of its crossing point and the angle θ (Fig. 1.4-1).

An optical system is a set of optical components placed between two transverse planes at z_1 and z_2, referred to as the input and output planes, respectively. The system is characterized completely by its effect on an incoming ray of arbitrary position and direction (y_1, θ_1). It steers the ray so that it has new position and direction (y_2, θ_2) at the output plane (Fig. 1.4-2).

28   CHAPTER 1   RAY OPTICS

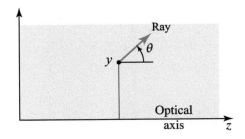

Figure 1.4-1   A ray is characterized by its coordinate y and its angle θ.

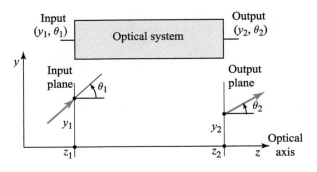

Figure 1.4-2   A ray enters an optical system at location z_1 with position y_1 and angle θ_1 and leaves at position y_2 and angle θ_2.

In the paraxial approximation, when all angles are sufficiently small so that $\sin\theta \approx \theta$, the relation between (y_2, θ_2) and (y_1, θ_1) is linear and can generally be written in the form

$$y_2 = Ay_1 + B\theta_1 \tag{1.4-1}$$

$$\theta_2 = Cy_1 + D\theta_1, \tag{1.4-2}$$

where A, B, C, and D are real numbers. Equations (1.4-1) and (1.4-2) may be conveniently written in matrix form as

$$\begin{bmatrix} y_2 \\ \theta_2 \end{bmatrix} = \begin{bmatrix} A & B \\ C & D \end{bmatrix} \begin{bmatrix} y_1 \\ \theta_1 \end{bmatrix}. \tag{1.4-3}$$

The matrix **M**, whose elements are A, B, C, and D, characterizes the optical system completely since it permits (y_2, θ_2) to be determined for any (y_1, θ_1). It is known as the **ray-transfer matrix**. As will be seen in the examples provided in Sec. 1.4B, angles that turn out to be negative point downward from the z axis in their direction of travel. Radii that turn out to be negative indicate concave surfaces whereas those that are positive indicate convex surfaces.

EXERCISE 1.4-1

Special Forms of the Ray-Transfer Matrix.   Consider the following situations in which one of the four elements of the ray-transfer matrix vanishes:

(a) Show that $A = 0$ represents a *focusing system*, in which all rays entering the system at a particular angle, whatever their position, leave at a single position.

(b) Show that $B = 0$ represents an *imaging system*, in which all rays entering the system at a particular position, whatever their angle, leave at a single position.

(c) What are the special features of a system for which $C = 0$ or $D = 0$?

B. Matrices of Simple Optical Components

Free-Space Propagation

Since rays travel along straight lines in a medium of uniform refractive index such as free space, a ray traversing a distance d is altered in accordance with $y_2 = y_1 + \theta_1 d$ and $\theta_2 = \theta_1$. The ray-transfer matrix is therefore

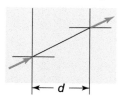

$$\mathbf{M} = \begin{bmatrix} 1 & d \\ 0 & 1 \end{bmatrix}. \quad (1.4\text{-}4)$$

Refraction at a Planar Boundary

At a planar boundary between two media of refractive indices n_1 and n_2, the ray angle changes in accordance with Snell's law $n_1 \sin\theta_1 = n_2 \sin\theta_2$. In the paraxial approximation, $n_1 \theta_1 \approx n_2 \theta_2$. The position of the ray is not altered, $y_2 = y_1$. The ray-transfer matrix is

$$\mathbf{M} = \begin{bmatrix} 1 & 0 \\ 0 & \frac{n_1}{n_2} \end{bmatrix}. \quad (1.4\text{-}5)$$

Refraction at a Spherical Boundary

The relation between θ_1 and θ_2 for paraxial rays refracted at a spherical boundary between two media is provided in (1.2-8). The ray height is not altered, $y_2 \approx y_1$. The ray-transfer matrix is

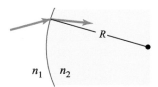

Convex: $R > 0$; concave: $R < 0$

$$\mathbf{M} = \begin{bmatrix} 1 & 0 \\ -\frac{(n_2-n_1)}{n_2 R} & \frac{n_1}{n_2} \end{bmatrix}. \quad (1.4\text{-}6)$$

Transmission Through a Thin Lens

The relation between θ_1 and θ_2 for paraxial rays transmitted through a thin lens of focal length f is given in (1.2-11). Since the height remains unchanged $(y_2 = y_1)$, we have

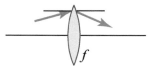

Convex: $f > 0$; concave: $f < 0$

$$\mathbf{M} = \begin{bmatrix} 1 & 0 \\ -\frac{1}{f} & 1 \end{bmatrix}. \quad (1.4\text{-}7)$$

Reflection from a Planar Mirror

Upon reflection from a planar mirror, the ray position is not altered, $y_2 = y_1$. Adopting the convention that the z axis points in the general direction of travel of the rays, i.e., toward the mirror for the incident rays and away from it for the reflected rays, we conclude that $\theta_2 = \theta_1$. The ray-transfer matrix is therefore the identity matrix

$$\mathbf{M} = \begin{bmatrix} 1 & 0 \\ 0 & 1 \end{bmatrix}. \quad (1.4\text{-}8)$$

Reflection from a Spherical Mirror
Using (1.2-1), and the convention that the z axis follows the general direction of the rays as they reflect from mirrors, we similarly obtain

$$\mathbf{M} = \begin{bmatrix} 1 & 0 \\ \frac{2}{R} & 1 \end{bmatrix}. \quad (1.4\text{-}9)$$

Concave: $R < 0$; convex: $R > 0$

Note the similarity between the ray-transfer matrices of a spherical mirror (1.4-9) and a thin lens (1.4-7). A mirror with radius of curvature R bends rays in a manner that is identical to that of a thin lens with focal length $f = -R/2$.

C. Matrices of Cascaded Optical Components
A cascade of N optical components or systems whose ray-transfer matrices are $\mathbf{M}_1, \mathbf{M}_2, \ldots, \mathbf{M}_N$ is equivalent to a single optical system of ray-transfer matrix

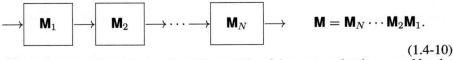

$$\mathbf{M} = \mathbf{M}_N \cdots \mathbf{M}_2 \mathbf{M}_1.$$

(1.4-10)

Note the order of matrix multiplication: The matrix of the system that is crossed by the rays is first placed to the right, so that it operates on the column matrix of the incident ray first. A sequence of matrix multiplications is not, in general, commutative, although it is associative.

EXERCISE 1.4-2

A Set of Parallel Transparent Plates. Consider a set of N parallel planar transparent plates of refractive indices $n_1, n_2, \ldots, n_N$ and thicknesses $d_1, d_2, \ldots, d_N$, placed in air ($n = 1$) normal to the z axis. Using induction, show that the ray-transfer matrix is

$$\mathbf{M} = \begin{bmatrix} 1 & \sum_{i=1}^{N} \frac{d_i}{n_i} \\ 0 & 1 \end{bmatrix}. \quad (1.4\text{-}11)$$

Note that the order in which the plates are placed does not affect the overall ray-transfer matrix. What is the ray-transfer matrix of an inhomogeneous transparent plate of thickness d_0 and refractive index $n(z)$?

EXERCISE 1.4-3

A Gap Followed by a Thin Lens. Show that the ray-transfer matrix of a distance d of free space followed by a lens of focal length f is

$$\mathbf{M} = \begin{bmatrix} 1 & d \\ -\frac{1}{f} & 1 - \frac{d}{f} \end{bmatrix}. \quad (1.4\text{-}12)$$

EXERCISE 1.4-4

Imaging with a Thin Lens. Derive an expression for the ray-transfer matrix of a system comprised of free space/thin lens/free space, as shown in Fig. 1.4-3. Show that if the imaging condition $(1/d_1 + 1/d_2 = 1/f)$ is satisfied, all rays originating from a single point in the input plane reach the output plane at the single point y_2, regardless of their angles. Also show that if $d_2 = f$, all parallel incident rays are focused by the lens onto a single point in the output plane.

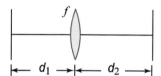

Figure 1.4-3 Single-lens imaging system.

Imaging with an Arbitrary Paraxial Optical System

A paraxial system comprising an arbitrary set of cascaded optical elements is characterized completely by the four elements A, B, C, D of its ray-transfer matrix **M**. Alternatively, the system may be characterized by the locations of four *cardinal points*: two focal points that determine the transmission of rays between its input and output planes. In accordance with (1.4-3), an incoming ray parallel to the optical axis ($\theta_1 = 0$) at height y_1 exits the system at height $y_2 = Ay_1$ and angle $\theta_2 = Cy_1$. This ray crosses the axis at a point F called the **back focal point**, which is located a distance $y_2/\theta_2 = A/C$ from the system's back vertex V, as shown in Fig. 1.4-4(a). The intersection of the extensions of the incoming and outgoing rays defines the **back principal point** H, which lies at a distance $f = y_1/\theta_2 = -1/C$ to the left of F, and is known as the **back focal length**. The back principal point H is thus located at a distance $h = -1/C + A/C$ to the left of the back vertex V. Note that the locations of the focal and principal points are independent of y_1 as long as the paraxial approximation is applicable.

Similarly, rays parallel to the axis but entering the system in the opposite direction (from right to left) are focused to the **front focal point** F' and define the **front principal point** H', which lies at a distance h' from the front vertex V'. The front focal point lies at a distance f' to the left of H', where f' is the **front focal length**. These distances may be expressed in terms of the elements of the inverse ray-transfer matrix

$$\mathbf{M}^{-1} = \begin{bmatrix} A' & B' \\ C' & D' \end{bmatrix} = \frac{1}{\det[\mathbf{M}]} \begin{bmatrix} D & -B \\ -C & DA \end{bmatrix} \quad (1.4\text{-}13)$$

via the relations $-f' = -1/C'$ and $-h' = -1/C' + A'/C'$. The determinant of **M**, denoted $\det[\mathbf{M}]$, is given by $AD - BC$.

In summary, the focal lengths and locations of the principal points may be determined from the $ABCD$ parameters by use of the following relations:

$$f = -1/C, \qquad h = (1-A)f \qquad (1.4\text{-}14)$$

$$f' = \det[\mathbf{M}]\,f, \qquad h' = -f' + Df. \qquad (1.4\text{-}15)$$

Negative signs indicate directions opposite to those denoted by the arrows in Fig. 1.4-4(a). The four distances may alternatively be established by tracing two rays, parallel to the optical axis but pointing in opposite directions, through the system. The $ABCD$ parameters may be determined from f, f', h, and h' by inverting (1.4-14) and (1.4-15).

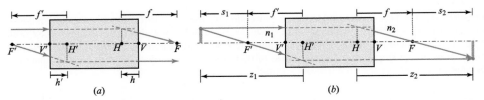

Figure 1.4-4 (a) Paraxial system representing an arbitrary set of cascaded optical elements. The designations F, V, and H represent the focal, vertex, and principal points, respectively, whereas f and h represent the focal length and distance from the principal point to the vertex, respectively. Primed quantities refer to the input plane while unprimed quantities refer to the output plane. (b) Imaging with this system. The refractive indices of the media in which the optical system is embedded are denoted n_1 and n_2, as shown.

The imaging condition is determined by considering the geometry portrayed in Fig. 1.4-4(b). Since $s_2/f = f'/s_1$, the imaging condition is simply $s_1 s_2 = f f'$, or equivalently $(z_1 - f')(z_2 - f) = f f'$, which leads to

$$\frac{f'}{z_1} + \frac{f}{z_2} = 1. \qquad (1.4\text{-}16)$$

If the refractive indices of the media within which the system is embedded are equal, then $\det[\mathbf{M}] = 1$ and, in accordance with (1.4-15), we have $f' = f$. The imaging condition in (1.4-16) then reduces to the familiar imaging equation $1/z_1 + 1/z_2 = 1/f$ [see (1.2-4)]; note, however, that here the distances z_1 and z_2 are measured from the principal points H' and H, respectively.

EXERCISE 1.4-5

Imaging with a Thick Lens. Consider a glass lens of refractive index n, thickness d, and two spherical surfaces of equal radii R. Determine the ray-transfer matrix of the lens assuming that it is placed in air (unity refractive index). Show that the back and front focal lengths are equal ($f' = f$) and that the principal points are located at equal distances from the vertices ($h' = h$), where

$$\frac{1}{f} = \frac{(n-1)}{R}\left[2 - \frac{n-1}{n}\frac{d}{R}\right] \qquad (1.4\text{-}17)$$

$$h = \frac{(n-1)fd}{nR}. \qquad (1.4\text{-}18)$$

Demonstrate that the transfer matrix of the system between two conjugate planes at distances z_1 and z_2 from the principal points of the lens (i.e., at distances $d_1 = z_1 - h'$ and $d_2 = z_2 - h$ from the vertices) that satisfies the imaging equation yields $B = 0$, indicating that it does indeed satisfy the imaging condition [see Exercise 1.4-1(b)].

D. Periodic Optical Systems

A periodic optical system is a cascade of identical unit systems. An example is a sequence of equally spaced identical relay lenses used to guide light, as shown in Fig. 1.2-19(a). Another example is the reflection of light between two mirrors that form an optical resonator (see Sec. 11.2A); in that case, the ray repeatedly traverses the same unit system (a round trip of reflections). Even a homogeneous medium, such as a glass fiber, may be considered as a periodic system if it is divided into contiguous identical segments of equal length. We proceed to formulate a general theory of ray propagation in periodic optical systems using matrix methods.

Difference Equation for the Ray Position

A periodic system is composed of a cascade of identical unit systems (stages), each with a ray-transfer matrix (A, B, C, D), as shown in Fig. 1.4-5. A ray enters the system with initial position y_0 and slope θ_0. To determine the position and slope (y_m, θ_m) of the ray at the exit of the mth stage, we apply the $ABCD$ matrix m times,

$$\begin{bmatrix} y_m \\ \theta_m \end{bmatrix} = \begin{bmatrix} A & B \\ C & D \end{bmatrix}^m \begin{bmatrix} y_0 \\ \theta_0 \end{bmatrix}. \tag{1.4-19}$$

We can also iteratively apply the relations

$$y_{m+1} = Ay_m + B\theta_m \tag{1.4-20}$$

$$\theta_{m+1} = Cy_m + D\theta_m \tag{1.4-21}$$

to determine (y_1, θ_1) from (y_0, θ_0), then (y_2, θ_2) from (y_1, θ_1), and so on, using a software routine.

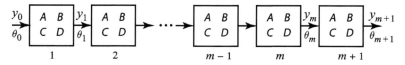

Figure 1.4-5 A cascade of identical optical systems.

It is of interest to derive equations that govern the dynamics of the position y_m, $m = 0, 1, \ldots$, irrespective of the angle θ_m. This is achieved by eliminating θ_m from (1.4-20) and (1.4-21). From (1.4-20)

$$\theta_m = \frac{y_{m+1} - Ay_m}{B}. \tag{1.4-22}$$

Replacing m with $m+1$ in (1.4-22) yields

$$\theta_{m+1} = \frac{y_{m+2} - Ay_{m+1}}{B}. \tag{1.4-23}$$

Substituting (1.4-22) and (1.4-23) into (1.4-21) gives

$$y_{m+2} = 2by_{m+1} - F^2 y_m, \tag{1.4-24}$$

Recurrence Relation for Ray Position

where
$$b = \frac{A + D}{2} \tag{1.4-25}$$

$$F^2 = AD - BC = \det[\mathbf{M}], \tag{1.4-26}$$

and $\det[\mathbf{M}]$ is the determinant of $\mathbf{M}$.

Equation (1.4-24) is a linear difference equation governing the ray position y_m. It can be solved iteratively by computing y_2 from y_0 and y_1, then y_3 from y_1 and y_2, and so on. The quantity y_1 may be computed from y_0 and θ_0 by use of (1.4-20) with $m = 0$.

It is useful, however, to derive an explicit expression for y_m by solving the difference equation (1.4-24). As with linear differential equations, a solution satisfying a linear difference equation and the initial conditions is a unique solution. It is therefore appropriate to make a judicious guess for the solution of (1.4-24). We use a trial solution of the geometric form

$$y_m = y_0 h^m, \tag{1.4-27}$$

where h is a constant. Substituting (1.4-27) into (1.4-24) immediately shows that the trial solution is suitable provided that h satisfies the quadratic algebraic equation

$$h^2 - 2bh + F^2 = 0, \tag{1.4-28}$$

from which

$$h = b \pm j\sqrt{F^2 - b^2}. \tag{1.4-29}$$

The results can be presented in a more compact form by defining the variable

$$\varphi = \cos^{-1}(b/F), \tag{1.4-30}$$

so that $b = F\cos\varphi$, $\sqrt{F^2 - b^2} = F\sin\varphi$, and therefore $h = F(\cos\varphi \pm j\sin\varphi) = F\exp(\pm j\varphi)$, whereupon (1.4-27) becomes $y_m = y_0 F^m \exp(\pm jm\varphi)$.

A general solution may be constructed from the two solutions with positive and negative signs by forming their linear combination. The sum of the two exponential functions can always be written as a harmonic (circular) function, so that

$$y_m = y_{\max} F^m \sin(m\varphi + \varphi_0), \tag{1.4-31}$$

where $y_{\max}$ and φ_0 are constants to be determined from the initial conditions y_0 and y_1. In particular, setting $m = 0$ we obtain $y_{\max} = y_0/\sin\varphi_0$.

The parameter F is related to the determinant of the ray-transfer matrix of the unit system by $F = \sqrt{\det[\mathbf{M}]}$. It can be shown that regardless of the unit system, $\det[\mathbf{M}] = n_1/n_2$, where n_1 and n_2 are the refractive indices of the initial and final sections of the unit system. This general result is easily verified for the ray-transfer matrices of all the optical components considered in this section. Since the determinant of a product of two matrices is the product of their determinants, it follows that the relation $\det[\mathbf{M}] = n_1/n_2$ is applicable to any cascade of these optical components. For example, if $\det[\mathbf{M}_1] = n_1/n_2$ and $\det[\mathbf{M}_2] = n_2/n_3$, then $\det[\mathbf{M}_2\mathbf{M}_1] = (n_2/n_3)(n_1/n_2) = n_1/n_3$. In most applications the first and last stages are air ($n = 1$) so that $n_1 = n_2$, which leads to $\det[\mathbf{M}] = 1$ and $F = 1$. In that case the solution for the ray position is

$$\boxed{y_m = y_{\max}\sin(m\varphi + \varphi_0).} \tag{1.4-32}$$

Ray Position
Periodic System

1.4 MATRIX OPTICS

We shall henceforth assume that $F = 1$. The corresponding solution for the ray angle is obtained by use of the relation $\theta_m = (y_{m+1} - Ay_m)/B$, which is derived from (1.4-20).

Condition for a Harmonic Trajectory

For y_m to be a harmonic (instead of hyperbolic) function, $\varphi = \cos^{-1} b$ must be real. This requires that

$$\boxed{|b| \leq 1 \quad \text{or} \quad \tfrac{1}{2}|A + D| \leq 1.} \quad (1.4\text{-}33)$$
Stability Condition

If, instead, $|b| > 1$, φ is then imaginary and the solution is a hyperbolic function (cosh or sinh), which increases without bound, as illustrated in Fig. 1.4-6(a). A harmonic solution ensures that y_m is bounded for all m, with a maximum value of $y_{\max}$. The bound $|b| \leq 1$ therefore provides a condition of **stability** (boundedness) of the ray trajectory.

Since y_m and y_{m+1} are both harmonic functions, so too is the ray angle corresponding to (1.4-32), by virtue of (1.4-22) and trigonometric identities. Thus, $\theta_m = \theta_{\max} \sin(m\varphi + \varphi_1)$, where the constants $\theta_{\max}$ and φ_1 are determined by the initial conditions. The maximum angle $\theta_{\max}$ must be sufficiently small so that the paraxial approximation, which underlies this analysis, is applicable.

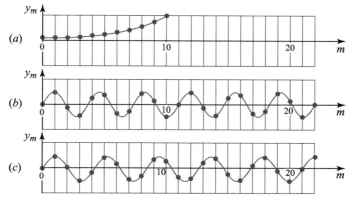

Figure 1.4-6 Examples of trajectories in periodic optical systems: (a) unstable trajectory ($b > 1$); (b) stable and periodic trajectory ($\varphi = 6\pi/11$; period = 11 stages); (c) stable but nonperiodic trajectory ($\varphi = 1.5$).

Condition for a Periodic Trajectory

The harmonic function (1.4-32) is periodic in m if it is possible to find an integer s such that $y_{m+s} = y_m$ for all m. The smallest integer is the period. The ray then retraces its path after s stages. This condition is satisfied if $s\varphi = 2\pi q$, where q is an integer. Thus, the necessary and sufficient condition for a periodic trajectory is that $\varphi/2\pi$ is a rational number q/s. If $\varphi = 6\pi/11$, for example, then $\varphi/2\pi = 3/11$ and the trajectory is periodic with period $s = 11$ stages. This case is illustrated in Fig. 1.4-6(b). Periodic optical systems will be revisited in Chapter 7.

Summary

A paraxial ray ($\theta_{\max} \ll 1$) traveling through a cascade of identical unit optical systems, each with a ray-transfer matrix with elements (A, B, C, D) such that $AD - BC = 1$, follows a harmonic (and therefore bounded) trajectory if the condition $|\frac{1}{2}(A+D)| \leq 1$, called the stability condition, is satisfied. The position at the mth stage is then $y_m = y_{\max} \sin(m\varphi + \varphi_0)$, $m = 0, 1, 2, \ldots$, where $\varphi = \cos^{-1}[\frac{1}{2}(A+D)]$. The constants $y_{\max}$ and φ_0 are determined from the initial positions y_0 and $y_1 = Ay_0 + B\theta_0$, where θ_0 is the initial ray inclination. The ray angles are related to the positions by $\theta_m = (y_{m+1} - Ay_m)/B$ and follow a harmonic function $\theta_m = \theta_{\max} \sin(m\varphi + \varphi_1)$. The ray trajectory is periodic with period s if $\varphi/2\pi$ is a rational number q/s.

EXAMPLE 1.4-1. A Sequence of Equally Spaced Identical Lenses. A set of identical lenses of focal length f separated by distance d, as shown in Fig. 1.4-7, may be used to relay light between two locations. The unit system, a distance of d of free space followed by a lens, has a ray-transfer matrix given by (1.4-12); $A = 1$, $B = d$, $C = -1/f$, $D = 1 - d/f$. The parameter $b = \frac{1}{2}(A + D) = 1 - d/2f$ and the determinant is unity. The condition for a stable ray trajectory, $|b| \leq 1$ or $-1 \leq b \leq 1$, is therefore

$$0 \leq d \leq 4f, \tag{1.4-34}$$

so that the spacing between the lenses must be smaller than four times the focal length. Under this condition the positions of paraxial rays obey the harmonic function

$$y_m = y_{\max} \sin(m\varphi + \varphi_0), \qquad \varphi = \cos^{-1}\left(1 - \frac{d}{2f}\right). \tag{1.4-35}$$

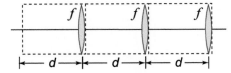

Figure 1.4-7 A periodic sequence of lenses.

When $d = 2f$, $\varphi = \pi/2$, and $\varphi/2\pi = 1/4$, so that the trajectory of an arbitrary ray is periodic with period equal to four stages. When $d = f$, $\varphi = \pi/3$, and $\varphi/2\pi = 1/6$, so that the ray trajectory is periodic and retraces itself each six stages. These cases are illustrated in Fig. 1.4-8.

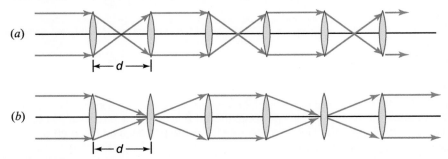

Figure 1.4-8 Examples of stable ray trajectories in a periodic lens system: (a) $d = 2f$; (b) $d = f$.

EXERCISE 1.4-6

A Periodic Set of Pairs of Different Lenses. Examine the trajectories of paraxial rays through a periodic system comprising a sequence of lens pairs with alternating focal lengths f_1 and f_2, as shown in Fig. 1.4-9. Show that the ray trajectory is bounded (stable) if

$$0 \le \left(1 - \frac{d}{2f_1}\right)\left(1 - \frac{d}{2f_2}\right) \le 1. \qquad (1.4\text{-}36)$$

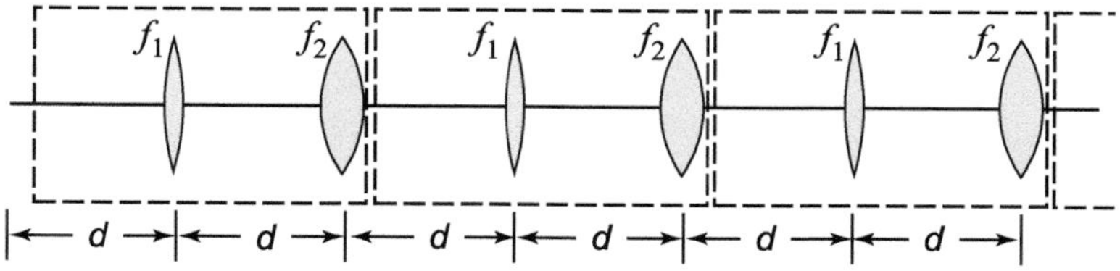

Figure 1.4-9 A periodic sequence of lens pairs.

EXERCISE 1.4-7

An Optical Resonator. Paraxial rays are reflected repeatedly between two spherical mirrors of radii R_1 and R_2 separated by a distance d (Fig. 1.4-10). Regarding this as a periodic system whose unit system is a single round trip between the mirrors, determine the condition of stability for the ray trajectory. Optical resonators will be studied in detail in Chapter 11.

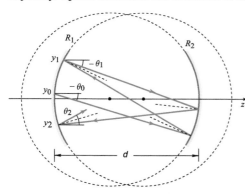

Figure 1.4-10 The optical resonator as a periodic optical system.

READING LIST

General Optics

F. L. Pedrotti, L. M. Pedrotti, and L. S. Pedrotti, *Introduction to Optics*, Cambridge University Press, 3rd ed. 2018.

T.-C. Poon and T. Kim, *Engineering Optics with MATLAB*, World Scientific, 2nd ed. 2018.

E. Hecht, *Optics*, Pearson, 5th ed. 2016.

S. D. Gupta, N. Ghosh, and A. Banerjee, *Wave Optics: Basic Concepts and Contemporary Trends*, CRC Press/Taylor & Francis, 2016.

B. D. Guenther, *Modern Optics*, Oxford University Press, 2nd ed. 2015.

C. A. DiMarzio, *Optics for Engineers*, CRC Press/Taylor & Francis, 2011.

M. Mansuripur, *Classical Optics and Its Applications*, Cambridge University Press, 2nd ed. 2009.

A. Walther, *The Ray and Wave Theory of Lenses*, Cambridge University Press, 1995, paperback ed. 2006.

A. Siciliano, *Optics: Problems and Solutions*, World Scientific, 2006.

M. Born and E. Wolf, *Principles of Optics*, Cambridge University Press, 7th expanded and corrected ed. 2002.

D. T. Moore, ed., *Selected Papers on Gradient-Index Optics*, SPIE Optical Engineering Press (Milestone Series Volume 67), 1993.

F. A. Jenkins and H. E. White, *Fundamentals of Optics*, McGraw–Hill, 1937, 4th revised ed. 1991.

R. W. Wood, *Physical Optics*, Macmillan, 3rd ed. 1934; Optical Society of America, 1988.

A. Sommerfeld, *Lectures on Theoretical Physics: Optics*, Academic Press, paperback ed. 1954.

Geometrical Optics

P. D. Lin, *Advanced Geometrical Optics*, Springer-Verlag, 2017.

V. N. Mahajan, *Fundamentals of Geometrical Optics*, SPIE Optical Engineering Press, 2014.

E. Dereniak and T. D. Dereniak, *Geometrical and Trigonometric Optics*, Cambridge University Press, 2008.

Yu. A. Kravtsov, *Geometrical Optics in Engineering Physics*, Alpha Science, 2005.

J. E. Greivenkamp, *Field Guide to Geometrical Optics*, SPIE Optical Engineering Press, 2004.

P. Mouroulis and J. Macdonald, *Geometrical Optics and Optical Design*, Oxford University Press, 1997.

Optical System Design

D. Malacara-Hernández and Z. Malacara-Hernández, *Handbook of Optical Design*, CRC Press/Taylor & Francis, 3rd ed. 2013.

J. Sasián, *Introduction to Aberrations in Optical Imaging Systems*, Cambridge University Press, 2013.

K. J. Kasunic, *Optical Systems Engineering*, McGraw–Hill, 2011.

R. E. Fischer, B. Tadic-Galeb, and P. R. Yoder, *Optical System Design*, McGraw–Hill, 2nd ed. 2008.

W. J. Smith, *Modern Optical Engineering*, McGraw–Hill, 1966, 4th ed. 2008.

D. C. O'Shea, *Elements of Modern Optical Design*, Wiley, 1985.

Matrix Optics

A. Gerrard and J. M. Burch, *Introduction to Matrix Methods in Optics*, Wiley, 1975; Dover, reprinted 2012.

J. W. Blaker, *Geometric Optics: The Matrix Theory*, CRC Press, 1971.

Popular and Historical

R. J. Weiss, *A Brief History of Light and Those that Lit the Way*, World Scientific, 1996.

A. R. Hall, *All was Light: An Introduction to Newton's Opticks*, Clarendon Press/Oxford University Press, 1993.

R. Kingslake, *A History of the Photographic Lens*, Academic Press, 1989.

M. I. Sobel, *Light*, University of Chicago Press, 1987.

A. I. Sabra, *Theories of Light from Descartes to Newton*, Cambridge University Press, 1981.

V. Ronchi, *The Nature of Light: An Historical Survey*, Harvard University Press, 1970.

W. H. Bragg, *Universe of Light*, Dover, paperback ed. 1959.

I. Newton, *Opticks: or A Treatise of the Reflections, Refractions, Inflections & Colours of Light*, 4th ed. 1704; Dover, reissued 1979.

PROBLEMS

1.1-2 **Fermat's Principle with Maximum Time.** Consider the elliptical mirror shown in Fig. P1.1-2(a), whose foci are denoted A and B. Geometrical properties of the ellipse dictate that the pathlength $\overline{APB}$ is identical to the pathlengths $\overline{AP'B}$ and $\overline{AP''B}$ for adjacent points on the ellipse.

(a) Now consider another mirror with a radius of curvature smaller than that of the elliptical mirror, but tangent to it at P, as displayed in Fig. P1.1-2(b). Show that the path $\overline{APB}$ followed by the light ray in traveling between points A and B is a path of *maximum* time, i.e., is greater than the adjacent paths $\overline{AQ'B}$ and $\overline{AQ''B}$.

(b) Finally, consider a mirror that crosses the ellipse, but is tangent to it at P, as illustrated in Fig. P1.1-2(c). Show that the possible ray paths $\overline{AQ'B}$, $\overline{APB}$, and $\overline{AQ''B}$ exhibit a point of inflection.

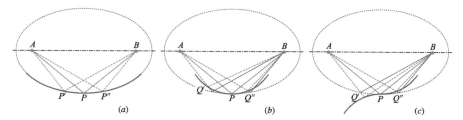

Figure P1.1-2 (a) Reflection from an elliptical mirror. (b) Reflection from an inscribed tangential mirror with greater curvature. (c) Reflection from a tangential mirror with curvature changing from concave to convex.

1.2-7 **Transmission through Planar Plates.**

(a) Use Snell's law to show that a ray entering a planar plate of thickness d and refractive index n_1 (placed in air; $n \approx 1$) emerges parallel to its initial direction. The ray need not be paraxial. Derive an expression for the lateral displacement of the ray as a function of the angle of incidence θ. Explain your results in terms of Fermat's principle.

(b) If the plate instead comprises a stack of N parallel layers stacked against each other with thicknesses $d_1, d_2, \ldots, d_N$ and refractive indices $n_1, n_2, \ldots, n_N$, show that the transmitted ray is parallel to the incident ray. If θ_m is the angle of the ray in the mth layer, show that $n_m \sin \theta_m = \sin \theta$, $m = 1, 2, \ldots$.

1.2-8 **Lens in Water.** Determine the focal length f of a biconvex lens with radii 20 cm and 30 cm and refractive index $n = 1.5$. What is the focal length when the lens is immersed in water ($n = 4/3$)?

1.2-9 **Numerical Aperture of a Cladless Fiber.** Determine the numerical aperture and the acceptance angle of an optical fiber if the refractive index of the core is $n_1 = 1.46$ and the cladding is stripped out (replaced with air $n_2 \approx 1$).

1.2-10 **Fiber Coupling Spheres.** Tiny glass balls are often used as lenses to couple light into and out of optical fibers. The fiber end is located at a distance f from the sphere. For a sphere of radius $a = 1$ mm and refractive index $n = 1.8$, determine f such that a ray parallel to the optical axis at a distance $y = 0.7$ mm is focused onto the fiber, as illustrated in Fig. P1.2-10.

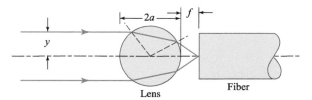

Figure P1.2-10 Focusing light into an optical fiber with a spherical glass ball.

1.2-11 **Extraction of Light from a High-Refractive-Index Medium.** Assume that light is generated isotropically in all directions inside a material of refractive index $n = 3.7$ cut in the shape of a parallelepiped and placed in air ($n = 1$) (see Exercise 1.2-6).

(a) If a reflective material acting as a perfect mirror is coated on all sides except the front side, determine the percentage of light that may be extracted from the front side.

(b) If another transparent material of refractive index $n = 1.4$ is placed on the front side, would that help extract some of the trapped light?

1.3-3 **Axially Graded Plate.** A plate of thickness d is oriented normal to the z axis. The refractive index $n(z)$ is graded in the z direction. Show that a ray entering the plate from air at an incidence angle θ_0 in the y–z plane makes an angle $\theta(z)$ at position z in the medium given by $n(z) \sin \theta(z) = \sin \theta_0$. Show that the ray emerges into air parallel to the original incident ray. *Hint:* You may use the results of Prob. 1.2-7. Show that the ray position $y(z)$ inside the plate obeys the differential equation $(dy/dz)^2 = (n^2/\sin^2 \theta - 1)^{-1}$.

1.3-4 **Ray Trajectories in GRIN Fibers.** Consider a graded-index optical fiber with cylindrical symmetry about the z axis and refractive index $n(\rho)$, $\rho = \sqrt{x^2 + y^2}$. Let (ρ, ϕ, z) be the position vector in a cylindrical coordinate system. Rewrite the paraxial ray equations, (1.3-4), in a cylindrical system and derive differential equations for ρ and ϕ as functions of z.

1.4-8 **Ray-Transfer Matrix of a Lens System.** Determine the ray-transfer matrix for an optical system made of a thin convex lens of focal length f and a thin concave lens of focal length $-f$ separated by a distance f. Discuss the imaging properties of this composite lens.

1.4-9 **Ray-Transfer Matrix of a GRIN Plate.** Determine the ray-transfer matrix of a SELFOC plate [i.e., a graded-index material with parabolic refractive index $n(y) \approx n_0(1 - \frac{1}{2}\alpha^2 y^2)$] of thickness d.

1.4-10 **The GRIN Plate as a Periodic System.** Consider the trajectories of paraxial rays inside a SELFOC plate normal to the z axis. This system may be regarded as a periodic system comprising a sequence of identical contiguous plates, each of thickness d. Using the result of Prob. 1.4-9, determine the stability condition of the ray trajectory. Is this condition dependent on the choice of d?

1.4-11 **Recurrence Relation for a Planar-Mirror Resonator.** Consider a planar-mirror optical resonator, with mirror separation d, as a periodic optical system. Determine the unit ray-transfer matrix for this system, demonstrating that $b = 1$ and $F = 1$. Show that there is then only a single root to the quadratic equation (1.4-28) so that the ray position must then take the form $\alpha + m\beta$, where α and β are constants.

1.4-12 **4×4 Ray-Transfer Matrix for Skewed Rays.** Matrix methods may be generalized to describe skewed paraxial rays in circularly symmetric systems, and to astigmatic (non-circularly symmetric) systems. A ray crossing the plane $z = 0$ is generally characterized by four variables — the coordinates (x, y) of its position in the plane, and the angles (θ_x, θ_y) that its projections in the x–z and y–z planes make with the z axis. The emerging ray is also characterized by four variables that are linearly related to the initial four variables. The optical system may then be characterized completely, within the paraxial approximation, by a 4×4 matrix.

(a) Determine the 4×4 ray-transfer matrix of a distance d in free space.

(b) Determine the 4×4 ray-transfer matrix of a thin cylindrical lens with focal length f oriented in the y direction. The cylindrical lens has focal length f for rays in the y–z plane, and no focusing power for rays in the x–z plane.

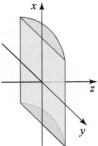

CHAPTER

2

WAVE OPTICS

2.1	POSTULATES OF WAVE OPTICS	43
2.2	MONOCHROMATIC WAVES	44
	A. Complex Representation and the Helmholtz Equation	
	B. Elementary Waves	
	C. Paraxial Waves	
*2.3	RELATION BETWEEN WAVE OPTICS AND RAY OPTICS	52
2.4	SIMPLE OPTICAL COMPONENTS	53
	A. Reflection and Refraction	
	B. Transmission Through Optical Components	
	C. Graded-Index Optical Components	
2.5	INTERFERENCE	61
	A. Interference of Two Waves	
	B. Multiple-Wave Interference	
2.6	POLYCHROMATIC AND PULSED LIGHT	71
	A. Temporal and Spectral Description	
	B. Light Beating	

Christiaan Huygens (1629–1695) advanced a number of novel concepts pertaining to the propagation of light waves.

Thomas Young (1773–1829) championed the wave theory of light and discovered the principle of optical interference.

Fundamentals of Photonics, Third Edition. Bahaa E. A. Saleh and Malvin Carl Teich.
©2019 John Wiley & Sons, Inc. Published 2019 by John Wiley & Sons, Inc.

Light propagates in the form of waves. In free space, light waves travel with a constant speed, $c_o = 3.0 \times 10^8$ m/s (30 cm/ns or 0.3 mm/ps or 0.3 μm/fs or 0.3 nm/as). As illustrated in Fig. 2.0-1, the range of optical wavelengths comprises three principal sub-regions: infrared (0.760 to 300 μm), visible (390 to 760 nm), and ultraviolet (10 to 390 nm). The corresponding range of optical frequencies stretches from 1 THz in the far-infrared to 30 PHz in the extreme ultraviolet.

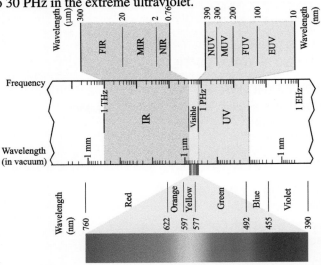

Figure 2.0-1 Optical frequencies and wavelengths. The infrared (**IR**) region of the spectrum comprises the near-infrared (**NIR**), mid-infrared (**MIR**), and far-infrared (**FIR**) bands. The **MWIR** and **LWIR** bands both lie within the MIR band; radiation in these regions can penetrate the atmosphere. The ultraviolet (**UV**) region comprises the near-ultraviolet (**NUV**), mid-ultraviolet (**MUV**) or deep-ultraviolet (**DUV**), far-ultraviolet (**FUV**), and extreme-ultraviolet (**EUV** or **XUV**) bands. The vacuum ultraviolet (**VUV**) consists of the FUV and EUV bands. The ultraviolet region is also divided into the **UVA**, **UVB**, and **UVC** bands, which have chemical and biological significance. The infrared, visible, and ultraviolet regions are gathered under the rubric "optical" since they make use of similar types of components (e.g., lenses and mirrors). The terahertz (**THz**) region occupies frequencies that stretch from 0.3 to 3 THz, corresponding to wavelengths that extend from 1 mm to 100 μm; the THz region partially overlaps the FIR band. For X-ray wavelengths, see Fig. 16.3-7.

The **wave theory** of light encompasses the ray theory (Fig. 2.0-2). Strictly speaking, ray optics is the limit of wave optics when the wavelength is infinitesimally short. However, the wavelength need not actually be zero for the ray-optics theory to be useful. As long as the light waves propagate through and around objects whose dimensions are much greater than the wavelength, the ray theory suffices for describing most optical phenomena. Because the wavelength of visible light is much smaller than the dimensions of the usual objects we encounter on a daily basis, the manifestations of the wave nature of light are usually not apparent without careful observation.

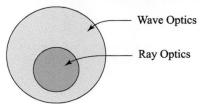

Figure 2.0-2 Wave optics encompasses ray optics. Ray optics is the limit of wave optics when the wavelength is very short.

This Chapter

In the context of wave optics, light is described by a scalar function, called the wavefunction, that obeys a second-order differential equation known as the wave equation. A discussion of the physical significance of the wavefunction is deferred to Chapter 5, where we consider electromagnetic optics; it will become apparent there that the wavefunction represents any of the components of the electric or magnetic fields. The wave equation, together with a relation between the optical power density and the wavefunction, constitute the postulates of the scalar-wave model of light known as **wave optics**. The consequences of these simple postulates are manifold and far reaching. Wave optics constitutes a basis for describing a host of optical phenomena that fall outside the confines of ray optics, including interference and diffraction, as will become clear in this and the following two chapters (Chapters 3 and 4).

Wave optics does have its limitations, however. It is not capable of providing a complete picture of the reflection and refraction of light at the boundaries between various media, nor can it accommodate optical phenomena that require a vector formulation, such as polarization effects. Those issues will be considered from a fundamental perspective in Chapters 5–8, as will the conditions under which scalar wave optics provides a good approximation to electromagnetic optics.

The chapter begins with the postulates of wave optics (Sec. 2.1). In Secs. 2.2–2.5 we consider monochromatic waves. Elementary waves, such as the plane wave, the spherical wave, and paraxial waves are introduced in Sec. 2.2. Section 2.3 establishes how ray optics is formally derived from wave optics. The interaction of optical waves with simple optical components such as mirrors, prisms, lenses, and various graded-index elements is examined in Sec. 2.4. Interference, an important manifestation of the wave nature of light, is the subject of Secs. 2.5 and 2.6, where polychromatic and pulsed light are discussed.

2.1 POSTULATES OF WAVE OPTICS

The Wave Equation

Light propagates in the form of waves. In free space, light waves travel with speed c_o. A homogeneous transparent medium such as glass is characterized by a single constant, its refractive index n (≥ 1). In a medium of refractive index n, light waves travel with a reduced speed

$$c = \frac{c_o}{n}.$$

(2.1-1)
Speed of Light
in a Medium

An optical wave is described mathematically by a real function of position $\mathbf{r} = (x, y, z)$ and time t, denoted $u(\mathbf{r}, t)$ and known as the **wavefunction**. It satisfies a partial differential equation called the **wave equation**,

$$\nabla^2 u - \frac{1}{c^2}\frac{\partial^2 u}{\partial t^2} = 0,$$

(2.1-2)
Wave Equation

where ∇^2 is the Laplacian operator, which is $\nabla^2 = \partial^2/\partial x^2 + \partial^2/\partial y^2 + \partial^2/\partial z^2$ in Cartesian coordinates. Any function that satisfies (2.1-2) represents a possible optical wave.

Because the wave equation is linear, the **principle of superposition** applies: if $u_1(\mathbf{r}, t)$ and $u_2(\mathbf{r}, t)$ represent possible optical waves, then $u(\mathbf{r}, t) = u_1(\mathbf{r}, t) + u_2(\mathbf{r}, t)$ also represents a possible optical wave.

At the boundary between two different media, the wavefunction changes in a way that depends on their refractive indices. However, the laws that govern this change depend on the physical significance assigned to the wavefunction which, as will be seen in Chapter 5, is an electromagnetic-field component. The underlying physical origin of the refractive index derives from electromagnetic optics (Sec. 5.5B).

The wave equation is also approximately applicable for media with refractive indices that are position dependent, provided that the variation is slow within distances of the order of a wavelength. The medium is then said to be locally homogeneous. For such media, n in (2.1-1) and c in (2.1-2) are simply replaced by the appropriate position-dependent functions $n(\mathbf{r})$ and $c(\mathbf{r})$, respectively.

Intensity, Power, and Energy

The optical **intensity** $I(\mathbf{r}, t)$, defined as the optical power per unit area (units of watts/cm^2), is proportional to the average of the squared wavefunction:

$$\boxed{I(\mathbf{r}, t) = 2\langle u^2(\mathbf{r}, t)\rangle\,.} \qquad (2.1\text{-}3)$$
Optical Intensity

The operation $\langle \cdot \rangle$ denotes averaging over a time interval much longer than the time of an optical cycle, but much shorter than any other time of interest (such as the duration of a pulse of light). The duration of an optical cycle is very short: $2 \times 10^{-15}\,\text{s} = 2\,\text{fs}$ for light of wavelength 600 nm, as an example. This concept is further elucidated in Sec. 2.6. The quantity $I(\mathbf{r}, t)$ is sometimes also called the **irradiance**.

Although the physical significance of the wavefunction $u(\mathbf{r}, t)$ has not been explicitly specified, (2.1-3) represents its connection with a physically measurable quantity — the optical intensity. There is some arbitrariness in the definition of the wavefunction and its relation to the intensity. For example, (2.1-3) could have been written without the factor 2 and the wavefunction scaled by a factor $\sqrt{2}$, in which case the intensity would remain the same. The choice of the factor 2 in (2.1-3) will later prove convenient, however.

The optical **power** $P(t)$ (units of watts) flowing into an area A normal to the direction of propagation of light is the integrated intensity

$$P(t) = \int_A I(\mathbf{r}, t)\, dA\,. \qquad (2.1\text{-}4)$$

The optical **energy** E (units of joules) collected in a given time interval is the integral of the optical power over the time interval.

2.2 MONOCHROMATIC WAVES

A monochromatic wave is represented by a wavefunction with harmonic time dependence,

$$u(\mathbf{r}, t) = a(\mathbf{r}) \cos[2\pi\nu t + \varphi(\mathbf{r})]\,, \qquad (2.2\text{-}1)$$

as illustrated in Fig. 2.2-1(a), where

$a(\mathbf{r})$ = amplitude
$\varphi(\mathbf{r})$ = phase
ν = frequency (cycles/s or Hz)
$\omega = 2\pi\nu$ = angular frequency (radians/s or s^{-1})
$T = 1/\nu = 2\pi/\omega$ = period (s).

Both the amplitude and phase are generally position dependent, but the wavefunction is a harmonic function of time with frequency ν at all positions. Optical waves have frequencies that lie in the range 3×10^{11} to 3×10^{16} Hz, as depicted in Fig. 2.0-1.

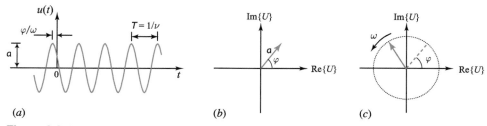

(a)   (b)   (c)

Figure 2.2-1 Representations of a monochromatic wave at a fixed position $\mathbf{r}$: (a) the wavefunction $u(t)$ is a harmonic function of time; (b) the complex amplitude $U = a\exp(j\varphi)$ is a fixed phasor; (c) the complex wavefunction $U(t) = U\exp(j2\pi\nu t)$ is a phasor rotating with angular velocity $\omega = 2\pi\nu$ radians/s.

A. Complex Representation and the Helmholtz Equation

Complex Wavefunction

It is convenient to represent the real wavefunction $u(\mathbf{r}, t)$ in (2.2-1) in terms of a complex function

$$U(\mathbf{r}, t) = a(\mathbf{r}) \exp[j\varphi(\mathbf{r})] \exp(j2\pi\nu t), \qquad (2.2\text{-}2)$$

so that

$$u(\mathbf{r}, t) = \text{Re}\{U(\mathbf{r}, t)\} = \tfrac{1}{2}[U(\mathbf{r}, t) + U^*(\mathbf{r}, t)], \qquad (2.2\text{-}3)$$

where the symbol * signifies complex conjugation. The function $U(\mathbf{r}, t)$, known as the **complex wavefunction**, describes the wave completely; the **wavefunction** $u(\mathbf{r}, t)$ is simply its real part. Like the wavefunction $u(\mathbf{r}, t)$, the complex wavefunction $U(\mathbf{r}, t)$ must also satisfy the wave equation

$$\boxed{\nabla^2 U - \frac{1}{c^2}\frac{\partial^2 U}{\partial t^2} = 0.} \qquad (2.2\text{-}4)$$
Wave Equation

The two functions satisfy the same boundary conditions.

Complex Amplitude

Equation (2.2-2) may be written in the form

$$U(\mathbf{r}, t) = U(\mathbf{r}) \exp(j2\pi\nu t), \tag{2.2-5}$$

where the time-independent factor $U(\mathbf{r}) = a(\mathbf{r}) \exp[j\varphi(\mathbf{r})]$ is referred to as the **complex amplitude** of the wave. The wavefunction $u(\mathbf{r}, t)$ is therefore related to the complex amplitude by

$$u(\mathbf{r}, t) = \text{Re}\{U(\mathbf{r}) \exp(j2\pi\nu t)\} = \tfrac{1}{2}[U(\mathbf{r}) \exp(j2\pi\nu t) + U^*(\mathbf{r}) \exp(-j2\pi\nu t)]. \tag{2.2-6}$$

At a given position $\mathbf{r}$, the complex amplitude $U(\mathbf{r})$ is a complex variable [depicted in Fig. 2.2-1(b)] whose magnitude $|U(\mathbf{r})| = a(\mathbf{r})$ is the amplitude of the wave and whose argument $\arg\{U(\mathbf{r})\} = \varphi(\mathbf{r})$ is the phase. The complex wavefunction $U(\mathbf{r}, t)$, shown in Fig. 2.2-1(c), is represented graphically by a phasor that rotates with angular velocity $\omega = 2\pi\nu$ radians/s. Its initial value at $t = 0$ is the complex amplitude $U(\mathbf{r})$.

The Helmholtz Equation

Substituting $U(\mathbf{r}, t) = U(\mathbf{r}) \exp(j2\pi\nu t)$ from (2.2-5) into the wave equation (2.2-4) leads to a differential equation for the complex amplitude $U(\mathbf{r})$:

$$\boxed{\nabla^2 U + k^2 U = 0,} \tag{2.2-7}$$
Helmholtz Equation

which is known as the **Helmholtz equation**, where

$$\boxed{k = \frac{2\pi\nu}{c} = \frac{\omega}{c}} \tag{2.2-8}$$
Wavenumber

is referred to as the **wavenumber**. Different solutions are obtained from different boundary conditions.

Optical Intensity

The optical intensity is determined by inserting (2.2-1) into (2.1-3):

$$\begin{aligned} 2u^2(\mathbf{r}, t) &= 2a^2(\mathbf{r}) \cos^2[2\pi\nu t + \varphi(\mathbf{r})] \\ &= |U(\mathbf{r})|^2 \{1 + \cos(2[2\pi\nu t + \varphi(\mathbf{r})])\}. \end{aligned} \tag{2.2-9}$$

Averaging (2.2-9) over a time longer than an optical period, $1/\nu$, causes the second term of (2.2-9) to vanish, whereupon

$$\boxed{I(\mathbf{r}) = |U(\mathbf{r})|^2.} \tag{2.2-10}$$
Optical Intensity

The optical intensity of a monochromatic wave is the absolute square of its complex amplitude.

The intensity of a monochromatic wave does *not* vary with time.

Wavefronts

The wavefronts are the surfaces of equal phase, $\varphi(\mathbf{r}) = $ constant. The constants are often taken to be multiples of 2π so that $\varphi(\mathbf{r}) = 2\pi q$, where q is an integer. The wavefront normal at position $\mathbf{r}$ is parallel to the gradient vector $\nabla\varphi(\mathbf{r})$ (a vector that has components $\partial\varphi/\partial x$, $\partial\varphi/\partial y$, and $\partial\varphi/\partial z$ in a Cartesian coordinate system). It represents the direction at which the rate of change of the phase is maximum.

Summary

- A monochromatic wave of frequency ν is described by a *complex wavefunction* $U(\mathbf{r},t) = U(\mathbf{r})\exp(j2\pi\nu t)$, which satisfies the wave equation.
- The *complex amplitude* $U(\mathbf{r})$ satisfies the Helmholtz equation; its magnitude $|U(\mathbf{r})|$ and argument $\arg\{U(\mathbf{r})\}$ are the *amplitude* and *phase* of the wave, respectively. The optical *intensity* is $I(\mathbf{r}) = |U(\mathbf{r})|^2$. The *wavefronts* are the surfaces of constant phase, $\varphi(\mathbf{r}) = \arg\{U(\mathbf{r})\} = 2\pi q$ ($q = $ integer).
- The *wavefunction* $u(\mathbf{r},t)$ is the real part of the complex wavefunction, $u(\mathbf{r},t) = \mathrm{Re}\{U(\mathbf{r},t)\}$. The wavefunction also satisfies the wave equation.

B. Elementary Waves

The simplest solutions of the Helmholtz equation in a homogeneous medium are the plane wave and the spherical wave.

The Plane Wave

The plane wave has complex amplitude

$$U(\mathbf{r}) = A\exp(-j\mathbf{k}\cdot\mathbf{r}) = A\exp\left[-j(k_x x + k_y y + k_z z)\right], \qquad (2.2\text{-}11)$$

where A is a complex constant called the **complex envelope** that represents the strength of the wave, and $\mathbf{k} = (k_x, k_y, k_z)$ is called the **wavevector**.[†] Substituting (2.2-11) into the Helmholtz equation (2.2-7) yields the relation $k_x^2 + k_y^2 + k_z^2 = k^2$, so that the magnitude of the wavevector $\mathbf{k}$ is the wavenumber k.

Since the phase of the wave is $\arg\{U(\mathbf{r})\} = \arg\{A\} - \mathbf{k}\cdot\mathbf{r}$, the surfaces of constant phase (wavefronts) obey $\mathbf{k}\cdot\mathbf{r} = k_x x + k_y y + k_z z = 2\pi q + \arg\{A\}$ with q integer. This is the equation describing parallel planes perpendicular to the wavevector $\mathbf{k}$ (hence the name "plane wave"). Consecutive planes are separated by a distance $\lambda = 2\pi/k$, so that

$$\boxed{\lambda = \frac{c}{\nu},} \qquad (2.2\text{-}12)$$
Wavelength

where λ is called the **wavelength**. The plane wave has a constant intensity $I(\mathbf{r}) = |A|^2$ everywhere in space so that it carries infinite power. This wave is clearly an idealization since it exists everywhere and at all times.

[†] The complex wavefunction for a monochromatic plane wave is written in a form commonly used in electrical engineering: $U(\mathbf{r},t) = A\exp[j(\omega t - \mathbf{k}\cdot\mathbf{r})]$. In the physics literature, this same wave is usually written as $U(\mathbf{r},t) = A\exp[-i(\omega t - \mathbf{k}\cdot\mathbf{r})]$; correspondence is attained by simply replacing i with $-j$, where $i = j = \sqrt{-1}$. This choice has no bearing on the final result, as is evidenced by observing that the wavefunction $u(\mathbf{r},t)$ in (2.2-13) takes the form of a cosine function, for which $\cos(x) = \cos(-x)$.

If the z axis is taken along the direction of the wavevector $\mathbf{k}$, then $U(\mathbf{r}) = A\exp(-jkz)$ and the corresponding wavefunction obtained from (2.2-6) is

$$u(\mathbf{r},t) = |A|\cos\left[2\pi\nu t - kz + \arg\{A\}\right] = |A|\cos\left[2\pi\nu(t - z/c) + \arg\{A\}\right]. \tag{2.2-13}$$

The wavefunction is therefore periodic in time with period $1/\nu$, and periodic in space with period $2\pi/k$, which is equal to the wavelength λ (see Fig. 2.2-2). Since the phase of the complex wavefunction, $\arg\{U(\mathbf{r},t)\} = 2\pi\nu(t - z/c) + \arg\{A\}$, varies with time and position as a function of the variable $t - z/c$ (see Fig. 2.2-2), c is called the **phase velocity** of the wave.

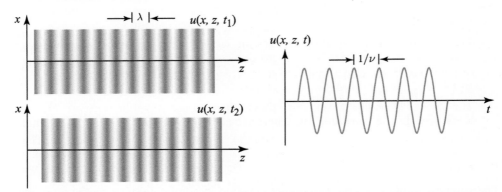

Figure 2.2-2 The wavefunction of a plane wave traveling in the z direction, schematically drawn as a graded red pattern, is a periodic function of z with spatial period λ, and a periodic function of t with temporal period $1/\nu$. The surfaces of constant phase (wavefronts) comprise a parallel set of planes normal to the z axis. The wavelengths displayed in Fig. 2.0-1 are in free space ($\lambda = \lambda_o$).

In a medium of refractive index n, the wave has phase velocity $c = c_o/n$ and wavelength $\lambda = c/\nu = c_o/n\nu$, so that $\lambda = \lambda_o/n$ where $\lambda_o = c_o/\nu$ is the wavelength in free space. Thus, for a given frequency ν, the wavelength in the medium is reduced relative to that in free space by the factor n. As a consequence, the wavenumber $k = 2\pi/\lambda$ is increased relative to that in free space ($k_o = 2\pi/\lambda_o$) by the factor n.

As a monochromatic wave propagates through media of different refractive indices its frequency remains the same, but its velocity, wavelength, and wavenumber are altered:

$$\boxed{c = \frac{c_o}{n}, \qquad \lambda = \frac{\lambda_o}{n}, \qquad k = nk_o.} \tag{2.2-14}$$

The Spherical Wave

Another simple solution of the Helmholtz equation (in spherical coordinates) is the spherical wave complex amplitude

$$U(\mathbf{r}) = \frac{A_0}{r}\exp(-jkr), \tag{2.2-15}$$

where r is the distance from the origin, $k = 2\pi\nu/c = \omega/c$ is the wavenumber, and A_0 is a constant. The intensity $I(\mathbf{r}) = |A_0|^2/r^2$ is inversely proportional to the square of the distance. Taking $\arg\{A_0\} = 0$ for simplicity, the wavefronts are the surfaces $kr = 2\pi q$ or $r = q\lambda$, where q is an integer. These are concentric spheres separated by a radial distance $\lambda = 2\pi/k$ that advance radially at the phase velocity c (Fig. 2.2-3).

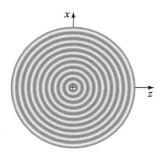

Figure 2.2-3 Cross section of the wavefunction of a spherical wave. The associated wavefronts are a set of concentric spheres.

A spherical wave originating at the position $\mathbf{r}_0$ has a complex amplitude $U(\mathbf{r}) = (A_0/|\mathbf{r} - \mathbf{r}_0|)\exp(-jk|\mathbf{r} - \mathbf{r}_0|)$. Its wavefronts are spheres centered about $\mathbf{r}_0$. A wave with complex amplitude $U(\mathbf{r}) = (A_0/r)\exp(+jkr)$ is a spherical wave traveling inwardly (toward the origin) instead of outwardly (away from the origin).

Fresnel Approximation of the Spherical Wave: The Paraboloidal Wave

Let us examine a spherical wave (originating at $\mathbf{r} = 0$) at points $\mathbf{r} = (x, y, z)$ that are sufficiently close to the z axis but far from the origin, so that $\sqrt{x^2 + y^2} \ll z$. The paraxial approximation of ray optics (Sec. 1.2) would be applicable were these points the endpoints of rays beginning at the origin. Denoting $\theta^2 = (x^2 + y^2)/z^2 \ll 1$, we use an approximation based on the Taylor-series expansion:

$$r = \sqrt{x^2 + y^2 + z^2} = z\sqrt{1 + \theta^2} = z\left(1 + \frac{\theta^2}{2} - \frac{\theta^4}{8} + \cdots\right)$$

$$\approx z\left(1 + \frac{\theta^2}{2}\right) = z + \frac{x^2 + y^2}{2z}. \qquad (2.2\text{-}16)$$

This expression, $r \approx z + (x^2 + y^2)/2z$, is now substituted into the phase of $U(\mathbf{r})$ in (2.2-15). A less accurate expression, $r \approx z$, can be substituted for the magnitude since it is less sensitive to errors than is the phase. The result is known as the **Fresnel approximation** of a spherical wave:

$$\boxed{U(\mathbf{r}) \approx \frac{A_0}{z}\exp(-jkz)\exp\left[-jk\frac{x^2 + y^2}{2z}\right].} \qquad (2.2\text{-}17)$$
Fresnel Approximation of a Spherical Wave

This approximation plays an important role in simplifying the theory of optical-wave transmission through apertures (**diffraction**), as discussed in Chapter 4.

The complex amplitude in (2.2-17) may be viewed as representing a plane wave $A_0 \exp(-jkz)$ modulated by the factor $(1/z)\exp[-jk(x^2 + y^2)/2z]$, which involves the phase $k(x^2 + y^2)/2z$. This phase factor serves to bend the planar wavefronts of the plane wave into paraboloidal surfaces (Fig. 2.2-4), since the equation of a paraboloid of revolution is $(x^2 + y^2)/z = \text{constant}$. In this region the spherical wave is well approximated by a **paraboloidal wave**. When z becomes very large, the paraboloidal phase factor in (2.2-17) approaches 0 so that the overall phase of the wave becomes kz. Since the magnitude A_0/z varies slowly with z, the spherical wave eventually approaches the plane wave $\exp(-jkz)$, as illustrated in Fig. 2.2-4.

The condition of validity for the Fresnel approximation is *not* simply that $\theta^2 \ll 1$, however. Although the third term of the series expansion, $\theta^4/8$, may be very small

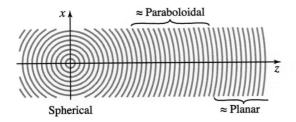

Figure 2.2-4 A spherical wave may be approximated at points near the z axis and sufficiently far from the origin by a paraboloidal wave. For points very far from the origin, the spherical wave approaches a plane wave.

in comparison with the second and first terms, when multiplied by kz it can become comparable to π. The approximation used in the foregoing is therefore valid when $kz\theta^4/8 \ll \pi$, or $(x^2 + y^2)^2 \ll 4z^3\lambda$. For points (x, y) lying within a circle of radius a centered about the z axis, the validity condition is thus $a^4 \ll 4z^3\lambda$ or

$$\frac{N_F \theta_m^2}{4} \ll 1, \qquad (2.2\text{-}18)$$

where $\theta_m = a/z$ is the maximum angle and

$$\boxed{N_F = \frac{a^2}{\lambda z}} \qquad (2.2\text{-}19)$$
Fresnel Number

is known as the **Fresnel number**.

EXERCISE 2.2-1

Validity of the Fresnel Approximation. Determine the radius of a circle within which a spherical wave of wavelength $\lambda = 633$ nm, originating at a distance 1 m away, may be approximated by a paraboloidal wave. Determine the maximum angle θ_m and the Fresnel number N_F.

C. Paraxial Waves

A wave is said to be paraxial if its wavefront normals are paraxial rays. One way of constructing a paraxial wave is to start with a plane wave $A \exp(-jkz)$, regard it as a "carrier" wave, and modify or "modulate" its complex envelope A, making it a slowly varying function of position, $A(\mathbf{r})$, so that the complex amplitude of the modulated wave becomes

$$U(\mathbf{r}) = A(\mathbf{r}) \exp(-jkz). \qquad (2.2\text{-}20)$$

The variation of the envelope $A(\mathbf{r})$ and its derivative with position z must be slow within the distance of a wavelength $\lambda = 2\pi/k$ so that the wave approximately maintains its underlying plane-wave nature.

The wavefunction of a paraxial wave, $u(\mathbf{r}, t) = |A(\mathbf{r})| \cos[2\pi\nu t - kz + \arg\{A(\mathbf{r})\}]$, is sketched in Fig. 2.2-5(a) as a function of z at $t = 0$ and $x = y = 0$. It is a sinusoidal function of z with amplitude $|A(0, 0, z)|$ and phase $\arg\{A(0, 0, z)\}$, both of which vary slowly with z. Since the phase $\arg\{A(x, y, z)\}$ changes little within the distance of a wavelength, the planar wavefronts $kz = 2\pi q$ of the carrier plane wave bend only slightly, so that their normals form paraxial rays [Fig. 2.2-5(b)].

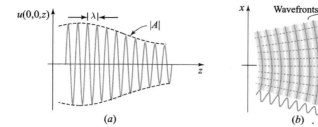

Figure 2.2-5 (a) Wavefunction of a paraxial wave at point on the z axis as a function of the axial distance z. (b) The wavefronts and wavefront normals of a paraxial wave in the x–z plane.

The Paraxial Helmholtz Equation

For the paraxial wave (2.2-20) to satisfy the Helmholtz equation (2.2-7), the complex envelope $A(\mathbf{r})$ must satisfy another partial differential equation that is obtained by substituting (2.2-20) into (2.2-7). The assumption that $A(\mathbf{r})$ varies slowly with respect to z signifies that within a distance $\Delta z = \lambda$, the change ΔA is much smaller than A itself, i.e., $\Delta A \ll A$. This inequality of complex variables applies to the magnitudes of the real and imaginary parts separately. Since $\Delta A = (\partial A/\partial z)\Delta z = (\partial A/\partial z)\lambda$, it follows that $\partial A/\partial z \ll A/\lambda = Ak/2\pi$, so that

$$\frac{\partial A}{\partial z} \ll kA. \qquad (2.2\text{-}21)$$

The derivative $\partial A/\partial z$ itself must also vary slowly within the distance λ, so that $\partial^2 A/\partial z^2 \ll k\,\partial A/\partial z$, which provides

$$\frac{\partial^2 A}{\partial z^2} \ll k^2 A. \qquad (2.2\text{-}22)$$

Substituting (2.2-20) into (2.2-7), and neglecting $\partial^2 A/\partial z^2$ in comparison with $k\,\partial A/\partial z$ or $k^2 A$, leads to a partial differential equation for the complex envelope $A(\mathbf{r})$:

$$\boxed{\nabla_T^2 A - j\,2k\frac{\partial A}{\partial z} = 0,} \qquad (2.2\text{-}23)$$
Paraxial Helmholtz Equation

where $\nabla_T^2 = \partial^2/\partial x^2 + \partial^2/\partial y^2$ is the transverse Laplacian operator.

Equation (2.2-23) is the **slowly varying envelope approximation** of the Helmholtz equation. We shall simply call it the **paraxial Helmholtz equation**. It bears some similarity to the Schrödinger equation of quantum physics [see (14.1-1)]. The simplest solution of the paraxial Helmholtz equation is the paraboloidal wave (Exercise 2.2-2), which is the paraxial approximation of a spherical wave. One of the most interesting and useful solutions, however, is the **Gaussian beam**, to which Chapter 3 is devoted.

EXERCISE 2.2-2

The Paraboloidal Wave and the Gaussian Beam. Verify that a paraboloidal wave with the complex envelope $A(\mathbf{r}) = (A_0/z)\exp[-jk(x^2 + y^2)/2z]$ [see (2.2-17)] satisfies the paraxial Helmholtz equation (2.2-23). Show that the wave whose complex envelope is given by $A(\mathbf{r}) =$

$[A_1/q(z)]\exp[-jk(x^2+y^2)/2q(z)]$, where $q(z) = z + jz_0$ and z_0 is a constant, also satisfies the paraxial Helmholtz equation. This wave, called the Gaussian beam, is the subject of Chapter 3. Sketch the intensity of the Gaussian beam in the plane $z = 0$.

*2.3 RELATION BETWEEN WAVE OPTICS AND RAY OPTICS

We proceed to show that ray optics emerges as the limit of wave optics when the wavelength $\lambda_o \to 0$. Consider a monochromatic wave of free-space wavelength λ_o in a medium with refractive index $n(\mathbf{r})$ that varies sufficiently slowly with position so that the medium may be regarded as locally homogeneous. We write the complex amplitude in (2.2-5) in the form

$$U(\mathbf{r}) = a(\mathbf{r})\exp[-jk_o S(\mathbf{r})], \qquad (2.3\text{-}1)$$

where $a(\mathbf{r})$ is its magnitude, $-k_o S(\mathbf{r})$ is its phase, and $k_o = 2\pi/\lambda_o$ is the free-space wavenumber. We assume that $a(\mathbf{r})$ varies sufficiently slowly with $\mathbf{r}$ that it may be regarded as constant within the distance of a wavelength λ_o.

The wavefronts are the surfaces $S(\mathbf{r}) = $ constant and the wavefront normals point in the direction of the gradient vector ∇S. In the neighborhood of a given position $\mathbf{r}_0$, the wave can be locally regarded as a plane wave with amplitude $a(\mathbf{r}_0)$ and wavevector $\mathbf{k}$ with magnitude $k = n(\mathbf{r}_0)k_o$ and direction parallel to the gradient vector ∇S at $\mathbf{r}_0$. A different neighborhood exhibits a local plane wave of different amplitude and different wavevector.

In ray optics it was shown that the optical rays are normal to the equilevel surfaces of a function $S(\mathbf{r})$ called the eikonal (see Sec. 1.3C). We therefore associate the local wavevectors (wavefront normals) in wave optics with the ray of ray optics and recognize that the function $S(\mathbf{r})$, which is proportional to the phase of the wave, is nothing but the eikonal of ray optics (Fig. 2.3-1). This association has a formal mathematical basis, as will be demonstrated shortly. With this analogy, ray optics can serve to determine the approximate effects of optical components on the wavefront normals, as illustrated in Fig. 2.3-1.

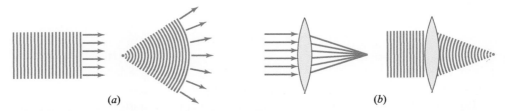

Figure 2.3-1 (*a*) The rays of ray optics are orthogonal to the wavefronts of wave optics (see also Fig. 1.3-10). (*b*) The effect of a lens on rays and wavefronts.

The Eikonal Equation

Substituting (2.3-1) into the Helmholtz equation (2.2-7) provides

$$k_o^2\left[n^2 - |\nabla S|^2\right]a + \nabla^2 a - jk_o\left[2\nabla S \cdot \nabla a + a\nabla^2 S\right] = 0, \qquad (2.3\text{-}2)$$

where $a = a(\mathbf{r})$ and $S = S(\mathbf{r})$. The real and imaginary parts of the left-hand side of (2.3-2) must both vanish. Equating the real part to zero and using $k_o = 2\pi/\lambda_o$, we obtain

$$|\nabla S|^2 = n^2 + \left(\frac{\lambda_o}{2\pi}\right)^2 \frac{\nabla^2 a}{a}. \tag{2.3-3}$$

The assumption that a varies slowly over the distance λ_o means that $\lambda_o^2 \nabla^2 a/a \ll 1$, so that the second term of the right-hand side may be neglected in the limit $\lambda_o \to 0$, whereupon

$$\boxed{|\nabla S|^2 \approx n^2.} \tag{2.3-4}$$
Eikonal Equation

This is the eikonal equation (1.3-20), which may be regarded as the main postulate of ray optics (Fermat's principle can be derived from the eikonal equation and *vice versa*).

Thus, the scalar function $S(\mathbf{r})$, which is proportional to the phase in wave optics, is the eikonal of ray optics. This is also consistent with the observation that in ray optics $S(\mathbf{r}_B) - S(\mathbf{r}_A)$ equals the optical pathlength between the points $\mathbf{r}_A$ and $\mathbf{r}_B$.

The eikonal equation is the limit of the Helmholtz equation when $\lambda_o \to 0$. Given $n(\mathbf{r})$ we may use the eikonal equation to determine $S(\mathbf{r})$. By equating the imaginary part of (2.3-2) to zero, we obtain a relation between a and S, thereby permitting us to determine the wavefunction.

2.4 SIMPLE OPTICAL COMPONENTS

In this section we examine the effects of optical components, such as mirrors, transparent plates, prisms, and lenses, on optical waves.

A. Reflection and Refraction

Reflection from a Planar Mirror

A plane wave of wavevector $\mathbf{k}_1$ is incident onto a planar mirror located in free space in the $z = 0$ plane. A reflected plane wave of wavevector $\mathbf{k}_2$ is created. The angles of incidence and reflection are θ_1 and θ_2, as illustrated in Fig. 2.4-1. The sum of the two waves satisfies the Helmholtz equation if the wavenumber is the same, i.e., if $k_1 = k_2 = k_o$. Certain boundary conditions must be satisfied at the surface of the mirror. Since these conditions are the same at all points (x, y), it is necessary that the phases of the two waves match, i.e.,

$$\mathbf{k}_1 \cdot \mathbf{r} = \mathbf{k}_2 \cdot \mathbf{r} \quad \text{for all } \mathbf{r} = (x, y, 0). \tag{2.4-1}$$

This *phase-matching condition* may also be regarded as matching of the tangential components of the two wavevectors in the mirror plane. Substituting $\mathbf{r} = (x, y, 0)$, $\mathbf{k}_1 = (k_o \sin \theta_1, 0, k_o \cos \theta_1)$, and $\mathbf{k}_2 = (k_o \sin \theta_2, 0, -k_o \cos \theta_2)$ into (2.4-1), we obtain $k_o x \sin \theta_1 = k_o x \sin \theta_2$, from which $\theta_1 = \theta_2$, so that the angles of incidence and reflection must be equal. Thus, the law of reflection of optical rays is applicable to the wavevectors of plane waves.

Reflection and Refraction at a Planar Dielectric Boundary

We now consider a plane wave of wavevector $\mathbf{k}_1$ incident on a planar boundary between two homogeneous media of refractive indices n_1 and n_2. The boundary lies in the

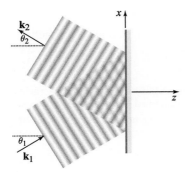

Figure 2.4-1 Reflection of a plane wave from a planar mirror. Phase matching at the surface of the mirror requires that the angles of incidence and reflection be equal.

$z = 0$ plane (Fig. 2.4-2). Refracted and reflected plane waves of wavevectors $\mathbf{k}_2$ and $\mathbf{k}_3$ emerge. The combination of the three waves satisfies the Helmholtz equation everywhere if each of the waves has the appropriate wavenumber in the medium in which it propagates ($k_1 = k_3 = n_1 k_o$ and $k_2 = n_2 k_0$).

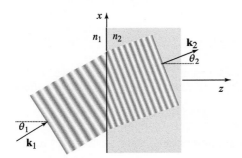

Figure 2.4-2 Refraction of a plane wave at a dielectric boundary. The wavefronts are matched at the boundary so that the distance between wavefronts for the incident wave, $\lambda_1 / \sin\theta_1 = \lambda_o/n_1 \sin\theta_1$, equals that for the refracted wave, $\lambda_2/\sin\theta_2 = \lambda_o/n_2 \sin\theta_2$, from which Snell's law follows.

Since the boundary conditions are invariant to x and y, it is necessary that the phases of the three waves match, i.e.,

$$\mathbf{k}_1 \cdot \mathbf{r} = \mathbf{k}_2 \cdot \mathbf{r} = \mathbf{k}_3 \cdot \mathbf{r} \qquad \text{for all } \mathbf{r} = (x, y, 0). \tag{2.4-2}$$

This *phase-matching condition* is tantamount to matching the tangential components of the three wavevectors at the boundary plane. Since $\mathbf{k}_1 = (n_1 k_o \sin\theta_1, 0, n_1 k_o \cos\theta_1)$, $\mathbf{k}_3 = (n_1 k_o \sin\theta_3, 0, -n_1 k_o \cos\theta_3)$, and $\mathbf{k}_2 = (n_2 k_o \sin\theta_2, 0, n_2 k_o \cos\theta_2)$, where θ_1, θ_2, and θ_3 are the angles of incidence, refraction, and reflection, respectively, it follows from (2.4-2) that $\theta_1 = \theta_3$ and $n_1 \sin\theta_1 = n_2 \sin\theta_2$. These are the laws of reflection and refraction (Snell's law) of ray optics, now applicable to the wavevectors.

It is not possible to determine the amplitudes of the reflected and refracted waves using scalar wave optics since the boundary conditions are not completely specified in this theory. This will be achieved in Sec. 6.2 using electromagnetic optics (Chapters 5 and 6).

B. Transmission Through Optical Components

We now proceed to examine the transmission of optical waves through transparent optical components such as plates, prisms, and lenses. The effect of reflection at the surfaces of these components will be ignored, since it cannot be properly accounted for using the scalar wave theory of light. Nor can the effect of absorption in the material, which is relegated to Sec. 5.5. The principal emphasis here is on the phase shift introduced by these components and on the associated wavefront bending.

Transparent Plate

Consider first the transmission of a plane wave through a transparent plate of refractive index n and thickness d surrounded by free space. The surfaces of the plate are the planes $z = 0$ and $z = d$ and the incident wave travels in the z direction (Fig. 2.4-3). Let $U(x, y, z)$ be the complex amplitude of the wave. Since external and internal reflections are ignored, $U(x, y, z)$ is assumed to be continuous at the boundaries. The ratio $t(x, y) = U(x, y, d)/U(x, y, 0)$ therefore represents the **complex amplitude transmittance** of the plate; it permits us to determine $U(x, y, d)$ for arbitrary $U(x, y, 0)$ at the input. The effect of reflection is considered in Sec. 6.2 and the effect of multiple internal reflections within the plate is examined in Sec. 11.1.

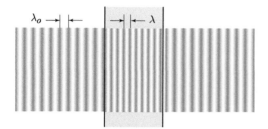

Figure 2.4-3 Transmission of a plane wave through a transparent plate.

Once inside the plate, the wave continues to propagate as a plane wave with wavenumber nk_o, so that $U(x, y, z)$ is proportional to $\exp(-jnk_o z)$. Thus, the ratio $U(x, y, d)/U(x, y, 0) = \exp(-jnk_o d)$, so that

$$t(x, y) = \exp(-jnk_o d). \qquad (2.4\text{-}3)$$

Transmittance
Transparent Plate

The plate is seen to introduce a phase shift $nk_o d = 2\pi(d/\lambda)$.

If the incident plane wave makes an angle θ with respect to the z axis and has wavevector $\mathbf{k}$ (Fig. 2.4-4), the refracted and transmitted waves are also plane waves with wavevectors $\mathbf{k}_1$ and $\mathbf{k}$ and angles θ_1 and θ, respectively, where θ_1 and θ are related by Snell's law: $\sin\theta = n\sin\theta_1$. The complex amplitude $U(x, y, z)$ inside the plate is now proportional to $\exp(-j\mathbf{k}_1 \cdot \mathbf{r}) = \exp[-jnk_o(z\cos\theta_1 + x\sin\theta_1)]$, so that the complex amplitude transmittance of the plate $U(x, y, d)/U(x, y, 0)$ is

$$t(x, y) = \exp\left(-jnk_o d \cos\theta_1\right). \qquad (2.4\text{-}4)$$

If the angle of incidence θ is small (i.e., if the incident wave is *paraxial*), then $\theta_1 \approx \theta/n$ is also small and the approximation $\cos\theta_1 \approx 1 - \frac{1}{2}\theta_1^2$ yields $t(x, y) \approx \exp(-jnk_o d)\exp(jk_o\theta^2 d/2n)$. If the plate is *sufficiently thin*, and the angle θ is *sufficiently small* such that $k_o\theta^2 d/2n \ll 2\pi$ [or $(d/\lambda_o)\theta^2/2n \ll 1$], then the transmittance of the plate may be approximated by (2.4-3). Under these conditions the transmittance of the plate is approximately independent of the angle θ.

Thin Transparent Plate of Varying Thickness

We now determine the amplitude transmittance of a thin transparent plate whose thickness $d(x, y)$ varies smoothly as a function of x and y, assuming that the incident wave is an arbitrary *paraxial* wave. The plate lies between the planes $z = 0$ and $z = d_0$, which are regarded as the boundaries encasing the optical component (Fig. 2.4-5).

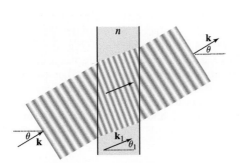

Figure 2.4-4 Transmission of an oblique plane wave through a thin transparent plate.

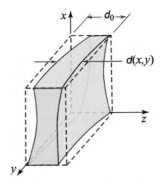

Figure 2.4-5 A transparent plate of varying thickness.

In the vicinity of the position $(x, y, 0)$ the incident paraxial wave may be regarded locally as a plane wave traveling along a direction that makes a small angle with the z axis. It crosses a thin plate of material of thickness $d(x, y)$ surrounded by thin layers of air of total thickness $d_0 - d(x, y)$. In accordance with the approximate relation (2.4-3), the local transmittance is the product of the transmittances of a thin layer of air of thickness $d_0 - d(x, y)$ and a thin layer of material of thickness $d(x, y)$, so that $\mathsf{t}(x, y) \approx \exp[-jnk_o d(x, y)] \exp[-jk_o(d_0 - d(x, y))]$, from which

$$\mathsf{t}(x, y) \approx h_0 \exp[-j(n-1)k_o d(x, y)], \qquad (2.4\text{-}5)$$

Transmittance
Variable-Thickness Plate

where $h_0 = \exp(-jk_o d_0)$ is a constant phase factor. This relation is valid in the paraxial approximation (where all angles θ are small) and when the thickness d_0 is sufficiently small so that $(d_0/\lambda_o)\theta^2/2n \ll 1$.

EXERCISE 2.4-1

Transmission Through a Prism. Use (2.4-5) to show that the complex amplitude transmittance of a thin inverted prism with small apex angle $\alpha \ll 1$ and thickness d_0 (Fig. 2.4-6) is $\mathsf{t}(x, y) = h_0 \exp[-j(n-1)\alpha k_o x]$, where $h_0 = \exp(-jk_o d_0)$. The transmittance is independent of y since the prism extends in the y direction. What is the effect of the prism on an incident plane wave traveling in the z direction? Compare your results with that obtained via the ray-optics model, as provided in (1.2-7).

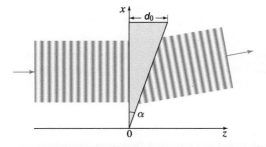

Figure 2.4-6 Transmission of a plane wave through a thin prism.

EXAMPLE 2.4-1. *Transmission Through a Biprism and an Axicon.* The biprism depicted in Fig. 1.2-12(a) comprises an inverted prism, such as that illustrated in Fig. 2.4-6, juxtaposed with an identical uninverted prism. Taking its thickness to be d_0 and its edge angle $\alpha \ll 1$, the results of Exercise 2.4-1 generalize to $t(x,y) = h_0\{\exp[-j(n-1)\alpha k_o x] + \exp[+j(n-1)\alpha k_o x]\} = 2h_0 \cos[(n-1)\alpha k_o x]$, with $h_0 = \exp(-jk_o d_0)$. The biprism thus converts an incident plane wave into a pair of waves that are tilted with respect to each other. The Fresnel biprism portrayed in Fig. 1.2-12(b) behaves in the same way.

The cone-shaped axicon shown in Fig. 1.2-12(c) is constructed by rotating the prism cross section depicted in Fig. 2.4-6 about a horizontal axis located at its top edge, from $\phi = -\pi$ to π. At any angle ϕ, the cross section of this device is an isosceles triangle of thickness d_0 and edge angle $\alpha \ll 1$. Using polar coordinates and integrating the results presented in Exercise 2.4-1 over ϕ provides $t(x,y) = h_0 \int_{-\pi}^{\pi} \exp[-j(n-1)\alpha(k_o \cos\phi)x - j(n-1)\alpha(k_o \sin\phi)y]\,d\phi = h_0 \int_{-\pi}^{\pi} \exp[-j(n-1)\alpha k_o \sqrt{x^2+y^2}\sin(\phi+\theta)]\,d\phi$. Since the integration is over 2π, the integral is independent of θ. Given that $\int_{-\pi}^{\pi} \exp(-ju\sin\phi)\,d\phi = 2\pi J_0(u)$, where $J_0(u)$ is the Bessel function of the first kind and zeroth order, the amplitude transmittance may be rewritten as $t(x,y) = 2\pi h_0 J_0[(n-1)\alpha k_o \sqrt{x^2+y^2}]$. The axicon thus converts an incident plane wave into an infinite number of plane waves, all directed toward its central axis in the form of a cone of half angle $(n-1)\alpha$. This device may be used to convert a plane wave into a Bessel beam (see Sec. 3.5A and Example 4.3-5).

Thin Lens

The general expression (2.4-5) for the complex amplitude transmittance of a thin transparent plate of variable thickness is now applied to the plano-convex thin lens shown in Fig. 2.4-7. Since the lens is the cap of a sphere of radius R, the thickness at the point (x,y) is $d(x,y) = d_0 - \overline{PQ} = d_0 - (R - \overline{QC})$, or

$$d(x,y) = d_0 - \left[R - \sqrt{R^2 - (x^2+y^2)}\right]. \quad (2.4\text{-}6)$$

This expression may be simplified by considering only points for which x and y are sufficiently small in comparison with R so that $x^2 + y^2 \ll R^2$. In that case

$$\sqrt{R^2 - (x^2+y^2)} = R\sqrt{1 - \frac{x^2+y^2}{R^2}} \approx R\left(1 - \frac{x^2+y^2}{2R^2}\right), \quad (2.4\text{-}7)$$

where we have used the same Taylor-series expansion that led to the Fresnel approximation of a spherical wave in (2.2-17). Using this approximation in (2.4-6) then provides

$$d(x,y) \approx d_0 - \frac{x^2+y^2}{2R}. \quad (2.4\text{-}8)$$

Finally, substitution into (2.4-5) yields

$$\boxed{t(x,y) \approx h_0 \exp\left[jk_o \frac{x^2+y^2}{2f}\right],} \quad (2.4\text{-}9)$$

Transmittance Thin Lens

where

$$f = \frac{R}{n-1} \quad (2.4\text{-}10)$$

is the focal length of the lens (see Sec. 1.2C) and $h_0 = \exp(-jnk_o d_0)$ is another constant phase factor that is usually of no significance.

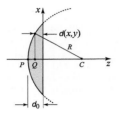

Figure 2.4-7 A plano-convex thin lens. The lens imparts a phase proportional to $x^2 + y^2$ to an incident plane wave, thereby transforming it into a paraboloidal wave centered at a distance f from the lens (see Exercise 2.4-3).

EXERCISE 2.4-2

Double-Convex Lens. Show that the complex amplitude transmittance of the double-convex lens (also called a spherical lens) shown in Fig. 2.4-8 is given by (2.4-9) with

$$\frac{1}{f} = (n-1)\left(\frac{1}{R_1} - \frac{1}{R_2}\right). \qquad (2.4\text{-}11)$$

You may prove this either by using the general formula (2.4-5) or by regarding the double-convex lens as a cascade of two plano-convex lenses. Recall that, by convention, the radius of a convex/concave surface is positive/negative, so that R_1 is positive and R_2 is negative for the lens displayed in Fig. 2.4-8. The parameter f is recognized as the focal length of the lens [see (1.2-12)].

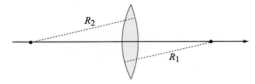

Figure 2.4-8 A double-convex lens.

EXERCISE 2.4-3

Focusing of a Plane Wave by a Thin Lens. Show that when a plane wave is transmitted through a thin lens of focal length f in a direction parallel to the axis of the lens, it is converted into a paraboloidal wave (the Fresnel approximation of a spherical wave) centered about a point at a distance f from the lens, as illustrated in Fig. 2.4-9. What is the effect of the lens on a plane wave incident at a small angle θ?

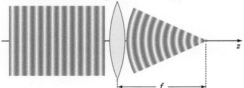

Figure 2.4-9 A thin lens transforms a plane wave into a paraboloidal wave.

EXERCISE 2.4-4

Imaging Property of a Lens. Show that a paraboloidal wave centered at the point P_1 (Fig. 2.4-10) is converted by a lens of focal length f into a paraboloidal wave centered at P_2, where $1/z_1 + 1/z_2 = 1/f$, a formula known as the **imaging equation**.

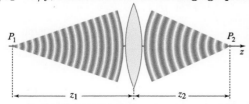

Figure 2.4-10 A lens transforms a paraboloidal wave into another paraboloidal wave. The two waves are centered at distances that satisfy the imaging equation.

Diffraction Gratings

A **diffraction grating** is an optical component that serves to periodically modulate the phase or amplitude of an incident wave. It can be made of a transparent plate with periodically varying thickness or periodically graded refractive index (see Sec. 2.4C). Repetitive arrays of diffracting elements such as apertures, obstacles, or absorbing elements (see Sec. 4.3) can also be used for this purpose. A reflection diffraction grating is often fabricated from a periodically ruled thin film of aluminum that has been evaporated onto a glass substrate.

Consider a diffraction grating made of a thin transparent plate placed in the $z = 0$ plane whose thickness varies periodically in the x direction with period Λ (Fig. 2.4-11). As will be demonstrated in Exercise 2.4-5, this plate converts an incident plane wave of wavelength $\lambda \ll \Lambda$, traveling at a small angle θ_i with respect to the z axis, into several plane waves at small angles with respect to the z axis:

$$\boxed{\theta_q \approx \theta_i + q\frac{\lambda}{\Lambda},} \qquad (2.4\text{-}12)$$

Grating Equation

where $q = 0, \pm 1, \pm 2, \ldots$, is called the diffraction order. Successive diffracted waves are separated by an angle $\theta = \lambda/\Lambda$, as shown schematically in Fig. 2.4-11.

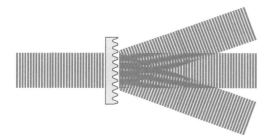

Figure 2.4-11 A thin transparent plate with periodically varying thickness serves as a diffraction grating. It splits an incident plane wave into multiple plane waves traveling in different directions.

EXERCISE 2.4-5

Transmission Through a Diffraction Grating.

(a) The thickness of a thin transparent plate varies sinusoidally in the x direction, $d(x,y) = \frac{1}{2}d_0[1+\cos(2\pi x/\Lambda)]$, as illustrated in Fig. 2.4-11. Show that the complex amplitude transmittance is $t(x,y) = h_0 \exp[-j\frac{1}{2}(n-1)k_o d_0 \cos(2\pi x/\Lambda)]$ where $h_0 = \exp[-j\frac{1}{2}(n+1)k_o d_0]$.
(b) Show that an incident plane wave traveling at a small angle θ_i with respect to the z direction is transmitted in the form of a sum of plane waves traveling at angles θ_q given by (2.4-12). *Hint:* Expand the periodic function $t(x,y)$ in a Fourier series.

Equation (2.4-12) is valid only in the paraxial approximation, when all angles are small, and when the period Λ is much greater than the wavelength λ. A more general analysis of a thin diffraction grating that does not rely on the paraxial approximation reveals that an incident plane wave at an angle θ_i gives rise to a collection of plane waves at angles θ_q that satisfy

$$\sin\theta_q = \sin\theta_i + q\frac{\lambda}{\Lambda}. \qquad (2.4\text{-}13)$$

This result may be derived by expanding the periodic transmittance $t(x,y)$ as a sum of Fourier components of the form $\exp(-jq2\pi x/\Lambda)$, where $q = 0, \pm 1, \pm 2, \ldots$ is the diffraction order. An incident plane wave $\exp(-jkx\sin\theta_i)$, modulated by the harmonic component $\exp(-jq2\pi x/\Lambda)$, generates a transmitted plane wave at the angle θ_q given by $\exp(-jkx\sin\theta_q) \propto \exp(-jkx\sin\theta_i)\exp(-jq2\pi x/\Lambda)$. This leads to the phase-matching condition $k\sin\theta_q = k\sin\theta_o + q2\pi/\Lambda$. Equation (2.4-13) follows since $k = 2\pi/\lambda$; this result is also applicable to waves reflected from the grating.

Diffraction gratings are used as filters and spectrum analyzers. Since the angles θ_q depend on the wavelength λ (and therefore on the frequency ν), an incident polychromatic wave is separated by the grating into its spectral components (Fig. 2.4-12). Diffraction gratings have found numerous applications in spectroscopy.

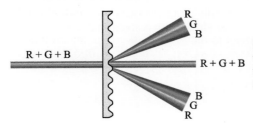

Figure 2.4-12 A diffraction grating directs two waves of different wavelengths, λ_1 and λ_2, into two different directions, θ_1 and θ_2. It therefore serves as a spectrum analyzer or a spectrometer.

C. Graded-Index Optical Components

The effect of a prism, lens, or diffraction grating on an incident optical wave lies in the phase shift it imparts, which serves to bend the wavefront in some prescribed manner. This phase shift is controlled by the variation in the thickness of the material with the transverse distance from the optical axis (linearly, quadratically, or periodically, in the cases of a prism, lens, and diffraction grating, respectively). The same phase shift may instead be introduced by a transparent planar plate of fixed thickness but with varying refractive index. This is a result of the fact that the thickness and refractive index appear as a product in (2.4-3).

The complex amplitude transmittance of a thin transparent planar plate of thickness d_0 and graded refractive index $n(x,y)$ is, from (2.4-3),

$$\boxed{t(x,y) = \exp\left[-jn(x,y)k_o d_0\right]}.$$
(2.4-14)
Transmittance
Graded-Index Thin Plate

By selecting the appropriate variation of $n(x,y)$ with x and y, the action of any constant-index thin optical component can be reproduced, as demonstrated in Exercise 2.4-6.

EXERCISE 2.4-6

Graded-Index Lens. Show that a thin plate of uniform thickness d_0 (Fig. 2.4-13) and quadratically graded refractive index $n(x,y) = n_0[1 - \frac{1}{2}\alpha^2(x^2+y^2)]$, with $\alpha d_0 \ll 1$, acts as a lens of focal length $f = 1/n_0 d_0 \alpha^2$ (see Exercise 1.3-1).

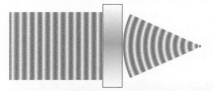

Figure 2.4-13 A graded-index plate acts as a lens.

2.5 INTERFERENCE

When two or more optical waves are simultaneously present in the same region of space and time, the total wavefunction is the sum of the individual wavefunctions. This basic principle of superposition follows from the linearity of the wave equation. For monochromatic waves of the same frequency, the superposition principle carries over to the complex amplitudes, which follows from the linearity of the Helmholtz equation.

The superposition principle does not apply to the optical intensity since the intensity of the sum of two or more waves is not necessarily the sum of their intensities. The disparity is associated with interference. The phenomenon of interference cannot be explained on the basis of ray optics since it is dependent on the phase relationship between the superposed waves.

In this section we examine the interference between two or more monochromatic waves of the same frequency. The interference of waves of different frequencies is discussed in Sec. 2.6.

A. Interference of Two Waves

When two monochromatic waves with complex amplitudes $U_1(\mathbf{r})$ and $U_2(\mathbf{r})$ are superposed, the result is a monochromatic wave of the same frequency that has a complex amplitude

$$U(\mathbf{r}) = U_1(\mathbf{r}) + U_2(\mathbf{r}). \qquad (2.5\text{-}1)$$

In accordance with (2.2-10), the intensities of the constituent waves are $I_1 = |U_1|^2$ and $I_2 = |U_2|^2$, while the intensity of the total wave is

$$I = |U|^2 = |U_1 + U_2|^2 = |U_1|^2 + |U_2|^2 + U_1^* U_2 + U_1 U_2^*. \qquad (2.5\text{-}2)$$

The explicit dependence on $\mathbf{r}$ has been omitted for convenience. Substituting

$$U_1 = \sqrt{I_1}\, \exp(j\varphi_1) \quad \text{and} \quad U_2 = \sqrt{I_2}\, \exp(j\varphi_2) \qquad (2.5\text{-}3)$$

into (2.5-2), where φ_1 and φ_2 are the phases of the two waves, we obtain

$$\boxed{I = I_1 + I_2 + 2\sqrt{I_1 I_2}\, \cos\varphi,} \qquad (2.5\text{-}4)$$
Interference Equation

with

$$\varphi = \varphi_2 - \varphi_1. \qquad (2.5\text{-}5)$$

This relation, called the **interference equation**, can also be understood in terms of the geometry of the phasor diagram displayed in Fig. 2.5-1(a), which demonstrates that the magnitude of the phasor U is sensitive not only to the magnitudes of the constituent phasors but also to the phase difference φ.

It is clear, therefore, that the intensity of the sum of the two waves is *not* the sum of their intensities [Fig. 2.5-1(b)]; an additional term, attributed to **interference** between the two waves, is present in (2.5-4). This term may be positive or negative, corresponding to constructive or destructive interference, respectively. If $I_1 = I_2 = I_0$, for example, then (2.5-4) yields $I = 2I_0(1 + \cos\varphi) = 4I_0 \cos^2(\varphi/2)$, so that for $\varphi = 0$, $I = 4I_0$ (i.e., the total intensity is four times the intensity of each of the

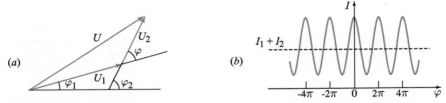

Figure 2.5-1 (a) Phasor diagram for the superposition of two waves of intensities I_1 and I_2 and phase difference $\varphi = \varphi_2 - \varphi_1$. (b) Dependence of the total intensity I on the phase difference φ.

superposed waves). For $\varphi = \pi$, on the other hand, the superposed waves cancel one another and the total intensity $I = 0$. Complete cancellation of the intensity in a region of space is generally not possible unless the intensities of the constituent superposed waves are equal. When $\varphi = \pi/2$ or $3\pi/2$, the interference term vanishes and $I = 2I_0$; for these special phase relationships the total intensity is the sum of the constituent intensities. The strong dependence of the intensity I on the phase difference φ permits us to measure phase differences by detecting light intensity. This principle is used in numerous optical systems.

Interference is accompanied by a spatial redistribution of the optical intensity without a violation of power conservation. For example, the two waves may have uniform intensities I_1 and I_2 in a particular plane, but as a result of a position-dependent phase difference φ, the total intensity can be smaller than $I_1 + I_2$ at some positions and larger at others, with the total power (integral of the intensity) conserved.

Interference is not observed under ordinary lighting conditions since the random fluctuations of the phases φ_1 and φ_2 cause the phase difference φ to assume random values that are uniformly distributed between 0 and 2π, so that $\cos\varphi$ averages to 0 and the interference term washes out. Light with such randomness is said to be *partially coherent* and Chapter 12 is devoted to its study. The analysis carried out here, and in subsequent chapters prior to Chapter 12, assume that the light is *coherent*, and therefore deterministic.

Interferometers

Consider the superposition of two plane waves, each of intensity I_0, propagating in the z direction, and assume that one wave is delayed by a distance d with respect to the other so that $U_1 = \sqrt{I_0}\exp(-jkz)$ and $U_2 = \sqrt{I_0}\exp[-jk(z-d)]$. The intensity I of the sum of these two waves can be determined by substituting $I_1 = I_2 = I_0$ and $\varphi = kd = 2\pi d/\lambda$ into the interference equation (2.5-4),

$$I = 2I_0\left[1 + \cos\left(2\pi\frac{d}{\lambda}\right)\right]. \qquad (2.5\text{-}6)$$

The dependence of I on the delay d is sketched in Fig. 2.5-2. When the delay is an integer multiple of λ, complete constructive interference occurs and the total intensity $I = 4I_0$. On the other hand, when d is an odd integer multiple of $\lambda/2$, complete destructive interference occurs and $I = 0$. The average intensity is the sum of the two intensities, i.e., $2I_0$.

An **interferometer** is an optical instrument that splits a wave into two waves using a beamsplitter, delays them by unequal distances, redirects them using mirrors, recombines them using another (or the same) beamsplitter, and detects the intensity of their superposition. Three important examples are illustrated in Fig. 2.5-3: the **Mach–Zehnder interferometer**, the **Michelson interferometer**, and the **Sagnac interferometer**.

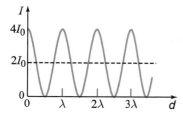

Figure 2.5-2 Dependence of the intensity I of the superposition of two waves, each of intensity I_0, on the delay distance d. When the delay distance is a multiple of λ, the interference is constructive; when it is an odd multiple of $\lambda/2$, the interference is destructive.

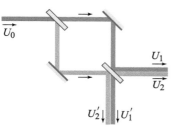

(a) Mach–Zehnder

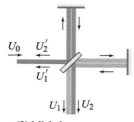

(b) Michelson

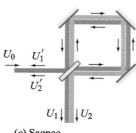

(c) Sagnac

Figure 2.5-3 Interferometers: A wave U_0 is split into two waves U_1 and U_2 (they are shown as shaded light and dark for ease of visualization but are actually congruent). After traveling through different paths, the waves are recombined into a superposition wave $U = U_1 + U_2$ whose intensity is recorded. The waves are split and recombined using beamsplitters. In the Sagnac interferometer the two waves travel through the same path, but in opposite directions.

Since the intensity I is sensitive to the phase $\varphi = 2\pi d/\lambda = 2\pi nd/\lambda_0 = 2\pi n\nu d/c_o$, where d is the difference between the distances traveled by the two waves, the interferometer can be used to measure small changes in the distance d, the refractive index n, or the wavelength λ_o (or frequency ν). For example, if $d/\lambda_o = 10^4$, a change of the refractive index of only $\Delta n = 10^{-4}$ corresponds to an easily observable phase change $\Delta\varphi = 2\pi$. The phase φ also changes by a full 2π if d changes by a wavelength λ. An incremental change of the frequency $\Delta\nu = c/d$ has the same effect.

Interferometers have numerous applications. These include the determination of distance in metrological applications such as strain measurement and surface profiling; refractive-index measurements; and spectrometry for the analysis of polychromatic light (see Sec. 12.2B). In the Sagnac interferometer the optical paths are identical but opposite in direction, so that rotation of the interferometer results in a phase shift φ proportional to the angular velocity of rotation. This system can therefore be used as a gyroscope. Because of its precision, optical interferometry is also being co-opted to detect the passage of gravitational waves, as discussed subsequently.

Finally, we demonstrate that energy conservation in an interferometer requires that the phases of the waves reflected and transmitted at a beamsplitter differ by $\pi/2$. Each of the interferometers considered in Fig. 2.5-3 has an output wave $U = U_1 + U_2$ that exits from one side of the beamsplitter and also another output wave $U' = U'_1 + U'_2$ that exits from the opposite side. Energy conservation dictates that the sum of the intensities of these two waves must equal the intensity of the incident wave, so that if one output wave has high intensity by virtue of constructive interference, the other must have low intensity by virtue of destructive interference. This complementarity can only be achieved if the phase differences φ and φ', associated with the components of output waves U and U', respectively, differ by π. Since the components of U and the components of U' experience the same pathlength differences, and the same numbers of reflections from mirrors, the π phase difference must be attributable to different phases introduced by the beamsplitter upon reflection and transmission. Examination

Interference of Two Oblique Plane Waves

Consider now the interference of two plane waves of equal intensities: one propagating in the z direction, $U_1 = \sqrt{I_0} \exp(-jkz)$; the other propagating at an angle θ with respect to the z axis, in the x–z plane, $U_2 = \sqrt{I_0} \exp[-j(k\cos\theta\ z + k\sin\theta\ x)]$, as illustrated in Fig. 2.5-4. At the $z = 0$ plane the two waves have a phase difference $\varphi = k\sin\theta\ x$, for which the interference equation (2.5-4) yields a total intensity

$$I = 2I_0\left[1 + \cos(k\sin\theta\ x)\right]. \tag{2.5-7}$$

This pattern varies sinusoidally with x, with period $2\pi/k\sin\theta = \lambda/\sin\theta$, as shown in Fig. 2.5-4. If $\theta = 30°$, for example, the period is 2λ. This suggests a method of printing a sinusoidal pattern of high resolution for use as a diffraction grating. It also suggests a method of monitoring the angle of arrival θ of a wave by mixing it with a reference wave and recording the resultant intensity distribution. As discussed in Sec. 4.5, this is the principle that lies behind holography.

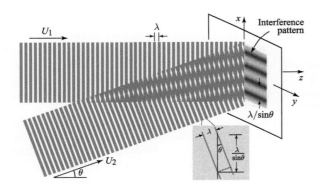

Figure 2.5-4 The interference of two plane waves traveling at an angle θ with respect to each other results in a sinusoidal intensity pattern in the x direction with period $\lambda/\sin\theta$.

EXERCISE 2.5-1

Interference of a Plane Wave and a Spherical Wave. A plane wave traveling along the z direction with complex amplitude $A_1 \exp(-jkz)$, and a spherical wave centered at $z = 0$ and approximated by the paraboloidal wave of complex amplitude $(A_2/z)\exp(-jkz)\exp[-jk(x^2 + y^2)/2z]$ [see (2.2-17)], interfere in the $z = d$ plane. Derive an expression for the total intensity $I(x, y, d)$. Assuming that the two waves have the same intensities at the $z = d$ plane, verify that the locus of points of zero intensity is a set of concentric rings, as illustrated in Fig. 2.5-5.

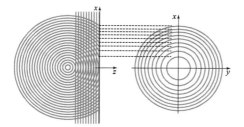

Figure 2.5-5 The interference of a plane wave and a spherical wave creates a pattern of concentric rings (illustrated at the plane $z = d$).

EXERCISE 2.5-2

Interference of Two Spherical Waves. Two spherical waves of equal intensity I_0, originating at the points $(-a, 0, 0)$ and $(a, 0, 0)$, interfere in the plane $z = d$ as illustrated in Fig. 2.5-6. This double-pinhole system is similar to that used by Thomas Young in his celebrated double-slit experiment in which he demonstrated interference. Use the paraboloidal approximation for the spherical waves to show that the intensity at the plane $z = d$ is

$$I(x, y, d) \approx 2I_0 \left(1 + \cos \frac{2\pi x \theta}{\lambda} \right), \qquad (2.5\text{-}8)$$

where the angle subtended by the centers of the two waves at the observation plane is $\theta \approx 2a/d$. The intensity pattern is periodic with period λ/θ.

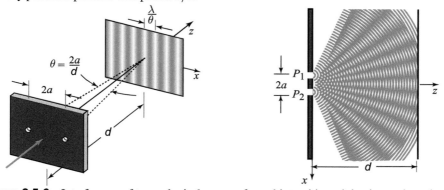

Figure 2.5-6 Interference of two spherical waves of equal intensities originating at the points P_1 and P_2. The two waves can be obtained by permitting a plane wave to impinge on two pinholes in a screen. The light intensity at an observation plane a large distance d from the pinholes takes the form of a sinusoidal interference pattern, with period $\approx \lambda/\theta$, along the direction of the line connecting the pinholes.

B. Multiple-Wave Interference

The superposition of M monochromatic waves of the same frequency, with complex amplitudes $U_1, U_2, \ldots, U_M$, gives rise to a wave whose frequency remains the same and whose complex amplitude is given by $U = U_1 + U_2 + \cdots + U_M$. Knowledge of the intensities of the individual waves, $I_1, I_2, \ldots, I_M$, is not sufficient to determine the total intensity $I = |U|^2$ since the relative phases must also be known. The role played by the phase is dramatically illustrated in the following examples.

Interference of M Waves with Equal Amplitudes and Equal Phase Differences

We first examine the interference of M waves with complex amplitudes

$$U_m = \sqrt{I_0}\, \exp[j(m-1)\varphi], \qquad m = 1, 2, \ldots, M. \tag{2.5-9}$$

The waves have equal intensities I_0, and phase difference φ between successive waves, as illustrated in Fig. 2.5-7(a). To derive an expression for the intensity of the superposition, it is convenient to introduce the quantity $h = \exp(j\varphi)$ whereupon $U_m = \sqrt{I_0}\, h^{m-1}$. The complex amplitude of the superposed wave is then

$$U = \sqrt{I_0}\,(1 + h + h^2 + \cdots + h^{M-1}) = \sqrt{I_0}\,\frac{1 - h^M}{1 - h}$$

$$= \sqrt{I_0}\,\frac{1 - \exp(jM\varphi)}{1 - \exp(j\varphi)}, \tag{2.5-10}$$

which has the corresponding intensity

$$I = |U|^2 = I_0 \left| \frac{\exp(-jM\varphi/2) - \exp(jM\varphi/2)}{\exp(-j\varphi/2) - \exp(j\varphi/2)} \right|^2, \tag{2.5-11}$$

whence

$$\boxed{I = I_0 \frac{\sin^2(M\varphi/2)}{\sin^2(\varphi/2)}.} \tag{2.5-12}$$

Interference of M Waves

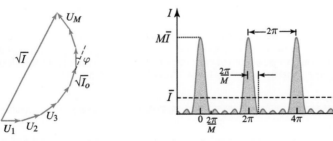

Figure 2.5-7 (a) The sum of M phasors of equal magnitudes and equal phase differences. (b) The intensity I as a function of φ. The peak intensity occurs when all the phasors are aligned; it is then M times greater than the mean intensity $\bar{I} = M I_0$. In this example $M = 5$.

The intensity I is evidently strongly dependent on the phase difference φ, as illustrated in Fig. 2.5-7(b) for $M = 5$. When $\varphi = 2\pi q$, where q is an integer, all the phasors are aligned so that the amplitude of the total wave is M times that of an individual component, and the intensity reaches its peak value of $M^2 I_0$. The mean intensity averaged over a uniform distribution of φ is $\bar{I} = (1/2\pi)\int_0^{2\pi} I\,d\varphi = M I_0$, which is the same as the result obtained in the absence of interference. The peak intensity is therefore M times greater than the mean intensity. The sensitivity of the intensity to the phase is therefore dramatic for large M. At its peak value, the intensity is magnified by a factor M over the mean but it decreases sharply as the phase difference φ deviates slightly from $2\pi q$. In particular, when $\varphi = 2\pi/M$ the intensity becomes zero. It is instructive to compare Fig. 2.5-7(b) for $M = 5$ with Fig. 2.5-2 for $M = 2$.

EXERCISE 2.5-3

Bragg Reflection. Consider light reflected at an angle θ from M parallel reflecting planes separated by a distance Λ, as shown in Fig. 2.5-8. Assume that only a small fraction of the light is reflected from each plane, so that the amplitudes of the M reflected waves are approximately equal. Show that the reflected waves have a phase difference $\varphi = k(2\Lambda \sin\theta)$ and that the angle θ at which the intensity of the total reflected light is maximum satisfies

$$\sin\theta_\mathcal{B} = \frac{\lambda}{2\Lambda}.$$

(2.5-13)
Bragg Angle

This equation defines the **Bragg angle** $\theta_\mathcal{B}$. Such reflections are encountered when light is reflected from a multilayer structure (see Sec. 7.1) or when X-ray waves are reflected from atomic planes in crystalline structures. It also occurs when light is reflected from a periodic structure created by an acoustic wave (see Chapter 20). An exact treatment of Bragg reflection is provided in Sec. 7.1C.

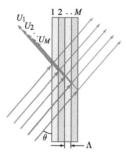

Figure 2.5-8 Reflection of a plane wave from M parallel planes separated from each other by a distance Λ. The reflected waves interfere constructively and yield maximum intensity when the angle θ is the Bragg angle $\theta_\mathcal{B}$. Note that θ is defined with respect to the parallel planes.

Fresnel Zone Plate

A **Fresnel zone plate** comprises a set of ring apertures of increasing radii, decreasing widths, and equal areas, as illustrated in Fig. 2.5-9.

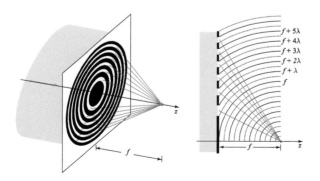

Figure 2.5-9 The Fresnel zone plate serves as a spherical lens with multiple focal lengths.

The structure serves as a spherical lens with multiple focal lengths, as may be understood from the perspective of interference. The center of the mth ring has a radius

ρ_m at the mth peak of the cosine function, i.e., $\pi \rho_m^2 / \lambda f = m 2\pi$ [see (2.4-9)]. At a focal point $z = f$, the distance R_m to the mth ring is given by $R_m^2 = f^2 + \rho_m^2$, so that $R_m = \sqrt{f^2 + 2m\lambda f}$. If f is sufficiently large so that the angles subtended by the rings are small, then a Taylor-series expansion provides $R_m \approx f + m\lambda$. Thus, the waves transmitted through consecutive rings have pathlengths differing by a wavelength, so that they interfere constructively at the focal point. A similar argument applies for the other foci. The operation of the Fresnel zone plate may also be understood from the perspective of Fourier optics, as explained in Sec. 4.1A.

Interference of an Infinite Number of Waves of Progressively Smaller Amplitudes and Equal Phase Differences

We now examine the superposition of an infinite number of waves with equal phase differences and with amplitudes that decrease at a geometric rate:

$$U_1 = \sqrt{I_0}, \quad U_2 = hU_1, \quad U_3 = hU_2 = h^2 U_1, \quad \ldots, \quad (2.5\text{-}14)$$

where $h = |h|e^{j\varphi}$, $|h| < 1$, and I_0 is the intensity of the initial wave. The amplitude of the mth wave is smaller than that of the $(m-1)$st wave by the factor $|h|$ and the phase differs by φ. The phasor diagram is shown in Fig. 2.5-10(a).

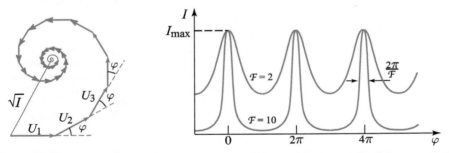

Figure 2.5-10 (a) The sum of an infinite number of phasors whose magnitudes are successively reduced at a geometric rate and whose phase differences φ are equal. (b) Dependence of the intensity I on the phase difference φ for two values of $\mathcal{F}$. Peak values occur at $\varphi = 2\pi q$. The full width at half maximum of each peak is approximately $2\pi / \mathcal{F}$ when $\mathcal{F} \gg 1$. The sharpness of the peaks increases with increasing $\mathcal{F}$.

The superposition wave has a complex amplitude

$$\begin{aligned} U &= U_1 + U_2 + U_3 + \cdots \\ &= \sqrt{I_0}\,(1 + h + h^2 + \cdots) \\ &= \frac{\sqrt{I_0}}{1-h} = \frac{\sqrt{I_0}}{1-|h|e^{j\varphi}}. \end{aligned} \quad (2.5\text{-}15)$$

The total intensity is then

$$I = |U|^2 = \frac{I_0}{|1-|h|e^{j\varphi}|^2} = \frac{I_0}{(1-|h|\cos\varphi)^2 + |h|^2 \sin^2\varphi}, \quad (2.5\text{-}16)$$

from which

$$I = \frac{I_0}{(1-|h|)^2 + 4|h|\sin^2(\varphi/2)}. \quad (2.5\text{-}17)$$

It is convenient to write this equation in the form

$$I = \frac{I_{\max}}{1 + (2\mathcal{F}/\pi)^2 \sin^2(\varphi/2)}, \qquad I_{\max} = \frac{I_0}{(1-|h|)^2}, \qquad (2.5\text{-}18)$$

Intensity of an Infinite Number of Waves

where the quantity

$$\mathcal{F} = \frac{\pi\sqrt{|h|}}{1-|h|} \qquad (2.5\text{-}19)$$

Finesse

is a parameter known as the **finesse**.

The intensity I is a periodic function of φ with period 2π, as illustrated in Fig. 2.5-10(b). It reaches its maximum value $I_{\max}$ when $\varphi = 2\pi q$, where q is an integer. This occurs when the phasors align to form a straight line. (This result is not unlike that displayed in Fig. 2.5-7(b) for the interference of M waves of equal amplitudes and equal phase differences.) When the finesse $\mathcal{F}$ is large (i.e., the factor $|h|$ is close to 1), I becomes a sharply peaked function of φ. Consider values of φ near the $\varphi = 0$ peak, as a representative example. For $|\varphi| \ll 1$, $\sin(\varphi/2) \approx \varphi/2$ whereupon (2.5-18) can be written as

$$I \approx \frac{I_{\max}}{1 + (\mathcal{F}/\pi)^2 \varphi^2}. \qquad (2.5\text{-}20)$$

The intensity I then decreases to half its peak value when $\varphi = \pi/\mathcal{F}$, so that the full-width at half-maximum (FWHM) of the peak becomes

$$\Delta\varphi \approx \frac{2\pi}{\mathcal{F}}. \qquad (2.5\text{-}21)$$

Width of Interference Pattern

In the regime $\mathcal{F} \gg 1$, we then have $\Delta\varphi \ll 2\pi$ and the assumption that $\varphi \ll 1$ is applicable. The finesse $\mathcal{F}$ is the ratio of the period 2π to the FWHM of the peaks in the interference pattern. It is therefore a measure of the sharpness of the interference function, i.e., the sensitivity of the intensity to deviations of φ from the values $2\pi q$ corresponding to the peaks.

A useful device based on this principle is the Fabry–Perot interferometer. It consists of two parallel mirrors within which light undergoes multiple reflections. In the course of each *round trip*, the light suffers a fixed amplitude reduction $|h| = |\mathrm{r}|$, arising from losses at the mirrors, and a phase shift $\varphi = k2d = 4\pi\nu d/c = 2\pi\nu/(c/2d)$ associated with the propagation, where d is the mirror separation. The total light intensity depends on the phase shift φ in accordance with (2.5-18), attaining maxima when $\varphi/2$ is an integer multiple of π. The proportionality of the phase shift φ to the optical frequency ν shows that the intensity transmission of the Fabry–Perot device will exhibit peaks separated in frequency by $c/2d$. The width of these peaks will be $(c/2d)/\mathcal{F}$, where the finesse $\mathcal{F}$ is governed by the loss via (2.5-19). The Fabry–Perot interferometer, which also serves as a spectrum analyzer, is considered further in Sec. 7.1B. It is commonly used as a resonator for lasers, as discussed in Secs. 11.1 and 16.1A.

EXAMPLE 2.5-1. *The LIGO Interferometer.* The LIGO interferometer[†] comprises a Michelson interferometer (MI) with a Fabry–Perot interferometer (FPI) embedded in each of its reflecting arms, as illustrated in Fig. 2.5-11. The MI is sensitive to the phase difference encountered by the optical waves that propagate through its arms; the FPIs serve to amplify the phases in each arm and thereby to significantly increase the sensitivity of the overall instrument.

If the phase shift encountered in a double pass within the FPI is denoted φ, the phase of the overall intracavity reflecting field U is, in accordance with (2.5-15),

$$\arg\{U\} = \arg\left\{\frac{1}{1 - |h|e^{j\varphi}}\right\} = \arctan\left\{\frac{|h|\sin\varphi}{1 - |h|\cos\varphi}\right\}. \qquad (2.5\text{-}22)$$

If φ is taken to be an integer multiple of 2π, to which is added a very small double-pass deviation $2\Delta\varphi \ll \pi$, a Taylor-series expansion of (2.5-22) yields $\arg\{U\} \approx 2\Delta\varphi|h|/(1 - |h|)$. This result is closely related to the finesse of the FPI, $\mathcal{F} = \pi\sqrt{|h|}/(1 - |h|)$, as provided in (2.5-19). When $|h| \approx 1$ and the finesse is high, we have $\arg\{U\} \approx 2\Delta\varphi \cdot 1/(1 - |h|)$ and $\mathcal{F} \approx \pi/(1 - |h|)$, so that $\arg\{U\} \approx (2\mathcal{F}/\pi)\Delta\varphi$. Thus, a very small phase deviation $\Delta\varphi$ imposed on the FPI is amplified by the factor $2\mathcal{F}/\pi$, which is large. This phase amplification results from the many reflections of the light between the mirrors of the FPI, which effectively increases its length and thus its sensitivity.

The interference pattern associated with the Michelson interferometer is characterized by the two-wave interference equation (2.5-4). If the light injected into both of its arms is of equal intensity, i.e., if $I_1 = I_2 = \frac{1}{2}I_0$, (2.5-4) becomes $I = I_0[1 + \cos(\varphi_2 - \varphi_1)]$. If the interferometer is then operated at a null and the phases for the two arms are taken to be $\varphi_{2,1} = (2\mathcal{F}/\pi)\Delta\varphi_{2,1}$, the LIGO interference pattern is given by

$$1 - \cos\tfrac{2\mathcal{F}}{\pi}(\Delta\varphi_2 - \Delta\varphi_1). \qquad (2.5\text{-}23)$$

The LIGO interferometer is thus a factor of $2\mathcal{F}/\pi$ more sensitive to the phase difference $\Delta\varphi_2 - \Delta\varphi_1$ than is a Michelson interferometer with the same arm lengths.

This increased sensitivity is the rationale for using the LIGO interferometer as a gravitational-wave detector. Generated by cataclysmic events in the distant universe, gravitational waves impose a dynamic strain on the fabric of space, which results in differential length variations in the orthogonal arms of the interferometer. This in turn modulates the phase difference $\Delta\varphi_2 - \Delta\varphi_1$, resulting in an overall light intensity whose magnitude is proportional to the gravitational-wave-induced strain. Gravitational waves were first detected by LIGO in 2015, a hundred years after Einstein first predicted their existence.[‡]

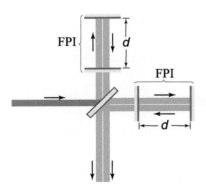

Figure 2.5-11 The LIGO interferometer is a Michelson interferometer (MI) with Fabry–Perot interferometers (FPIs) nested in each of its arms. Each FPI in the advanced-LIGO instrument has a length $d \approx 4$ km and a finesse $\mathcal{F} \approx 450$, so that the enhancement in sensitivity with respect to an ordinary MI is $2\mathcal{F}/\pi \approx 286$. The gravitational wave observed in 2015 imparted to the LIGO interferometer a differential spatial strain $(\Delta d_2 - \Delta d_1)/d \equiv \Delta d/d$ with a magnitude of roughly 5×10^{-22}, which corresponds to a differential length deviation Δd of about 2 am (some 400 times smaller than the radius of a proton). The light source was a 20-W Nd:YAG laser operated at $\lambda_o = c_o/\nu = 1.064$ μm. The corresponding phase difference $\Delta\varphi_2 - \Delta\varphi_1$ thus had a magnitude of $2\pi\nu\Delta d/c_o \approx 1.8 \times 10^{-11}$ rad; its oscillations were in the audio-frequency range.

[†] LIGO is an acronym for Laser Interferometer Gravitational-Wave Observatory, a facility with dual sites in Livingston, Louisiana and Hanford, Washington.

[‡] B. P. Abbott *et al.*, Observation of Gravitational Waves from a Binary Black Hole Merger, *Physical Review Letters*, vol. 116, 061102, 2016. The near-simultaneous detections at both LIGO sites, which are separated by a distance of ≈ 3000 km and a time of ≈ 10 msec, unequivocally confirmed the cosmological origin of the waves.

2.6 POLYCHROMATIC AND PULSED LIGHT

Since the wavefunction of monochromatic light is a harmonic function of time extending over all time (from $-\infty$ to ∞), it is an idealization that cannot be met in reality. This section is devoted to waves of arbitrary time dependence, including optical pulses of finite time duration. Such waves are polychromatic rather than monochromatic. A more detailed introduction to the optics of pulsed light is provided in Chapter 23.

A. Temporal and Spectral Description

Although a polychromatic wave is described by a wavefunction $u(\mathbf{r}, t)$ with nonharmonic time dependence, it may be expanded as a superposition of harmonic functions, each of which represents a monochromatic wave. Since we already know how monochromatic waves propagate in free space and through various optical components, we can determine the effect of optical systems on polychromatic light by using the principle of superposition.

Fourier methods permit the expansion of an arbitrary function of time $u(t)$, representing the wavefunction $u(\mathbf{r}, t)$ at a fixed position $\mathbf{r}$, as a superposition integral of harmonic functions of different frequencies, amplitudes, and phases:

$$u(t) = \int_{-\infty}^{\infty} v(\nu) \exp(j2\pi\nu t) \, d\nu, \qquad (2.6\text{-}1)$$

where $v(\nu)$ is determined by carrying out the **Fourier transform**

$$v(\nu) = \int_{-\infty}^{\infty} u(t) \exp(-j2\pi\nu t) \, dt. \qquad (2.6\text{-}2)$$

A review of the Fourier transform and its properties is presented in Sec. A.1 of Appendix A. The expansion in (2.6-1) extends over positive and negative frequencies. However, since $u(t)$ is real, $v(-\nu) = v^*(\nu)$ (see Sec. A.1). Thus, the negative-frequency components are not independent; they are simply conjugated versions of the corresponding positive-frequency components.

Complex Representation

It is convenient to represent the real function $u(t)$ in (2.6-1) by a complex function

$$U(t) = 2\int_0^{\infty} v(\nu) \exp(j2\pi\nu t) \, d\nu \qquad (2.6\text{-}3)$$

that includes only the positive-frequency components (multiplied by a factor of 2), and suppresses all the negative frequencies. The Fourier transform of $U(t)$ is therefore a function $V(\nu) = 2v(\nu)$ for $\nu \geq 0$, and 0 for $\nu < 0$.

The real function $u(t)$ can be determined from its complex representation $U(t)$ by simply taking the real part,

$$u(t) = \text{Re}\{U(t)\} = \tfrac{1}{2}[U(t) + U^*(t)]. \qquad (2.6\text{-}4)$$

The complex function $U(t)$ is known as the **complex analytic signal**. The validity of (2.6-4) can be verified by breaking the integral in (2.6-1) into two parts, with limits from 0 to $+\infty$ and from $-\infty$ to 0. The first integral equals $\tfrac{1}{2}U(t)$ by virtue of (2.6-3), whereas the second is given by

$$\int_{-\infty}^{0} v(\nu)\exp(j2\pi\nu t)\,d\nu = \int_{0}^{\infty} v(-\nu)\exp(-j2\pi\nu t)\,d\nu$$
$$= \int_{0}^{\infty} v^*(\nu)\exp(-j2\pi\nu t)\,d\nu = \tfrac{1}{2}U^*(t). \quad (2.6\text{-}5)$$

The first step above reflects a simple change of variable from ν to $-\nu$, while the second step uses the symmetry relation $v(-\nu) = v^*(\nu)$. The net result is that $u(t)$ can be expressed as a sum of the complex function $\tfrac{1}{2}U(t)$ and its conjugate, confirming (2.6-4).

As a simple example, the complex representation of the real harmonic function $u(t) = \cos(\omega t)$ is the complex harmonic function $U(t) = \exp(j\omega t)$. This is the complex representation introduced in Sec. 2.2A for monochromatic waves. In fact, the complex representation of a polychromatic wave, as described in this section, is simply a superposition of the complex representations of each of its monochromatic Fourier components.

The complex analytic signal corresponding to the wavefunction $u(\mathbf{r}, t)$ is called the **complex wavefunction** $U(\mathbf{r}, t)$. Since each of its Fourier components satisfies the wave equation, so too does the complex wavefunction $U(\mathbf{r}, t)$,

$$\boxed{\nabla^2 U - \frac{1}{c^2}\frac{\partial^2 U}{\partial t^2} = 0.} \quad \begin{array}{l}(2.6\text{-}6)\\ \text{Wave Equation}\end{array}$$

Figure 2.6-1 shows the magnitudes of the Fourier transforms of the wavefunction $u(\mathbf{r}, t)$ and the complex wavefunction $U(\mathbf{r}, t)$. In this illustration the optical wave is **quasi-monochromatic**, i.e., it has Fourier components with frequencies confined within a narrow band of width $\Delta\nu$ surrounding a central frequency ν_0, such that $\Delta\nu \ll \nu_0$.

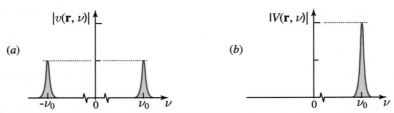

Figure 2.6-1 (a) The magnitude $|v(\mathbf{r}, \nu)|$ of the Fourier transform of the wavefunction $u(\mathbf{r}, t)$. (b) The magnitude $|V(\mathbf{r}, \nu)|$ of the Fourier transform of the corresponding complex wavefunction $U(\mathbf{r}, t)$.

Intensity of a Polychromatic Wave

The optical intensity is related to the wavefunction by (2.1-3):

$$I(\mathbf{r}, t) = 2\langle u^2(\mathbf{r}, t)\rangle$$
$$= 2\left\langle \left\{\tfrac{1}{2}[U(\mathbf{r}, t) + U^*(\mathbf{r}, t)]\right\}^2 \right\rangle$$
$$= \tfrac{1}{2}\langle U^2(\mathbf{r}, t)\rangle + \tfrac{1}{2}\langle U^{*2}(\mathbf{r}, t)\rangle + \langle U(\mathbf{r}, t)U^*(\mathbf{r}, t)\rangle. \quad (2.6\text{-}7)$$

For a quasi-monochromatic wave with central frequency ν_0 and spectral width $\Delta\nu \ll \nu_0$, the average $\langle \cdot \rangle$ is taken over a time interval much longer than the time of an optical cycle $1/\nu_0$ but much shorter than $1/\Delta\nu$ (see Sec. 2.1). Since $U(\mathbf{r},t)$ is given by (2.6-4), the term U^2 in (2.6-7) has components oscillating at frequencies $\approx 2\nu_0$. Similarly, the components of U^{*2} oscillate at frequencies $\approx -2\nu_0$. These terms are therefore washed out by the averaging operation. The third term, however, contains only frequency differences, which are of the order of $\Delta\nu \ll \nu_0$. It therefore varies slowly and is unaffected by the time-averaging operation. Thus, the third term in (2.6-7) survives and the light intensity becomes

$$I(\mathbf{r},t) = |U(\mathbf{r},t)|^2 . \qquad (2.6\text{-}8)$$
Optical Intensity

The optical intensity of a quasi-monochromatic wave is the absolute square of its complex wavefunction.

The simplicity of this result is, in fact, the rationale for introducing the concept of the complex wavefunction.

Pulsed Plane Wave

The simplest example of pulsed light is a pulsed plane wave. The complex wavefunction has the form

$$U(\mathbf{r},t) = \mathcal{A}\left(t - \frac{z}{c}\right) \exp\left[j2\pi\nu_0\left(t - \frac{z}{c}\right)\right], \qquad (2.6\text{-}9)$$

where the **complex envelope** $\mathcal{A}(t)$ is a time-varying function and ν_0 is the central optical frequency. The monochromatic plane wave is a special case of (2.6-9) for which $\mathcal{A}(t)$ is constant, i.e., $U(\mathbf{r},t) = \mathcal{A}\exp[j2\pi\nu_0(t-z/c)] = \mathcal{A}\exp(-jk_0z)\exp(j\omega_0t)$, where $k_0 = \omega_0/c$ and $\omega_0 = 2\pi\nu_0$.

Since $U(\mathbf{r},t)$ in (2.6-9) is a function of $t - z/c$ it satisfies the wave equation (2.6-6) regardless of the form of the function $\mathcal{A}(\cdot)$ (provided that $d^2\mathcal{A}/dt^2$ exists). This can be verified by direct substitution.

If $\mathcal{A}(t)$ is of finite duration τ, then at any fixed position z the wave lasts for a time period τ, and at any fixed time t it extends over a distance $c\tau$. It is therefore a **wavepacket** of fixed extent traveling in the z direction (Fig. 2.6-2). As an example, a pulse of duration $\tau = 1$ ps extends over a distance $c\tau = 0.3$ mm in free space.

The Fourier transform of the complex wavefunction in (2.6-9) is

$$V(\mathbf{r},\nu) = A(\nu - \nu_0)\exp(-j2\pi\nu z/c), \qquad (2.6\text{-}10)$$

where $A(\nu)$ is the Fourier transform of $\mathcal{A}(t)$. This may be shown by use of the frequency translation property of the Fourier transform (see Sec. A.1 of Appendix A). The complex envelope $\mathcal{A}(t)$ is often slowly varying in comparison with an optical cycle, so that its Fourier transform $A(\nu)$ has a spectral width $\Delta\nu$ much smaller than the central frequency ν_0. The spectral width $\Delta\nu$ is inversely proportional to the temporal width τ. In particular, if $\mathcal{A}(t)$ is Gaussian, then its Fourier transform $A(\nu)$ is also Gaussian. If the temporal and spectral widths are defined as the power-RMS widths, then their product equals $1/4\pi$ (see Sec. A.2 of Appendix A). For example, if $\tau = 1$ ps, then $\Delta\nu = 80$ GHz. If the central frequency ν_0 is 5×10^{14} Hz (corresponding to $\lambda_o = 0.6$ μm), then $\Delta\nu/\nu_0 = 1.6 \times 10^{-4}$, so that the light is quasi-monochromatic. Fig. 2.6-2 illustrates the temporal, spatial, and spectral characteristics of the pulsed plane wave in terms of the wavefunction.

74   CHAPTER 2 WAVE OPTICS

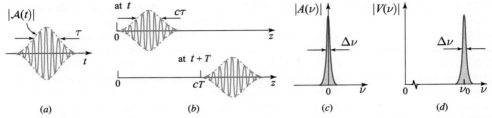

Figure 2.6-2 Temporal, spatial, and spectral characteristics of a pulsed plane wave. (*a*) The wavefunction at a fixed position has duration τ. (*b*) The wavefunction as a function of position at times t and $t + T$. The pulse travels with speed c and occupies a distance $c\tau$. (*c*) The magnitude $|A(\nu)|$ of the Fourier transform of the complex envelope. (*d*) The magnitude $|V(\nu)|$ of the Fourier transform of the complex wavefunction is centered at ν_0.

The propagation of a pulsed plane wave through a medium with frequency-dependent refractive index (i.e., with a frequency-dependent speed of light $c = c_o/n$) is discussed in Sec. 5.7 while other aspects of pulsed optics are considered in Chapter 23.

B. Light Beating

The dependence of the intensity of a polychromatic wave on time may be attributed to interference among the monochromatic components that constitute the wave. This concept is now demonstrated by means of two examples: interference between two monochromatic waves and interference among a finite number of monochromatic waves.

Interference of Two Monochromatic Waves of Different Frequencies

An optical wave composed of two monochromatic waves of frequencies ν_1 and ν_2 and intensities I_1 and I_2 has a complex wavefunction at some location in space

$$U(t) = \sqrt{I_1}\, \exp(j2\pi\nu_1 t) + \sqrt{I_2}\, \exp(j2\pi\nu_2 t), \qquad (2.6\text{-}11)$$

where the phases are taken to be zero and the **r** dependence has been suppressed for convenience. The intensity of the total wave is determined by use of the interference equation (2.5-4),

$$I(t) = I_1 + I_2 + 2\sqrt{I_1 I_2}\, \cos\left[2\pi(\nu_2 - \nu_1)t\right]. \qquad (2.6\text{-}12)$$

The intensity therefore varies sinusoidally at the difference frequency $|\nu_2 - \nu_1|$, which is known as the **beat frequency**. This phenomenon goes by a number of names: **light beating, optical mixing, photomixing, optical heterodyning,** and **coherent detection**.

Equation (2.6-12) is analogous to (2.5-7), which describes the *spatial* interference of two waves of the same frequency traveling in different directions. This can be understood in terms of the phasor diagram in Fig. 2.5-1. The two phasors U_1 and U_2 rotate at angular frequencies $\omega_1 = 2\pi\nu_1$ and $\omega_2 = 2\pi\nu_2$, so that the difference angle is $\varphi = \varphi_2 - \varphi_1 = 2\pi(\nu_2 - \nu_1)t$, in accord with (2.6-12). Waves of different frequencies traveling in different directions exhibit spatiotemporal interference.

In electronics, beating or mixing is said to occur when the sum of two sinusoidal signals is detected by a nonlinear (e.g., quadratic) device called a mixer, producing signals at the difference and sum frequencies. This device is used in heterodyne radio receivers. In optics, photodetectors are responsive to the optical (Sec. 19.1B), or optical intensity which, in accordance with (2.6-8), is proportional to the absolute square of the complex wavefunction. Optical detectors are therefore sensitive only to the difference frequency.

Much as (2.5-7) provides the basis for determining the direction of a wave via the spatial interference pattern at a screen, (2.6-12) provides a way of determining the frequency of an optical wave by measuring the temporal interference pattern at the output of a photodetector. The use of optical beating in optical heterodyne receivers is discussed in Sec. 25.4. Other forms of optical mixing make use of nonlinear media to generate optical-frequency differences and sums, as described in Chapter 22.

EXERCISE 2.6-1

Optical Doppler Radar. As a result of the **Doppler effect**, a monochromatic optical wave of frequency ν, reflected from an object moving with a velocity component v along the line of sight from an observer, undergoes a frequency shift $\Delta\nu = \pm(2v/c)\nu$, depending on whether the object is moving toward (+) or away (−) from the observer. Assuming that the original and reflected waves are superimposed, derive an expression for the intensity of the resultant wave. Suggest a method for measuring the velocity of a target using such an arrangement. If one of the mirrors of a Michelson interferometer [Fig. 2.5-3(b)] moves with velocity $\pm v$, use (2.5-6) to show that the beat frequency is $\pm(2v/c)\nu$.

Interference of M Monochromatic Waves with Equal Intensities and Equally Spaced Frequencies

The interference of a large number of monochromatic waves with equal intensities, equal phases, and equally spaced frequencies can result in the generation of brief pulses of light. Consider an odd number of waves, $M = 2L + 1$, each with intensity I_0 and zero phase, and with frequencies

$$\nu_q = \nu_0 + q\nu_F, \qquad q = -L, \ldots, 0, \ldots L, \tag{2.6-13}$$

centered about frequency ν_0 and spaced by frequency $\nu_F \ll \nu_0$. At a given position, the total wave has a complex wavefunction

$$U(t) = \sqrt{I_0} \sum_{q=-L}^{L} \exp\left[j2\pi(\nu_0 + q\nu_F)t\right]. \tag{2.6-14}$$

This represents the sum of M phasors of equal magnitudes and successive phases that differ by $\varphi = 2\pi\nu_F t$. Results for the intensity are immediately available from the analysis carried out in Sec. 2.5B, which is mathematically identical to the case at hand. Referring to (2.5-12) and Fig. 2.5-7, and using the substitution $\varphi = 2\pi t/T_F$ with $T_F = 1/\nu_F$, the total intensity is

$$I(t) = |U(t)|^2 = I_0 \frac{\sin^2(M\pi t/T_F)}{\sin^2(\pi t/T_F)}. \tag{2.6-15}$$

As illustrated in Fig. 2.6-3, the intensity $I(t)$ is a periodic sequence of optical pulses with period T_F, peak intensity $M^2 I_0$, and mean intensity $\bar{I} = MI_0$. The peak intensity is therefore M times greater than the mean intensity. The duration of each pulse is approximately T_F/M so that the pulses become very short when M is large. If $\nu_F = 1$ GHz, for example, then $T_F = 1$ ns; for $M = 1000$, pulses of 1-ps duration are generated.

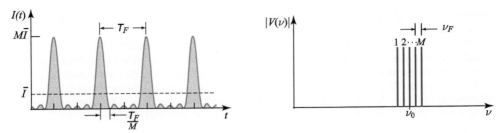

Figure 2.6-3 Time dependence of the optical intensity $I(t)$ of a polychromatic wave comprising M monochromatic waves of equal intensities, equal phases, and successive frequencies that differ by ν_F. The intensity $I(t)$ is a periodic train of pulses of period $T_F = 1/\nu_F$ with a peak that is M times greater than the mean $\bar{I}$. The duration of each pulse is M times shorter than the period. In this example $M = 5$. These graphs should be compared with those in Fig. 2.5-7. The magnitude of the Fourier transform $|V(\nu)|$ is shown in the lower graph.

This example provides a dramatic demonstration of how M monochromatic waves can conspire to produce a train of very short optical pulses. We shall see in Sec. 16.4D that the modes of a laser can be *mode-locked* in the fashion described above to produce a sequence of ultrashort laser pulses.

READING LIST

Wave Optics and Interferometry
See also the reading list on general optics in Chapter 1.

D. Fleisch and L. Kinnaman, *A Student's Guide to Waves*, Cambridge University Press, 2015.

M. Mansuripur, *Classical Optics and Its Applications*, Cambridge University Press, 2nd ed. 2009.

P. Hariharan, *Basics of Interferometry*, Academic Press, 2nd ed. 2006.

J. R. Pierce, *Almost All About Waves*, MIT Press, 1974; Dover, reissued 2006.

H. J. Pain, *The Physics of Vibrations and Waves*, Wiley, 6th ed. 2005.

R. H. Webb, *Elementary Wave Optics*, Academic Press, 1969; Dover, reissued 2005.

E. Hecht and A. Zajac, *Optics*, Addison–Wesley, 2nd ed. 1990.

J. M. Vaughan, *The Fabry–Perot Interferometer*, CRC Press, 1989.

H. D. Young, *Fundamentals of Waves, Optics, and Modern Physics*, McGraw–Hill, paperback 2nd ed. 1976.

M. Françon, N. Krauzman, J. P. Matieu, and M. May, *Experiments in Physical Optics*, CRC Press, 1970.

M. Françon, *Optical Interferometry*, Academic Press, 1966.

Spectroscopy

D. L. Pavia, G. M. Lampman, G. S. Kriz and J. A. Vyvyan, *Introduction to Spectroscopy*, Brooks/Cole, 5th ed. 2014.

B. C. Smith, *Fundamentals of Fourier Transform Infrared Spectroscopy*, CRC Press/Taylor & Francis, 2nd ed. 2011.

J. M. Hollas, *Modern Spectroscopy*, Wiley, paperback 4th ed. 2010.

P. R. Griffiths and J. A. de Haseth, *Fourier Transform Infrared Spectrometry*, Wiley, 2nd ed. 2007.

Diffraction Gratings

C. Palmer, *Diffraction Grating Handbook*, Richardson Gratings (Rochester, NY), 7th ed. 2014.

E. G. Loewen and E. Popov, *Diffraction Gratings and Applications*, CRC Press, 1997.

Interferometry for Gravitational-Wave Detection

S. Wills, Gravitational Waves: The Road Ahead, *Optics & Photonics News*, vol. 29, no. 5, pp. 44–51, 2018.

B. P. Abbott *et al.* (LIGO Scientific Collaboration and Virgo Collaboration), GW170817: Observation of Gravitational Waves from a Binary Neutron Star Inspiral, *Physical Review Letters*, vol. 119, 161101, 2017.

B. P. Abbott *et al.* (LIGO Scientific Collaboration and Virgo Collaboration), Observation of Gravitational Waves from a Binary Black Hole Merger, *Physical Review Letters*, vol. 116, 061102, 2016.

B. P. Abbott *et al.*, Astrophysical Implications of the Binary Black Hole Merger GW150914, *The Astrophysical Journal Letters*, vol. 818, L22, 2016.

R. W. P. Drever, Fabry–Perot Cavity Gravity-Wave Detectors, in D. G. Blair, ed., *The Detection of Gravitational Waves*, Cambridge University Press, 1991, Chapter 12, pp. 306–328.

A. Brillet, J. Gea-Banacloche, G. Leuchs, C. N. Man, and J. Y. Vinet, Advanced Techniques: Recycling and Squeezing, in D. G. Blair, ed., *The Detection of Gravitational Waves*, Cambridge University Press, 1991, Chapter 15, pp. 369–405.

D. G. Blair, Gravitational Waves in General Relativity, in D. G. Blair, ed., *The Detection of Gravitational Waves*, Cambridge University Press, 1991, Chapter 1, pp. 3–15.

A. Einstein, Die Feldgleichungen der Gravitation (The Field Equations of Gravitation), Sitzungsberichte der Königlich Preussische Akademie der Wissenschaften (Berlin), pp. 844–847 (part 2), 1915.

Popular and Historical

P. Daukantas, 200 Years of Fresnel's Legacy, *Optics & Photonics News*, vol. 26, no. 9, pp. 40–47, 2015.

T. Levitt, *A Short Bright Flash: Augustin Fresnel and the Birth of the Modern Lighthouse*, Norton, 2013.

F. J. Dijksterhuis, *Lenses and Waves: Christiaan Huygens and the Mathematical Science of Optics in the Seventeenth Century*, 2004, Springer-Verlag, paperback ed. 2011.

J. Z. Buchwald, *The Rise of the Wave Theory of Light: Optical Theory and Experiment in the Early Nineteenth Century*, University of Chicago Press, paperback ed. 1989.

W. E. Kock, *Sound Waves and Light Waves: The Fundamentals of Wave Motion*, Doubleday/Anchor, 1965.

C. Huygens, *Treatise on Light*, 1690, University of Chicago Press, 1945; Echo Library, reprinted 2007.

Seminal Articles

G. W. Kamerman, ed., *Selected Papers on Laser Radar*, SPIE Optical Engineering Press (Milestone Series Volume 133), 1997.

P. Hariharan and D. Malacara-Hernandez, eds., *Selected Papers on Interference, Interferometry, and Interferometric Metrology*, SPIE Optical Engineering Press (Milestone Series Volume 110), 1995.

D. Maystre, ed., *Selected Papers on Diffraction Gratings*, SPIE Optical Engineering Press (Milestone Series Volume 83), 1993.

P. Hariharan, ed., *Selected Papers on Interferometry*, SPIE Optical Engineering Press (Milestone Series Volume 28), 1991.

PROBLEMS

2.2-3 **Spherical Waves.** Use a spherical coordinate system to verify that the complex amplitude of the spherical wave (2.2-15) satisfies the Helmholtz equation (2.2-7).

2.2-4 **Intensity of a Spherical Wave.** Derive an expression for the intensity I of a spherical wave at a distance r from its center in terms of the optical power P. What is the intensity at $r = 1$ m for $P = 100$ W?

78   CHAPTER 2   WAVE OPTICS

2.2-5   **Cylindrical Waves.** Derive expressions for the complex amplitude and intensity of a monochromatic wave whose wavefronts are cylinders centered about the y axis.

2.2-6   **Paraxial Helmholtz Equation.** Derive the paraxial Helmholtz equation (2.2-23) using the approximations in (2.2-21) and (2.2-22).

2.2-7   **Conjugate Waves.** Compare a monochromatic wave with complex amplitude $U(\mathbf{r})$ to a monochromatic wave of the same frequency but with complex amplitude $U^*(\mathbf{r})$, with respect to intensity, wavefronts, and wavefront normals. Use the plane wave $U(\mathbf{r}) = A\exp[-jk(x+y)/\sqrt{2}]$ and the spherical wave $U(\mathbf{r}) = (A/r)\exp(-jkr)$ as examples.

2.3-1   **Wave in a GRIN Slab.** Sketch the wavefronts of a wave traveling in the graded-index SELFOC slab described in Example 1.3-1.

2.4-7   **Reflection of a Spherical Wave from a Planar Mirror.** A spherical wave is reflected from a planar mirror sufficiently far from the wave origin so that the Fresnel approximation is satisfied. By regarding the spherical wave locally as a plane wave with slowly varying direction, use the law of reflection of plane waves to determine the nature of the reflected wave.

2.4-8   **Optical Pathlength.** A plane wave travels in a direction normal to a thin plate made of N thin parallel layers of thicknesses d_q and refractive indices n_q, $q = 1, 2, \ldots, N$. If all reflections are ignored, determine the complex amplitude transmittance of the plate. If the plate is replaced with a distance d of free space, what should d be so that the same complex amplitude transmittance is obtained? Show that this distance is the optical pathlength defined in Sec. 1.1.

2.4-9   **Diffraction Grating.** Repeat Exercise 2.4-5 for a thin transparent plate whose thickness $d(x, y)$ is a square (instead of sinusoidal) periodic function of x of period $\Lambda \gg \lambda$. Show that the angle θ between the diffracted waves is still given by $\theta \approx \lambda/\Lambda$. If a plane wave is incident in a direction normal to the grating, determine the amplitudes of the different diffracted plane waves.

2.4-10 **Reflectance of a Spherical Mirror.** Show that the complex amplitude reflectance $\mathfrak{r}(x, y)$ (the ratio of the complex amplitudes of the reflected and incident waves) of a thin spherical mirror of radius R is given by $\mathfrak{r}(x, y) = h_0 \exp[-jk_o(x^2 + y^2)/R]$, where h_0 is a constant. Compare this to the complex amplitude transmittance of a lens of focal length $f = -R/2$.

2.5-4   **Standing Waves.** Derive an expression for the intensity I of the superposition of two plane waves of wavelength λ traveling in opposite directions along the z axis. Sketch I versus z.

2.5-5   **Fringe Visibility.** The visibility of an interference pattern such as that described by (2.5-4) and plotted in Fig. 2.5-1 is defined as the ratio $\mathcal{V} = (I_{\max} - I_{\min})/(I_{\max} + I_{\min})$, where $I_{\max}$ and $I_{\min}$ are the maximum and minimum values of I. Derive an expression for $\mathcal{V}$ as a function of the ratio I_1/I_2 of the two interfering waves and determine the ratio I_1/I_2 for which the visibility is maximum.

2.5-6   **Michelson Interferometer.** If one of the mirrors of the Michelson interferometer [Fig. 2.5-3(b)] is misaligned by a small angle $\Delta\theta$, describe the shape of the interference pattern in the detector plane. What happens to this pattern as the other mirror moves?

2.6-2   **Pulsed Spherical Wave.**
   (a) Show that a pulsed spherical wave has a complex wavefunction of the form $U(\mathbf{r}, t) = (1/r)\mathfrak{a}(t - r/c)$, where $\mathfrak{a}(t)$ is an arbitrary function.
   (b) An ultrashort optical pulse has a complex wavefunction with central frequency corresponding to a wavelength $\lambda_o = 585$ nm and a Gaussian envelope of RMS width of $\sigma_t = 6$ fs (1 fs$= 10^{-15}$ s). How many optical cycles are contained within the pulse width? If the pulse propagates in free space as a spherical wave initiated at the origin at $t = 0$, describe the spatial distribution of the intensity as a function of the radial distance at time $t = 1$ ps.

CHAPTER 3

BEAM OPTICS

3.1	THE GAUSSIAN BEAM	80
	A. Complex Amplitude	
	B. Properties	
	C. Beam Quality	
3.2	TRANSMISSION THROUGH OPTICAL COMPONENTS	91
	A. Transmission Through a Thin Lens	
	B. Beam Shaping	
	C. Reflection from a Spherical Mirror	
*D.	Transmission Through an Arbitrary Optical System	
3.3	HERMITE–GAUSSIAN BEAMS	99
3.4	LAGUERRE–GAUSSIAN BEAMS	102
3.5	NONDIFFRACTING BEAMS	105
	A. Bessel Beams	
*B.	Airy Beams	

The Gaussian beam, named after the German mathematician **Carl Friedrich Gauss (1777–1855)**, is circularly symmetric and has a radial intensity that follows the form of a Gaussian distribution.

Edmond Nicolas Laguerre (1834–1886), a French mathematician, devised a set of polynomials useful for describing circularly symmetric light beams with helical wavefronts and orbital angular momentum.

Friedrich Wilhelm Bessel (1784–1846), a noted German astronomer, established a set of functions that characterize the radial intensity of circularly symmetric, planar-wavefront, non-diffracting optical beams.

Fundamentals of Photonics, Third Edition. Bahaa E. A. Saleh and Malvin Carl Teich.
©2019 John Wiley & Sons, Inc. Published 2019 by John Wiley & Sons, Inc.

Can light be spatially confined and transported in free space without angular spread? Although the wave nature of light precludes the possibility of such idealized transport, light can, in fact, be confined in the form of beams that come as close as possible to waves that are spatially localized and nondiverging.

The two extremes of angular and spatial confinement are the plane wave and the spherical wave, respectively. The wavefront normals (rays) of a plane wave coincide with the direction of travel of the wave so that there is no angular spread, but its energy extends spatially over all space. The spherical wave, in contrast, originates from a single spatial point, but its wavefront normals (rays) diverge in all angular directions.

Waves whose wavefront normals make small angles with the z axis are called paraxial waves. They must satisfy the paraxial Helmholtz equation, which was derived in Sec. 2.2C. The **Gaussian beam** is an important solution of this equation that exhibits the characteristics of an optical beam, as attested to by a number of its properties. The beam power is principally concentrated within a small cylinder that surrounds the beam axis. The intensity distribution in any transverse plane is a circularly symmetric Gaussian function centered about the beam axis. The width of this function is minimum at the beam waist and gradually becomes larger as the distance from the waist increases in both directions. The wavefronts are approximately planar near the beam waist, then gradually curve as the distance from the waist increases, and ultimately become approximately spherical far from the beam waist. The angular divergence of the wavefront normals assumes the minimum value permitted by the wave equation for a given beam width. The wavefront normals are therefore much like a thin pencil of rays. Under ideal conditions, the light from many types of lasers takes the form of a Gaussian beam.

This Chapter

An expression for the complex amplitude of the Gaussian beam is set forth in Sec. 3.1 and a detailed discussion of its physical properties (intensity, power, beam width, beam divergence, depth of focus, and phase) is provided. The shaping of Gaussian beams (focusing, relaying, collimating, and expanding) via the use of various optical components is the subject of Sec. 3.2. In Secs. 3.3 and 3.4 we introduce more general families of optical beams, known as Hermite–Gaussian and Laguerre–Gaussian beams, respectively, of which the simple Gaussian beam is a member. Finally, in Sec. 3.5 we discuss nondiffracting beams, including Bessel, Bessel–Gaussian, and Airy beams.

3.1 THE GAUSSIAN BEAM

A. Complex Amplitude

The concept of paraxial waves was introduced in Sec. 2.2C. A monochromatic paraxial wave is a plane wave traveling along the z direction e^{-jkz} (with wavenumber $k = 2\pi/\lambda$ and wavelength λ), modulated by a complex envelope $A(\mathbf{r})$ that is a slowly varying function of position (see Fig. 2.2-5), so that its complex amplitude is

$$U(\mathbf{r}) = A(\mathbf{r}) \exp(-jkz). \qquad (3.1\text{-}1)$$

The envelope is taken to be approximately constant within a neighborhood of size λ, so that the wave locally maintains its plane-wave nature but exhibits wavefront normals that are paraxial rays.

3.1 THE GAUSSIAN BEAM

In order that the complex amplitude $U(\mathbf{r})$ satisfy the Helmholtz equation, $\nabla^2 U + k^2 U = 0$, the complex envelope $A(\mathbf{r})$ must satisfy the paraxial Helmholtz equation (2.2-23)

$$\nabla_T^2 A - j\, 2k \frac{\partial A}{\partial z} = 0, \qquad (3.1\text{-}2)$$

where $\nabla_T^2 = \partial^2/\partial x^2 + \partial^2/\partial y^2$ is the transverse Laplacian operator. A simple solution to the paraxial Helmholtz equation yields the paraboloidal wave (see Exercise 2.2-2), for which

$$A(\mathbf{r}) = \frac{A_1}{z} \exp\left(-jk\frac{\rho^2}{2z}\right), \qquad \rho^2 = x^2 + y^2, \qquad (3.1\text{-}3)$$

where A_1 is a constant. The paraboloidal wave is the paraxial approximation of the spherical wave $U(r) = (A_1/r)\exp(-jkr)$ when x and y are much smaller than z (see Sec. 2.2B).

Another solution of the paraxial Helmholtz equation leads to the Gaussian beam. It is obtained from the paraboloidal wave by use of a simple transformation. Since the complex envelope of the paraboloidal wave (3.1-3) is a solution of the paraxial Helmholtz equation (3.1-2), so too is a shifted version of it, with $z - \xi$ replacing z where ξ is a constant:

$$A(\mathbf{r}) = \frac{A_1}{q(z)} \exp\left[-jk\frac{\rho^2}{2q(z)}\right], \qquad q(z) = z - \xi. \qquad (3.1\text{-}4)$$

This represents a paraboloidal wave centered about the point $z = \xi$ instead of about $z = 0$. Equation (3.1-4) remains a solution of (3.1-2) even when ξ is complex, but the solution acquires dramatically different properties. In particular, when ξ is purely imaginary, say $\xi = -jz_0$ where z_0 is real, (3.1-4) yields the complex envelope of the Gaussian beam

$$\boxed{A(\mathbf{r}) = \frac{A_1}{q(z)} \exp\left[-jk\frac{\rho^2}{2q(z)}\right], \qquad q(z) = z + jz_0.} \qquad (3.1\text{-}5)$$
Complex Envelope

The quantity $q(z)$ is called the **q-parameter** of the beam and the parameter z_0 is known as the **Rayleigh range**.

To separate the amplitude and phase of this complex envelope, we write the complex function $1/q(z) = 1/(z+jz_0)$ in terms of its real and imaginary parts by defining two new real functions, $R(z)$ and $W(z)$, such that

$$\boxed{\frac{1}{q(z)} = \frac{1}{R(z)} - j\frac{\lambda}{\pi W^2(z)}.} \qquad (3.1\text{-}6)$$

It will be shown subsequently that $W(z)$ and $R(z)$ are measures of the beam width and wavefront radius of curvature, respectively. Expressions for $W(z)$ and $R(z)$ as functions of z and z_0 are provided in (3.1-8) and (3.1-9). Substituting (3.1-6) into (3.1-5) and using (3.1-1) leads directly to an expression for the complex amplitude $U(\mathbf{r})$ of the Gaussian beam:

$$U(\mathbf{r}) = A_0 \frac{W_0}{W(z)} \exp\left[-\frac{\rho^2}{W^2(z)}\right] \exp\left[-jkz - jk\frac{\rho^2}{2R(z)} + j\zeta(z)\right] \quad (3.1\text{-}7)$$

Complex Amplitude

$$W(z) = W_0 \sqrt{1 + \left(\frac{z}{z_0}\right)^2} \quad (3.1\text{-}8)$$

$$R(z) = z\left[1 + \left(\frac{z_0}{z}\right)^2\right] \quad (3.1\text{-}9)$$

$$\zeta(z) = \tan^{-1}\frac{z}{z_0} \quad (3.1\text{-}10)$$

$$W_0 = \sqrt{\frac{\lambda z_0}{\pi}}. \quad (3.1\text{-}11)$$

Beam Parameters

A new constant $A_0 = A_1/jz_0$ has been defined for convenience.

The expression for the complex amplitude of the Gaussian beam provided above is central to this chapter. It is described by two independent parameters, A_0 and z_0, which are determined from the boundary conditions. All other parameters are related to the z_0 and the wavelength λ by (3.1-8) to (3.1-11). The significance of these parameters will become clear in the sequel.

B. Properties

Equations (3.1-7)–(3.1-11) will now be used to determine the properties of the Gaussian beam.

Intensity

The optical intensity $I(\mathbf{r}) = |U(\mathbf{r})|^2$ is a function of the axial and radial positions, z and $\rho = \sqrt{x^2 + y^2}$, respectively

$$I(\rho, z) = I_0 \left[\frac{W_0}{W(z)}\right]^2 \exp\left[-\frac{2\rho^2}{W^2(z)}\right], \quad (3.1\text{-}12)$$

where $I_0 = |A_0|^2$. At any value of z the intensity is a Gaussian function of the radial distance ρ — hence the appellation "Gaussian beam." The Gaussian function has its peak on the z axis, at $\rho = 0$, and decreases monotonically as ρ increases. The beam width $W(z)$ of the Gaussian distribution increases with the axial distance z as illustrated in Fig. 3.1-1.

On the beam axis ($\rho = 0$) the intensity in (3.1-12) reduces to

$$I(0, z) = I_0 \left[\frac{W_0}{W(z)}\right]^2 = \frac{I_0}{1 + (z/z_0)^2}, \quad (3.1\text{-}13)$$

3.1 THE GAUSSIAN BEAM

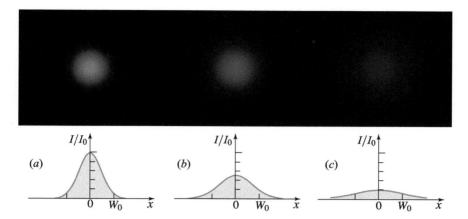

Figure 3.1-1 Normalized Gaussian beam intensity I/I_0 as a function of the radial distance ρ at different axial distances: (a) $z = 0$; (b) $z = z_0$; (c) $z = 2z_0$.

which has its maximum value I_0 at $z = 0$ and decays gradually with increasing z, reaching half its peak value at $z = \pm z_0$ (Fig. 3.1-2). When $|z| \gg z_0$, $I(0, z) \approx I_0 z_0^2/z^2$, so that the intensity decreases with distance in accordance with an inverse-square law, as for spherical and paraboloidal waves. Overall, the beam center ($z = 0, \rho = 0$) is the location of the greatest intensity: $I(0,0) = I_0$.

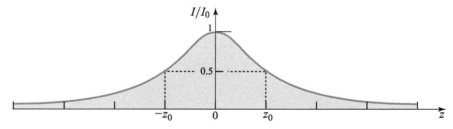

Figure 3.1-2 The normalized beam intensity I/I_0 at points on the beam axis ($\rho = 0$) as a function of distance along the beam axis, z.

Power

The total optical power carried by the beam is the integral of the optical intensity over any transverse plane (say at position z),

$$P = \int_0^\infty I(\rho, z)\, 2\pi\rho\, d\rho, \tag{3.1-14}$$

which yields

$$P = \tfrac{1}{2} I_0 \left(\pi W_0^2\right). \tag{3.1-15}$$

The beam power is thus half the peak intensity multiplied by the beam area. The result is independent of z, as expected. Since optical beams are often described by their power

P, it is useful to express I_0 in terms of P via (3.1-15), whereupon (3.1-12) can be rewritten in the form

$$I(\rho, z) = \frac{2P}{\pi W^2(z)} \exp\left[-\frac{2\rho^2}{W^2(z)}\right].$$

(3.1-16)
Beam Intensity

The ratio of the power carried within a circle of radius ρ_0 in the transverse plane to the total power, at position z, is

$$\frac{1}{P} \int_0^{\rho_0} I(\rho, z) \, 2\pi\rho \, d\rho = 1 - \exp\left[-\frac{2\rho_0^2}{W^2(z)}\right].$$

(3.1-17)

The power contained within a circle of radius $\rho_0 = W(z)$ is therefore approximately 86% of the total power. About 99% of the power is contained within a circle of radius $1.5\,W(z)$.

Beam Width

At any transverse plane, the beam intensity assumes its peak value on the beam axis, and decreases by the factor $1/e^2 \approx 0.135$ at the radial distance $\rho = W(z)$. Since 86% of the power is carried within a circle of radius $W(z)$, we regard $W(z)$ as the beam radius (or beam width). The RMS width of the intensity distribution, on the other hand, is $\sigma = \frac{1}{2}W(z)$ (see Appendix A, Sec. A.2, for the different definitions of width).

The dependence of the beam width on z is governed by (3.1-8),

$$W(z) = W_0 \sqrt{1 + \left(\frac{z}{z_0}\right)^2}.$$

(3.1-18)
Beam Width
(Beam Radius)

It assumes its minimum value, W_0, at the plane $z = 0$. This is the beam waist and W_0 is thus known as the **waist radius**. The waist diameter $2W_0$ is also called the **spot size**. The beam width increases monotonically with z, and assumes the value $\sqrt{2}W_0$ at $z = \pm z_0$ (Fig. 3.1-3).

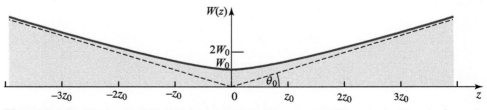

Figure 3.1-3 The beam width $W(z)$ assumes its minimum value W_0 at the beam waist ($z = 0$), reaches $\sqrt{2}W_0$ at $z = \pm z_0$, and increases linearly with z for large z.

Beam Divergence

For $z \gg z_0$ the first term of (3.1-18) may be neglected, which results in the linear relation

$$W(z) \approx \frac{W_0}{z_0} z = \theta_0 z. \tag{3.1-19}$$

As illustrated in Fig. 3.1-3, the beam then diverges as a cone of half-angle

$$\theta_0 = \frac{W_0}{z_0} = \frac{\lambda}{\pi W_0}, \tag{3.1-20}$$

where we have made use of (3.1-11). Approximately 86% of the beam power is confined within this cone, as indicated following (3.1-17).

Rewriting (3.1-20) in terms of the spot size, the angular divergence of the beam becomes

$$\boxed{2\theta_0 = \frac{4}{\pi} \frac{\lambda}{2W_0}.} \tag{3.1-21}$$
Divergence Angle

The divergence angle is directly proportional to the wavelength λ and inversely proportional to the spot size $2W_0$. Squeezing the spot size (beam-waist diameter) therefore leads to increased beam divergence. It is clear that a highly directional beam is constructed by making use of a short wavelength and a thick beam waist.

Depth of Focus

Since the beam has its minimum width at $z = 0$, as shown in Fig. 3.1-3, it achieves its best focus at the plane $z = 0$. In either direction, the beam gradually grows "out of focus." The axial distance within which the beam width is no greater than a factor $\sqrt{2}$ times its minimum value, so that its area is within a factor of 2 of the minimum, is known as the **depth-of-focus** or **confocal parameter** (Fig. 3.1-4). It is evident from (3.1-18) and (3.1-11) that the actual depth of focus is twice the Rayleigh range:

$$\boxed{2z_0 = \frac{2\pi W_0^2}{\lambda}.} \tag{3.1-22}$$
Depth of Focus

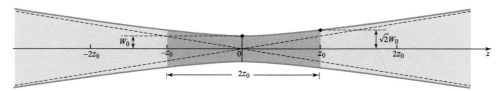

Figure 3.1-4 Depth of focus of a Gaussian beam.

The depth of focus is therefore directly proportional to the area of the beam at its waist, πW_0^2, and inversely proportional to the wavelength, λ. A beam focused to a small spot size thus has a short depth of focus; locating the plane of focus thus requires increased accuracy. Small spot size and long depth of focus can be simultaneously attained only for short wavelengths. As an example, at $\lambda_o = 633$ nm (a common He–Ne laser-line wavelength), a spot size $2W_0 = 2$ cm corresponds to a depth of focus $2z_0 \approx 1$ km. A much smaller spot size of 20 μm corresponds to a much shorter depth of focus of 1 mm.

Phase

The phase of the Gaussian beam is, from (3.1-7),[†]

$$\varphi(\rho, z) = kz - \zeta(z) + \frac{k\rho^2}{2R(z)}. \tag{3.1-23}$$

On the beam axis ($\rho = 0$) the phase comprises two components:

$$\varphi(0, z) = kz - \zeta(z). \tag{3.1-24}$$

The first, kz, is the phase of a plane wave. The second represents a phase retardation $\zeta(z)$ given by (3.1-10), which ranges from $-\pi/2$ at $z = -\infty$ to $+\pi/2$ at $z = \infty$, as illustrated in Fig. 3.1-5. This phase retardation corresponds to an excess delay of the wavefront in relation to a plane wave (see also Fig. 3.1-8). The total accumulated excess retardation as the wave travels from $z = -\infty$ to $z = \infty$ is π. This phenomenon is known as the **Gouy effect**. It arises from the transverse spatial confinement of the beam, which is accompanied by a spread in its transverse wavevector components by virtue of the Fourier transform. This results in a reduction in the axial component of the wavevector k_z from its plane-wave value $k_z = \sqrt{k^2 - k_x^2 - k_y^2}$ (see Sec. 2.2B).[‡]

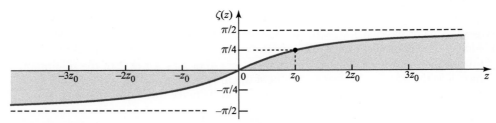

Figure 3.1-5 The function $\zeta(z)$ represents the phase retardation of the Gaussian beam relative to a uniform plane wave at points on the beam axis.

Wavefronts

The third component in (3.1-23) is responsible for wavefront bending. It represents the deviation of the phase at off-axis points in a given transverse plane from that at the axial point. The surfaces of constant phase satisfy $k[z + \rho^2/2R(z)] - \zeta(z) = 2\pi q$. Since $\zeta(z)$ and $R(z)$ are relatively slowly varying functions, they are effectively constant at points within the beam width on each wavefront. We may therefore write $z + \rho^2/2R \approx q\lambda + \zeta\lambda/2\pi$, where $R = R(z)$ and $\zeta = \zeta(z)$. This is the equation of a paraboloidal surface with **radius of curvature** R. Thus, $R(z)$, plotted in Fig. 3.1-6, is the radius of curvature of the wavefront at position z along the beam axis.

[†] The phase $\varphi(\rho, z)$ in (3.1-23), and throughout this chapter, is related to the phase factor specified in (3.1-7) by $\exp(-j\varphi)$.

[‡] See S. Feng and H. G. Winful, Physical Origin of the Gouy Phase Shift, *Optics Letters*, vol. 26, pp. 485–487, 2001.

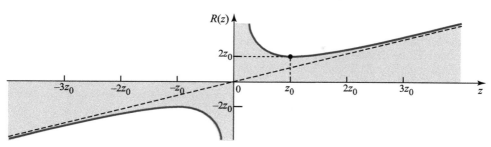

Figure 3.1-6 The radius of curvature $R(z)$ of the wavefronts of a Gaussian beam as a function of position along the beam axis. The dashed line is the radius of curvature of a spherical wave.

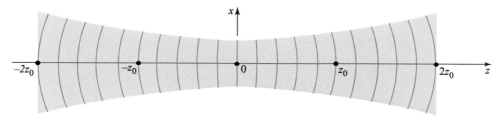

Figure 3.1-7 Wavefronts of a Gaussian beam.

As illustrated in Fig. 3.1-6, the radius of curvature $R(z)$ is infinite at $z = 0$, so that the wavefronts are planar, i.e., they have no curvature. The radius decreases to a minimum value of $2z_0$ at $z = z_0$, where the wavefront has the greatest curvature (Fig. 3.1-7). The radius of curvature subsequently increases as z increases further until $R(z) \approx z$ for $z \gg z_0$. The wavefronts are then approximately the same as those of a spherical wave. The pattern of the wavefronts is identical for negative z, except for a change in sign (Fig. 3.1-8). We have adopted the convention that a diverging wavefront has a positive radius of curvature whereas a converging wavefront has a negative radius of curvature.

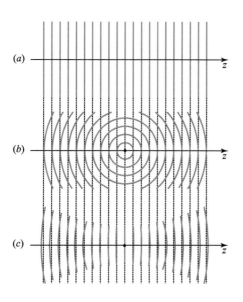

Figure 3.1-8 Wavefronts of (*a*) a uniform plane wave; (*b*) a spherical wave; (*c*) a Gaussian beam. At points near the beam center, the Gaussian beam resembles a plane wave. At large z the beam behaves like a spherical wave except that its phase is retarded by $\pi/2$ (a quarter of the distance between two adjacent wavefronts).

Parameters Required to Characterize a Gaussian Beam

Assuming that the wavelength λ is known, how many parameters are required to describe a plane wave, a spherical wave, and a Gaussian beam? The plane wave is completely specified by its complex amplitude and direction. The spherical wave is specified by its complex amplitude and the location of its origin. The Gaussian beam, in contrast, requires more parameters for its characterization — its peak amplitude [determined by A_0 in (3.1-7)], its direction (the beam axis), the location of its waist, *and* one additional parameter, such as the waist radius W_0 *or* the Rayleigh range z_0. Thus, if the beam peak amplitude and the axis are known, two additional parameters are required for full specification.

If the complex q-parameter, $q(z) = z + jz_0$, is known, the distance to the beam waist z and the Rayleigh range z_0 are readily identified as the real and imaginary parts thereof. As an example, if $q(z)$ is $3 + j4$ cm at some point on the beam axis, we infer that the beam waist lies at a distance $z = 3$cm to the left of that point and that the depth of focus is $2z_0 = 8$ cm. The waist radius W_0 may then be determined via (3.1-11). The quantity $q(z)$ is therefore sufficient for characterizing a Gaussian beam of known peak amplitude and beam axis. Given $q(z)$ at a single point, the linear dependence of q on z permits it to be determined at all points: if $q(z) = q_1$ and $q(z + d) = q_2$, then $q_2 = q_1 + d$. Using the example provided immediately above, at $z = 13$ cm it is evident that $q = 13 + j4$.

If the beam width $W(z)$ and the radius of curvature $R(z)$ are known at an arbitrary point on the beam axis, the beam can be fully identified by solving (3.1-8), (3.1-9), and (3.1-11) for z, z_0, and W_0. Alternatively, the beam can be identified by determining $q(z)$ from $W(z)$ and $R(z)$ using (3.1-6).

Summary: Properties of the Gaussian Beam at Special Locations

- *At the location $z = z_0$.* At an axial distance z_0 from the beam waist, the wave has the following properties:
  - The intensity on the beam axis is $\frac{1}{2}$ the peak intensity.
  - The beam width is a factor of $\sqrt{2}$ greater than the width at the beam waist, and the beam area is larger by a factor of 2.
  - The phase on the beam axis is retarded by an angle $\pi/4$ relative to the phase of a plane wave.
  - The radius of curvature of the wavefront achieves its minimum value, $R = 2z_0$, so that the wavefront has the greatest curvature.

- *Near the beam center.* At locations for which $|z| \ll z_0$ and $\rho \ll W_0$, the quantity $\exp[-\rho^2/W^2(z)] \approx \exp(-\rho^2/W_0^2) \approx 1$, so that the beam intensity, which is proportional to the square of this quantity, is approximately constant. Also, $R(z) \approx z_0^2/z$ and $\zeta(z) \approx 0$, so that the phase $k[z + \rho^2/2R(z)] \approx kz(1 + \rho^2/2z_0^2) \approx kz$, by virtue of (3.1-11) when $z_0 \gg \lambda$. The Gaussian beam may therefore be approximated near its center by a plane wave.

- *Far from the beam waist.* At transverse locations within the waist radius ($\rho < W_0$), but far from the beam waist ($z \gg z_0$), the wave behaves approximately like a spherical wave. In this domain $W(z) \approx W_0 z/z_0 \gg W_0$ and $\rho < W_0$, so that $\exp[-\rho^2/W^2(z)] \approx 1$ and the beam intensity is approximately uniform. Since $R(z) \approx z$ in this regime, the wavefronts are approximately spherical. Thus, except for the Gouy phase retardation $\zeta(z) \approx \pi/2$, the complex amplitude of the Gaussian beam approaches that of the paraboloidal wave, which in turn approaches that of the spherical wave in the paraxial approximation.

EXERCISE 3.1-1

Parameters of a Gaussian Laser Beam. A 1-mW He–Ne laser produces a Gaussian beam at a wavelength of $\lambda = 633$ nm with a spot size $2W_0 = 0.1$ mm.

(a) Determine the angular divergence of the beam, its depth of focus, and its diameter at $z = 3.5 \times 10^5$ km (approximately the distance to the moon).
(b) What is the radius of curvature of the wavefront at $z = 0$, $z = z_0$, and $z = 2z_0$?
(c) What is the optical intensity (in W/cm^2) at the beam center ($z = 0$, $\rho = 0$) and at the axial point $z = z_0$? Compare this with the intensity at $z = z_0$ of a 100-W spherical wave produced by a small isotropically emitting light source located at $z = 0$.

EXERCISE 3.1-2

Validity of the Paraxial Approximation for a Gaussian Beam. The complex envelope $A(\mathbf{r})$ of a Gaussian beam is an exact solution of the paraxial Helmholtz equation (3.1-2), but its corresponding complex amplitude $U(\mathbf{r}) = A(\mathbf{r})\exp(-jkz)$ is only an approximate solution of the Helmholtz equation (2.2-7). This is because the paraxial Helmholtz equation is itself approximate. The approximation is satisfactory if the condition (2.2-21) is satisfied. Show that if the divergence angle θ_0 of a Gaussian beam is small ($\theta_0 \ll 1$), the necessary condition (2.2-21) for the validity of the paraxial Helmholtz equation is indeed satisfied.

EXERCISE 3.1-3

Determination of a Beam with Given Width and Curvature. Consider a Gaussian beam whose width W and radius of curvature R are known at a particular point on the beam axis (Fig. 3.1-9). Show that the beam waist is located to the left at a distance

$$z = \frac{R}{1 + (\lambda R/\pi W^2)^2} \tag{3.1-25}$$

and that the waist radius is

$$W_0 = \frac{W}{\sqrt{1 + (\pi W^2/\lambda R)^2}}. \tag{3.1-26}$$

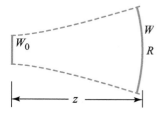

Figure 3.1-9 Given W and R, determine z and W_0.

EXERCISE 3.1-4

Determination of the Width and Curvature at One Point Given the Width and Curvature at Another Point. Assume that the width and radius of curvature of a Gaussian beam of wavelength $\lambda = 1$ μm at some point on the beam axis are $W_1 = 1$ mm and $R_1 = 1$ m, respectively (Fig. 3.1-10). Determine the beam width W_2 and radius of curvature R_2 at a distance $d = 10$ cm to the right.

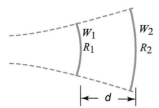

Figure 3.1-10 Given W_1, R_1 and d, determine W_2 and R_2.

EXERCISE 3.1-5

Identification of a Beam with Known Curvatures at Two Points. A Gaussian beam has radii of curvature R_1 and R_2 at two points on the beam axis separated by a distance d, as illustrated in Fig. 3.1-11. Verify that the location of the beam center and its depth of focus may be determined from the relations

$$z_1 = \frac{-d(R_2 - d)}{R_2 - R_1 - 2d} \tag{3.1-27}$$

$$z_0^2 = \frac{-d(R_1 + d)(R_2 - d)(R_2 - R_1 - d)}{(R_2 - R_1 - 2d)^2} \tag{3.1-28}$$

$$W_0 = \sqrt{\frac{\lambda z_0}{\pi}}.$$

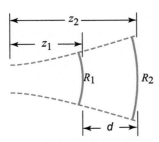

Figure 3.1-11 Given R_1, R_2, and d, determine z_1, z_2, z_0, and W_0.

C. Beam Quality

The Gaussian beam is an idealization that is only approximately met, even in well-designed laser systems. A measure of the quality of an optical beam is the deviation of its profile from Gaussian form. For a beam of waist diameter $2W_m$ and angular divergence $2\theta_m$, a useful numerical measure of the beam quality is provided by the M^2-factor, which is defined as the ratio of the waist-diameter–divergence product, $2W_m \cdot 2\theta_m$ (usually measured in units of mm·mrad), to that expected for a Gaussian beam, which is $2W_0 \cdot 2\theta_0 = 4\lambda/\pi$. Thus,

$$\mathrm{M}^2 = \frac{2W_m \cdot 2\theta_m}{4\lambda/\pi}. \tag{3.1-29}$$

If the two beams have the same waist diameter, the M^2-factor is simply the ratio of their angular divergences,

$$\mathrm{M}^2 = \theta_m/\theta_0, \tag{3.1-30}$$

where $\theta_0 = \lambda/\pi W_0 = \lambda/\pi W_m$ [see (3.1-21)]. Since the Gaussian beam enjoys the smallest possible divergence angle of all beams with the same waist diameter, $\mathrm{M}^2 \geq 1$. The specification of the M^2-factor of an optical beam thus signifies a divergence angle that is M^2 times greater than that of a Gaussian beam of the same waist diameter.

Optical beams produced by commonly available Helium–Neon lasers usually exhibit $\mathrm{M}^2 < 1.1$. For ion lasers, M^2 is typically in the range 1.1–1.3. Collimated TEM_{00} diode-laser beams usually exhibit $\mathrm{M}^2 \approx 1.1$–1.7, whereas high-energy multimode lasers display M^2 factors as high as 3 or 4.

For an optical beam that is approximately Gaussian, the M^2-factor may be determined by making use of a charge-coupled device (CCD) camera to measure the intensity profile of the beam at various locations along the axis of the beam. The beam is focused, by a high-quality lens with a long focal length and large $F_\#$, to a size that is roughly the same as that of the CCD array [see (3.2-17)]. First, the beam center is located by finding the plane at which the spot size is minimized; the waist diameter $2W_m$ is then measured. The axial distance from the beam center to the plane at which the beam diameter increases by a factor of $\sqrt{2}$ provides the Rayleigh range z_m. An estimate of the angular divergence $2\theta_m$ is obtained by using the Gaussian-beam relation $\theta_m = \sqrt{\lambda/\pi z_m}$, which is obtained from (3.1-11) and (3.1-20). Finally, the $\mathbb{M}^2$-factor is computed by means of (3.1-29).

3.2 TRANSMISSION THROUGH OPTICAL COMPONENTS

We proceed now to a discussion of the effects of various optical components on a Gaussian beam. We demonstrate that if a Gaussian beam is transmitted through a set of circularly symmetric optical components aligned with the beam axis, *the Gaussian beam remains a Gaussian beam*, provided that the overall system maintains the paraxial nature of the wave. The beam is reshaped, however — its waist and curvature are altered. The results of this section are of importance in the design of optical instruments that rely on Gaussian beams.

A. Transmission Through a Thin Lens

The complex amplitude transmittance of a thin lens of focal length f is proportional to $\exp(jk\rho^2/2f)$ [see (2.4-9)]. When a Gaussian beam traverses such a component, its complex amplitude, given in (3.1-7), is multiplied by this phase factor. As a result, although the beam width is not altered ($W' = W$), the wavefront is.

To be specific consider a Gaussian beam centered at $z = 0$, with waist radius W_0, transmitted through a thin lens located at position z, as illustrated in Fig. 3.2-1. The phase of the incident wave at the plane of the lens is $kz + k\rho^2/2R - \zeta$, as prescribed by (3.1-23), where $R = R(z)$ and $\zeta = \zeta(z)$ are given in (3.1-9) and (3.1-10), respectively. The phase of the emerging wave therefore becomes

$$kz + k\frac{\rho^2}{2R} - \zeta - k\frac{\rho^2}{2f} = kz + k\frac{\rho^2}{2R'} - \zeta, \qquad (3.2\text{-}1)$$

where

$$\frac{1}{R'} = \frac{1}{R} - \frac{1}{f}. \qquad (3.2\text{-}2)$$

We conclude that the transmitted wave is itself a Gaussian beam with width $W' = W$ and radius of curvature R', where R' satisfies the imaging equation $1/R - 1/R' = 1/f$. The sign of R is positive since the wavefront of the incident beam is diverging whereas the opposite is true of R'.

The parameters of the emerging beam are determined by referring to the outcome of Exercise 3.1-3, in which the parameters of a Gaussian beam are determined from its width and curvature at a given point. Equation (3.1-26) provides that the waist radius is

$$W_0' = \frac{W}{\sqrt{1 + (\pi W^2/\lambda R')^2}} \qquad (3.2\text{-}3)$$

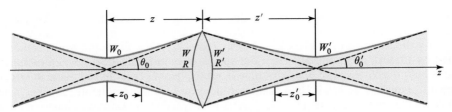

Figure 3.2-1 Transmission of a Gaussian beam through a thin lens.

whereas (3.1-25) provides that the beam center is located at a distance from the lens given by

$$-z' = \frac{R'}{1 + (\lambda R'/\pi W^2)^2}. \tag{3.2-4}$$

The minus sign in (3.2-4) indicates that the beam waist lies to the right of the lens. Substituting $W = W_0\sqrt{1 + (z/z_0)^2}$ and $R = z[1 + (z_0/z)^2]$ from (3.1-8) and (3.1-9) into (3.2-2) to (3.2-4) yields a set of formulas that relate the unprimed parameters of the Gaussian beam incident on the lens to the primed parameters of the Gaussian beam that emerges from the lens, as represented in Fig. 3.2-1:

Waist radius	$W_0' = MW_0$	(3.2-5)
Waist location	$(z' - f) = M^2(z - f)$	(3.2-6)
Depth of focus	$2z_0' = M^2(2z_0)$	(3.2-7)
Divergence angle	$2\theta_0' = \dfrac{2\theta_0}{M}$	(3.2-8)
Magnification	$M = \dfrac{M_r}{\sqrt{1 + r^2}}$	(3.2-9)
$r = \dfrac{z_0}{z - f},$	$M_r = \left\| \dfrac{f}{z - f} \right\|.$	(3.2-9a)

Parameter Transformation by a Lens

The magnification factor M evidently plays an important role. The waist radius is magnified by M, the depth of focus is magnified by M^2, and the divergence angle is minified by M.

Limit of Ray Optics

Consider the limiting case in which $(z - f) \gg z_0$, so that the lens is well outside the depth of focus of the incident beam (Fig. 3.2-2). The beam may then be approximated by a spherical wave, and, in accordance with (3.2-9) and (3.2-9a), $r \ll 1$ so that $M \approx M_r$. In this case (3.2-5)–(3.2-9a) reduce to

$$W_0' \approx MW_0 \tag{3.2-10}$$

$$\frac{1}{z'} + \frac{1}{z} \approx \frac{1}{f} \tag{3.2-11}$$

$$M \approx M_r = \left| \frac{f}{z-f} \right|. \qquad (3.2\text{-}12)$$

Equations (3.2-10)–(3.2-12) are precisely the relations provided by ray optics for the location and size of a patch of light of diameter $2W_0$ located at a distance z to the left of a thin lens (see Sec. 1.2C). Indeed, the magnification factor M_r is identically that based on ray optics. Since (3.2-9) provides that $M < M_r$, the maximum Gaussian-beam magnification attainable is the ray-optics limit M_r. As r^2 increases, the magnification is reduced and the deviation from ray optics widens. Equations (3.2-10)–(3.2-12) also correspond to the results obtained from wave optics for the focusing of a spherical wave in the paraxial approximation (see Sec. 2.4B).

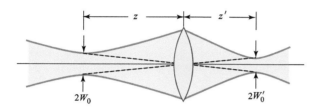

Figure 3.2-2 Beam imaging in the ray-optics limit.

B. Beam Shaping

A lens, or sequence of lenses, may be used to reshape a Gaussian beam without compromising its Gaussian nature. Of course, graded-index components can serve this purpose as well.

Beam Focusing

For a lens placed at the waist of a Gaussian beam, as illustrated in Fig. 3.2-3, the appropriate parameter-transformation formulas are obtained by simply substituting $z = 0$ in (3.2-5) to (3.2-9a). The transmitted beam is then focused to a waist radius W_0' at a distance z' given by

$$W_0' = \frac{W_0}{\sqrt{1 + (z_0/f)^2}} \qquad (3.2\text{-}13)$$

$$z' = \frac{f}{1 + (f/z_0)^2}. \qquad (3.2\text{-}14)$$

In the special case when the depth of focus of the incident beam $2z_0$ is much longer than the focal length f of the lens, as illustrated in Fig. 3.2-4, (3.2-13) reduces to $W_0' \approx (f/z_0)W_0$. Using $z_0 = \pi W_0^2/\lambda$ from (3.1-11), along with (3.1-20), then leads to the simple result

$$W_0' \approx \frac{\lambda}{\pi W_0} f = \theta_0 f \qquad (3.2\text{-}15)$$

$$z' \approx f. \qquad (3.2\text{-}16)$$

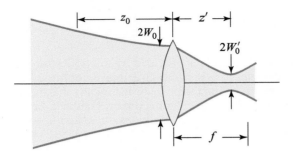

Figure 3.2-3 Focusing a Gaussian beam with a lens at the beam waist.

The transmitted beam is then focused in the focal plane of the lens as would be expected for a collimated beam of parallel rays impinging on the lens. This result emerges because, at its waist, the incident Gaussian beam is well approximated by a plane wave. Wave optics provides that the focused waist radius W_0' is directly proportional to the wavelength and the focal length, and inversely proportional to the radius of the incident beam. The spot size expected from ray optics is, of course, zero, a result that is indeed obtained from the wave-optics formulas as $\lambda \to 0$.

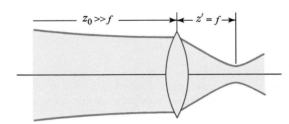

Figure 3.2-4 Focusing a collimated beam.

In many applications, such as laser scanning, laser printing, compact-disc (CD) burning, and laser fusion, it is desired to generate the smallest possible spot size. It is clear from (3.2-15) that this is achieved by making use of the shortest possible wavelength, the thickest incident beam, and the shortest focal-length lens. Since the lens must intercept the incident beam, its diameter D should be at least $2W_0$. Taking $D = 2W_0$, and making use of (3.2-15), the diameter of the focused spot is given by

$$\boxed{2W_0' \approx \frac{4}{\pi}\lambda F_\#} \qquad F_\# = \frac{f}{D}, \qquad (3.2\text{-}17)$$
Focused Spot Size

where the F-number of the lens is denoted $F_\#$. A microscope objective with small F-number is often used for this purpose. A caveat is in order: since (3.2-15) and (3.2-16) are approximate their validity must always be confirmed before use.

EXERCISE 3.2-1

Beam Relaying. A Gaussian beam of radius W_0 and wavelength λ is repeatedly focused by a sequence of identical lenses, each of focal length f and separated by a distance d (Fig. 3.2-5). The focused waist radius is equal to the incident waist radius, i.e., $W_0' = W_0$. Using (3.2-6), (3.2-9), and (3.2-9a) show that this condition can arise only if the inequality $d \leq 4f$ is satisfied. Note that this is the same as the ray-confinement condition for a sequence of lenses derived in Example 1.4-1 using ray optics.

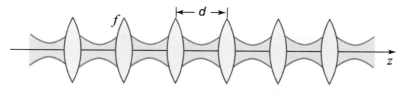

Figure 3.2-5 Beam relaying.

EXERCISE 3.2-2

Beam Collimation. A Gaussian beam is transmitted through a thin lens of focal length f.

(a) Show that the locations of the waists of the incident and transmitted beams, z and z', respectively, are related by

$$\frac{z'}{f} - 1 = \frac{z/f - 1}{(z/f - 1)^2 + (z_0/f)^2}. \tag{3.2-18}$$

This relation is plotted in Fig. 3.2-6.

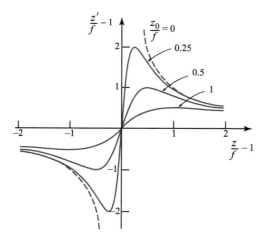

Figure 3.2-6 Relation between the waist locations of the incident and transmitted beams.

(b) The beam is collimated by making the location of the new waist z' as distant as possible from the lens. This is achieved by using the smallest possible ratio z_0/f (short depth of focus and long focal length). For a given ratio z_0/f, show that the optimal value of z for collimation is $z = f + z_0$.

(c) Given $\lambda = 1$ μm, $z_0 = 1$ cm, and $f = 50$ cm, determine the optimal value of z for collimation, and the corresponding magnification M, distance z', and width W_0'' of the collimated beam.

EXERCISE 3.2-3

Beam Expansion. A Gaussian beam may be expanded and collimated by using two lenses of focal lengths f_1 and f_2, as illustrated in Fig. 3.2-7. Parameters of the initial beam (W_0, z_0) are modified by the first lens to (W_0'', z_0'') and subsequently altered by the second lens to (W_0', z_0'). The first lens, which has a short focal length, serves to reduce the depth of focus $2z_0''$ of the beam. This prepares it for collimation by the second lens, which has a long focal length. The system functions as an inverse Keplerian telescope.

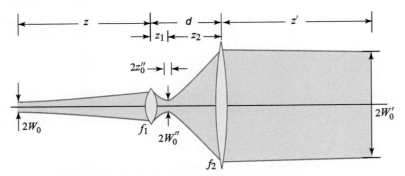

Figure 3.2-7 Beam expansion using a two-lens system.

(a) Assuming that $f_1 \ll z$ and $z - f_1 \gg z_0$, use the results of Exercise 3.2-2 to determine the optical distance d between the lenses such that the distance z' to the waist of the final beam is as large as possible.
(b) Determine an expression for the overall magnification $M = W_0'/W_0$ of the system.

C. Reflection from a Spherical Mirror

We now examine the reflection of a Gaussian beam from a spherical mirror. The complex amplitude reflectance of the mirror is proportional to $\exp(-jk\rho^2/R)$ (see Prob. 2.4-10), where by convention $R > 0$ for convex mirrors and $R < 0$ for concave mirrors. The action of the mirror on a Gaussian beam of width W_1 and radius of curvature R_1 is therefore to reflect the beam and to modify its phase by the factor $-k\rho^2/R$, while leaving the beam width unaltered. The reflected beam therefore remains Gaussian, with parameters W_2 and R_2 given by

$$W_2 = W_1 \tag{3.2-19}$$

$$\frac{1}{R_2} = \frac{1}{R_1} + \frac{2}{R}. \tag{3.2-20}$$

Equation (3.2-20) is identical to (3.2-2) provided $f = -R/2$. Thus, the Gaussian beam is modified in precisely the same way as it is by a lens, except for a reversal of the direction of propagation.

Three special cases, illustrated in Fig. 3.2-8, are of interest:

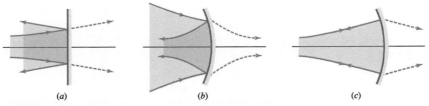

Figure 3.2-8 Reflection of a Gaussian beam with radius of curvature R_1 from a mirror with radius of curvature R: (a) $R = \infty$; (b) $R_1 = \infty$; (c) $R_1 = -R$. The dashed curves show the effects of replacing the mirror by a lens of focal length $f = -R/2$.

- If the *mirror is planar*, i.e., $R = \infty$, then $R_2 = R_1$, so that the mirror reverses the direction of the beam without altering its curvature, as illustrated in Fig. 3.2-8(a).
- If $R_1 = \infty$, i.e., if the *beam waist lies on the mirror*, then $R_2 = R/2$. If the mirror is concave ($R < 0$), $R_2 < 0$ so that the reflected beam acquires a negative curvature and the wavefronts converge. The mirror then focuses the beam to a smaller spot size, as illustrated in Fig. 3.2-8(b).
- If $R_1 = -R$, i.e., if the incident *beam has the same curvature as the mirror*, then $R_2 = R$. The wavefronts of both the incident and reflected waves then coincide with the mirror and the wave retraces its path as shown in Fig. 3.2-8(c). This is expected since the wavefront normals are also normal to the mirror so that the mirror reflects the wave back onto itself. In the illustration in Fig. 3.2-8(c) the mirror is concave ($R < 0$); the incident wave is diverging ($R_1 > 0$) and the reflected wave is converging ($R_2 < 0$).

EXERCISE 3.2-4

Variable-Reflectance Mirrors. A spherical mirror of radius R has a variable power reflectance characterized by $\mathcal{R}(\rho) = |\mathrm{r}(\rho)|^2 = \exp(-2\rho^2/W_m^2)$, which is a Gaussian function of the radial distance ρ. The reflectance is unity on axis and falls by a factor $1/e^2$ when $\rho = W_m$. Determine the effect of the mirror on a Gaussian beam with radius of curvature R_1 and beam width W_1 at the mirror.

---

*D. Transmission Through an Arbitrary Optical System

In the paraxial ray-optics approximation, an optical system is completely characterized by the 2×2 ray-transfer matrix relating the position and inclination of the transmitted ray to those of the incident ray (see Sec. 1.4). We now consider how an arbitrary paraxial optical system, characterized by a matrix **M** of elements (A, B, C, D), modifies a Gaussian beam (Fig. 3.2-9).

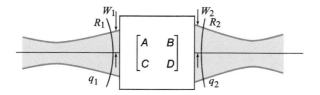

Figure 3.2-9 Modification of a Gaussian beam by an arbitrary paraxial system described by an *ABCD* matrix.

The ABCD Law

The q-parameters, q_1 and q_2, of the incident and transmitted Gaussian beams at the input and output planes of a paraxial optical system described by the (A, B, C, D) matrix are related by

$$q_2 = \frac{Aq_1 + B}{Cq_1 + D}. \qquad (3.2\text{-}21)$$

The *ABCD* Law

Because the complex q-parameter identifies the width W and radius of curvature R of the Gaussian beam (see Exercise 3.1-3), this simple expression, called the *ABCD* **law**, governs the effect of an arbitrary paraxial system on a Gaussian beam. The *ABCD* law will be established by verification in special cases; its generality will ultimately be proved by induction.

Transmission Through Free Space

When the optical system is a distance d of free space (or of any homogeneous medium), the elements of the ray-transfer matrix **M** are $A = 1$, $B = d$, $C = 0$, $D = 1$ [see (1.4-4)]. Since it has been established earlier that $q = z + jz_0$ in free space, the q-parameter is modified by the optical system in accordance with $q_2 = q_1 + d$. This is, in fact, is equal to $(1 \cdot q_1 + d)/(0 \cdot q_1 + 1)$ so that the *ABCD* law is seen to apply.

Transmission Through a Thin Optical Component

An arbitrary thin optical component does not affect the ray position so that

$$y_2 = y_1, \tag{3.2-22}$$

but does alter the inclination angle in accordance with

$$\theta_2 = Cy_1 + D\theta_1, \tag{3.2-23}$$

as illustrated in Fig. 3.2-10. Thus, $A = 1$ and $B = 0$, but C and D are arbitrary. However, in all of the thin optical components described in Sec. 1.4B, $D = n_1/n_2$. By virtue of the vanishing thickness of the component, the beam width does not change, i.e.,

$$W_2 = W_1. \tag{3.2-24}$$

Moreover, if the beams at the input and output planes of the component are approximated by spherical waves of radii R_1 and R_2, respectively, then in the paraxial approximation, when θ_1 and θ_2 are small, $\theta_1 \approx y_1/R_1$ and $\theta_2 \approx y_2/R_2$. Substituting these expressions into (3.2-23), with the help of (3.2-22) we obtain

$$\frac{1}{R_2} = C + \frac{D}{R_1}. \tag{3.2-25}$$

Using (3.1-6), which is the expression for q as a function of R and W, and noting that $D = n_1/n_2 = \lambda_2/\lambda_1$, (3.2-24) and (3.2-25) can be combined into a single equation,

$$\frac{1}{q_2} = C + \frac{D}{q_1}, \tag{3.2-26}$$

from which $q_2 = (1 \cdot q_1 + 0)/(Cq_1 + D)$, so that the *ABCD* law again applies.

Invariance of the ABCD Law to Cascading

If the *ABCD* law is applicable to each of two optical systems with matrices $\mathbf{M}_i = (A_i, B_i, C_i, D_i)$, $i = 1, 2$, it must also apply to a system comprising their cascade (a system with matrix $\mathbf{M} = \mathbf{M}_2\mathbf{M}_1$). This may be shown by straightforward substitution.

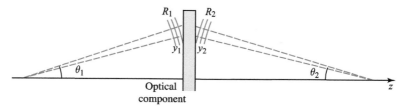

Figure 3.2-10 Modification of a Gaussian beam by a thin optical component.

Generality of the ABCD Law

Since the $ABCD$ law applies to thin optical components as well as to propagation in a homogeneous medium, it also applies to any combination thereof. All of the paraxial optical systems of interest are combinations of propagation in homogeneous media and thin optical components such as thin lenses and mirrors. It is therefore apparent that the $ABCD$ law is applicable to all of these systems. Furthermore, since an inhomogeneous continuously varying medium may be regarded as a cascade of incremental thin elements followed by incremental distances, we conclude that the $ABCD$ law applies to these systems as well, provided that all rays (wavefront normals) remain paraxial.

EXERCISE 3.2-5

Transmission of a Gaussian Beam Through a Transparent Plate. Use the $ABCD$ law to examine the transmission of a Gaussian beam from air, through a transparent plate of refractive index n and thickness d, and again into air. Assume that the beam axis is normal to the plate.

3.3 HERMITE–GAUSSIAN BEAMS

The Gaussian beam is not the only beam-like solution of the paraxial Helmholtz equation (3.1-2). Of particular interest are solutions that exhibit non-Gaussian intensity distributions but share the wavefronts of the Gaussian beam. Such beams have the salutary feature of being able to match the curvatures of spherical mirrors of large radius, such as those that form an optical resonator, and reflect between them without being altered. Such self-reproducing waves are called the **modes** of the resonator (see Appendix C). The optics of resonators is discussed in Chapter 11.

Consider a Gaussian beam of complex envelope [see (3.1-5)]

$$A_G(x, y, z) = \frac{A_1}{q(z)} \exp\left[-jk\frac{x^2 + y^2}{2q(z)}\right], \qquad (3.3\text{-}1)$$

where $q(z) = z + jz_0$. Expressions for the beam width $W(z)$ and the wavefront radius of curvature $R(z)$ are provided in (3.1-8) and (3.1-9), respectively. Now consider a second wave whose complex envelope is a modulated version of the Gaussian beam,

$$A(x, y, z) = \mathcal{X}\left[\sqrt{2}\frac{x}{W(z)}\right] \mathcal{Y}\left[\sqrt{2}\frac{y}{W(z)}\right] \exp[j\mathcal{Z}(z)]\, A_G(x, y, z), \qquad (3.3\text{-}2)$$

where $\mathcal{X}(\cdot)$, $\mathcal{Y}(\cdot)$, and $\mathcal{Z}(\cdot)$ are real functions. This wave, should it be shown to exist, has the following two properties:

1. The phase is the same as that of the underlying Gaussian wave, except for an excess phase $\mathcal{Z}(z)$ that is independent of x and y. If $\mathcal{Z}(z)$ is a slowly varying function of z, both waves have wavefronts with the same radius of curvature $R(z)$. These two waves are therefore focused by thin lenses and mirrors in precisely the same manner.
2. The magnitude

$$A_0 \mathcal{X}\left[\sqrt{2}\frac{x}{W(z)}\right] \mathcal{Y}\left[\sqrt{2}\frac{y}{W(z)}\right] \left[\frac{W_0}{W(z)}\right] \exp\left[-\frac{x^2+y^2}{W^2(z)}\right], \quad (3.3\text{-}3)$$

where $A_0 = A_1/jz_0$, is a function of $x/W(z)$ and $y/W(z)$ whose widths in the x and y directions vary with z in accordance with the same scaling factor $W(z)$. As z increases, the intensity distribution in the transverse plane remains fixed, except for a magnification factor $W(z)$. This distribution is a Gaussian function modulated in the x and y directions by the functions $\mathcal{X}^2(\cdot)$ and $\mathcal{Y}^2(\cdot)$, respectively.

The modulated wave therefore represents a beam of non-Gaussian intensity distribution, but it shares the same wavefronts and angular divergence as the underlying Gaussian wave.

The existence of this wave is assured if three real functions $\mathcal{X}(\cdot)$, $\mathcal{Y}(\cdot)$, and $\mathcal{Z}(z)$ can be found such that (3.3-2) satisfies the paraxial Helmholtz equation (3.1-2). Substituting (3.3-2) into (3.1-2), using the fact that A_G itself satisfies (3.1-2), and defining two new variables $u = \sqrt{2}\,x/W(z)$ and $v = \sqrt{2}\,y/W(z)$, we obtain

$$\frac{1}{\mathcal{X}}\left(\frac{\partial^2 \mathcal{X}}{\partial u^2} - 2u\frac{\partial \mathcal{X}}{\partial u}\right) + \frac{1}{\mathcal{Y}}\left(\frac{\partial^2 \mathcal{Y}}{\partial v^2} - 2v\frac{\partial \mathcal{Y}}{\partial v}\right) + kW^2(z)\frac{\partial \mathcal{Z}}{\partial z} = 0. \quad (3.3\text{-}4)$$

Since the left-hand side of this equation is the sum of three terms, each of which is a function of a single independent variable, u, v, and z, respectively, each of these terms must be constant. Equating the first term to the constant $-2\mu_1$ and the second to $-2\mu_2$, the third must be equal to $2(\mu_1 + \mu_2)$. This technique of "separation of variables" permits us to reduce the partial differential equation (3.3-4) into three ordinary differential equations, for $\mathcal{X}(u)$, $\mathcal{Y}(v)$, and $\mathcal{Z}(z)$, respectively:

$$-\frac{1}{2}\frac{d^2\mathcal{X}}{du^2} + u\frac{d\mathcal{X}}{du} = \mu_1 \mathcal{X} \quad (3.3\text{-}5\text{a})$$

$$-\frac{1}{2}\frac{d^2\mathcal{Y}}{dv^2} + v\frac{d\mathcal{Y}}{dv} = \mu_2 \mathcal{Y} \quad (3.3\text{-}5\text{b})$$

$$z_0\left[1 + \left(\frac{z}{z_0}\right)^2\right]\frac{d\mathcal{Z}}{dz} = \mu_1 + \mu_2, \quad (3.3\text{-}5\text{c})$$

where we have made use of the expression for $W(z)$ given in (3.1-8) and (3.1-11).

Equation (3.3-5a) represents an eigenvalue problem (see Appendix C) whose eigenvalues are $\mu_l = l$, where $l = 0, 1, 2, \ldots$ and whose eigenfunctions are the **Hermite polynomials** $\mathcal{X}(u) = \mathbb{H}_l(u)$, $l = 0, 1, 2, \ldots$. These polynomials are defined by the recurrence relation

$$\mathbb{H}_{l+1}(u) = 2u\,\mathbb{H}_l(u) - 2l\,\mathbb{H}_{l-1}(u) \quad (3.3\text{-}6)$$

with
$$\mathbb{H}_0(u) = 1, \qquad \mathbb{H}_1(u) = 2u. \tag{3.3-7}$$

Thus,
$$\mathbb{H}_2(u) = 4u^2 - 2, \qquad \mathbb{H}_3(u) = 8u^3 - 12u, \quad \ldots. \tag{3.3-8}$$

Similarly, the solutions of (3.3-5b) are $\mu_2 = m$ and $\mathcal{Y}(v) = \mathbb{H}_m(v)$, where $m = 0, 1, 2, \ldots$. There is therefore a family of solutions labeled by the indices (l, m). Substituting $\mu_1 = l$ and $\mu_2 = m$ in (3.3-5c), and integrating, we obtain

$$\mathcal{Z}(z) = (l+m)\,\zeta(z), \tag{3.3-9}$$

where $\zeta(z) = \tan^{-1}(z/z_0)$. The excess phase $\mathcal{Z}(z)$ thus varies slowly between $-(l+m)\,\pi/2$ and $+(l+m)\,\pi/2$, as z varies between $-\infty$ and ∞ (see (3.1-10) and Fig. 3.1-5).

Complex Amplitude

Finally, substitution into (3.3-2) yields an expression for the complex envelope of the beam labeled by the indices (l,m). Rearranging terms and multiplying by $\exp(-jkz)$ provides the complex amplitude

$$\boxed{\begin{aligned}U_{l,m}(x,y,z) = A_{l,m}& \left[\frac{W_0}{W(z)}\right] \mathbb{G}_l\!\left[\frac{\sqrt{2}\,x}{W(z)}\right] \mathbb{G}_m\!\left[\frac{\sqrt{2}\,y}{W(z)}\right] \\ & \times \exp\left[-jkz - jk\frac{x^2+y^2}{2R(z)} + j(l+m+1)\zeta(z)\right]\end{aligned}} \tag{3.3-10}$$

Hermite–Gaussian Beam

where

$$\mathbb{G}_l(u) = \mathbb{H}_l(u)\,\exp\!\left(\frac{-u^2}{2}\right), \qquad l = 0, 1, 2, \ldots \tag{3.3-11}$$

is known as the **Hermite–Gaussian function** of order l, and $A_{l,m}$ is a constant.

Since $\mathbb{H}_0(u) = 1$, the Hermite–Gaussian function of order 0 is simply the Gaussian function. Continuing to higher order, $\mathbb{G}_1(u) = 2u\exp(-u^2/2)$ is an odd function, $\mathbb{G}_2(u) = (4u^2 - 2)\exp(-u^2/2)$ is even, $\mathbb{G}_3(u) = (8u^3 - 12u)\exp(-u^2/2)$ is odd, and so on. These functions are displayed schematically in Fig. 3.3-1.

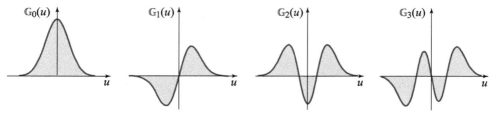

Figure 3.3-1 Low-order Hermite–Gaussian functions: (a) $\mathbb{G}_0(u)$; (b) $\mathbb{G}_1(u)$; (c) $\mathbb{G}_2(u)$; (d) $\mathbb{G}_3(u)$.

An optical wave with complex amplitude given by (3.3-10) is known as a **Hermite–Gaussian beam** of order (l, m), which is often denoted HG_{lm}. The Hermite–Gaussian beam of order $(0, 0)$, namely HG_{00}, is the simple Gaussian beam.

Intensity Distribution

The optical intensity of the HG_{lm} Hermite–Gaussian beam, $I_{l,m} = |U_{l,m}|^2$, is given by

$$I_{l,m}(x, y, z) = |A_{l,m}|^2 \left[\frac{W_0}{W(z)}\right]^2 \mathbb{G}_l^2\left[\frac{\sqrt{2}\,x}{W(z)}\right] \mathbb{G}_m^2\left[\frac{\sqrt{2}\,y}{W(z)}\right]. \qquad (3.3\text{-}12)$$

Figure 3.3-2 illustrates the dependence of the intensity on the normalized transverse distances $u = \sqrt{2}\,x/W(z)$ and $v = \sqrt{2}\,y/W(z)$ for several values of l and m. Beams of higher order have larger widths than those of lower order, as is evident in Fig. 3.3-1. Regardless of the order, however, the width of the beam is proportional to $W(z)$; thus, as z increases, the transverse spatial extent of the intensity pattern is magnified by the factor $W(z)/W_0$ but otherwise maintains its profile. The only circularly symmetric member of the family of Hermite–Gaussian beams is the elementary Gaussian beam itself.

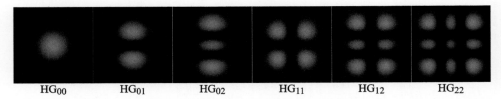

HG_{00} HG_{01} HG_{02} HG_{11} HG_{12} HG_{22}

Figure 3.3-2 Intensity distributions of several low-order Hermite–Gaussian beams, HG_{lm}, in the transverse plane. The HG_{00} beam is the elementary Gaussian beam displayed in Fig. 3.1-1.

The Hermite–Gaussian beam defined in (3.3-10) may be generalized by ascribing different beam widths to its x and y components, $W_x(z)$ and $W_y(z)$, respectively, thereby defining the **elliptic Hermite–Gaussian beam**. Because (3.3-10) is a separable function of x and y, this constitutes yet another exact solution of the paraxial Helmholtz equation. A special case is the **elliptic Gaussian beam** that appears in Prob. 3.1-8; it exhibits elliptical, rather than circular, contours of constant intensity.

3.4 LAGUERRE–GAUSSIAN BEAMS

Laguerre–Gaussian Beams

The Hermite–Gaussian beams form a complete set of solutions to the paraxial Helmholtz equation. Any other solution can be written as a superposition of these beams. An alternate complete set of solutions, known as **Laguerre–Gaussian beams**, is obtained by writing the paraxial Helmholtz equation in cylindrical coordinates (ρ, ϕ, z) and then using the separation-of-variables technique in ρ and ϕ, rather than in x and y. The complex amplitude of the Laguerre–Gaussian beam, denoted LG_{lm}, can be expressed as

$$U_{l,m}(\rho, \phi, z) = A_{l,m} \left[\frac{W_0}{W(z)}\right] \left(\frac{\rho}{W(z)}\right)^l \mathbb{L}_m^l\left(\frac{2\rho^2}{W^2(z)}\right) \exp\left(-\frac{\rho^2}{W^2(z)}\right)$$

$$\times \exp\left[-jkz - jk\frac{\rho^2}{2R(z)} \mp jl\phi + j(l + 2m + 1)\zeta(z)\right], \qquad (3.4\text{-}1)$$

where the $\mathbb{L}_m^l(\cdot)$ represent generalized Laguerre polynomials,[†] and where $W(z)$, $R(z)$, $\zeta(z)$, and W_0 are given by (3.1-8)–(3.1-11). The integers $l = 0, 1, 2, \ldots$ and $m = 0, 1, 2, \ldots$ are azimuthal and radial indices, respectively. The lowest-order Laguerre–Gaussian beam LG_{00}, like the lowest-order Hermite–Gaussian beam HG_{00}, is the simple Gaussian beam.

EXERCISE 3.4-1

Laguerre–Gaussian Beam as a Superposition of Hermite–Gaussian Beams. Demonstrate that the Laguerre–Gaussian beam LG_{10} is equivalent to the superposition of two Hermite–Gaussian beams, HG_{01} and HG_{10}, with equal amplitudes and a phase shift of $\pi/2$, i.e., $LG_{10} \equiv \frac{1}{\sqrt{2}}(HG_{01} + jHG_{10})$.

The intensity of the Laguerre–Gaussian beam, which is proportional to the absolute square of (3.4-1), is a function of ρ and z, but not of ϕ, so that it is circularly symmetric. As illustrated in Fig. 3.4-1(a), the transverse intensity distribution for the LG_{10} beam assumes a toroidal shape. Its peak value is attained at a radius of $\rho = \sqrt{1/2}\,W(z)$, which increases with the distance z from the beam center (much as for the Gaussian beam). Beams of any order $l \neq 0$ are also toroidal when $m = 0$, and attain their peak values at radii $\sqrt{l/2}\,W(z)$. All beams with $l \neq 0$ have zero intensity at the beam center ($\rho = 0$); those with a radial index $m > 0$ take the form of multiple rings.

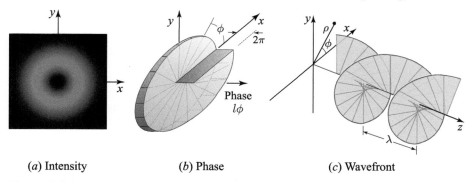

(a) Intensity   (b) Phase   (c) Wavefront

Figure 3.4-1 The Laguerre–Gaussian beam LG_{10}. (a) The transverse intensity distribution takes the form of a toroid. (b) The phase component $l\phi$, plotted for $l = 1$, is a linear function of the azimuthal angle ϕ. (c) The wavefront is a left-handed helical surface that undergoes corkscrew-like motion as it travels in the z direction.

The phase behavior of the Laguerre–Gaussian beam has the same dependence on ρ and z as does the Gaussian beam [see (3.1-7)], with two notable exceptions: (1) the Gouy phase is enhanced by the factor $(l + 2m + 1)$, and (2) there is an additional phase factor $e^{\mp jl\phi}$ that is proportional to the azimuthal angle ϕ. The phase component $l\phi$, illustrated in Fig. 3.4-1(b) for $l = 1$, is associated with the phase factor $\exp(-jl\phi)$ [see (3.1-23) and associated footnote]. It results in the wavefront assuming the form of a left-handed helix that undergoes corkscrew-like motion as the wave advances in the z direction, as shown in Fig. 3.4-1(c). Beams with $l > 1$ have wavefronts comprising l distinct but intertwined helices. The pitch of each helix is $l\lambda$ and the $\mp$ sign determines its handedness.

[†] The generalized Laguerre polynomials are expressible as $\mathbb{L}_m^l(x) = (m+l)! \sum_{i=0}^m (-x)^i / [i!(m-i)!(l+i)!]$. A few elementary examples are $\mathbb{L}_0^l(x) = 1$; $\mathbb{L}_1^l(x) = 1 - x + l$; $\mathbb{L}_2^l(x) = \frac{1}{2}[x^2 - 2(l+2)x + (l+1)(l+2)]$. The generalized Laguerre polynomials $\mathbb{L}_m^l(x)$ reduce to the simple Laguerre polynomials $\mathbb{L}_m(x)$ when $l = 0$.

As discussed in Sec. 2.5A, the phase of an optical beam may be determined by detecting its interference with an auxiliary optical field of known form (e.g., a plane wave). The phase of a Laguerre–Gaussian beam can be readily observed by detecting its superposition with another Laguerre–Gaussian beam of the same order but opposite handedness. Such a superposition, which constitutes a form of standing wave, has an intensity proportional to $|\exp(-jl\phi) + \exp(jl\phi)|^2 = 4\cos^2(l\phi)$, explicitly illustrating that the resulting intensity is sensitive to $l\phi$ as shown in Fig. 3.4-2. The number of angular interference fringes is equal to $2l$.

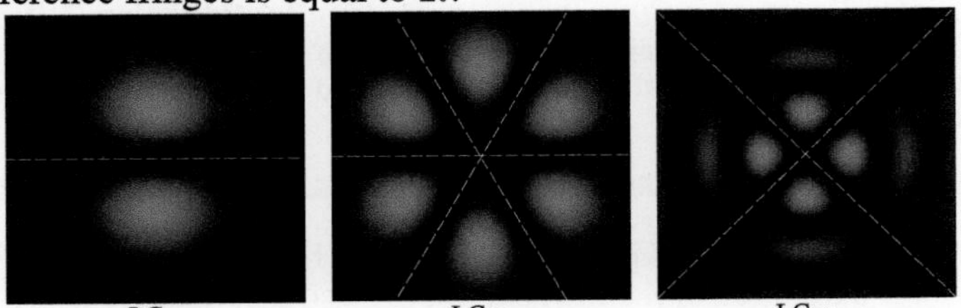

Figure 3.4-2 Transverse intensity distributions of the superposition of two Laguerre–Gaussian beams of the same order LG_{lm} but opposite handedness. The dashed white lines signify the loci of zero intensity (the nodes of the standing waves). The number of such lines is equal to the azimuthal order l. As the azimuthal angle ϕ moves from one node to the next, the phase changes by 2π.

Laguerre–Gaussian beams may be directly generated as laser modes or as combinations of Hermite–Gaussian laser modes, as discussed in Exercise 3.4-1. A Gaussian beam may be converted into an Laguerre–Gaussian beam by imparting to it the phase factor $\exp(-jl\phi)$ with the help of a spiral phase plate, a dielectric slab whose optical thickness increases linearly with ϕ [see Fig. 3.4-1(b)]. However, the method of choice for converting a Gaussian beam to a Laguerre–Gaussian beam is to make use of a diffractive optical element, or hologram, endowed with a fork dislocation centered on the beam axis, as exhibited in Example 4.5-3.

Beams with spiral phase carry orbital angular momentum. This may be understood by observing that an optical wave carries linear momentum that points along the direction orthogonal to its wavefronts (see Secs. 5.1 and 13.1D), which is also the direction of the optical rays. Since rays orthogonal to the helical wavefront of a Laguerre–Gaussian beam have azimuthal components that revolve about the beam axis, their linear momentum is accompanied by orbital angular momentum. This can also be visualized by considering that refracted optical rays incident on the surface of a spiral phase plate acquire azimuthal components [see Fig. 3.4-1(b)]. By virtue of their orbital angular momentum, Laguerre–Gaussian beams can exert a mechanical torque on micro-objects and can thus be used to manipulate microparticles.

Optical Vortices

An **optical vortex** is an optical field that exhibits a *line* of zero optical intensity, such as the line along the axis of a Laguerre–Gaussian beam with $l \neq 0$. It is also called a *screw dislocation* since the phase of the field is twisted like a corkscrew about the axis of travel. An optical vortex in a plane is a *point* at which the optical field vanishes; it is also called a *phase singularity*. An example of the latter is the point $(x, y) = (0, 0)$ in the transverse plane of the Laguerre–Gaussian beam illustrated in Fig. 3.4-1(a).

The strength of a vortex is indicated by its **topological charge**, which is determined by the number of full twists that the phase undergoes in a distance of one wavelength. For the Laguerre–Gaussian beam, the topological charge is the azimuthal index l, which is indicated by the number of lines of zero intensity that appear in the standing wave generated by the combination of two beams of the same order but opposite handedness, as illustrated in Fig. 3.4-2. This number also determines the orbital angular momentum of the associated photon, as will be discussed in Sec. 13.1D.

Optical vortex beams can assume forms that are far more complex than the simple Laguerre–Gaussian beam. Interference among three or more randomly directed plane-wave components of similar intensities always results in a field cross-section that contains many vortices. Such beams often exhibit unusual and dramatic properties — the field surrounding a vortex can, for example, tangle and form links and knots.[†]

Ince–Gaussian Beams

As discussed in Sec. 3.3 and at the beginning of this section, Hermite–Gaussian and Laguerre–Gaussian beams form complete sets of exact solutions to the paraxial Helmholtz equation, in Cartesian and cylindrical coordinates, respectively. A third complete set of exact solutions, known as Ince–Gaussian (IG) beams,[‡] exists in elliptic cylindrical coordinates, another three-dimensional orthogonal coordinate system. The transverse structure of these beams is characterized by Ince polynomials, which have an intrinsic elliptical character. Laguerre–Gaussian and Hermite–Gaussian beams are limiting forms of Ince–Gaussian beams when the ellipticity parameter is 0 and ∞, respectively.

3.5 NONDIFFRACTING BEAMS

A. Bessel Beams

In the search for beam-like waves, it is natural to attempt to construct waves whose wavefronts are planar but whose intensity distributions are nonuniform in the transverse plane. Consider, for example, a wave with complex amplitude

$$U(\mathbf{r}) = A(x,y)\, e^{-j\beta z}. \tag{3.5-1}$$

In order that this wave satisfy the Helmholtz equation (2.2-7), $\nabla^2 U + k^2 U = 0$, the quantity $A(x,y)$ must satisfy

$$\nabla_T^2 A + k_T^2 A = 0, \tag{3.5-2}$$

where $k_T^2 + \beta^2 = k^2$ and $\nabla_T^2 = \partial^2/\partial x^2 + \partial^2/\partial y^2$ is the transverse Laplacian operator. Equation (3.5-2), known as the **two-dimensional Helmholtz equation**, may be solved by employing the method of separation of variables. Using polar coordinates ($x = \rho\cos\phi$, $y = \rho\sin\phi$), the result turns out to be

$$A(x,y) = A_m\, J_m(k_T \rho)\, e^{jm\phi}, \qquad m = 0, \pm 1, \pm 2, \ldots, \tag{3.5-3}$$

where $J_m(\cdot)$ is the Bessel function of the first kind and mth order, and A_m is a constant. Solutions of (3.5-3) that are singular at $\rho = 0$ are not included.

For $m = 0$, the wave has complex amplitude

$$U(\mathbf{r}) = A_0\, J_0(k_T \rho)\, e^{-j\beta z}, \tag{3.5-4}$$

and therefore has planar wavefronts — the wavefront normals (rays) are all parallel to the z axis. The intensity distribution $I(\rho,\phi,z) = |A_0|^2 J_0^2(k_T \rho)$ is circularly symmetric

---

[†] See M. R. Dennis, R. P. King, B. Jack, K. O'Holleran, and M. J. Padgett, Isolated Optical Vortex Knots, *Nature Physics*, vol. 6, pp. 118–121, 2010.

[‡] See M. A. Bandres and J. C. Gutiérrez-Vega, Ince–Gaussian Beams, *Optics Letters*, vol. 29, pp. 144–146, 2004.

and varies with ρ as illustrated in Fig. 3.5-1(a); it is independent of z so that there is no spread of the optical power. This wave is known as the **Bessel beam**.

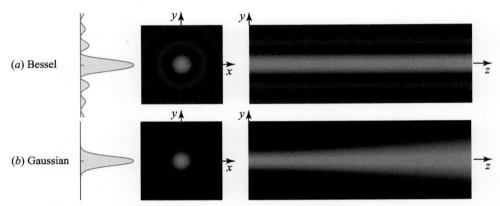

Figure 3.5-1 (a) The intensity distribution of the Bessel beam in the transverse plane is independent of z. The beam is nondiffracting and therefore does not diverge. (b) Transverse intensity distribution of a Gaussian beam for comparison with the Bessel beam. Parameters are selected such that the peak intensities and $1/e^2$ widths are identical in both cases.

It is useful to compare the Bessel beam with the Gaussian beam. Whereas the complex amplitude of the Bessel beam is an *exact* solution of the Helmholtz equation, the complex amplitude of the Gaussian beam is only an *approximate* solution thereof (since its complex envelope is an exact solution of the paraxial Helmholtz equation). The intensity distributions of these two beams are compared graphically in Fig. 3.5-1.

It is apparent that the asymptotic behavior of these distributions in the limit of large radial distances is significantly different. The intensity of the Gaussian beam decreases exponentially with ρ as $\exp[-2\rho^2/W^2(z)]$. The intensity of the Bessel beam, on the other hand, decreases as $J_0^2(k_T\rho) \approx (2/\pi k_T\rho)\cos^2(k_T\rho - \pi/4)$, which is an oscillatory function superimposed on a slow inverse-power-law decay with ρ. As a consequence, the transverse RMS width of the Gaussian beam, $\sigma = \frac{1}{2}W(z)$, is finite, while the transverse RMS width of the ideal Bessel beam is infinite for all z (see Appendix A, Sec. A.2 for the definition of RMS width). This is a manifestation of the tradeoff between beam size and divergence: the RMS width of the ideal Bessel beam is infinite and its divergence is zero, just as for the ideal plane wave.

As shown in Examples 2.4-1 and 4.3-5, the Bessel beam is associated with a continuum of plane waves whose directions form a cone of fixed half angle with respect to the propagation direction. It can be implemented by use of an axicon [Fig. 1.2-12(c)]. A derivation of the complex amplitude for the Bessel beam along with a general discussion of nondiffracting beams from the perspective of Fourier optics is provided in Sec. 4.3C.

A hybrid beam, called a **Bessel–Gaussian beam**, is a Bessel beam modulated by a Gaussian function of the radial coordinate ρ. The Gaussian serves as a window function that accelerates the slow radial decay of the Bessel beam (see Fig. 3.5-1). The Bessel–Gaussian beam can be generated by illuminating an axicon with a Gaussian beam.

*B. Airy Beams

In analogy with the Bessel beam, the Airy beam arises as a diffraction-free exact solution to the paraxial Helmholtz equation (2.2-23). Although the shape of its transverse intensity distribution is maintained, the beam center is transversally displaced in an accelerated manner as it propagates along the axial direction,[†] as shown in Fig. 3.5-2.

The complex envelope of the Airy beam in one dimension is expressed as

$$A(x, z) = \text{Ai}\left(\frac{x}{W_0} - \frac{z^2}{(4z_0)^2}\right) \exp\left(-j\frac{x}{W_0}\frac{z}{2z_0}\right) \exp\left(j\frac{z^3}{(24z_0)^3}\right), \quad (3.5\text{-}5)$$

where $\text{Ai}(x)$ is the Airy function, a special function that is the solution of the Airy differential equation $d^2y/dx^2 = xy$. The parameters W_0 and z_0 are, respectively, transverse and axial scaling factors that obey the relation $W_0^2 = \lambda z_0/\pi$, which also applies to the Gaussian beam [see (3.1-11)]. At $z = 0$ the transverse intensity of the Airy beam, $I(x, 0) = \text{Ai}^2(x/W_0)$, is distinctly asymmetrical as illustrated on the left-hand side of Fig. 3.5-2. At an arbitrary value of z, the intensity has the same transverse distribution except that it exhibits an axially dependent transverse shift $x = W_0 z^2/(4z_0)^2$ that follows a parabolic trajectory, $x = z^2/4a$ with $a = 4z_0^2/W_0$, thereby mimicking the path of a ballistic projectile. At $z = 4z_0$, for example, the transverse shift is W_0 while at $z = 20\, z_0$ it grows to $25\, W_0$, which provides the rationale for the appellation **accelerating beam**.

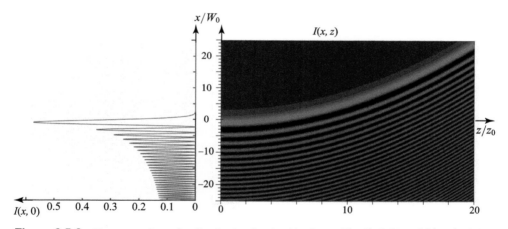

Figure 3.5-2 Transverse intensity distribution for the Airy beam $I(x, 0)$ (left) and $I(x, z)$ (right).

The Airy beam may be generated by making use of an optical Fourier-transform system, as described in Prob. 4.2-6. Applications of the Airy beam include microscopy and prodding small particles along curved trajectories.

Other Bessel-like and Airy-like beams with main-lobe intensity distributions that remain nearly invariant and symmetrical as they travel can be engineered to propagate along arbitrary trajectories in free space (including 3D spirals).[‡] These nondiffracting beams, which can be partially obstructed and yet recover further down the beam axis (so-called "self-healing"), are useful for applications such as optical trapping and precision drilling.

[†] See G. A. Siviloglou, J. Broky, A. Dogariu, and D. N. Christodoulides, Observation of Accelerating Airy Beams, *Physical Review Letters*, vol. 99, 213901, 2007.

[‡] See J. Zhao, P. Zhang, D. Deng, J. Liu, Y. Gao, I. D. Chremmos, N. K. Efremidis, D. N. Christodoulides, and Z. Chen, Observation of Self-Accelerating Bessel-Like Optical Beams Along Arbitrary Trajectories, *Optics Letters*, vol. 38, pp. 498–500, 2013.

READING LIST

Books

See also the reading list on lasers in Chapter 16.

J. Secor, R. Alfano, and S. Ashrafi, *Complex Light*, IOP Publishing, 2017.

F. M. Dickey, ed., *Laser Beam Shaping: Theory and Techniques*, CRC Press/Taylor & Francis, 2nd ed. 2014.

A. N. Oraevsky, *Gaussian Beams and Optical Resonators*, Nova Science, 1996.

Articles

M. J. Padgett, Orbital Angular Momentum 25 Years On, *Optics Express*, vol. 25, pp. 11265–11274, 2017.

J. Zhao, P. Zhang, D. Deng, J. Liu, Y. Gao, I. D. Chremmos, N. K. Efremidis, D. N. Christodoulides, and Z. Chen, Observation of Self-Accelerating Bessel-Like Optical Beams Along Arbitrary Trajectories, *Optics Letters*, vol. 38, pp. 498–500, 2013.

A. Dudley, M. Lavery, M. Padgett, and A. Forbes, Unraveling Bessel Beams, *Optics & Photonics News*, vol. 24, no. 6, pp. 22–29, 2013.

A. M. Yao and M. J. Padgett, Orbital Angular Momentum: Origins, Behavior and Applications, *Advances in Optics and Photonics*, vol. 3, pp. 161–204, 2011.

M. R. Dennis, R. P. King, B. Jack, K. O'Holleran, and M. J. Padgett, Isolated Optical Vortex Knots, *Nature Physics*, vol. 6, pp. 118–121, 2010.

M. R. Dennis, K. O'Holleran, and M. J. Padgett, Singular Optics: Optical Vortices and Polarization Singularities, in E. Wolf, ed., *Progress in Optics*, Elsevier, 2009, vol. 53, pp. 293–363.

J. Vickers, M. Burch, R. Vyas, and S. Singh, Phase and Interference Properties of Optical Vortex Beams, *Journal of the Optical Society of America A*, vol. 25, pp. 823–827, 2008.

M. Martinelli and P. Martelli, Laguerre Mathematics in Optical Communications, *Optics & Photonics News*, vol. 19, no. 2, pp. 30–35, 2008.

G. A. Siviloglou, J. Broky, A. Dogariu, and D. N. Christodoulides, Observation of Accelerating Airy Beams, *Physical Review Letters*, vol. 99, 213901, 2007.

M. A. Bandres and J. C. Gutiérrez-Vega, Ince–Gaussian Modes of the Paraxial Wave Equation and Stable Resonators, *Journal of the Optical Society of America*, vol. 21, pp. 873–880, 2004.

F. Gori, G. Guattari, and C. Padovani, Bessel–Gauss Beams, *Optics Communications*, vol. 64, pp. 491–495, 1987.

J. Durnin, J. J. Miceli, Jr., and J. H. Eberly, Diffraction-Free Beams, *Physical Review Letters*, vol. 58, pp. 1499–1501, 1987.

Special issue on propagation and scattering of beam fields, *Journal of the Optical Society of America A*, vol. 3, no. 4, 1986.

H. Kogelnik and T. Li, Laser Beams and Resonators, *Proceedings of the IEEE*, vol. 54, pp. 1312–1329, 1966.

G. D. Boyd and J. P. Gordon, Confocal Multimode Resonator for Millimeter Through Optical Wavelength Masers, *Bell System Technical Journal*, vol. 40, pp. 489–508, 1961.

A. G. Fox and T. Li, Resonant Modes in a Maser Interferometer, *Bell System Technical Journal*, vol. 40, pp. 453–488, 1961.

PROBLEMS

3.1-6 **Beam Parameters.** The light emitted from a Nd:YAG laser at a wavelength of 1.06 μm is a Gaussian beam of 1-W optical power and beam divergence $2\theta_0 = 1$ mrad. Determine the beam waist radius, the depth of focus, the maximum intensity, and the intensity on the beam axis at a distance $z = 100$ cm from the beam waist.

3.1-7 **Beam Identification by Two Widths.** A Gaussian beam of wavelength $\lambda_o = 10.6$ μm (emitted by a CO_2 laser) has widths $W_1 = 1.699$ mm and $W_2 = 3.380$ mm at two points separated by a distance $d = 10$ cm. Determine the location of the waist and the waist radius.

3.1-8 **The Elliptic Gaussian Beam.** The paraxial Helmholtz equation admits a Gaussian beam with intensity $I(x,y,0) = |A_0|^2 \exp[-2(x^2/W_{0x}^2 + y^2/W_{0y}^2)]$ in the $z=0$ plane, with the beam waist radii W_{0x} and W_{0y} in the x and y directions, respectively. The contours of constant intensity are therefore ellipses instead of circles. Write expressions for the beam depth of focus, angular divergence, and radii of curvature in the x and y directions, as functions of W_{0x}, W_{0y}, and the wavelength λ. If $W_{0x} = 2W_{0y}$, sketch the shape of the beam spot in the $z=0$ plane and in the far field (z much greater than the depths of focus in both transverse directions).

3.2-6 **Beam Focusing.** An argon-ion laser produces a Gaussian beam of wavelength $\lambda = 488$ nm with waist radius $W_0 = 0.5$ mm. Design a single-lens optical system for focusing the light to a spot of diameter 100 μm. What is the shortest focal-length lens that may be used?

3.2-7 **Spot Size.** A Gaussian beam of Rayleigh range $z_0 = 50$ cm and wavelength $\lambda = 488$ nm is converted into a Gaussian beam of waist radius W_0' using a lens of focal length $f = 5$ cm at a distance z from its waist, as illustrated in Fig. 3.2-2. Plot W_0' as a function of z. Verify that in the limit $z - f \gg z_0$, (3.2-10) and (3.2-12) hold; and that in the limit $z \ll z_0$, (3.2-13) holds.

3.2-8 **Beam Refraction.** A Gaussian beam is incident from air ($n=1$) into a medium with a planar boundary and refractive index $n = 1.5$. The beam axis is normal to the boundary and the beam waist lies at the boundary. Sketch the transmitted beam. If the angular divergence of the beam in air is 1 mrad, what is the angular divergence in the medium?

*3.2-9 **Transmission of a Gaussian Beam Through a Graded-Index Slab.** The $ABCD$ matrix of a SELFOC graded-index slab with quadratic refractive index $n(y) \approx n_0(1 - \frac{1}{2}\alpha^2 y^2)$ (see Sec. 1.3B) and length d is $A = \cos\alpha d$, $B = (1/\alpha)\sin\alpha d$, $C = -\alpha\sin\alpha d$, $D = \cos\alpha d$ for paraxial rays along the z direction. A Gaussian beam of wavelength λ_o, waist radius W_0 in free space, and axis in the z direction enters the slab at its waist. Use the $ABCD$ law to determine an expression for the beam width in the y direction as a function of d. Sketch the shape of the beam as it travels through the medium.

3.3-2 **Power Confinement in Hermite–Gaussian Beams.** Determine the ratio of the power contained within a circle of radius $W(z)$ in the transverse plane, to the total power, for the Hermite–Gaussian beams HG_{00}, HG_{01}, HG_{10}, and HG_{11}. What is the ratio of the power contained within a circle of radius $\frac{1}{10}W(z)$ to the total power for the HG_{00} and HG_{11} beams?

3.3-3 **The Donut Beam.** Consider a wave that is a superposition of two Hermite–Gaussian beams, HG_{01} and HG_{10}, with equal intensities. The two beams have independent and random phases so that their intensities add with no interference. Show that the total intensity is described by a donut-shaped (toroidal) circularly symmetric function. Assuming that $W_0 = 1$ mm, determine the radius of the circle of peak intensity and the radii of the two circles of $1/e^2$ times the peak intensity at the beam waist.

3.3-4 **Axial Phase.** Consider the Hermite–Gaussian beams of all orders, HG_{lm}, with Rayleigh range $z_0 = 30$ cm in a medium of refractive index $n = 1$. Determine the frequencies within the band $\nu = 10^{14} \pm 2 \times 10^9$ Hz for which the phase retardation between the planes $z = -z_0$ and $z = z_0$ is an integer multiple of π along the beam axis. These frequencies are the modes of a resonator comprising two spherical mirrors placed at the $z = \pm z_0$ planes, as described in Sec. 11.2D.

CHAPTER

4

FOURIER OPTICS

4.1	PROPAGATION OF LIGHT IN FREE SPACE	113
	A. Spatial Harmonic Functions and Plane Waves	
	B. Transfer Function of Free Space	
	C. Impulse Response Function of Free Space	
	D. Huygens–Fresnel Principle	
4.2	OPTICAL FOURIER TRANSFORM	124
	A. Fourier Transform in the Far Field	
	B. Fourier Transform Using a Lens	
4.3	DIFFRACTION OF LIGHT	129
	A. Fraunhofer Diffraction	
	*B. Fresnel Diffraction	
	*C. Nondiffracting Waves	
4.4	IMAGE FORMATION	137
	A. Ray Optics of a Single-Lens Imaging System	
	B. Wave Optics of a 4-f Imaging System	
	C. Wave Optics of a Single-Lens Imaging System	
	D. Near-Field Imaging	
4.5	HOLOGRAPHY	147

Josef von Fraunhofer (1787–1826) developed the diffraction grating and contributed to our understanding of diffraction. His epitaph reads *Approximavit sidera* (he brought the stars closer).

Jean-Baptiste Joseph Fourier (1768–1830) demonstrated that periodic functions could be constructed from sums of sinusoids. Harmonic analysis is the basis of Fourier optics; it has many applications.

Dennis Gabor (1900–1979) invented holography and contributed to its development. He made the first hologram in 1947 and received the Nobel Prize in 1971 for carrying out this body of work.

Fundamentals of Photonics, Third Edition. Bahaa E. A. Saleh and Malvin Carl Teich.
©2019 John Wiley & Sons, Inc. Published 2019 by John Wiley & Sons, Inc.

Fourier optics provides a description of the propagation of light waves based on harmonic analysis (the Fourier transform) and linear systems. The methods of harmonic analysis have proved useful for describing signals and systems in many disciplines. Harmonic analysis is based on the expansion of an arbitrary function of time $f(t)$ in terms of a superposition (a sum or integral) of harmonic functions of time of different frequencies (see Appendix A, Sec. A.1). The harmonic function $F(\nu)\exp(j2\pi\nu t)$, which has frequency ν and complex amplitude $F(\nu)$, is the building block of the theory. Several of these functions, each with its own amplitude $F(\nu)$, are added to construct the function $f(t)$, as illustrated in Fig. 4.0-1. The complex amplitude $F(\nu)$, as a function of frequency, is called the Fourier transform of $f(t)$. This approach is highly useful for analyzing linear systems (see Appendix B, Sec. B.1). If the response of the system to each harmonic function is known, the response to an arbitrary input function is readily determined by the use of harmonic analysis at the input of the system and superposition at the output.

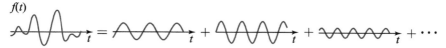

Figure 4.0-1 An arbitrary function $f(t)$ may be analyzed as a sum of harmonic functions of different frequencies and complex amplitudes.

An arbitrary complex function $f(x,y)$ of two variables that represent spatial coordinates in a plane, say x and y, may similarly be written as a superposition of harmonic functions of x and y, each of the form $F(\nu_x,\nu_y)\exp[-j2\pi(\nu_x x + \nu_y y)]$, where $F(\nu_x,\nu_y)$ is the complex amplitude and ν_x and ν_y are the **spatial frequencies** (cycles per unit length; typically cycles/mm) in the x and y directions, respectively.[†] The harmonic function $F(\nu_x,\nu_y)\exp[-j2\pi(\nu_x x+\nu_y y)]$ is the two-dimensional building block of the theory. It can be used to generate an arbitrary function of two variables $f(x,y)$, as depicted in Fig. 4.0-2 and explained in Appendix A, Sec. A.3.

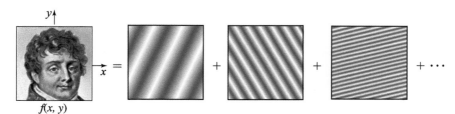

Figure 4.0-2 An arbitrary function $f(x,y)$ may be analyzed in terms of a sum of harmonic functions of different spatial frequencies and complex amplitudes, drawn here schematically as graded blue lines.

The monochromatic plane wave $U(x,y,z) = A\exp[-j(k_x x + k_y y + k_z z)]$ plays an important role in wave optics. The coefficients (k_x, k_y, k_z) are the components of the wavevector **k**, and A is a complex constant. $U(x,y,z)$ reduces to a spatial harmonic function of the points in an arbitrary plane. At the $z = 0$ plane, for example, $U(x,y,0)$ becomes the harmonic function $f(x,y) = A\exp[-j2\pi(\nu_x x + \nu_y y)]$,

---

[†] The spatial harmonic function is defined with a minus sign in the exponent, in contrast to the plus sign used in the definition of the temporal harmonic function (compare (A.1-1) and (A.3-1) in Appendix A). This sign convention is chosen to match that for the forward-traveling plane wave set forth in (2.2-11).

where $\nu_x = k_x/2\pi$ and $\nu_y = k_y/2\pi$ are the spatial frequencies (cycles/mm), and k_x and k_y are the spatial angular frequencies (radians/mm). There is a one-to-one correspondence between the plane wave $U(x, y, z)$ and the spatial harmonic function $f(x, y) = U(x, y, 0)$ since knowledge of k_x and k_y is sufficient to determine k_z via the relation $k_x^2 + k_y^2 + k_z^2 = \omega^2/c^2 = (2\pi/\lambda)^2$. As will be explained subsequently, k_x and k_y may not exceed ω/c under usual circumstances; i.e., the spatial frequencies ν_x and ν_y may not exceed the inverse wavelength $1/\lambda$.

Since an arbitrary function $f(x, y)$ can be analyzed as a superposition of harmonic functions, an arbitrary traveling wave $U(x, y, z)$ may be analyzed in terms of a sum of plane waves (Fig. 4.0-3). The plane wave is thus the building block used to construct a wave of arbitrary complexity. Furthermore, if it can be determined how a linear optical system modifies plane waves, the principle of superposition can be used to establish the effect of the system on an arbitrary wave.

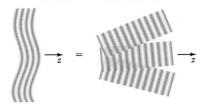

Figure 4.0-3 The principle of Fourier optics: An arbitrary wave in free space can be analyzed in terms of a superposition of plane waves.

Because of the important role that Fourier analysis plays in describing linear systems, it is useful to consider the propagation of light through linear optical components, including free space, in terms of a linear-systems approach. The complex amplitudes at two planes normal to the optic (z) axis are regarded as the input and output of the system (Fig. 4.0-4). A linear system may be characterized by either its **impulse response function**, which is the response of the system to a point (i.e., an impulse) at its input, or by its **transfer function**, which is the response of the system to a set of spatial harmonic functions (as described in Appendix B).

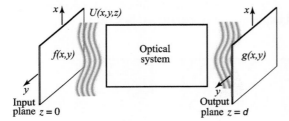

Figure 4.0-4 The transmission of an optical wave $U(x, y, z)$ through an optical system located between an input plane $z = 0$ and an output plane $z = d$. This configuration is regarded as a linear system whose input and output are the functions of $f(x, y) = U(x, y, 0)$ and $g(x, y) = U(x, y, d)$, respectively.

This Chapter

The chapter begins with a Fourier description of the propagation of monochromatic light in free space (Sec. 4.1). The transfer function and impulse response function of the free-space propagation system are determined. In Sec. 4.2 we show that a lens may be used to carry out the spatial Fourier-transform operation. The transmission of light through apertures is discussed in Sec. 4.3; this section comprises a Fourier-optics approach to the diffraction of light, a subject usually presented in introductory textbooks from the perspective of the Huygens principle. Section 4.4 is devoted to image formation and spatial filtering in the context of both ray and wave optics. Subwavelength imaging, in the form of near-field optical microscopy, is also considered. Finally, an introduction to holography, the recording and reconstruction of optical waves, is presented in Sec. 4.5. It is important to understand the basic properties of Fourier transforms and linear systems in one and two dimensions (as reviewed in Appendices A and B, respectively) to follow this chapter.

4.1 PROPAGATION OF LIGHT IN FREE SPACE

A. Spatial Harmonic Functions and Plane Waves

A *monochromatic* plane wave of complex amplitude $U(x,y,z) = A\exp[-j(k_x x + k_y y + k_z z)]$ has wavevector $\mathbf{k} = (k_x, k_y, k_z)$, wavelength λ, wavenumber $k = \sqrt{k_x^2 + k_y^2 + k_z^2} = 2\pi/\lambda$, and complex envelope A. The vector $\mathbf{k}$ makes angles $\theta_x = \sin^{-1}(k_x/k)$ and $\theta_y = \sin^{-1}(k_y/k)$ with the y–z and x–z planes, respectively, as illustrated in Fig. 4.1-1. Thus, if $\theta_x = 0$, there is no component of $\mathbf{k}$ in the x direction. The complex amplitude at the $z = 0$ plane, $U(x,y,0)$, is a spatial harmonic function $f(x,y) = A\exp[-j2\pi(\nu_x x + \nu_y y)]$ with spatial frequencies $\nu_x = k_x/2\pi$ and $\nu_y = k_y/2\pi$. The angles of the wavevector are therefore related to the spatial frequencies of the harmonic function by

$$\theta_x = \sin^{-1}\lambda\nu_x, \quad \theta_y = \sin^{-1}\lambda\nu_y. \tag{4.1-1}$$

Spatial Frequencies and Angles

The spatial frequency $\nu = k/2\pi$ is specified in cycles/mm, whereas the optical frequency $\nu = kc/2\pi = c/\lambda$ is specified in cycles/sec or Hz, as shown in Sec. 2.2.

Recognizing $\Lambda_x = 1/\nu_x$ and $\Lambda_y = 1/\nu_y$ as the periods of the harmonic functions in the x and y directions (mm/cycle), we see that the angles $\theta_x = \sin^{-1}(\lambda/\Lambda_x)$ and $\theta_y = \sin^{-1}(\lambda/\Lambda_y)$ are governed by the ratios of the wavelength of light to the period of the harmonic function in each direction. These geometrical relations follow from matching the wavefronts of the wave to the periodic pattern of the harmonic function in the $z = 0$ plane, as illustrated in Fig. 4.1-1.

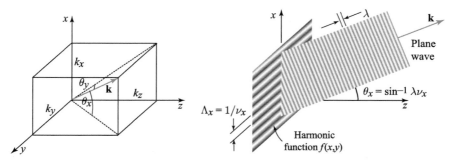

Figure 4.1-1 A harmonic function of spatial frequencies ν_x and ν_y at the plane $z = 0$ is consistent with a plane wave traveling at angles $\theta_x = \sin^{-1}\lambda\nu_x$ and $\theta_y = \sin^{-1}\lambda\nu_y$.

If $k_x \ll k$ and $k_y \ll k$, so that the wavevector $\mathbf{k}$ is paraxial, the angles θ_x and θ_y are small ($\sin\theta_x \approx \theta_x$ and $\sin\theta_y \approx \theta_y$) and

$$\theta_x \approx \lambda\nu_x, \quad \theta_y \approx \lambda\nu_y. \tag{4.1-2}$$

Spatial Frequencies and Angles
(Paraxial Approximation)

The angles of inclination of the wavevector are then directly proportional to the spatial frequencies of the corresponding harmonic function. Apparently, there is a one-to-one correspondence between the plane wave $U(x, y, z)$ and the harmonic function $f(x, y)$. Given one, the other can be readily determined, provided the wavelength λ is known: the harmonic function $f(x, y)$ is obtained by sampling at the $z = 0$ plane, $f(x, y) = U(x, y, 0)$. Given the harmonic function $f(x, y)$, on the other hand, the wave $U(x, y, z)$ is constructed by using the relation $U(x, y, z) = f(x, y) \exp(-jk_z z)$ with

$$k_z = \pm\sqrt{k^2 - k_x^2 - k_y^2}, \qquad k = 2\pi/\lambda. \qquad (4.1\text{-}3)$$

A condition for the validity of this correspondence is that $k_x^2 + k_y^2 < k^2$, so that k_z is real. This condition implies that $\lambda\nu_x < 1$ and $\lambda\nu_y < 1$, so that the angles θ_x and θ_y defined by (4.1-1) exist. The $+$ and $-$ signs in (4.1-3) represent waves traveling in the forward and backward directions, respectively. We shall be concerned with forward waves only.

Spatial Spectral Analysis

When a plane wave of unity amplitude traveling in the z direction is transmitted through a thin optical element with complex amplitude transmittance $f(x, y) = \exp[-j2\pi(\nu_x x + \nu_y y)]$ the wave is modulated by the harmonic function, so that $U(x, y, 0) = f(x, y)$. The incident wave is then converted into a plane wave with a wavevector at angles $\theta_x = \sin^{-1} \lambda\nu_x$ and $\theta_y = \sin^{-1} \lambda\nu_y$ (see Fig. 4.1-2). The element thus acts much as a prism, bending the wave upward in this illustration. If the complex amplitude transmittance is $f(x, y) = \exp[+j2\pi(\nu_x x + \nu_y y)]$, the wave is converted into a plane wave whose wavevector makes angles $-\theta_x$ and $-\theta_y$ with the z axis, so the wave is bent downward instead.

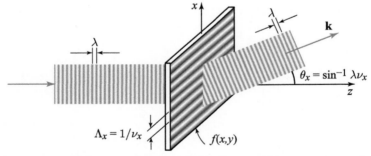

Figure 4.1-2 A thin element whose complex amplitude transmittance is a harmonic function of spatial frequency ν_x (period $\Lambda_x = 1/\nu_x$) bends a plane wave of wavelength λ by an angle $\theta_x = \sin^{-1} \lambda\nu_x = \sin^{-1}(\lambda/\Lambda_x)$. The dark blue and white stripes are used to indicate that the element is a *phase* grating (changing only the phase of the wave).

The wave-deflection property of an optical element with harmonic-function transmittance may be understood as an interference phenomenon. In a direction making an angle θ_x, two points on the element separated by a the period $\Lambda = 1/\nu_x$, have a relative pathlength difference of $\Lambda \sin \theta_x = (1/\nu_x)\lambda\nu_x = \lambda$, i.e., equal to a wavelength. Hence, all segments separated by a period interfere constructively in this direction.

If the transmittance of the optical element $f(x, y)$ is the sum of several harmonic functions of different spatial frequencies, the transmitted optical wave is also the sum of an equal number of plane waves dispersed into different directions; each spatial frequency is mapped into a corresponding direction, in accordance with (4.1-1). The amplitude of each wave is proportional to the amplitude of the corresponding harmonic component of $f(x, y)$.

Examples.

- A complex amplitude transmittance of the form $f(x, y) = \cos(2\pi\nu_x x) = \frac{1}{2}\{\exp(-j2\pi\nu_x x) + \exp(+j2\pi\nu_x x)\}$ bends an incident plane wave into components traveling at angles $\pm \sin^{-1}(\lambda\nu_x)$, namely in both the upward and downward directions.

- An element with a transmittance that varies as $1 + \cos(2\pi\nu_y y)$ behaves as a diffraction grating (see Exercise 2.4-5); the incident wave is bent into components that travel to the right and left, while a portion travels straight through.

- An element with transmittance $f(x, y) = \mathcal{U}[\cos(2\pi\nu_x x)]$, where $\mathcal{U}(x)$ is the unit step function [$\mathcal{U}(x) = 1$ if $x > 0$, and $\mathcal{U}(x) = 0$ if $x < 0$], represents a periodic set of slits with $f(x, y) = 1$ set in an opaque screen [$f(x, y) = 0$]. This periodic function may be analyzed via a Fourier series as a sum of harmonic functions of spatial frequencies $0, \pm\nu_x, \pm 2\nu_x, \ldots$, corresponding to waves traveling at angles $0, \pm \sin^{-1} \lambda\nu_x, \pm \sin^{-1} 2\lambda\nu_x, \ldots$, with amplitudes proportional to the coefficients of the Fourier series (in the case at hand, these vanish for even harmonics). At these angles, the waves transmitted through the slits interfere constructively.

More generally, if $f(x, y)$ is a superposition integral of harmonic functions,

$$f(x, y) = \iint_{-\infty}^{\infty} F(\nu_x, \nu_y) \exp\left[-j2\pi(\nu_x x + \nu_y y)\right] d\nu_x \, d\nu_y, \quad (4.1\text{-}4)$$

with frequencies (ν_x, ν_y) and amplitudes $F(\nu_x, \nu_y)$, the transmitted wave $U(x, y, z)$ is the superposition of plane waves,

$$U(x, y, z) = \iint_{-\infty}^{\infty} F(\nu_x, \nu_y) \exp\left[-j(2\pi\nu_x x + 2\pi\nu_y y)\right] \exp(-jk_z z) \, d\nu_x \, d\nu_y,$$

$$(4.1\text{-}5)$$

with complex envelopes $F(\nu_x, \nu_y)$ where $k_z = \sqrt{k^2 - k_x^2 - k_y^2} = 2\pi\sqrt{\lambda^{-2} - \nu_x^2 - \nu_y^2}$.
Note that $F(\nu_x, \nu_y)$ is the Fourier transform of $f(x, y)$ [see (A.3-2) in Appendix A].

Since an arbitrary function may be Fourier analyzed as a superposition integral of the form (4.1-4), the light transmitted through a thin optical element of arbitrary transmittance may be written as a superposition of plane waves (see Fig. 4.1-3), provided that $\nu_x^2 + \nu_y^2 < \lambda^{-2}$.

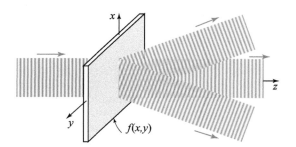

Figure 4.1-3 A thin optical element of amplitude transmittance $f(x, y)$ decomposes an incident plane wave into many plane waves. The plane wave traveling at the angles $\theta_x = \sin^{-1} \lambda\nu_x$ and $\theta_y = \sin^{-1} \lambda\nu_y$ has a complex envelope $F(\nu_x, \nu_y)$, the Fourier transform of $f(x, y)$.

This process of "spatial spectral analysis" is akin to the angular dispersion of different temporal-frequency components (wavelengths) provided by a prism. Free-space

Amplitude Modulation

Consider a transparency with complex amplitude transmittance $f_0(x, y)$. If the Fourier transform $F_0(\nu_x, \nu_y)$ extends over widths $\pm\Delta\nu_x$ and $\pm\Delta\nu_y$ in the x and y directions, the transparency will deflect an incident plane wave by angles θ_x and θ_y in the range $\pm\sin^{-1}(\lambda\Delta\nu_x)$ and $\pm\sin^{-1}(\lambda\Delta\nu_y)$, respectively.

Consider a second transparency of complex amplitude transmittance $f(x, y) = f_0(x, y)\exp[-j2\pi(\nu_{x0}x + \nu_{y0}y)]$, where $f_0(x, y)$ is slowly varying compared to $\exp[-j2\pi(\nu_{x0}x + \nu_{y0}y)]$ so that $\Delta\nu_x \ll \nu_{x0}$ and $\Delta\nu_y \ll \nu_{y0}$. We may regard $f(x, y)$ as an amplitude-modulated function with a carrier frequency ν_{x0} and ν_{y0} and modulation function $f_0(x, y)$. The Fourier transform of $f(x, y)$ is $F_0(\nu_x - \nu_{x0}, \nu_y - \nu_{y0})$, in accordance with the frequency-shifting property of the Fourier transform (see Appendix A). The transparency will deflect a plane wave to directions centered about the angles $\theta_{x0} = \sin^{-1}\lambda\nu_{x0}$ and $\theta_{y0} = \sin^{-1}\lambda\nu_{y0}$ (Fig. 4.1-4). This can also be readily seen by regarding $f(x, y)$ as a transparency of transmittance $f_0(x, y)$ in contact with a grating or prism of transmittance $\exp[-j2\pi(\nu_{x0}x + \nu_{y0}y)]$ that provides the angular deflection θ_{x0} and θ_{y0}.

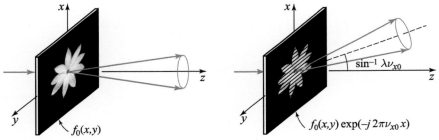

Figure 4.1-4 Deflection of light by the transparencies $f_0(x, y)$ and $f_0(x, y)\exp(-j2\pi\nu_{x0}x)$. The "carrier" harmonic function $\exp(-j2\pi\nu_{x0}x)$ acts as a prism that deflects the wave by an angle $\theta_{x0} = \sin^{-1}\lambda\nu_{x0}$.

This idea may be used to record two images $f_1(x, y)$ and $f_2(x, y)$ on the same transparency using the *spatial-frequency multiplexing* scheme $f(x, y) = f_1(x, y)\exp[-j2\pi(\nu_{x1}x + \nu_{y1}y)] + f_2(x, y)\exp[-j2\pi(\nu_{x2}x + \nu_{y2}y)]$. The two images may be easily separated by illuminating the transparency with a plane wave, whereupon the two images are deflected at different angles and are thus separated. This principle will prove useful in holography (Sec. 4.5), where it is often desired to separate two images recorded on the same transparency.

Frequency Modulation

The foregoing examples relate to the transmittance of plane waves through transparencies endowed with one or more 2D harmonic functions that extend over the entire region of the transparency. We now examine the transmission of a plane wave through a transparency comprising a "collage" of several regions, the transmittance of each of which is a harmonic function of some spatial frequency, as illustrated in Fig. 4.1-5. If the dimensions of each region are much greater than the period, each region acts as a grating or prism that deflects the wave in a particular direction, so that different portions of the incident wavefront are deflected into different directions. This principle may be used to create maps of optical interconnections, as described in Sec. 24.1A.

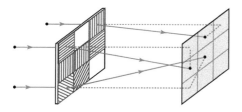

Figure 4.1-5 Deflection of light by a transparency made of several harmonic functions (phase gratings) of different spatial frequencies.

A transparency may also have a harmonic transmittance with a spatial frequency that varies continuously and slowly with position (in comparison with λ), much as the frequency of a frequency-modulated (FM) signal varies slowly with time. Consider, for example, the phase function $f(x,y) = \exp[-j2\pi\varphi(x,y)]$, where $\varphi(x,y)$ is a continuous slowly varying function of x and y. In the neighborhood of a point (x_0, y_0), we may use the Taylor-series expansion $\varphi(x,y) \approx \varphi(x_0, y_0) + (x-x_0)\nu_x + (y-y_0)\nu_y$, where the derivatives $\nu_x = \partial\varphi/\partial x$ and $\nu_y = \partial\varphi/\partial y$ are evaluated at the position (x_0, y_0). The local variation of $f(x,y)$ with x and y is therefore proportional to the quantity $\exp[-j2\pi(\nu_x x + \nu_y y)]$, which is a harmonic function with spatial frequencies $\nu_x = \partial\varphi/\partial x$ and $\nu_y = \partial\varphi/\partial y$. Since these derivatives vary with x and y, so do the spatial frequencies. The transparency $f(x,y) = \exp[-j2\pi\varphi(x,y)]$ therefore deflects the portion of the wave at the position (x,y) by the position-dependent angles $\theta_x = \sin^{-1}(\lambda\partial\varphi/\partial x)$ and $\theta_y = \sin^{-1}(\lambda\partial\varphi/\partial y)$.

EXAMPLE 4.1-1. *Scanning.* A thin transparency with complex amplitude transmittance $f(x,y) = \exp(j\pi x^2/\lambda f)$ introduces a phase shift $2\pi\varphi(x,y)$ where $\varphi(x,y) = -x^2/2\lambda f$, so that the wave is deflected at the position (x,y) by the angles $\theta_x = \sin^{-1}(\lambda\partial\varphi/\partial x) = \sin^{-1}(-x/f)$ and $\theta_y = 0$. If $|x/f| \ll 1$, $\theta_x \approx -x/f$ and the deflection angle θ_x is directly proportional to the transverse distance x. This transparency may be used to deflect a narrow beam of light. Moreover, if the transparency is moved at a uniform speed, the beam is deflected by a linearly increasing angle as illustrated in Fig. 4.1-6.

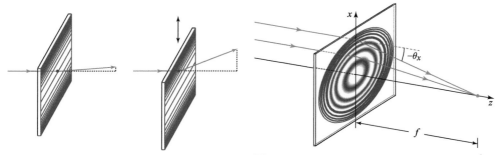

Figure 4.1-6 Making use of a frequency-modulated transparency to scan an optical beam.

Figure 4.1-7 A transparency with transmittance $f(x,y) = \exp[j\pi(x^2+y^2)/\lambda f]$ acts as a spherical lens with focal length f.

EXAMPLE 4.1-2. *Imaging.* If the transparency illustrated in Example 4.1-1 is illuminated by a plane wave, each strip of the wave at a given value of x is deflected by a different angle and as a result the wavefront is altered. The local wavevector at position x bends by an angle $-x/f$ so that all wavevectors meet at a single line on the optical axis a distance f from the transparency. The transparency then acts as a cylindrical lens with a focal length f. Similarly, a transparency with transmittance $f(x,y) = \exp[j\pi(x^2+y^2)/\lambda f]$ acts as a spherical lens with focal length f, as illustrated in Fig. 4.1-7. Indeed, this is the expression for the amplitude transmittance of a thin lens provided in (2.4-9).

EXERCISE 4.1-1

Binary-Plate Cylindrical Lens. Use harmonic analysis near the position x to show that a transparency with complex amplitude transmittance equal to the binary function

$$f(x,y) = \mathcal{U}\left[\cos\left(\pi\frac{x^2}{\lambda f}\right)\right], \tag{4.1-6}$$

where $\mathcal{U}(x)$ is the unit step function [$\mathcal{U}(x) = 1$ if $x \geq 0$, and $\mathcal{U}(x) = 0$ if $x < 0$], acts as a cylindrical lens with multiple focal lengths equal to $\infty, \pm f, \pm f/3, \pm f/5, \ldots$.

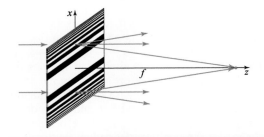

Figure 4.1-8 Binary plate as a cylindrical lens with multiple foci.

Fresnel Zone Plate

A two-dimensional generalization of the binary plate in Exercise 4.1-1 is a circularly symmetric transparency of complex amplitude transmittance

$$f(x,y) = \mathcal{U}\left[\cos\left(\pi\frac{x^2+y^2}{\lambda f}\right)\right], \tag{4.1-7}$$

known as the **Fresnel zone plate**. It is a set of ring apertures of increasing radii, decreasing widths, and equal areas (see Fig. 4.1-9). The structure serves as a spherical lens with multiple focal lengths. A ray incident at each point is split into multiple rays, and the transmitted rays meet at multiple foci with focal lengths $\infty, \pm f, \pm f/3, \pm f/5, \ldots$, together with a component transmitted without deflection.

The operation of the Fresnel zone plate may also be described in terms of interference (see Sec. 2.5B). The center of the mth ring has a radius ρ_m at the mth peak of the cosine function, i.e., $\pi\rho_m^2/\lambda f = m2\pi$. At a focal point $z = f$, the distance R_m to the mth ring is given by $R_m^2 = f^2 + \rho_m^2$, so that $R_m = \sqrt{f^2 + 2m\lambda f}$. If f is sufficiently large so that the angles subtended by the rings are small, then $R_m \approx f + m\lambda$. Thus, the waves transmitted through consecutive rings have pathlengths differing by a wavelength, so that they interfere constructively at the focal point. A similar argument applies for the other foci.

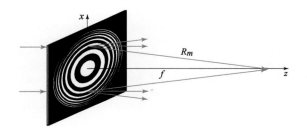

Figure 4.1-9 The Fresnel zone plate.

B. Transfer Function of Free Space

We now examine the propagation of a monochromatic optical wave of wavelength λ and complex amplitude $U(x, y, z)$ in the free space between the planes $z = 0$ and $z = d$, called the input and output planes, respectively (see Fig. 4.1-10). Given the complex amplitude of the wave at the input plane, $f(x,y) = U(x,y,0)$, we shall determine the complex amplitude at the output plane, $g(x,y) = U(x,y,d)$.

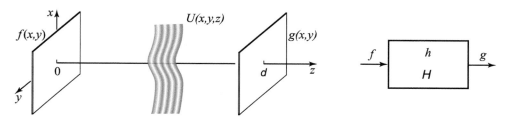

Figure 4.1-10 Propagation of light between two planes is regarded as a linear system whose input and output are the complex amplitudes of the wave in the two planes.

We regard $f(x, y)$ and $g(x, y)$ as the input and output of a linear system. The system is linear since the Helmholtz equation, which $U(x, y, z)$ must satisfy, is linear. The system is shift-invariant because of the invariance of free space to displacement of the coordinate system. A linear shift-invariant system is characterized by its impulse response function $h(x, y)$ or by its transfer function $H(\nu_x, \nu_y)$, as explained in Appendix B, Sec. B.2. We now proceed to determine expressions for these functions.

The transfer function $H(\nu_x, \nu_y)$ is the factor by which an input spatial harmonic function of frequencies ν_x and ν_y is multiplied to yield the output harmonic function. We therefore consider a harmonic input function $f(x, y) = A \exp[-j2\pi(\nu_x x + \nu_y y)]$. As explained earlier, this corresponds to a plane wave $U(x, y, z) = A \exp[-j(k_x x + k_y y + k_z z)]$ where $k_x = 2\pi\nu_x$, $k_y = 2\pi\nu_y$, and

$$k_z = \sqrt{k^2 - k_x^2 - k_y^2} = 2\pi\sqrt{\lambda^{-2} - \nu_x^2 - \nu_y^2}. \qquad (4.1\text{-}8)$$

The output $g(x, y) = A \exp[-j(k_x x + k_y y + k_z d)]$, so that we can write $H(\nu_x, \nu_y) = g(x,y)/f(x,y) = \exp(-jk_z d)$, from which

$$H(\nu_x, \nu_y) = \exp\left(-j2\pi d\sqrt{\lambda^{-2} - \nu_x^2 - \nu_y^2}\right). \qquad (4.1\text{-}9)$$

Transfer Function
of Free Space

The transfer function $H(\nu_x, \nu_y)$ is therefore a circularly symmetric complex function of the spatial frequencies ν_x and ν_y. Its magnitude and phase are sketched in Fig. 4.1-11.

For spatial frequencies for which $\nu_x^2 + \nu_y^2 \leq \lambda^{-2}$ (i.e., frequencies lying within a circle of radius $1/\lambda$) the magnitude $|H(\nu_x, \nu_y)| = 1$ and the phase $\arg\{H(\nu_x, \nu_y)\}$ is a function of ν_x and ν_y. A harmonic function with such frequencies therefore undergoes a spatial phase shift as it propagates, but its magnitude is not altered.

At higher spatial frequencies, $\nu_x^2 + \nu_y^2 > \lambda^{-2}$, the quantity under the square root in (4.1-9) is negative so that the exponent is real and the transfer function $\exp[-2\pi d(\nu_x^2 + \nu_y^2 - \lambda^{-2})^{1/2}]$ represents an attenuation factor; the wave is then called an **evanescent**

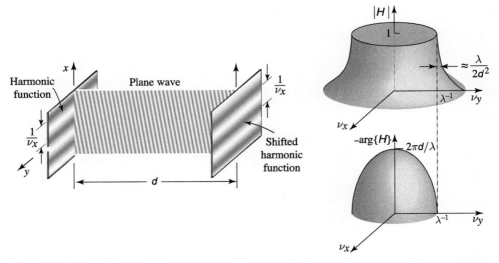

Figure 4.1-11 Magnitude and phase of the transfer function $H(\nu_x, \nu_y)$ for free-space propagation between two planes separated by a distance d.

wave.[†] When $\nu_\rho = (\nu_x^2 + \nu_y^2)^{1/2}$ exceeds λ^{-1} slightly, i.e., $\nu_\rho \approx \lambda^{-1}$, the attenuation factor is $\exp[-2\pi d(\nu_\rho^2 - \lambda^{-2})^{1/2}] = \exp[-2\pi d(\nu_\rho - \lambda^{-1})^{1/2}(\nu_\rho + \lambda^{-1})^{1/2}] \approx \exp[-2\pi d(\nu_\rho - \lambda^{-1})^{1/2}(2\lambda^{-1})^{1/2}]$, which equals $\exp(-2\pi)$ when $(\nu_\rho - \lambda^{-1}) \approx \lambda/2d^2$, or $(\nu_\rho - 1/\lambda)/(1/\lambda) \approx \frac{1}{2}(\lambda/d)^2$. For $d \gg \lambda$ the attenuation factor decreases sharply when the spatial frequency slightly exceeds λ^{-1}, as illustrated in Fig. 4.1-11. We may therefore regard λ^{-1} as the cutoff spatial frequency (the spatial bandwidth) of the system. Thus,

> *The spatial bandwidth of light propagation in free space is approximately λ^{-1} cycles/mm.*

Features contained in spatial frequencies greater than λ^{-1} (corresponding to details of size finer than λ) cannot be transmitted by an optical wave of wavelength λ over distances much greater than λ.

Fresnel Approximation

The expression for the transfer function in (4.1-9) may be simplified if the input function $f(x, y)$ contains only spatial frequencies that are much smaller than the cutoff frequency λ^{-1}, so that $\nu_x^2 + \nu_y^2 \ll \lambda^{-2}$. The plane-wave components of the propagating light then make small angles $\theta_x \approx \lambda\nu_x$ and $\theta_y \approx \lambda\nu_y$ corresponding to paraxial rays.

Denoting $\theta^2 = \theta_x^2 + \theta_y^2 \approx \lambda^2(\nu_x^2 + \nu_y^2)$, where θ is the angle with the optical axis, the phase factor in (4.1-9) is

$$2\pi d\sqrt{\lambda^{-2} - \nu_x^2 - \nu_y^2} = 2\pi \frac{d}{\lambda}\sqrt{1 - \theta^2}$$

[†] Evanescent waves are neither forward nor backward propagating; rather, they propagate in the transverse plane where they are generated. We select the negative sign before the square root in (4.1-3) because such waves must attenuate, rather than grow, in the positive z direction absent a gain mechanism enabling them to grow.

$$= 2\pi \frac{d}{\lambda}\left(1 - \frac{\theta^2}{2} + \frac{\theta^4}{8} - \cdots\right). \tag{4.1-10}$$

Neglecting the third and higher terms of this expansion, (4.1-9) may be approximated by

$$\boxed{H(\nu_x, \nu_y) \approx H_0 \exp\left[j\pi\lambda d\left(\nu_x^2 + \nu_y^2\right)\right],} \tag{4.1-11}$$
Transfer Function of Free Space
(Fresnel Approximation)

where $H_0 = \exp(-jkd)$. In this approximation, the phase is a quadratic function of ν_x and ν_y, as illustrated in Fig. 4.1-12. This approximation is known as the **Fresnel approximation**.

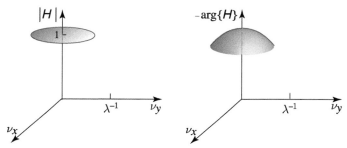

Figure 4.1-12 The transfer function of free-space propagation for low spatial frequencies (much less than $1/\lambda$ cycles/mm) has a constant magnitude and a quadratic phase.

The condition of validity of the Fresnel approximation is that the third term in (4.1-10) is much smaller than π for all θ. This is equivalent to

$$\frac{\theta^4 d}{4\lambda} \ll 1. \tag{4.1-12}$$

If a is the largest radial distance at the output plane, the largest angle $\theta_m \approx a/d$, and (4.1-12) may be written in the form [see (2.2-18)]

$$\boxed{N_F \frac{\theta_m^2}{4} \ll 1,} \tag{4.1-13}$$
Fresnel Approximation
Condition of Validity

$$\boxed{N_F = \frac{a^2}{\lambda d},} \tag{4.1-14}$$
Fresnel Number

where N_F is the Fresnel number. For example, if $a = 1$ cm, $d = 100$ cm, and $\lambda = 0.5$ μm, then $\theta_m = 10^{-2}$ radian, $N_F = 200$, and $N_F \theta^2/4 = 5 \times 10^{-3}$. In this case the Fresnel approximation is applicable.

Input–Output Relation

Given the input function $f(x, y)$, the output function $g(x, y)$ may be determined as follows: (1) we determine the Fourier transform

$$F(\nu_x, \nu_y) = \iint_{-\infty}^{\infty} f(x, y) \exp\left[j2\pi(\nu_x x + \nu_y y)\right] \, dx \, dy, \qquad (4.1\text{-}15)$$

which represents the complex envelopes of the plane-wave components at the input plane; (2) the product $H(\nu_x, \nu_y)F(\nu_x, \nu_y)$ gives the complex envelopes of the plane-wave components at the output plane; and (3) the complex amplitude at the output plane is the sum of the contributions of these plane waves,

$$g(x, y) = \iint_{-\infty}^{\infty} H(\nu_x, \nu_y) F(\nu_x, \nu_y) \exp[-j2\pi(\nu_x x + \nu_y y)] \, d\nu_x d\nu_y. \qquad (4.1\text{-}16)$$

Using the Fresnel approximation for $H(\nu_x, \nu_y)$, which is given by (4.1-11), we have

$$\boxed{g(x, y) = H_0 \iint_{-\infty}^{\infty} F(\nu_x, \nu_y) \exp\left[j\pi\lambda d\left(\nu_x^2 + \nu_y^2\right)\right] \exp\left[-j2\pi(\nu_x x + \nu_y y)\right] d\nu_x \, d\nu_y.}$$

(4.1-17)

Equations (4.1-17) and (4.1-15) serve to relate the output function $g(x, y)$ to the input function $f(x, y)$.

C. Impulse Response Function of Free Space

The impulse response function $h(x, y)$ of the system of free-space propagation is the response $g(x, y)$ when the input $f(x, y)$ is a point at the origin $(0, 0)$. It is the inverse Fourier transform of the transfer function $H(\nu_x, \nu_y)$. Using the results of Sec. A.3 and Table A.1-1 of Appendix A, together with $k = 2\pi/\lambda$, the inverse Fourier transform of (4.1-11) turns out to be

$$\boxed{h(x, y) \approx h_0 \exp\left[-jk\frac{x^2 + y^2}{2d}\right],}$$

(4.1-18)
Impulse Response Function
Free Space (Fresnel Approximation)

where $h_0 = (j/\lambda d) \exp(-jkd)$. This function is proportional to the complex amplitude at the $z = d$ plane of a paraboloidal wave centered about the origin $(0, 0)$ [see (2.2-17)]. Thus, each point at the input plane generates a paraboloidal wave; all such waves are superimposed at the output plane.

Free-Space Propagation as a Convolution

An alternative procedure for relating complex amplitudes $f(x, y)$ and $g(x, y)$ is to regard $f(x, y)$ as a superposition of different points (delta functions), each producing a paraboloidal wave. The wave originating at the point (x', y') has an amplitude $f(x', y')$

and is centered about (x', y') so that it generates a wave with amplitude $f(x', y')h(x - x', y - y')$ at the point (x, y) at the output plane. The sum of these contributions is the two-dimensional convolution

$$g(x,y) = \iint_{-\infty}^{\infty} f(x',y')h(x-x',y-y')\,dx'\,dy', \qquad (4.1\text{-}19)$$

which, in the Fresnel approximation, becomes

$$g(x,y) = h_0 \iint_{-\infty}^{\infty} f(x',y') \exp\left[-j\pi \frac{(x-x')^2 + (y-y')^2}{\lambda d}\right] dx'\,dy', \qquad (4.1\text{-}20)$$

where $h_0 = (j/\lambda d) \exp(-jkd)$.

In summary: within the Fresnel approximation, there are two approaches to determining the complex amplitude $g(x, y)$ at the output plane, given the complex amplitude $f(x, y)$ at the input plane: (1) Equation (4.1-20) is based on a space-domain approach in which the input wave is expanded in terms of paraboloidal elementary waves; and (2) Equation (4.1-17) is a frequency-domain approach in which the input wave is expanded as a sum of plane waves.

EXERCISE 4.1-2

Gaussian Beams Revisited. If the function $f(x,y) = A\exp[-(x^2+y^2)/W_0^2]$ represents the complex amplitude of an optical wave $U(x, y, z)$ in the plane $z = 0$, show that $U(x, y, z)$ is the Gaussian beam displayed in (3.1-7). Use both space- and frequency-domain methods.

D. Huygens–Fresnel Principle

The **Huygens–Fresnel principle** states that each point on a wavefront generates a spherical wave (Fig. 4.1-13). The envelope of these secondary waves constitutes a new wavefront. Their superposition constitutes the wave in another plane. The system's impulse response function for propagation between the planes $z = 0$ and $z = d$ is

$$h(x,y) \propto \frac{1}{r}\exp(-jkr), \qquad r = \sqrt{x^2+y^2+d^2}. \qquad (4.1\text{-}21)$$

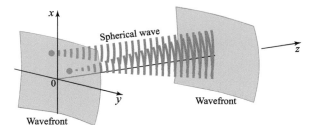

Figure 4.1-13 The Huygens–Fresnel principle. Each point on a wavefront generates a spherical wave.

In the paraxial approximation, the spherical wave given by (4.1-21) is approximated by the paraboloidal wave in (4.1-18) (see Sec. 2.2B). Our derivation of the impulse response function is therefore consistent with the Huygens–Fresnel principle.

4.2 OPTICAL FOURIER TRANSFORM

As has been shown in Sec. 4.1, the propagation of light in free space is described conveniently by Fourier analysis. If the complex amplitude of a monochromatic wave of wavelength λ in the $z = 0$ plane is a function $f(x, y)$ composed of harmonic components of different spatial frequencies, each harmonic component corresponds to a plane wave: the plane wave traveling at angles $\theta_x = \sin^{-1} \lambda \nu_x$, $\theta_y = \sin^{-1} \lambda \nu_y$ corresponds to the components with spatial frequencies ν_x and ν_y and has an amplitude $F(\nu_x, \nu_y)$, the Fourier transform of $f(x, y)$. This suggests that light can be used to compute the Fourier transform of a two-dimensional function $f(x, y)$, simply by making a transparency with amplitude transmittance $f(x, y)$ through which a uniform plane wave of unity magnitude is transmitted.

Because each of the plane waves has an infinite extent and therefore overlaps with the other plane waves, however, it is necessary to find a method of separating these waves. It will be shown that at a sufficiently large distance, only a single plane wave contributes to the total amplitude at each point at the output plane, so that the Fourier components are eventually separated naturally. A more practical approach is to use a lens to focus each of the plane waves into a single point, as described subsequently.

A. Fourier Transform in the Far Field

We now proceed to show that if the propagation distance d is sufficiently long, the only plane wave that contributes to the complex amplitude at a point (x, y) at the output plane is the wave with direction making angles $\theta_x \approx x/d$ and $\theta_y \approx y/d$ with the optical axis (see Fig. 4.2-1). This is the wave with wavevector components $k_x \approx (x/d)k$ and $k_y \approx (y/d)k$ and amplitude $F(\nu_x, \nu_y)$ with $\nu_x = x/\lambda d$ and $\nu_y = x/\lambda d$. The complex amplitudes $g(x, y)$ and $f(x, y)$ of the wave at the $z = d$ and $z = 0$ planes are related by

$$g(x,y) \approx h_0 \, F\!\left(\frac{x}{\lambda d}, \frac{y}{\lambda d}\right). \qquad (4.2\text{-}1)$$

Free-Space Propagation as Fourier Transform (Fraunhofer Approximation)

where $F(\nu_x, \nu_y)$ is the Fourier transform of $f(x, y)$ and $h_0 = (j/\lambda d) \exp(-jkd)$. Contributions of all other waves cancel out as a result of destructive interference. This approximation is known as the **Fraunhofer approximation**.

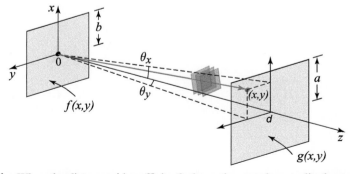

Figure 4.2-1 When the distance d is sufficiently long, the complex amplitude at point (x, y) in the $z = d$ plane is proportional to the complex amplitude of the plane-wave component with angles $\theta_x \approx x/d \approx \lambda \nu_x$ and $\theta_y \approx y/d \approx \lambda \nu_y$, i.e., to the Fourier transform $F(\nu_x, \nu_y)$ of $f(x, y)$, with $\nu_x = x/\lambda d$ and $\nu_y = y/\lambda d$.

As noted in the following proofs, the conditions of validity of Fraunhofer approximation are:

$$N_F \ll 1 \quad \text{and} \quad N_F' \ll 1. \qquad (4.2\text{-}2)$$

Fraunhofer Approximation
Condition of Validity
$N_F = a^2/\lambda d, \; N_F' = b^2/\lambda d$

The Fraunhofer approximation is therefore valid whenever the Fresnel numbers N_F and N_F' are small. The Fraunhofer approximation is more difficult to satisfy than the Fresnel approximation, which requires that $N_F \theta_m^2/4 \ll 1$ [see (4.1-13)]. Since $\theta_m \ll 1$ in the paraxial approximation, it is possible to satisfy the Fresnel condition $N_F \theta_m^2/4 \ll 1$ for Fresnel numbers N_F not necessarily $\ll 1$.

☐ **Proofs of the Fourier Transform Property of Free-Space Propagation in the Fraunhofer Approximation.** We begin with the relation between $g(x,y)$ and $f(x,y)$ in (4.1-20). The phase in the argument of the exponent is $(\pi/\lambda d)[(x-x')^2+(y-y')^2] = (\pi/\lambda d)[(x^2+y^2)+(x'^2+y'^2)-2(xx'+yy')]$. If $f(x,y)$ is confined to a small area of radius b, and if the distance d is sufficiently large so that the Fresnel number $N_F' = b^2/\lambda d$ is small, then the phase factor $(\pi/\lambda d)(x'^2+y'^2) \leq \pi(b^2/\lambda d)$ is negligible and (4.1-20) may be approximated by

$$g(x,y) = h_0 \exp\left(-j\pi \frac{x^2+y^2}{\lambda d}\right) \iint_{-\infty}^{\infty} f(x',y') \exp\left(j2\pi \frac{xx'+yy'}{\lambda d}\right) dx'\, dy'. \qquad (4.2\text{-}3)$$

The factors $x/\lambda d$ and $y/\lambda d$ may be regarded as the frequencies $\nu_x = x/\lambda d$ and $\nu_y = y/\lambda d$, so that

$$g(x,y) = h_0 \exp\left(-j\pi \frac{x^2+y^2}{\lambda d}\right) F\left(\frac{x}{\lambda d}, \frac{y}{\lambda d}\right), \qquad (4.2\text{-}4)$$

where $F(\nu_x, \nu_y)$ is the Fourier transform of $f(x,y)$. The phase factor given by $\exp[-j\pi(x^2+y^2)/\lambda d]$ in (4.2-4) may also be neglected and (4.2-1) obtained if we also limit our interest to points at the output plane within a circle of radius a centered about the z-axis so that $\pi(x^2+y^2)/\lambda d \leq \pi a^2/\lambda d \ll \pi$. This is applicable when the Fresnel number $N_F = a^2/\lambda d \ll 1$.

Another proof is based on (4.1-17), which expresses the complex amplitude $g(x,y)$ as an integral of plane waves of different frequencies. If d is sufficiently large so that the phase in the integrand is much greater than 2π, it can be shown using the method of stationary phase[†] that only one value of ν_x contributes to the integral. This is the value for which the derivative of the phase $\pi \lambda d \nu_x^2 - 2\pi \nu_x x$ with respect to ν_x vanishes; i.e., $\nu_x = x/\lambda d$. Similarly, the only value of ν_y that contributes to the integral is $\nu_y = y/\lambda d$. This proves the assertion that only one plane wave contributes to the far field at a given point. ■

EXERCISE 4.2-1

Conditions of Validity of the Fresnel and Fraunhofer Approximations: A Comparison.
Demonstrate that the Fraunhofer approximation is more restrictive than the Fresnel approximation by taking $\lambda = 0.5\ \mu m$, and assuming that the object points lie within a circular aperture of radius $b = 1$ cm and the observation points lie within a circular aperture of radius $a = 2$ cm. Determine the range of distances d between the object plane and the observation plane for which each of these approximations is applicable.

[†] See, e.g., M. Born and E. Wolf, *Principles of Optics*, Cambridge University Press, 7th expanded and corrected ed. 2002, Appendix III.

Summary

In the Fraunhofer approximation, the complex amplitude $g(x, y)$ of a wave of wavelength λ in the $z = d$ plane is proportional to the Fourier transform $F(\nu_x, \nu_y)$ of the complex amplitude $f(x, y)$ in the $z = 0$ plane, evaluated at the spatial frequencies $\nu_x = x/\lambda d$ and $\nu_y = y/\lambda d$. The approximation is valid if $f(x, y)$ at the input plane is confined to a circle of radius b satisfying $b^2/\lambda d \ll 1$, and at points at the output plane within a circle of radius a satisfying $a^2/\lambda d \ll 1$.

B. Fourier Transform Using a Lens

The plane-wave components that constitute a wave may also be separated by use of a lens. A thin spherical lens transforms a plane wave into a paraboloidal wave focused to a point in the lens focal plane (see Sec. 2.4 and Exercise 2.4-3). If the plane wave arrives at small angles θ_x and θ_y, the paraboloidal wave is centered about the point $(\theta_x f, \theta_y f)$, where f is the focal length (see Fig. 4.2-2). The lens therefore maps each direction (θ_x, θ_y) into a single point $(\theta_x f, \theta_y f)$ in the focal plane and thus separates the contributions of the different plane waves.

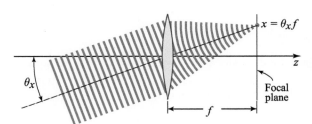

Figure 4.2-2 Focusing of a plane wave into a point. A direction (θ_x, θ_y) is mapped into a point $(x, y) = (\theta_x f, \theta_y f)$. (see Exercise 2.4-3.)

In reference to the optical system shown in Fig. 4.2-3, let $f(x, y)$ be the complex amplitude of the optical wave in the $z = 0$ plane. Light is decomposed into plane waves, with the wave traveling at small angles $\theta_x = \lambda \nu_x$ and $\theta_y = \lambda \nu_y$ having a complex amplitude proportional to the Fourier transform $F(\nu_x, \nu_y)$. This wave is focused by the lens into a point (x, y) in the focal plane where $x = \theta_x f = \lambda f \nu_x$ and $y = \theta_y f = \lambda f \nu_y$. The complex amplitude at point (x, y) at the output plane is therefore proportional to the Fourier transform of $f(x, y)$ evaluated at $\nu_x = x/\lambda f$ and $\nu_y = y/\lambda f$, so that

$$g(x, y) \propto F\left(\frac{x}{\lambda f}, \frac{y}{\lambda f}\right). \tag{4.2-5}$$

To determine the proportionality factor in (4.2-5), we analyze the input function $f(x, y)$ into its Fourier components and trace the plane wave corresponding to each component through the optical system. We then superpose the contributions of these waves at the output plane to obtain $g(x, y)$. Assuming that these waves are paraxial and using the Fresnel approximation, we obtain:

$$g(x, y) = h_l \exp\left[j\pi \frac{(x^2 + y^2)(d - f)}{\lambda f^2}\right] F\left(\frac{x}{\lambda f}, \frac{y}{\lambda f}\right), \tag{4.2-6}$$

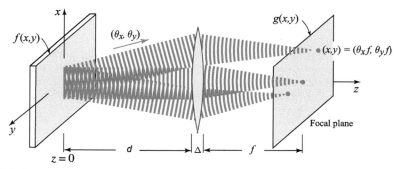

Figure 4.2-3 Focusing of the plane waves associated with the harmonic Fourier components of the input function $f(x, y)$ into points in the focal plane. The amplitude of the plane wave with direction $(\theta_x, \theta_y) = (\lambda\nu_x, \lambda\nu_y)$ is proportional to the Fourier transform $F(\nu_x, \nu_y)$ and is focused at the point $(x, y) = (\theta_x f, \theta_y f) = (\lambda f \nu_x, \lambda f \nu_y)$.

where $h_l = H_0 h_0 = (j/\lambda f) \exp[-jk(d + f)]$. Thus, the coefficient of proportionality in (4.2-5) contains a phase factor that is a quadratic function of x and y.

Since $|h_l| = 1/\lambda f$ it follows from (4.2-6) that the optical intensity at the output plane is

$$I(x, y) = \frac{1}{(\lambda f)^2} \left| F\left(\frac{x}{\lambda f}, \frac{y}{\lambda f}\right) \right|^2. \quad (4.2\text{-}7)$$

The intensity of light at the output plane (the back focal plane of the lens) is therefore proportional to the absolute-squared value of the Fourier transform of the complex amplitude of the wave at the input plane, regardless of the distance d.

The phase factor in (4.2-6) vanishes if $d = f$, so that

$$g(x, y) = h_l F\left(\frac{x}{\lambda f}, \frac{y}{\lambda f}\right), \quad (4.2\text{-}8)$$

Fourier-Transform Property of a Lens

where $h_l = (j/\lambda f) \exp(-j2kf)$. In this geometry, known as the **2-f system** (see Fig. 4.2-4), the complex amplitudes at the front and back focal planes of the lens are related by a Fourier transform, both magnitude and phase.

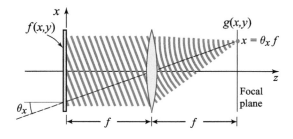

Figure 4.2-4 The 2-f system. The Fourier component of $f(x, y)$ with spatial frequencies ν_x and ν_y generates a plane wave at angles $\theta_x = \lambda\nu_x$ and $\theta_y = \lambda\nu_y$ and is focused by the lens to the point $(x, y) = (f\theta_x, f\theta_y) = (\lambda f \nu_x, \lambda f \nu_y)$ so that $g(x, y)$ is proportional to the Fourier transform $F(x/\lambda f, y/\lambda f)$.

Summary

The complex amplitude of light at a point (x, y) in the back focal plane of a lens of focal length f is proportional to the Fourier transform of the complex amplitude in the front focal plane evaluated at the frequencies $\nu_x = x/\lambda f, \nu_y/\lambda f$. This relation is valid in the Fresnel approximation. Without the lens, the Fourier transformation is obtained only in the Fraunhofer approximation, which is more restrictive.

☐ ***Proof of the Fourier Transform Property of the Lens in the Fresnel Approximation.** The proof takes the following four steps.

1. The plane wave with angles $\theta_x = \lambda\nu_x$ and $\theta_y = \lambda\nu_y$ has a complex amplitude $U(x,y,0) = F(\nu_x, \nu_y)\exp[-j2\pi(\nu_x x + \nu_y y)]$ in the $z = 0$ plane and $U(x, y, d) = H(\nu_x, \nu_y) F(\nu_x, \nu_y)\exp[-j2\pi(\nu_x x + \nu_y y)]$ in the $z = d$ plane, immediately before crossing the lens, where $H(\nu_x, \nu_y) = H_0 \exp[j\pi\lambda d(\nu_x^2 + \nu_y^2)]$ is the transfer function of a distance d of free space and $H_0 = \exp(-jkd)$.

2. Upon crossing the lens, the complex amplitude is multiplied by the lens phase factor $\exp[j\pi(x^2 + y^2)/\lambda f]$ [the phase factor $\exp(-jk\Delta)$, where Δ is the width of the lens, has been ignored]. Thus,

$$U(x, y, d + \Delta) = H_0 \exp\left[j\pi(x^2+y^2)/\lambda f\right] \\ \times \exp\left[j\pi\lambda d \left(\nu_x^2 + \nu_y^2\right)\right] F(\nu_x, \nu_y) \exp\left[-j2\pi(\nu_x x + \nu_y y)\right]. \quad (4.2\text{-}9)$$

This expression is simplified by writing $-2\nu_x x + x^2/\lambda f = (x^2 - 2\nu_x\lambda f x)/\lambda f = [(x-x_0)^2 - x_0^2]/\lambda f$, with $x_0 = \lambda\nu_x f$; a similar relation for y is written with $y_0 = \lambda\nu_y f$, so that

$$U(x, y, d + \Delta) = A(\nu_x, \nu_y)\exp\left[j\pi\frac{(x-x_0)^2 + (y-y_0)^2}{\lambda f}\right], \quad (4.2\text{-}10)$$

where

$$A(\nu_x, \nu_y) = H_0 \exp\left[j\pi\lambda(d-f)\left(\nu_x^2 + \nu_y^2\right)\right] F(\nu_x, \nu_y). \quad (4.2\text{-}11)$$

Equation (4.2-10) is recognized as the complex amplitude of a paraboloidal wave converging toward the point (x_0, y_0) in the lens focal plane, $z = d + \Delta + f$.

3. We now examine the propagation in the free space between the lens and the output plane to determine $U(x, y, d + \Delta + f)$. We apply (4.1-20) to (4.2-10), use the relation $\int \exp[j2\pi(x - x_0)x'/\lambda f]\, dx' = \lambda f\delta(x - x_0)$, and obtain

$$U(x, y, d + \Delta + f) = h_0(\lambda f)^2 A(\nu_x, \nu_y)\delta(x - x_0)\delta(y - y_0), \quad (4.2\text{-}12)$$

where $h_0 = (j/\lambda f)\exp(-jkf)$. Indeed, the plane wave is focused into a single point at $x_0 = \lambda\nu_x f$ and $y_0 = \lambda\nu_y f$.

4. The last step is to integrate over all the plane waves (all ν_x and ν_y). By virtue of the sifting property of the delta function, $\delta(x - x_0) = \delta(x - \lambda f\nu_x) = (1/\lambda f)\delta(\nu_x - x/\lambda f)$, this integral gives $g(x, y) = h_0 A(x/\lambda f, y/\lambda f)$. Substituting from (4.2-11) we finally obtain (4.2-6). ∎

EXERCISE 4.2-2

The Inverse Fourier Transform. In the single-lens optical system depicted in Fig. 4.2-4, the field distribution in the front focal plane ($z = 2f$) is a scaled version of the Fourier transform of the field distribution in the back focal plane ($z = 0$). Verify that if the coordinate system in the front focal plane is inverted, i.e., $(x, y) \to (-x, -y)$, then the resultant field distribution yields the inverse Fourier transform.

4.3 DIFFRACTION OF LIGHT

When an optical wave is transmitted through an aperture in an opaque screen and travels some distance in free space, its intensity distribution is called the diffraction pattern. If light were treated as rays, the diffraction pattern would be a shadow of the aperture. Because of the wave nature of light, however, the diffraction pattern may deviate slightly or substantially from the aperture shadow, depending on the distance between the aperture and observation plane, the wavelength, and the dimensions of the aperture. An example is illustrated in Fig. 4.3-1. It is difficult to determine exactly the manner in which the screen modifies the incident wave, but the propagation in free space beyond the aperture is always governed by the laws described earlier in this chapter.

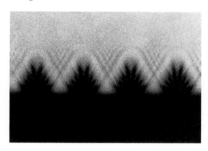

Figure 4.3-1 Diffraction pattern of the teeth of a saw. (Adapted from M. Cagnet, M. Françon, and J. C. Thrierr, *Atlas of Optical Phenomena*, Springer-Verlag, 1962.)

The simplest theory of diffraction is based on the *assumption* that the incident wave is transmitted without change at points within the aperture, but is reduced to zero at points on the back side of the opaque part of the screen. If $U(x,y)$ and $f(x,y)$ are the complex amplitudes of the wave immediately to the left and right of the screen (Fig. 4.3-2), respectively, then in accordance with this assumption,

$$f(x,y) = U(x,y)\, p(x,y), \tag{4.3-1}$$

where

$$p(x,y) = \begin{cases} 1 & \text{inside the aperture} \\ 0, & \text{outside the aperture} \end{cases} \tag{4.3-2}$$

is called the **aperture function**.

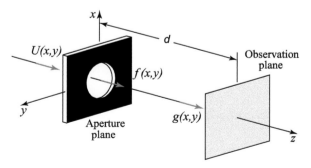

Figure 4.3-2 A wave $U(x,y)$ is transmitted through an aperture of amplitude transmittance $p(x,y)$, generating a wave of complex amplitude $f(x,y) = U(x,y)p(x,y)$. After propagation a distance d in free space, the complex amplitude is $g(x,y)$ and the diffraction pattern is the intensity $I(x,y) = |g(x,y)|^2$.

Given $f(x,y)$, the complex amplitude $g(x,y)$ at an observation plane a distance d from the screen may be determined using the methods described in Secs. 4.1 and 4.2. The diffraction pattern $I(x,y) = |g(x,y)|^2$ is known as **Fraunhofer diffraction** or

Fresnel diffraction, depending on whether free-space propagation is described using the Fraunhofer approximation or the Fresnel approximation, respectively.

Although this approach gives reasonably accurate results in most cases, it is not exact. The validity and self-consistency of the assumption that the complex amplitude $f(x, y)$ vanishes at points outside the aperture on the back of the screen are questionable since the transmitted wave propagates in all directions and therefore reaches those points as well. A theory of diffraction based on the exact solution of the Helmholtz equation under the boundary conditions imposed by the aperture is mathematically difficult. Only a few geometrical structures have yielded exact solutions. However, different theories of diffraction have been developed using a variety of assumptions, leading to results with varying accuracies. Rigorous diffraction theory is beyond the scope of this book.

A. Fraunhofer Diffraction

Fraunhofer diffraction is the theory of transmission of light through apertures, assuming that the incident wave is multiplied by the aperture function and that the Fraunhofer approximation determines the propagation of light in the free space beyond the aperture. The Fraunhofer approximation is valid if the propagation distance d between the aperture and observation planes is sufficiently large so that the Fresnel number $N_F' = b^2/\lambda d \ll 1$, where b is the largest radial distance within the aperture.

Assuming that the incident wave is a plane wave of intensity I_i traveling in the z direction so that $U(x, y) = \sqrt{I_i}$, then $f(x,y) = \sqrt{I_i}\, p(x, y)$. In the Fraunhofer approximation [see (4.2-1)],

$$g(x, y) \approx \sqrt{I_i}\, h_0\, P\left(\frac{x}{\lambda d}, \frac{y}{\lambda d}\right), \qquad (4.3\text{-}3)$$

where

$$P(\nu_x, \nu_y) = \iint_{-\infty}^{\infty} p(x, y) \exp\left[j 2\pi (\nu_x x + \nu_y y)\right] dx\, dy \qquad (4.3\text{-}4)$$

is the Fourier transform of $p(x, y)$ and $h_0 = (j/\lambda d) \exp(-jkd)$. The diffraction pattern is therefore

$$\boxed{\,I(x, y) = \frac{I_i}{(\lambda d)^2}\left|P\left(\frac{x}{\lambda d}, \frac{y}{\lambda d}\right)\right|^2.\,} \qquad (4.3\text{-}5)$$

In summary: the Fraunhofer diffraction pattern at the point (x, y) is proportional to the squared magnitude of the Fourier transform of the aperture function $p(x, y)$ evaluated at the spatial frequencies $\nu_x = x/\lambda d$ and $\nu_y = y/\lambda d$.

EXERCISE 4.3-1

Fraunhofer Diffraction from a Rectangular Aperture. Verify that the Fraunhofer diffraction pattern from a rectangular aperture, of height and width D_x and D_y respectively, observed at a distance d is

$$I(x, y) = I_o\, \text{sinc}^2\left(\frac{D_x x}{\lambda d}\right) \text{sinc}^2\left(\frac{D_y y}{\lambda d}\right), \qquad (4.3\text{-}6)$$

where $I_o = (D_x D_y/\lambda d)^2\, I_i$ is the peak intensity and $\text{sinc}(x) \equiv \sin(\pi x)/(\pi x)$. Verify that the first zeros of this pattern occur at $x = \pm \lambda d/D_x$ and $y = \pm \lambda d/D_y$, so that the angular divergence of the diffracted light is given by

$$\theta_x = \frac{\lambda}{D_x}, \qquad \theta_y = \frac{\lambda}{D_y}. \tag{4.3-7}$$

If $Dy < D_x$, the diffraction pattern is wider in the y direction than in the x direction, as illustrated in Fig. 4.3-3.

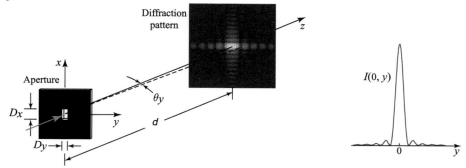

Figure 4.3-3 Fraunhofer diffraction from a rectangular aperture. The central lobe of the pattern has half-angular widths $\theta_x = \lambda/D_x$ and $\theta_y = \lambda/D_y$.

EXERCISE 4.3-2

Fraunhofer Diffraction from a Circular Aperture. Verify that the Fraunhofer diffraction pattern from a circular aperture of diameter D (Fig. 4.3-4) is

$$I(x,y) = I_o \left[\frac{2J_1(\pi D\rho/\lambda d)}{\pi D\rho/\lambda d} \right]^2, \qquad \rho = \sqrt{x^2 + y^2}, \tag{4.3-8}$$

where $I_o = (\pi D^2/4\lambda d)^2 I_i$ is the peak intensity and $J_1(\cdot)$ is the Bessel function of order 1. The Fourier transform of circularly symmetric functions is discussed in Appendix A, Sec. A.3. The circularly symmetric pattern (4.3-8), known as the **Airy pattern**, consists of a central disk surrounded by rings. Verify that the radius of the central disk, known as the **Airy disk**, is $\rho_s = 1.22\lambda d/D$ and subtends an angle

$$\theta = 1.22 \frac{\lambda}{D}. \tag{4.3-9}$$

Airy Disk Half Angle

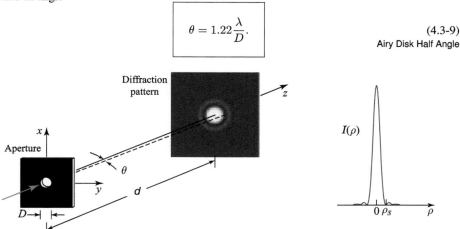

Figure 4.3-4 The Fraunhofer diffraction pattern from a circular aperture produces the Airy pattern with the radius of the central disk subtending an angle $\theta = 1.22\lambda/D$.

The Fraunhofer approximation is valid for distances d that are usually extremely large. It is satisfied, for example, in applications of long-distance free-space optical communications such as laser radar (lidar) and satellite communications. However, as shown in Sec. 4.2B, if a lens of focal length f is used to focus the diffracted light, the intensity pattern in the focal plane is proportional to the squared magnitude of the Fourier transform of $p(x,y)$ evaluated at $\nu_x = x/\lambda f$ and $\nu_y = y/\lambda f$. The observed pattern is therefore identical to that obtained from (4.3-5), with the distance d replaced by the focal length f.

EXERCISE 4.3-3

Spot Size of a Focused Optical Beam. A beam of light is focused using a lens of focal length f with a circular aperture of diameter D (Fig. 4.3-5). If the beam is approximated by a plane wave at points within the aperture, verify that the pattern of the focused spot is

$$I(x,y) = I_o \left[\frac{2J_1(\pi D\rho/\lambda f)}{\pi D\rho/\lambda f} \right]^2, \qquad \rho = \sqrt{x^2 + y^2}, \qquad (4.3\text{-}10)$$

where I_o is the peak intensity. Compare the radius of the focused spot,

$$\rho_s = 1.22 \lambda \frac{f}{D}, \qquad (4.3\text{-}11)$$

to the spot size obtained when a Gaussian beam of waist radius W_0 is focused by an ideal lens of infinite aperture [see (3.2-15)].

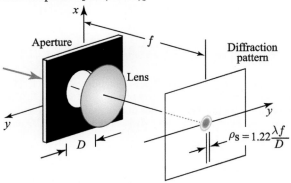

Figure 4.3-5 Focusing of a plane wave transmitted through a circular aperture of diameter D.

---

*B. Fresnel Diffraction

The theory of Fresnel diffraction is based on the assumption that the incident wave is multiplied by the aperture function $p(x,y)$ and propagates in free space in accordance with the Fresnel approximation. If the incident wave is a plane wave traveling in the z-direction with intensity I_i, the complex amplitude immediately after the aperture is $f(x,y) = \sqrt{I_i}\, p(x,y)$. Using (4.1-20), the diffraction pattern $I(x,y) = |g(x,y)|^2$ at a distance d is

$$I(x,y) = \frac{I_i}{(\lambda d)^2} \left| \iint_{-\infty}^{\infty} p(x',y') \exp\left[-j\pi \frac{(x-x')^2 + (y-y')^2}{\lambda d} \right] dx'\, dy' \right|^2 .$$

$$(4.3\text{-}12)$$

It is convenient to normalize all distances using $\sqrt{\lambda d}$ as a unit of distance, so that $X = x/\sqrt{\lambda d}$ and $X' = x'/\sqrt{\lambda d}$ are the normalized distances (and similarly for y and y'). Equation (4.3-12) then gives

$$I(X,Y) = I_i \left| \int\!\!\!\int_{-\infty}^{\infty} p(X',Y') \exp\left\{-j\pi\left[(X-X')^2 + (Y-Y')^2\right]\right\} dX'\,dY' \right|^2.$$
(4.3-13)

The integral in (4.3-13) is the convolution of $p(X,Y)$ and $\exp[-j\pi(X^2+Y^2)]$. The real and imaginary parts of $\exp(-j\pi X^2)$, $\cos \pi X^2$ and $\sin \pi X^2$, respectively, are plotted in Fig. 4.3-6. They oscillate at an increasing frequency and their first lobes lie in the intervals $|X| < 1/\sqrt{2}$ and $|X| < 1$, respectively. The total area under the function $\exp(-j\pi X^2)$ is 1, with the main contribution to the area coming from the first few lobes, since subsequent lobes cancel out. If a is the radius of the aperture, the radius of the normalized function $p(X,Y)$ is $a/\sqrt{\lambda d}$. The result of the convolution, which depends on the relative size of the two functions, is therefore governed by the Fresnel number $N_F = a^2/\lambda d$.

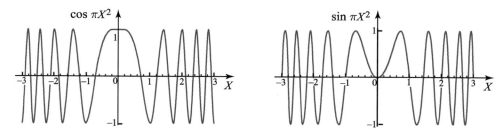

Figure 4.3-6 The real and imaginary parts of $\exp(-j\pi X^2)$.

If the Fresnel number is large, the normalized width of the aperture $a/\sqrt{\lambda d}$ is much greater than the width of the main lobe, and the convolution yields approximately the wider function $p(X,Y)$. Under this condition the Fresnel diffraction pattern is a shadow of the aperture, as would be expected from ray optics. Note that ray optics is applicable in the limit $\lambda \to 0$, which corresponds to the limit $N_F \to \infty$. In the opposite limit, when N_F is small, the Fraunhofer approximation becomes applicable and the Fraunhofer diffraction pattern is obtained.

EXAMPLE 4.3-1. *Fresnel Diffraction from a Slit.* Assume that the aperture is a slit of width $D = 2a$, so that $p(x,y) = 1$ when $|x| \leq a$, and 0 elsewhere. The normalized coordinate $X = x/\sqrt{\lambda d}$ and

$$P(X,Y) = \begin{cases} 1, & |X| \leq \dfrac{a}{\sqrt{\lambda d}} = \sqrt{N_F} \\ 0, & \text{elsewhere}, \end{cases}$$
(4.3-14)

where $N_F = a^2/\lambda d$ is the Fresnel number. Substituting into (4.3-13), we obtain $I(X,Y) = I_i|g(X)|^2$, where

$$g(X) = \int_{-\sqrt{N_F}}^{\sqrt{N_F}} \exp\left[-j\pi(X-X')^2\right] dX' = \int_{X-\sqrt{N_F}}^{X+\sqrt{N_F}} \exp(-j\pi X'^2)\,dX'.$$
(4.3-15)

This integral is usually written in terms of the Fresnel integrals

$$C(x) = \int_0^x \cos\frac{\pi\alpha^2}{2}\, d\alpha, \qquad S(x) = \int_0^x \sin\frac{\pi\alpha^2}{2}\, d\alpha, \qquad (4.3\text{-}16)$$

which are available in the standard computer mathematical libraries.

The complex function $g(X)$ may also be evaluated using Fourier-transform techniques. Since $g(x)$ is the convolution of a rectangular function of width $\sqrt{N_F}$ and $\exp(-j\pi X^2)$, its Fourier transform $G(\nu_x) \propto \mathrm{sinc}(\sqrt{N_F}\,\nu_x)\exp(j\pi\nu_x^2)$ (see Table A.1-1 in Appendix A). Thus, $g(X)$ may be computed by determining the inverse Fourier transform of $G(\nu_x)$. If $N_F \gg 1$, the width of $\mathrm{sinc}(\sqrt{N_F}\,\nu_x)$ is much narrower than the width of the first lobe of $\exp(j\pi\nu_x^2)$ (see Fig. 4.3-6) so that $G(\nu_x) \approx \mathrm{sinc}(\sqrt{N_F}\,\nu_x)$ and $g(X)$ is the rectangular function representing the aperture shadow.

The diffraction pattern from a slit is plotted in Fig. 4.3-7 for different Fresnel numbers corresponding to different distances d from the aperture. At very small distances (very large N_F), the diffraction pattern is a perfect shadow of the slit. As the distance increases (N_F decreases), the wave nature of light is exhibited in the form of small oscillations around the edges of the aperture (see also the diffraction pattern in Fig. 4.3-1). For very small N_F, the Fraunhofer pattern described by (4.3-6) is obtained. This is a sinc function with the first zero subtending an angle $\lambda/D = \lambda/2a$.

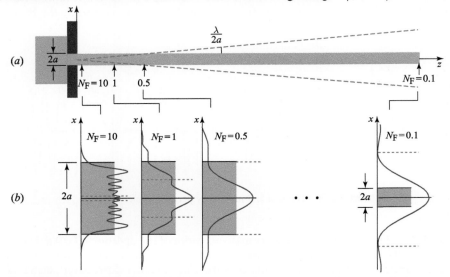

Figure 4.3-7 Fresnel diffraction from a slit of width $D = 2a$. (a) Shaded area is the geometrical shadow of the aperture. The dashed line is the width of the Fraunhofer diffracted beam. (b) Diffraction pattern at four axial positions marked by the arrows in (a) and corresponding to the Fresnel numbers $N_F = 10, 1, 0.5$, and 0.1. The shaded area represents the geometrical shadow of the slit. The dashed lines at $|x| = (\lambda/D)d$ represent the width of the Fraunhofer pattern in the far field. Where the dashed lines coincide with the edges of the geometrical shadow, the Fresnel number $N_F = a^2/\lambda d = 0.5$.

EXAMPLE 4.3-2. *Fresnel Diffraction from a Gaussian Aperture.* If the aperture function $p(x,y)$ is the Gaussian function $p(x,y) = \exp[-(x^2+y^2)/W_0^2]$, the Fresnel diffraction equation (4.3-12) may be evaluated exactly by finding the convolution of $\exp[-(x^2+y^2)/W_0^2]$ with $h_0 \exp[-j\pi(x^2+y^2)/\lambda d]$ using, for example, Fourier transform techniques (see Appendix A). The resultant diffraction pattern is

$$I(x,y) = I_i \left[\frac{W_0}{W(d)}\right]^2 \exp\left[-2\frac{x^2+y^2}{W^2(d)}\right], \qquad (4.3\text{-}17)$$

where $W^2(d) = W_0^2 + \theta_0^2 d^2$ and $\theta_0 = \lambda/\pi W_0$.

The diffraction pattern is a Gaussian function of $1/e^2$ half-width $W(d)$. For small d, $W(d) \approx W_0$; but as d increases, $W(d)$ increases and approaches $W(d) \approx \theta_0 d$ when d is sufficiently large for the Fraunhofer approximation to be applicable, so that the angle subtended by the Fraunhofer diffraction pattern is θ_0. These results are illustrated in Fig. 4.3-8, which is analogous to the illustration in Fig. 4.3-7 for diffraction from a slit. The wave diffracted from a Gaussian aperture is the Gaussian beam described in detail in Chapter 3.

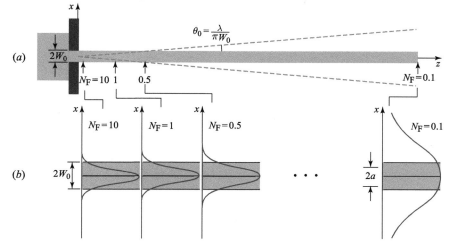

Figure 4.3-8 Fresnel diffraction pattern for a Gaussian aperture of radius W_0 at distances d such that the parameter $(\pi/2)W_0^2/\lambda d$, which is analogous to the Fresnel number N_F in Fig. 4.3-7, is 10, 1, 0.5, and 0.1. These values correspond to $W(d)/W_0 = 1.001, 1.118, 1.414$, and 5.099, respectively. The diffraction pattern is Gaussian at all distances.

Summary

In the order of increasing distance from the aperture, the diffraction pattern is:

1. A shadow of the aperture.
2. A Fresnel diffraction pattern, which is the convolution of the normalized aperture function with $\exp[-j\pi(X^2 + Y^2)]$.
3. A Fraunhofer diffraction pattern, which is the absolute-squared value of the Fourier transform of the aperture function. The far field has an angular divergence proportional to λ/D, where D is the diameter of the aperture.

Fresnel Diffraction from a Periodic Aperture: The Talbot Effect

Fresnel diffraction from a one-dimensional periodic aperture is best described in the Fourier domain by expanding the aperture function $p(x)$ in a Fourier series. If Λ is its period, then the Fourier expansion has frequencies $\nu_x = m/\Lambda$, where $m = 0, \pm 1, \pm 2, \ldots$. The transfer function of free space (4.1-11), at a distance z, is then

$$H_0 \exp\left(j\pi\lambda z \nu_x^2\right) = H_0 \exp\left(j\pi\lambda z m^2/\Lambda^2\right) = H_0 \exp\left(j 2\pi m^2 z/z_T\right), \quad (4.3\text{-}18)$$

where $z_T = 2\Lambda^2/\lambda$. At $z = z_T$, or multiples thereof, the transfer function is simply a constant H_0, independent of the harmonic order m. At these specific distances, then, each of the harmonic functions comprising the aperture function $p(x)$ is multiplied by the same factor so that the function $p(x)$ is reproduced. This process of self imaging is known as the **Talbot effect**, and the distance z_T is called the *Talbot* *distance*. At distances z unequal to multiples of z_T, the field is given by $U(x,z) = H_0 \sum_{m=-\infty}^{\infty} c_m \exp(jmx/\Lambda) \exp\left(j2\pi m^2 z/z_\text{T}\right)$, where the c_m are coefficients of the Fourier series expansion of $p(x)$. The corresponding intensity $I(x,z) = |U(x,z)|^2$ for an opaque screen with parallel slits exhibits a carpet-like pattern, as illustrated in Fig. 4.3-9.

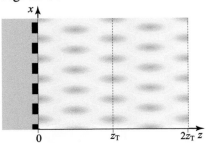

Figure 4.3-9 Talbot effect. Fresnel diffraction pattern from a periodic aperture that takes the form of parallel slits in an otherwise opaque screen. The pattern is reproduced at distances that are multiples of the Talbot distance z_T. The result has the appearance of a carpet with periodic patterns in x and z.

The Talbot effect is also observed for two-dimensional periodic apertures provided that the period is the same in the x and y directions.

*C. Nondiffracting Waves

In accordance with (4.1-8) and (4.1-9), the transfer function of free-space is $\exp(-jk_z z)$, where $k_z = \sqrt{k_o^2 - k_x^2 - k_y^2}$ is a circularly symmetric complex function of $k_T^2 = k_x^2 + k_y^2$. Any two input harmonic functions with spatial frequencies for which k_T^2 is the same thus have the same value of the transfer function. It follows that a function $f(x,y)$ that is a superposition of harmonic functions, all with the same value of k_T and therefore the same k_z, creates a stationary wave $U(x,y,z) = f(x,y)\exp(-jk_z z)$ that maintains its transverse distribution, and is therefore nondiffracting, as it travels through free space, regardless of the distance z. The wavefronts of such a wave are planes orthogonal to the z axis, and the propagation constant is k_z. Nondiffracting optical beams were considered in Sec. 3.5.

EXAMPLE 4.3-3. *Two Plane Waves.* The function $f(x,y) = \cos(\alpha x) = \frac{1}{2}[\exp(-j\alpha x) + \exp(+j\alpha x)]$ comprises two harmonic components with spatial angular frequencies $k_x = \pm\alpha$. On propagation through free space, each of these components is modified by the same factor $\exp(-jk_z z)$, where $k_z = \sqrt{k_o^2 - \alpha^2}$. The result is a stationary wave that takes the form $U(x,y,z) = \cos(\alpha x)\exp(-jk_z z)$, which has a sinusoidal transverse distribution representing the interference between two oblique plane waves at angles $\pm\sin^{-1}(\alpha/k_o)$, as provided by (2.5-7).

EXAMPLE 4.3-4. *Four Plane Waves.* A plane wave traveling in the z direction that is modulated by the function $f(x,y) = \cos(\alpha_x x)\cos(\alpha_y y)$ results in four waves, at angles $\pm\sin^{-1}(\alpha_x/k_o)$ and $\pm\sin^{-1}(\alpha_y/k_o)$ with respect to the x and y axes, respectively. Since the quantity $k_T^2 = \alpha_x^2 + \alpha_y^2$ is the same for all four waves, so too is $k_z = \sqrt{k_o^2 - k_T^2}$. The outcome at z is thus the stationary wave $U(x,y,z) = \cos(\alpha_x x)\cos(\alpha_y y)\exp(-jk_z z)$, as shown.

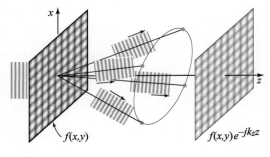

EXAMPLE 4.3-5. *Infinite Number of Plane Waves.* Consider now a function $f(x,y)$ composed of several harmonic functions of angular frequencies $k_x = k_T \cos\phi$ and $k_y = k_T \sin\phi$, with

fixed k_T but different ϕ. The quantities $k_x^2 + k_y^2 = k_T^2$ and $k_z = \sqrt{k_o^2 - k_T^2}$ are thus the same for each of these functions. This superposition of waves therefore corresponds to a stationary wave $U(x,y,z) = f(x,y)\exp(-jk_z z)$, no matter how many values of ϕ are included. A limiting case is the superposition in which a continuum of harmonic functions extends over all angles ϕ, which yields $f(x,y) = \int_{-\pi}^{\pi} \exp(-j\,k_T\cos\phi\,x - j\,k_T\sin\phi\,y)\,d\phi$.

The result is a continuum of plane waves whose directions form a cone of half-angle $\sin^{-1}(k_T/k_o)$. This superposition wave is nothing other than the Bessel beam $U(x,y,z) = 2\pi J_0\left(k_T\sqrt{x^2+y^2}\right)\exp(-jk_z z)$ described in (3.5-4) and illustrated at right. The connection is explicitly forged via the identity $\int_{-\pi}^{\pi} \exp(-ju\sin\phi)\,d\phi = 2\pi J_0(u)$, where $J_0(u)$ is the Bessel function of the first kind and zeroth order.

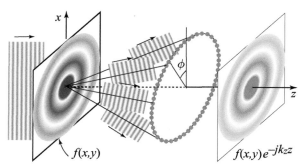

The plane-wave superposition associated with the Bessel beam may be implemented by making use of an axicon (see Example 2.4-1 and Sec. 3.5A).

4.4 IMAGE FORMATION

An ideal image formation system is an optical system that replicates the distribution of light in one plane, the object plane, into another, the image plane. Since the optical transmission process is never perfect, the image is never an exact replica of the object. Aside from image magnification, there is also blur resulting from imperfect focusing and from the diffraction of optical waves. This section is devoted to the description of image formation systems and their fidelity. Methods of linear systems, such as the impulse response function and the transfer function (Appendix B), are used to characterize image formation. A simple ray-optics approach is presented first, then a treatment based on wave optics is subsequently developed.

A. Ray Optics of a Single-Lens Imaging System

Consider an imaging system using a lens of focal length f at distances d_1 and d_2 from the object and image planes, respectively, as shown in Fig. 4.4-1. When $1/d_1 + 1/d_2 = 1/f$, the system is focused so that paraxial rays emitted from each point in the object plane reach a single corresponding point in the image plane. Within the ray theory of light, the imaging is "ideal," with each point of the object producing a single point of the image. The impulse response function of the system is an impulse function.

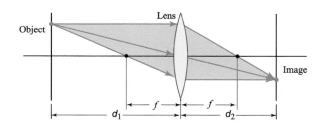

Figure 4.4-1 Rays in a focused imaging system.

Suppose now that the system is not in focus, as illustrated in Fig. 4.4-2, and assume that the focusing error is

$$\epsilon = \frac{1}{d_1} + \frac{1}{d_2} - \frac{1}{f}.$$
(4.4-1)
Focusing Error

A point in the object plane generates a patch of light in the image plane that is a shadow of the lens aperture. The distribution of this patch is the system's impulse response function. For simplicity, we shall consider an object point lying on the optical axis and determine the distribution of light $h(x, y)$ it generates in the image plane.

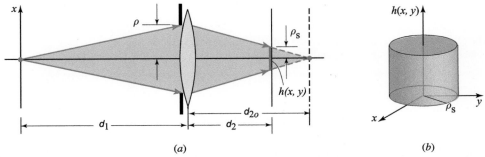

Figure 4.4-2 (a) Rays in a defocused imaging system. (b) The impulse response function of an imaging system with a circular aperture of diameter D is a circle of radius $\rho_s = \epsilon d_2 D/2$, where ϵ is the focusing error.

Assume that the plane of the focused image lies at a distance d_{2o} satisfying the imaging equation $1/d_{2o} + 1/d_1 = 1/f$. The shadow of a point on the edge of the aperture at a radial distance ρ is a point in the image plane with radial distance ρ_s where the ratio $\rho_s/\rho = (d_{2o} - d_2)/d_{2o} = 1 - d_2/d_{2o} = 1 - d_2(1/f - 1/d_1) = 1 - d_2(1/d_2 - \epsilon) = \epsilon d_2$. If $p(x, y)$ is the aperture function, also called the **pupil function** [$p(x, y) = 1$ for points inside the aperture, and 0 elsewhere], then $h(x, y)$ is a scaled version of $p(x, y)$ magnified by a factor $\rho_s/\rho = \epsilon d_2$, so that

$$h(x, y) \propto p\left(\frac{x}{\epsilon d_2}, \frac{y}{\epsilon d_2}\right).$$
(4.4-2)
Impulse Response Function
(Ray-Optics)

As an example, a circular aperture of diameter D corresponds to an impulse response function confined to a circle of radius

$$\rho_s = \tfrac{1}{2}\epsilon d_2 D,$$
(4.4-3)
Blur Spot Radius

as illustrated in Fig. 4.4-2. The radius ρ_s of this "blur spot" is an inverse measure of resolving power and image quality. A small value of ρ_s means that the system is capable of resolving fine details. Since ρ_s is proportional to the aperture diameter D, the image quality may be improved by use of a small aperture. A small aperture corresponds to a reduced sensitivity of the system to focusing errors, so that it corresponds to an increased "depth of focus."

B. Wave Optics of a 4-*f* Imaging System

Consider now the two-lens imaging system illustrated in Fig. 4.4-3. This system, called the **4-*f* system**, serves as a focused imaging system with unity magnification, as can be easily verified by ray tracing.

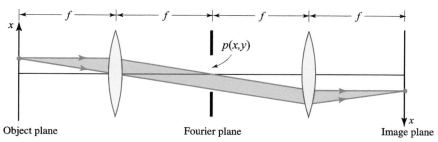

Figure 4.4-3 The 4-*f* imaging system. If an inverted coordinate system is used in the image plane, the magnification is unity.

The analysis of wave propagation through this system becomes simple if we recognize it as a cascade of two Fourier-transforming subsystems. The first subsystem (between the object plane and the Fourier plane) performs a Fourier transform, and the second (between the Fourier plane and the image plane) performs an inverse Fourier transform since the coordinate system in the image plane is inverted (see Exercise 4.2-2). As a result, in the absence of an aperture the image is a perfect replica of the object.

Let $f(x, y)$ be the complex amplitude transmittance of a transparency placed in the object plane and illuminated by a plane wave $\exp(-jkz)$ traveling in the z direction, as illustrated in Fig. 4.4-4, and let $g(x, y)$ be the complex amplitude in the image plane. The first lens system analyzes $f(x, y)$ into its spatial Fourier transform and separates its Fourier components so that each point in the Fourier plane corresponds to a single spatial frequency. These components are then recombined by the second lens system and the object distribution is perfectly reconstructed.

The 4-f imaging system can be used as a spatial filter in which the image $g(x, y)$ is a filtered version of the object $f(x, y)$. Since the Fourier components of $f(x, y)$ are available in the Fourier plane, a mask may be used to adjust them selectively, blocking some components and transmitting others, as illustrated in Fig. 4.4-5. The Fourier component of $f(x, y)$ at the spatial frequency (ν_x, ν_y) is located in the Fourier plane at the point $x = \lambda f \nu_x, y = \lambda f \nu_y$. To implement a filter of transfer function $H(\nu_x, \nu_y)$, the complex amplitude transmittance $p(x, y)$ of the mask must be proportional to $H(x/\lambda f, y/\lambda f)$. Thus, the transfer function of the filter realized by a mask of transmittance $p(x, y)$ is

$$H(\nu_x, \nu_y) = p(\lambda f \nu_x, \lambda f \nu_y), \tag{4.4-4}$$

Transfer Function 4-*f* System

where we have ignored the phase factor $j \exp(-j2kf)$ associated with each Fourier transform operation [the argument of h_l in (4.2-8)]. The Fourier transforms $G(\nu_x, \nu_y)$ and $F(\nu_x, \nu_y)$ of $g(x, y)$ and $f(x, y)$ are related by $G(\nu_x, \nu_y) = H(\nu_x, \nu_y) F(\nu_x, \nu_y)$.

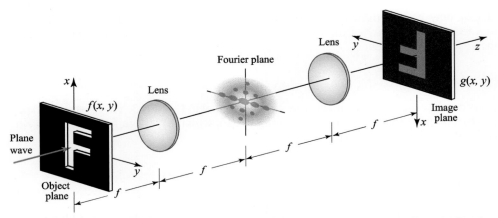

Figure 4.4-4 The 4-f imaging system performs a Fourier transform followed by an inverse Fourier transform, so that the image is a perfect replica of the object.

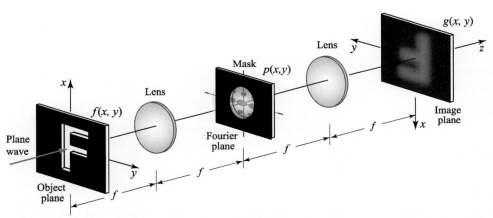

Figure 4.4-5 Spatial filtering. The transparencies in the object and Fourier planes have complex amplitude transmittances $f(x, y)$ and $p(x, y)$. A plane wave traveling in the z direction is modulated by the object transparency, Fourier transformed by the first lens, multiplied by the transmittance of the mask in the Fourier plane, and inverse Fourier transformed by the second lens. As a result, the complex amplitude in the image plane $g(x, y)$ is a filtered version of $f(x, y)$. The system has a transfer function $H(\nu_x, \nu_y) = p(\lambda f \nu_x, \lambda f \nu_y)$.

This is a rather simple result. *The transfer function has the same shape as the pupil function.* The corresponding impulse response function $h(x, y)$ is the inverse Fourier transform of $H(\nu_x, \nu_y)$,

$$h(x, y) = \frac{1}{(\lambda f)^2} P\left(\frac{x}{\lambda f}, \frac{y}{\lambda f}\right),$$

(4.4-5)
Impulse Response Function
4-f System

where $P(\nu_x, \nu_y)$ is the Fourier transform of $p(x, y)$.

Examples of Spatial Filters

- The ideal circularly symmetric *low-pass filter* has a transfer function $H(\nu_x, \nu_y) = 1$, $\nu_x^2 + \nu_y^2 < \nu_s^2$ and $H(\nu_x, \nu_y) = 0$, otherwise. It passes spatial frequencies that are smaller than the cutoff frequency ν_s and blocks higher frequencies. This filter is implemented by a mask in the form of a circular aperture of diameter D, with $D/2 = \nu_s \lambda f$. For example, if $D = 2$ cm, $\lambda = 1$ μm, and $f = 100$ cm, the cutoff frequency (spatial bandwidth) $\nu_s = D/2\lambda f = 10$ lines/mm. This filter eliminates spatial frequencies that are greater than 10 lines/mm, so that the smallest size of discernible detail in the filtered image is approximately 0.1 mm.

- The *high-pass filter* is the complement of the low-pass filter. It blocks low frequencies and transmits high frequencies. The mask is a clear transparency with an opaque central circle. The filter output is high at regions of large rate of change and small at regions of smooth or slow variation of the object. The filter is therefore useful for edge enhancement in image-processing applications.

- The *vertical-pass filter* blocks horizontal frequencies and transmits vertical frequencies. Only variations in the x direction are transmitted. If the mask is a vertical slit of width D, the highest transmitted frequency is $\nu_y = (D/2)/\lambda f$.

Examples of these filters and their effects on images are illustrated in Fig. 4.4-6.

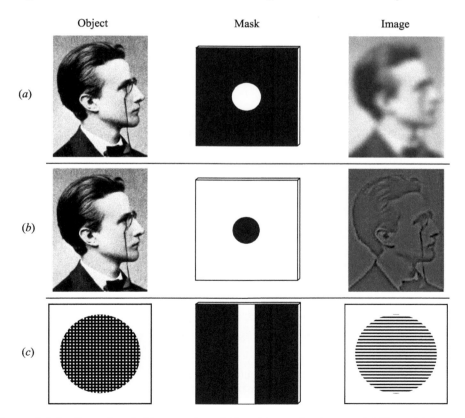

Figure 4.4-6 Examples of object, mask, and filtered image for three spatial filters: (*a*) low-pass filter; (*b*) high-pass filter; (*c*) vertical-pass filter. Black means the transmittance is zero and white means the transmittance is unity.

C. Wave Optics of a Single-Lens Imaging System

We now consider image formation in the single-lens imaging system illustrated in Fig. 4.4-7, using a wave-optics approach. We first determine the impulse response function, and then derive the transfer function. These functions are determined by the defocusing error ϵ, given by (4.4-1), and by the pupil function $p(x, y)$ (the transmittance of the aperture in the lens plane). The pupil function in this single-lens imaging system plays the same role of the mask function in the 4-f imaging system described in the previous section.

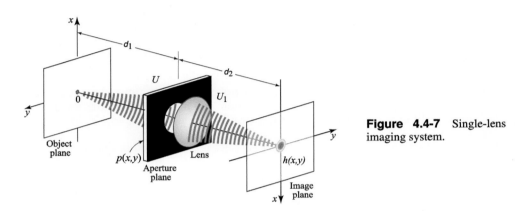

Figure 4.4-7 Single-lens imaging system.

Impulse Response Function

To determine the impulse response function we consider an object composed of a single point (an impulse) on the optical axis at the point $(0, 0)$, and follow the emitted optical wave as it travels to the image plane. The resultant complex amplitude is the impulse response function $h(x, y)$.

An impulse in the object plane produces in the aperture plane a spherical wave approximated by [see (4.1-18)]

$$U(x, y) \approx h_1 \exp\left[-jk\frac{x^2 + y^2}{2d_1}\right], \tag{4.4-6}$$

where $h_1 = (j/\lambda d_1) \exp(-jkd_1)$. Upon crossing the aperture and the lens, $U(x, y)$ is multiplied by the pupil function $p(x, y)$ and the lens quadratic phase factor $\exp[jk(x^2 + y^2)/2f]$, becoming

$$U_1(x, y) = U(x, y) \exp\left(jk\frac{x^2 + y^2}{2f}\right) p(x, y). \tag{4.4-7}$$

The resultant field $U_1(x, y)$ then propagates in free space a distance d_2. In accordance with (4.1-20) it produces the impulse response function

$$h(x, y) = h_2 \iint_{-\infty}^{\infty} U_1(x', y') \exp\left[-j\pi\frac{(x - x')^2 + (y - y')^2}{\lambda d_2}\right] dx' \, dy', \tag{4.4-8}$$

where $h_2 = (j/\lambda d_2) \exp(-jkd_2)$. Substituting from (4.4-6) and (4.4-7) into (4.4-8) and casting the integrals as a Fourier transform, we obtain

$$h(x,y) = h_1 h_2 \exp\left(-j\pi \frac{x^2 + y^2}{\lambda d_2}\right) P_1\left(\frac{x}{\lambda d_2}, \frac{y}{\lambda d_2}\right), \qquad (4.4\text{-}9)$$

where $P_1(\nu_x, \nu_y)$ is the Fourier transform of the function

$$p_1(x,y) = p(x,y) \exp\left(-j\pi\epsilon \frac{x^2 + y^2}{\lambda}\right), \qquad (4.4\text{-}10)$$
Generalized Pupil Function

known as the **generalized pupil function**. The factor ϵ is the focusing error given by (4.4-1).

For a high-quality imaging system, the impulse response function is a narrow function, extending only over a small range of values of x and y. If the phase factor $\pi(x^2 + y^2)/\lambda d_2$ in (4.4-9) is much smaller than 1 for all x and y within this range, it can be neglected, so that

$$h(x,y) = h_0 P_1\left(\frac{x}{\lambda d_2}, \frac{y}{\lambda d_2}\right), \qquad (4.4\text{-}11)$$
Impulse Response Function

where $h_0 = h_1 h_2$ is a constant of magnitude $(1/\lambda d_1)(1/\lambda d_2)$. It follows that the system's impulse response function is proportional to the Fourier transform of the generalized pupil function $p_1(x,y)$ evaluated at $\nu_x = x/\lambda d_2$ and $\nu_y = y/\lambda d_2$.

If the system is focused ($\epsilon = 0$), then $p_1(x,y) = p(x,y)$, and

$$h(x,y) \approx h_0 P\left(\frac{x}{\lambda d_2}, \frac{y}{\lambda d_2}\right), \qquad (4.4\text{-}12)$$

where $P(\nu_x, \nu_y)$ is the Fourier transform of $p(x,y)$. This result is similar to the corresponding result in (4.4-5) for the 4-f system.

EXAMPLE 4.4-1. *Impulse Response Function of a Focused Imaging System with a Circular Aperture.* If the aperture is a circle of diameter D so that $p(x,y) = 1$ if $\rho = \sqrt{x^2 + y^2} \leq D/2$, and zero otherwise, then the impulse response function is

$$h(x,y) = h(0,0) \frac{2J_1(\pi D \rho/\lambda d_2)}{\pi D \rho / \lambda d_2}, \qquad \rho = \sqrt{x^2 + y^2}, \qquad (4.4\text{-}13)$$

and $|h(0,0)| = (\pi D^2/4\lambda^2 d_1 d_2)$. This is a circularly symmetric function whose cross section is shown in Fig. 4.4-8. It drops to zero at a radius

$$\rho_s = 1.22\lambda \frac{d_2}{D} \qquad (4.4\text{-}14)$$

and oscillates slightly before it vanishes. The radius ρ_s is therefore a measure of the size of the blur circle. If the system is focused at ∞, then $d_1 = \infty$ and $d_2 = f$, so that

$$\rho_s = 1.22\lambda F_\# , \qquad (4.4\text{-}15)$$
Spot Radius

where $F_\# = f/D$ is the lens F-number. Thus, systems of smaller $F_\#$ (larger apertures) have better image quality. This assumes, of course, that the larger lens does not introduce geometrical aberrations.

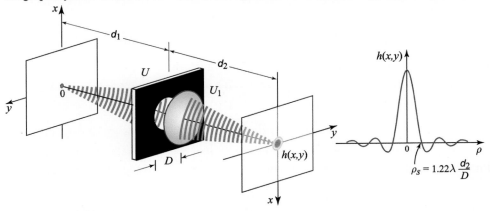

Figure 4.4-8 Impulse response function of an imaging system with a circular aperture.

Transfer Function

The transfer function of a linear system can only be defined when the system is shift invariant (see Appendix B). Evidently, the single-lens imaging system is not shift invariant since a shift Δ of a point in the object plane is accompanied by a *different* shift $M\Delta$ in the image plane, where $M = -d_2/d_1$ is the magnification.

The image is different from the object in two ways. First, the image is a magnified replica of the object, i.e., the point (x, y) of the object is located at a new point (Mx, My) in the image. Second, every point is smeared into a patch as a result of defocusing or diffraction. We can therefore think of image formation as a cascade of two systems — a system of ideal magnification followed by a system of blur, as depicted in Fig. 4.4-9. By its nature, the magnification system is shift-variant. For points near the optical axis, the blur system is approximately shift invariant and therefore can be described by a transfer function.

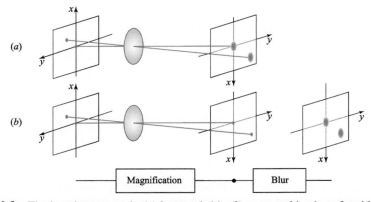

Figure 4.4-9 The imaging system in (a) is regarded in (b) as a combination of an ideal imaging system with only magnification, followed by shift-invariant blur in which each point is blurred into a patch with a distribution equal to the impulse response function.

The transfer function $H(\nu_x, \nu_y)$ of the blur system is determined by obtaining the Fourier transform of the impulse response function $h(x, y)$ in (4.4-11). The result is

$$H(\nu_x, \nu_y) \approx p_1(\lambda d_2 \nu_x, \lambda d_2 \nu_y),$$

(4.4-16)
Transfer Function

where $p_1(x, y)$ is the generalized pupil function and we have ignored a constant phase factor $\exp(-jkd_1) \exp(-jkd_2)$. If the system is focused, then

$$H(\nu_x, \nu_y) \approx p(\lambda d_2 \nu_x, \lambda d_2 \nu_y),$$

(4.4-17)

where $p(x, y)$ is the pupil function. This result is identical to that obtained for the 4-f imaging system [see (4.4-4)]. If the aperture is a circle of diameter D, for example, then the transfer function is constant within a circle of radius ν_s, where

$$\nu_s = \frac{D}{2\lambda d_2},$$

(4.4-18)

and vanishes elsewhere, as illustrated in Fig. 4.4-10.

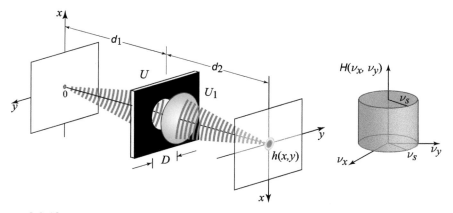

Figure 4.4-10 Transfer function of a focused imaging system with a circular aperture of diameter D. The system has a spatial bandwidth $\nu_s = D/2\lambda d_2$.

If the lens is focused at infinity, i.e., $d_2 = f$,

$$\nu_s = \frac{1}{2\lambda F_\#},$$

(4.4-19)
Spatial Bandwidth

where $F_\# = f/D$ is the lens F-number. For example, for an F-2 lens ($F_\# = f/D = 2$) and for $\lambda = 0.5$ μm, $\nu_s = 500$ lines/mm. The frequency ν_s is the spatial bandwidth, i.e., the highest spatial frequency that the imaging system can transmit.

D. Near-Field Imaging

It was shown in Sec. 4.1B that the spatial bandwidth of light propagating in free space at a wavelength λ is λ^{-1} cycles/mm. Fourier components of the object distribution with spatial frequencies greater than λ^{-1} lead to evanescent waves that decay rapidly and diminish at distances from the object plane of the order of a wavelength, so that object features smaller than a wavelength cannot be transmitted. Moreover, it was shown in Sec. 4.4C that an imaging system using a lens with a specified $F_\#$ has an impulse response function whose radius is $1.22\lambda F_\#$, so that points separated by a distance smaller than $1.22\lambda F_\#$ cannot be discriminated [see Fig. 4.4-11(a)]. Another imaging modality that makes use of a laser beam focused by a lens to scan an object, as depicted in Fig. 4.4-11(b), behaves similarly. The resolution of this system is dictated by the size of the focused spot, which has a radius of $1.22\lambda F_\#$, as was shown in Example 4.4-1. In both of these cases, therefore, object details with dimensions much smaller than a wavelength are washed out in the scanned image. This fundamental limit on the resolution of image-formation systems is referred to as the **diffraction limit**.

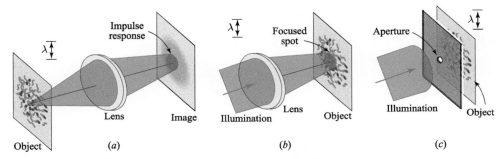

Figure 4.4-11 In a single-lens imaging system, the subwavelength spatial details of an object are washed out (a) in an image formed by a single lens, or (b) by making use of a focused laser-scanning system. (c) On the other hand, a scanning imaging system that makes use of illumination transmitted through a subwavelength aperture preserves the subwavelength details of the object, provided that the object plane is placed at a subwavelength distance from the aperture plane.

The diffraction limit may be circumvented, however. *Light can be localized to a spot with dimensions much smaller than a wavelength, within a single plane.* The difficulty is that the evanescent waves fully decay a short distance away from that plane, whereupon the spot diverges and acquires a size that exceeds the wavelength. At yet greater distances, the wave ultimately becomes a spherical wave. Hence, the diffraction limit can be circumvented if the object is illuminated in the very vicinity of the subwavelength spot, before the beam waist has an opportunity to grow. This may be implemented in a scanning configuration by passing the illumination beam through an aperture of diameter much smaller than a wavelength, as depicted in Fig. 4.4-11(c). The object is placed at a distance from the aperture that is usually less than half the diameter of the aperture so that the beam illuminates a subwavelength-size area of the object. Upon transmission through the object, the traveling components of the wave form a spherical wave whose amplitude is proportional to the object transmittance at the location of the spot illumination. The resolution of this imaging system is therefore of the order of the aperture size, which is much smaller than the wavelength. An image is constructed by raster-scanning the illuminated aperture across the object and recording the optical response via a conventional far-field imaging system. This technique is known as near-field optical imaging or **scanning near-field optical microscopy (SNOM)**. Subwavelength imaging falls in the domain of **nanophotonics** since the imaging takes place over a subwavelength (nanometer) spatial scale. Other approaches for achieving subwavelength imaging make use of negative-index and hyperbolic materials, as considered in Sec. 8.1.

SNOM is typically implemented by sending the illumination light through an optical fiber with an aluminum-coated tapered tip, as illustrated in Fig. 4.4-12. The light is guided through the fiber by total internal reflection. As the diameter of the fiber decreases, the light is guided by reflection from the metallic surface, which acts like a conical mirror. As the fiber diameter grows yet smaller in the region of the tip, the wave can no longer be guided (see Sec. 9.1) and becomes evanescent. The distribution of the illumination wave at, and beyond, the end of the tip is complex and is typically determined numerically. Aperture diameters and spatial resolutions of the order of tens of nanometers are usually achieved in SNOM with visible light. Since the tip of the fiber scans the object at a distance of only a few nanometers, an elaborate feedback system must be employed to maintain the distance for a specimen of arbitrary topography. Applications of SNOM include the non-destructive characterization of inorganic, organic, composite, and biological materials and nanostructures.

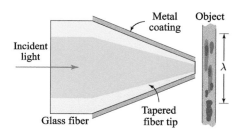

Figure 4.4-12 An optical fiber with a tapered metal-coated tip for near-field imaging.

4.5 HOLOGRAPHY

Holography involves the recording and reconstruction of optical waves. A hologram is a 2D transparency that contains a coded record of the amplitude and phase of an optical wave originating from a 3D object. Consider a monochromatic optical wave whose complex amplitude in some plane, say the $z = 0$ plane, is $U_o(x, y)$. If, somehow, a thin optical element (call it a transparency) with complex amplitude transmittance $t(x, y)$ equal to $U_o(x, y)$ were able to be made, it would provide a complete record of the wave. The wave could then be reconstructed simply by illuminating the transparency with a uniform plane wave of unit amplitude traveling in the z direction. The transmitted wave would have a complex amplitude in the $z = 0$ plane $U(x, y) = 1 \cdot t(x, y) = U_o(x, y)$. The original wave would then be reproduced at all points in the $z = 0$ plane, and therefore reconstructed everywhere in the space $z > 0$.

As an example, we know that a uniform plane wave traveling at an angle θ with respect to the z axis in the x–z plane has a complex amplitude $U_o(x, y) = \exp[-jkx \sin \theta]$. A record of this wave would be a transparency with complex amplitude transmittance $t(x, y) = \exp[-jkx \sin \theta]$. Such a transparency acts as a prism that bends an incident plane wave $\exp(-jkz)$ by an angle θ (see Sec. 2.4B), thus reproducing the original wave.

The question is how to make a transparency $t(x, y)$ from the original wave $U_o(x, y)$. One key impediment is that optical detectors, including the photographic emulsions used to make transparencies, are responsive to the optical intensity, $|U_o(x, y)|^2$, and are therefore insensitive to the phase $\arg\{U_o(x, y)\}$. Phase information is obviously important and cannot be disregarded, however. For example, if the phase of the oblique wave $U_o(x, y) = \exp[-jkx \sin \theta]$ were not recorded, neither would the direction of travel of the wave. To record the phase of $U_o(x,y)$, a code must be found that transforms phase into intensity. The recorded information could then be optically decoded in order to reconstruct the wave.

The Holographic Code

The holographic code is based on mixing the original wave (hereafter called the **object wave**) U_o with a known **reference wave** U_r and recording their interference pattern in the $z = 0$ plane. The intensity of the sum of the two waves is photographically recorded and a transparency of complex amplitude transmittance t, proportional to the intensity, is made [Fig. 4.5-1(a)]. The transmittance is therefore given by

$$\begin{aligned} t &\propto |U_o + U_r|^2 = |U_r|^2 + |U_o|^2 + U_r^* U_o + U_r U_o^*, \\ &= I_r + I_o + U_r^* U_o + U_r U_o^*, \\ &= I_r + I_o + 2\sqrt{I_r I_o}\cos\left[\arg\{U_r\} - \arg\{U_o\}\right], \end{aligned} \quad (4.5\text{-}1)$$

where I_r and I_o are, respectively, the intensities of the reference wave and the object wave at the $z = 0$ plane.

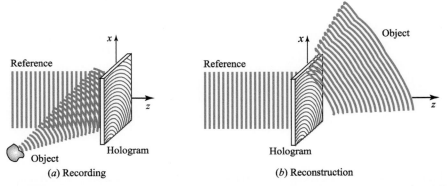

Figure 4.5-1 (a) A hologram is a transparency on which the interference pattern between the original wave (object wave) and a reference wave is recorded. (b) The original wave is reconstructed by illuminating the hologram with the reference wave.

The transparency, called a **hologram**, clearly carries coded information pertinent to the magnitude and phase of the wave U_o. In fact, as an interference pattern the transmittance t is highly sensitive to the difference between the phases of the two waves, as was shown in Sec. 2.5 (the temporal analog to holography is heterodyning, discussed in Sec. 2.6). As indicated above, ordinary photography is responsive only to the intensity of the incident wave and therefore records no phase information.

To decode the information in the hologram and reconstruct the object wave, the reference wave U_r is again used to illuminate the hologram [Fig. 4.5-1(b)]. The result is a wave with complex amplitude

$$U = tU_r \propto U_r I_r + U_r I_o + I_r U_o + U_r^2 U_o^* \quad (4.5\text{-}2)$$

in the hologram plane $z = 0$. The third term on the right-hand side is the original wave multiplied by the intensity I_r of the reference wave. If I_r is uniform (independent of x and y), this term constitutes the desired reconstructed wave. But it must be separated from the other three terms. The fourth term is a conjugated version of the original wave modulated by U_r^2. The first two terms represent the reference wave, modulated by the sum of the intensities of the two waves.

If the reference wave is selected to be a uniform plane wave propagating along the z axis $\sqrt{I_r}\exp(-jkz)$, then in the $z=0$ plane $U_r(x,y) = \sqrt{I_r}$ is a constant independent of x and y. Dividing (4.5-2) by $U_r = \sqrt{I_r}$ gives

$$U(x,y) \propto I_r + I_o(x,y) + \sqrt{I_r}\,U_o(x,y) + \sqrt{I_r}\,U_o^*(x,y). \qquad (4.5\text{-}3)$$

Reconstructed Wave in Plane of Hologram

The significance of the various terms in (4.5-3), and the methods of extracting the original wave (the third term), are clarified by means of a number of examples.

EXAMPLE 4.5-1. *Hologram of an Oblique Plane Wave.* If the object wave is an oblique plane wave at angle θ [Fig. 4.5-2(a)], $U_o(x,y) = \sqrt{I_o}\exp(-jkx\sin\theta)$, then (4.5-3) gives $U(x,y) \propto I_r + I_o + \sqrt{I_r I_o}\exp(-jkx\sin\theta) + \sqrt{I_r I_o}\exp(-jkx\sin\theta)$. Since the first two terms are constant, they correspond to a wave propagating in the z direction (the continuance of the reference wave). The third term corresponds to the original object wave, whereas the fourth term represents the **conjugate wave**, a plane wave traveling at an angle $-\theta$. The object wave is therefore separable from the other waves. In fact, this hologram is nothing but a recording of the interference pattern formed from two oblique plane waves at an angle θ (Sec. 2.5A). It serves as a sinusoidal diffraction grating that splits an incident reference wave into three waves at angles 0, θ, and $-\theta$ [see Fig. 4.5-2(b) and Sec. 2.4B].

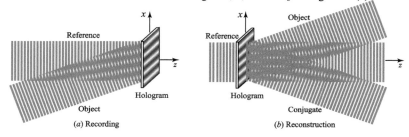

Figure 4.5-2 The hologram of an oblique plane wave is a sinusoidal diffraction grating.

EXAMPLE 4.5-2. *Hologram of a Point Source.* Here the object wave is a spherical wave originating at the point $\mathbf{r}_0 = (0,0,-d)$, as illustrated in Fig. 4.5-3, so that $U_o(x,y) \propto \exp(-jk|\mathbf{r}-\mathbf{r}_0|)/|\mathbf{r}-\mathbf{r}_0|$, where $\mathbf{r} = (x,y,0)$. The first term of (4.5-3) corresponds to a plane wave traveling in the z direction, whereas the third is proportional to the amplitude of the original spherical wave originating at $(0,0,-d)$. The fourth term is proportional to the amplitude of the conjugate wave $U_o^*(x,y) \propto \exp(jk|\mathbf{r}-\mathbf{r}_0|)/|\mathbf{r}-\mathbf{r}_0|$, which is a converging spherical wave centered at the point $(0,0,d)$. The second term is proportional to $1/|\mathbf{r}-\mathbf{r}_0|^2$ and its corresponding wave therefore travels in the z direction with very small angular spread since its intensity varies slowly in the transverse plane.

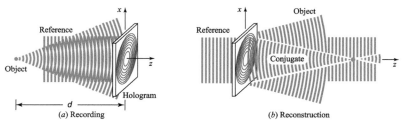

Figure 4.5-3 Hologram of a spherical wave originating from a point source. The conjugate wave forms a real image of the point.

Off-Axis Holography

One means of separating the four components of the reconstructed wave is to ensure that they vary at well-separated spatial frequencies, so that they have well-separated directions. This form of spatial-frequency multiplexing (see Sec. 4.1A) is assured if the object and reference waves are offset so that they arrive from well-separated directions.

Let us consider the case when the object wave has a complex amplitude $U_o(x,y) = f(x,y)\exp(-jkx\sin\theta)$. This is a wave of complex envelope $f(x,y)$ modulated by a phase factor equal to that introduced by a prism with deflection angle θ. It is assumed that $f(x,y)$ varies slowly so that its maximum spatial frequency ν_s corresponds to an angle $\theta_s = \sin^{-1}\lambda\nu_s$ much smaller than θ. The object wave therefore has directions centered about the angle θ, as illustrated in Fig. 4.5-4. Equation (4.5-3) gives

$$U(x,y) \propto I_r + |f(x,y)|^2 + \sqrt{I_r}\,f(x,y)\exp(-jkx\sin\theta)$$
$$+ \sqrt{I_r}\,f^*(x,y)\exp(+jkx\sin\theta). \qquad (4.5\text{-}4)$$

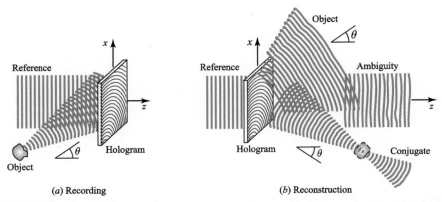

Figure 4.5-4 Hologram of an off-axis object wave. The object wave is separated from both the reference and conjugate waves.

The third term is evidently a replica of the object wave, which arrives from a direction at angle θ. The presence of the phase factor $\exp(+jkx\sin\theta)$ in the fourth term indicates that it is deflected in the $-\theta$ direction. The first term corresponds to a plane wave traveling in the z direction. The second term, usually known as the **ambiguity term**, corresponds to a nonuniform plane wave in directions within a cone of small angle $2\theta_s$ around the z direction. The offset of the directions of the object and reference waves results in a natural angular separation of the object and conjugate waves from each other and from the other two waves if $\theta > 3\theta_s$, thus allowing the original wave to be recovered unambiguously. An alternative method of reducing the effect of the ambiguity wave is to make the intensity of the reference wave much greater than that of the object wave. The ambiguity wave [second term of (4.5-3)] is then much smaller than the other terms since it involves only object waves; it is therefore relatively negligible.

Fourier-Transform Holography

The Fourier transform $F(\nu_x, \nu_y)$ of a function $f(x,y)$ may be computed optically by use of a lens (see Sec. 4.2). If $f(x,y)$ is the complex amplitude in one focal plane of the lens, then $F(x/\lambda f, y/\lambda f)$ is the complex amplitude in the other focal plane, where

f is the focal length of the lens and λ is the wavelength. Since the Fourier transform is usually a complex-valued function, it cannot be recorded directly.

The Fourier transform $F(x/\lambda f, y/\lambda f)$ may be recorded holographically by regarding it as an object wave, $U_o(x,y) = f(x/\lambda f, y/\lambda f)$, mixing it with a reference wave $U_f(x,y)$, and recording the superposition as a hologram [Fig. 4.5-5(a)]. Reconstruction is achieved by illumination of the hologram with the reference wave as usual. The reconstructed wave may be inverse Fourier transformed using a lens so that the original function $f(x,y)$ is recovered [Fig. 4.5-5(b)].

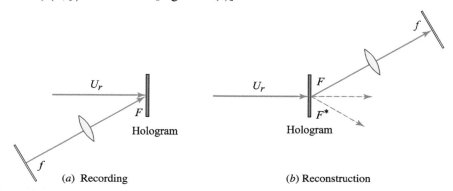

Figure 4.5-5 (a) Hologram of a wave whose complex amplitude represents the Fourier transform of a function $f(x,y)$. (b) Reconstruction of $f(x,y)$ by use of a Fourier-transform lens.

Holographic Spatial Filters

A spatial filter of transfer function $H(\nu_x, \nu_y)$ may be implemented by use of a 4-f optical system with a mask of complex amplitude transmittance $p(x,y) = H(x/\lambda f, y/\lambda f)$ placed in the Fourier plane (see Sec. 4.4B). Since the transfer function $H(\nu_x, \nu_y)$ is usually complex-valued, the mask transmittance $p(x,y)$ has a phase component and is difficult to fabricate using conventional printing techniques. If the filter impulse response function $h(x,y)$ is real-valued, however, a Fourier-transform hologram of $h(x,y)$ may be created by holographically recording the Fourier transform $U_o(x,y) = H(x/\lambda f, y/\lambda f)$, as depicted in Fig. 4.5-6(a).

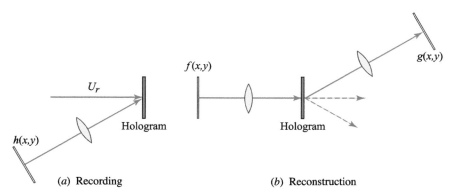

Figure 4.5-6 The VanderLugt holographic filter. (a) A hologram of the Fourier transform of $h(x,y)$ is recorded. (b) The Fourier transform of $f(x,y)$ is transmitted through the hologram and inverse Fourier transformed by a lens. The result is a function $g(x,y)$ proportional to the convolution of $f(x,y)$ and $h(x,y)$. The overall process provides a spatial filter with impulse response function $h(x,y)$.

Using the Fourier transform of the input $f(x,y)$ as a reference, $U_r(x,y) = F(x/\lambda f, y/\lambda f)$, the hologram constructs the wave

$$U_r(x,y)U_o(x,y) = F(x/\lambda f, y/\lambda f)\, H(x/\lambda f, y/\lambda f). \qquad (4.5\text{-}5)$$

The inverse Fourier transform of the reconstructed object wave, obtained with a lens of focal length f as illustrated in Fig. 4.5-6(b), therefore yields a complex amplitude $g(x,y)$ with a Fourier transform $G(\nu_x, \nu_y) = H(\nu_x, \nu_y)F(\nu_x, \nu_y)$. Thus, $g(x,y)$ is the convolution of $f(x,y)$ with $h(x,y)$. The overall system, known as the **VanderLugt filter**, performs the operation of convolution, which is the basis of spatial filtering.

If the conjugate wave $U_r(x,y)U_o^*(x,y) = F(x/\lambda f, y/\lambda f)H^*(x/\lambda f, y/\lambda f)$ is, instead, inverse Fourier transformed, the correlation, instead of the convolution, of the functions $f(x,y)$ and $h(x,y)$ is obtained. The operation of correlation is useful in image-processing applications, including pattern recognition.

The Holographic Apparatus

An essential condition for the successful fabrication of a hologram is the availability of a monochromatic light source with minimal phase fluctuations. The presence of phase fluctuations results in the random shifting of the interference pattern and the washing out of the hologram. For this reason, a coherent light source (usually a laser) is a necessary part of the apparatus. The coherence requirements for the interference of light waves are discussed in Chapter 12.

Figure 4.5-7 illustrates a typical experimental configuration used to record a hologram and reconstruct the optical wave scattered from the surface of a physical object. Using a beamsplitter, laser light is split into two portions; one is used as the reference wave, whereas the other is scattered from the object to form the object wave. The optical path difference between the two waves should be as small as possible to ensure that the two beams maintain a nonrandom phase difference [the term $\arg\{U_r\} - \arg\{U_o\}$ in (4.5-1)].

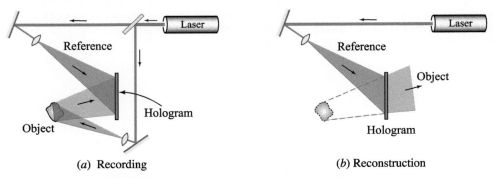

Figure 4.5-7 Holographic recording and reconstruction.

Since the interference pattern forming the hologram is composed of fine lines separated by distances of the order of $\lambda/\sin\theta$, where θ is the angular offset between the reference and object waves, the photographic film must be of high resolution and the system must not vibrate during the exposure. The larger θ, the smaller the distances between the hologram lines, and the more stringent these requirements are. The object wave is reconstructed when the recorded hologram is illuminated with the reference wave, so that a viewer see the object as if it were actually there, with its three-dimensional character preserved.

Volume Holography

It has been assumed so far that the hologram is a thin planar transparency on which the interference pattern of the object and reference waves is recorded. We now consider recording the hologram in a relatively thick medium and show that this offers an advantage. Consider the simple case when the object and reference waves are plane waves with wavevectors $\mathbf{k}_r$ and $\mathbf{k}_o$. The recording medium extends between the planes $z = 0$ and $z = \Delta$, as illustrated in Fig. 4.5-8. The interference pattern is now a function of x, y, and z:

$$I(x,y,z) = \left|\sqrt{I_r}\exp(-j\mathbf{k}_r \cdot \mathbf{r}) + \sqrt{I_o}\exp(-j\mathbf{k}_o \cdot \mathbf{r})\right|^2$$
$$= I_r + I_o + 2\sqrt{I_r I_o}\cos(\mathbf{k}_o \cdot \mathbf{r} - \mathbf{k}_r \cdot \mathbf{r})$$
$$= I_r + I_o + 2\sqrt{I_r I_o}\cos(\mathbf{k}_g \cdot \mathbf{r}), \qquad (4.5\text{-}6)$$

where $\mathbf{k}_g = \mathbf{k}_o - \mathbf{k}_r$. This is a sinusoidal pattern of period $\Lambda = 2\pi/|\mathbf{k}_g|$ and with the surfaces of constant intensity normal to the vector $\mathbf{k}_g$.

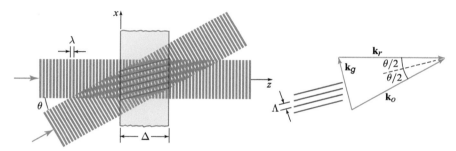

Figure 4.5-8 Interference pattern when the reference and object waves are plane waves. Since $|\mathbf{k}_r| = |\mathbf{k}_o| = 2\pi/\lambda$ and $|\mathbf{k}_g| = 2\pi/\Lambda$, from the geometry of the vector diagram $2\pi/\Lambda = 2(2\pi/\lambda)\sin(\theta/2)$, so that $\Lambda = \lambda/2\sin(\theta/2)$.

For example, if the reference wave points in the z direction and the object wave makes an angle θ with the z axis, $|\mathbf{k}_g| = 2k\sin(\theta/2)$ and the period is

$$\Lambda = \frac{\lambda}{2\sin(\theta/2)} \qquad (4.5\text{-}7)$$

as illustrated in Fig. 4.5-8.

If recorded in emulsion, this pattern serves as a thick diffraction grating, a **volume hologram**. The vector $\mathbf{k}_g$ is called the **grating vector**. When illuminated with the reference wave as illustrated in Fig. 4.5-9, the parallel planes of the grating reflect the wave only when the Bragg condition $\sin\phi = \lambda/2\Lambda$ is satisfied, where ϕ is the angle between the planes of the grating and the incident reference wave (Exercise 2.5-3). In our case $\phi = \theta/2$, so that $\sin(\theta/2) = \lambda/2\Lambda$. In view of (4.5-7), the Bragg condition is indeed satisfied, so that the reference wave is indeed reflected. As evident from the geometry, the reflected wave is an extension of the object wave, so that the reconstruction process is successful.

Suppose now that the hologram is illuminated with a reference wave of different wavelength λ'. Evidently, the Bragg condition, $\sin(\theta/2) = \lambda'/2\Lambda$, will not be satisfied and the wave will not be reflected. It follows that the object wave is reconstructed only

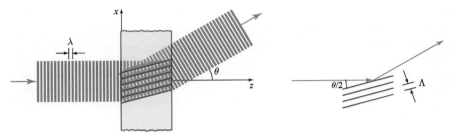

Figure 4.5-9 The reference wave is Bragg reflected from the thick hologram and the object wave is reconstructed.

if the wavelength of the reconstruction source matches that of the recording source. If light with a broad spectrum (white light) is used as a reconstruction source, only the "correct" wavelength would be reflected and the reconstruction process would be successful.

Although the recording process must be done with monochromatic light, the reconstruction can be achieved with white light. This provides a clear advantage in many applications of holography. Other geometries for recording a reconstruction of a volume hologram are illustrated in Fig. 4.5-10.

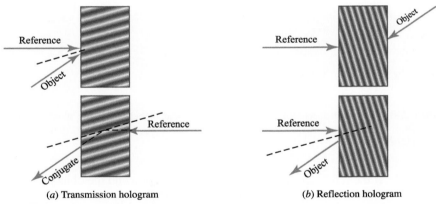

(a) Transmission hologram   (b) Reflection hologram

Figure 4.5-10 Two geometries for recording and reconstruction of a volume hologram. (a) This hologram is recorded with the reference and object waves arriving from the same side, and is reconstructed by use of a reversed reference wave; the reconstructed wave is a conjugate wave traveling in a direction opposite to the original object wave. (b) A reflection hologram is recorded with the reference and object waves arriving from opposite sides; the object wave is reconstructed by reflection from the grating.

Another type of hologram that may be viewed with white light is the **rainbow hologram**. This hologram is recorded through a narrow slit so that the reconstructed image, of course, also appears as if seen through a slit. However, if the wavelength of reconstruction differs from the recording wavelength, the reconstructed wave will appear to be coming from a displaced slit since a magnification effect will be introduced. If white light is used for reconstruction, the reconstructed wave appears as the object seen through many displaced slits, each with a different wavelength (color). The result is a rainbow of images seen through parallel slits. Each slit displays the object with parallax effect in the direction of the slit, but not in the orthogonal direction. Rainbow holograms enjoy wide commercial use as displays.

Computer-Generated Holography

A **computer-generated hologram** is a hologram of an object that does not physically exist. The hologram is generated by computing, and then digitally recording, the interference pattern of a reference wave with a mathematically defined wave that represents light scattered from a particular virtual object. The hologram may take the form of a mask, film, or spatial light modulator; when illuminated by the reference wave it generates the desired object wave. Computer-generated holography is principally geared toward 3D visualization, including applications in CAD (computer-aided design), gaming, and video displays.

An important application is the generation of **holographic optical elements** (HOEs). One example is the hologram of a point source, described in Example 4.5-2, which functions as a lens. A HOE that converts a planar wave into another optical beam with a mathematically defined complex amplitude, such as a Hermite–Gaussian, Laguerre–Gaussian, Bessel, or Airy beam (see Secs. 3.4 and 3.5), may be created by computing and digitally recording the interference pattern of the desired beam with a planar wave, as described in Example 4.5-3.

EXAMPLE 4.5-3.   *Holographic Optical Element for Generating a Spiral-Phased Wave.*
The task at hand is to create a HOE that converts a reference planar wave into a spiral-phased object wave with complex amplitude $U_o = \exp(-jl\phi)$ at the $z = 0$ plane. Here $\phi = \arctan(y/x)$ is the azimuthal angle [see Fig. 3.4-1(b)] and $l = 1, 2, \ldots$ is the topological charge of the associated optical vortex, as discussed in Sec. 3.4. Choosing a reference planar wave that propagates in the x–z plane, at an angle θ with respect to the z axis (see Fig. 2.5-4), gives rise to $U_r = \exp(-jk_o \sin\theta\, x)$ at the $z = 0$ plane. Making use of (4.5-1) thus yields an interference pattern given by

$$I(x, y) = 2 + 2\cos(2\pi x/\Lambda - l\phi), \quad \phi = \arctan(y/x), \tag{4.5-8}$$

where $\Lambda = \lambda_o / \sin\theta$.

The resultant holograms take the form of vertical sinusoidal fringes of period Λ with dislocations near $x = 0$, as displayed in Fig. 4.5-11. Illuminating the recorded hologram with the reference wave will result in the generation of a spiral-phased wave with a helical wavefront and the associated value of l.

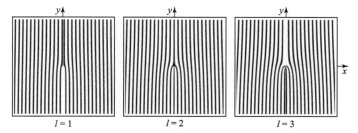

Figure 4.5-11 Computer-generated holographic optical elements for introducing a spiral phase into a planar wave, for three values of l.

READING LIST

Fourier Optics and Optical Signal Processing
J. W. Goodman, *Introduction to Fourier Optics*, Freeman, 4th ed. 2017.
K. Khare, *Fourier Optics and Computational Imaging*, Wiley, 2016.

R. K. Tyson, *Principles and Applications of Fourier Optics*, IOP Publishing, 2014.

D. Voelz, *Computational Fourier Optics: A MATLAB Tutorial*, SPIE Optical Engineering Press, 2011.

O. K. Ersoy, *Diffraction, Fourier Optics, and Imaging*, Wiley, 2006.

E. G. Steward, *Fourier Optics: An Introduction*, Halsted Press, 2nd ed. 1987; Dover, reissued 2004.

W. Lauterborn and T. Kurz, *Coherent Optics: Fundamentals and Applications*, Springer-Verlag, 2nd ed. 2003.

E. L. O'Neill, *Introduction to Statistical Optics*, Addison–Wesley, 1963; Dover, reissued 2003.

M. A. Abushagur and H. Caulfield, eds., *Selected Papers on Fourier Optics*, SPIE Optical Engineering Press (Milestone Series Volume 105), 1995.

P. W. Hooijmans, *Coherent Optical System Design*, Wiley, 1994.

A. VanderLugt, *Optical Signal Processing*, Wiley, 1992.

F. T. Yu and S. Yin, eds., *Selected Papers on Coherent Optical Processing*, SPIE Optical Engineering Press (Milestone Series Volume 52), 1992.

G. Reynolds, J. B. DeVelis, G. B. Parrent, and B. J. Thompson, *The New Physical Optics Notebook: Tutorials in Fourier Optics*, SPIE Optical Engineering Press, 1989.

G. Harburn, C. A. Taylor, and T. R. Welberry, *Atlas of Optical Transforms*, Cornell University Press, 1975.

M. Cagnet, M. Françon, and S. Mallick, *Atlas of Optical Phenomena*, Springer-Verlag, reprinted with supplement 1971.

M. Cagnet, M. Françon, and J. C. Thrierr, *Atlas of Optical Phenomena*, Springer-Verlag, 1962.

Diffraction

P. Ya. Ufimtsev, *Fundamentals of the Physical Theory of Diffraction*, Wiley, 2nd ed. 2014.

M. Nieto-Vesperinas, *Scattering and Diffraction in Physical Optics*, World Scientific, 2nd ed. 2006.

H. M. Nussenzveig, *Diffraction Effects in Semiclassical Scattering*, Cambridge University Press, 1992, paperback ed. 2006.

A. Sommerfeld, *Mathematical Theory of Diffraction*, Mathematische Annalen, 1896; Birkhäuser, 2004.

D. C. O'Shea, T. J. Suleski, A. D. Kathman, and D. W. Prather, *Diffractive Optics: Design, Fabrication, and Test*, SPIE Optical Engineering Press, 2003.

J. M. Cowley, *Diffraction Physics*, Elsevier, 3rd revised ed. 1995.

K. E. Oughstun, ed., *Selected Papers on Scalar Wave Diffraction*, SPIE Optical Engineering Press (Milestone Series Volume 51), 1992.

M. Françon, *Diffraction: Coherence in Optics*, Pergamon, 1966.

Imaging

G. de Villiers and E. R. Pike, *The Limits of Resolution*, CRC Press/Taylor & Francis, 2017.

D. F. Buscher, *Practical Optical Interferometry: Imaging at Visible and Infrared Wavelengths*, Cambridge University Press, 2015.

G. Saxby, *The Science of Imaging*, CRC Press/Taylor & Francis, 2nd ed. 2011.

B. E. A. Saleh, *Introduction to Subsurface Imaging*, Cambridge University Press, 2011.

V. N. Mahajan, *Optical Imaging and Aberrations, Part II. Wave Diffraction Optics*, SPIE Optical Engineering Press, 2nd ed. 2011.

R. L. Easton, Jr., *Fourier Methods in Imaging*, Wiley, 2010.

D. J. Brady, *Optical Imaging and Spectroscopy*, Wiley, 2009.

H. H. Barrett and K. J. Myers, *Foundations of Image Science*, Wiley, 2004.

S. Jutamulia, ed., *Selected Papers on Near-Field Optics*, SPIE Optical Engineering Press (Milestone Series Volume 172), 2002.

C. S. Williams and O. A. Becklund, *Introduction to the Optical Transfer Function*, SPIE Optical Engineering Press, 2002.

G. D. Boreman, *Modulation Transfer Function in Optical and Electro-Optical Systems*, SPIE Optical Engineering Press, 2001.

M. Françon, *Optical Image Formation and Processing*, Academic Press, 1979.

Holography

U. Schnars, C. Falldorf, J. Watson, and W. Jüptner, *Digital Holography and Wavefront Sensing: Principles, Techniques and Applications*, Springer-Verlag, 2nd ed. 2015.

T.-C. Poon and J.-P. Liu, *Introduction to Modern Digital Holography: With Matlab*, Cambridge University Press, 2014.

P.-A. Blanche, *Field Guide to Holography*, SPIE Optical Engineering Press, 2014.

V. Toal, *Introduction to Holography*, CRC Press/Taylor & Francis, 2012.

F. Unterseher, J. Hansen, and B. Schlesinger, *Holography Handbook: Making Holograms the Easy Way*, Ross, paperback 3rd ed. 2010.

L. Yaroslavsky, *Digital Holography and Digital Image Processing: Principles, Methods, Algorithms*, Kluwer, 2004, paperback ed. 2010.

G. Saxby, *Practical Holography*, Institute of Physics, 3rd ed. 2004.

P. Hariharan, *Basics of Holography*, Cambridge University Press, 2002.

H. I. Bjelkhagen and H. J. Caulfield, eds., *Selected Papers on Fundamental Techniques in Holography*, SPIE Optical Engineering Press (Milestone Series Volume 171), 2001.

J. E. Kasper and S. A. Feller, *Complete Book of Holograms: How They Work and How to Make Them*, Wiley, 1987; Dover, reissued 2001.

R. S. Sirohi and K. D. Hinsch, eds., *Selected Papers on Holographic Interferometry Principles and Techniques*, SPIE Optical Engineering Press (Milestone Series Volume 144), 1998.

M. Françon, *Holography*, Academic Press, 1974.

D. Gabor, Holography, 1948–1971 (Nobel Lecture in Physics, 1971), in S. Lundqvist, ed., *Nobel Lectures in Physics 1971–1980*, World Scientific, 1992.

PROBLEMS

4.1-3 **Correspondence Between Harmonic Functions and Plane Waves.** The complex amplitudes of a monochromatic wave of wavelength λ in the $z = 0$ and $z = d$ planes are $f(x,y)$ and $g(x,y)$, respectively. Assuming that $d = 10^4 \lambda$, use harmonic analysis to determine $g(x,y)$ in the following cases:

(a) $f(x,y) = 1$;

(b) $f(x,y) = \exp[(-j\pi/\lambda)(x+y)]$;

(c) $f(x,y) = \cos(\pi x/2\lambda)$;

(d) $f(x,y) = \cos^2(\pi y/2\lambda)$;

(e) $f(x,y) = \sum_m \text{rect}[(x/10\lambda) - 2m]$, $m = 0, \pm 1, \pm 2, \ldots$, where $\text{rect}(x) = 1$ if $|x| \leq 1/2$ and 0, otherwise.

Describe the physical nature of the wave in each case.

4.1-4 **Conical Confinement Angle.** In Prob. 4.1-3, if $f(x,y)$ is a circularly symmetric function with a maximum spatial frequency of 200 lines/mm, determine the angle of the cone within which the wave directions are confined. Assume that $\lambda = 633$ nm.

4.1-5 **Logarithmic Interconnection Map.** A transparency of amplitude transmittance $t(x,y) = \exp[-j2\pi\phi(x)]$ is illuminated with a uniform plane wave of wavelength $\lambda = 1$ μm. The transmitted light is focused by an adjacent lens of focal length $f = 100$ cm. What must $\phi(x)$ be so that the ray that hits the transparency at position x is deflected and focused to a position $x' = \ln(x)$ for all $x > 0$? (Note that x and x' are measured in millimeters.) If the lens is removed, how should $\phi(x)$ be modified so that the system performs the same function? This system may be used to perform a logarithmic coordinate transformation, as discussed in Chapter 24 [see Exercise 24.1-2].

4.2-3 **Proof of the Lens Fourier-Transform Property.**

(a) Show that the convolution of $f(x)$ and $\exp(-j\pi x^2/\lambda d)$ may be obtained via three steps: multiply $f(x)$ by $\exp(-j\pi x^2/\lambda d)$; evaluate the Fourier transform of the product at the frequency $\nu_x = x/\lambda d$; and multiply the result by $\exp(-j\pi x^2/\lambda d)$.

(b) The Fourier transform system in Fig. 4.2-4 is a cascade of three systems — propagation a distance f in free space, transmission through a lens of focal length f, and propagation a distance f in free space. Noting that propagation a distance d in free space is equivalent to convolution with $\exp(-j\pi x^2/\lambda d)$ [see (4.1-20)], and using the result in (a), derive the lens' Fourier-transform equation (4.2-8). For simplicity ignore the y dependence.

4.2-4 **Fourier Transform of the Line Functions.** A transparency of amplitude transmittance $t(x, y)$ is illuminated with a plane wave of wavelength $\lambda = 1$ μm and focused with a lens of focal length $f = 100$ cm. Sketch the intensity distribution in the plane of the transparency and in the lens focal plane in the following cases (all distances are measured in mm):
(a) $t(x, y) = \delta(x - y)$;
(b) $t(x, y) = \delta(x + a) + \delta(x - a)$, $a = 1$ mm;
(c) $t(x, y) = \delta(x + a) + j\delta(x - a)$, $a = 1$ mm;
where $\delta(\cdot)$ is the delta function (see Appendix A, Sec. A.1).

4.2-5 **Design of an Optical Fourier-Transform System.** Consider a lens used to display the Fourier transform of a two-dimensional function with spatial frequencies between 20 and 200 lines/mm. If the wavelength of light is $\lambda = 488$ nm, what should be the focal length of the lens so that the highest and lowest spatial frequencies are separated by a distance of 9 cm in the Fourier plane?

*4.2-6 **Generation of the Airy Beam by Use of an Optical Fourier-Transform System.** As described in Sec. 3.5B, the Airy beam has an amplitude $A(x, 0) = \mathrm{Ai}(x/W_0)$ in the $z = 0$ plane, where $\mathrm{Ai}(x)$ is the Airy function and W_0 is a measure of the beam width. Given that the Fourier transform of the phase function $\exp(jx^3/3)$ is equal to $2\pi \mathrm{Ai}(2\pi \nu_x)$, design an optical Fourier-transform system that generates the Airy beam using a lens of focal length f and a mask whose amplitude transmittance is $\exp(jx^3/3)$. Determine an expression for W_0 of the beam generated in terms of f and the wavelength λ.

4.3-4 **Fraunhofer Diffraction from a Diffraction Grating.** Derive an expression for the Fraunhofer diffraction pattern for an aperture made of $M = 2L + 1$ parallel slits of infinitesimal widths separated by equal distances $a = 10\lambda$,

$$p(x, y) = \sum_{m=-L}^{L} \delta(x - ma).$$

Sketch the pattern as a function of the observation angle $\theta = x/d$, where d is the observation distance.

4.3-5 **Fraunhofer Diffraction with an Oblique Incident Wave.** The diffraction pattern from an aperture with aperture function $p(x, y)$ is proportional to $|P(x/\lambda d, y/\lambda d)|^2$, where $P(\nu_x, \nu_y)$ is the Fourier transform of $p(x, y)$ and d is the distance between the aperture and observation planes. What is the diffraction pattern when the direction of the incident wave makes a small angle $\theta_x \ll 1$, with the z-axis in the x–z plane?

*4.3-6 **Fresnel Diffraction from Two Pinholes.** Show that the Fresnel diffraction pattern from two pinholes separated by a distance $2a$, i.e., $p(x, y) = [\delta(x-a)+\delta(x+a)]\delta(y)$, at an observation distance d is the periodic pattern, $I(x, y) = (2/\lambda d)^2 \cos^2(2\pi ax/\lambda d)$.

*4.3-7 **Relation Between Fresnel and Fraunhofer Diffraction.** Show that the Fresnel diffraction pattern of the aperture function $p(x, y)$ is equal to the Fraunhofer diffraction pattern of the aperture function $p(x, y) \exp[-j\pi(x^2 + y^2)/\lambda d]$.

4.4-1 **Blurring a Sinusoidal Grating.** An object $f(x, y) = \cos^2(2\pi x/a)$ is imaged by a defocused single-lens imaging system whose impulse response function $h(x, y) = 1$ within a square of width D, and is 0 elsewhere. Derive an expression for the distribution of the image $g(x, 0)$ in the x direction. Derive an expression for the contrast of the image in terms of the ratio D/a. The contrast is defined as $(\max - \min)/(\max + \min)$, where max and min are the maximum and minimum values of $g(x, 0)$, respectively.

4.4-2 **Image of a Phase Object.** An imaging system has an impulse response function $h(x, y) = \mathrm{rect}(x)\,\delta(y)$. If the input wave is

$$f(x, y) = \begin{cases} \exp\left(j\dfrac{\pi}{2}\right) & \text{for } x > 0 \\ \exp\left(-j\dfrac{\pi}{2}\right) & \text{for } x \leq 0, \end{cases}$$

determine and sketch the intensity $|g(x, y)|^2$ of the output wave $g(x, y)$. Verify that even though the intensity of the input wave $|f(x, y)|^2 = 1$, the intensity of the output wave is not uniform.

4.4-3 **Optical Spatial Filtering.** Consider the spatial filtering system shown in Fig. 4.4-5 with $f = 1000$ mm. The system is illuminated with a uniform plane wave of unit amplitude and wavelength $\lambda = 10^{-3}$ mm. The input transparency has amplitude transmittance $f(x, y)$ and the mask has amplitude transmittance $p(x, y)$. Write an expression relating the complex amplitude $g(x, y)$ of light in the image plane to $f(x, y)$ and $p(x, y)$. Assuming that all distances are measured in mm, sketch $g(x, 0)$ in the following cases:
(a) $f(x, y) = \delta(x - 5)$ and $p(x, y) = \text{rect}(x)$;
(b) $f(x, y) = \text{rect}(x)$ and $p(x, y) = \text{sinc}(x)$.
Determine $p(x, y)$ such that $g(x, y) = \nabla_T^2 f(x, y)$, where $\nabla_T^2 = \partial^2/\partial x^2 + \partial^2/\partial y^2$ is the transverse Laplacian operator.

4.4-4 **Optical Correlation.** Show how a spatial filter may be used to perform the operation of optical correlation (see Appendix A) between two images described by the real-valued functions $f_1(x, y)$ and $f_2(x, y)$. Under what conditions would the complex amplitude transmittances of the masks and transparencies used be real-valued?

*4.4-5 **Impulse Response Function of a Severely Defocused System.** Using wave optics, show that the impulse response function of a severely defocused imaging system (where the defocusing error ϵ is very large) may be approximated by $h(x, y) = p(x/\epsilon d_2, y/\epsilon d_2)$, where $p(x, y)$ is the pupil function. *Hint:* Use the method of stationary phase described on page 125 (second proof) to evaluate the integral resulting from the use of (4.4-11) and (4.4-10). Note that this is the same result as that predicted by the ray theory of light [see (4.4-2)].

4.4-6 **Two-Point Resolution.**
(a) Consider the single-lens imaging system discussed in Sec. 4.4C. Assuming a square aperture of width D, unit magnification, and perfect focus, write an expression for the impulse response function $h(x, y)$.
(b) Determine the response of the system to an object consisting of two points separated by a distance b, i.e.,
$$f(x, y) = \delta(x)\,\delta(y) + \delta(x - b)\,\delta(y).$$
(c) If $\lambda d_2/D = 0.1$ mm, sketch the magnitude of the image $g(x, 0)$ as a function of x when the points are separated by a distance $b = 0.5, 1,$ and 2 mm. What is the minimum separation between the two points such that the image remains discernible as two spots instead of a single spot, i.e., has two peaks?

4.4-7 **Ring Aperture.**
(a) A focused single-lens imaging system, with magnification $M = 1$ and focal length $f = 100$ cm has an aperture in the form of a ring
$$p(x, y) = \begin{cases} 1, & a \leq \sqrt{x^2 + y^2} \leq b, \\ 0, & \text{otherwise,} \end{cases}$$
where $a = 5$ mm and $b = 6$ mm. Determine the transfer function $H(\nu_x, \nu_y)$ of the system and sketch its cross section $H(\nu_x, 0)$. Assume that the wavelength $\lambda = 1$ μm.
(b) If the image plane is now moved closer to the lens so that its distance from the lens becomes $d_2 = 25$ cm, with the distance between the object plane and the lens d_1 as in (a), use the ray-optics approximation to determine the impulse response function of the imaging system $h(x, y)$ and sketch $h(x, 0)$.

4.5-1 **Holography with a Spherical Reference Wave.** The choice of a uniform plane wave as a reference wave is not essential to holography; other waves can be used. Assuming that the reference wave is a spherical wave centered about the point $(0, 0, -d)$, determine the hologram pattern and examine the reconstructed wave when:
(a) the object wave is a plane wave traveling at an angle θ_x;
(b) the object wave is a spherical wave centered at $(-x_0, 0, -d_1)$.
Approximate spherical waves by paraboloidal waves.

4.5-2 **Optical Correlation via Holography.** A transparency with an amplitude transmittance given by $f(x, y) = f_1(x - a, y) + f_2(x + a, y)$ is Fourier transformed by a lens and the intensity is recorded on a transparency (hologram). The hologram is subsequently illuminated with a reference wave and the reconstructed wave is Fourier transformed with a lens to generate the function $g(x, y)$. Derive an expression relating $g(x, y)$ to $f_1(x, y)$ and $f_2(x, y)$. Show how the correlation of $f_1(x, y)$ and $f_2(x, y)$ may be determined with this system.

CHAPTER

5

ELECTROMAGNETIC OPTICS

5.1	ELECTROMAGNETIC THEORY OF LIGHT	162
5.2	ELECTROMAGNETIC WAVES IN DIELECTRIC MEDIA	166
	A. Linear, Nondispersive, Homogeneous, and Isotropic Media	
	B. Nonlinear, Dispersive, Inhomogeneous, or Anisotropic Media	
5.3	MONOCHROMATIC ELECTROMAGNETIC WAVES	172
5.4	ELEMENTARY ELECTROMAGNETIC WAVES	175
	A. Plane, Dipole, and Gaussian Electromagnetic Waves	
	B. Relation Between Electromagnetic Optics and Scalar Wave Optics	
	C. Vector Beams	
5.5	ABSORPTION AND DISPERSION	181
	A. Absorption	
	B. Dispersion	
	C. The Resonant Medium	
5.6	SCATTERING OF ELECTROMAGNETIC WAVES	192
	A. Born Approximation	
	B. Rayleigh Scattering	
	C. Mie Scattering	
	D. Attenuation in a Medium with Scatterers	
5.7	PULSE PROPAGATION IN DISPERSIVE MEDIA	199

James Clerk Maxwell (1831–1879) advanced the theory that light is an electromagnetic wave phenomenon. He formulated a set of fundamental equations of enormous importance that bear his name.

Lord Rayleigh (John William Strutt) (1842–1919) contributed extensively to many areas of optics, including blackbody radiation, image formation, and scattering. He was awarded the Nobel Prize in 1904.

Fundamentals of Photonics, Third Edition. Bahaa E. A. Saleh and Malvin Carl Teich.
©2019 John Wiley & Sons, Inc. Published 2019 by John Wiley & Sons, Inc.

It is apparent from the results presented in Chapters 2–4 that wave optics has a far greater reach than ray optics. Remarkably, both approaches provide similar results for many simple optical phenomena involving paraxial waves, such as the focusing of light by a lens and the behavior of light in graded-index media and periodic systems. But it is also clear that wave optics offers something that ray optics cannot: the ability to explain phenomena such as interference and diffraction, which involve phase, and therefore lie hopelessly beyond the reach of a simple construct like ray optics. In spite of its many successes, however, wave optics, like ray optics, is unable to quantitatively account for some simple observations in an optics experiment, such as the division of light at a beamsplitter. The fraction of light reflected (and transmitted) turns out to depend on the polarization of the incident light, which means that the light must be treated in the context of a vector, rather than a scalar, theory. That's where electromagnetic optics enters the picture. In common with radio waves and X-rays, as shown in Fig. 5.0-1, light is an electromagnetic phenomenon that is described by a *vector* wave theory. Electromagnetic radiation propagates in the form of two mutually coupled vector waves, an electric-field wave and a magnetic-field wave. From this perspective, the wave-optics approach set forth in Chapter 2, and developed in Chapters 3 and 4, is merely a *scalar* approximation to the more complete electromagnetic theory.

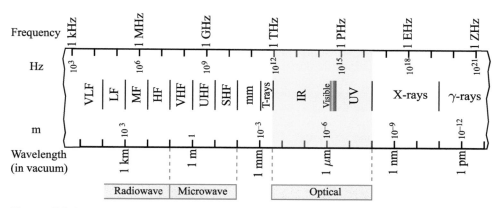

Figure 5.0-1 The electromagnetic spectrum from low frequencies (long wavelengths) to high frequencies (short wavelengths). The optical region, shown as shaded, is displayed in greater detail in Fig. 2.0-1.

Electromagnetic optics thus encompasses wave optics, which in turn reduces to ray optics in the limit of short wavelengths, as shown in Chapter 2. This hierarchy is displayed in Fig. 5.0-2.

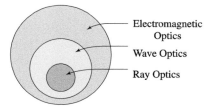

Figure 5.0-2 Electromagnetic optics is a *vector* theory comprising an electric field and a magnetic field that vary in time and space. Wave optics is an approximation to electromagnetic optics that relies on the wavefunction, a *scalar* function of time and space. Ray optics is the limit of wave optics when the wavelength is very short.

Optical frequencies occupy a band of the electromagnetic spectrum that extends from the infrared through the visible to the ultraviolet, as shown in Fig. 5.0-1. The range of wavelengths that is generally considered to lie in the optical domain extends from 10 nm to 300 μm (as is shown in greater detail in Fig. 2.0-1). Because these wavelengths are substantially shorter than those of radiowaves, or even microwaves, the techniques involved in their generation, transmission, and detection have traditionally been rather distinct. In recent years, however, the march toward miniaturization has served to blur these differences: it is now commonplace to encounter wavelength- and subwavelength-size resonators, antennas, waveguides, lasers, and other structures.

This Chapter

This chapter offers a brief review of those aspects of electromagnetic theory that are of paramount importance in optics. The fundamental theoretical construct — Maxwell's equations — is set forth in Sec. 5.1. The behavior of optical electromagnetic waves in dielectric media is examined in Sec. 5.2. Together, these sections lay out the fundamentals of electromagnetic optics and provide the set of laws that govern the remaining sections of the chapter. These rules simplify considerably for the special case of monochromatic light, as discussed in Sec. 5.3. Elementary electromagnetic waves (plane waves, dipole waves, and Gaussian beams), introduced in Sec. 5.4, provide important examples that are often encountered in practice. Section 5.5 is devoted to a study of the propagation of light in dispersive media, which exhibit wavelength-dependent absorption and refraction, as do real media. The scattering of electromagnetic waves, considered in Sec. 5.6, plays an important role in optics and plasmonics, as discussed in Chapter 8. Finally, in Sec. 5.7, we consider pulse propagation in dispersive media, which provides a basic underpinning for Chapters 10, 23, and 25.

Chapter 6, which is based on the theory of electromagnetic optics presented in this chapter, deals explicitly with the polarization of light and the interaction of polarized light with dielectric and anisotropic media such as liquid crystals. The material set forth here also forms the basis for the expositions provided in Chapters 7–11, which deal, respectively, with the optics of layered and periodic media, metals and metamaterials, guided waves, fibers, and resonators. Chapters 12 and 22, devoted to statistical optics and nonlinear optics, respectively, are also based on electromagnetic optics.

5.1 ELECTROMAGNETIC THEORY OF LIGHT

An electromagnetic field is described by two related *vector* fields that are functions of position and time: the **electric field** $\mathcal{E}(\mathbf{r}, t)$ and the **magnetic field** $\mathcal{H}(\mathbf{r}, t)$. In general, therefore, six scalar functions of position and time are required to describe light in free space. Fortunately, these six functions are interrelated since they must satisfy the celebrated set of coupled partial differential equations known as Maxwell's equations.

Maxwell's Equations in Free Space

The electric- and magnetic-field vectors in free space satisfy **Maxwell's equations**:

$$\nabla \times \mathcal{H} = \epsilon_o \frac{\partial \mathcal{E}}{\partial t} \qquad (5.1\text{-}1)$$

$$\nabla \times \mathcal{E} = -\mu_o \frac{\partial \mathcal{H}}{\partial t} \qquad (5.1\text{-}2)$$

$$\nabla \cdot \mathcal{E} = 0 \qquad (5.1\text{-}3)$$

$$\nabla \cdot \mathcal{H} = 0, \qquad (5.1\text{-}4)$$

Maxwell's Equations
(Free Space)

where the constants $\epsilon_o \approx (1/36\pi) \times 10^{-9}$ F/m and $\mu_o = 4\pi \times 10^{-7}$ H/m (MKS units) are, respectively, the **electric permittivity** and the **magnetic permeability** of free space. The vector operators $\nabla\cdot$ and $\nabla\times$ represent the divergence and curl, respectively.[†]

The Wave Equation

A necessary condition for $\mathcal{E}$ and $\mathcal{H}$ to satisfy Maxwell's equations is that each of their components satisfy the wave equation

$$\nabla^2 u - \frac{1}{c_o^2}\frac{\partial^2 u}{\partial t^2} = 0.$$

(5.1-5)
Wave Equation
(Free Space)

Here

$$c_o = \frac{1}{\sqrt{\epsilon_o \mu_o}} \quad \approx 3 \times 10^8 \text{ m/s}$$

(5.1-6)
Speed of Light
(Free Space)

is the speed of light in vacuum, and the scalar function $u(\mathbf{r}, t)$ represents any of the three components $(\mathcal{E}_x, \mathcal{E}_y, \mathcal{E}_z)$ of $\mathcal{E}$ or the three components $(\mathcal{H}_x, \mathcal{H}_y, \mathcal{H}_z)$ of $\mathcal{H}$.

The wave equation may be derived from Maxwell's equations by applying the curl operation $\nabla\times$ to (5.1-2), making use of the vector identity $\nabla \times (\nabla \times \mathcal{E}) = \nabla(\nabla \cdot \mathcal{E}) - \nabla^2 \mathcal{E}$, and then using (5.1-1) and (5.1-3) to show that each component of $\mathcal{E}$ satisfies the wave equation. A similar procedure is followed for $\mathcal{H}$. Since Maxwell's equations and the wave equation are linear, the principle of superposition applies: if two sets of electric and magnetic fields are solutions to these equations separately, their sum is also a solution.

The connection between electromagnetic optics and wave optics is now evident. The wave equation (2.1-2), which is the basis of wave optics, is embedded in the structure of electromagnetic theory; the speed of light is related to the electromagnetic constants ϵ_o and μ_o by (5.1-6); and the scalar wavefunction $u(\mathbf{r}, t)$ in Chapter 2 represents any of the six components of the electric- and magnetic-field vectors. Electromagnetic optics reduces to wave optics in problems for which the vector nature of the electromagnetic fields is not of essence. As we shall see in this and the following chapters, the vector character of light underlies polarization phenomena and governs the amount of light reflected or transmitted through boundaries between different media, and therefore determines the characteristics of light propagation in waveguides, layered media, and optical resonators.

Maxwell's Equations in a Medium

In a medium devoid of free electric charges and currents, two additional vector fields are required — the **electric flux density** (also called the **electric displacement**) $\mathcal{D}(\mathbf{r}, t)$ and the **magnetic flux density** $\mathcal{B}(\mathbf{r}, t)$. The four fields, $\mathcal{E}$, $\mathcal{H}$, $\mathcal{D}$, and $\mathcal{B}$, are related by Maxwell's equations in a source-free medium:

[†] In a Cartesian coordinate system $\nabla \cdot \mathcal{E} = \partial \mathcal{E}_x/\partial x + \partial \mathcal{E}_y/\partial y + \partial \mathcal{E}_z/\partial z$ whereas $\nabla \times \mathcal{E}$ is a vector with Cartesian components $(\partial \mathcal{E}_z/\partial y - \partial \mathcal{E}_y/\partial z)$, $(\partial \mathcal{E}_x/\partial z - \partial \mathcal{E}_z/\partial x)$, and $(\partial \mathcal{E}_y/\partial x - \partial \mathcal{E}_x/\partial y)$.

$$\boxed{\begin{aligned} \nabla \times \mathcal{H} &= \frac{\partial \mathcal{D}}{\partial t} \\ \nabla \times \mathcal{E} &= -\frac{\partial \mathcal{B}}{\partial t} \\ \nabla \cdot \mathcal{D} &= 0 \\ \nabla \cdot \mathcal{B} &= 0. \end{aligned}} \quad \begin{aligned} &(5.1\text{-}7) \\ &(5.1\text{-}8) \\ &(5.1\text{-}9) \\ &(5.1\text{-}10) \end{aligned}$$

Maxwell's Equations
(Source-Free Medium)

Conductive media such as metals have free electric charges, requiring the addition of an associated current density $\mathcal{J}$ to the right-hand side of (5.1-7), as discussed in Sec. 8.2A. Maxwell's original formulation in 1865 comprised 20 simultaneous equations with 20 variables; these were condensed into their present form by Oliver Heaviside in 1885.

The relationship between the electric flux density $\mathcal{D}$ and the electric field $\mathcal{E}$ depends on the electric properties of the medium, which are characterized by the **polarization density** $\mathcal{P}$. In a dielectric medium, the polarization density is the macroscopic sum of the electric dipole moments induced by the electric field. Similarly, the relation between the magnetic flux density $\mathcal{B}$ and the magnetic field $\mathcal{H}$ depends on the magnetic properties of the medium, embodied in the **magnetization density** $\mathcal{M}$, which is defined analogously to the polarization density. The equations relating the flux densities and the fields are

$$\mathcal{D} = \epsilon_o \mathcal{E} + \mathcal{P} \qquad (5.1\text{-}11)$$
$$\mathcal{B} = \mu_o \mathcal{H} + \mu_o \mathcal{M}. \qquad (5.1\text{-}12)$$

The vector fields $\mathcal{P}$ and $\mathcal{M}$ are in turn related to the externally applied electric and magnetic fields $\mathcal{E}$ and $\mathcal{H}$ by relationships that depend on the electric and magnetic character of the medium, respectively, as will be described in Sec. 5.2. Equations relating $\mathcal{P}$ and $\mathcal{E}$, as well as $\mathcal{M}$ and $\mathcal{H}$, are established once the medium is specified. When these latter equations are substituted into Maxwell's equations in a source-free medium, the flux densities disappear.

In free space, $\mathcal{P} = \mathcal{M} = 0$, so that $\mathcal{D} = \epsilon_o \mathcal{E}$ and $\mathcal{B} = \mu_o \mathcal{H}$ whereupon (5.1-7)–(5.1-10) reduce to the free-space Maxwell's equations, (5.1-1)–(5.1-4).

Boundary Conditions

In a homogeneous medium, all components of the fields $\mathcal{E}$, $\mathcal{H}$, $\mathcal{D}$, and $\mathcal{B}$ are continuous functions of position. At the boundary between two dielectric media, in the absence of free electric charges and currents, the tangential components of the electric and magnetic fields $\mathcal{E}$ and $\mathcal{H}$, and the normal components of the electric and magnetic flux densities $\mathcal{D}$ and $\mathcal{B}$, must be continuous (Fig. 5.1-1).

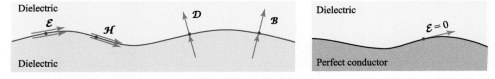

Figure 5.1-1 Boundary conditions at: (*a*) the interface between two dielectric media; (*b*) the interface between a perfect conductor and a dielectric material.

At the boundary between a dielectric medium and a perfectly conductive medium, the tangential components of the electric-field vector must vanish. Since a perfect mirror is made of a perfectly conductive material (a metal), the component of the electric field parallel to the surface of the mirror must be zero. This requires that at normal incidence the electric fields of the reflected and incident waves must have equal magnitudes and a phase shift of π so that their sum adds up to zero.

These boundary conditions are an integral part of Maxwell's equations. They are used to determine the reflectance and transmittance of waves at various boundaries (see Sec. 6.2), and the propagation of waves in periodic dielectric structures (see Sec. 7.1) and waveguides (see Sec. 9.2).

Intensity, Power, and Energy

The flow of electromagnetic power is governed by the vector

$$\mathcal{S} = \mathcal{E} \times \mathcal{H}, \qquad (5.1\text{-}13)$$

which is known as the **Poynting vector**. The direction of power flow is along the direction of the Poynting vector, i.e., orthogonal to both $\mathcal{E}$ and $\mathcal{H}$. The **optical intensity** $I(\mathbf{r}, t)$ (power flow across a unit area normal to the vector $\mathcal{S}$)[†] is the magnitude of the time-averaged Poynting vector $\langle \mathcal{S} \rangle$. The average is taken over times that are long in comparison with an optical cycle, but short compared to other times of interest. The wave-optics equivalent is given in (2.1-3).

Using the vector identity $\nabla \cdot (\mathcal{E} \times \mathcal{H}) = (\nabla \times \mathcal{E}) \cdot \mathcal{H} - (\nabla \times \mathcal{H}) \cdot \mathcal{E}$, together with Maxwell's equations (5.1-7)–(5.1-8) and (5.1-11)–(5.1-12), we obtain

$$\nabla \cdot \mathcal{S} = -\frac{\partial}{\partial t}\left(\frac{1}{2}\epsilon_o \mathcal{E}^2 + \frac{1}{2}\mu_o \mathcal{H}^2\right) - \mathcal{E} \cdot \frac{\partial \mathcal{P}}{\partial t} - \mu_o \mathcal{H} \cdot \frac{\partial \mathcal{M}}{\partial t}. \qquad (5.1\text{-}14)$$

The first and second terms in parentheses in (5.1-14) represent the energy densities (per unit volume) stored in the electric and magnetic fields, respectively. The third and fourth terms represent the power densities associated with the material's electric and magnetic dipoles. Equation (5.1-14), known as the **Poynting theorem**, therefore represents conservation of energy: the power flow escaping from the surface of an incremental volume equals the time rate of change of the energy stored inside the volume.

Momentum

An electromagnetic wave carries linear momentum, which results in radiation pressure on objects from which the wave reflects or scatters. In free space, the linear momentum density (per unit volume) is a vector

$$\epsilon_o \mathcal{E} \times \mathcal{B} = \frac{1}{c^2}\mathcal{S} \qquad (5.1\text{-}15)$$
Linear Momentum Density

proportional to the Poynting vector $\mathcal{S}$. The average momentum in a cylinder of length c and unit area is $(\langle \mathcal{S} \rangle / c^2) \cdot c = \langle \mathcal{S} \rangle / c$. This momentum crosses the unit area in a unit time, so that the average rate (per unit time) of momentum flow across a unit area oriented perpendicular to the direction of $\mathcal{S}$ is $\langle \mathcal{S} \rangle / c$.

An electromagnetic wave may also carry angular momentum and may therefore exert torque on an object. The average rate of angular momentum transported by an electromagnetic field is $\mathbf{r} \times \langle \mathcal{S} \rangle / c$. For example, the Laguerre–Gaussian beams introduced in Sec. 3.4 have helical wavefronts; the Poynting vector has an azimuthal component that leads to an orbital angular momentum.

[†] For a discussion of this interpretation, see M. Born and E. Wolf, *Principles of Optics*, Cambridge University Press, 7th expanded and corrected ed. 2002, pp. 7–10.

5.2 ELECTROMAGNETIC WAVES IN DIELECTRIC MEDIA

The character of the medium is embodied in the relation between the polarization and magnetization densities, $\mathcal{P}$ and $\mathcal{M}$, on the one hand, and the electric and magnetic fields, $\mathcal{E}$ and $\mathcal{H}$, on the other; these are known as the **constitutive relation**. In most media, the constitutive relation separates into a pair of constitutive relations, one between $\mathcal{P}$ and $\mathcal{E}$, and another between $\mathcal{M}$ and $\mathcal{H}$. The former describes the dielectric properties of the medium, whereas the latter describes its magnetic properties. With the notable exceptions of magnetic materials, optically active materials, and metamaterials, the principal emphasis in this book is on the dielectric properties. We therefore direct our attention to the $\mathcal{P}$-$\mathcal{E}$ relations for various dielectric media; the $\mathcal{M}$-$\mathcal{H}$ relations for magnetic media obey similar relations under similar conditions.

It is useful to regard the $\mathcal{P}$-$\mathcal{E}$ constitutive relation as arising from a system in which $\mathcal{E}$ is the input and $\mathcal{P}$ is the output or response (Fig. 5.2-1). Note that $\mathcal{E} = \mathcal{E}(\mathbf{r}, t)$ and $\mathcal{P} = \mathcal{P}(\mathbf{r}, t)$ are functions of both position and time.

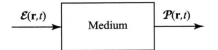

Figure 5.2-1 In response to an applied electric field $\mathcal{E}$, the dielectric medium creates a polarization density $\mathcal{P}$.

Definitions

- A dielectric medium is said to be *linear* if the vector field $\mathcal{P}(\mathbf{r}, t)$ is linearly related to the vector field $\mathcal{E}(\mathbf{r}, t)$. The principle of superposition then applies.
- The medium is said to be *nondispersive* if its response is instantaneous, i.e., if $\mathcal{P}$ at time t is determined by $\mathcal{E}$ at the same time t and not by prior values of $\mathcal{E}$. Nondispersiveness is clearly an idealization since all physical systems, no matter how rapidly they may respond, do have a response time that is finite.
- The medium is said to be *homogeneous* if the relation between $\mathcal{P}$ and $\mathcal{E}$ is independent of the position $\mathbf{r}$.
- The medium is said to be *isotropic* if the relation between the vectors $\mathcal{P}$ and $\mathcal{E}$ is independent of the direction of the vector $\mathcal{E}$, so that the medium exhibits the same behavior from all directions. The vectors $\mathcal{P}$ and $\mathcal{E}$ must then be parallel.
- The medium is said to be *spatially nondispersive* if the relation between $\mathcal{P}$ and $\mathcal{E}$ is local, i.e., if $\mathcal{P}$ at each position $\mathbf{r}$ is influenced only by $\mathcal{E}$ at the same position $\mathbf{r}$. The medium is assumed to be spatially nondispersive throughout this chapter (optically active media, considered in Sec. 6.4A, are spatially dispersive).

A. Linear, Nondispersive, Homogeneous, and Isotropic Media

Let us first consider the simplest case of *linear*, *nondispersive*, *homogeneous*, and *isotropic* dielectric media. The vectors $\mathcal{P}$ and $\mathcal{E}$ at every position and time are then parallel and proportional, so that

$$\mathcal{P} = \epsilon_o \chi \mathcal{E}, \qquad (5.2\text{-}1)$$

where the scalar constant χ is called the **electric susceptibility** (Fig. 5.2-2).

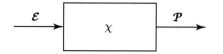

Figure 5.2-2 A linear, nondispersive, homogeneous, and isotropic medium is fully characterized by a single constant, the electric susceptibility χ.

Substituting (5.2-1) in (5.1-11) shows that $\mathcal{D}$ and $\mathcal{E}$ are also parallel and proportional,

$$\mathcal{D} = \epsilon \mathcal{E}, \qquad (5.2\text{-}2)$$

where the scalar quantity

$$\epsilon = \epsilon_o(1+\chi) \qquad (5.2\text{-}3)$$

is defined as the **electric permittivity** of the medium. The **relative permittivity** $\epsilon/\epsilon_o = 1 + \chi$ is also called the **dielectric constant** of the medium.

Under similar conditions, the magnetic relation can be written in the form

$$\mathcal{B} = \mu \mathcal{H}, \qquad (5.2\text{-}4)$$

where μ is the **magnetic permeability** of the medium.

With the relations (5.2-2) and (5.2-4), Maxwell's equations in (5.1-7)–(5.1-10) relate only the two vector fields $\mathcal{E}(\mathbf{r}, t)$ and $\mathcal{H}(\mathbf{r}, t)$, simplifying to

$$\nabla \times \mathcal{H} = \epsilon \frac{\partial \mathcal{E}}{\partial t} \qquad (5.2\text{-}5)$$

$$\nabla \times \mathcal{E} = -\mu \frac{\partial \mathcal{H}}{\partial t} \qquad (5.2\text{-}6)$$

$$\nabla \cdot \mathcal{E} = 0 \qquad (5.2\text{-}7)$$

$$\nabla \cdot \mathcal{H} = 0. \qquad (5.2\text{-}8)$$

Maxwell's Equations
(Linear, Nondispersive, Homogeneous,
Isotropic, Source-Free Medium)

It is apparent that (5.2-5)–(5.2-8) are identical in form to the free-space Maxwell's equations in (5.1-1)–(5.1-4) except that ϵ replaces ϵ_o and μ replaces μ_o. Each component of $\mathcal{E}$ and $\mathcal{H}$ therefore satisfies the wave equation

$$\nabla^2 u - \frac{1}{c^2} \frac{\partial^2 u}{\partial t^2} = 0, \qquad (5.2\text{-}9)$$

**Wave Equation
(in a Medium)**

where the speed of light in the medium is denoted c:

$$c = \frac{1}{\sqrt{\epsilon \mu}}. \qquad (5.2\text{-}10)$$

**Speed of Light
(in a Medium)**

The ratio of the speed of light in free space to that in the medium, c_o/c, is defined as the *refractive index* n:

$$n = \frac{c_o}{c} = \sqrt{\frac{\epsilon}{\epsilon_o}\frac{\mu}{\mu_o}}, \qquad (5.2\text{-}11)$$

Refractive Index

where (5.1-6) provides

$$c_o = \frac{1}{\sqrt{\epsilon_o \mu_o}}. \qquad (5.2\text{-}12)$$

For a nonmagnetic material, $\mu = \mu_o$ and

$$n = \sqrt{\frac{\epsilon}{\epsilon_o}} = \sqrt{1+\chi}, \qquad (5.2\text{-}13)$$

Refractive Index (Nonmagnetic Media)

so that the refractive index is the square root of the relative permittivity. These relations provide another point of connection with scalar wave optics (Sec. 2.1), as discussed further in Sec. 5.4B.

Finally, the Poynting theorem (5.1-14) based on Maxwell's equations (5.2-5) and (5.2-6) takes the form of a continuity equation

$$\nabla \cdot \mathcal{S} = -\frac{\partial \mathcal{W}}{\partial t} \qquad (5.2\text{-}14)$$

where

$$\mathcal{W} = \tfrac{1}{2}\epsilon \mathcal{E}^2 + \tfrac{1}{2}\mu \mathcal{H}^2 \qquad (5.2\text{-}15)$$

is the energy density stored in the medium.

B. Nonlinear, Dispersive, Inhomogeneous, or Anisotropic Media

We now consider nonmagnetic dielectric media for which one or more of the properties of linearity, nondispersiveness, homogeneity, and isotropy are not satisfied.

Inhomogeneous Media

We first consider an inhomogeneous dielectric (such as a graded-index medium) that is linear, nondispersive, and isotropic. The simple proportionalities, $\mathcal{P} = \epsilon_o \chi \mathcal{E}$ and $\mathcal{D} = \epsilon \mathcal{E}$, remain intact, but the coefficients χ and ϵ become functions of position: $\chi = \chi(\mathbf{r})$ and $\epsilon = \epsilon(\mathbf{r})$ (Fig. 5.2-3). The refractive index therefore also becomes position dependent so that $n = n(\mathbf{r})$.

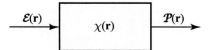

Figure 5.2-3 An inhomogeneous (but linear, nondispersive, and isotropic) medium is characterized by a position dependent susceptibility $\chi(\mathbf{r})$.

Beginning with Maxwell's equations, (5.1-7)–(5.1-10), and noting that $\epsilon = \epsilon(\mathbf{r})$ is position dependent, we apply the curl operation $\nabla \times$ to both sides of (5.1-8). We then use (5.1-7) to write

$$\frac{\epsilon_o}{\epsilon} \nabla \times (\nabla \times \mathcal{E}) = -\frac{1}{c_o^2} \frac{\partial^2 \mathcal{E}}{\partial t^2}. \qquad (5.2\text{-}16)$$

Wave Equation
(Inhomogeneous Medium)

The magnetic field satisfies a different equation:

$$\nabla \times \left(\frac{\epsilon_o}{\epsilon} \nabla \times \mathcal{H} \right) = -\frac{1}{c_o^2} \frac{\partial^2 \mathcal{H}}{\partial t^2}. \qquad (5.2\text{-}17)$$

Wave Equation
(Inhomogeneous Medium)

Equation (5.2-16) may also be written in the form

$$\nabla^2 \mathcal{E} + \nabla \left(\frac{1}{\epsilon} \nabla \epsilon \cdot \mathcal{E} \right) - \mu_o \epsilon \frac{\partial^2 \mathcal{E}}{\partial t^2} = 0. \qquad (5.2\text{-}18)$$

The validity of (5.2-18) can be demonstrated by employing the following procedure. Use the identity $\nabla \times (\nabla \times \mathcal{E}) = \nabla(\nabla \cdot \mathcal{E}) - \nabla^2 \mathcal{E}$, valid for a rectilinear coordinate system. Invoke (5.1-9), which yields $\nabla \cdot \epsilon \mathcal{E} = 0$, together with the identity $\nabla \cdot \epsilon \mathcal{E} = \epsilon \nabla \cdot \mathcal{E} + \nabla \epsilon \cdot \mathcal{E}$, which provides $\nabla \cdot \mathcal{E} = -(1/\epsilon)\nabla \epsilon \cdot \mathcal{E}$. Finally, substitute in (5.2-16) to obtain (5.2-18).

For media with gradually varying dielectric properties, i.e., when $\epsilon(\mathbf{r})$ varies sufficiently slowly so that it can be assumed constant within distances of the order of a wavelength, the second term on the left-hand side of (5.2-18) is negligible in comparison with the first, so that

$$\nabla^2 \mathcal{E} - \frac{1}{c^2(\mathbf{r})} \frac{\partial^2 \mathcal{E}}{\partial t^2} \approx 0, \qquad (5.2\text{-}19)$$

where $c(\mathbf{r}) = 1/\sqrt{\mu_o \epsilon} = c_o/n(\mathbf{r})$ is spatially varying and $n(\mathbf{r}) = \sqrt{\epsilon(\mathbf{r})/\epsilon_o}$ is the refractive index at position $\mathbf{r}$. This relation was invoked without proof in Sec. 2.1, but it is clearly an approximate consequence of Maxwell's equations.

For a homogeneous dielectric medium of refractive index n perturbed by a slowly varying spatially dependent change Δn, it is often useful to write (5.2-19) in the form

$$\nabla^2 \mathcal{E} - \frac{1}{c^2} \frac{\partial^2 \mathcal{E}}{\partial t^2} \approx -\mathcal{S}, \qquad \mathcal{S} = -\mu_o \frac{\partial^2 \Delta \mathcal{P}}{\partial t^2}, \qquad \Delta \mathcal{P} = 2\epsilon_o n \, \Delta n \, \mathcal{E}, \qquad (5.2\text{-}20)$$

where $c = c_o/n$ is the speed of light in the homogenous medium. Thus, $\mathcal{E}$ satisfies the wave equation with a radiation source $\mathcal{S}$ created by a perturbation of the polarization density $\Delta \mathcal{P}$, which in turn is proportional to Δn and $\mathcal{E}$ itself. These equations may be verified by expanding the term $1/c^2(\mathbf{r})$ in (5.2-19) as $(n+\Delta n)^2/c_o^2 \approx (n^2+2n\Delta n)/c_o^2$ and bringing the perturbation term to the right-hand side of the equation. The term $\Delta \mathcal{P}$ is the perturbation in $\mathcal{P}$, as can be shown by noting that $\mathcal{P} = \epsilon_o \chi \mathcal{E} = \epsilon_o(\epsilon/\epsilon_o - 1)\mathcal{E} = \epsilon_o(n^2 - 1)\mathcal{E}$, so that $\Delta \mathcal{P} = \epsilon_o \Delta(n^2 - 1)\mathcal{E} = 2\epsilon_o n \Delta n \mathcal{E}$.

Anisotropic Media

The relation between the vectors $\mathcal{P}$ and $\mathcal{E}$ in an anisotropic dielectric medium depends on the direction of the vector $\mathcal{E}$; the requirement that the two vectors remain parallel is not maintained. If the medium is linear, nondispersive, and homogeneous, each component of $\mathcal{P}$ is a linear combination of the three components of $\mathcal{E}$:

$$\mathcal{P}_i = \sum_j \epsilon_o \chi_{ij} \mathcal{E}_j, \tag{5.2-21}$$

where the indices $i, j = 1, 2, 3$ denote the x, y, and z components, respectively.

The dielectric properties of the medium are then described by a 3×3 array of constants $\{\chi_{ij}\}$, which are elements of what is called the **electric susceptibility tensor** χ (Fig. 5.2-4). A similar relation between $\mathcal{D}$ and $\mathcal{E}$ applies:

$$\mathcal{D}_i = \sum_j \epsilon_{ij} \mathcal{E}_j, \tag{5.2-22}$$

where $\{\epsilon_{ij}\}$ are the elements of the **electric permittivity tensor** ϵ.

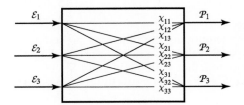

Figure 5.2-4 An anisotropic (but linear, homogeneous, and nondispersive) medium is characterized by nine constants, the components of the electric susceptibility tensor χ_{ij}. Each component of $\mathcal{P}$ is a weighted superposition of the three components of $\mathcal{E}$.

The optical properties of anisotropic media are examined in Chapter 6. The relation between $\mathcal{B}(t)$ and $\mathcal{H}(t)$ for anisotropic magnetic media takes a form similar to that of (5.2-22), under similar assumptions.

Dispersive Media

The relation between the vectors $\mathcal{P}$ and $\mathcal{E}$ in a dispersive dielectric medium is dynamic rather than instantaneous. The vector $\mathcal{E}(t)$ may be thought of as an input that induces the bound electrons in the atoms of the medium to oscillate, which then collectively give rise to the polarization-density vector $\mathcal{P}(t)$ as the output. The presence of a time delay between the output and the input indicates that the system possesses memory. Only when this time is short in comparison with other times of interest can the response be regarded as instantaneous, in which case the medium is approximately nondispersive.

For dispersive media that are linear, homogeneous, and isotropic, the dynamic relation between $\mathcal{P}(t)$ and $\mathcal{E}(t)$ may be described by a linear differential equation such as that associated with a driven harmonic oscillator: $a_1 \, d^2\mathcal{P}/dt^2 + a_2 \, d\mathcal{P}/dt + a_3 \, \mathcal{P} = \mathcal{E}$, where a_1, a_2, and a_3 are constants. A simple analysis along these lines (see Sec. 5.5C) provides a physical rationale for the presence of dispersion (and absorption).

More generally, the linear-systems approach provided in Appendix B may be used to investigate an arbitrary linear system, which is characterized by its response to an impulse (impulse response function). An electric-field impulse of magnitude $\delta(t)$ applied at time $t = 0$ induces a time-dispersed polarization density of magnitude $\epsilon_o \mathsf{x}(t)$, where $\mathsf{x}(t)$ is a scalar function of time with finite duration that begins at $t = 0$. Since the medium is linear, an arbitrary electric field $\mathcal{E}(t)$ then induces a polarization density

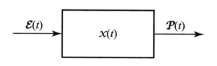

Figure 5.2-5 In a dispersive (but linear, homogeneous, and isotropic) medium the relation between $\mathcal{P}(t)$ and $\mathcal{E}(t)$ is governed by a dynamic linear system described by an impulse response function $\epsilon_o \mathrm{x}(t)$ that corresponds to a frequency dependent susceptibility $\chi(\nu)$.

that is a superposition of the effects of $\mathcal{E}(t')$ for all $t' \le t$, so that the polarization density can be expressed as a convolution, as defined in Appendix A:

$$\mathcal{P}(t) = \epsilon_o \int_{-\infty}^{\infty} \mathrm{x}(t-t')\,\mathcal{E}(t')\,dt'. \qquad (5.2\text{-}23)$$

This dielectric medium is completely described by its impulse response function $\epsilon_o \mathrm{x}(t)$.

Alternatively, a dynamic linear system may be described by its transfer function, which governs the response to harmonic inputs. The transfer function is the Fourier transform of the impulse response function (see Appendix B). In the example at hand, the transfer function at frequency ν is $\epsilon_o \chi(\nu)$, where $\chi(\nu)$ is the Fourier transform of $\mathrm{x}(t)$ so that it is a frequency-dependent susceptibility (Fig. 5.2-5). This concept is discussed further in Secs. 5.3 and 5.5.

For magnetic media under similar assumptions, the relation between $\mathcal{M}(t)$ and $\mathcal{H}(t)$ is analogous to (5.2-23).

Nonlinear Media

A nonlinear dielectric medium is defined as one in which the relation between $\mathcal{P}$ and $\mathcal{E}$ is nonlinear, in which case the wave equation as written in (5.2-9) is not applicable. Rather, Maxwell's equations can be used to derive a nonlinear wave equation that electromagnetic waves obey in a such a medium.

We first derive a general wave equation valid for homogeneous and isotropic nonmagnetic media. Operating on Maxwell's equation (5.1-8) with the curl operator $\nabla \times$, and using the relation $\mathcal{B} = \mu_o \mathcal{H}$ from (5.2-4) together with (5.1-7), we obtain $\nabla \times (\nabla \times \mathcal{E}) = -\mu_o \partial^2 \mathcal{D}/\partial t^2$. Making use of the vector identity $\nabla \times (\nabla \times \mathcal{E}) = \nabla(\nabla \cdot \mathcal{E}) - \nabla^2 \mathcal{E}$ and the relation $\mathcal{D} = \epsilon_o \mathcal{E} + \mathcal{P}$ from (5.1-11) then yields

$$\nabla(\nabla \cdot \mathcal{E}) - \nabla^2 \mathcal{E} = -\epsilon_o \mu_o \frac{\partial^2 \mathcal{E}}{\partial t^2} - \mu_o \frac{\partial^2 \mathcal{P}}{\partial t^2}. \qquad (5.2\text{-}24)$$

For homogeneous and isotropic media $\mathcal{D} = \epsilon \mathcal{E}$; thus $\nabla \cdot \mathcal{D} = 0$ from (5.1-9) is equivalent to $\nabla \cdot \mathcal{E} = 0$. Substituting this, along with $\epsilon_o \mu_o = 1/c_o^2$ from (5.1-6), into (5.2-24) therefore provides

$$\boxed{\nabla^2 \mathcal{E} - \frac{1}{c_o^2}\frac{\partial^2 \mathcal{E}}{\partial t^2} = \mu_o \frac{\partial^2 \mathcal{P}}{\partial t^2}.} \qquad (5.2\text{-}25)$$
Wave Equation
(Homogeneous and Isotropic Medium)

Equation (5.2-25) is applicable for all homogeneous and isotropic dielectric media: nonlinear or linear, nondispersive or dispersive.

Now, if the medium is nonlinear, nondispersive, and nonmagnetic, the polarization density $\mathcal{P}$ can be written as a memoryless nonlinear function of $\mathcal{E}$, say $\mathcal{P} = \Psi(\mathcal{E})$, valid at every position and time. (The simplest example of such a function is $\mathcal{P} =$

$a_1 \mathcal{E} + a_2 \mathcal{E}^2$, where a_1 and a_2 are constants.) Under these conditions (5.2-25) becomes a nonlinear partial differential equation for the electric-field vector $\mathcal{E}(\mathbf{r}, t)$:

$$\nabla^2 \mathcal{E} - \frac{1}{c_o^2} \frac{\partial^2 \mathcal{E}}{\partial t^2} = \mu_o \frac{\partial^2 \Psi(\mathcal{E})}{\partial t^2}. \tag{5.2-26}$$

The principle of superposition is no longer applicable by virtue of the nonlinear nature of this wave equation. Nonlinear magnetic materials may be similarly described.

Most dielectric media are approximately linear unless the optical intensity is substantial, as in the case of focused laser beams. Nonlinear optics is discussed in Chapter 22.

5.3 MONOCHROMATIC ELECTROMAGNETIC WAVES

For the special case of *monochromatic* electromagnetic waves in an optical medium, all components of the electric and magnetic fields are harmonic functions of time with the same frequency ν and corresponding angular frequency $\omega = 2\pi\nu$. Adopting the complex representation used in Sec. 2.2A, these six real field components may be expressed as

$$\mathcal{E}(\mathbf{r}, t) = \text{Re}\{\mathbf{E}(\mathbf{r}) \exp(j\omega t)\}$$
$$\mathcal{H}(\mathbf{r}, t) = \text{Re}\{\mathbf{H}(\mathbf{r}) \exp(j\omega t)\}, \tag{5.3-1}$$

where $\mathbf{E}(\mathbf{r})$ and $\mathbf{H}(\mathbf{r})$ represent electric- and magnetic-field complex-amplitude vectors, respectively. Analogous complex-amplitude vectors $\mathbf{P}, \mathbf{D}, \mathbf{M},$ and $\mathbf{B}$ are similarly associated with the real vectors $\mathcal{P}, \mathcal{D}, \mathcal{M},$ and $\mathcal{B}$, respectively.

Maxwell's Equations in a Medium

Inserting (5.3-1) into Maxwell's equations (5.1-7)–(5.1-10), and using the relation $(\partial/\partial t)\, e^{j\omega t} = j\omega\, e^{j\omega t}$ for monochromatic waves of angular frequency ω, yields a set of equations obeyed by the field complex-amplitude vectors:

$$\nabla \times \mathbf{H} = j\omega \mathbf{D} \tag{5.3-2}$$
$$\nabla \times \mathbf{E} = -j\omega \mathbf{B} \tag{5.3-3}$$
$$\nabla \cdot \mathbf{D} = 0 \tag{5.3-4}$$
$$\nabla \cdot \mathbf{B} = 0. \tag{5.3-5}$$

Maxwell's Equations
(Source-Free Medium;
Monochromatic Fields)

Similarly, (5.1-11) and (5.1-12) give rise to

$$\mathbf{D} = \epsilon_o \mathbf{E} + \mathbf{P} \tag{5.3-6}$$
$$\mathbf{B} = \mu_o \mathbf{H} + \mu_o \mathbf{M}. \tag{5.3-7}$$

Intensity and Power

As indicated in Sec. 5.1, the flow of electromagnetic power is governed by the time average of the Poynting vector $\mathbf{S} = \mathcal{E} \times \mathcal{H}$. Casting this expression in terms of complex amplitudes yields

$$\mathcal{S} = \text{Re}\left\{\mathbf{E}e^{j\omega t}\right\} \times \text{Re}\left\{\mathbf{H}e^{j\omega t}\right\} = \tfrac{1}{2}\left(\mathbf{E}e^{j\omega t} + \mathbf{E}^*e^{-j\omega t}\right) \times \tfrac{1}{2}\left(\mathbf{H}e^{j\omega t} + \mathbf{H}^*e^{-j\omega t}\right)$$
$$= \tfrac{1}{4}\left(\mathbf{E} \times \mathbf{H}^* + \mathbf{E}^* \times \mathbf{H} + e^{j2\omega t}\mathbf{E} \times \mathbf{H} + e^{-j2\omega t}\mathbf{E}^* \times \mathbf{H}^*\right). \qquad (5.3\text{-}8)$$

The terms containing the factors $e^{j2\omega t}$ and $e^{-j2\omega t}$ oscillate at optical frequencies and are therefore washed out by the averaging process, which is slow in comparison with an optical cycle. Thus,

$$\langle \mathcal{S} \rangle = \tfrac{1}{4}(\mathbf{E} \times \mathbf{H}^* + \mathbf{E}^* \times \mathbf{H}) = \tfrac{1}{2}(\mathbf{S} + \mathbf{S}^*) = \text{Re}\{\mathbf{S}\}, \qquad (5.3\text{-}9)$$

where the vector

$$\boxed{\mathbf{S} = \tfrac{1}{2}\mathbf{E} \times \mathbf{H}^*} \qquad (5.3\text{-}10)$$
Complex Poynting Vector

may be regarded as a complex Poynting vector. The optical intensity is the magnitude of the vector $\text{Re}\{\mathbf{S}\}$.

Linear, Nondispersive, Homogeneous, and Isotropic Media

For monochromatic waves, the relations provided in (5.2-2) and (5.2-4) become the **material equations**

$$\mathbf{D} = \epsilon \mathbf{E} \quad \text{and} \quad \mathbf{B} = \mu \mathbf{H}, \qquad (5.3\text{-}11)$$

so that Maxwell's equations, (5.3-2)–(5.3-5), depend solely on the interrelated complex-amplitude vectors $\mathbf{E}$ and $\mathbf{H}$:

$$\boxed{\begin{aligned} \nabla \times \mathbf{H} &= j\omega\epsilon\mathbf{E} \\ \nabla \times \mathbf{E} &= -j\omega\mu\mathbf{H} \\ \nabla \cdot \mathbf{E} &= 0 \\ \nabla \cdot \mathbf{H} &= 0. \end{aligned}} \qquad \begin{aligned}(5.3\text{-}12)\\(5.3\text{-}13)\\(5.3\text{-}14)\\(5.3\text{-}15)\end{aligned}$$
Maxwell's Equations
(Linear, Nondispersive, Homogeneous,
Isotropic, Source-Free Medium;
Monochromatic Light)

Substituting the electric and magnetic fields $\mathcal{E}$ and $\mathcal{H}$ given in (5.3-1) into the wave equation (5.2-9) yields the Helmholtz equation

$$\boxed{\nabla^2 U + k^2 U = 0,} \quad k = nk_o = \omega\sqrt{\epsilon\mu} \qquad (5.3\text{-}16)$$
Helmholtz Equation

where the scalar function $U = U(\mathbf{r})$ represents the complex amplitude of any of the three components (E_x, E_y, E_z) of $\mathbf{E}$ or three components (H_x, H_y, H_z) of $\mathbf{H}$; and where $n = \sqrt{(\epsilon/\epsilon_o)(\mu/\mu_o)}$, $k_o = \omega/c_o$, and $c = c_o/n$. In the context of wave optics, the Helmholtz equation in (2.2-7) was written in terms of the complex amplitude $U(\mathbf{r})$ of the real wavefunction $u(\mathbf{r}, t)$.

Inhomogeneous Media

In an inhomogeneous nonmagnetic medium, Maxwell's equations (5.3-12)–(5.3-15) remain applicable, but the electric permittivity of the medium becomes position dependent, $\epsilon = \epsilon(\mathbf{r})$. For locally homogeneous media in which $\epsilon(\mathbf{r})$ varies slowly with respect to the wavelength, the Helmholtz equation (5.3-16) remains approximately valid, subject to the substitutions $k = n(\mathbf{r})k_o$ and $n(\mathbf{r}) = \sqrt{\epsilon(\mathbf{r})/\epsilon_o}$.

Dispersive Media

In a dispersive dielectric medium, $\mathcal{P}(t)$ and $\mathcal{E}(t)$ are connected by the dynamic relation provided in (5.2-23). To determine the corresponding relation between the complex-amplitude vectors $\mathbf{P}$ and $\mathbf{E}$, we substitute (5.3-1) into (5.2-23), which gives rise to

$$\mathbf{P} = \epsilon_o \chi(\nu) \mathbf{E} \qquad (5.3\text{-}17)$$

where

$$\chi(\nu) = \int_{-\infty}^{\infty} \chi(t) \exp(-j2\pi\nu t)\, dt \qquad (5.3\text{-}18)$$

is the Fourier transform of $\chi(t)$.

Equation (5.3-17) can also be directly inferred from (5.2-23) by invoking the convolution theorem: convolution in the time domain corresponds to multiplication in the frequency domain (see Secs. A.1 and B.1 of Appendices A and B, respectively), and recognizing $\mathbf{E}$ and $\mathbf{P}$ as the components of $\mathcal{E}$ and $\mathcal{P}$ of frequency ν. The function $\epsilon_o \chi(\nu)$ may therefore be regarded as the transfer function of the linear system that relates $\mathcal{P}(t)$ to $\mathcal{E}(t)$.

The relation between $\mathcal{D}$ and $\mathcal{E}$ is similar,

$$\mathbf{D} = \epsilon(\nu) \mathbf{E} \qquad (5.3\text{-}19)$$

where

$$\epsilon(\nu) = \epsilon_o[1 + \chi(\nu)]. \qquad (5.3\text{-}20)$$

Therefore, in dispersive media the susceptibility χ and the permittivity ϵ are frequency-dependent and, in general, complex-valued quantities. The Helmholtz equation (5.3-16) is thus readily adapted for use in dispersive nonmagnetic media by taking

$$k = \omega\sqrt{\epsilon(\nu)\mu_o}. \qquad (5.3\text{-}21)$$

When $\chi(\nu)$ and $\epsilon(\nu)$ are approximately constant within the frequency band of interest, the medium may be treated as approximately nondispersive. The implications of the complex-valued nature of χ and k in dispersive media are discussed further in Sec. 5.5.

5.4 ELEMENTARY ELECTROMAGNETIC WAVES

A. Plane, Dipole, and Gaussian Electromagnetic Waves

We now examine three elementary solutions to Maxwell's equations that are of substantial importance in optics: plane waves and spherical (dipole) waves, which were discussed in Sec. 2.2B in the context of wave optics, and the Gaussian beam, which was studied in Chapter 3 using the wave-optics formalism. The medium is assumed to be linear, homogeneous, nondispersive, and isotropic, and the waves are assumed to be monochromatic.

The Transverse Electromagnetic (TEM) Plane Wave

Consider a monochromatic electromagnetic wave whose magnetic- and electric-field complex-amplitude vectors are plane waves with wavevector $\mathbf{k}$ (see Sec. 2.2B) so that

$$\mathbf{H}(\mathbf{r}) = \mathbf{H}_0 \exp(-j\mathbf{k} \cdot \mathbf{r}) \tag{5.4-1}$$

$$\mathbf{E}(\mathbf{r}) = \mathbf{E}_0 \exp(-j\mathbf{k} \cdot \mathbf{r}), \tag{5.4-2}$$

where the complex envelopes $\mathbf{H}_0$ and $\mathbf{E}_0$ are constant vectors. All components of $\mathbf{H}(\mathbf{r})$ and $\mathbf{E}(\mathbf{r})$ satisfy the Helmholtz equation provided that the magnitude of $\mathbf{k}$ is $k = nk_o$, where n is the refractive index of the medium.

We now examine the conditions that must be obeyed by $\mathbf{H}_0$ and $\mathbf{E}_0$ in order that Maxwell's equations be satisfied. Substituting (5.4-1) and (5.4-2) into Maxwell's equations (5.3-12) and (5.3-13), respectively, leads to

$$\mathbf{k} \times \mathbf{H}_0 = -\omega \epsilon \mathbf{E}_0 \tag{5.4-3}$$

$$\mathbf{k} \times \mathbf{E}_0 = \omega \mu \mathbf{H}_0. \tag{5.4-4}$$

The other two Maxwell's equations, (5.3-14) and (5.3-15), are satisfied identically since the divergence of a uniform plane wave is zero.

It follows from (5.4-3) that $\mathbf{E}$ must be perpendicular to both $\mathbf{k}$ and $\mathbf{H}$ and from (5.4-4) that $\mathbf{H}$ must be perpendicular to both $\mathbf{k}$ and $\mathbf{E}$. Thus, $\mathbf{E}$, $\mathbf{H}$, and $\mathbf{k}$ are mutually orthogonal, as illustrated in Fig. 5.4-1. Since $\mathbf{E}$ and $\mathbf{H}$ lie in a plane normal to the direction of propagation $\mathbf{k}$, the wave is called a **transverse electromagnetic (TEM)** wave.

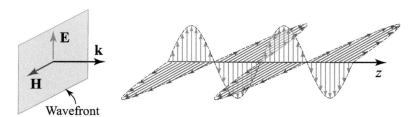

Figure 5.4-1 The TEM plane wave. The vectors $\mathbf{E}$, $\mathbf{H}$, and $\mathbf{k}$ are mutually orthogonal. The wavefronts (surfaces of constant phase) are normal to the wavevector $\mathbf{k}$.

In accordance with (5.4-3), the magnitudes H_0 and E_0 are related by $H_0 = (\omega\epsilon/k)E_0$. Similarly, (5.4-4) yields $H_0 = (k/\omega\mu)E_0$. For these two equations to be consistent, we must have $\omega\epsilon/k = k/\omega\mu$, or $k = \omega\sqrt{\epsilon\mu} = \omega/c = n\omega/c_o = nk_o$.

This is, in fact, the same condition required in order that the wave satisfy the Helmholtz equation.

The ratio between the amplitudes of the electric and magnetic fields is $E_0/H_0 = \omega\mu/k = c\mu = \sqrt{\mu/\epsilon}$. This quantity is known as the **impedance** of the medium,

$$\eta = \frac{E_0}{H_0} = \sqrt{\frac{\mu}{\epsilon}}. \tag{5.4-5}$$
Impedance

For nonmagnetic media $\mu = \mu_o$, whereupon $\eta = \sqrt{\mu_o/\epsilon}$ may be defined in terms of the impedance of free space η_o via

$$\eta = \frac{\eta_o}{n}, \tag{5.4-6}$$
Impedance
(Nonmagnetic Media)

where

$$\eta_o = \sqrt{\frac{\mu_o}{\epsilon_o}} \approx 120\pi \approx 377\,\Omega. \tag{5.4-7}$$

The complex Poynting vector $\mathbf{S} = \frac{1}{2}\mathbf{E} \times \mathbf{H}^*$ [see (5.3-10)] is parallel to the wavevector $\mathbf{k}$, so that the power flows along a direction normal to the wavefronts. Its magnitude is $\frac{1}{2}E_0 H_0^* = |E_0|^2/2\eta$, and the intensity I is therefore given by

$$I = \frac{|E_0|^2}{2\eta}. \tag{5.4-8}$$
Intensity

The intensity of a TEM wave is thus seen to be proportional to the absolute square of the complex envelope of the electric field. As an example, an intensity of 10 W/cm^2 in free space corresponds to an electric field of ≈ 87 V/cm. Note the similarity between (5.4-8) and the relation $I = |U|^2$, which was defined for scalar waves in Sec. 2.2A.

Equation (5.2-15) provides that the time-averaged energy density $W = \langle \mathcal{W} \rangle$ of the plane wave is

$$W = \tfrac{1}{2}\epsilon|E_0|^2, \tag{5.4-9}$$

since the electric and magnetic contributions are equal, i.e., $\frac{1}{2}\epsilon|E_0|^2/2 = \frac{1}{2}\mu|H_0|^2/2$. The intensity in (5.4-8) and the time-averaged energy density in (5.4-9) are therefore related by

$$I = cW, \tag{5.4-10}$$

indicating that the time-averaged power density flow I results from the transport of the time-averaged energy density at the velocity of light c. This is readily visualized by considering a cylinder of area A and length c whose axis lies parallel to the direction of propagation. The energy stored in the cylinder, cAW, is transported across the area in one second, confirming that the intensity (power per unit area) is $I = cW$.

The linear momentum density (per unit volume) transported by a plane wave is $(1/c^2)\mathbf{S} = (1/c^2)I\,\widehat{k} = (W/c)\,\widehat{k}$.

The Dipole Wave

An oscillating **electric dipole** radiates a wave with features that resemble the scalar spherical wave discussed in Sec. 2.2B. The radiation frequency is determined by the frequency at which the dipole oscillates. This electromagnetic wave is readily constructed from an auxiliary vector field $\mathbf{A}(\mathbf{r})$, known as the **vector potential**, which is often used to facilitate the solution of Maxwell's equations in electromagnetics. For the case at hand we set

$$\mathbf{A}(\mathbf{r}) = a_0 \, U(\mathbf{r}) \, \hat{\mathbf{x}}, \tag{5.4-11}$$

where a_0 is a constant and $\hat{\mathbf{x}}$ is a unit vector in the direction of the dipole (the x direction). The quantity $U(\mathbf{r})$ represents a scalar spherical wave with the origin at $r = 0$:

$$U(\mathbf{r}) = \frac{1}{4\pi r} \exp(-jkr). \tag{5.4-12}$$

Because $U(\mathbf{r})$ satisfies the Helmholtz equation, as was established in Sec. 2.2B, $\mathbf{A}(\mathbf{r})$ will also satisfy the Helmholtz equation $\nabla^2 \mathbf{A} + k^2 \mathbf{A} = 0$.

We now define the magnetic field in terms of the curl of this vector

$$\mathbf{H} = \frac{1}{\mu} \nabla \times \mathbf{A}, \tag{5.4-13}$$

and determine the corresponding electric field from Maxwell's equation (5.3-12):

$$\mathbf{E} = \frac{1}{j\omega\epsilon} \nabla \times \mathbf{H}. \tag{5.4-14}$$

The form of (5.4-13) and (5.4-14) ensures that $\nabla \cdot \mathbf{E} = 0$ and $\nabla \cdot \mathbf{H} = 0$, as required by (5.3-14) and (5.3-15), since the divergence of the curl of any vector field vanishes. Because $\mathbf{A}(\mathbf{r})$ satisfies the Helmholtz equation, it can readily be shown that the remaining Maxwell's equation, $\nabla \times \mathbf{E} = -j\omega\mu\mathbf{H}$, is also satisfied. It is therefore clear that (5.4-11)–(5.4-14) define a valid electromagnetic wave that satisfies Maxwell's equations.

Explicit expressions for $\mathbf{E}$ and $\mathbf{H}$ are obtained by carrying out the curl operations prescribed in (5.4-13) and (5.4-14). This is conveniently accomplished by making use of the spherical coordinate system (r, θ, ϕ) defined in Fig. 5.4-2(a), with unit vectors $(\hat{\mathbf{r}}, \hat{\boldsymbol{\theta}}, \hat{\boldsymbol{\phi}})$. The exact results for $\mathbf{E}$ and $\mathbf{H}$ turn out to be

$$\mathbf{E}(\mathbf{r}) = 2e_0 \cos\theta \left[\frac{1}{jkr} + \frac{1}{(jkr)^2} \right] U(\mathbf{r}) \, \hat{\mathbf{r}} + e_0 \sin\theta \left[1 + \frac{1}{jkr} + \frac{1}{(jkr)^2} \right] U(\mathbf{r}) \, \hat{\boldsymbol{\theta}}, \tag{5.4-15}$$

$$\mathbf{H}(\mathbf{r}) = h_0 \sin\theta \left[1 + \frac{1}{jkr} \right] U(\mathbf{r}) \, \hat{\boldsymbol{\phi}}, \tag{5.4-16}$$

where $h_0 = (jk/\mu)A_0$ and $e_0 = \eta H_0$. It can be shown that an electric dipole moment p pointing in the x direction radiates the wave described in (5.4-15) and (5.4-16) with $a_0 = j\mu\omega\text{p}$, so that $h_0 = (-\omega^2/c)\text{p}$ and $e_0 = -\mu\omega^2\text{p}$.

For points at distances from the origin that are much greater than a wavelength ($kr = 2\pi r/\lambda \gg 1$), the complex-amplitude vectors in (5.4-15) and (5.4-16) may be approximated by

$$\mathbf{E}(\mathbf{r}) \approx e_0 \sin\theta \, U(\mathbf{r}) \, \widehat{\boldsymbol{\theta}}, \tag{5.4-17}$$

$$\mathbf{H}(\mathbf{r}) \approx h_0 \sin\theta \, U(\mathbf{r}) \, \widehat{\boldsymbol{\phi}}. \tag{5.4-18}$$

The wavefronts are then spherical (as for the scalar spherical wave) and, as illustrated in Fig. 5.4-2(b), the electric and magnetic fields are orthogonal to one another and to the radial direction $\widehat{\mathbf{r}}$, with the electric field pointing in the polar direction and the magnetic field pointing in the azimuthal direction. Since the field strength is proportional to $\sin\theta$, the radiation pattern is toroidal with a null in the direction of the dipole, as shown in 5.4-2(c). In the paraxial approximation, at points near the z axis and far from the origin, such that $\theta \approx \pi/2$ and $\phi \approx \pi/2$, the wavefront normals are nearly parallel to the z axis (corresponding to paraxial rays), and $\sin\theta \approx 1$.

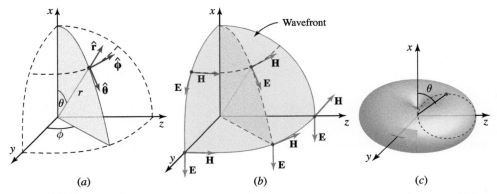

Figure 5.4-2 (a) Spherical coordinate system. (b) Electric- and magnetic-field vectors and wavefronts of the electromagnetic field, at distances $r \gg \lambda/2\pi$, radiated by an oscillating electric dipole. (c) The radiation pattern (field magnitude versus polar angle θ) is toroidal.

In a Cartesian coordinate system, $\widehat{\boldsymbol{\theta}} = -\sin\theta \, \widehat{\mathbf{x}} + \cos\theta \cos\phi \, \widehat{\mathbf{y}} + \cos\theta \sin\phi \, \widehat{\mathbf{z}} \approx -\widehat{\mathbf{x}} + (x/z)(y/z) \, \widehat{\mathbf{y}} + (x/z) \, \widehat{\mathbf{z}} \approx -\widehat{\mathbf{x}} + (x/z) \, \widehat{\mathbf{z}}$, so that

$$\mathbf{E}(\mathbf{r}) \approx e_0 \left(-\widehat{\mathbf{x}} + \frac{x}{z} \widehat{\mathbf{z}} \right) U(\mathbf{r}), \tag{5.4-19}$$

where $U(\mathbf{r})$ is the paraxial approximation of the spherical wave, i.e., the paraboloidal wave discussed in Sec. 2.2B. For sufficiently large values of z, the term (x/z) in (5.4-19) may also be neglected, whereupon

$$\mathbf{E}(\mathbf{r}) \approx -e_0 \, U(\mathbf{r}) \, \widehat{\mathbf{x}}, \tag{5.4-20}$$

$$\mathbf{H}(\mathbf{r}) \approx h_0 \, U(\mathbf{r}) \, \widehat{\mathbf{y}}. \tag{5.4-21}$$

In this approximation $U(\mathbf{r})$ approaches $(1/4\pi z) \, e^{-jkz}$, so that a TEM plane wave ultimately emerges, as portrayed in Fig. 2.2-4.

An electromagnetic wave that is dual to the electric-dipole wave discussed above is radiated by a **magnetic dipole** with the magnetic dipole moment $\mathfrak{m}$ pointing in the x direction. In the far field ($kr \gg 1$), it has an electric field pointing in the azimuthal direction and an orthogonal magnetic field pointing in the polar direction, with complex-amplitude vectors given by

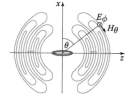

$$\mathbf{H}(\mathbf{r}) \approx h_0 \sin\theta\, U(\mathbf{r})\, \widehat{\boldsymbol{\theta}}, \qquad (5.4\text{-}22)$$

$$\mathbf{E}(\mathbf{r}) \approx e_0 \sin\theta\, U(\mathbf{r})\, \widehat{\boldsymbol{\phi}}, \qquad (5.4\text{-}23)$$

where $h_0 = (\omega^2/c^2)m$ and $e_0 = \mu(\omega^2/c)m$. At radio frequencies, this type of wave is radiated by electric current flowing in a loop antenna placed in a plane orthogonal to the x axis. At optical frequencies, tiny metal loops serve as optical antennas (Sec. 8.2D) and as important components in metamaterials (Sec. 8.3A).

The Gaussian Beam

It was demonstrated in Sec. 3.1 that a scalar Gaussian beam is readily obtained from a paraboloidal wave (the paraxial approximation to a spherical wave) by replacing the coordinate z by $z + jz_0$, where z_0 is a real constant.

The same transformation applied to the corresponding electromagnetic wave leads to the electromagnetic **vector Gaussian beam**. Replacing z in (5.4-19) by $z + jz_0$ yields

$$\mathbf{E}(\mathbf{r}) = e_0 \left(-\widehat{\mathbf{x}} + \frac{x}{z + jz_0}\, \widehat{\mathbf{z}} \right) U(\mathbf{r}), \qquad (5.4\text{-}24)$$

where $U(\mathbf{r})$ now represents the scalar complex amplitude of a Gaussian beam provided in (3.1-7). The wavefronts of the Gaussian beam are illustrated in Fig. 5.4-3(a) (these are also shown in Fig. 3.1-7), while the E-field lines determined from (5.4-24) are displayed in Fig. 5.4-3(b). In this case, the direction of the **E** field is not spatially uniform.

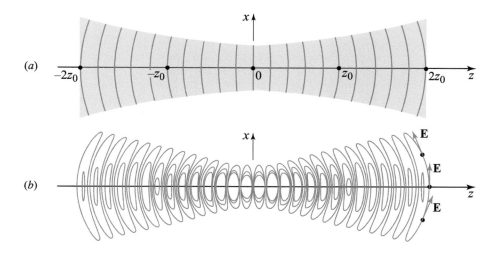

Figure 5.4-3 (a) Wavefronts of the scalar Gaussian beam $U(\mathbf{r})$ in the x–z plane. (b) Electric-field lines of the electromagnetic Gaussian beam in the x–z plane. (Adapted from H. A. Haus, *Waves and Fields in Optoelectronics*, Prentice Hall, 1984, Fig. 5.3a.)

B. Relation Between Electromagnetic Optics and Scalar Wave Optics

The paraxial scalar wave, defined in Sec. 2.2C, has wavefront normals that form small angles with respect to the axial coordinate z. The wavefronts behave locally as plane waves while the complex envelope and direction of propagation vary slowly with z.

This notion is also applicable to electromagnetic waves in linear isotropic media. A paraxial electromagnetic wave is locally approximated by a TEM plane wave. At each point, the vectors **E** and **H** lie in a plane that is tangential to the wavefront surfaces and normal to the wavevector **k** (Fig. 5.4-4). The optical power flows along the direction **E** × **H**, which is parallel to **k** and approximately parallel to the coordinate z.

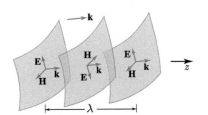

Figure 5.4-4 The paraxial electromagnetic wave. The vectors **E** and **H** reverse directions after propagation a distance of a half wavelength.

A paraxial scalar wave of intensity $I = |U|^2$ [see (2.2-10)] may be associated with a paraxial electromagnetic wave of the same intensity $I = |E|^2/2\eta$ [see (5.4-8)] by setting the complex amplitude to $U = E/\sqrt{2\eta}$ and matching the wavefronts. As attested to by the extensive development provided in Chapters 2–4, the scalar-wave description of light provides a very good approximation for solving a great many problems involving the interference, diffraction, propagation, and imaging of paraxial waves. The Gaussian beam with small divergence angle, considered in Chapter 3, provides a case in point. Most features of these beams, such as their intensity, focusing by a lens, reflection from a mirror, and interference, are addressed satisfactorily within the context of scalar wave optics. Of course, when polarization comes into play, wave optics is mute and we must appeal to electromagnetic optics.

It is of interest to note that U (as defined above) and E do not satisfy the same boundary conditions. For an electric field tangential to the boundary between two dielectric media, for example, E is continuous (Fig. 5.1-1), but $U = E/\sqrt{2\eta}$ is discontinuous since η changes value at the boundary. Thus, problems involving reflection and refraction at boundaries cannot be addressed completely within the scalar wave theory, although the matching of phase that leads to the law of reflection and Snell's law is adequately carried out within its confines (Sec. 2.4). Indeed, calculations of reflectance and transmittance at a boundary depend on the polarization state of the light and therefore require electromagnetic optics (see Sec. 6.2). Similarly, problems involving the transmission of light through dielectric waveguides require an analysis based on electromagnetic theory, as discussed in Chapters 9 and 10.

C. Vector Beams

Maxwell's equations in the paraxial approximation admit other cylindrically symmetric beam solutions for which the direction of the electric-field vector is spatially nonuniform. One example is a beam in which the electric field is aligned in an azimuthal orientation with respect to the beam axis, as illustrated in Fig. 5.4-5(a), i.e.,

$$\mathbf{E}(\mathbf{r}) = U(\rho, z)\exp(-jkz)\widehat{\boldsymbol{\phi}}. \qquad (5.4\text{-}25)$$

The scalar function $U(\rho, z)$ turns out to be the Bessel–Gauss solution to the Helmholtz equation, as discussed in Sec. 3.5A. This beam vanishes on-axis ($\rho = 0$) and has a toroidal transverse spatial distribution. The beam diverges in the axial direction and the spot size increases, much like the Gaussian beam.

Yet another cylindrically-symmetric beam has an azimuthally oriented magnetic-field vector, so that the electric-field vector is radial, as illustrated schematically in Fig. 5.4-5(b). It also has a spatial distribution with an on-axis null. The distribution of the vector field of this beam bears some resemblance to the electromagnetic field radiated by a dipole oriented along the beam axis (see Fig. 5.4-2).

It has been shown that a vector beam with radial electric-field vector may be focused by a lens of large numerical aperture to a spot of significantly smaller size than is possible with a conventional scalar Gaussian beam. Clearly, applications for such beams find a place in high-resolution microscopy. There are, it turns out, many variations on the theme of optical vector beams and their uses.

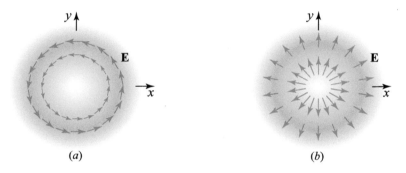

Figure 5.4-5 Vector beams with cylindrical symmetry. (a) Electric-field vectors oriented in the azimuthal direction. (b) Electric-field vectors oriented in the radial direction. The shading indicates the spatial distribution of the optical intensity in the transverse plane.

5.5 ABSORPTION AND DISPERSION

In this section, we consider absorption and dispersion in nonmagnetic media.

A. Absorption

The dielectric media considered thus far have been assumed to be fully transparent, i.e., not to absorb light. Glass is such a material in the visible region of the optical spectrum but it is, in fact, absorptive in the ultraviolet and infrared regions. Transmissive optical components in those bands are fabricated from other materials: examples are quartz and magnesium fluoride in the ultraviolet; and germanium and barium fluoride in the infrared. Figure 5.5-1 illustrates the spectral windows within which some commonly encountered optical materials are transparent (see Sec. 14.1D for further discussion).

In this section, we adopt a phenomenological approach to the absorption of light in linear media. Consider a complex electric susceptibility

$$\chi = \chi' + j\chi'', \qquad (5.5\text{-}1)$$

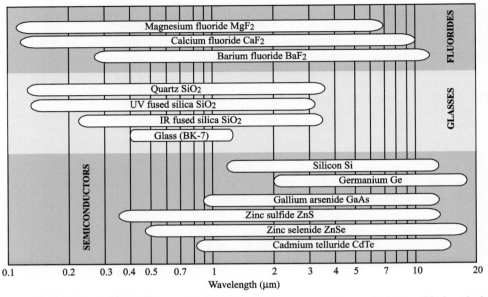

Figure 5.5-1 The white regions indicate the spectral bands within which the specified optical materials transmit light. Selected fluorides, glasses, and semiconductors are displayed.

corresponding to a complex electric permittivity $\epsilon = \epsilon_o(1+\chi)$ and a complex relative permittivity $\epsilon/\epsilon_o = (1+\chi)$. For monochromatic light, the Helmholtz equation (5.3-16) for the complex amplitude $U(\mathbf{r})$ remains valid, $\nabla^2 U + k^2 U = 0$, *but the wavenumber k itself becomes complex-valued*:

$$k = \omega\sqrt{\epsilon\mu_o} = k_o\sqrt{1+\chi} = k_o\sqrt{1+\chi'+j\chi''}, \qquad (5.5\text{-}2)$$

where $k_o = \omega/c_o$ is the wavenumber in free space.

Writing k in terms of real and imaginary parts, $k = \beta - j\frac{1}{2}\alpha$, allows β and α to be related to the susceptibility components χ' and χ'':

$$k = \beta - j\tfrac{1}{2}\alpha = k_o\sqrt{1+\chi'+j\chi''}. \qquad (5.5\text{-}3)$$

As a result of the imaginary part of k, a plane wave with complex amplitude $U = A\exp(-jkz)$ traveling through such a medium in the z-direction undergoes a change in magnitude (as well as the usual change in phase). Substituting $k = \beta - j\frac{1}{2}\alpha$ into the exponent of this plane wave yields $U = A\exp(-\frac{1}{2}\alpha z)\exp(-j\beta z)$. For $\alpha > 0$, which corresponds to absorption in the medium, the envelope A of the original plane wave is attenuated by the factor $\exp(-\frac{1}{2}\alpha z)$ so that the intensity, which is proportional to $|U|^2$, is attenuated by $|\exp(-\frac{1}{2}\alpha z)|^2 = \exp(-\alpha z)$. The coefficient α is therefore recognized as the **absorption coefficient** (also called the **attenuation coefficient**) of the medium. This simple exponential decay formula for the intensity provides the rationale for writing the imaginary part of k as $-\frac{1}{2}\alpha$. It will be seen in Sec. 15.1A that certain media, such as those used in lasers, can exhibit $\alpha < 0$, in which case $\gamma \equiv -\alpha$ is called the *gain coefficient* and the medium amplifies rather than attenuates light.

Since the parameter β is the rate at which the phase changes with z, it represents the propagation constant of the wave. The medium therefore has an effective refractive index n defined by

$$\beta = nk_o, \tag{5.5-4}$$

and the wave travels with a phase velocity $c = c_o/n$.

Substituting (5.5-4) into (5.5-3) thus relates the refractive index n and the absorption coefficient α to the real and imaginary parts of the susceptibility χ' and χ'':

$$\boxed{n - j\frac{1}{2}\frac{\alpha}{k_o} = \sqrt{\epsilon/\epsilon_o} = \sqrt{1 + \chi' + j\chi''}.} \tag{5.5-5}$$
Absorption Coefficient and Refractive Index

Note that the square root in (5.5-5) provides two complex numbers with opposite signs (phase difference of π). The sign is selected such that if χ'' is negative, i.e., the medium is absorbing, then α is positive, i.e., the wave is attenuated. If $(1 + \chi')$ is positive, then the complex number $1 + \chi' + j\chi''$ is in the fourth quadrant, and its square root can be in either the second or the fourth quadrant. By selecting the value in the fourth quadrant, we ensure that α is positive, and n is then also positive. Similarly, if $(1+\chi')$ is negative, then $1 + \chi' + j\chi''$ is in the third quadrant, and its square root is selected to be in the fourth quadrant so that both α and n are positive. The impedance associated with the complex susceptibility χ, which is also complex, is given by

$$\boxed{\eta = \sqrt{\frac{\mu_o}{\epsilon}} = \frac{\eta_o}{\sqrt{1+\chi}}.} \tag{5.5-6}$$
Impedance

Hence, in the context of our formulation, χ, k, ϵ, and η are complex quantities while α, β, and n are real.

Weakly Absorbing Media

In a weakly absorbing medium, we have the condition $\chi'' \ll 1 + \chi'$, so that $\sqrt{1+\chi'+j\chi''} = \sqrt{1+\chi'}\sqrt{1+j\delta} \approx \sqrt{1+\chi'}(1+j\frac{1}{2}\delta)$, where $\delta = \chi''/(1+\chi')$. It follows from (5.5-5) that

$$\boxed{\begin{aligned} n &\approx \sqrt{1+\chi'} \\ \alpha &\approx -\frac{k_o}{n}\chi''. \end{aligned}}$$
(5.5-7)
(5.5-8)
Weakly Absorbing Medium

Under these circumstances, the refractive index is determined by the real part of the susceptibility and the absorption coefficient is proportional to the imaginary part thereof. In an absorptive medium χ'' is negative so that α is positive whereas in an amplifying medium χ'' is positive and α is negative.

EXERCISE 5.5-1

Dilute Absorbing Medium. A nonabsorptive medium of refractive index n_0 serves as host to a dilute suspension of impurities characterized by susceptibility $\chi = \chi' + j\chi''$, where $\chi' \ll 1$ and

$\chi'' \ll 1$. Determine the overall susceptibility of the medium and demonstrate that the refractive index and absorption coefficient are given approximately by

$$n \approx n_0 + \frac{\chi'}{2n_0} \tag{5.5-9}$$

$$\alpha \approx -\frac{k_o \chi''}{n_0}. \tag{5.5-10}$$

Strongly Absorbing Media

In a strongly absorbing medium, $|\chi''| \gg |1 + \chi'|$, so that (5.5-5) yields $n - j\alpha/2k_o \approx \sqrt{j\chi''} = \sqrt{-j}\sqrt{(-\chi'')} = \pm\frac{1}{\sqrt{2}}(1-j)\sqrt{(-\chi'')}$, whereupon

$$n \approx \sqrt{(-\chi'')/2} \tag{5.5-11}$$

$$\alpha \approx 2k_o\sqrt{(-\chi'')/2}. \tag{5.5-12}$$

Strongly Absorbing Medium

Since χ'' is negative for an absorbing medium, the plus sign of the square root was selected to ensure that α is positive, and this yields a positive value for n as well.

B. Dispersion

Dispersive media are characterized by a frequency-dependent (and thus wavelength-dependent) susceptibility $\chi(\nu)$, electric permittivity $\epsilon(\nu)$, refractive index $n(\nu)$, and speed $c_o/n(\nu)$. Since the angle of refraction in Snell's law depends on refractive index, which is wavelength dependent, optical components fabricated from dispersive materials, such as prisms and lenses, bend light of different wavelengths by different angles. This accounts for the wavelength-resolving capabilities of refracting surfaces and for the wavelength-dependent focusing power of lenses (and the attendant chromatic aberration in imaging systems). Polychromatic light is therefore refracted into a range of directions. These effects are illustrated schematically in Fig. 5.5-2.

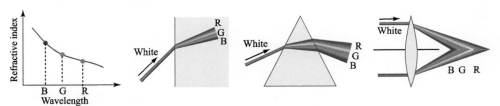

Figure 5.5-2 Optical components fabricated from dispersive materials refract waves of different wavelengths by different angles (B = blue, G = green, R = red).

Moreover, by virtue of the frequency-dependent speed of light in a dispersive medium, each of the frequency components comprising a short pulse of light experiences a different time delay. If the propagation distance through a medium is substantial, as is often the case in an optical fiber, for example, a brief light pulse at the input will be substantially dispersed in time so that its width at the output is increased, as illustrated in Fig. 5.5-3.

The wavelength dependence of the refractive index of some common optical materials is displayed in Fig. 5.5-4.

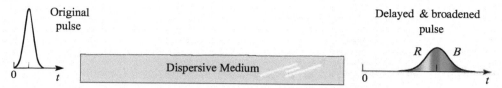

Figure 5.5-3 A dispersive medium serves to broaden a pulse of light because the different frequency components that constitute the pulse travel at different velocities. In this illustration, the low-frequency component (long wavelength, denoted R) travels faster than the high-frequency component (short wavelength, denoted B) and therefore arrives earlier.

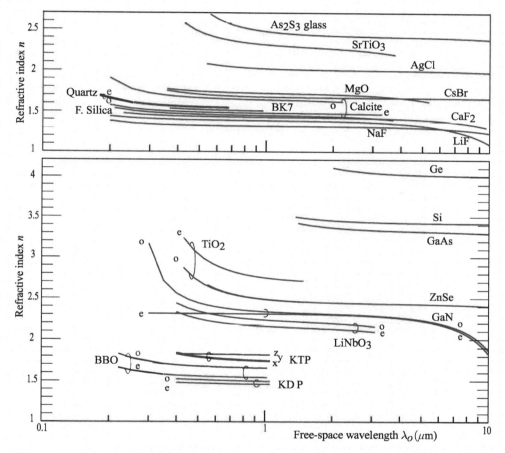

Figure 5.5-4 Wavelength dependence of the refractive index of selected optical materials, including glasses, crystals, and semiconductors. The designations 'e' and 'o' represent ordinary and extraordinary refraction, respectively, for anisotropic materials (see Sec. 6.3).

Measures of Dispersion

Material dispersion can be quantified in a number of different ways. For glass optical components and broad-spectrum light that covers the visible band (white light), a commonly used measure is the Abbe number $\mathbb{V} = (n_d - 1)/(n_F - n_C)$, where n_F, n_d, and n_C are the refractive indices of the glass at three standard wavelengths: blue at 486.1 nm, yellow at 587.6 nm, and red at 656.3 nm, respectively. For flint glass $\mathbb{V} \approx 38$

whereas for fused silica $\mathbb{V} \approx 68$.

On the other hand, if dispersion in the vicinity of a particular wavelength λ_o is of interest, an often used measure is the magnitude of the derivative $dn/d\lambda_o$ at that wavelength. This measure is appropriate for prisms, for example, in which the ray deflection angle θ_d is a function of n [see (1.2-6)]. The angular dispersion $d\theta_d/d\lambda_o = (d\theta_d/dn)(dn/d\lambda_o)$ is then a product of the material dispersion factor, $dn/d\lambda_o$, and another factor, $d\theta_d/dn$, that depends on the geometry of the prism and the refractive index of the material of which it is made.

The effect of material dispersion on the propagation of brief pulses of light is governed not only by the refractive index n and its first derivative $dn/d\lambda_o$, but also by the second derivative $d^2n/d\lambda_o^2$, as will be elucidated in Sec. 5.7 and Sec. 23.3.

Absorption and Dispersion: The Kramers–Kronig Relations

Absorption and dispersion are intimately related. Indeed, a dispersive material, i.e., a material whose refractive index is wavelength dependent, *must* be absorptive and must exhibit an absorption coefficient that is also wavelength dependent. The relation between the absorption coefficient and the refractive index is a result of the Kramers–Kronig relations, which relate the real and imaginary parts of the susceptibility of a medium, $\chi'(\nu)$ and $\chi''(\nu)$:

$$\chi'(\nu) = \frac{2}{\pi} \int_0^\infty \frac{s\chi''(s)}{s^2 - \nu^2} \, ds \qquad (5.5\text{-}13)$$

$$\chi''(\nu) = \frac{2}{\pi} \int_0^\infty \frac{\nu\chi'(s)}{\nu^2 - s^2} \, ds. \qquad (5.5\text{-}14)$$

Kramers–Kronig Relations

Given the real or the imaginary component of $\chi(\nu)$ for all ν, these powerful formulas allow the complementary component to be determined for all ν. The Kramers–Kronig relations connecting $\chi''(\nu)$ and $\chi'(\nu)$ translate into relations between the absorption coefficient $\alpha(\nu)$ and the refractive index $n(\nu)$ by virtue of (5.5-5), which relates α and n to χ'' and χ'.

The Kramers–Kronig relations are a special Hilbert-transform pair, as can be understood from linear systems theory (see Sec. B.1 of Appendix B). They are applicable for all linear, shift-invariant, causal systems with real impulse response functions. The linear system at hand is the polarization-density response of a medium $\mathcal{P}(t)$ to an applied electric field $\mathcal{E}(t)$ set forth in (5.2-23). Since $\mathcal{E}(t)$ and $\mathcal{P}(t)$ are real, so too is the impulse response function $\epsilon_o\chi(t)$. As a consequence, its Fourier transform, the transfer function $\epsilon_o\chi(\nu)$, exhibits Hermitian symmetry: $\chi(-\nu) = \chi^*(\nu)$ [see Sec. A.1 of Appendix A]. This system therefore obeys all of the conditions required for the Kramers–Kronig relations to apply. The real and imaginary parts of the transfer function $\epsilon_o\chi(\nu)$ are therefore related by (B.1-6) and (B.1-7) and, in particular, by (5.5-13) and (5.5-14).

C. The Resonant Medium

We now set forth a simple classical microscopic theory that leads to a complex susceptibility and provides an underlying rationale for the presence of frequency-dependent absorption and dispersion in an optical medium. The approach is known as the **Lorentz oscillator model**. A more thorough discussion of the interaction of light and matter is provided in Chapter 14.

5.5 ABSORPTION AND DISPERSION

Consider a dielectric medium such as a collection of resonant atoms, in which the dynamic relation between the polarization density $\mathcal{P}(t)$ and the electric field $\mathcal{E}(t)$, considered for a single polarization, is described by a linear second-order ordinary differential equation of the form

$$\frac{d^2\mathcal{P}}{dt^2} + \zeta\frac{d\mathcal{P}}{dt} + \omega_0^2\mathcal{P} = \omega_0^2\epsilon_o\chi_0\,\mathcal{E}, \qquad (5.5\text{-}15)$$

Resonant Dielectric Medium

where ζ, ω_0, and χ_0 are constants.

An equation of this form emerges when the motion of a bound charge associated with a resonant atom is modeled phenomenologically as a classical harmonic oscillator, in which the displacement of the charge $x(t)$ and the applied force $\mathcal{F}(t)$ are related by

$$\frac{d^2x}{dt^2} + \zeta\frac{dx}{dt} + \omega_0^2 x = \frac{\mathcal{F}}{m}. \qquad (5.5\text{-}16)$$

Here m is the mass of the bound charge, $\omega_0 = \sqrt{\kappa/m}$ is its resonance angular frequency, κ is the elastic constant of the restoring force, and ζ is the damping coefficient.

If the dipole moment associated with each individual atom is $\mathrm{p} = -ex$, the polarization density of the medium as a whole is related to the displacement by $\mathcal{P} = N\mathrm{p} = -Nex$, where $-e$ is the electronic charge and N is the number of atoms per unit volume of the medium. The electric field and force are related by $\mathcal{E} = \mathcal{F}/(-e)$. The quantities $\mathcal{P}$ and $\mathcal{E}$ are therefore proportional to x and $\mathcal{F}$, respectively, and comparison of (5.5-15) and (5.5-16) provides

$$\chi_0 = \frac{Ne^2}{\epsilon_o m \omega_0^2}. \qquad (5.5\text{-}17)$$

The applied electric field can thus be thought of as inducing a time-dependent electric dipole moment in each atom, as portrayed in Fig. 5.5-5, and hence a time-dependent polarization density in the medium as a whole.

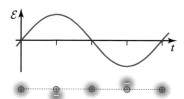

Figure 5.5-5 A time-varying electric field $\mathcal{E}$ applied to a Lorentz-oscillator atom induces a time-varying dipole moment p that contributes to the overall polarization density $\mathcal{P}$.

The medium is completely characterized by its impulse response function $\epsilon_o\chi(t)$, an exponentially decaying harmonic function, or equivalently by its transfer function $\epsilon_o\chi(\nu)$, which is obtained by solving (5.5-15) one frequency at a time, as follows. Substituting $\mathcal{E}(t) = \mathrm{Re}\{E\exp(j\omega t)\}$ and $\mathcal{P}(t) = \mathrm{Re}\{P\exp(j\omega t)\}$ into (5.5-15) yields

$$(-\omega^2 + j\zeta\omega + \omega_0^2)P = \omega_0^2\epsilon_o\chi_0 E, \qquad (5.5\text{-}18)$$

from which $P = \epsilon_o[\chi_0\omega_0^2/(\omega_0^2 - \omega^2 + j\zeta\omega)]E$. Writing this relation in the form $P = \epsilon_o\chi(\nu)E$, and substituting $\omega = 2\pi\nu$, yields an expression for the frequency-dependent susceptibility,

$$\chi(\nu) = \chi_0 \frac{\nu_0^2}{\nu_0^2 - \nu^2 + j\nu\,\Delta\nu},$$

(5.5-19)
Susceptibility
(Resonant Medium)

where $\nu_0 = \omega_0/2\pi$ is the resonance frequency and $\Delta\nu = \zeta/2\pi$.

The real and imaginary parts of $\chi(\nu)$, denoted $\chi'(\nu)$ and $\chi''(\nu)$ respectively, are therefore given by

$$\chi'(\nu) = \chi_0 \frac{\nu_0^2 \left(\nu_0^2 - \nu^2\right)}{\left(\nu_0^2 - \nu^2\right)^2 + (\nu\,\Delta\nu)^2} \qquad (5.5\text{-}20)$$

$$\chi''(\nu) = -\chi_0 \frac{\nu_0^2 \nu\,\Delta\nu}{\left(\nu_0^2 - \nu^2\right)^2 + (\nu\,\Delta\nu)^2}. \qquad (5.5\text{-}21)$$

These equations are plotted in Fig. 5.5-6.

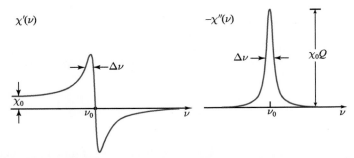

Figure 5.5-6 Real and imaginary parts of the susceptibility of a resonant dielectric medium. The real part $\chi'(\nu)$ is positive below resonance, zero at resonance, and negative above resonance. The imaginary part $\chi''(\nu)$ is negative so that $-\chi''(\nu)$ is positive everywhere and has a peak value $\chi_0 Q$ at $\nu = \nu_0$, where $Q = \nu_0/\Delta\nu$. The illustration portrays results for $Q = 10$.

At frequencies well below resonance ($\nu \ll \nu_0$), $\chi'(\nu) \approx \chi_0$ and $\chi''(\nu) \approx 0$, so that the low-frequency susceptibility is simply χ_0. At frequencies well above resonance ($\nu \gg \nu_0$), $\chi'(\nu) \approx \chi''(\nu) \approx 0$ so that the medium behaves like free space. Precisely at resonance ($\nu = \nu_0$), $\chi'(\nu_0) = 0$ and $-\chi''(\nu_0)$ reaches its peak value of $\chi_0 Q$, where $Q = \nu_0/\Delta\nu$. The resonance frequency ν_0 is usually much greater than $\Delta\nu$ so that $Q \gg 1$. Thus, the magnitude of the peak value of $-\chi''(\nu)$, which is $\chi_0 Q$, is much larger than the magnitude of the low-frequency value of $\chi'(\nu)$, which is χ_0. The maximum and minimum values of $\chi'(\nu)$ are $\pm \chi_0\, Q/(2 \mp 1/Q)$ and occur at frequencies $\nu_0 \sqrt{1 \mp 1/Q}$, respectively. For large Q, χ' swings between positive and negative values with a magnitude approximately equal to $\chi_0 Q/2$, i.e., one half of the peak value of χ''. The signs of χ' and χ'' determine the phase of χ, which simply determines the angle between the phasors P and E.

The behavior of $\chi(\nu)$ in the vicinity of resonance ($\nu \sim \nu_0$) is often of particular interest. In this region, we may use the approximation $(\nu_0^2 - \nu^2) = (\nu_0 + \nu)(\nu_0 - \nu) \approx$

$2\nu_0(\nu_0 - \nu)$ in the real part of the denominator of (5.5-19), and replace ν with ν_0 in the imaginary part thereof, to obtain

$$\chi(\nu \sim \nu_0) \approx \chi_0 \frac{\nu_0/2}{(\nu_0 - \nu) + j\Delta\nu/2}, \tag{5.5-22}$$

from which

$$\chi''(\nu) \approx -\chi_0 \frac{\nu_0 \Delta\nu}{4} \frac{1}{(\nu_0 - \nu)^2 + (\Delta\nu/2)^2} \tag{5.5-23}$$

$$\chi'(\nu) \approx 2 \frac{\nu - \nu_0}{\Delta\nu} \chi''(\nu). \tag{5.5-24}$$

Susceptibility (Near Resonance)

The function $\chi''(\nu)$ in (5.5-23), known as the **Lorentzian function**, decreases to half its peak value when $|\nu - \nu_0| = \Delta\nu/2$. The parameter $\Delta\nu$ therefore represents the full-width at half-maximum (FWHM) value of $\chi''(\nu)$.

The behavior of $\chi(\nu)$ far from resonance is also of interest. In the limit $|(\nu - \nu_0)| \gg \Delta\nu$, the susceptibility given in (5.5-19) is approximately real,

$$\chi(\nu) \approx \chi_0 \frac{\nu_0^2}{\nu_0^2 - \nu^2}, \tag{5.5-25}$$

Susceptibility (Far from Resonance)

so that the medium exhibits negligible absorption.

The absorption coefficient and the refractive index of a resonant medium may be determined by substituting the expressions for $\chi'(\nu)$ and $\chi''(\nu)$, e.g., (5.5-23) and (5.5-24) into (5.5-5). Each of these parameters generally depends on both $\chi'(\nu)$ and $\chi''(\nu)$. However, in the special case for which the resonant atoms are embedded in a nondispersive host medium of refractive index n_0, and are sufficiently dilute so that $\chi''(\nu)$ and $\chi'(\nu)$ are both $\ll 1$, this dependence is much simpler, namely, the refractive index and the absorption coefficient are dependent on χ' and χ'', respectively. Using the results of Exercise 5.5-1, it can be shown that these parameters are related by:

$$\alpha(\nu) \approx -\left(\frac{2\pi\nu}{n_0 c_o}\right) \chi''(\nu) \tag{5.5-26}$$

$$n(\nu) \approx n_0 + \frac{\chi'(\nu)}{2n_0}. \tag{5.5-27}$$

The dependence of these quantities on ν is illustrated in Fig. 5.5-7.

Media with Multiple Resonances

A typical dielectric medium contains multiple resonances corresponding to different lattice and electronic vibrations. The overall susceptibility arises from a superposition of contributions from these resonances. Whereas the imaginary part of the susceptibility is confined to frequencies near the resonance, the real part contributes at all frequencies near and *below* resonance, as shown in Fig. 5.5-6. This is exhibited in the frequency dependence of the absorption coefficient and the refractive index, as illustrated in Fig. 5.5-8. Absorption and dispersion are strongest near the resonance

190 CHAPTER 5 ELECTROMAGNETIC OPTICS

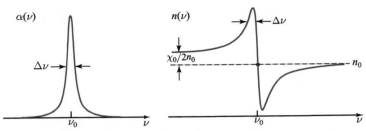

Figure 5.5-7 Absorption coefficient $\alpha(\nu)$ and refractive index $n(\nu)$ of a dielectric medium of refractive index n_0 containing a dilute concentration of atoms of resonance frequency ν_0.

frequencies. Away from the resonance frequencies, the refractive index is constant and the medium is approximately nondispersive and nonabsorptive. Each resonance does, however, contribute a constant value to the refractive index at all frequencies below its resonance frequency.

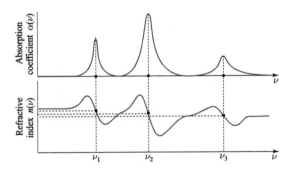

Figure 5.5-8 Frequency dependence of the absorption coefficient $\alpha(\nu)$ and the refractive index $n(\nu)$ for a medium with three resonances.

Other complex processes can also contribute to the absorption coefficient and the refractive index of a material, so that different patterns of frequency dependence emerge. Figure 5.5-9 shows an example of the *wavelength* dependence of the absorption coefficient and refractive index for a dielectric material that is essentially transparent at visible wavelengths. The illustration shows a decreasing refractive index with increasing wavelength in the visible region by virtue of a nearby ultraviolet resonance. The material is therefore more dispersive at shorter visible wavelengths where the rate of decrease of the index is greatest. This behavior is not unlike that exhibited in Fig. 5.5-1 and Fig. 5.5-4 for various real dielectric materials.

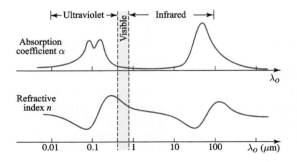

Figure 5.5-9 Typical wavelength dependence of the absorption coefficient and refractive index for a dielectric medium exhibiting resonant absorption in the ultraviolet and infrared bands, concomitant with low absorption in the visible band. In this diagram the abscissa is wavelength rather than frequency.

The Sellmeier Equation

In a medium with multiple resonances, labeled $i = 1, 2, \ldots$, the susceptibility is approximately given by a sum of terms, each of the form of (5.5-25), for frequencies far from any of the resonances. Using the relation between the refractive index and the real susceptibility provided in (5.2-13), $n^2 = 1 + \chi$, the dependence of n on frequency and wavelength assumes a form known as the **Sellmeier equation**:

$$n^2 \approx 1 + \sum_i \chi_{0i} \frac{\nu_i^2}{\nu_i^2 - \nu^2} = 1 + \sum_i \chi_{0i} \frac{\lambda^2}{\lambda^2 - \lambda_i^2}. \qquad (5.5\text{-}28)$$
Sellmeier Equation

The Sellmeier equation provides a good description of the refractive index for most optically transparent materials. At wavelengths for which $\lambda \ll \lambda_i$ the ith term becomes approximately proportional to λ^2, and for $\lambda \gg \lambda_i$ it becomes approximately constant. As an example, the dispersion in fused silica, illustrated in Example 5.7-1, is well described by three resonances. For some materials the Sellmeier equation is conveniently approximated by a power series.

The Sellmeier equations for a few selected materials, extracted from measured data using a least-squares fitting algorithm, are provided in Table 5.5-1.

Table 5.5-1 Sellmeier equations for the wavelength dependence of the refractive indices for selected materials at room temperature. The quantities n_o and n_e indicate the ordinary and extraordinary indices of refraction, respectively, for anisotropic materials (see Sec. 6.3). The range of wavelengths where the results are valid is indicated in the rightmost column.

Material	Sellmeier Equation (Wavelength λ in μm)	Wavelength Range (μm)
Fused silica	$n^2 = 1 + \dfrac{0.6962\lambda^2}{\lambda^2 - (0.06840)^2} + \dfrac{0.4079\lambda^2}{\lambda^2 - (0.1162)^2} + \dfrac{0.8975\lambda^2}{\lambda^2 - (9.8962)^2}$	0.21–3.71
Si	$n^2 = 1 + \dfrac{10.6684\lambda^2}{\lambda^2 - (0.3015)^2} + \dfrac{0.0030\lambda^2}{\lambda^2 - (1.1347)^2} + \dfrac{1.5413\lambda^2}{\lambda^2 - (1104.0)^2}$	1.36–11
GaAs	$n^2 = 3.5 + \dfrac{7.4969\lambda^2}{\lambda^2 - (0.4082)^2} + \dfrac{1.9347\lambda^2}{\lambda^2 - (37.17)^2}$	1.4–11
BBO	$n_o^2 = 2.7359 + \dfrac{0.01878}{\lambda^2 - 0.01822} - 0.01354\lambda^2$	0.22–1.06
	$n_e^2 = 2.3753 + \dfrac{0.01224}{\lambda^2 - 0.01667} - 0.01516\lambda^2$	
KDP	$n_o^2 = 1 + \dfrac{1.2566\lambda^2}{\lambda^2 - (0.09191)^2} + \dfrac{33.8991\lambda^2}{\lambda^2 - (33.3752)^2}$	0.4–1.06
	$n_e^2 = 1 + \dfrac{1.1311\lambda^2}{\lambda^2 - (0.09026)^2} + \dfrac{5.7568\lambda^2}{\lambda^2 - (28.4913)^2}$	
LiNbO$_3$	$n_o^2 = 2.3920 + \dfrac{2.5112\lambda^2}{\lambda^2 - (0.217)^2} + \dfrac{7.1333\lambda^2}{\lambda^2 - (16.502)^2}$	0.4–3.1
	$n_e^2 = 2.3247 + \dfrac{2.2565\lambda^2}{\lambda^2 - (0.210)^2} + \dfrac{14.503\lambda^2}{\lambda^2 - (25.915)^2}$	

5.6 SCATTERING OF ELECTROMAGNETIC WAVES

Previous chapters have described the propagation of optical waves through homogeneous media, the reflection and refraction of light at dielectric boundaries, wave transmission through optical components, and diffraction through apertures. In Sec. 5.5, we considered the absorption and dispersion of light. We turn now to the scattering of light, which plays an important role in various domains of optics, including nanophotonics.

In particular, we examine light scattering from a homogeneous medium containing localized inhomogeneities, irregularities, material defects, grains, or suspended particles. Both the medium and the scatterers are assumed to be dielectrics with linear and isotropic optical properties. The scattering from a small metal sphere is considered in Sec. 8.2C and various forms of light scattering are discussed in Sec. 14.5C.

A. Born Approximation

When an optical wave traveling in a given direction in a homogeneous medium encounters an object with different optical properties, the wave is scattered into other directions. This effect may be analyzed by solving Maxwell's equations and applying the appropriate boundary conditions. However, analytical solutions of this problem exist only in few ideal cases. We therefore resort to a commonly used approximate approach for solving such problems, known as the **Born approximation**. It is applicable for *weak scattering*, i.e., when the scattering object may be regarded as a small perturbation to the relative permittivity (or other optical properties) of the medium.

To introduce the Born approximation, it is convenient to first address the scattering of a scalar wave and then to subsequently consider an electromagnetic wave. The scalar complex amplitude $U(\mathbf{r})$ obeys the Helmholtz equation (2.2-7),

$$\left[\nabla^2 + k^2(\mathbf{r})\right] U = 0, \tag{5.6-1}$$

where inside the scattering object the wavenumber is $k(\mathbf{r}) = k_s(\mathbf{r})$ and in the host medium, which is taken to be uniform, the wavenumber is $k(\mathbf{r}) = k$. By writing $k^2(\mathbf{r}) = k^2 + [k^2(\mathbf{r}) - k^2]$, (5.6-1) may be rewritten as the Helmholtz equation for the scattered complex amplitude $U_s(\mathbf{r})$,

$$\left(\nabla^2 + k^2\right) U_s = -S, \tag{5.6-2}$$

with a source

$$S(\mathbf{r}) = [k_s^2(\mathbf{r}) - k^2] U_s(\mathbf{r}) \tag{5.6-3}$$

that is localized within the volume V of the scatterer, and is zero outside of it. As will be justified shortly, the solution to (5.6-2) is

$$U_s(\mathbf{r}) = \int_V S(\mathbf{r}') \frac{e^{-jk|\mathbf{r}-\mathbf{r}'|}}{4\pi |\mathbf{r}-\mathbf{r}'|} \, d\mathbf{r}' \tag{5.6-4}$$

at positions $\mathbf{r}$ outside the volume V. However, the integral in (5.6-4) cannot be readily evaluated to determine $U_s(\mathbf{r})$ since, in accordance with (5.6-3), the source $S(\mathbf{r}')$ itself depends on the wave $U_s(\mathbf{r})$, which is unknown.

If the scattering is weak, however, it is safe to assume that the incident wave $U_0(\mathbf{r})$ is essentially unaffected by the process of scattering within the volume V, in which case the complex amplitude $U_s(\mathbf{r})$ in the expression for the scattering source (5.6-3) may be approximated by the incident complex amplitude $U_0(\mathbf{r})$, whereupon

$$S(\mathbf{r}) \approx [k_s^2(\mathbf{r}) - k^2] U_0(\mathbf{r}). \tag{5.6-5}$$

This expression may then be used in (5.6-4) to determine the scattered complex amplitude $U_s(\mathbf{r})$. Implicit in the assumption of weak scattering is the condition that a wave scattered from one point in the scattering volume V is not subsequently scattered from another point, i.e., multiple scattering is a negligible second-order effect.

It is evident from (5.6-4) that the scattered wave $U_s(\mathbf{r})$ is then approximately a superposition of spherical waves generated by a continuum of point sources within the scatterer, as schematized in Fig. 5.6-1. Each point at position $\mathbf{r}'$ creates a spherical wave with amplitude $S(\mathbf{r}')$ given by the approximate expression (5.6-5). The concept is similar to that of the Huygens–Fresnel principle of diffraction described in Sec. 4.1D (see Fig. 4.1-13). In this type of scattering, known as **elastic scattering**, the frequency of the scattered light remains the same as that of the incident light.

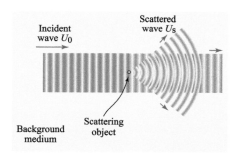

Figure 5.6-1 Under the Born approximation, the scattered wave $U_s(\mathbf{r})$ is a superposition of spherical waves, each generated by a point in the scatterer.

B. Rayleigh Scattering

Rayleigh scattering involves *small scatterers*. It is engendered by variations in a medium that are introduced, for example, by the presence of particles whose sizes are much smaller than a wavelength or by random inhomogeneities at a scale much finer than a wavelength.

Weak Scattering: Scalar Waves

If the contrast between the optical properties of the scattering and surrounding media is low, i.e., if the scattering is weak, then the Born approximation is applicable.

If we consider a single scattering object, much smaller than the wavelength of light and located at $\mathbf{r} = 0$, the source distribution in (5.6-5) may be approximated as $S(\mathbf{r}) \approx (k_s^2 - k^2) U_0 V \delta(\mathbf{r})$, where $\delta(\mathbf{r})$ is the delta function and k_s is the wavenumber within the small scatterer. Substituting this in the integral provided in (5.6-4) yields

$$U_s \approx (k_s^2 - k^2) V U_0 \frac{e^{-jkr}}{4\pi r}, \tag{5.6-6}$$

which represents a single spherical wave centered about $\mathbf{r} = 0$ (the location of the scatterer), with an amplitude proportional to that of the incident wave U_0.

In accordance with (2.2-10), the intensity of the scattered wave is therefore

$$I_s = |U_s|^2 \approx (k_s^2 - k^2)^2 \frac{V^2}{(4\pi r)^2} I_0, \tag{5.6-7}$$

where $I_0 = |U_0|^2$. Since the scalar scattered wave is isotropic, the total scattered power $P_s = 4\pi r^2 I_s$ becomes

$$P_s \approx \frac{1}{4\pi}\left(k_s^2 - k^2\right)^2 V^2 I_0, \qquad (5.6\text{-}8)$$

which reveals that the scattered power is proportional to the square of the scatterer volume V.

Since k_s and k are both proportional to ω, it is clear from (5.6-8) that the scattered power is proportional to ω^4 or, in terms of wavelength, to $1/\lambda_o^4$. Known as the **Rayleigh inverse fourth-power law**, this indicates that incident waves of short wavelength undergo greater scattering than those of long wavelength. As an example, the Rayleigh scattering of light at a wavelength of $\lambda_o = 400$ nm exceeds that of light at a wavelength of $\lambda_o = 800$ nm by the factor $2^4 = 16$. Rayleigh scattering from the density fluctuations of air, which are finer than the wavelengths of light in the visible spectral band, is responsible for the blue color of the sky. The short-wavelength (blue) light is preferentially scattered over a large range of angles, whereas the light arriving directly from the sun is reduced in blue and therefore appears to have a yellowish tint. In silica-glass optical fibers, Rayleigh scattering is responsible for the greater attenuation of visible than infrared light, as discussed in Sec. 10.3A.

Weak Scattering: Electromagnetic Waves

The derivation of the scattered wave considered above was predicated on a scalar complex amplitude that obeys the Helmholtz equation (5.6-1). The scattering of an electromagnetic wave may be formulated in a similar manner by beginning with the vector potential $\mathbf{A}$, which also satisfies the Helmholtz equation. Applying the Born approximation, the vector potential of the scattered wave may be expressed as a superposition of dipole waves centered at points within the scatterer, in analogy with (5.6-4). The vector potential $\mathbf{A}$ for an oscillating dipole has the distribution of a spherical wave, with the associated electric and magnetic complex amplitudes $\mathbf{E}$ and $\mathbf{H}$ described in Sec. 5.4A.

From an electromagnetic point-of-view, scattering can thus be viewed as the creation, by the incident field, of a collection of oscillating electric dipoles at all points within the scatterer, each radiating a dipole wave.

For a single small scatterer at the origin, the scattered electromagnetic wave is identical to that radiated by a single electric dipole pointing along the direction of the electric field $\mathbf{E}_0$ of the incident wave, as illustrated in Fig. 5.6-2. In the far zone ($r \gg \lambda$), the electric and magnetic fields of the scattered wave point in the polar and azimuthal directions, respectively, as provided in (5.4-17) and (5.4-18), as well as in Fig. 5.4-2. The electric-field complex amplitude of the scattered wave is thus given by

$$\mathbf{E}_s \approx E_{s0} \sin\theta \, \frac{e^{-jkr}}{4\pi r} \, \widehat{\boldsymbol{\theta}}, \qquad E_{s0} = -(k_s^2 - k^2)V E_0, \qquad (5.6\text{-}9)$$

so that the scattered-wave intensity is

$$I_s \approx \left|k_s^2 - k^2\right|^2 \frac{V^2}{(4\pi r)^2} I_0 \sin^2\theta, \qquad (5.6\text{-}10)$$

where I_0 is given by (5.4-8). The angular distribution of the scattered wave is thus independent of ϕ and assumes the toroidal pattern illustrated in Fig. 5.6-2. The scattering is at a maximum when $\theta = \pi/2$, i.e., when the direction of the scattered wave is orthogonal to the direction of the electric field of the incident wave. In particular, back-scattering has the same intensity as forward-scattering.

The expression for the electromagnetic intensity given in (5.6-10) differs from that for the scalar-wave intensity provided in (5.6-7) by the factor $\sin^2 \theta$. This distinction arises because the oscillating dipole radiates a transverse electromagnetic wave, which precludes scattering in a direction parallel to the incident electric field, whereas scalar wave optics does not take polarization into account.

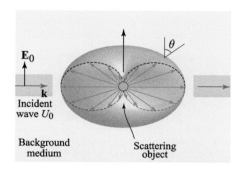

Figure 5.6-2 A transverse electromagnetic plane wave with electric field $\mathbf{E}_0$ scattered from a point object (blue circle at center) creates a scattered electric-dipole wave $\mathbf{E}_s$ with a toroidal directional pattern. The scattered intensity $I_s \propto \sin^2 \theta$, where θ is the scattering angle.

The total scattered power is calculated by integrating (5.6-10) over the surface of a sphere. Using the incremental integration area in spherical coordinates, $r^2 \sin \theta \, d\theta \, d\phi$, and noting that $\int_0^\pi \sin^3 \theta \, d\theta = 4/3$, leads to

$$P_s \approx \frac{1}{6\pi} \left| k_s^2 - k^2 \right|^2 V^2 I_0. \tag{5.6-11}$$

The electromagnetic scattered power is thus 2/3 of that obtained in the scalar-wave case, as provided in (5.6-8). The results differ because of the distinction in the integration over θ in the two cases. In the isotropic case the appropriate integration is $\int_0^\pi \sin \theta \, d\theta = 2$, whereas in the electromagnetic case the integration yields 4/3, as indicated above, which is a factor of 2/3 smaller.

It is commonplace to characterize the strength of scattering in terms of a **scattering cross section** σ_s. Writing the total scattered power P_s as the product

$$P_s = \sigma_s I_0, \tag{5.6-12}$$

where I_0 is the incident light intensity [W/m^2], it is evident that σ_s may be regarded as the area of an aperture [m^2] that intercepts the incident wave and collects an amount of power equal to the actual scattered power. Based on (5.6-11), the scattering cross section under the Born approximation (weak scattering) and the small-scatterer approximation (Rayleigh scattering) is therefore

$$\sigma_s = \frac{1}{6\pi} \left| k_s^2 - k^2 \right|^2 V^2. \tag{5.6-13}$$

Let us consider a specific example: the scattering cross section of a spherical dielectric scatterer of radius a and permittivity ϵ_s embedded in a dielectric medium of permittivity ϵ, under the assumption that both media have the same magnetic permeability μ. Substituting $k = \omega\sqrt{\epsilon\mu} = 2\pi/\lambda$, $k_s = \omega\sqrt{\epsilon_s\mu} = \sqrt{\epsilon_s/\epsilon} \cdot 2\pi/\lambda$, and $V = \frac{4}{3}\pi a^3$ into (5.6-13), we obtain

$$\sigma_s = \pi a^2 Q_s, \qquad Q_s = \frac{8}{3} \left| \frac{\epsilon_s - \epsilon}{3\epsilon} \right|^2 \left(2\pi \frac{a}{\lambda} \right)^4. \tag{5.6-14}$$

The scattering cross section of the spherical scatterer is thus given by the product of its geometrical area, πa^2, and a small dimensionless factor Q_s, known as the **scattering efficiency**. The quantity Q_s is proportional to the fourth power of the ratio a/λ, where λ is the wavelength of light in the background medium, and to the square of the contrast factor $(\epsilon_s - \epsilon)/\epsilon = (n_s^2 - n^2)/n^2$, where n_s and n are the refractive indices of the scatterer and the medium, respectively. Rayleigh scattering is evidently highly dependent on the size of the scatterer; the scattered power is proportional to the sixth power of the radius of a spherical scatterer. Of course the validity of these results requires that the radius of the scatterer be small in comparison with a wavelength.

EXAMPLE 5.6-1. *Rayleigh Scattering from a Dielectric Nanosphere.* Light of wavelength $\lambda = 600$ nm is scattered from a spherical nanoparticle of radius $a = 60$ nm and a relative permittivity that is 10% greater than the background value. Since the $a/\lambda = 0.1$, the small-scatterer condition is satisfied. Also, since the contrast $(\epsilon_s - \epsilon)/\epsilon = 0.1$, the weak-scattering condition is satisfied. In accordance with (5.6-14), the scattering efficiency is $Q_s \approx 4.6 \times 10^{-4}$ and the scattering cross section is $\sigma_s \approx 5.2$ nm^2. Hence, if the intensity of the incident light is $I_0 \approx 10^5$ W/m^2 (corresponding to a 3-mW laser beam of 100-μm radius), the scattered power is $P_s \approx 0.52$ pW.

Strong Scattering: Nanosphere

The Born approximation is not applicable in the case of strong scattering, i.e., when the contrast $(\epsilon_s - \epsilon)/\epsilon$ between the relative permittivities of the scatterer and the background is not small. However, an alternative method, known as the **quasi-static approximation**, may be used to determine the Rayleigh scattered field if the scatterer is spherical and its a radius much smaller than the optical wavelength, i.e., a nanosphere.

Again, the scattered electric-field complex amplitude $\mathbf{E}_s$ is that radiated by an electric dipole, as in (5.4-15) and (5.4-16). As explained below, in the far zone the field is described by

$$\mathbf{E}_s \approx E_{s0} \sin\theta \frac{e^{-jkr}}{4\pi r} \widehat{\boldsymbol{\theta}}, \qquad E_{s0} = -4\pi \left(\frac{\epsilon_s - \epsilon}{\epsilon_s + 2\epsilon}\right) k^2 a^3 E_0, \qquad (5.6\text{-}15)$$

and the associated scattering cross section turns out to be approximately given by

$$\boxed{\sigma_s = \pi a^2 Q_s, \qquad Q_s = \frac{8}{3}\left|\frac{\epsilon_s - \epsilon}{\epsilon_s + 2\epsilon}\right|^2 \left(2\pi\frac{a}{\lambda}\right)^4.} \qquad (5.6\text{-}16)$$
Nanosphere
Cross Section

If $\epsilon_s \approx \epsilon$, then $\epsilon_s + 2\epsilon \approx 3\epsilon$, whereupon the weak-scattering results are recovered from the above equations, i.e., (5.6-15) reproduces (5.6-9), and (5.6-16) reproduces (5.6-14).

These results may be confirmed by applying appropriate boundary conditions at the surface of the scattering sphere ($r = a$), namely, matching the tangential components of the external and internal electric fields $\mathbf{E}$, as well as the normal components of the displacement fields $\mathbf{D}$, which are products of the permittivities and the electric fields in each medium (see Fig. 5.1-1). The *internal electric field* $\mathbf{E}_i$ within the scattering sphere is uniformly distributed, with amplitude

$$\boxed{E_i = \frac{3\epsilon}{\epsilon_s + 2\epsilon} E_0,} \qquad (5.6\text{-}17)$$
Nanosphere Internal Field

and with a direction that is parallel to the electric field of the incident wave, as shown in Fig. 5.6-3.

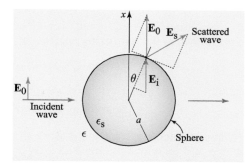

Figure 5.6-3 Scattering of a plane wave with electric field $\mathbf{E}_0$ from a dielectric nanosphere of radius $a \ll \lambda$. The scattered wave $\mathbf{E}_s$ is identical to the wave radiated by an electric dipole, and the internal field $\mathbf{E}_i$ is uniform within the sphere. Boundary conditions dictate that the polar components of $\mathbf{E}_0 + \mathbf{E}_s$ and $\mathbf{E}_i$ are equal, and the radial components of $\epsilon(\mathbf{E}_0 + \mathbf{E}_s)$ and $\epsilon_s \mathbf{E}_i$ are also equal. Scattering from a metal nanosphere is considered in Sec. 8.2C.

The external field is the sum of the incident field $\mathbf{E}_0$ and the scattered field $\mathbf{E}_s$, which is a dipole wave. Since the radius of the sphere is taken to be much smaller than the wavelength of the light ($r \ll \lambda$), at the boundary $r = a$ we have $kr \ll 1$. It follows that points on the sphere lie in the near-field zone of the dipole wave. As a consequence, the full expression for the electric field of the dipole wave provided in (5.4-15) may be approximated by the $1/(jkr)^2$ terms. At $r = a$, the radial and polar components of $\mathbf{E}_s$ are therefore $2(jka)^{-2}(E_{s0}\cos\theta)(4\pi a)^{-1}e^{-jka}$ and $(jka)^{-2}(E_{s0}\sin\theta)(4\pi a)^{-1}e^{-jka}$, respectively. Inserting these expressions in the boundary conditions results in (5.6-16) and (5.6-17). This solution, which is valid for long wavelengths ($\lambda \gg a$), i.e., low frequencies, may also be obtained by solving the electrostatic problem of a dielectric sphere in an applied steady electric field, which explains the appellation *quasi-static approximation*.

C. Mie Scattering

For weak scattering, the Born approximation is applicable for scatterers of all sizes, including those with dimensions comparable to, or larger than, the wavelength of the incident light. The scattered wave is formulated as an integral of dipole waves centered at points within the scatterer, and with amplitudes proportional to the local value of $k_s^2(\mathbf{r}) - k^2$, as set forth in Sec. 5.6A. The resultant scattering pattern is sensitive to the size and shape of the scatterer.

If the Born approximation cannot be used because the scattering is not sufficiently weak, the problem can be solved analytically for a few special shapes, such as spheres. This is known as **Mie scattering**. Quadrupole solutions, which are terms of order higher than the dipole solutions to the Helmholtz equation that we have considered thus far, become important for large spheres, which renders the mathematical analysis more complicated. The directional pattern of the scattering assumes complex, and often asymmetric, shapes so that scattering in the forward direction can become stronger than that in the backward direction. For spheres that are large in comparison with the wavelength, the scattered power turns out to be proportional to the square of the particle diameter, rather than to the sixth power as for Rayleigh scattering.

Moveover, the strength of Mie scattering is roughly independent of wavelength, in contrast to Rayleigh scattering, so all wavelengths in white light are scattered approximately equally. Mie scattering from the water droplets suspended in clouds, which are comparable in size to the visible wavelengths comprising sunlight, is responsible for their white (or gray) color. It is also responsible for the white glare around light sources (such as automobile headlights) in the presence of mist and fog.

D. Attenuation in a Medium with Scatterers

Although the intensity that is Rayleigh scattered from a single scatterer is very small, the cumulative effect of a large number of scatterers distributed within a medium can result in significant attenuation. A wave propagating through a homogeneous medium with an average of N_s identical scatterers per unit volume, each with scattering cross section σ_s, is attenuated exponentially at a rate α_s, known as the **scattering coefficient**:

$$\boxed{\alpha_s = N_s \sigma_s.} \qquad (5.6\text{-}18)$$
Scattering Coefficient

This result is derived by considering a plane wave of intensity I traveling along the z axis of a cylinder with unit cross-sectional area and incremental length Δz, as illustrated in Fig. 5.6-4. The incremental slice contains $N_s \Delta z$ scatterers, each of which scatters a small amount of power $\sigma_s I$ away from the z direction. In passing through this slice, the intensity therefore decreases by the increment $\Delta I = -(N_s \Delta z)\sigma_s I$. In the limit as $\Delta z \to 0$, this yields $dI/dz = -\alpha_s I$, where $\alpha_s = N_s \sigma_s$, so that the intensity of the wave decays exponentially at the rate α_s, thereby decreasing by the factor $\exp(-\alpha_s z)$ upon traveling a distance z.

Medium with Absorbing Scatterers

If the scatterers are absorptive as well, then additional attenuation is encountered as the wave passes through the medium. The overall attenuation coefficient, also called the intensity **extinction coefficient**,[†] is the sum of the **absorption coefficient** α_a and the **scattering coefficient** α_s, i.e., $\alpha = \alpha_a + \alpha_s$.

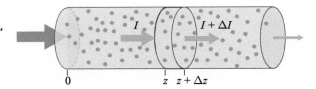

Figure 5.6-4 Scattering and absorption from scatterers embedded in a nonabsorbing homogenous medium results in wave extinction.

We proceed to derive an expression for the absorption coefficient α_a for a nonabsorbing homogeneous medium of real permittivity ϵ in which a concentration of N_s spherical scatterers per unit volume, each of complex permittivity ϵ_s and volume V, is embedded. The absorptive nature of the scatterers is embodied in the imaginary part of ϵ_s. The complex effective permittivity of the composite medium (the host medium and the embedded scatterers) is denoted ϵ_e. The wavenumbers of the composite and host media are $k_e = \omega\sqrt{\epsilon_e \mu}$ and $k = \omega\sqrt{\epsilon \mu}$, which are complex and real, respectively. Based on (5.5-3) we can therefore write $\alpha_a/2 = -\text{Im}\{k_e\} = -k\,\text{Im}\{\sqrt{\epsilon_e/\epsilon}\,\}$.

An approximate expression for ϵ_e is provided by the weighted average based on the volume fraction occupied by the scatterers, $\text{f} = N_s V$:

$$\epsilon_e \approx (1 - \text{f})\epsilon + \text{f}\epsilon_s = \epsilon + (\epsilon_s - \epsilon)\text{f}. \qquad (5.6\text{-}19)$$

This expression is valid for a dilute medium ($\text{f} \ll 1$) with weak scattering ($\epsilon_s \approx \epsilon$) and small scatterers. For small spherical scatterers of arbitrary concentration, and for

---

[†] The field extinction coefficient, in contrast, is the rate at which the field, rather than the intensity, decreases.

arbitrary values of ϵ_s and ϵ, the Maxwell-Garnett mixing rule is applicable:[‡]

$$\epsilon_e \approx \epsilon + 3f\epsilon \frac{\epsilon_s - \epsilon}{\epsilon_s + 2\epsilon - (\epsilon_s - \epsilon)f} = \epsilon \frac{2(1-f)\epsilon + (1+2f)\epsilon_s}{(2+f)\epsilon + (1-f)\epsilon_s}. \qquad (5.6\text{-}20)$$

This relation may be derived by noting that the average electric field is $\overline{E} = fE_i + (1-f)E_0$ and the average displacement is $\overline{D} = f\epsilon_s E_i + (1-f)\epsilon E_0$, where E_i and E_0 are the internal and external fields, respectively. Using (5.6-17), which provides $E_0/E_i = (\epsilon_s + 2\epsilon)/3\epsilon$, and forming the ratio $\epsilon_e \approx \overline{D}/\overline{E}$, yields the desired result. In a dilute medium of small spherical scatterers, for which $f \ll 1$, (5.6-20) takes the simpler form

$$\epsilon_e \approx \epsilon + 3f\epsilon \frac{\epsilon_s - \epsilon}{\epsilon_s + 2\epsilon}. \qquad (5.6\text{-}21)$$

If, additionally, the scattering is weak ($\epsilon_s \approx \epsilon$), then (5.6-21) reduces to the weighted-average formula (5.6-19).

If the scatterers are small and dilute, but not necessarily weak, then $\epsilon_e = \epsilon + \Delta\epsilon$, where $\Delta\epsilon \ll \epsilon$. In this case, $\alpha_a/2 = -k\,\text{Im}\{\sqrt{\epsilon_e/\epsilon}\} = -k\,\text{Im}\{(1+\Delta\epsilon/\epsilon)^{1/2}\} \approx -k\,\text{Im}\{1 + \Delta\epsilon/2\epsilon\} = -k\,\text{Im}\{\Delta\epsilon/2\epsilon\}$ so that $\alpha_a \approx -k\,\text{Im}\{(\epsilon_e - \epsilon)/\epsilon\}$. With $k = 2\pi/\lambda$ and $f = N_s V = N_s 4\pi a^3/3$ for spherical scatterers of radius a, use of (5.6-21) leads to the approximate result

$$\boxed{\alpha_a = N_s \sigma_a, \quad \sigma_a = \pi a^2 Q_a, \quad Q_a \approx -4\,\text{Im}\left\{\frac{\epsilon_s - \epsilon}{\epsilon_s + 2\epsilon}\right\}\left(2\pi\frac{a}{\lambda}\right),} \qquad (5.6\text{-}22)$$
Absorption Coefficient

where σ_a is the absorption cross section and the dimensionless factor Q_a is the **absorption efficiency**. Note that if we use the weighted-average formula (5.6-19) for ϵ_e, which is valid for dilute, weak, and small scatterers, we obtain an expression for α_a identical to that provided in (5.6-22) except that the factor $\epsilon_s + 2\epsilon$ in Q_a is replaced by 3ϵ. The overall attenuation coefficient $\alpha = \alpha_a + \alpha_s$ is obtained by combining the results provided in (5.6-22), (5.6-16), and (5.6-18).

5.7 PULSE PROPAGATION IN DISPERSIVE MEDIA

The propagation of pulses of light in dispersive media is important in many applications including optical fiber communication systems, as will be discussed in detail in Chapters 10, 23, and 25. As indicated above, a dispersive medium is characterized by a frequency-dependent refractive index and absorption coefficient, so that monochromatic waves of different frequencies travel through the medium with different velocities and undergo different attenuations. Since a pulse of light comprises a sum of many monochromatic waves, each of which is modified differently, the pulse is delayed and broadened (dispersed in time); in general its shape is also altered. In this section we provide a simplified analysis of these effects; a detailed description is deferred to Chapter 23.

[‡] See J. C. Maxwell Garnett, XII. Colours in Metal Glasses and in Metallic Films, *Philosophical Transactions of the Royal Society A*, vol. 203, pp. 385–420, 1904; M. Born and E. Wolf, *Principles of Optics*, Cambridge University Press, 7th expanded and corrected ed. 2002.

Group Velocity

Consider a pulsed plane wave traveling in the z direction through a lossless dispersive medium with refractive index $n(\omega)$. Following the example set forth in Sec. 2.6, assume that the initial complex wavefunction at $z = 0$ is $U(0,t) = \mathcal{A}(t)\exp(j\omega_0 t)$, where ω_0 is the central angular frequency and $\mathcal{A}(t)$ is the complex envelope of the wave. It will be shown below that if the dispersion is weak, i.e., if n varies slowly within the spectral bandwidth of the wave, then the complex wavefunction at a distance z is approximately $U(z,t) = \mathcal{A}(t-z/v)\exp[j\omega_0(t-z/c)]$, where $c = c_o/n(\omega_0)$ is the speed of light in the medium at the central frequency, and v is the velocity at which the envelope travels (see Fig. 5.7-1). The parameter v, called the **group velocity**, is given by

$$\boxed{\frac{1}{v} = \beta' = \frac{d\beta}{d\omega}}, \qquad (5.7\text{-}1)$$
Group Velocity

where $\beta = \omega n(\omega)/c_o$ is the frequency-dependent propagation constant and the derivative in (5.7-1), which is often denoted β', is evaluated at the central frequency ω_0. The group velocity is a characteristic of the dispersive medium, and generally varies with the central frequency. The corresponding time delay $\tau_d = z/v$ is called the **group delay**.

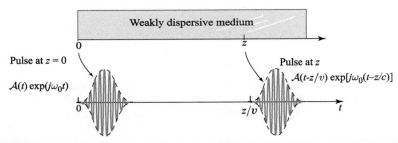

Figure 5.7-1 An optical pulse traveling in a dispersive medium that is weak enough so that its group velocity is frequency independent. The envelope travels with group velocity v while the underlying wave travels with phase velocity c.

Since the phase factor $\exp[j\omega_0(t-z/c)]$ is a function of $t-z/c$, the speed of light c, given by $1/c = \beta(\omega_0)/\omega_0$, is often called the **phase velocity**. In an ideal (nondispersive) medium, $\beta(\omega) = \omega/c$ whereupon $v = c$ and the group and phase velocities are identical.

□ **Derivation of the Formula for the Group Velocity.** The proof of (5.7-1) relies on a Fourier decomposition of the envelope $\mathcal{A}(t)$ into its constituent harmonic functions. A component of frequency Ω, assumed to have a Fourier amplitude $A(\Omega)$, corresponds to a monochromatic wave of frequency $\omega = \omega_0 + \Omega$ traveling with propagation constant $\beta(\omega_0 + \Omega)$. This component of the pulsed plane wave therefore travels as $A(\Omega)\exp\{-j[\beta(\omega_0 + \Omega)]z\}\exp[j(\omega_0 + \Omega)t]$. If $\beta(\omega)$ varies slowly near the central frequency ω_0, it may be approximately linearized via a two-term Taylor series expansion: $\beta(\omega_0 + \Omega) \approx \beta(\omega_0) + \Omega\, d\beta/d\omega = \omega_0/c + \Omega/v$. The Ω component of the complex wavefunction may therefore be approximated by $A(\Omega)\exp[j\Omega(t-z/v)]\exp[j\omega_0(t-z/c)]$. It follows that, upon traveling a distance z, the envelope of the Fourier component $A(\Omega)\exp(j\Omega t)$ becomes $A(\Omega)\exp[j\Omega(t-z/v)]$ for every value of Ω; thus the pulse envelope $\mathcal{A}(t)$ becomes $\mathcal{A}(t-z/v)$. The pulse therefore travels at the group velocity $v = 1/(d\beta/d\omega)$, in accordance with (5.7-1). ■

Since the index of refraction of most materials is typically measured and tabulated as a function of optical wavelength rather than frequency, it is convenient to express the group velocity v in terms of $n(\lambda_o)$. Using the relations $\beta = \omega n(\lambda_o)/c_o = 2\pi n(\lambda_o)/\lambda_o$ and $\lambda_o = 2\pi c_o/\omega$ in (5.7-1), along with the chain rule $d\beta/d\omega = (d\beta/d\lambda_o)(d\lambda_o/d\omega)$, provides

$$v = \frac{c_o}{N} \qquad N = n(\lambda_o) - \lambda_o \frac{dn}{d\lambda_o}. \qquad (5.7\text{-}2)$$
Group Velocity and Group Index

The derivative $dn/d\lambda_o$ in (5.7-2) is evaluated at the central wavelength λ_0. The parameter N is often called the **group index**.

Group Velocity Dispersion (GVD)

Since the group velocity $v = 1/(d\beta/d\omega)$ is itself often frequency dependent, different frequency components of the pulse undergo different delays $\tau_d = z/v$. As a result, the pulse spreads in time. This phenomenon is known as **group velocity dispersion** (GVD). To estimate the spread associated with GVD it suffices to note that, upon traveling a distance z, two identical pulses of central frequencies ν and $\nu + \delta\nu$ suffer a differential group delay

$$\delta\tau = \frac{d\tau_d}{d\nu}\delta\nu = \frac{d}{d\nu}\left(\frac{z}{v}\right)\delta\nu = D_\nu\, z\, \delta\nu, \qquad (5.7\text{-}3)$$

where the quantity

$$D_\nu = \frac{d}{d\nu}\left(\frac{1}{v}\right) = 2\pi\beta'' \qquad (5.7\text{-}4)$$
Dispersion Coefficient

is called the **dispersion coefficient** and $\beta'' \equiv d^2\beta/d\omega^2|_{\omega_0}$. This effect is actually associated with the higher-order terms in the Taylor series expansion of $\beta(\omega)$ that were neglected in the derivation of the group velocity carried out above; a more complete treatment will be provided in Chapter 23.

If the pulse has an initial spectral width σ_ν (Hz), in accordance with (5.7-3) a good estimate of its temporal spread is then provided by

$$\sigma_\tau = |D_\nu|\sigma_\nu z. \qquad (5.7\text{-}5)$$
Pulse Spread

The dispersion coefficient D_ν is a measure of the pulse-time broadening per unit distance per unit spectral width (s/m-Hz). This temporal broadening process is illustrated schematically in Fig. 5.7-2. If the refractive index is specified in terms of the wavelength, $n(\lambda_o)$, then (5.7-2) and (5.7-4) give

$$D_\nu = \frac{\lambda_o^3}{c_o^2}\frac{d^2n}{d\lambda_o^2}. \qquad (5.7\text{-}6)$$
Dispersion Coefficient (s/m-Hz)

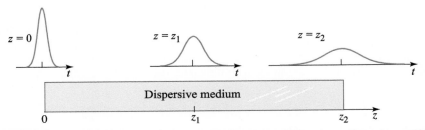

Figure 5.7-2 The temporal spread of an optical pulse traveling in a dispersive medium is proportional to the product of the dispersion coefficient D_ν, the spectral width σ_ν, and the distance traveled z.

It is also common to define a dispersion coefficient D_λ in terms of the wavelength[†] instead of the frequency. Using $D_\lambda\, d\lambda = D_\nu\, d\nu$ yields $D_\lambda = D_\nu\, d\nu/d\lambda_o = D_\nu\,(-c_o/\lambda_o^2)$, which leads directly to

$$D_\lambda = -\frac{\lambda_o}{c_o}\frac{d^2 n}{d\lambda_o^2}.$$
(5.7-7)
Dispersion Coefficient
(s/m-nm)

In analogy with (5.7-5), for a source of spectral width σ_λ the temporal broadening of a pulse of light is

$$\sigma_\tau = |D_\lambda|\sigma_\lambda z.$$
(5.7-8)
Pulse Spread

As discussed in Secs. 10.3, 23.3, and 25.1, D_λ is usually specified in units of ps/km-nm in fiber-optics applications: the pulse broadening is measured in picoseconds, the length of the medium in kilometers, and the source spectral width in nanometers.

Normal and Anomalous Dispersion

Although the sign of the dispersion coefficient D_ν (or D_λ) does not affect the pulse-broadening rate, it does affect the phase of the complex envelope of the optical pulse. As such, the sign can play an important role in pulse propagation through media consisting of cascades of materials with different dispersion properties, as examined in Chapter 23. If $D_\nu > 0$ ($D_\lambda < 0$), the medium is said to exhibit **normal dispersion**. In this case, the travel time for higher-frequency components is greater than the travel time for lower-frequency components so that shorter-wavelength components of the pulse arrive later than longer-wavelength components, as illustrated schematically in Fig. 5.7-3. If $D_\nu < 0$ ($D_\lambda > 0$), the medium is said to exhibit **anomalous dispersion**, in which case the shorter-wavelength components travel faster and arrive earlier. Most glasses exhibit normal dispersion in the visible region of the spectrum; at longer wavelengths, however, the dispersion often becomes anomalous.

Single-Resonance Medium

The group velocity and dispersion coefficient for an optical pulse propagating through a single-resonance medium is determined by substituting (5.5-20) and (5.5-21) into

---

[†] An alternative definition of the dispersion coefficient, $M = -D_\lambda$, is also widely used in the literature.

5.7 PULSE PROPAGATION IN DISPERSIVE MEDIA

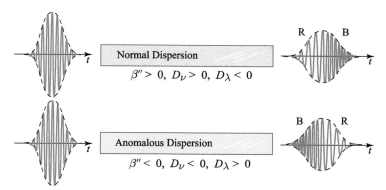

Figure 5.7-3 Propagation of an optical pulse through media with normal and anomalous dispersion. In a medium with normal dispersion the shorter-wavelength components of the pulse (B) arrive later that those with longer wavelength (R). A medium with anomalous dispersion exhibits the opposite behavior. The pulses are said to be chirped since the instantaneous frequency is time varying.

(5.5-5) and making use of (5.7-2) and (5.7-7). To illustrate the behavior of the pulse in this medium, the wavelength dependence of the refractive index n, the group index N, and the dispersion coefficient D_λ, are plotted in Fig. 5.7-4 as a function of normalized wavelength λ/λ_0, for a medium with parameters $\chi_0 = 0.05$ and $\Delta\nu/\nu_0 = 0.1$.

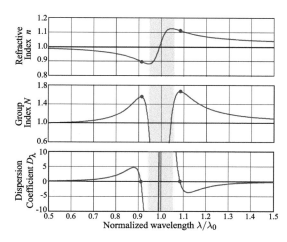

Figure 5.7-4 Wavelength dependence of the optical parameters associated with a single-resonance medium plotted as a function of the wavelength normalized to the wavelength at the resonance frequency, λ/λ_0: the refractive index $n = c_o/c$ (dots indicate points of inflection), the group index $N = c_o/v$ (dots indicate maxima), and the dispersion coefficient D_λ (dots indicate zeros). The parameters N and D_λ are not meaningful near resonance (shaded area).

In the vicinity of the resonance (shaded area in figure), n varies sufficiently rapidly with wavelength that the parameters N and D_λ, which are defined on the basis of a Taylor-series approximation comprising a few terms, cease to be meaningful. Away from the resonance, on both sides thereof, the refractive index decreases monotonically with increasing wavelength and exhibits points of inflection (indicated by dots). The first derivative of the refractive index achieves local maxima at these locations so that the group index N attains its maximum values there. Moreover, the second derivative vanishes at these points so that the dispersion coefficient changes sign. As the wavelength approaches the resonance wavelength from below, the dispersion changes from anomalous to normal; the reverse is true as the wavelength approaches resonance from above, as is evident in Fig. 5.7-4.

Fast and Slow Light in Resonant Media

As is evident in Fig. 5.7-4, in a resonant medium the refractive index n and the group index N undergo rapid changes near the resonance frequency, and may be substantially greater or smaller than unity. Consequently, the phase velocity $c = c_o/n$ and the group velocity $v = c_o/N$ may be significantly less than, or greater than, the velocity of light in free space, c_o. The group index, and hence the group velocity, may even be negative. This raises the question of a potential conflict with causality and the special theory of relativity, which provides that information cannot be transmitted at a velocity greater than c_o. It turns out that there is no such conflict since neither the group velocity nor the phase velocity corresponds to the **information velocity**, which is the speed at which information is transmitted between two points. The information velocity may be determined by tracing the propagation of a nonanalytic point on the pulse, for example, the onset of a rectangular pulse. It cannot exceed c_o.

The concepts of phase and group velocity were considered earlier in the context of an optical pulse traveling in a weakly dispersive medium, i.e., a medium with propagation constant $\beta(\omega)$ that is approximately linear in the vicinity of the central frequency of the pulse, ω_0. After traveling a distance z, the pulse is delayed by a time z/v and is modulated by a phase factor $\exp(-j\omega_0 z/c)$. This phase, which travels at the phase velocity c, carries no information. It is the group velocity v that governs the time of "arrival" of the pulse. Since, in this approximation, the pulse envelope maintains its shape as it travels (Fig. 5.7-1), the group velocity is a good approximation of the information velocity. In the resonant medium, this occurs at wavelengths far from resonance, where the group index is greater than unity and the group velocity is less than c_o.

At frequencies closer to resonance, higher-order dispersion terms become appreciable. In the presence of second-order dispersion (GVD), but negligible higher-order dispersion, a Gaussian pulse, for example, remains Gaussian, albeit with an increased width; its peak travels at the group velocity v. However, since the Gaussian pulse has a continuous profile and infinite support (i.e., extends over all time), the velocity at which the peak travels is not necessarily the information velocity; indeed, it can be greater than the free-space speed of light.

In the immediate vicinity of resonance, where the group velocity can be significantly greater than c_o and can even be negative (Fig. 5.7-4), higher-order dispersion terms must be considered. The pulse shape can then be significantly altered and the group velocity can no longer be considered as a possible information velocity. For sufficiently short distances, however, the pulse may travel without a significant alteration in shape, and this may occur at a group velocity significantly higher than c_o. The pulse can also travel at a negative group velocity, signifying that a point on the pulse, identified by a peak for example, arrives at the end of the medium before the corresponding point on the input pulse even enters the medium! In the opposite limit of slow light, certain special resonance media permit the group velocity of light to be made exceedingly small so that a light pulse may be substantially slowed or even halted. It should be emphasized, however, that in none of these situations does the information velocity exceed c_o.

Since fast- and slow-light phenomena can only be observed near resonance, where the absorption coefficient is large (and frequency dependent), a mechanism for optical amplification is necessary, and nonlinear effects are often exploited to enhance this phenomenon.

EXAMPLE 5.7-1. *Dispersion in a Multi-Resonance Medium: Fused Silica.* In the region between 0.21 and 3.71 μm, the wavelength dependence of the refractive index n for fused silica at room temperature is well characterized by the Sellmeier equation (5.5-28). Expressing all wavelengths in μm, this is achieved using three resonance wavelengths at $\lambda_1 = 0.06840$ μm, $\lambda_2 =$

0.1162 μm, and $\lambda_3 = 9.8962$ μm, with weights $\chi_{01} = 0.6962$, $\chi_{02} = 0.4079$, and $\chi_{03} = 0.8975$, respectively. Expressions for the group index N and the dispersion coefficient D_λ are readily derived from this equation by means of (5.7-2) and (5.7-7). The results of this calculation in the 600–1600-nm wavelength range are presented in Fig. 5.7-5.

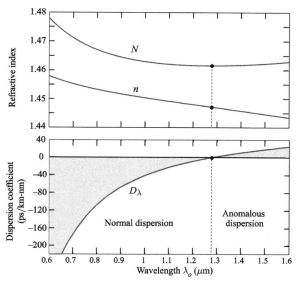

Figure 5.7-5 Wavelength dependence of optical parameters for fused silica calculated on the basis of the Sellmeier equation (5.5-28): the refractive index $n = c_o/c$ (dot indicates point of inflection), the group index $N = c_o/v$ (dot indicates minimum), and the dispersion coefficient D_λ (dot indicates zero).

The refractive index n is seen to decrease monotonically with increasing wavelength, and to exhibit a point of inflection at $\lambda_o = 1.276$ μm. At this wavelength, the group index N is minimum so that the group velocity $v = c_o/N$ is maximum. Since the dispersion coefficient D_λ is proportional to the second derivative of n with respect to λ_o, it vanishes at this wavelength. Zero dispersion coefficient signifies minimal pulse broadening. At wavelengths shorter than 1.276 μm, $D_\lambda < 0$ and the medium exhibits normal dispersion whereas at longer wavelengths, $D_\lambda > 0$ and the dispersion is anomalous. The presence of a zero-dispersion wavelength offers significant advantages in the design of optical fiber communication systems in which optical pulses carry information, as will become evident in Secs. 10.3, 25.1, and 23.3. The silica-glass fibers used in such systems are doped and exhibit zero dispersion close to 1.312 μm. The group index N is seen to be larger than the refractive index n by $\sim 2\%$, as is common in many materials.

READING LIST

Electromagnetics
See also the reading list on general optics in Chapter 1.

J.-M. Liu, *Principles of Photonics*, Cambridge University Press, 2016.

N. Ida, *Engineering Electromagnetics*, Springer-Verlag, 3rd ed. 2015.

J. C. Rautio, The Long Road to Maxwell's Equations, *IEEE Spectrum*, vol. 51, no. 12, pp. 36–40 & 54–56, 2014.

F. T. Ulaby and U. Ravaioli, *Fundamentals of Applied Electromagnetics*, Prentice Hall, 7th ed. 2014.

M. N. O. Sadiku, *Elements of Electromagnetics*, Oxford University Press, 6th ed. 2014.

U. S. Inan, A. Inan, and R. Said, *Engineering Electromagnetics and Waves*, Prentice Hall, 2nd ed. 2014.

A. Zangwill, *Modern Electrodynamics*, Cambridge University Press, 2013.

D. Fleisch, *A Student's Guide to Maxwell's Equations*, Cambridge University Press, 2013.

A. Balanis, *Advanced Engineering Electromagnetics*, Wiley, 2nd ed. 2012.

V. Lucarini, J. J. Saarinen, K.-E. Peiponen, and E. M. Vartiainen, *Kramers–Kronig Relations in Optical Materials Research*, Springer-Verlag, 2005.

S. A. Akhmanov and S. Yu. Nikitin, *Physical Optics*, Oxford University Press, 1997.

Electromagnetics Classics

J. D. Jackson, *Classical Electrodynamics*, Wiley, 3rd ed. 1999.

S. Ramo, J. R. Whinnery, and T. Van Duzer, *Fields and Waves in Communication Electronics*, Wiley, 3rd ed. 1994.

H. A. Haus and J. R. Melcher, *Electromagnetic Fields and Energy*, Prentice Hall, 1989.

H. A. Haus, *Waves and Fields in Optoelectronics*, Prentice Hall, 1984.

L. D. Landau, E. M. Lifshitz, and L. P. Pitaevskii, *Electrodynamics of Continuous Media*, Nauka (Moscow), 2nd ed. 1982; Butterworth–Heinemann, 2nd English ed. 1984, reprinted 2004.

L. D. Landau and E. M. Lifshitz, *The Classical Theory of Fields*, Nauka (Moscow), 6th revised ed. 1973; Butterworth–Heinemann, 4th revised English ed. 1975, reprinted with corrections 2000.

J. A. Stratton, *Electromagnetic Theory*, McGraw–Hill, 1941; Wiley–IEEE Classic reissue 2007.

H. A. Lorentz, *The Theory of Electrons and Its Applications to the Phenomena of Light and Radiant Heat*, Teubner, 1906; Dover, reissued 2004.

O. Heaviside, *Electromagnetic Theory*, "The Electrician" Printing and Publishing (London), 1893.

J. C. Maxwell, *A Treatise on Electricity and Magnetism*, Macmillan/Clarendon Press (Oxford), 1873.

Vector Beams

Q. Zhan, ed., *Vectorial Optical Fields*, World Scientific, 2014.

R. Dorn, S. Quabis, and G. Leuchs, Sharper Focus for a Radially Polarized Light Beam, *Physical Review Letters*, vol. 91, 233901, 2003.

D. G. Hall, Vector-Beam Solutions of Maxwell's Wave Equation, *Optics Letters*, vol. 21, pp. 9–11, 1996.

Optical Constants

P. Hartmann, *Optical Glass*, SPIE Optical Engineering Press, 2014.

M. Bass and V. N. Mahajan, eds., *Handbook of Optics*, McGraw–Hill, 3rd ed. 2010.

A. R. Hilton, Sr., *Chalcogenide Glasses for Infrared Optics*, McGraw–Hill, 2010.

E. D. Palik, ed., *Handbook of Optical Constants of Solids III*, Academic Press, 1998.

Y.-M. Chiang, D. Birnie III, and W. D. Kingery, *Physical Ceramics: Principles for Ceramic Science and Engineering*, Wiley, 1997.

W. L. Wolfe and G. J. Zissis, eds., *The Infrared Handbook*, Environmental Research Institute of Michigan, revised ed. 1993.

Fast and Slow Light

J. B. Khurgin and R. S. Tucker, eds., *Slow Light: Science and Applications*, CRC Press/Taylor & Francis, 2009.

P. W. Milonni, *Fast Light, Slow Light and Left-Handed Light*, Taylor & Francis, 2005.

L. Brillouin, *Wave Propagation and Group Velocity*, Academic Press, 1960.

Light Scattering

J. A. Adam, *Rays, Waves, and Scattering: Topics in Classical Mathematical Physics*, Princeton University Press, 2017.

G. Gouesbet and G. Gréhan, *Generalized Lorenz-Mie Theories*, Springer-Verlag, 2nd ed. 2017.

B. Chu, *Laser Light Scattering: Basic Principles and Practice*, Dover, paperback 2nd ed. 2007.

M. Nieto-Vesperinas, *Scattering and Diffraction in Physical Optics*, World Scientific, 2nd ed. 2006.

H. M. Nussenzveig, *Diffraction Effects in Semiclassical Scattering*, Cambridge University Press, 1992, paperback ed. 2006.

L. Tsang, J. A. Kong and K-H. Ding, *Scattering of Electromagnetic Waves: Theories and Applications*, Wiley, 2000.

B. J. Berne and R. Pecora, *Dynamic Light Scattering: With Applications to Chemistry, Biology, and Physics*, Wiley, 1976; Dover, reissued 2000.

M. I. Mishchenko, J. W. Hovenier, and L. D. Travis, eds., *Light Scattering by Nonspherical Particles: Theory, Measurements, and Applications*, Academic Press, 2000.

M. Kerker, ed., *Selected Papers on Light Scattering*, SPIE Optical Engineering Press (Milestone Series Volume 4), 1988.

C. F. Bohren and D. R. Huffman, *Absorption and Scattering of Light by Small Particles*, Wiley, 1983, paperback ed. 1998.

H. C. van de Hulst, *Light Scattering by Small Particles*, Wiley, 1957; Dover, reissued 1981.

Optical Pulse Propagation

K. E. Oughstun, *Electromagnetic and Optical Pulse Propagation 1: Spectral Representations in Temporally Dispersive Media*, Springer-Verlag, 2007, paperback ed. 2010.

PROBLEMS

5.1-1 **An Electromagnetic Wave.** An electromagnetic wave in free space has an electric-field vector $\mathcal{E} = f(t - z/c_o)\hat{\mathbf{x}}$, where $\hat{\mathbf{x}}$ is a unit vector in the x direction, and $f(t) = \exp(-t^2/\tau^2)\exp(j2\pi\nu_0 t)$, where τ is a constant. Describe the physical nature of this wave and determine an expression for the magnetic-field vector.

5.2-1 **Dielectric Media.** Identify the media described by the following equations, with respect to linearity, dispersiveness, spatial dispersiveness, and homogeneity. Assume that all media are isotropic.

(a) $\mathcal{P} = \epsilon_o \chi \mathcal{E} - a\nabla \times \mathcal{E}$,

(b) $\mathcal{P} + a\mathcal{P}^2 = \epsilon_o \mathcal{E}$,

(c) $a_1 \partial^2 \mathcal{P}/\partial t^2 + a_2 \partial \mathcal{P}/\partial t + \mathcal{P} = \epsilon_o \chi \mathcal{E}$,

(d) $\mathcal{P} = \epsilon_o \{a_1 + a_2 \exp[-(x^2 + y^2)]\}\mathcal{E}$,

where χ, a, a_1, and a_2 are constants.

5.3-1 **Traveling Standing Wave.** The electric-field complex-amplitude vector for a monochromatic wave of wavelength λ_o traveling in free space is $\mathbf{E}(\mathbf{r}) = E_0 \sin(\beta y) \exp(-j\beta z)\hat{\mathbf{x}}$.

(a) Determine a relation between β and λ_o.

(b) Derive an expression for the magnetic-field complex-amplitude vector $\mathbf{H}(\mathbf{r})$.

(c) Determine the direction of the flow of optical power.

(d) This wave may be regarded as the sum of two TEM plane waves. Determine their directions of propagation.

5.4-1 **Electric Field of Focused Light.**

(a) 1 W of optical power is focused uniformly on a flat target of size 0.1×0.1 mm^2 placed in free space. Determine the peak value of the electric field E_0 (V/m). Assume that the optical wave is approximated as a TEM plane wave within the area of the target.

(b) Determine the electric field at the center of a Gaussian beam (the point on the beam axis located at the beam waist) if the beam power is 1 W and the beam waist radius $W_0 = 0.1$ mm (refer to Sec. 3.1).

5.5-2 **Amplitude-Modulated Wave in a Dispersive Medium.** An amplitude-modulated wave whose complex wavefunction takes the form $\mathcal{A}(t) = [1 + m\cos(2\pi f_s t)]\exp(j2\pi\nu_0 t)$ at $z = 0$, where $f_s \ll \nu_0$, travels a distance z through a dispersive medium of propagation constant $\beta(\nu)$ and negligible attenuation. If $\beta(\nu_0) = \beta_0$, $\beta(\nu_0 - f_s) = \beta_1$, and $\beta(\nu_0 + f_s) = \beta_2$, derive an expression for the complex envelope of the transmitted wave as a function of β_0, β_1, β_2, and z. Show that at certain distances z the wave is amplitude modulated with no phase modulation.

5.7-1 **Group Velocity Dispersion in a Medium Described by the Sellmeier Equation.**

(a) Derive expressions for the group index N and the group velocity dispersion coefficient D_λ for a medium whose refractive index is described by the Sellmeier equation (5.5-28).

(b) Plot the wavelength dependence of n, N, and D_λ for fused silica in the region between 0.25 and 3.5 μm. Make use of the parameters provided in Table 5.5-1 (and in Example 5.7-1). Verify the curves provided in Fig. 5.7-5.

(c) Construct a similar collection of plots for GaAs in the region between 1.5 and 10.5 μm. As indicated in Table 5.5-1, in the wavelength region between 1.4 and 11 μm, at room temperature, GaAs is characterized by a 3-term Sellmeier equation with resonance wavelengths at 0 μm, 0.4082 μm, and 37.17 μm, with associated weights given by 3.5, 7.4969, and 1.9347, respectively. Compare and contrast the behavior of the dispersion properties of fused silica with those of GaAs.

5.7-2 **Refractive Index of Air.** The refractive index of air can be precisely measured with the help of a Michelson interferometer and a tunable light source. At atmospheric pressure, and a temperature of 20° C, the refractive index of air differs from unity by $n - 1 = 2.672 \times 10^{-4}$ at a wavelength of 0.76 μm, by $n - 1 = 2.669 \times 10^{-4}$ at a wavelength of 0.81 μm, and by $n - 1 = 2.665 \times 10^{-4}$ at a wavelength of 0.86 μm.

(a) Using a quadratic fit to these data, determine the wavelength dependence of the group velocity.

(b) Obtain an expression for the dispersion coefficient D_λ in ps/km-nm and compare your result with that for a silica optical fiber.

CHAPTER 6

POLARIZATION OPTICS

6.1	POLARIZATION OF LIGHT	211
	A. Polarization	
	B. Matrix Representation	
6.2	REFLECTION AND REFRACTION	221
6.3	OPTICS OF ANISOTROPIC MEDIA	227
	A. Refractive Indices	
	B. Propagation Along a Principal Axis	
	C. Propagation in an Arbitrary Direction	
	D. Dispersion Relation, Rays, Wavefronts, and Energy Transport	
	E. Double Refraction	
6.4	OPTICAL ACTIVITY AND MAGNETO-OPTICS	240
	A. Optical Activity	
	B. Magneto-Optics: The Faraday Effect	
6.5	OPTICS OF LIQUID CRYSTALS	244
6.6	POLARIZATION DEVICES	247
	A. Polarizers	
	B. Wave Retarders	
	C. Polarization Rotators	
	D. Nonreciprocal Polarization Devices	

The French physicist **Augustin-Jean Fresnel (1788–1827)** put forth a transverse wave theory of light. Equations describing the partial reflection and refraction of light are named in his honor. Fresnel also made important contributions to the theory of light diffraction.

Sir George Gabriel Stokes (1819–1903), an Irish mathematician and physicist, developed a description of light that encompasses intensity as well as state of polarization. He also made seminal contributions to wave optics, fluorescence, and optical aberrations.

Fundamentals of Photonics, Third Edition. Bahaa E. A. Saleh and Malvin Carl Teich.
©2019 John Wiley & Sons, Inc. Published 2019 by John Wiley & Sons, Inc.

The polarization of light at a given position in space is determined by the path taken by its electric-field vector $\mathcal{E}(\mathbf{r}, t)$ in time. In a simple medium, this vector lies in a plane tangential to the wavefront at that position. For monochromatic light, any two orthogonal components of the complex-amplitude vector $\mathbf{E}(\mathbf{r})$ in that plane vary sinusoidally in time, with amplitudes and phases that generally differ, so that the endpoint of the vector $\mathbf{E}(\mathbf{r})$ traces out an ellipse. Since the directions of the wavefront normals are generally position-dependent, so too are the tangential planes, along with the orientations and shapes of the ellipses, as illustrated in Fig. 6.0-1(a).

For a plane wave, however, the wavefronts are parallel transverse planes and the polarization ellipses are the same everywhere, as illustrated in Fig. 6.0-1(b), although the field vectors need not be parallel at any given time. A plane wave is therefore described by a single ellipse and is said to be **elliptically polarized**. The orientation and ellipticity of the polarization ellipse determine the state of polarization of the plane wave, while its size is established by the optical intensity. When the ellipse reduces to a straight line or becomes a circle, the wave is said to be **linearly polarized** or **circularly polarized**, respectively.

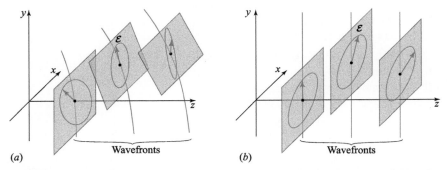

Figure 6.0-1 Trace of the time course of the electric-field vector endpoint for monochromatic light at several positions. (a) Arbitrary wave. (b) Plane or paraxial wave traveling in the z direction.

In paraxial optics, light propagates along directions that lie within a narrow cone centered about the optical axis (the z axis). Waves are approximately transverse electromagnetic (TEM) in character and the electric-field vectors therefore lie approximately in transverse planes, with negligible axial components. From the perspective of polarization, paraxial waves may be approximated by plane waves and described by a single polarization ellipse (or line or circle).

Polarization plays an important role in the interaction of light with matter as attested to by the following examples:
- The proportion of light reflected at the boundary between two materials depends on the polarization of the incident wave.
- The proportion of light absorbed by certain materials is polarization-dependent.
- Light scattering from matter is generally polarization-sensitive.
- The refractive index of anisotropic materials depends on the polarization. Waves with different polarizations travel at different velocities and undergo different phase shifts, so that the polarization ellipse is modified as the wave advances (linearly polarized light can be transformed into circularly polarized light, as a simple example). This feature finds use in the design of many optical devices.

- The polarization plane of linearly polarized light is rotated by passage through certain materials, including optically active media, liquid crystals, and some substances in the presence of an external magnetic field.

This Chapter

This chapter is devoted to a description of elementary polarization phenomena and a number of their applications. Elliptically polarized light is introduced in Sec. 6.1 using a matrix formalism that is convenient for describing polarization devices. Section 6.2 describes the effect of polarization on the reflection and refraction of light at boundaries between dielectric media. The propagation of light through anisotropic media (crystals), optically active media, and liquid crystals are the topics of Secs. 6.3, 6.4, and 6.5, respectively. Finally, elementary polarization devices (such as polarizers, retarders, rotators, and isolators) are discussed in Sec. 6.6. Polarization is important for understanding how light behaves in photonic crystals, metals, and metamaterials (Chapters 7 and 8), and how light is guided (Chapters 9 and 10) and stored (Chapter 11). The polarization properties of random light are considered in Sec. 12.4. Polarization clearly plays a central role in many areas of optics and photonics.

6.1 POLARIZATION OF LIGHT

A. Polarization

Consider a monochromatic plane wave of frequency ν and angular frequency $\omega = 2\pi\nu$ traveling in the z direction with velocity c. The electric field lies in the x–y plane and is generally described by

$$\mathcal{E}(z,t) = \mathrm{Re}\left\{\mathbf{A}\exp\left[j\omega\left(t - \frac{z}{c}\right)\right]\right\}, \tag{6.1-1}$$

where the complex envelope

$$\mathbf{A} = A_x\widehat{\mathbf{x}} + A_y\widehat{\mathbf{y}}, \tag{6.1-2}$$

is a vector with complex components A_x and A_y. To describe the polarization of this wave, we trace the endpoint of the vector $\mathcal{E}(z,t)$ at each position z as a function of time.

Polarization Ellipse

Expressing A_x and A_y in terms of their magnitudes and phases, $A_x = a_x \exp(j\varphi_x)$ and $A_y = a_y \exp(j\varphi_y)$, and substituting into (6.1-2) and (6.1-1) we obtain

$$\mathcal{E}(z,t) = \mathcal{E}_x\widehat{\mathbf{x}} + \mathcal{E}_y\widehat{\mathbf{y}}, \tag{6.1-3}$$

where

$$\mathcal{E}_x = a_x \cos\left[\omega\left(t - \frac{z}{c}\right) + \varphi_x\right] \tag{6.1-4a}$$

$$\mathcal{E}_y = a_y \cos\left[\omega\left(t - \frac{z}{c}\right) + \varphi_y\right] \tag{6.1-4b}$$

are the x and y components of the electric-field vector $\mathcal{E}(z,t)$. The components $\mathcal{E}_x$ and $\mathcal{E}_y$ are periodic functions of $(t - z/c)$ that oscillate at frequency ν. Equations (6.1-4) are the parametric equations of the ellipse

$$\frac{\mathcal{E}_x^2}{a_x^2} + \frac{\mathcal{E}_y^2}{a_y^2} - 2\cos\varphi\,\frac{\mathcal{E}_x\mathcal{E}_y}{a_x a_y} = \sin^2\varphi, \tag{6.1-5}$$

where $\varphi = \varphi_y - \varphi_x$ is the phase difference.

At a fixed value of z, the tip of the electric-field vector rotates periodically in the x–y plane, tracing out this ellipse. At a fixed time t, the locus of the tip of the electric-field vector traces out a helical trajectory in space that lies on the surface of an elliptical cylinder (see Fig. 6.1-1). The electric field rotates as the wave advances, repeating its motion periodically for each distance corresponding to a wavelength $\lambda = c/\nu$.

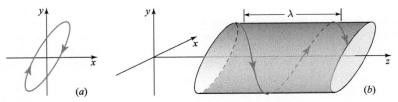

Figure 6.1-1 (a) Rotation of the endpoint of the electric-field vector in the x–y plane at a fixed position z. (b) Trajectory of the endpoint of the electric-field vector as the wave advances.

The state of polarization of the wave is determined by the orientation and shape of the **polarization ellipse**, which is characterized by the two angles defined in Fig. 6.1-2: the angle ψ determines the direction of the major axis, whereas the angle χ determines the ellipticity, namely the ratio of the minor to major axes of the ellipse b/a. These angles depend on the ratio of the complex-envelope magnitudes $R = a_y/a_x$, and on the phase difference $\varphi = \varphi_y - \varphi_x$, in accordance with the following relations:

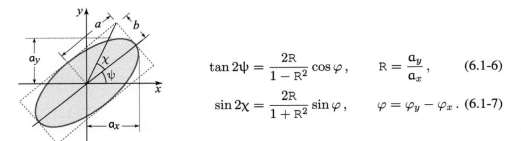

$$\tan 2\psi = \frac{2R}{1 - R^2} \cos \varphi, \qquad R = \frac{a_y}{a_x}, \qquad (6.1\text{-}6)$$

$$\sin 2\chi = \frac{2R}{1 + R^2} \sin \varphi, \qquad \varphi = \varphi_y - \varphi_x. \qquad (6.1\text{-}7)$$

Figure 6.1-2
Polarization ellipse.

Equations (6.1-6) and (6.1-7) may be derived by finding the angle ψ that achieves a transformation of the coordinate system of $\mathcal{E}_x$ and $\mathcal{E}_y$ in (6.1-5) such that the rotated ellipse has no cross term. The size of the ellipse is determined by the intensity of the wave, which is proportional to $|A_x|^2 + |A_y|^2 = a_x^2 + a_y^2$.

Linearly Polarized Light

If one of the components vanishes ($a_x = 0$, for example), the light is **linearly polarized (LP)** in the direction of the other component (the y direction). The wave is also linearly polarized if the phase difference $\varphi = 0$ or π, since (6.1-5) then yields $\mathcal{E}_y = \pm(a_y/a_x)\mathcal{E}_x$, which is the equation of a straight line of slope $\pm a_y/a_x$ (the $+$ and $-$ signs correspond to $\varphi = 0$ and π, respectively). In these cases, the elliptical cylinder in Fig. 6.1-1(b) collapses to a plane, as illustrated in Fig. 6.1-3. The wave is therefore also said to have **planar polarization**. As an example, if $a_x = a_y$, the plane of polarization makes an angle of $45°$ with respect to the x axis. If $a_x = 0$, on the other hand, the plane of polarization is the y–z plane.

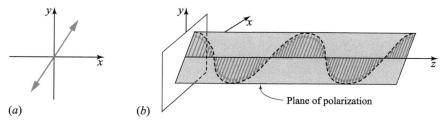

(a) (b)

Figure 6.1-3 Linearly polarized light (also called plane polarized light). (a) Time course of the field at a fixed position z. (b) Endpoint of the electric-field vector at position z at a fixed time t.

Circularly Polarized Light

If $\varphi = \pm\pi/2$ and $a_x = a_y = a_0$, (6.1-4) gives $\mathcal{E}_x = a_0 \cos[\omega(t - z/c) + \varphi_x]$ and $\mathcal{E}_y = \mp a_0 \sin[\omega(t - z/c) + \varphi_x]$, from which $\mathcal{E}_x^2 + \mathcal{E}_y^2 = a_0^2$, which is the equation of a circle. The elliptical cylinder in Fig. 6.1-1(b) becomes a circular cylinder and the wave is said to be circularly polarized. In the case $\varphi = -\pi/2$, the electric field at a fixed position z rotates in a counterclockwise direction when viewed from the direction toward which the wave is approaching. The light is then said to be **left circularly polarized (LCP)** [Fig. 6.1-4(a)]. The case $\varphi = +\pi/2$ corresponds to clockwise rotation and **right circularly polarized (RCP)** light.[†] In the left circular case, the locus traced by the endpoint of the electric-field vector at different positions is a left-handed helix, as illustrated in Fig. 6.1-4(b). For right circular polarization, it is a right-handed helix. The helical path of the *endpoint* of the electric-field vector for the plane-wave circularly polarized light considered here is to be distinguished from the helical *wavefront* of the Laguerre–Gaussian beam discussed in Sec. 3.4 [Fig. 3.4-1(c)].

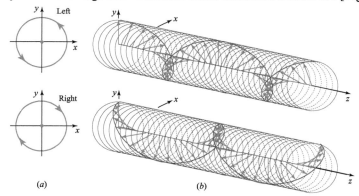

(a) (b)

Figure 6.1-4 Motion of the endpoints of the electric-field vectors for left and right circularly polarized plane waves. (a) Time course at a fixed position z. (b) Trajectories of the endpoints.

Poincaré Sphere and Stokes Parameters

As indicated above, the state of polarization of a light wave can be described by two real parameters: the magnitude ratio $R = a_y/a_x$ and the phase difference $\varphi = \varphi_y - \varphi_x$. These are sometimes lumped into a single complex number $R\exp(j\varphi)$, called the **complex polarization ratio**. Alternatively, we may characterize the state of polarization by the two angles ψ and χ, which represent the orientation and ellipticity of the polarization ellipse, respectively, as defined in Fig. 6.1-2.

[†] This convention is used in most optics textbooks. The opposite designation is often used in the engineering literature: in the case of right (left) circularly polarized light, the electric-field vector at a fixed position rotates counterclockwise (clockwise) when viewed from the direction toward which the wave is approaching.

The **Poincaré sphere** (see Fig. 6.1-5) is a geometrical construct in which the state of polarization is represented by a point on the surface of a sphere of unit radius, with coordinates ($r = 1$, $\theta = 90° - 2\chi$, $\phi = 2\psi$) in a spherical coordinate system. Each point on the sphere represents a polarization state. For example, points on the equator ($\chi = 0°$) represent states of linear polarization, with the two points $2\psi = 0°$ and $2\psi = 180°$ representing linear polarization along the x and y axes, respectively. The north and south poles ($2\chi = \pm 90°$) represent right-handed and left-handed circular polarization, respectively. Other points on the surface of the sphere represent states of elliptical polarization.

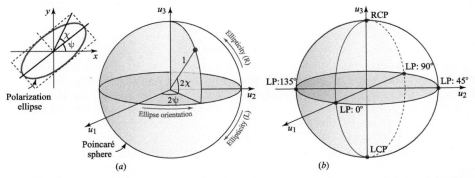

Figure 6.1-5 (a) The orientation and ellipticity of the polarization ellipse are represented geometrically as a point on the surface of the Poincaré sphere. (b) Points on the Poincaré sphere representing linearly polarized (LP) light at various angles with the x direction, as well as right-circularly polarized (RCP) and left-circularly polarized (LCP) light. Points in the interior of the sphere represent partially polarized light (the point at the origin of the sphere represents unpolarized light), as illustrated in Fig. 12.4-1.

The two real quantities (R, φ), or equivalently the angles (χ, ψ), describe the state of polarization but contain no information about the intensity of the wave. Another representation that does contain such information is the **Stokes vector**. This is a set of four real numbers (S_0, S_1, S_2, S_3), called the **Stokes parameters**. The first of these, $S_0 = a_x^2 + a_y^2$, is proportional to the optical intensity whereas the other three, (S_1, S_2, S_3), are the Cartesian coordinates of the point on the Poincaré sphere, (u_1, u_2, u_3) = ($\cos 2\chi \cos 2\psi$, $\cos 2\chi \sin 2\psi$, $\sin 2\chi$), multiplied by S_0, so that

$$S_1 = S_0 \cos 2\chi \cos 2\psi \qquad (6.1\text{-}8a)$$

$$S_2 = S_0 \cos 2\chi \sin 2\psi \qquad (6.1\text{-}8b)$$

$$S_3 = S_0 \sin 2\chi. \qquad (6.1\text{-}8c)$$

Using (6.1-6) and (6.1-7), together with a few trigonometric identities, the Stokes parameters in (6.1-8) may be expressed in terms of the field parameters (a_x, a_y, φ), and in terms of the components of the complex envelope (A_x, A_y), as:

$$S_0 = a_x^2 + a_y^2 \quad = |A_x|^2 + |A_y|^2 \qquad (6.1\text{-}9a)$$

$$S_1 = a_x^2 - a_y^2 \quad = |A_x|^2 - |A_y|^2 \qquad (6.1\text{-}9b)$$

$$S_2 = 2a_x a_y \cos \varphi = 2 \operatorname{Re}\{A_x^* A_y\} \qquad (6.1\text{-}9c)$$

$$S_3 = 2a_x a_y \sin \varphi = 2 \operatorname{Im}\{A_x^* A_y\}. \qquad (6.1\text{-}9d)$$

Stokes Parameters

Since $S_1^2 + S_2^2 + S_3^2 = S_0^2$, only three of the four components of the Stokes vector are independent; they completely define the intensity and the state of polarization of the light. A generalization of the Stokes parameters suitable for describing partially coherent light is presented in Sec. 12.4.

We conclude that there are a number of equivalent representations for describing the state of polarization of an optical field: (1) the polarization ellipse; (2) the Poincaré sphere; and (3) the Stokes vector. Yet another equivalent representation, the Jones vector, is introduced in the following section.

B. Matrix Representation

The Jones Vector

As indicated above, a monochromatic plane wave of frequency ν traveling in the z direction is completely characterized by the complex envelopes $A_x = a_x \exp(j\varphi_x)$ and $A_y = a_y \exp(j\varphi_y)$ of the x and y components of the electric-field vector. These complex quantities may be written in the form of a column matrix known as the **Jones vector**:

$$\mathbf{J} = \begin{bmatrix} A_x \\ A_y \end{bmatrix}. \tag{6.1-10}$$

Given $\mathbf{J}$, we can determine the total intensity of the wave, $I = (|A_x|^2 + |A_y|^2)/2\eta$, and use the ratio $R = a_y/a_x = |A_y|/|A_x|$ and the phase difference $\varphi = \varphi_y - \varphi_x = \arg\{A_y\} - \arg\{A_x\}$ to determine the orientation and shape of the polarization ellipse, as well as the Poincaré sphere and the Stokes parameters.

The Jones vectors for some special polarization states are provided in Table 6.1-1. The intensity in each case has been normalized so that $|A_x|^2 + |A_y|^2 = 1$ and the phase of the x component is taken to be $\varphi_x = 0$.

Table 6.1-1 Jones vectors of linearly polarized (LP) and right- and left-circularly polarized (RCP, LCP) light.

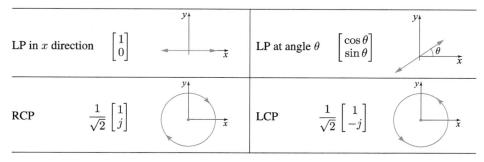

Orthogonal Polarizations

Two polarization states represented by the Jones vectors $\mathbf{J}_1$ and $\mathbf{J}_2$ are said to be orthogonal if the inner product between $\mathbf{J}_1$ and $\mathbf{J}_2$ is zero. The inner product is defined by

$$(\mathbf{J}_1, \mathbf{J}_2) = A_{1x} A_{2x}^* + A_{1y} A_{2y}^*, \tag{6.1-11}$$

where A_{1x} and A_{1y} are the elements of $\mathbf{J}_1$ and A_{2x} and A_{2y} are the elements of $\mathbf{J}_2$. An example of orthogonal Jones vectors are the linearly polarized waves in the x and

y directions, or any other pair of orthogonal directions. Another example is provided by right and left circularly polarized waves.

Expansion of Arbitrary Polarization as a Superposition of Two Orthogonal Polarizations

An arbitrary Jones vector $\mathbf{J}$ can always be analyzed as a weighted superposition of two orthogonal Jones vectors, say $\mathbf{J}_1$ and $\mathbf{J}_2$, that form the expansion basis; thus $\mathbf{J} = \alpha_1 \mathbf{J}_1 + \alpha_2 \mathbf{J}_2$. If $\mathbf{J}_1$ and $\mathbf{J}_2$ are normalized such that $(\mathbf{J}_1, \mathbf{J}_1) = (\mathbf{J}_2, \mathbf{J}_2) = 1$, the expansion coefficients are the inner products $\alpha_1 = (\mathbf{J}, \mathbf{J}_1)$ and $\alpha_2 = (\mathbf{J}, \mathbf{J}_2)$.

EXAMPLE 6.1-1. *Expansions in Linearly Polarized and Circularly Polarized Bases.*
Using the x and y linearly polarized vectors $\begin{bmatrix} 1 \\ 0 \end{bmatrix}$ and $\begin{bmatrix} 0 \\ 1 \end{bmatrix}$ as an expansion basis, the expansion coefficients for a Jones vector of components A_x and A_y with $|A_x|^2 + |A_y|^2 = 1$ are, by definition, $\alpha_1 = A_x$ and $\alpha_2 = A_y$. The same polarization state may be expanded in other bases.

- In a basis of linearly polarized vectors at angles $45°$ and $135°$, i.e., $\mathbf{J}_1 = \frac{1}{\sqrt{2}} \begin{bmatrix} 1 \\ 1 \end{bmatrix}$ and $\mathbf{J}_2 = \frac{1}{\sqrt{2}} \begin{bmatrix} -1 \\ 1 \end{bmatrix}$, the expansion coefficients α_1 and α_2 are:

$$A_{45} = \frac{1}{\sqrt{2}}(A_x + A_y), \quad A_{135} = \frac{1}{\sqrt{2}}(A_y - A_x). \tag{6.1-12}$$

- Similarly, if the right and left circularly polarized waves $\frac{1}{\sqrt{2}} \begin{bmatrix} 1 \\ j \end{bmatrix}$ and $\frac{1}{\sqrt{2}} \begin{bmatrix} 1 \\ -j \end{bmatrix}$ are used as an expansion basis, the coefficients α_1 and α_2 are:

$$A_R = \frac{1}{\sqrt{2}}(A_x - jA_y), \quad A_L = \frac{1}{\sqrt{2}}(A_x + jA_y). \tag{6.1-13}$$

For example, a linearly polarized wave with a plane of polarization that makes an angle θ with the x axis (i.e., $A_x = \cos\theta$ and $A_y = \sin\theta$) is equivalent to a superposition of right and left circularly polarized waves with coefficients $\frac{1}{\sqrt{2}} e^{-j\theta}$ and $\frac{1}{\sqrt{2}} e^{j\theta}$, respectively. A linearly polarized wave therefore equals a weighted sum of right and left circularly polarized waves.

EXERCISE 6.1-1

Measurement of the Stokes Parameters. Show that the Stokes parameters defined in (6.1-9) for light with Jones vector components A_x and A_y are given by

$$S_0 = |A_x|^2 + |A_y|^2 \tag{6.1-14a}$$

$$S_1 = |A_x|^2 - |A_y|^2 \tag{6.1-14b}$$

$$S_2 = |A_{45}|^2 - |A_{135}|^2 \tag{6.1-14c}$$

$$S_3 = |A_R|^2 - |A_L|^2, \tag{6.1-14d}$$

where A_{45} and A_{135} are the coefficients of expansion in a basis of linearly polarized vectors at angles $45°$ and $135°$ as in (6.1-12), and A_R and A_L are the coefficients of expansion in a basis of the right and left circularly polarized waves set forth in (6.1-13). Suggest a method of measuring the Stokes parameters for light of arbitrary polarization.

Matrix Representation of Polarization Devices

Consider the transmission of a plane wave of arbitrary polarization through an optical system that maintains the plane-wave nature of the wave, but alters its polarization, as illustrated schematically in Fig. 6.1-6. The system is assumed to be linear, so that the principle of superposition of optical fields is obeyed. Two examples of such systems are the reflection of light from a planar boundary between two media, and the transmission of light through a plate with anisotropic optical properties.

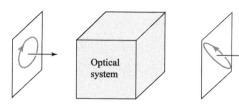

Figure 6.1-6 An optical system that alters the polarization of a plane wave.

The complex envelopes of the two electric-field components of the input (incident) wave, A_{1x} and A_{1y}, and those of the output (transmitted or reflected) wave, A_{2x} and A_{2y}, are in general related by the weighted superpositions

$$A_{2x} = T_{11} A_{1x} + T_{12} A_{1y}$$
$$A_{2y} = T_{21} A_{1x} + T_{22} A_{1y}, \qquad (6.1\text{-}15)$$

where T_{11}, T_{12}, T_{21}, and T_{22} are constants describing the device. Equations (6.1-15) are general relations that all linear optical polarization devices must satisfy.

The linear relations in (6.1-15) may conveniently be written in matrix notation by defining a 2×2 matrix **T** with elements T_{11}, T_{12}, T_{21}, and T_{22} so that

$$\begin{bmatrix} A_{2x} \\ A_{2y} \end{bmatrix} = \begin{bmatrix} T_{11} & T_{12} \\ T_{21} & T_{22} \end{bmatrix} \begin{bmatrix} A_{1x} \\ A_{1y} \end{bmatrix}. \qquad (6.1\text{-}16)$$

If the input and output waves are described by the Jones vectors $\mathbf{J}_1$ and $\mathbf{J}_2$, respectively, then (6.1-16) may be written in the compact matrix form

$$\mathbf{J}_2 = \mathbf{T}\mathbf{J}_1. \qquad (6.1\text{-}17)$$

The matrix **T**, called the **Jones matrix**, describes the optical system, whereas the vectors $\mathbf{J}_1$ and $\mathbf{J}_2$ describe the input and output waves.

The structure of the Jones matrix **T** of a given optical system determines its effect on the polarization state and intensity of the wave. The following is a compilation of the Jones matrices of some systems with simple characteristics. Physical devices that have such characteristics will be discussed subsequently in this chapter.

Linear polarizers. The system represented by the Jones matrix

$$\mathbf{T} = \begin{bmatrix} 1 & 0 \\ 0 & 0 \end{bmatrix} \qquad (6.1\text{-}18)$$
Linear Polarizer
Along x Direction

transforms a wave of components (A_{1x}, A_{1y}) into a wave of components $(A_{1x}, 0)$ by eliminating the y component, thereby yielding a wave polarized along the x direction, as illustrated in Fig. 6.1-7. The system is a **linear polarizer** with its transmission axis pointing in the x direction.

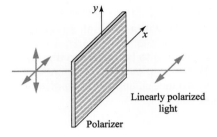

Figure 6.1-7 The linear polarizer. The lines in the polarizer represent the field direction that is permitted to pass.

Wave retarders. The system represented by the matrix

$$\mathbf{T} = \begin{bmatrix} 1 & 0 \\ 0 & e^{-j\Gamma} \end{bmatrix} \qquad (6.1\text{-}19)$$
Wave-Retarder
(Fast Axis Along x Direction)

transforms a wave with field components (A_{1x}, A_{1y}) into another with components $(A_{1x}, e^{-j\Gamma}A_{1y})$, thereby delaying the y component by a phase Γ while leaving the x component unchanged. It is therefore called a **wave retarder**. The x and y axes are called the fast and slow axes of the retarder, respectively.

The simple application of matrix algebra permits the results illustrated in Fig. 6.1-8 to be understood:

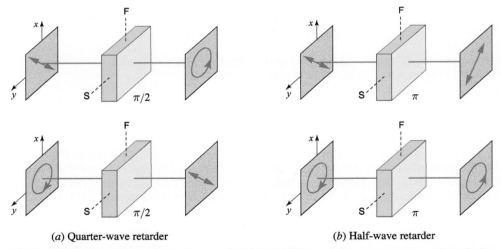

(a) Quarter-wave retarder (b) Half-wave retarder

Figure 6.1-8 Operations of quarter-wave ($\pi/2$) and half-wave (π) retarders on several particular states of polarization are shown in (a) and (b), respectively. F and S represent the fast and slow axes of the retarder, respectively.

- When $\Gamma = \pi/2$, the retarder (called a **quarter-wave retarder** and described by the Jones matrix $\begin{bmatrix} 1 & 0 \\ 0 & -j \end{bmatrix}$) converts the linearly polarized wave $\begin{bmatrix} 1 \\ 1 \end{bmatrix}$ into the left circularly polarized wave $\begin{bmatrix} 1 \\ -j \end{bmatrix}$, and converts the right circularly polarized wave $\begin{bmatrix} 1 \\ j \end{bmatrix}$ into the linearly polarized wave $\begin{bmatrix} 1 \\ 1 \end{bmatrix}$.

- When $\Gamma = \pi$, the retarder (called a **half-wave retarder** and described by the Jones matrix $\begin{bmatrix} 1 & 0 \\ 0 & -1 \end{bmatrix}$) converts the linearly polarized wave $\begin{bmatrix} 1 \\ 1 \end{bmatrix}$ into the linearly polarized wave $\begin{bmatrix} 1 \\ -1 \end{bmatrix}$, thereby rotating the plane of polarization by 90°. The half-wave retarder converts the right circularly polarized wave $\begin{bmatrix} 1 \\ j \end{bmatrix}$ into the left circularly polarized wave $\begin{bmatrix} 1 \\ -j \end{bmatrix}$.

Polarization rotators. While a wave retarder can transform a wave with one form of polarization into another, a **polarization rotator** always maintains the linear polarization of a wave but rotates the plane of polarization by a particular angle. The Jones matrix

$$\mathbf{T} = \begin{bmatrix} \cos\theta & -\sin\theta \\ \sin\theta & \cos\theta \end{bmatrix} \qquad (6.1\text{-}20)$$
Polarization Rotator

represents a device that converts a linearly polarized wave $\begin{bmatrix} \cos\theta_1 \\ \sin\theta_1 \end{bmatrix}$ into another linearly polarized wave $\begin{bmatrix} \cos\theta_2 \\ \sin\theta_2 \end{bmatrix}$, where $\theta_2 = \theta_1 + \theta$. It therefore rotates the plane of polarization of a linearly polarized wave by an angle θ.

Cascaded Polarization Devices

The action of cascaded optical systems on polarized light may be conveniently determined by using conventional matrix multiplication formulas. A system characterized by the Jones matrix $\mathbf{T}_1$ followed by another characterized by $\mathbf{T}_2$ are equivalent to a single system characterized by the product matrix $\mathbf{T} = \mathbf{T}_2\mathbf{T}_1$. The matrix of the system through which light is first transmitted must stand to the right in the matrix product since it is the first to affect the input Jones vector.

EXERCISE 6.1-2

Cascaded Wave Retarders. Show that two cascaded quarter-wave retarders with parallel fast axes are equivalent to a half-wave retarder. What is the result if the fast axes are orthogonal?

Coordinate Transformation

The elements of the Jones vectors and Jones matrices are dependent on the choice of the coordinate system. However, if these elements are known in one coordinate system, they can be determined in another coordinate system by using matrix methods. If $\mathbf{J}$ is the Jones vector in the x–y coordinate system, then in a new coordinate system x'–y', with the x' direction making an angle θ with the x direction, the Jones vector $\mathbf{J}'$ is given by

$$\mathbf{J}' = \mathbf{R}(\theta)\,\mathbf{J}, \qquad (6.1\text{-}21)$$

where $\mathbf{R}(\theta)$ is the coordinate-transformation matrix

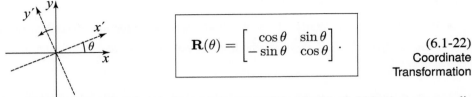

$$\mathbf{R}(\theta) = \begin{bmatrix} \cos\theta & \sin\theta \\ -\sin\theta & \cos\theta \end{bmatrix}. \qquad (6.1\text{-}22)$$
Coordinate Transformation

This can be verified by relating the components of the electric field in the two coordinate systems.

The Jones matrix $\mathbf{T}$, which represents an optical system, is similarly transformed into $\mathbf{T}'$, in accordance with the matrix relations

$$\mathbf{T}' = \mathbf{R}(\theta)\,\mathbf{T}\,\mathbf{R}(-\theta) \qquad (6.1\text{-}23)$$

$$\mathbf{T} = \mathbf{R}(-\theta)\,\mathbf{T}'\,\mathbf{R}(\theta), \qquad (6.1\text{-}24)$$

where $\mathbf{R}(-\theta)$ is given by (6.1-22) with $-\theta$ replacing θ. The matrix $\mathbf{R}(-\theta)$ is the inverse of $\mathbf{R}(\theta)$, so that $\mathbf{R}(-\theta)\,\mathbf{R}(\theta)$ is a unit matrix. Equation (6.1-23) can be obtained by using the relation $\mathbf{J}_2 = \mathbf{T}\mathbf{J}_1$ and the transformation $\mathbf{J}_2' = \mathbf{R}(\theta)\,\mathbf{J}_2 = \mathbf{R}(\theta)\,\mathbf{T}\,\mathbf{J}_1$. Since $\mathbf{J}_1 = \mathbf{R}(-\theta)\,\mathbf{J}_1'$, we have $\mathbf{J}_2' = \mathbf{R}(\theta)\,\mathbf{T}\,\mathbf{R}(-\theta)\,\mathbf{J}_1'$; since $\mathbf{J}_2' = \mathbf{T}'\,\mathbf{J}_1'$, (6.1-23) follows.

EXERCISE 6.1-3

Jones Matrix of a Rotated Half-Wave Retarder. Show that the Jones matrix of a half-wave retarder whose fast axis makes an angle θ with the x axis is

$$\mathbf{T} = \begin{bmatrix} \cos 2\theta & \sin 2\theta \\ \sin 2\theta & -\cos 2\theta \end{bmatrix}. \qquad (6.1\text{-}25)$$
Half-Wave Retarder at Angle θ

Derive (6.1-25) using (6.1-19), (6.1-22), and (6.1-24). Demonstrate that if $\theta = 22.5°$, the output waves are proportional to the sum and difference of the input waves.

Normal Modes

The normal modes of a polarization system are the states of polarization that are not changed when the wave is transmitted through the system (see Appendix C). These states have Jones vectors satisfying

$$\mathbf{T}\mathbf{J} = \mu\mathbf{J}, \qquad (6.1\text{-}26)$$

where μ is constant. The normal modes are therefore the eigenvectors of the Jones matrix $\mathbf{T}$, and the values of μ are the corresponding eigenvalues. Since the matrix $\mathbf{T}$ is of size 2×2 there are only two independent normal modes, $\mathbf{T}\mathbf{J}_1 = \mu_1\mathbf{J}_1$ and $\mathbf{T}\mathbf{J}_2 = \mu_2\mathbf{J}_2$. If the matrix $\mathbf{T}$ is a Hermitian, i.e., if $T_{12} = T_{21}^*$, the normal modes are orthogonal: $(\mathbf{J}_1, \mathbf{J}_2) = 0$. The normal modes are usually used as an expansion basis, so that an arbitrary input wave $\mathbf{J}$ may be expanded as a superposition of normal modes: $\mathbf{J} = \alpha_1\mathbf{J}_1 + \alpha_2\mathbf{J}_2$. The response of the system may then be easily evaluated since $\mathbf{T}\mathbf{J} = \mathbf{T}(\alpha_1\mathbf{J}_1 + \alpha_2\mathbf{J}_2) = \alpha_1\mathbf{T}\mathbf{J}_1 + \alpha_2\mathbf{T}\mathbf{J}_2 = \alpha_1\mu_1\mathbf{J}_1 + \alpha_2\mu_2\mathbf{J}_2$ (see Appendix C).

EXERCISE 6.1-4

Normal Modes of Simple Polarization Systems.

(a) Show that the normal modes of the linear polarizer are linearly polarized waves.
(b) Show that the normal modes of the wave retarder are linearly polarized waves.
(c) Show that the normal modes of the polarization rotator are right and left circularly polarized waves.

What are the eigenvalues of the systems described above?

6.2 REFLECTION AND REFRACTION

In this section we examine the reflection and refraction of a monochromatic plane wave of arbitrary polarization incident at a planar boundary between two dielectric media. The media are assumed to be linear, homogeneous, and isotropic with impedances η_1 and η_2, and refractive indices n_1 and n_2. The incident, refracted, and reflected waves are labeled with the subscripts 1, 2, and 3, respectively, as illustrated in Fig. 6.2-1.

As shown in Sec. 2.4A, the wavefronts of these waves are matched at the boundary if the angles of reflection and incidence are equal, $\theta_3 = \theta_1$, and if the angles of refraction and incidence satisfy Snell's law,

$$n_1 \sin \theta_1 = n_2 \sin \theta_2. \tag{6.2-1}$$

To relate the amplitudes and polarizations of the three waves, we associate with each wave an x–y coordinate system in a plane normal to the direction of propagation (Fig. 6.2-1). The electric-field envelopes of these waves are described by the Jones vectors

$$\mathbf{J}_1 = \begin{bmatrix} A_{1x} \\ A_{1y} \end{bmatrix}, \qquad \mathbf{J}_2 = \begin{bmatrix} A_{2x} \\ A_{2y} \end{bmatrix}, \qquad \mathbf{J}_3 = \begin{bmatrix} A_{3x} \\ A_{3y} \end{bmatrix}. \tag{6.2-2}$$

We proceed to determine the relations between $\mathbf{J}_2$ and $\mathbf{J}_1$ and between $\mathbf{J}_3$ and $\mathbf{J}_1$. These relations are written in the form of matrices $\mathbf{J}_2 = \mathbf{t}\mathbf{J}_1$ and $\mathbf{J}_3 = \mathbf{r}\mathbf{J}_1$, where $\mathbf{t}$ and $\mathbf{r}$ are 2×2 Jones matrices describing the transmission and reflection of the wave, respectively.

The elements of the transmission and reflection matrices may be determined by imposing the boundary conditions required by electromagnetic theory, namely the continuity at the boundary of the tangential components of $\mathbf{E}$ and $\mathbf{H}$ and the normal components of $\mathbf{D}$ and $\mathbf{B}$. The electric field associated with each wave is orthogonal to the magnetic field; the ratio of their envelopes is the characteristic impedance, which is η_1 for the incident and reflected waves and η_2 for the transmitted wave. The result is a set of equations that are solved to obtain relations between the components of the electric fields of the three waves.

The algebra involved is reduced substantially if we observe that the two normal modes for this system are linearly polarized waves with polarizations along the x and y directions. This may be proved if we show that an incident, a reflected, and a refracted wave with their electric-field vectors pointing in the x direction are self-consistent with the boundary conditions, and similarly for three waves linearly polarized in the y direction. This is indeed the case. The x and y polarized waves are therefore uncoupled.

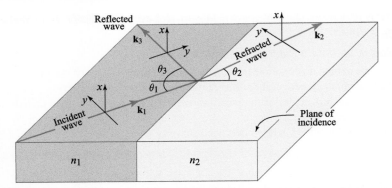

Figure 6.2-1 Reflection and refraction at the boundary between two dielectric media.

The x-polarized mode is called the **transverse electric (TE)** polarization or the **orthogonal polarization**, since the electric fields are orthogonal to the plane of incidence. The y-polarized mode is called the **transverse magnetic (TM)** polarization since the magnetic field is orthogonal to the plane of incidence, or the **parallel polarization** since the electric fields are parallel to the plane of incidence. The orthogonal and parallel polarizations are also called the s (for the German *senkrecht*, meaning "perpendicular") and p (for "parallel") polarizations, respectively. The y axes in Fig. 6.2-1 are arbitrarily defined such that their components parallel to the boundary between the dielectrics all point in the same direction.

The independence of the x and y polarizations implies that the Jones matrices **t** and **r** are diagonal,

$$\mathbf{t} = \begin{bmatrix} t_x & 0 \\ 0 & t_y \end{bmatrix}, \qquad \mathbf{r} = \begin{bmatrix} r_x & 0 \\ 0 & r_y \end{bmatrix} \tag{6.2-3}$$

so that

$$E_{2x} = t_x E_{1x}, \qquad E_{2y} = t_y E_{1y} \tag{6.2-4}$$

$$E_{3x} = r_x E_{1x}, \qquad E_{3y} = r_y E_{1y}. \tag{6.2-5}$$

The coefficients t_x and t_y are the complex amplitude transmittances for the TE and TM polarizations, respectively; r_x and r_y are the analogous complex amplitude reflectances.

Applying the boundary conditions (i.e., equating the tangential components of the electric fields and the tangential components of the magnetic fields at both sides of the boundary) in each of the TE and TM cases, we obtain the following expressions for the complex-amplitude reflectances and transmittances:

$$r_x = \frac{\eta_2 \sec\theta_2 - \eta_1 \sec\theta_1}{\eta_2 \sec\theta_2 + \eta_1 \sec\theta_1}, \qquad t_x = 1 + r_x, \tag{6.2-6}$$
TE Polarization

$$r_y = \frac{\eta_2 \cos\theta_2 - \eta_1 \cos\theta_1}{\eta_2 \cos\theta_2 + \eta_1 \cos\theta_1}, \qquad t_y = (1 + r_y)\frac{\cos\theta_1}{\cos\theta_2}. \tag{6.2-7}$$
TM Polarization
Reflection & Transmission

The characteristic impedance $\eta = \sqrt{\mu/\epsilon}$ is complex if ϵ and/or μ are complex, as is the case for lossy or conductive media. For nonlossy, nonmagnetic, dielectric media, $\eta = \eta_o/n$ is real, where $\eta_o = \sqrt{\mu_o/\epsilon_o}$ and n is the refractive index. In this case, the reflection and transmission coefficients in (6.2-6) and (6.2-7) yield the **Fresnel equations**:

$$r_x = \frac{n_1 \cos\theta_1 - n_2 \cos\theta_2}{n_1 \cos\theta_1 + n_2 \cos\theta_2}, \quad t_x = 1 + r_x,$$ (6.2-8)
TE Polarization

$$r_y = \frac{n_1 \sec\theta_1 - n_2 \sec\theta_2}{n_1 \sec\theta_1 + n_2 \sec\theta_2}, \quad t_y = (1 + r_y)\frac{\cos\theta_1}{\cos\theta_2}.$$ (6.2-9)
TM Polarization
Fresnel Equations

Given n_1, n_2, and θ_1, the reflection coefficients can be determined from the Fresnel equations by first determining θ_2 using Snell's law, (6.2-1), from which

$$\cos\theta_2 = \sqrt{1 - \sin^2\theta_2} = \sqrt{1 - (n_1/n_2)^2 \sin^2\theta_1}.$$ (6.2-10)

Since the quantities under the square-root signs in (6.2-10) can be negative, the reflection and transmission coefficients are in general complex. The magnitudes $|r_x|$ and $|r_y|$, and the phase shifts $\varphi_x = \arg\{r_x\}$ and $\varphi_y = \arg\{r_y\}$, are plotted in Figs. 6.2-2 to 6.2-5 for the two polarizations, as functions of the angle of incidence θ_1. Plots are provided for external reflection ($n_1 < n_2$) as well as for internal reflection ($n_1 > n_2$).

TE Polarization

The dependence of the reflection coefficient r_x on θ_1 for the TE-polarized wave is given by (6.2-8):

External reflection ($n_1 < n_2$). The reflection coefficient r_x is always real and negative, corresponding to a phase shift $\varphi_x = \pi$. The magnitude $|r_x| = (n_2 - n_1)/(n_1 + n_2)$ at $\theta_1 = 0$ (normal incidence) and increases to unity at $\theta_1 = 90°$ (grazing incidence), as shown in Fig. 6.2-2.

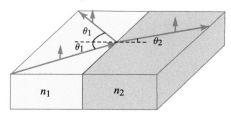

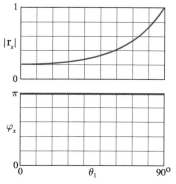

Figure 6.2-2 Magnitude and phase of the reflection coefficient as a function of the angle of incidence for *external reflection* of the *TE-polarized* wave ($n_2/n_1 = 1.5$).

Internal reflection ($n_1 > n_2$). For small θ_1 the reflection coefficient is real and positive. Its magnitude is $(n_1 - n_2)/(n_1 + n_2)$ when $\theta_1 = 0°$, and increases gradually to a value of unity, which is attained when θ_1 equals the critical angle $\theta_c = \sin^{-1}(n_2/n_1)$. For $\theta_1 > \theta_c$, the magnitude of r_x remains at unity, which corresponds to total internal reflection. This may be shown by using (6.2-10) to write[†] $\cos\theta_2 = -\sqrt{1 - \sin^2\theta_1/\sin^2\theta_c} = -j\sqrt{\sin^2\theta_1/\sin^2\theta_c - 1}$, and substituting into (6.2-8). Total internal reflection is accompanied by a phase shift $\varphi_x = \arg\{r_x\}$ given by

---

[†] The choice of the minus sign for the square root is consistent with the derivation that leads to the Fresnel equation.

$$\tan\frac{\varphi_x}{2} = \sqrt{\frac{\cos^2\theta_c}{\cos^2\theta_1} - 1}$$

(6.2-11)
TE-Reflection
Phase Shift

The phase shift φ_x increases from 0 at $\theta_1 = \theta_c$ to π at $\theta_1 = 90°$, as illustrated in Fig. 6.2-3. This phase plays an important role in dielectric waveguides (see Sec. 9.2). An evanescent wave is created in the vicinity of the boundary when total internal reflection occurs.

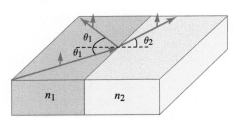

Figure 6.2-3 Magnitude and phase of the reflection coefficient as a function of the angle of incidence for *internal reflection* of the *TE-polarized* wave ($n_1/n_2 = 1.5$).

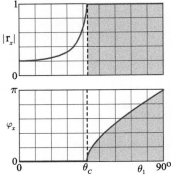

TM Polarization

Similarly, the dependence of the reflection coefficient r_y on θ_1 for the TM-polarized wave is provided by (6.2-9):

External reflection ($n_1 < n_2$). The reflection coefficient r_y is always real, as shown in Fig. 6.2-4. It assumes a negative value of $(n_1 - n_2)/(n_1 + n_2)$ at $\theta_1 = 0$ (normal incidence). Its magnitude then decreases until it vanishes when $n_1 \sec\theta_1 = n_2 \sec\theta_2$, at an angle $\theta_1 = \theta_B$, known as the **Brewster angle**:

$$\theta_B = \tan^{-1}(n_2/n_1)$$

(6.2-12)
Brewster Angle

(see Prob. 6.2-4 for other properties of the Brewster angle). For $\theta_1 > \theta_B$, r_y reverses sign (φ_y goes from π to 0) and its magnitude gradually increases until it approaches unity at $\theta_1 = 90°$. The absence of reflection of the TM wave at the Brewster angle is useful for making polarizers (see Sec. 6.6).

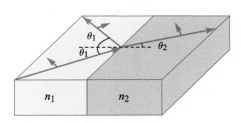

Figure 6.2-4 Magnitude and phase of the reflection coefficient as a function of the angle of incidence for *external reflection* of the *TM-polarized* wave ($n_2/n_1 = 1.5$).

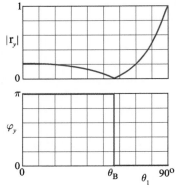

Internal reflection $(n_1 > n_2)$. At $\theta_1 = 0°$, the reflection coefficient r_y is positive and has magnitude $(n_1 - n_2)/(n_1 + n_2)$, as illustrated in Fig. 6.2-5. As θ_1 increases, the magnitude decreases until it vanishes at the Brewster angle $\theta_B = \tan^{-1}(n_2/n_1)$. As θ_1 increases beyond θ_B, r_y becomes negative and its magnitude increases until it reaches unity at the critical angle θ_c. For $\theta_1 > \theta_c$ the wave undergoes total internal reflection accompanied by a phase shift $\varphi_y = \arg\{r_y\}$ given by

$$\tan\frac{\varphi_y}{2} = \frac{-1}{\sin^2\theta_c}\sqrt{\frac{\cos^2\theta_c}{\cos^2\theta_1} - 1}.$$

(6.2-13)
TM-Reflection
Phase Shift

At normal incidence, evidently, the reflection coefficient is $r = (n_1 - n_2)/(n_1 + n_2)$, whether the reflection is TE or TM, or external or internal.

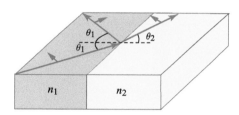
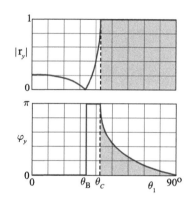

Figure 6.2-5 Magnitude and phase of the reflection coefficient as a function of the angle of incidence for *internal reflection* of the *TM-polarized wave* $(n_1/n_2 = 1.5)$.

EXERCISE 6.2-1

Brewster Windows. At what angle is a TM-polarized beam of light transmitted through a glass plate of refractive index $n = 1.5$ placed in air $(n = 1)$ without suffering reflection losses at either surface? Such plates, known as Brewster windows (Fig. 6.2-6), are used in lasers, as described in Sec. 16.2D.

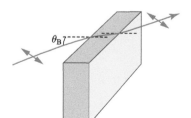

Figure 6.2-6 The Brewster window transmits TM-polarized light with no reflection loss.

Power Reflectance and Transmittance

The reflection and transmission coefficients r and t represent ratios of complex amplitudes. The power reflectance $\mathcal{R}$ and power transmittance $\mathcal{T}$ are defined as the ratios of power flow (along a direction normal to the boundary) of the reflected and transmitted waves to that of the incident wave. Because the reflected and incident waves propagate in the same medium and make the same angle with the normal to the surface, it follows that

$$\mathcal{R} = |\mathrm{r}|^2. \tag{6.2-14}$$

For both TE and TM polarizations, and for both external and internal reflection, the power reflectance at normal incidence is therefore

$$\boxed{\mathcal{R} = \left(\frac{n_1 - n_2}{n_1 + n_2}\right)^2.} \tag{6.2-15}$$

Power Reflectance at Normal Incidence

At the boundary between glass ($n = 1.5$) and air ($n = 1$), for example, $\mathcal{R} = 0.04$, so that 4% of the light is reflected at normal incidence. At the boundary between GaAs ($n = 3.6$) and air ($n = 1$), $\mathcal{R} \approx 0.32$, so that 32% of the light is reflected at normal incidence. However, at oblique angles the reflectance can be much greater or much smaller than 32%, as illustrated in Fig. 6.2-7.

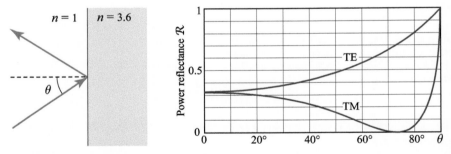

Figure 6.2-7 Power reflectance of TE- and TM-polarization plane waves at the boundary between air ($n = 1$) and GaAs ($n = 3.6$), as a function of the angle of incidence θ.

The power transmittance $\mathcal{T}$ is determined by invoking the conservation of power, so that in the absence of absorption loss the transmittance is simply

$$\mathcal{T} = 1 - \mathcal{R}. \tag{6.2-16}$$

It is important to note, however, that $\mathcal{T}$ is generally *not* equal to $|\mathrm{t}|^2$ since the power travels at different angles and with different impedances in the two media. For a wave traveling at an angle θ in a medium of refractive index n, the power flow in the direction normal to the boundary is $(|\mathcal{E}|^2/2\eta)\cos\theta = (|\mathcal{E}|^2/2\eta_o)\,n\cos\theta$. It follows that

$$\mathcal{T} = \frac{n_2 \cos\theta_2}{n_1 \cos\theta_1} |\mathrm{t}|^2. \tag{6.2-17}$$

Reflectance from a plate. The power reflectance at normal incidence from a plate with two surfaces is described by $\mathcal{R}(1 + \mathcal{T}^2)$ since the power reflected from the far surface involves a double transmission through the near surface. For a glass plate in air, the overall reflectance is $\mathcal{R}(1 + \mathcal{T}^2) = 0.04[1 + (0.96)^2] \approx 0.077$, so that about 7.7% of the incident light power is reflected. However, this calculation does not include interference effects, which are washed out when the light is incoherent (see Sec. 12.2), nor does it account for multiple reflections inside the plate. Optical transmission and reflectance from multiple boundaries in layered media are described in detail in Sec. 7.1.

6.3 OPTICS OF ANISOTROPIC MEDIA

A dielectric medium is said to be anisotropic if its macroscopic optical properties depend on direction. The macroscopic properties of a material are, of course, ultimately governed by its microscopic properties: the shape and orientation of the individual molecules and the organization of their centers in space. Optical materials have different kinds of positional and orientational types of order, which may be described as follows (see Fig. 6.3-1):

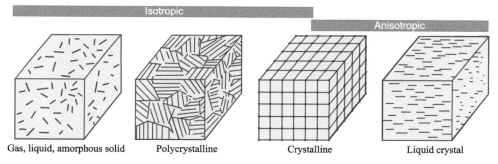

Figure 6.3-1 Positional and orientational order in different types of materials.

- If the molecules are located at totally random positions in space, and are themselves isotropic or oriented along random directions, the medium is isotropic. *Gases*, *liquids*, and *amorphous solids* follow this prescription.

- If the structure takes the form of disjointed crystalline grains that are randomly oriented with respect to each other, the material is said to be *polycrystalline*. The individual grains are, in general, anisotropic, but their averaged macroscopic behavior is isotropic.

- If the molecules are organized in space according to a regular periodic pattern and they are oriented in the same direction, as in *crystals*, the medium is, in general, anisotropic.

- If the molecules are anisotropic and their orientations are not totally random, the medium is anisotropic, even if their positions are totally random. This is the case for *liquid crystals*, which have orientational order but lack complete positional order.

A. Refractive Indices

Permittivity Tensor

In a linear anisotropic dielectric medium (a crystal, for example), each component of the electric flux density **D** is a linear combination of the three components of the electric field,

$$D_i = \sum_j \epsilon_{ij} E_j. \tag{6.3-1}$$

The indices $i, j = 1, 2, 3$ refer to the x, y, and z components, respectively, as described in Sec. 5.2B. The dielectric properties of the medium are therefore characterized by a 3×3 array of nine coefficients, $\{\epsilon_{ij}\}$, that form the **electric permittivity tensor** $\boldsymbol{\epsilon}$, which is a tensor of second rank. The **material equation** (6.3-1) is usually written in the symbolic form

$$\mathbf{D} = \boldsymbol{\epsilon}\mathbf{E}. \tag{6.3-2}$$

For most dielectric media, the electric permittivity tensor is symmetric, i.e., $\epsilon_{ij} = \epsilon_{ji}$. This means that the relation between the vectors **D** and **E** is reciprocal, i.e., their ratio remains the same if their directions are exchanged. This symmetry is obeyed for dielectric nonmagnetic materials that do not exhibit optical activity, and in the absence of an external magnetic field (Sec. 6.4). With this symmetry, the medium is characterized by only six independent numbers in an arbitrary coordinate system. For crystals of certain symmetries, even fewer coefficients suffice since some vanish and some are related.

Geometrical Representation of Vectors and Tensors

A *vector*, such as the electric field **E**, for example, describes a physical variable with magnitude and direction. It is represented *geometrically* by an arrow pointing in that particular direction, whose length is proportional to the magnitude of the vector [Fig. 6.3-2(a)]. A vector, which is a tensor of first rank, is represented *numerically* by three numbers: its projections on the three axes of a particular coordinate system. Though these components depend on the choice of the coordinate system, the magnitude and direction of the vector in physical space are independent of the choice of the coordinate system. A scalar, which is described by a single number, is a tensor of zero rank.

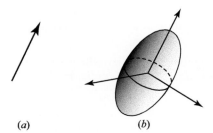

Figure 6.3-2 Geometrical representation of: (a) a vector; (b) a symmetric second-rank tensor.

A second-rank *tensor* is a rule that relates two vectors. In a given coordinate system, it is represented *numerically* by nine numbers. Changing the coordinate system yields a different set of nine numbers, but the physical nature of the rule is unchanged. A useful *geometrical* representation [Fig. 6.3-2(b)] of a *symmetric* second-rank tensor (the permittivity tensor ϵ, for example), is a quadratic surface (an ellipsoid) defined by

$$\sum_{ij} \epsilon_{ij}\, x_i x_j = 1, \qquad (6.3\text{-}3)$$

which is known as the **quadric representation**. This surface is invariant to the choice of the coordinate system; if the coordinate system is rotated, both x_i and ϵ_{ij} are altered but the ellipsoid remains intact in physical space. The ellipsoid has six degrees of freedom and carries all information about the symmetric second-rank tensor. In the principal coordinate system, ϵ_{ij} is diagonal and the ellipsoid assumes a particularly simple form:

$$\epsilon_1 x_1^2 + \epsilon_2 x_2^2 + \epsilon_3 x_3^2 = 1. \qquad (6.3\text{-}4)$$

Its principal axes are those of the tensor, and its axes have half-lengths $1/\sqrt{\epsilon_1}$, $1/\sqrt{\epsilon_2}$, and $1/\sqrt{\epsilon_3}$.

Principal Axes and Principal Refractive Indices

The elements of the permittivity tensor depend on how the coordinate system is chosen relative to the crystal structure. However, a coordinate system can always be found for which the off-diagonal elements of ϵ_{ij} vanish, so that

$$D_1 = \epsilon_1 E_1, \qquad D_2 = \epsilon_2 E_2, \qquad D_3 = \epsilon_3 E_3, \qquad (6.3\text{-}5)$$

where $\epsilon_1 = \epsilon_{11}$, $\epsilon_2 = \epsilon_{22}$, and $\epsilon_3 = \epsilon_{33}$. According to (6.3-1), **E** and **D** are parallel along these particular directions so that if, for example, **E** points in the x direction, then so too must **D**. This coordinate system defines the **principal axes** and principal planes of the crystal. Throughout the remainder of this chapter, the coordinate system x, y, z, which is equivalently denoted x_1, x_2, x_3, is assumed to lie along the principal axes of the crystal. This choice simplifies all analyses without loss of generality. The permittivities ϵ_1, ϵ_2, and ϵ_3 correspond to refractive indices

$$n_1 = \sqrt{\epsilon_1/\epsilon_o}, \qquad n_2 = \sqrt{\epsilon_2/\epsilon_o}, \qquad n_3 = \sqrt{\epsilon_3/\epsilon_o}, \qquad (6.3\text{-}6)$$

respectively, where ϵ_o is the permittivity of free space; these are known as the **principal refractive indices**.

Biaxial, Uniaxial, and Isotropic Crystals

Crystals in which the three principal refractive indices are different are termed **biaxial**. For crystals with certain symmetries, namely a single axis of threefold, fourfold, or sixfold symmetry, two of the refractive indices are equal ($n_1 = n_2$) and the crystal is called **uniaxial**. In this case, the indices are usually denoted $n_1 = n_2 = n_o$ and $n_3 = n_e$, which are known as the **ordinary** and **extraordinary** indices, respectively, for reasons that will become clear shortly. The crystal is said to be **positive uniaxial** if $n_e > n_o$, and **negative uniaxial** if $n_e < n_o$. The z axis of a uniaxial crystal is called the **optic axis**. In certain crystals with even greater symmetry (those with cubic unit cells, for example), all three indices are equal and the medium is optically isotropic.

Impermeability Tensor

The relation $\mathbf{D} = \boldsymbol{\epsilon}\mathbf{E}$ can be inverted and written in the form $\mathbf{E} = \boldsymbol{\epsilon}^{-1}\mathbf{D}$, where $\boldsymbol{\epsilon}^{-1}$ is the inverse of the tensor $\boldsymbol{\epsilon}$. It is also useful to define the **electric impermeability tensor** $\boldsymbol{\eta} = \epsilon_o \boldsymbol{\epsilon}^{-1}$ (not to be confused with the impedance of the medium η), so that $\epsilon_o \mathbf{E} = \boldsymbol{\eta}\mathbf{D}$. Since $\boldsymbol{\epsilon}$ is symmetric, so too is $\boldsymbol{\eta}$. Both tensors, $\boldsymbol{\epsilon}$ and $\boldsymbol{\eta}$, share the same principal axes. In the principal coordinate system, $\boldsymbol{\eta}$ is diagonal with principal values $\epsilon_o/\epsilon_1 = 1/n_1^2$, $\epsilon_o/\epsilon_2 = 1/n_2^2$, and $\epsilon_o/\epsilon_3 = 1/n_3^2$. Either tensor, $\boldsymbol{\epsilon}$ or $\boldsymbol{\eta}$, fully describes the optical properties of the crystal.

Index Ellipsoid

The **index ellipsoid** (also called the **optical indicatrix**) is the quadric representation of the electric impermeability tensor $\boldsymbol{\eta} = \epsilon_o \boldsymbol{\epsilon}^{-1}$:

$$\sum_{ij} \eta_{ij}\, x_i x_j = 1, \qquad i,j = 1,2,3. \tag{6.3-7}$$

If the principal axes were to be used as the coordinate system, we would obtain

$$\boxed{\frac{x_1^2}{n_1^2} + \frac{x_2^2}{n_2^2} + \frac{x_3^2}{n_3^2} = 1,} \tag{6.3-8}$$
Index Ellipsoid

with principal values $1/n_1^2$, $1/n_2^2$, and $1/n_3^2$, and axes of half-lengths n_1, n_2, and n_3.

The optical properties of the crystal (the directions of the principal axes and the values of the principal refractive indices) are therefore completely described by the index ellipsoid (Fig. 6.3-3). For a uniaxial crystal, the index ellipsoid reduces to an ellipsoid of revolution; for an isotropic medium it becomes a sphere.

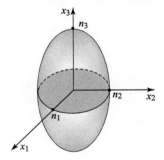

Figure 6.3-3 The index ellipsoid. The coordinates (x_1, x_2, x_3) are the principal axes while (n_1, n_2, n_3) are the principal refractive indices of the crystal.

B. Propagation Along a Principal Axis

The rules that govern the propagation of light in crystals under general conditions are rather complex. However, they become relatively simple if the light is a plane wave traveling along one of the principal axes of the crystal. We begin with this case.

Normal Modes

Let x–y–z be a coordinate system that coincides with the principal axes of a crystal. A plane wave traveling in the z direction and linearly polarized along the x direction [Fig. 6.3-4(a)] travels with phase velocity c_o/n_1 (wavenumber $k = n_1 k_o$) without changing its polarization. The reason for this is that the electric field has only one component, E_1 pointed along the x direction, so that **D** is also in the x direction with $D_1 = \epsilon_1 E_1$; the wave equation derived from Maxwell's equations therefore provides a velocity of light given by $1/\sqrt{\mu_o \epsilon_1} = c_o/n_1$. Similarly, a plane wave traveling in the z direction and linearly polarized along the y direction [Fig. 6.3-4(b)] travels with phase velocity c_o/n_2, thereby experiencing a refractive index n_2. Thus, the normal modes for propagation in the z direction are linearly polarized waves in the x and y directions. These waves are said to be normal modes because their velocities and polarizations are maintained as they propagate (see Appendix C). Other cases in which the wave propagates along one of the principal axes and is linearly polarized along another are treated similarly [Fig. 6.3-4(c)].

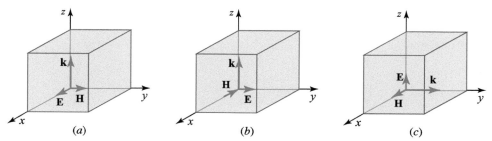

Figure 6.3-4 A wave traveling along a principal axis and polarized along another principal axis has phase velocity c_o/n_1, c_o/n_2, or c_o/n_3, when the electric-field vector points in the x, y, or z directions, respectively. (a) $k = n_1 k_o$; (b) $k = n_2 k_o$; (c) $k = n_3 k_o$.

Polarization Along an Arbitrary Direction

We now consider a wave traveling along one principal axis (the z axis, for example) that is linearly polarized along an arbitrary direction in the x–y plane. This case is addressed by analyzing the wave as a sum of the normal modes, namely the linearly polarized waves in the x and y directions. These two components travel with different phase velocities, c_o/n_1 and c_o/n_2, respectively. They therefore undergo different phase shifts, $\varphi_x = n_1 k_o d$ and $\varphi_y = n_2 k_o d$, respectively, after propagating a distance d. Their phase retardation is thus $\varphi = \varphi_y - \varphi_x = (n_2 - n_1)k_o d$. Recombination of the two components yields an elliptically polarized wave, as explained in Sec. 6.1 and illustrated in Fig. 6.3-5. Such a crystal can therefore serve as a **wave retarder**, a device in which two orthogonal polarizations travel at different phase velocities so that one is retarded with respect to the other (see Fig. 6.1-8).

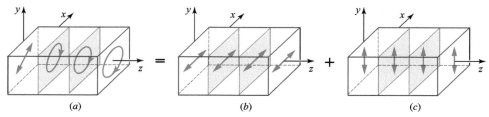

Figure 6.3-5 A linearly polarized wave at $45°$ in the $z = 0$ plane (a) is analyzed as a superposition of two linearly polarized components in the x and y directions (normal modes), which travel at velocities c_o/n_1 and c_o/n_2 [(b) and (c), respectively]. As a result of phase retardation, the wave is converted from plane polarization to elliptical polarization (a). It is therefore clear that the initial linearly polarized wave is not a normal mode of the system.

C. Propagation in an Arbitrary Direction

We now consider the general case of a plane wave traveling in an anisotropic crystal in an arbitrary direction defined by the unit vector $\hat{\mathbf{u}}$. We demonstrate that the two normal modes are linearly polarized waves. The refractive indices n_a and n_b, and the directions of polarization of these modes, may be determined by use of a procedure based on the index ellipsoid:

Index-Ellipsoid Construction for Determining Normal Modes

Figure 6.3-6 illustrates a geometrical construction for determining the polarizations and refractive indices n_a and n_b of the normal modes of a wave traveling in the direction of the unit vector $\hat{\mathbf{u}}$ in an anisotropic material characterized by the index ellipsoid:

$$\frac{x_1^2}{n_1^2} + \frac{x_2^2}{n_2^2} + \frac{x_3^2}{n_3^2} = 1.$$

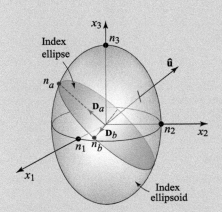

Figure 6.3-6 Determination of the normal modes from the index ellipsoid.

- Draw a plane passing through the origin of the index ellipsoid, normal to $\hat{\mathbf{u}}$. The intersection of the plane with the ellipsoid is an ellipse called the **index ellipse**.
- The half-lengths of the major and minor axes of the index ellipse are the refractive indices n_a and n_b of the two normal modes.
- The directions of the major and minor axes of the index ellipse are the directions of the vectors $\mathbf{D}_a$ and $\mathbf{D}_b$ for the normal modes. These directions are orthogonal.
- The vectors $\mathbf{E}_a$ and $\mathbf{E}_b$ may be determined from $\mathbf{D}_a$ and $\mathbf{D}_b$ with the help of (6.3-5).

□ **Proof of the Index-Ellipsoid Construction for Determining the Normal Modes.** To determine the normal modes (see Sec. 6.1B) for a plane wave traveling in the direction $\hat{\mathbf{u}}$, we cast Maxwell's equations (5.3-2)–(5.3-5), and the material equation $\mathbf{D} = \epsilon \mathbf{E}$ given in (6.3-2), as an eigenvalue problem. Since all fields are assumed to vary with the position $\mathbf{r}$ as $\exp(-j\mathbf{k} \cdot \mathbf{r})$, where $\mathbf{k} = k\hat{\mathbf{u}}$, Maxwell's equations (5.4-3) and (5.4-4) reduce to

$$\mathbf{k} \times \mathbf{H} = -\omega \mathbf{D} \qquad (6.3\text{-}9)$$

$$\mathbf{k} \times \mathbf{E} = \omega \mu_o \mathbf{H} \qquad (6.3\text{-}10)$$

Substituting (6.3-10) into (6.3-9) leads to

$$\mathbf{k} \times (\mathbf{k} \times \mathbf{E}) = -\omega^2 \mu_o \mathbf{D}. \qquad (6.3\text{-}11)$$

Using $\mathbf{E} = \boldsymbol{\epsilon}^{-1} \mathbf{D}$, we obtain

$$\mathbf{k} \times (\mathbf{k} \times \boldsymbol{\epsilon}^{-1} \mathbf{D}) = -\omega^2 \mu_o \mathbf{D}. \qquad (6.3\text{-}12)$$

This is an eigenvalue equation that $\mathbf{D}$ must satisfy. Working with $\mathbf{D}$ is convenient since we know that it lies in a plane normal to the wave direction $\hat{\mathbf{u}}$.

We now simplify (6.3-12) by using $\boldsymbol{\eta} = \epsilon_o \boldsymbol{\epsilon}^{-1}$, $\mathbf{k} = k\hat{\mathbf{u}}$, $n = k/k_o$, and $k_o^2 = \omega^2 \mu_o \epsilon_o$ to obtain

$$-\hat{\mathbf{u}} \times (\hat{\mathbf{u}} \times \boldsymbol{\eta} \mathbf{D}) = \frac{1}{n^2} \mathbf{D}. \qquad (6.3\text{-}13)$$

The operation $-\hat{\mathbf{u}} \times (\hat{\mathbf{u}} \times \boldsymbol{\eta} \mathbf{D})$ may be interpreted as a projection of the vector $\boldsymbol{\eta}\mathbf{D}$ onto a plane normal to $\hat{\mathbf{u}}$. We may therefore rewrite (6.3-13) in the form

$$\mathbf{P}_u \boldsymbol{\eta} \mathbf{D} = \frac{1}{n^2} \mathbf{D}, \qquad (6.3\text{-}14)$$

where $\mathbf{P}_u$ is an operator representing projection. Equation (6.3-14) is an eigenvalue equation for the operator $\mathbf{P}_u \boldsymbol{\eta}$, with eigenvalue $1/n^2$ and eigenvector $\mathbf{D}$. The two eigenvalues, $1/n_a^2$ and $1/n_b^2$, and two corresponding eigenvectors, $\mathbf{D}_a$ and $\mathbf{D}_b$, which are orthogonal, represent the two normal modes.

The eigenvalue problem (6.3-14) has a simple geometrical interpretation. The tensor $\boldsymbol{\eta}$ is represented geometrically by its quadric representation, the index ellipsoid. The operator $\mathbf{P}_u \boldsymbol{\eta}$ represents projection onto a plane normal to $\hat{\mathbf{u}}$. Solving the eigenvalue problem in (6.3-14) is thus equivalent to finding the principal axes of the ellipse formed by the intersection of the plane normal to $\hat{\mathbf{u}}$ with the index ellipsoid. This is precisely the construction set forth in Fig. 6.3-6 for determining the normal modes. ∎

Special Case: Uniaxial Crystals

In uniaxial crystals ($n_1 = n_2 = n_o$ and $n_3 = n_e$) the index ellipsoid of Fig. 6.3-6 is an ellipsoid of revolution. For a wave whose direction of travel $\hat{\mathbf{u}}$ forms an angle θ with the optic axis, the index ellipse has half-lengths n_o and $n(\theta)$, where $n(\theta)$ is determined from the index-ellipsoid equation by making the substitutions $x_1 = n(\theta) \cos\theta$, $x_2 = 0$, and $x_3 = -n(\theta) \sin\theta$. The result is

$$\boxed{\frac{1}{n^2(\theta)} = \frac{\cos^2 \theta}{n_o^2} + \frac{\sin^2 \theta}{n_e^2},} \qquad (6.3\text{-}15)$$
Refractive Index of Extraordinary Wave

so that the normal modes have refractive indices $n_b = n_o$ and $n_a = n(\theta)$. The first mode, called the **ordinary wave**, has a refractive index n_o regardless of θ. In accordance with the ellipse shown in Fig. 6.3-7, the second mode, called the **extraordinary wave**, has a refractive index $n(\theta)$ that varies from n_o when $\theta = 0°$, to n_e when $\theta = 90°$. The vector $\mathbf{D}$ of the ordinary wave is normal to the plane defined by the optic axis (z axis) and the direction of wave propagation $\mathbf{k}$, and the vectors $\mathbf{E}$ and $\mathbf{D}$ are parallel. The extraordinary wave, on the other hand, has a vector $\mathbf{D}$ that is normal to $\mathbf{k}$ and lies in the k–z plane, and $\mathbf{E}$ is not parallel to $\mathbf{D}$, as shown in Fig. 6.3-7.

234 CHAPTER 6 POLARIZATION OPTICS

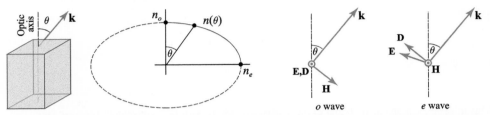

Figure 6.3-7 Variation of the refractive index $n(\theta)$ of the extraordinary wave with θ (the angle between the direction of propagation and the optic axis) in a uniaxial crystal, and directions of the electromagnetic fields of the ordinary (o) and extraordinary (e) waves. The circle with a dot at the center located at the origin signifies that the direction of the vector is out of the plane of the paper, toward the reader.

D. Dispersion Relation, Rays, Wavefronts, and Energy Transport

We now examine other properties of waves in anisotropic media including the dispersion relation (the relation between ω and $\mathbf{k}$).

The optical wave is characterized by the wavevector $\mathbf{k}$, the field vectors $\mathbf{E}$, $\mathbf{D}$, $\mathbf{H}$, and $\mathbf{B}$, and the complex Poynting vector $\mathbf{S} = \frac{1}{2}\mathbf{E} \times \mathbf{H}^*$ (direction of power flow). These vectors are related by (6.3-9) and (6.3-10). It follows from (6.3-9) that $\mathbf{D}$ is normal to both $\mathbf{k}$ and $\mathbf{H}$. Equation (6.3-10) similarly indicates that $\mathbf{H}$ is normal to both $\mathbf{k}$ and $\mathbf{E}$. These geometrical conditions are illustrated in Fig. 6.3-8, which also shows the complex Poynting vector $\mathbf{S}$, which is orthogonal to both $\mathbf{E}$ and $\mathbf{H}$. Thus, $\mathbf{D}$, $\mathbf{E}$, $\mathbf{k}$, and $\mathbf{S}$ lie in one plane to which $\mathbf{H}$ and $\mathbf{B}$ are normal. In this plane $\mathbf{D} \perp \mathbf{k}$ and $\mathbf{S} \perp \mathbf{E}$; but $\mathbf{D}$ is not necessarily parallel to $\mathbf{E}$, and $\mathbf{S}$ is not necessarily parallel to $\mathbf{k}$.

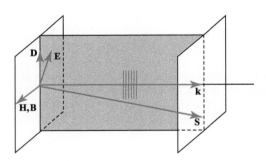

Figure 6.3-8 The vectors $\mathbf{D}$, $\mathbf{E}$, $\mathbf{k}$, and $\mathbf{S}$ all lie in a single plane, to which $\mathbf{H}$ and $\mathbf{B}$ are normal. Also $\mathbf{D} \perp \mathbf{k}$ and $\mathbf{E} \perp \mathbf{S}$. The wavefronts are orthogonal to $\mathbf{k}$.

Dispersion Relation: The $\mathbf{k}$ Surface

Using the relation $\mathbf{D} = \epsilon \mathbf{E}$ in (6.3-11), we obtain

$$\mathbf{k} \times (\mathbf{k} \times \mathbf{E}) + \omega^2 \mu_o \epsilon \mathbf{E} = 0. \qquad (6.3\text{-}16)$$

This vector equation, which $\mathbf{E}$ must satisfy, translates to three linear homogeneous equations for the components E_1, E_2, and E_3 along the principal axes, written in the matrix form

$$\begin{bmatrix} n_1^2 k_o^2 - k_2^2 - k_3^2 & k_1 k_2 & k_1 k_3 \\ k_2 k_1 & n_2^2 k_o^2 - k_1^2 - k_3^2 & k_2 k_3 \\ k_3 k_1 & k_3 k_2 & n_3^2 k_o^2 - k_1^2 - k_2^2 \end{bmatrix} \begin{bmatrix} E_1 \\ E_2 \\ E_3 \end{bmatrix} = \begin{bmatrix} 0 \\ 0 \\ 0 \end{bmatrix}, \qquad (6.3\text{-}17)$$

where (k_1, k_2, k_3) are the components of $\mathbf{k}$, $k_o = \omega/c_o$, and (n_1, n_2, n_3) are the principal refractive indices given by (6.3-6). A nontrivial solution to these equations obtains when the determinant of the matrix is set to zero, which yields

$$\sum_{j=1,2,3} \frac{k_j^2}{k^2 - n_j^2 k_o^2} = 1,$$

(6.3-18)
Dispersion Relation
k Surface

where $k^2 = k_1^2 + k_2^2 + k_3^2$ and $k_o = \omega/c_o$. This relation is known as the **dispersion relation**. It is the equation of a surface $\omega = \omega(k_1, k_2, k_3)$ in the k_1, k_2, k_3 space, known as the **normal surface** or the **k** surface.

The **k** surface is a centrosymmetric surface comprising two sheets, each corresponding to a solution (a normal mode). It can be shown that the **k** surface intersects each of the principal planes in an ellipse and a circle, as illustrated in Fig. 6.3-9. For biaxial crystals ($n_1 < n_2 < n_3$), the two sheets meet at four points, defining two optic axes. In the uniaxial case ($n_1 = n_2 = n_o$, $n_3 = n_e$), the two sheets become a sphere and an ellipsoid of revolution that meet at only two points, thereby defining a single optic axis (the z axis). In the isotropic case ($n_1 = n_2 = n_3 = n$), the two sheets degenerate into a single sphere.

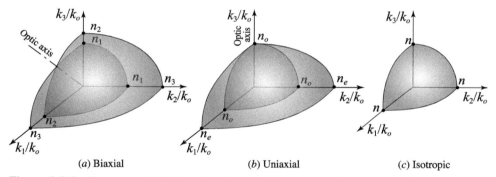

(a) Biaxial   (b) Uniaxial   (c) Isotropic

Figure 6.3-9 One octant of the **k** surface for (a) a biaxial crystal ($n_1 < n_2 < n_3$); (b) a uniaxial crystal ($n_1 = n_2 = n_o$, $n_3 = n_e$); and (c) an isotropic crystal ($n_1 = n_2 = n_3 = n$).

Determining Properties of the Normal Modes from the k Surface

Much like the index ellipsoid, the **k** surface can be used to determine the normal modes for waves propagating in any prescribed direction, but it also provides physical insight into a collection of other properties of these modes, as considered below:

Refractive indices. The intersection of the prescribed direction $\hat{\mathbf{u}} = (u_1, u_2, u_3)$ of the wave with the **k** surface is a point at a distance from the origin equal to the wavenumber $k = n\omega/c_o$, from which the refractive index n and the corresponding phase velocity $c = \omega/k$ may be determined. In fact, associated with each direction are two intersections corresponding to two wavenumbers and two refractive indices for the two normal modes. Substituting $(k_1, k_2, k_3) = (u_1 k, u_2 k, u_3 k)$ into (6.3-18) yields

$$\sum_{j=1,2,3} \frac{u_j^2 k^2}{k^2 - n_j^2 k_o^2} = 1. \qquad (6.3\text{-}19)$$

This is a fourth-order equation in k (or second order in k^2). It has four solutions, $\pm k_a$ and $\pm k_b$, of which only the two positive values are meaningful, since the negative values represent a reversed direction of propagation. The problem is therefore solved: the wavenumbers of the normal modes are k_a and k_b and the refractive indices are $n_a = k_a/k_o$ and $n_b = k_b/k_o$.

Polarization. To find the directions of polarization of the two normal modes, we determine the components $(k_1, k_2, k_3) = (ku_1, ku_2, ku_3)$ and the elements of the matrix in (6.3-17) for each of the two wavenumbers $k = k_a$ and $k = k_b$. We then solve two of the three equations in (6.3-17) to establish the ratios E_1/E_3 and E_2/E_3, from which we find the directions of the corresponding fields $\mathbf{E}_a$ and $\mathbf{E}_b$. Note that $\mathbf{E}_a$ and $\mathbf{E}_b$ for the two modes are not necessarily orthogonal, whereas $\mathbf{D}_a$ and $\mathbf{D}_b$ are.

Group velocity. The group velocity may also be determined from the k surface. In analogy with the group velocity $v = d\omega/dk$ that governs the propagation of light pulses (wavepackets), as discussed in Sec. 5.7, the group velocity for *rays* (localized beams or spatial wavepackets) is the vector $\mathbf{v} = \nabla_k \omega(\mathbf{k})$, the gradient of ω with respect to k. Since the k surface is the surface $\omega(k_1, k_2, k_3) = $ constant, $\mathbf{v}$ must be normal to the k surface. Thus, rays travel along directions normal to the k surface. The wavefronts are perpendicular to the wavevector k since the phase of the wave is $\mathbf{k} \cdot \mathbf{r}$. The wavefront normals are therefore parallel to the wavevector k.

Energy transport. The complex Poynting vector $\mathbf{S} = \frac{1}{2}\mathbf{E} \times \mathbf{H}^*$ is also normal to the k surface. This can be demonstrated by choosing a value for ω and considering two vectors k and $\mathbf{k} + \Delta\mathbf{k}$ that lie on the k surface. Taking the differentials of (6.3-9) and (6.3-10), and using certain vector identities, shows that $\Delta\mathbf{k} \cdot \mathbf{S} = 0$, so that $\mathbf{S}$ is normal to the k surface. Consequently, $\mathbf{S}$ is also parallel to the group velocity vector $\mathbf{v}$.

Optical rays. If the k surface is a sphere, as it is for isotropic media, the vectors k, S, and v are all parallel, indicating that rays are parallel to the wavevector k and energy flows in the same direction, as illustrated in Fig. 6.3-10(*a*). On the other hand, if the k surface is not normal to the wavevector k, as illustrated in Fig. 6.3-10(*b*), the rays and the direction of energy transport are not orthogonal to the wavefronts. Rays then have the "extraordinary" property of traveling at an oblique angle to their wavefronts [Fig. 6.3-10(*b*)].

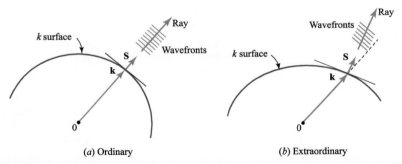

(*a*) Ordinary (*b*) Extraordinary

Figure 6.3-10 Rays and wavefronts for (*a*) a spherical k surface, and (*b*) a nonspherical k surface.

Special Case: Uniaxial Crystals

In uniaxial crystals ($n_1 = n_2 = n_o$ and $n_3 = n_e$), the equation of the k surface $\omega = \omega(k_1, k_2, k_3)$ simplifies to

$$(k^2 - n_o^2 k_o^2)\left(\frac{k_1^2 + k_2^2}{n_e^2} + \frac{k_3^2}{n_o^2} - k_o^2\right) = 0. \qquad (6.3\text{-}20)$$

This equation has two solutions: a sphere, corresponding to the leftmost factor vanishing:

$$k = n_o k_o, \qquad (6.3\text{-}21)$$

and an ellipsoid of revolution, corresponding to the rightmost factor vanishing:

$$\frac{k_1^2 + k_2^2}{n_e^2} + \frac{k_3^2}{n_o^2} = k_o^2. \tag{6.3-22}$$

Because of symmetry about the z axis (optic axis), there is no loss of generality in assuming that the vector **k** lies in the y–z plane. Its direction is then characterized by the angle θ it makes with the optic axis. It is thus convenient to draw the k-surfaces only in the y–z plane, as a circle and an ellipse, as shown in Fig. 6.3-11.

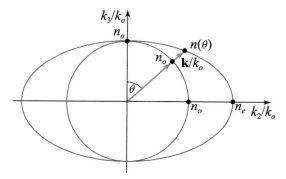

Figure 6.3-11 Intersection of the **k** surfaces with the y–z plane for a positive uniaxial crystal ($n_e > n_o$).

Given the direction $\hat{\mathbf{u}}$ of the vector **k**, the wavenumber k is determined by finding the intersection with the **k** surfaces. The two solutions define the two normal modes, the ordinary and extraordinary waves. The ordinary wave has wavenumber $k = n_o k_o$ regardless of the direction of $\hat{\mathbf{u}}$, whereas the extraordinary wave has wavenumber $n(\theta)k_o$, where $n(\theta)$ is given by (6.3-15), thereby confirming earlier results obtained from the index-ellipsoid geometrical construction. The directions of the rays, wavefronts, energy flow, and field vectors **E** and **D** for the ordinary and extraordinary waves in a uniaxial crystal are illustrated in Fig. 6.3-12.

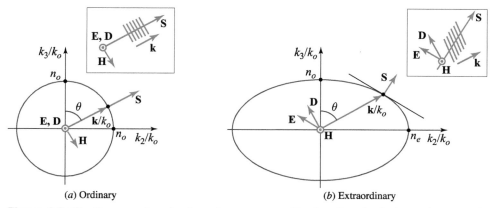

(a) Ordinary (b) Extraordinary

Figure 6.3-12 The normal modes for a plane wave traveling in a direction **k** that makes an angle θ with the optic axis z of a uniaxial crystal are: (a) An ordinary wave of refractive index n_o polarized in a direction normal to the k–z plane. (b) An extraordinary wave of refractive index $n(\theta)$ [given by (6.3-15)] polarized in the k–z plane along a direction tangential to the ellipse (the **k** surface) at the point of its intersection with **k**. This wave is "extraordinary" in the following ways: **D** is not parallel to **E** but both lie in the k–z plane, and **S** is not parallel to **k** so that power does not flow along the direction of **k**; the rays are therefore not normal to the wavefronts so that the wave travels "sideways."

E. Double Refraction

Refraction of Plane Waves

We now examine the refraction of a plane wave at the boundary between an isotropic medium (say air, $n = 1$) and an anisotropic medium (a crystal). The key principle that governs the refraction of waves for this configuration is that the wavefronts of the incident and refracted waves must be matched at the boundary. Because the anisotropic medium supports two modes with distinctly different phase velocities, and therefore different indices of refraction, an incident wave gives rise to two refracted waves with different directions and different polarizations. The effect is known as **double refraction** or **birefringence**.

The phase-matching condition requires that Snell's law be obeyed, i.e.,

$$k_o \sin \theta_1 = k \sin \theta, \qquad (6.3\text{-}23)$$

where θ_1 and θ are the angles of incidence and refraction, respectively. In an anisotropic medium, however, the wavenumber $k = n(\theta)k_o$ is itself a function of θ, so that

$$\sin \theta_1 = n(\theta_a + \theta) \sin \theta, \qquad (6.3\text{-}24)$$

where θ_a is the angle between the optic axis and the normal to the surface, so that $\theta_a + \theta$ is the angle the refracted ray makes with the optic axis. Equation (6.3-24) is a modified version of Snell's law. To solve (6.3-23), we draw the intersection of the **k** surface with the plane of incidence and search for an angle θ for which (6.3-23) is satisfied. Two solutions, corresponding to the two normal modes, are expected. The polarization state of the incident light governs the distribution of energy among the two refracted waves.

Take, for example, a uniaxial crystal and a plane of incidence parallel to the optic axis. The **k** surfaces intersect the plane of incidence in a circle and an ellipse (Fig. 6.3-13). The two refracted waves that satisfy the phase-matching condition are determined by satisfying (6.3-24):

- An ordinary wave of orthogonal polarization (TE) at an angle $\theta = \theta_o$, for which

$$\sin \theta_1 = n_o \sin \theta_o; \qquad (6.3\text{-}25)$$

- An extraordinary wave of parallel polarization (TM) at an angle $\theta = \theta_e$, for which

$$\sin \theta_1 = n(\theta_a + \theta_e) \sin \theta_e, \qquad (6.3\text{-}26)$$

where $n(\theta)$ is given by (6.3-15).

If the incident wave carries the two polarizations, the two refracted waves will emerge, as shown in Fig. 6.3-13.

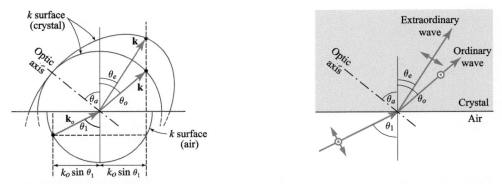

Figure 6.3-13 Determination of the angles of refraction by matching projections of the **k** vectors in air and in a uniaxial crystal.

Refraction of Rays

The analysis immediately above dealt with the refraction of plane waves. The refraction of rays is different in an anisotropic medium, since rays do not necessarily travel in directions normal to the wavefronts. In air, before entering the crystal, the wavefronts are normal to the rays. The refracted wave must have a wavevector that satisfies the phase-matching condition, so that Snell's law (6.3-24) is applicable, with the angle of refraction θ determining the direction of **k**. However, since the direction of **k** is not the direction of the ray, Snell's law is not applicable to rays in anisotropic media.

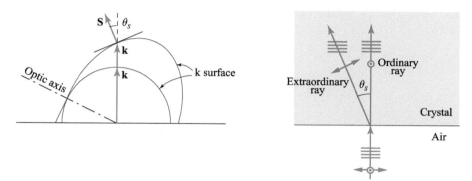

Figure 6.3-14 Double refraction at normal incidence.

An example that dramatizes the deviation from Snell's law is that of normal incidence into a uniaxial crystal whose optic axis is neither parallel nor perpendicular to the crystal boundary. The incident wave has a **k** vector normal to the boundary. To ensure phase matching, the refracted waves must also have wavevectors in the same direction. Intersections with the **k** surface yield two points corresponding to two waves. The ordinary ray is parallel to **k**. But the extraordinary ray points in the direction of the normal to the **k** surface, at an angle θ_s with the normal to the crystal boundary, as illustrated in Fig. 6.3-14. Thus, normal incidence creates oblique refraction. The principle of phase matching is maintained, however: wavefronts of both refracted rays are parallel to the crystal boundary and to the wavefront of the incident ray.

When light rays are transmitted through a plate of anisotropic material as described above, the two rays refracted at the first surface refract again at the second surface, creating two laterally separated rays with orthogonal polarizations, as illustrated in Fig. 6.3-15.

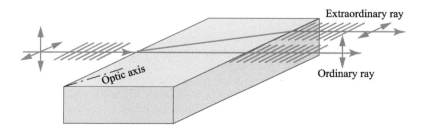

Figure 6.3-15 Double refraction through an anisotropic plate. The plate serves as a polarizing beamsplitter.

6.4 OPTICAL ACTIVITY AND MAGNETO-OPTICS

A. Optical Activity

Certain materials act as natural polarization rotators, a property known as **optical activity**. Their normal modes are waves that are circularly, rather than linearly polarized; waves with right- and left-circular polarizations travel at different phase velocities.

We demonstrate below that an optically active medium with right- and left-circular-polarization phase velocities c_o/n_+ and c_o/n_- acts as a polarization rotator with an angle of rotation $\pi(n_- - n_+)d/\lambda_o$ that is proportional to the thickness of the medium d. The rotatory power (rotation angle per unit length) of the optically active medium is therefore

$$\rho = \frac{\pi}{\lambda_o}(n_- - n_+). \qquad (6.4\text{-}1)$$
Rotatory Power

The direction in which the polarization plane rotates is the same as that of the circularly polarized component with the greater phase velocity (smaller refractive index). If $n_+ < n_-$, ρ is positive and the rotation is in the same direction as the electric-field vector of the right circularly polarized wave [clockwise when viewed from the direction toward which the wave is approaching, as illustrated in Fig. 6.4-1(a)]. Such materials are said to be **dextrorotatory**, whereas those for which $n_+ > n_-$ are termed **levorotatory**.

☐ **Derivation of the Rotatory Power.** Equation (6.4-1) may be derived by decomposing the incident linearly polarized wave into a sum of right and left circularly polarized components of equal amplitudes (see Example 6.1-1),

$$\begin{bmatrix} \cos\theta \\ \sin\theta \end{bmatrix} = \tfrac{1}{2}e^{-j\theta}\begin{bmatrix} 1 \\ j \end{bmatrix} + \tfrac{1}{2}e^{j\theta}\begin{bmatrix} 1 \\ -j \end{bmatrix}, \qquad (6.4\text{-}2)$$

where θ is the initial angle of the plane of polarization. After propagating a distance d through the medium, the phase shifts encountered by the right and left circularly polarized waves are $\varphi_+ = 2\pi n_+ d/\lambda_o$ and $\varphi_- = 2\pi n_- d/\lambda_o$, respectively, resulting in a Jones vector

$$\tfrac{1}{2}e^{-j\theta}e^{-j\varphi_+}\begin{bmatrix} 1 \\ j \end{bmatrix} + \tfrac{1}{2}e^{j\theta}e^{-j\varphi_-}\begin{bmatrix} 1 \\ -j \end{bmatrix} = e^{-j\varphi_o}\begin{bmatrix} \cos(\theta - \varphi/2) \\ \sin(\theta - \varphi/2) \end{bmatrix}, \qquad (6.4\text{-}3)$$

where $\varphi_o = \tfrac{1}{2}(\varphi_+ + \varphi_-)$ and $\varphi = \varphi_- - \varphi_+ = 2\pi(n_- - n_+)d/\lambda_o$. This Jones vector represents a linearly polarized wave with the plane of polarization rotated by an angle $\varphi/2 = \pi(n_- - n_+)d/\lambda_o$, as provided in (6.4-1). ■

Optical activity occurs in materials with an intrinsically helical structure. Examples include selenium, tellurium, tellurium oxide (TeO$_2$), quartz (α-SiO$_2$), and cinnabar (HgS). Optically active liquids consist of so-called chiral molecules, which come in distinct left- and right-handed mirror-image forms. Many organic compounds, such as amino acids and sugars, exhibit optical activity. Almost all amino acids are levorotatory, whereas common sugars come in both forms: dextrose (d-glucose) and levulose (fructose) are dextrorotatory and levorotatory, respectively, as their names imply. The rotatory power and sense of rotation for solutions of such substances are therefore sensitive to both the concentration and structure of the solute. A saccharimeter is used to determine the optical activity of sugar solutions, from which the sugar concentration is calculated.

6.4 OPTICAL ACTIVITY AND MAGNETO-OPTICS

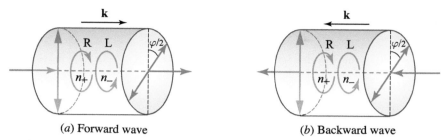

(a) Forward wave   (b) Backward wave

Figure 6.4-1 (a) The rotation of the plane of polarization by an optically active medium results from the difference in the velocities for the two circular polarizations. In this illustration, the right circularly polarized wave (R) is faster than the left circularly polarized wave (L), i.e., $n_+ < n_-$, so that ρ is positive and the material is dextrorotatory. (b) If the wave in (a) is reflected after traversing the medium, the plane of polarization rotates in the opposite direction so that the wave retraces itself.

Material Equations

A time-varying magnetic flux density **B** applied to an optically active structure induces a circulating current, by virtue of its helical character, that sets up an electric dipole moment (and hence a polarization) proportional to $j\omega \mathbf{B} = -\nabla \times \mathbf{E}$. The optically active medium is therefore spatially dispersive; i.e., the relation between $\mathbf{D}(\mathbf{r})$ and $\mathbf{E}(\mathbf{r})$ is not local. $\mathbf{D}(\mathbf{r})$ at position $\mathbf{r}$ is determined not only by $\mathbf{E}(\mathbf{r})$, but also by $\mathbf{E}(\mathbf{r}')$ at points $\mathbf{r}'$ in the immediate vicinity of $\mathbf{r}$, since it is dependent on the spatial derivatives contained in $\nabla \times \mathbf{E}(\mathbf{r})$. For a plane wave, we have $\mathbf{E}(\mathbf{r}) = \mathbf{E}\exp(-j\mathbf{k}\cdot\mathbf{r})$ and $\nabla \times \mathbf{E} = -j\mathbf{k} \times \mathbf{E}$, so that the electric permittivity tensor is dependent on the wavevector **k**. Spatial dispersiveness is analogous to temporal dispersiveness, which has its origin in the noninstantaneous response of the medium (see Sec. 5.2). While the permittivity of a medium exhibiting temporal dispersion depends on the frequency ω, that of a medium exhibiting spatial dispersion depends on the wavevector **k**.

An optically active medium is described by the **k**-dependent material equation

$$\mathbf{D} = \epsilon\mathbf{E} + j\epsilon_o \xi \mathbf{k} \times \mathbf{E}, \qquad (6.4\text{-}4)$$

where ξ is a pseudoscalar whose sign depends on the handedness of the coordinate system. This relation is a first-order approximation of the **k** dependence of the permittivity tensor, under appropriate symmetry conditions.[†] The first term represents the response of an isotropic dielectric medium whereas the second term accounts for the optical activity, as will be shown subsequently. This **D**–**E** relation is often written in the form

$$\mathbf{D} = \epsilon\mathbf{E} + j\epsilon_o \mathbf{G} \times \mathbf{E}, \qquad (6.4\text{-}5)$$

where $\mathbf{G} = \xi\mathbf{k}$ is a pseudovector known as the **gyration vector**. In such media the vector **D** is clearly not parallel to **E** since the vector $\mathbf{G} \times \mathbf{E}$ in (6.4-5) is perpendicular to **E**.

Normal Modes of the Optically Active Medium

We proceed to show that the two normal modes of the medium described by (6.4-5) are circularly polarized waves, and we determine the velocities c_o/n_+ and c_o/n_- in terms of the constant $G = \xi k$.

[†] See, e.g., L. D. Landau, E. M. Lifshitz, and L. P. Pitaevskii, *Electrodynamics of Continuous Media*, Butterworth–Heinemann, 2nd English ed. 1984, reprinted with corrections 2004, Chapter 12.

We assume that the wave propagates in the z direction, so that $\mathbf{k} = (0, 0, k)$ and thus $\mathbf{G} = (0, 0, G)$. Equation (6.4-5) may then be written in matrix form as

$$\begin{bmatrix} D_1 \\ D_2 \\ D_3 \end{bmatrix} = \epsilon_o \begin{bmatrix} n^2 & -jG & 0 \\ jG & n^2 & 0 \\ 0 & 0 & n^2 \end{bmatrix} \begin{bmatrix} E_1 \\ E_2 \\ E_3 \end{bmatrix}, \qquad (6.4\text{-}6)$$

where $n^2 = \epsilon/\epsilon_o$. The diagonal elements in (6.4-6) correspond to propagation in an isotropic medium with refractive index n, whereas the off-diagonal elements, proportional to G, represent the optical activity.

To prove that the normal modes are circularly polarized, consider the two circularly polarized waves with electric-field vectors $\mathbf{E} = (E_0, \pm jE_0, 0)$. The $+$ and $-$ signs correspond to right and left circularly polarized waves, respectively. Substitution in (6.4-6) yields $\mathbf{D} = (D_0, \pm jD_0, 0)$, where $D_0 = \epsilon_o(n^2 \pm G)E_0$. It follows that $\mathbf{D} = \epsilon_o n_\pm^2 \mathbf{E}$, where

$$\boxed{n_\pm = \sqrt{n^2 \pm G}\,.} \qquad (6.4\text{-}7)$$

Hence, for either of the two circularly polarized waves the vector $\mathbf{D}$ is parallel to the vector $\mathbf{E}$. Equation (6.3-11) is satisfied if the wavenumber $k = n_\pm k_o$. Thus, the right and left circularly polarized waves propagate without changing their state of polarization, with refractive indices n_+ and n_-, respectively. They are therefore the normal modes for this medium.

EXERCISE 6.4-1

Rotatory Power of an Optically Active Medium. Show that if $G \ll n$, the rotatory power of an optically active medium (rotation of the polarization plane per unit length) is approximately given by

$$\rho \approx -\frac{\pi G}{\lambda_o n}\,. \qquad (6.4\text{-}8)$$

The rotatory power is strongly dependent on the wavelength. Since G is proportional to k, as indicated by (6.4-5), it is inversely proportional to the wavelength λ_o. Thus, the rotatory power in (6.4-8) is inversely proportional to λ_o^2. Moreover, the refractive index n is itself wavelength dependent. By way of example, the rotatory power ρ of quartz is ≈ 31 deg/mm at $\lambda_o = 500$ nm and ≈ 22 deg/mm at $\lambda_o = 600$ nm; for silver thiogallate ($AgGaS_2$), ρ is ≈ 700 deg/mm at 490 nm and ≈ 500 deg/mm at 500 nm.

B. Magneto-Optics: The Faraday Effect

Many materials act as polarization rotators in the presence of a static magnetic field, a property known as the **Faraday effect**. The angle of rotation is then proportional to the thickness of the material, and the rotatory power ρ (rotation angle per unit length) is proportional to the component of the magnetic flux density B in the direction of the wave propagation,

$$\rho = \mathcal{V}B, \qquad (6.4\text{-}9)$$

where V is called the **Verdet constant**.

The sense of rotation is governed by the direction of the magnetic field: for $V > 0$, the rotation is in the direction of a right-handed screw pointing in the direction of the magnetic field [Fig. 6.4-2(a)]. In contrast to optical activity, however, the sense of rotation does not reverse with the reversal of the direction of propagation of the wave. Thus, when a wave travels through a Faraday rotator and then reflects back onto itself, traveling once more through the rotator in the opposite direction, it undergoes twice the rotation [Fig. 6.4-2(b)]. Materials that exhibit the Faraday effect include glasses as well

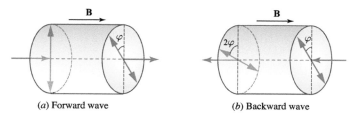

(a) Forward wave   (b) Backward wave

Figure 6.4-2 (a) Polarization rotation in a medium exhibiting the Faraday effect. (b) The sense of rotation is invariant to the direction of travel of the wave.

as yttrium iron garnet (YIG), terbium gallium garnet (TGG), and terbium aluminum garnet (TbAlG). The Verdet constant of TbAlG is $V \approx -1.16$ min/Oe-cm at $\lambda_o = 500$ nm. Thin films of these ferrimagnetic materials are used to make compact devices.

Material Equations

In magneto-optic materials, the electric permittivity tensor ϵ is altered by the application of a *static* magnetic field **H**, so that $\epsilon = \epsilon(\mathbf{H})$. This effect originates from the interaction of the static magnetic field with the motion of the electrons in the material in response to an *optical* electric field **E**. For the Faraday effect, in particular, the material equation is

$$\mathbf{D} = \epsilon \mathbf{E} + j\epsilon_o \mathbf{G} \times \mathbf{E} \tag{6.4-10}$$

with

$$\mathbf{G} = \gamma_B \mathbf{B}. \tag{6.4-11}$$

Here, $\mathbf{B} = \mu \mathbf{H}$ is the static magnetic flux density, and γ_B is a constant of the medium known as the **magnetogyration coefficient**.

Equation (6.4-10) is identical to (6.4-5) so that the vector $\mathbf{G} = \gamma_B \mathbf{B}$ in Faraday rotators plays the role of the gyration vector $\mathbf{G} = \xi \mathbf{k}$ in optically active media. For the Faraday effect, however, **G** does not depend on **k**, so that reversing the direction of propagation does not reverse the sense of rotation of the plane of polarization. This property is useful for constructing optical isolators, as explained in Sec. 6.6D.

With this analogy, and using (6.4-8), we conclude that the rotatory power of the Faraday medium is $\rho \approx -\pi G/\lambda_o n = -\pi \gamma_B B/\lambda_o n$, from which the Verdet constant (rotatory power per unit magnetic flux density) is seen to be

$$\boxed{V \approx -\frac{\pi \gamma_B}{\lambda_o n}.} \tag{6.4-12}$$

The Verdet constant is clearly a function of the wavelength λ_o.

6.5 OPTICS OF LIQUID CRYSTALS

Liquid Crystals

A liquid crystal comprises a collection of elongated organic molecules that are typically cigar-shaped. The molecules lack positional order (like liquids) but possess orientational order (like crystals). There are three types (phases) of liquid crystals, as illustrated in Fig. 6.5-1:

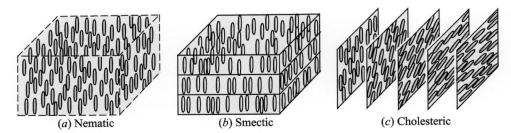

(a) Nematic   (b) Smectic   (c) Cholesteric

Figure 6.5-1 Molecular organizations of different types of liquid crystals.

- In **nematic liquid crystals** the orientations of the molecules tend to be the same but their positions are totally random.
- In **smectic liquid crystals** the orientations of the molecules are the same, but their centers are stacked in parallel layers within which they have random positions; they therefore have positional order only in one dimension.
- The **cholesteric liquid crystal** is a distorted form of its nematic cousin in which the orientations undergo helical rotation about an axis.

Liquid crystallinity, which was discovered in 1888, is a *fluid* state of matter; it is intermediate between liquid and solid. The molecules are able to change orientation when subjected to a force. When a thin layer of liquid crystal is placed between two parallel glass plates that are rubbed together, for example, the molecules orient themselves along the direction of rubbing.

Twisted nematic liquid crystals are nematic liquid crystals on which a twist (similar to the twist that exists naturally in the cholesteric phase) is externally imposed. This can be achieved, for example, by placing a thin layer of nematic liquid crystal between two glass plates that are polished in perpendicular directions, as schematized in Fig. 6.5-2. This section is devoted to a discussion of the optical properties of twisted nematic liquid crystals, which are widely used in photonics, e.g., for liquid-crystal displays. The electro-optic properties of twisted nematic liquid crystals, and their use as optical modulators and switches, are described in Sec. 21.3.

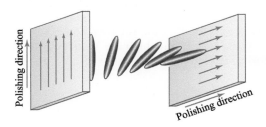

Figure 6.5-2 Molecular orientations of the twisted nematic liquid crystal.

Optical Properties of Twisted Nematic Liquid Crystals

The twisted nematic liquid crystal is an optically *inhomogeneous* and *anisotropic* medium that acts locally as a uniaxial crystal, with the optic axis parallel to the elongated direction. The optical properties are conveniently analyzed by considering the material to be divided into thin layers perpendicular to the axis of twist, each of which acts as a uniaxial crystal; the optic axis is taken to rotate gradually, in a helical fashion, along the axis of twist (Fig. 6.5-3). The cumulative effects of these layers on the transmitted wave is then calculated. We show that, under certain conditions, the twisted nematic liquid crystal acts as a polarization rotator in which the plane of polarization rotates in alignment with the molecular twist.

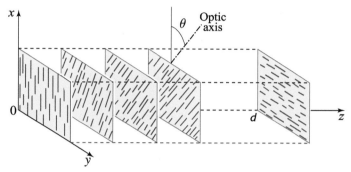

Figure 6.5-3 Propagation of light in a twisted nematic liquid crystal. In this diagram the angle of twist is 90°.

Consider the propagation of light along the axis of twist (the z axis) of a twisted nematic liquid crystal and assume that the twist angle θ varies linearly with z,

$$\theta = \alpha z, \qquad (6.5\text{-}1)$$

where α is the twist coefficient (degrees per unit length). The optic axis is therefore parallel to the x–y plane and makes an angle θ with the x direction. The ordinary and extraordinary refractive indices are n_o and n_e, respectively (typically, $n_e > n_o$), and the phase-retardation coefficient (retardation per unit length) is

$$\beta = (n_e - n_o)k_o. \qquad (6.5\text{-}2)$$

The liquid-crystal cell is completely characterized by the twist coefficient α and the retardation coefficient β.

In practice, $\beta \gg \alpha$ so that many cycles of phase retardation are introduced before the optic axis rotates appreciably. We show below that if this condition is satisfied, and the incident wave at $z = 0$ is linearly polarized in the x direction, then the wave maintains its linearly polarized state but the plane of polarization rotates in alignment with the molecular twist, so that the angle of rotation is $\theta = \alpha z$ and the total rotation in a crystal of length d is the angle of twist αd. The liquid-crystal cell then serves as a polarization rotator with rotatory power α. The polarization-rotation property of the twisted nematic liquid crystal is useful for making display devices, as explained in Sec. 21.3.

☐ **Proof that the Twisted Nematic Liquid Crystal Acts as a Polarization Rotator.** We proceed to show that the twisted nematic liquid crystal acts as a polarization rotator if $\beta \gg \alpha$. We divide the overall width of the cell d into N incremental layers of equal widths $\Delta z = d/N$. The mth layer, located at the distance $z = z_m = m\Delta z$, $m = 1, 2, \ldots, N$, is a wave retarder whose slow axis (the optic axis) makes an angle $\theta_m = m\Delta\theta$ with the x axis, where $\Delta\theta = \alpha\Delta z$. It therefore has a Jones matrix [see (6.1-24)]

$$\mathbf{T}_m = \mathbf{R}(-\theta_m)\,\mathbf{T}_r\,\mathbf{R}(\theta_m), \tag{6.5-3}$$

where

$$\mathbf{T}_r = \begin{bmatrix} \exp(-jn_e k_o \Delta z) & 0 \\ 0 & \exp(-jn_o k_o \Delta z) \end{bmatrix} \tag{6.5-4}$$

is the Jones matrix of a wave retarder whose axis is along the x direction and $\mathbf{R}(\theta)$ is the coordinate rotation matrix in (6.1-22).

It is convenient to rewrite $\mathbf{T}_r$ in terms of the phase-retardation coefficient $\beta = (n_e - n_o)k_o$,

$$\mathbf{T}_r = \exp(-j\varphi\Delta z)\begin{bmatrix} \exp(-j\beta\Delta z/2) & 0 \\ 0 & \exp(j\beta\Delta z/2) \end{bmatrix}, \tag{6.5-5}$$

where $\varphi = (n_o + n_e)k_o/2$. Since multiplying the Jones vector by a constant phase factor does not affect the state of polarization, we simply ignore the prefactor $\exp(-j\varphi\Delta z)$ in (6.5-5).

The overall Jones matrix of the device is the product

$$\mathbf{T} = \prod_{m=N}^{1} \mathbf{T}_m = \prod_{m=N}^{1} \mathbf{R}(-\theta_m)\,\mathbf{T}_r\,\mathbf{R}(\theta_m). \tag{6.5-6}$$

Using (6.5-3) and noting that $\mathbf{R}(\theta_m)\,\mathbf{R}(-\theta_{m-1}) = \mathbf{R}(\theta_m - \theta_{m-1}) = \mathbf{R}(\Delta\theta)$, we obtain

$$\mathbf{T} = \mathbf{R}(-\theta_N)\,[\mathbf{T}_r \mathbf{R}(\Delta\theta)]^{N-1}\,\mathbf{T}_r\,\mathbf{R}(\theta_1). \tag{6.5-7}$$

Substituting from (6.5-5) and (6.1-22), we obtain

$$\mathbf{T}_r\,\mathbf{R}(\Delta\theta) = \begin{bmatrix} \exp(-j\beta\Delta z/2) & 0 \\ 0 & \exp(j\beta\Delta z/2) \end{bmatrix}\begin{bmatrix} \cos\alpha\Delta z & \sin\alpha\Delta z \\ -\sin\alpha\Delta z & \cos\alpha\Delta z \end{bmatrix}. \tag{6.5-8}$$

Using (6.5-7) and (6.5-8), the Jones matrix $\mathbf{T}$ of the device can, in principle, be determined in terms of the parameters α, β, and $d = N\Delta z$.

For $\alpha \ll \beta$, we may assume that the incremental rotation matrix $\mathbf{R}(\Delta\theta)$ is approximately the identity matrix, whereupon

$$\mathbf{T} \approx \mathbf{R}(-\theta_N)\,[\mathbf{T}_r]^N\,\mathbf{R}(\theta_1) = \mathbf{R}(-\alpha N\Delta z)\begin{bmatrix} \exp(-j\beta\Delta z/2) & 0 \\ 0 & \exp(j\beta\Delta z/2) \end{bmatrix}^N$$

$$= \mathbf{R}(-\alpha N\Delta z)\begin{bmatrix} \exp(-j\beta N\Delta z/2) & 0 \\ 0 & \exp(j\beta N\Delta z/2) \end{bmatrix}, \tag{6.5-9}$$

so that

$$\mathbf{T} = \mathbf{R}(-\alpha d)\begin{bmatrix} \exp(-j\beta d/2) & 0 \\ 0 & \exp(j\beta d/2) \end{bmatrix}. \tag{6.5-10}$$

This Jones matrix represents a wave retarder of retardation βd with the slow axis along the x direction, followed by a polarization rotator with rotation angle αd. If the original wave is linearly polarized along the x direction, the wave retarder imparts only a phase shift; the device then simply rotates the polarization by an angle αd equal to the twist angle. A wave linearly polarized along the y direction is rotated by the same angle. ■

6.6 POLARIZATION DEVICES

This section offers a brief description of a number of devices that are used to modify the state of polarization of light. The basic principles underlying the operation of these devices have been set forth earlier in this chapter.

A. Polarizers

A linear polarizer is a device that transmits the component of the electric field that lies along the direction of its transmission axis while blocking the orthogonal component. The blocking action may be achieved by selective absorption, selective reflection from isotropic media, or selective reflection/refraction in anisotropic media.

Polarization by Selective Absorption (Dichroism)

The absorption of light by certain anisotropic media, called **dichroic materials**, depends on the direction of the incident electric field (Fig. 6.6-1). These materials generally have anisotropic molecular structures whose response is sensitive to the direction of the electric field. The most common dichroic material is **Polaroid H-sheet**, invented in 1938 and still in common use. It is fabricated from a sheet of iodine-impregnated polyvinyl alcohol that is heated and stretched in a particular direction. The analogous device in the infrared is the **wire-grid polarizer**, which comprises a planar configuration of closely spaced fine wires stretched in a single direction. The component of the incident electric field in the direction of the wires is absorbed whereas the component perpendicular to the wires passes through.

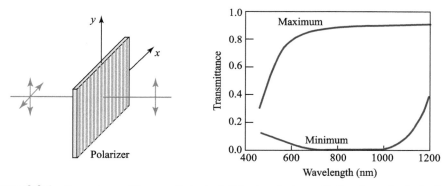

Figure 6.6-1 Power transmittances of a typical dichroic polarizer with the plane of polarization of the light aligned for maximum and minimum transmittance, as indicated.

Polarization by Selective Reflection

The reflectance of light at the boundary between two isotropic dielectric materials is dependent on its polarization, as discussed in Sec. 6.2. At the Brewster angle of incidence, in particular, the reflectance of TM-polarized light vanishes so that it is totally refracted (Fig. 6.2-4). At this angle, therefore, only TE-polarized light is reflected, so that the reflector serves as a polarizer, as depicted in Fig. 6.6-2.

Polarization by Selective Refraction (Polarizing Beamsplitters)

When light enters an anisotropic crystal, the ordinary and extraordinary waves refract at different angles and gradually separate from each other (see Sec. 6.3E and Fig. 6.3-15). This provides an effective means for obtaining polarized light from unpolarized

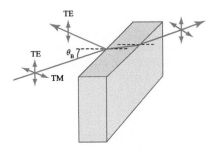

Figure 6.6-2 The Brewster-angle polarizer.

light, and it is commonly used. These devices usually consist of two cemented prisms comprising anisotropic (uniaxial) materials, often with different orientations, as illustrated by the examples in Fig. 6.6-3. These prisms therefore serve as **polarizing beamsplitters**.

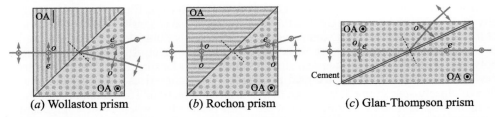

(a) Wollaston prism   (b) Rochon prism   (c) Glan-Thompson prism

Figure 6.6-3 Examples of polarizing beamsplitters. The parallel (p) and orthogonal (s) polarization components of a beam are separated by refraction or reflection at the boundary between two uniaxial crystals whose optic axes (OA) have different orientations. In this illustration, the crystals are negative uniaxial ($n_o > n_e$), such as calcite. (a) In the Wollaston prism the p component is extraordinary in the first crystal and ordinary in the second, while the opposite is true for the s component, so that they undergo different angles of refraction. (b) In the Rochon prism, the p component is transmitted without refraction since it is ordinary in both crystals, while the s component is refracted since it is ordinary in the first crystal and extraordinary in the second. (c) The operation of the Glan-Thompson prism is based on total internal reflection at the boundary between the first crystal and the cement layer. This occurs for only the p polarization since it is an ordinary wave with the higher refractive index. The s component is transmitted. The Glan–Thompson device has the merit of providing a large angular separation between the emerging waves.

B. Wave Retarders

A wave retarder serves to convert a wave with one form of polarization into another form. It is characterized by its retardation Γ and its fast and slow axes (see Sec. 6.1B). The normal modes are linearly polarized waves polarized along the directions of the axes. The velocities of the two waves differ so that transmission through the retarder imparts a relative phase shift Γ to these modes.

Wave retarders are often constructed from anisotropic crystals in the form of plates. As explained in Sec. 6.3B, when light travels along a principal axis of a crystal (say the z axis), the normal modes are linearly polarized waves pointing along the two other principal axes (the x and y axes). These modes experience the principal refractive indices n_1 and n_2, and thus travel at velocities c_o/n_1 and c_o/n_2, respectively. If $n_1 < n_2$, the x axis is the fast axis. If the plate has thickness d, the phase retardation is $\Gamma = (n_2 - n_1)k_o d = 2\pi(n_2 - n_1)d/\lambda_o$. The retardation is thus directly proportional to the thickness d of the plate and inversely proportional to the wavelength λ_o (note, however, that $n_2 - n_1$ is itself wavelength dependent).

The refractive indices of a thin sheet of mica, for example, are 1.599 and 1.594 at $\lambda_o = 633$ nm, so that $\Gamma/d \approx 15.8\pi$ rad/mm. A sheet of thickness 63.3 μm yields $\Gamma \approx \pi$ and thus serves as a half-wave retarder.

Light Intensity Control via a Wave Retarder and Two Polarizers

Consider a wave retarder of retardation Γ placed between two crossed polarizers whose axes are oriented at 45° with respect to the axes of the retarder, as illustrated in Fig. 6.6-4. The power (or intensity) transmittance of this system is

$$\mathcal{T} = \sin^2(\Gamma/2), \tag{6.6-1}$$

which may be established by making use of Jones matrices or by examining the polarization ellipse of the retarded light as a function of Γ, and then determining the component that lies in the direction of the output polarizer, as illustrated in Fig. 6.6-4. If $\Gamma = 0$ no light is transmitted through the system since the polarizers are orthogonal. On the other hand, if $\Gamma = \pi$ all of the light is transmitted since the retarder then rotates the plane of polarization 90° whereupon it matches the transmission axis of the second polarizer.

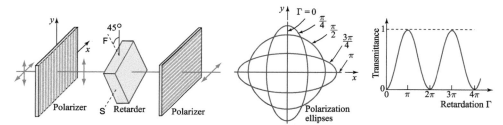

Figure 6.6-4 Controlling light intensity by means of a wave retarder with variable retardation Γ placed between two crossed polarizers.

The intensity of the transmitted light is thus readily controlled by altering the retardation Γ. This can be achieved, for example, by deliberately changing the indices n_1 and n_2 by application of an external DC electric field to the retarder. This is the basic principle that underlies the operation of electro-optic modulators, as discussed in Chapter 21.

Furthermore, since Γ depends on d, slight variations in the thickness of a sample can be monitored by examining the pattern of the transmitted light. Moreover, since Γ is wavelength dependent, the transmittance of the system is frequency sensitive. Though it can be used as a filter, the selectivity is not sharp. Other configurations using wave retarders and polarizers can be used to construct narrowband transmission filters.

C. Polarization Rotators

A polarization rotator serves to rotate the plane of polarization of linearly polarized light by a fixed angle, while maintaining its linearly polarized nature. Optically active media and materials exhibiting the Faraday effect act as polarization rotators, as discussed in Sec. 6.4. The twisted nematic liquid crystal also acts as a polarization rotator under certain conditions, as shown in Sec. 6.5.

If a polarization rotator is placed between two polarizers, the amount of light transmitted depends on the rotation angle. The intensity of the light can therefore be controlled (modulated) if the angle of rotation is externally changed (e.g., by varying the magnetic flux density applied to a Faraday rotator or by changing the molecular orientation of a liquid crystal by means of an applied electric field). Electro-optic modulation of light and liquid-crystal display devices are discussed in Chapter 21.

D. Nonreciprocal Polarization Devices

A device whose effect on the polarization state is invariant to reversal of the direction of propagation is said to be **reciprocal**. If a wave is transmitted through such a device in one direction and the emerging wave is retransmitted in the opposite direction, then it retraces the changes in the polarization state and arrives at the input in the very same initial polarization state. Devices that do not have this directional invariance are termed **nonreciprocal**. All of the polarization systems described in this chapter are reciprocal, with the exception of the Faraday rotator (Sec. 6.4B). A number of useful nonreciprocal polarization devices may be implemented by combining the Faraday rotator with other reciprocal polarization components (see Sec. 24.1C).

Optical Isolator

An **optical isolator** is a device that transmits light in only one direction, thereby acting as a "one-way valve." Optical isolators are useful for preventing reflected light from returning back to the source. Such feedback can have deleterious effects on the operation of certain devices, such as laser diodes.

An optical isolator is constructed by placing a Faraday rotator between two polarizers whose axes make a 45° angle with respect to each other. The magnetic flux density applied to the rotator is adjusted so that it rotates the polarization by 45° in the direction of a right-handed screw pointing in the z direction [Fig. 6.6-5(a)]. Light traveling through the system in the forward direction (from left to right) thus crosses polarizer A, rotates 45°, and is thence transmitted through polarizer B. Linearly polarized light with the polarization plane at 45° but traveling through the system in the backward direction [from right to left in Fig. 6.6-5(b)] successfully crosses polarizer B. However, on passing through the Faraday rotator, the plane of polarization rotates an additional 45° and is therefore blocked by polarizer A. Since the backward light might be generated by reflection of the forward wave from subsequent surfaces, the isolator serves to protect its source from reflected light.

Note that the Faraday rotator is a necessary component of the optical isolator. An optically active, or liquid-crystal, polarization rotator cannot be used in its place. In those *reciprocal* components, the sense of rotation is such that the polarization of the reflected wave retraces that of the incident wave so that the light would be transmitted back through the polarizers to the source.

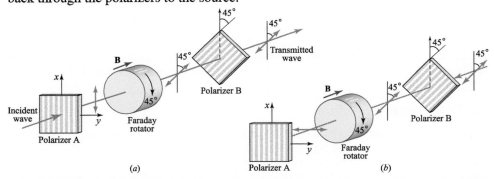

Figure 6.6-5 An optical isolator that makes use of a Faraday rotator transmits light in one direction. (a) A wave traveling in the forward direction is transmitted. (b) A wave traveling in the backward (or reverse) direction is blocked.

Faraday-rotator isolators constructed from yttrium iron garnet (YIG) or terbium gallium garnet (TGG) offer attenuations of the backward wave of up to 90 dB over a relatively wide wavelength range. Thin films of these materials placed in permanent magnetic fields are used to make compact optical isolators.

Nonreciprocal Polarization Rotation

The combination of a 45° Faraday rotator and a half-wave retarder is another useful nonreciprocal device. As illustrated in Fig. 6.6-6(a), the state of polarization of a forward-traveling linearly polarized wave, with its plane of polarization oriented at 22.5° with the fast axis of the retarder, maintains its state of polarization upon transmission through the device (since it undergoes 45° rotation by the Faraday rotator, followed by −45° rotation by the retarder). However, for a wave traveling in the reverse direction, the plane of polarization is rotated by 45° + 45° = 90°, as can be readily seen in Fig. 6.6-6(b). This device may therefore be used in combination with a polarizing beamsplitter to deflect the backward-traveling wave away from the source of the forward-traveling wave so that it can be accessed independently. A system of this kind can be useful for implementing nonreciprocal interconnects, such as **optical circulators**, as portrayed in Fig. 24.1-10(b).

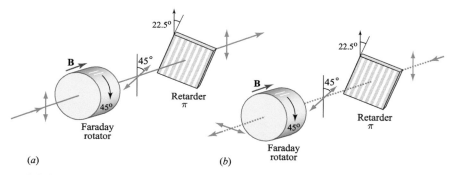

Figure 6.6-6 A Faraday rotator followed by a half-wave (π) retarder is a nonreciprocal device that: (a) maintains the polarization state of a linearly polarized forward-traveling wave, but (b) rotates the plane of polarization of the backward-traveling wave by 90°.

READING LIST

Polarization

See also the reading list on general optics in Chapter 1 and the reading list in Chapter 5.

A. Kumar and A. Ghatak, *Polarization of Light with Applications in Optical Fibers*, SPIE Optical Engineering Press, 2011.

D. H. Goldstein, *Polarized Light*, CRC Press/Taylor & Francis, 3rd ed. 2010.

L. Nikolova and P. S. Ramanujam, *Polarization Holography*, Cambridge University Press, 2009.

R. Martínez-Herrero, P. M. Mejías, and G. Piquero, *Characterization of Partially Polarized Light Fields*, Springer-Verlag, 2009.

E. Collett, *Field Guide to Polarization*, SPIE Optical Engineering Press, 2005.

J. N. Damask, *Polarization Optics in Telecommunications*, Springer-Verlag, 2004.

S. Sugano and N. Kojima, eds., *Magneto-Optics*, Springer-Verlag, 2000.

C. Brosseau, *Fundamentals of Polarized Light: A Statistical Optics Approach*, Wiley, 1998.

S. Huard, *Polarization of Light*, Wiley, 1996.

E. Collett, *Polarized Light: Fundamentals and Applications*, CRC Press, 1993.

A. Yariv and P. Yeh, *Optical Waves in Crystals: Propagation and Control of Laser Radiation*, Wiley, 1985, paperback reprint 2002.

R. M. A. Azzam and N. M. Bashara, *Ellipsometry and Polarized Light*, North-Holland, 1977, reprinted 1989.

B. A. Robson, *The Theory of Polarization Phenomena*, Clarendon, 1974.

L. Velluz, M. Le Grand, and M. Grosjean, *Optical Circular Dichroism: Principles, Measurements, and Applications*, Academic Press, 1965.

W. A. Shurcliff and S. S. Ballard, *Polarized Light*, Van Nostrand, 1964.

W. A. Shurcliff, *Polarized Light: Production and Use*, Harvard University Press, 1962, reprinted 1966.

Crystals and Tensor Analysis

C. Malgrange, C. Ricolleau, and M. Schlenker, *Symmetry and Physical Properties of Crystals*, Springer-Verlag, 2014.

D. Fleisch, *A Student's Guide to Vectors and Tensors*, Cambridge University Press, 2011.

S. Haussühl, *Physical Properties of Crystals*, Wiley–VCH, 2007.

R. J. D. Tilley, *Crystals and Crystal Structures*, Wiley, 2006.

E. A. Wood, *Crystals and Light: An Introduction to Optical Crystallography*, Dover, 2nd ed. 1977.

J. F. Nye, *Physical Properties of Crystals: Their Representation by Tensors and Matrices*, Oxford University Press, 1985.

Liquid Crystals

See also the reading list on liquid crystal devices and displays in Chapter 21.

J. V. Selinger, *Introduction to the Theory of Soft Matter: From Ideal Gases to Liquid Crystals*, Springer-Verlag, 2016.

I.-C. Khoo, *Liquid Crystals*, Wiley, 2nd ed. 2007.

T. Scharf, *Polarized Light in Liquid Crystals and Polymers*, Wiley, 2006.

P. Oswald and P. Pieranski, *Nematic and Cholesteric Liquid Crystals: Concepts and Physical Properties Illustrated by Experiments*, CRC Press/Taylor & Francis, 2005.

P. J. Collings, *Liquid Crystals: Nature's Delicate Phase of Matter*, Princeton University Press, 2nd ed. 2002.

M. E. Lines and A. M. Glass, *Principles and Applications of Ferroelectrics and Related Materials*, Clarendon, 1977; Oxford University Press, paperback 2nd ed. 2001.

P. G. de Gennes and J. Prost, *The Physics of Liquid Crystals*, Oxford University Press, 2nd ed. 1993.

S. Chandrasekhar, *Liquid Crystals*, Cambridge University Press, 2nd ed. 1992.

L. M. Blinov, *Electro-Optical and Magneto-Optical Properties of Liquid Crystals*, Wiley, 1983.

Articles

L. A. Whitehead and M. A. Mossman, Reflections on Total Internal Reflection, *Optics & Photonics News*, vol. 20, no. 2, pp. 28–34, 2009.

M. Mansuripur, The Faraday Effect, *Optics & Photonics News*, vol. 10, no. 11, pp. 32–36, 1999.

B. H. Billings, ed., *Selected Papers on Applications of Polarized Light*, SPIE Optical Engineering Press (Milestone Series Volume 57), 1992.

S. D. Jacobs, ed., *Selected Papers on Liquid Crystals for Optics*, SPIE Optical Engineering Press (Milestone Series Volume 46), 1992.

B. H. Billings, ed., *Selected Papers on Polarization*, SPIE Optical Engineering Press (Milestone Series Volume 23), 1990.

A. Lakhtakia, ed., *Selected Papers on Natural Optical Activity*, SPIE Optical Engineering Press (Milestone Series Volume 15), 1990.

W. Swindell, ed., *Benchmark Papers in Optics: Polarized Light*, Dowden, Hutchinson & Ross, 1975.

Sir David Brewster's Scientific Work, *Nature*, vol. 25, pp. 157–159, 15 December 1881.

PROBLEMS

6.1-5 **Orthogonal Polarizations.** Show that if two elliptically polarized states are orthogonal, the major axes of their ellipses are perpendicular and the senses of rotation are opposite.

6.1-6 **Rotating a Polarization Rotator.** Show that the Jones matrix of a polarization rotator is invariant to rotation of the coordinate system.

6.1-7 **Jones Matrix of a Polarizer.** Show that the Jones matrix of a linear polarizer whose transmission axis makes an angle θ with the x axis is

$$\mathbf{T} = \begin{bmatrix} \cos^2\theta & \sin\theta\cos\theta \\ \sin\theta\cos\theta & \sin^2\theta \end{bmatrix}.$$

Derive this result using (6.1-18), (6.1-22), and (6.1-24).

6.1-8 **Half-Wave Retarder.** Consider linearly polarized light passed through a half-wave retarder. If the polarization plane makes an angle θ with the fast axis of the retarder, show that the transmitted light is linearly polarized at an angle $-\theta$, i.e., it is rotated by an angle 2θ. Why is the half-wave retarder not equivalent to a polarization rotator?

6.1-9 **Wave Retarders in Tandem.** Write the Jones matrices for:

(a) A $\pi/2$ wave retarder with the fast axis along the x direction.

(b) A π wave retarder with the fast axis at $45°$ to the x direction.

(c) A $\pi/2$ wave retarder with the fast axis along the y direction.

If these three retarders are cascaded (placed in tandem), with (c) following (b) following (a), show that the resulting device introduces a $90°$ rotation. What happens if the order of the three retarders is reversed?

6.1-10 **Reflection of Circularly Polarized Light.** Show that circularly polarized light changes handedness (right becomes left, and *vice versa*), upon reflection from a mirror.

6.1-11 **Anti-Glare Screen.** A self-luminous object is viewed through a glass window. An antiglare screen is used to eliminate glare caused by the reflection of background light from the surfaces of the window. Show that such a screen may be constructed from a combination of a linear polarizer and a quarter-wave retarder whose axes are at $45°$ with respect to the transmission axis of the polarizer. Can the screen be regarded as an optical isolator?

6.2-2 **Derivation of Fresnel Equations.** Derive the reflection equation (6.2-6), which is used to derive the Fresnel equation (6.2-8) for TE polarization. How would you go about obtaining the reflection coefficient if the incident light took the form of a beam rather than a plane wave?

6.2-3 **Reflectance of Glass.** A plane wave is incident from air ($n = 1$) onto a glass plate ($n = 1.5$) at an angle of incidence of $45°$. Determine the power reflectances of the TE and TM waves. What is the average reflectance for unpolarized light (light carrying TE and TM waves of equal intensities)?

6.2-4 **Refraction at the Brewster Angle.** Use the condition $n_1 \sec\theta_1 = n_2 \sec\theta_2$ and Snell's law, $n_1 \sin\theta_1 = n_2 \sin\theta_2$, to derive (6.2-12) for the Brewster angle. Also show that at the Brewster angle, $\theta_1 + \theta_2 = 90°$, so that the directions of the reflected and refracted waves are orthogonal, and hence the electric field of the refracted TM wave is parallel to the direction of the reflected wave. The reflection of light may be regarded as a scattering process in which the refracted wave acts as a source of radiation generating the reflected wave. At the Brewster angle, this source oscillates in a direction parallel to the direction of propagation of the reflected wave, so that radiation cannot occur and no TM light is reflected.

6.2-5 **Retardation Associated with Total Internal Reflection.** Determine the phase retardation between the TE and TM waves that is introduced by total internal reflection at the boundary between glass ($n = 1.5$) and air ($n = 1$) at an angle of incidence $\theta = 1.2\,\theta_c$, where θ_c is the critical angle.

6.2-6 **Goos–Hänchen Shift.** Consider two TE plane waves undergoing total internal reflection at angles θ and $\theta + d\theta$, where $d\theta$ is an incremental angle. If the phase retardation introduced between the reflected waves is written in the form $d\varphi = \xi\,d\theta$, find an expression for the coefficient ξ. Sketch the interference patterns of the two incident waves and the two reflected waves and verify that they are shifted by a lateral distance proportional to ξ. When the incident wave is a beam (composed of many plane-wave components), the reflected beam is displaced laterally by a distance proportional to ξ. This is known as the Goos–Hänchen effect.

6.2-7 **Reflection from an Absorptive Medium.** Use Maxwell's equations and appropriate boundary conditions to show that the complex amplitude reflectance at the boundary between free space and a medium with refractive index n and absorption coefficient α, at normal incidence, is $r = [(n - j\alpha c/2\omega) - 1]/[(n - j\alpha c/2\omega) + 1]$.

6.3-1 **Maximum Retardation in Quartz.** Quartz is a positive uniaxial crystal with $n_e = 1.553$ and $n_o = 1.544$. (a) Determine the retardation per mm at $\lambda_o = 633$ nm when the crystal is oriented such that retardation is maximized. (b) At what thickness(es) does the crystal act as a quarter-wave retarder?

6.3-2 **Maximum Extraordinary Effect.** Determine the direction of propagation in quartz ($n_e = 1.553$ and $n_o = 1.544$) at which the angle between the wavevector **k** and the Poynting vector **S** (which is also the direction of ray propagation) is maximum.

6.3-3 **Double Refraction.** An unpolarized plane wave is incident from free space onto a quartz crystal ($n_e = 1.553$ and $n_o = 1.544$) at an angle of incidence 30°. The optic axis lies in the plane of incidence and is perpendicular to the direction of the incident wave before it enters the crystal. Determine the directions of the wavevectors and the rays of the two refracted components.

6.3-4 **Lateral Shift in Double Refraction.** What is the optimum geometry for maximizing the lateral shift between the refracted ordinary and extraordinary beams in a positive uniaxial crystal? Indicate all pertinent angles and directions.

6.3-5 **Transmission Through a LiNbO$_3$ Plate.** Examine the transmission of an unpolarized He–Ne laser beam ($\lambda_o = 633$ nm) normally incident on a LiNbO$_3$ plate ($n_e = 2.29$, $n_o = 2.20$) of thickness 1 cm, cut such that its optic axis makes an angle 45° with the normal to the plate. Determine the lateral shift at the output of the plate and the retardation between the ordinary and extraordinary beams.

*6.3-6 **Conical Refraction.** When the wavevector **k** points along an optic axis of a biaxial crystal an unusual situation occurs. The two sheets of the **k** surface meet and the surface can be approximated by a conical surface. Consider a ray normally incident on the surface of a biaxial crystal for which one of its optic axes is also normal to the surface. Show that multiple refraction occurs with the refracted rays forming a cone. This effect is known as **conical refraction**. What happens when the conical rays refract from the parallel surface of the crystal into air?

6.6-1 **Circular Dichroism.** Certain materials have different absorption coefficients for right and left circularly polarized light, a property known as **circular dichroism**. Determine the Jones matrix for a device that converts light with any state of polarization into right circularly polarized light.

6.6-2 **Polarization Rotation by a Sequence of Linear Polarizers.** A wave that is linearly polarized in the x direction is transmitted through a sequence of N linear polarizers whose transmission axes are inclined by angles $m\theta$ ($m = 1, 2, \ldots, N$; $\theta = \pi/2N$) with respect to the x axis. Show that the transmitted light is linearly polarized in the y direction but its amplitude is reduced by the factor $\cos^N \theta$. What happens in the limit as $N \to \infty$? *Hint:* Use Jones matrices and note that

$$\mathbf{R}[(m+1)\theta]\,\mathbf{R}(-m\theta) = \mathbf{R}(\theta),$$

where $\mathbf{R}(\theta)$ is the coordinate transformation matrix.

CHAPTER 7

PHOTONIC-CRYSTAL OPTICS

7.1	OPTICS OF DIELECTRIC LAYERED MEDIA	258
	A. Matrix Theory of Multilayer Optics	
	B. Fabry–Perot Etalon	
	C. Bragg Grating	
7.2	ONE-DIMENSIONAL PHOTONIC CRYSTALS	277
	A. Bloch Modes	
	B. Matrix Optics of Periodic Media	
	C. Fourier Optics of Periodic Media	
	D. Boundaries Between Periodic and Homogeneous Media	
7.3	TWO- AND THREE-DIMENSIONAL PHOTONIC CRYSTALS	291
	A. Two-Dimensional Photonic Crystals	
	*B. Three-Dimensional Photonic Crystals	

Felix Bloch (1905–1983) developed a theory that describes electron waves in the periodic structure of solids.

Eli Yablonovitch (born 1946) coinvented the concept of the photonic bandgap; he made the first photonic-bandgap crystal.

Sajeev John (born 1957) invoked the notion of photon localization and coinvented the photonic-bandgap concept.

Fundamentals of Photonics, Third Edition. Bahaa E. A. Saleh and Malvin Carl Teich.
©2019 John Wiley & Sons, Inc. Published 2019 by John Wiley & Sons, Inc.

The propagation of light in homogeneous media and its reflection and refraction at the boundaries between different media are a principal concern of optics, as described in the earlier chapters of this book. Photonic devices often comprise multiple layers of different materials arranged, for example, to suppress or enhance reflectance or to alter the spectral or the polarization characteristics of light. Multilayered and stratified media are also found in natural physical and biological systems and are responsible for the distinct colors of some insects and butterfly wings. Multilayered media can also be periodic, i.e., comprise identical dielectric structures replicated in a one-, two-, or three-dimensional periodic arrangement, as illustrated in Fig. 7.0-1. One-dimensional periodic structures include stacks of identical parallel planar multilayer segments. These are often used as gratings that reflect optical waves incident at certain angles, or as filters that selectively reflect waves of certain frequencies. Two-dimensional periodic structures include sets of parallel rods as well as sets of parallel cylindrical holes, such as those used to modify the characteristics of optical fibers known as holey fibers (see Sec. 10.4). Three-dimensional periodic structures comprise arrays of cubes, spheres, or holes of various shapes, organized in lattice structures much like those found in natural crystals. Photonic crystals are a special class of optical metamaterials, which are considered in Chapter 8.

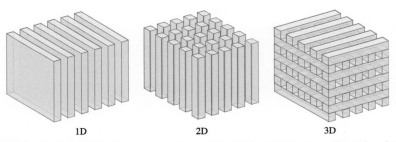

Figure 7.0-1 Periodic photonic structures in one-dimensional (1D), two-dimensional (2D), and three-dimensional (3D) configurations.

Optical waves, which are inherently periodic, interact with periodic media in a unique way, particularly when the scale of the periodicity is of the same order as that of the wavelength. For example, spectral bands emerge in which light waves cannot propagate through the medium without severe attenuation. Waves with frequencies lying within these forbidden bands, called **photonic bandgaps**, behave in a manner akin to total internal reflection, but are applicable for all directions. The dissolution of the transmitted wave is a result of destructive interference among the waves scattered by elements of the periodic structure in the forward direction. Remarkably, this effect extends over finite spectral bands, rather than occurring for just single frequencies.

This phenomenon is analogous to the electronic properties of crystalline solids such as semiconductors. In that case, the periodic wave associated with an electron travels in a periodic crystal lattice, and energy bandgaps often materialize. Because of this analogy, the photonic periodic structures have come to be called **photonic crystals**. Photonic crystals enjoy a whole raft of applications, including use as waveguides, fibers, resonators, lasers, filters, routers, switches, gates, and sensors; other applications are in the offing.

An electromagnetic-optics analysis is usually required to describe the optical properties of inhomogeneous media such as multilayered and periodic materials. For inhomogeneous dielectric media, as we know from Sec. 5.2B, the permittivity $\epsilon(\mathbf{r})$ is spatially varying and the wave equation takes the general forms of (5.2-16) and (5.2-17). For a harmonic wave of angular frequency ω, this leads to generalized Helmholtz equations for the electric and the magnetic fields expressible as

$$\eta(\mathbf{r}) \, \nabla \times (\nabla \times \mathbf{E}) = \frac{\omega^2}{c_o^2} \mathbf{E}, \qquad (7.0\text{-}1)$$

$$\nabla \times [\eta(\mathbf{r}) \, \nabla \times \mathbf{H}] = \frac{\omega^2}{c_o^2} \mathbf{H}, \qquad (7.0\text{-}2)$$

Generalized Helmholtz Equations

where $\eta(\mathbf{r}) = \epsilon_o/\epsilon(\mathbf{r})$ is the electric impermeability (see Sec. 6.3A). One of these equations may be solved for either the electric or the magnetic field, and the other field may be directly determined by use of Maxwell's equations. Note that (7.0-1) and (7.0-2) are cast in the form of an eigenvalue problem: a differential operator applied on the field function equals a constant multiplied by the field function. The eigenvalues are ω^2/c_o^2 and the eigenfunctions provide the spatial distributions of the modes of the propagating field (see Appendix C). For reasons to be explained in Secs. 7.2C and 7.3, we work with the magnetic-field equation (7.0-2) rather than the electric-field equation (7.0-1).

For multilayered media, $\epsilon(\mathbf{r})$ is piecewise constant, i.e., it is uniform within any given layer but changes from one layer to another. Wave propagation can then be studied by using the known properties of optical waves in homogeneous media, together with the appropriate boundary conditions that dictate the laws of reflection and transmission.

Periodic dielectric media are characterized by periodic values of $\epsilon(\mathbf{r})$ and $\eta(\mathbf{r})$. This periodicity imposes certain conditions on the optical wave. For example, the propagation constant deviates from simply proportionality to the angular frequency ω, as for a homogeneous medium. While the modes of propagation in a homogeneous medium are plane waves of the form $\exp(-j\mathbf{k} \cdot \mathbf{r})$, the modes of the periodic medium, known as **Bloch modes**, are traveling waves modulated by standing waves.

This Chapter

Previous chapters have focused on the optics of thin optical components that are well separated, such as thin lenses, planar gratings, and image-bearing films across which the light travels. This chapter addresses the optics of *bulk* media comprising multiple dielectric layers and periodic 1D, 2D, and 3D photonic structures. Section 7.1, in which 1D layered media are considered, serves as a prelude to periodic media and photonic crystals. A matrix approach offers a systematic treatment of the multiple reflections that occur at the multiple boundaries of the medium. Section 7.2 introduces photonic crystals in their simplest form — 1D periodic structures. Matrix methods are adopted to determine the dispersion relation and the band structure. An alternate approach, based on a Fourier-series representation of the periodic functions associated with the medium and the wave, is also presented. These results are generalized in Sec. 7.3 to two- and three-dimensional photonic crystals.

Throughout this chapter, the various media are assumed to be isotropic, and therefore described by a scalar permittivity ϵ, although reflection and refraction at boundaries have inherent polarization-sensitive characteristics.

Photonic Crystals in Other Chapters

By virtue of their omnidirectional reflection property, photonic crystals can be used as "perfect" dielectric mirrors. A slab of homogeneous medium embedded in a photonic crystal may be used to guide light by multiple reflections from the boundaries. Applications to optical waveguides are described in Sec. 9.5. Similarly, light may be guided through an optical fiber with a homogeneous core embedded in a cladding of the same material, but with cylindrical holes drilled parallel to the fiber axis. Such "holey" fibers, described in Sec. 10.4, offer a number of salutary features not present in ordinary optical fibers. A cavity burrowed in a photonic crystal may function as an optical resonator since it has perfectly reflecting walls at frequencies within the photonic bandgap. Photonic-crystal microresonators and lasers will be described briefly in Secs. 11.4D and 18.5C, respectively.

7.1 OPTICS OF DIELECTRIC LAYERED MEDIA

A. Matrix Theory of Multilayer Optics

A plane wave normally incident on a layered medium undergoes reflections and transmissions at the layer boundaries, which in turn undergo their own reflections and transmissions in an unending process, as illustrated in Fig. 7.1-1(a). The complex amplitudes of the transmitted and reflected waves may be determined by use of the Fresnel equations at each boundary (see Sec. 6.2); the overall transmittance and reflectance of the medium can, in principle, be calculated by superposition of these individual waves. This technique was used in Sec. 2.5B to determine the transmittance of the Fabry–Perot interferometer.

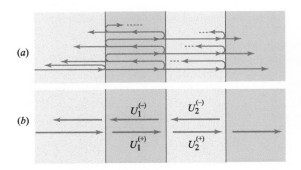

Figure 7.1-1 (a) Reflections of a single wave at the boundaries of a multilayered medium. (b) In each layer, the forward waves are lumped into a forward collected wave $U^{(+)}$ while the backward waves are lumped into a backward collected wave $U^{(-)}$.

When the number of layers is large, tracking the infinite number of "micro" reflections and transmissions can be tedious. An alternative "macro" approach is based on the recognition that within each layer there are two types of waves: forward waves traveling to the right, and backward waves traveling to the left. The sums of these waves add up to a single forward collected wave $U^{(+)}$ and a single backward collected wave $U^{(-)}$ at any point, as illustrated in Fig. 7.1-1(b). Determining the wave propagation in a layered medium is then equivalent to determining the amplitudes of this pair of waves everywhere. The complex amplitudes of the four waves on the two sides of a boundary may be related by imposing the appropriate boundary conditions, or by simply using the Fresnel equations of reflection and transmission.

Wave-Transfer Matrix

Tracking the complex amplitudes of the forward and backward waves through the boundaries of a multilayered medium is facilitated by use of matrix methods. Consider two arbitrary planes within a given optical system, denoted plane 1 and plane 2. The amplitudes of the forward and backward collected waves at plane 1, $U_1^{(+)}$ and $U_1^{(-)}$, respectively, are represented by a column matrix of dimension 2, and similarly for plane 2. These two column matrices are related by the matrix equation

$$\begin{bmatrix} U_2^{(+)} \\ U_2^{(-)} \end{bmatrix} = \begin{bmatrix} A & B \\ C & D \end{bmatrix} \begin{bmatrix} U_1^{(+)} \\ U_1^{(-)} \end{bmatrix}. \qquad (7.1\text{-}1)$$

The matrix **M**, whose elements are A, B, C, and D, is called the **wave-transfer matrix** (or transmission matrix). It depends on the optical properties of the layered medium between the two planes.

A multilayered medium is conveniently divided into a concatenation of basic elements described by known wave-transfer matrices, say $\mathbf{M}_1, \mathbf{M}_2, \ldots, \mathbf{M}_N$. The amplitudes of the forward and backward collected waves at the two ends of the overall medium are then related by a single matrix that is the matrix product

$$\mathbf{M} = \mathbf{M}_N \ldots \mathbf{M}_2 \mathbf{M}_1, \qquad (7.1\text{-}2)$$

where the elements $1, 2, \ldots, N$ are numbered from left to right as shown in the figure. The wave-transfer matrix cascade formula provided in (7.1-2) is identical to the ray-transfer matrix cascade formula given in (1.4-10), and it proves equally useful.

Scattering Matrix

An alternative to the wave-transfer matrix that relates the four complex amplitudes $U_{1,2}^{(\pm)}$ at the two edges of a layered medium is the scattering matrix, or **S** matrix. It is often used to describe transmission lines, microwave circuits, and scattering systems. In this case, the outgoing waves are expressed in terms of the incoming waves,

$$\begin{bmatrix} U_2^{(+)} \\ U_1^{(-)} \end{bmatrix} = \begin{bmatrix} t_{12} & r_{21} \\ r_{12} & t_{21} \end{bmatrix} \begin{bmatrix} U_1^{(+)} \\ U_2^{(-)} \end{bmatrix}, \qquad (7.1\text{-}3)$$

where the elements of the **S** matrix are denoted t_{12}, r_{21}, r_{12}, and t_{21}. Unlike the wave-transfer matrix, these elements have direct physical significance. The quantities t_{12} and r_{12} are the forward amplitude transmittance and reflectance (i.e., the transmittance and reflectance of a wave incident from the left), respectively, while t_{21} and r_{21} are the amplitude transmittance and reflectance in the backward direction (i.e., a wave coming from the right), respectively. The subscript 12, for example, signifies that the light is incident from medium 1 into medium 2. This can be easily verified by noting that if there is no backward wave at plane 2, so that $U_2^{(-)} = 0$, we obtain $U_2^{(+)} = t_{12} U_1^{(+)}$ and $U_1^{(-)} = r_{12} U_1^{(+)}$. Similarly, if there is no forward wave at plane 1, so that $U_1^{(+)} = 0$, we obtain $U_2^{(+)} = r_{21} U_2^{(-)}$ and $U_1^{(-)} = t_{21} U_2^{(-)}$.

A distinct advantage of the **S**-matrix formalism is that its elements are directly related to the physical parameters of the system. On the other hand, a disadvantage is that the **S** matrix of a cascade of elements is not the product of the **S** matrices of the constituent elements. A useful systematic procedure for analyzing a cascaded system therefore draws on both the wave-transfer and scattering matrix approaches: we use the handy multiplication formula of the **M** matrices and then convert to the **S** matrix to determine the overall transmittance and reflectance of the cascaded system.

EXAMPLE 7.1-1. *Propagation Through a Homogeneous Medium.* For a homogeneous layer of width d and refractive index n, the complex amplitudes of the collected waves at the planes indicated by the arrows are related by $U_2^{(+)} = e^{-j\varphi}U_1^{(+)}$ and $U_1^{(-)} = e^{-j\varphi}U_2^{(-)}$, where $\varphi = nk_o d$, so that in this case the wave-transfer matrix and the scattering matrix are:

$$\mathbf{M} = \begin{bmatrix} \exp(-j\varphi) & 0 \\ 0 & \exp(j\varphi) \end{bmatrix}, \quad \mathbf{S} = \begin{bmatrix} \exp(-j\varphi) & 0 \\ 0 & \exp(-j\varphi) \end{bmatrix}, \quad \varphi = nk_o d. \tag{7.1-4}$$

Relation between Scattering Matrix and Wave-Transfer Matrix

The elements of the **M** and **S** matrices are related by manipulating the defining equations (7.1-1) and (7.1-3), whereupon the following conversion equations emerge:

$$\mathbf{M} = \begin{bmatrix} A & B \\ C & D \end{bmatrix} = \frac{1}{t_{21}} \begin{bmatrix} t_{12}t_{21} - r_{12}r_{21} & r_{21} \\ -r_{12} & 1 \end{bmatrix}, \tag{7.1-5}$$

$$\mathbf{S} = \begin{bmatrix} t_{12} & r_{21} \\ r_{12} & t_{21} \end{bmatrix} = \frac{1}{D} \begin{bmatrix} AD - BC & B \\ -C & 1 \end{bmatrix}. \tag{7.1-6}$$

Matrix Conversion Relations

These equations are not valid in the limiting cases when $t_{21} = 0$ or $D = 0$.

Summary

Matrix wave optics offers a systematic procedure for determining the amplitude transmittance and reflectance of a stack of dielectric layers with prescribed thicknesses and refractive indices:

- The stack is divided into a cascade of elements encompassing boundaries with homogeneous layers between them.
- The **M** matrix is determined for each element. This may be achieved by using the Fresnel formulas for transmission and reflection to determine its **S** matrix, and then using the conversion relation (7.1-5) to calculate the corresponding **M** matrix.
- The **M** matrix for the full stack of elements is obtained by simply multiplying the **M** matrices for the individual elements, in accordance with the wave-transfer matrix formula provided in (7.1-2).
- Finally, the **S** matrix for the full stack is determined by conversion from the overall **M** matrix via (7.1-6). The elements of the **S** matrix then directly yield the amplitude transmittance and reflectance for the full stack of dielectric layers.

Two Cascaded Systems: Airy Formulas

Matrix methods may be used to derive explicit expressions for elements of the scattering matrix of a composite system in terms of elements of the scattering matrices of the constituent systems. Consider a wave transmitted through a system described by an **S** matrix with elements t_{12}, r_{21}, r_{12}, and t_{21}, followed by another system with **S** matrix elements t_{23}, r_{32}, r_{23}, and t_{32}. By multiplying the two associated **M** matrices, and then converting the result to an **S** matrix, the following formulas for the overall forward transmittance and reflectance can be derived:

$$t_{13} = \frac{t_{12} t_{23}}{1 - r_{21} r_{23}}, \quad r_{13} = r_{12} + \frac{t_{12} t_{21} r_{23}}{1 - r_{21} r_{23}}. \tag{7.1-7}$$

If the two cascaded systems are mediated by propagation through a homogeneous medium, as illustrated in Fig. 7.1-2, then by use of the wave-transfer matrix in (7.1-4), with the phase $\varphi = nk_o d$, where d is the propagation distance and n is the refractive index of the medium, the following formulas for the overall transmittance and reflectance, known as the **Airy formulas**, may be derived:

$$\boxed{t_{13} = \frac{t_{12} t_{23} \exp(-j\varphi)}{1 - r_{21} r_{23} \exp(-j2\varphi)}, \quad r_{13} = r_{12} + \frac{t_{12} t_{21} r_{23} \exp(-j2\varphi)}{1 - r_{21} r_{23} \exp(-j2\varphi)}.} \tag{7.1-8}$$
Airy Formulas

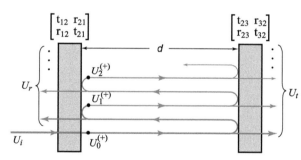

Figure 7.1-2 Transmission of a plane wave through a cascade of two separated systems.

The Airy formulas may also be derived by tracking the multiple transmissions and reflections experienced by an incident wave between the two systems and adding up their amplitudes, as portrayed in Fig. 7.1-2. A plane wave of complex amplitude U_i incident on the first system produces an initial internal wave of amplitude $U_0^{(+)} = t_{12} U_i$, which reflects back and forth between the two subsystems producing additional internal waves $U_1^{(+)}, U_2^{(+)}, \ldots$, all traveling in the forward direction. The amplitude of the overall transmitted wave U_t is related to the total internal amplitude $U^{(+)} = U_0^{(+)} + U_1^{(+)} + \cdots$ by $U_t = t_{23} \exp(-j\phi) U^{(+)}$, where $\phi = nk_o d$. The overall amplitude transmittance is therefore $t_{13} = U_t/U_i = t_{12} t_{23} \exp(-j\phi)(U^{(+)}/U_0^{(+)})$. Since $U^{(+)} = U_0^{(+)}(1 + h + h^2 + \cdots) = U_0^{(+)}/(1-h)$, where $h = r_{21} r_{23} \exp(-j2\phi)$ is the round-trip multiplication factor, the overall amplitude transmittance t_{13} yields the Airy formula in (7.1-8).

Conservation Relations for Lossless Media

If the medium between planes 1 and 2 is nonlossy, then the incoming and outgoing optical powers must be equal. Furthermore, if the media at the input and output planes have the same impedance and refractive index, then these powers are represented by the squared magnitudes of the complex amplitudes $|U_{1,2}^{(\pm)}|^2$. In this case, conservation of power dictates that $|U_1^{(+)}|^2 + |U_2^{(-)}|^2 = |U_2^{(+)}|^2 + |U_1^{(-)}|^2$ for any combination of incoming amplitudes. By choosing the incoming amplitudes $U_1^{(+)}$ and $U_2^{(-)}$ to be (1,0), (0,1), and (1,1), the conservation formula above yields three equations that relate the elements of the **S** matrix. These equations can be used to prove the following formulas:

$$|t_{12}| = |t_{21}| \equiv |t|, \quad |r_{12}| = |r_{21}| \equiv |r|, \quad |t|^2 + |r|^2 = 1, \qquad (7.1\text{-}9)$$

$$t_{12}/t_{21}^* = -r_{12}/r_{21}^*. \qquad (7.1\text{-}10)$$

Equations (7.1-9) relate the magnitudes of the elements of the **S** matrix for lossless media whose input and output planes see the same refractive index, whereas (7.1-10) relates their arguments.

The formulas in (7.1-9) and (7.1-10) translate to the following relations among the elements of the **M** matrix:

$$|D| = |A|, \quad |C| = |B|, \quad |A|^2 - |B|^2 = 1, \qquad (7.1\text{-}11)$$

$$\det \mathbf{M} = C/B^* = A/D^* = t_{12}/t_{21}, \quad |\det \mathbf{M}| = 1. \qquad (7.1\text{-}12)$$

These results can be derived by substituting the conservation relations for lossless media, (7.1-9) and (7.1-10), into the conversion relations between the wave-transfer and scattering matrices, (7.1-5) and (7.1-6).

EXAMPLE 7.1-2. *Single Dielectric Boundary.* Consider a system comprising a single boundary. In accordance with the Fresnel equations (Sec. 6.2), the transmittance and reflectance at a boundary between two media of refractive indices n_1 and n_2 are governed by the **S** matrix

$$\mathbf{S} = \begin{bmatrix} t_{12} & r_{21} \\ r_{12} & t_{21} \end{bmatrix} = \frac{1}{n_1 + n_2} \begin{bmatrix} 2n_1 & n_2 - n_1 \\ n_1 - n_2 & 2n_2 \end{bmatrix}. \qquad (7.1\text{-}13)$$

Substituting (7.1-13) into (7.1-5) yields the **M** matrix

$$\mathbf{M} = \frac{1}{2n_2} \begin{bmatrix} n_2 + n_1 & n_2 - n_1 \\ n_2 - n_1 & n_2 + n_1 \end{bmatrix}. \qquad (7.1\text{-}14)$$

Lossless Symmetric Systems

For lossless systems with reciprocal symmetry, namely systems whose transmission/reflection in the forward and backward directions are identical, we have $t_{21} = t_{12} \equiv t$ and $r_{21} = r_{12} \equiv r$. In this case, (7.1-9) and (7.1-10) yield

$$|t|^2 + |r|^2 = 1, \quad t/r = -(t/r)^*, \quad \arg\{r\} - \arg\{t\} = \pm\pi/2, \qquad (7.1\text{-}15)$$

indicating that the phases associated with transmission and reflection differ by $\pi/2$. Under these conditions, the elements of the **M** matrix satisfy the following relations:

$$A = D^*, \quad B = C^*, \quad |A|^2 - |B|^2 = 1, \quad \det \mathbf{M} = 1. \qquad (7.1\text{-}16)$$

7.1 OPTICS OF DIELECTRIC LAYERED MEDIA

The **S** and **M** matrices then take the simple form

$$\mathbf{S} = \begin{bmatrix} t & r \\ r & t \end{bmatrix}, \quad \mathbf{M} = \begin{bmatrix} 1/t^* & r/t \\ r^*/t^* & 1/t \end{bmatrix}, \qquad (7.1\text{-}17)$$
Lossless Symmetric System

and the system is described by two complex numbers t and r, related by (7.1-15).

Only relative phases are significant in such systems, so we may assume without loss of generality that $\arg\{t\} = 0$. It then follows from (7.1-15) that $\arg\{r\} = \pm\pi/2$ so that $r = \pm j|r|$, whereupon the matrices in (7.1-17) take the simpler forms

$$\mathbf{S} = \begin{bmatrix} |t| & j|r| \\ j|r| & |t| \end{bmatrix}, \quad \mathbf{M} = \frac{1}{|t|}\begin{bmatrix} 1 & j|r| \\ -j|r| & 1 \end{bmatrix}, \quad |t|^2 + |r|^2 = 1.. \qquad (7.1\text{-}18)$$

These equations are commonly used to describe lossless symmetric systems such as beamsplitters (e.g., cube and pellicle beamsplitters) and integrated-optic couplers. Moreover, if the system is balanced, i.e., if $|r| = |t| = \frac{1}{\sqrt{2}}$, we have $\mathbf{S} = \frac{1}{\sqrt{2}}\begin{bmatrix} 1 & j \\ j & 1 \end{bmatrix}$.

EXAMPLE 7.1-3. *Dielectric Slab.* Consider a lossless symmetric system comprising a cascade of three subsystems: a boundary between media of refractive indices n_1 and n_2, followed by travel through a medium of index n_2, followed in turn by a boundary between media with indices n_2 and n_1. By virtue of the results provided in (7.1-2) and Example 7.1-2, the overall **M** matrix is then a product of the three constituent **M** matrices, with the matrix multiplication taking place in reverse order:

$$\mathbf{M} = \frac{1}{4n_1 n_2}\begin{bmatrix} n_1+n_2 & n_1-n_2 \\ n_1-n_2 & n_1+n_2 \end{bmatrix}\begin{bmatrix} e^{-j\varphi} & 0 \\ 0 & e^{j\varphi} \end{bmatrix}\begin{bmatrix} n_2+n_1 & n_2-n_1 \\ n_2-n_1 & n_2+n_1 \end{bmatrix}. \qquad (7.1\text{-}19)$$

Here $\varphi = n_2 k_o d$ where d is the width of the slab. The elements of this matrix **M**, which are given by

$$A = D^* = \frac{1}{t^*} = \frac{1}{4n_1 n_2}\left[(n_1+n_2)^2 e^{-j\varphi} - (n_2-n_1)^2 e^{j\varphi}\right], \qquad (7.1\text{-}20)$$

$$B = C^* = \frac{r}{t} = -j\frac{(n_2^2 - n_1^2)}{4n_1 n_2}\sin\varphi, \qquad (7.1\text{-}21)$$

satisfy the properties of a lossless symmetric system, as described by (7.1-16). Expressions for t and r are determined directly from (7.1-20) and (7.1-21):

$$t = \frac{4n_1 n_2 \exp(-j\varphi)}{(n_1+n_2)^2 - (n_1-n_2)^2 \exp(-j2\varphi)}, \quad r = -j\left[\frac{n_2^2 - n_1^2}{4n_1 n_2}\sin\varphi\right]t. \qquad (7.1\text{-}22)$$

The intensity transmittance $|t|^2$, and the intensity reflectance $|r|^2 = 1-|t|^2$, are periodic functions of the phase φ, with period π. The magnitude of the phase difference between r and t is always maintained at $\pi/2$, regardless of the value of φ, but its sign changes as $\sin\varphi$ switches its sign every π interval. It is worthy of note that the expressions provided in (7.1-20) and (7.1-21) can also be directly derived by regarding the system as a combination of two boundaries mediated by propagation through a distance in a medium, and using the Airy formula (7.1-8) with $t_{12} = t_{32} = 2n_1/(n_1+n_2)$, $t_{21} = t_{23} = 2n_2/(n_1+n_2)$, and $r_{12} = r_{32} = -r_{21} = -r_{23} = (n_1-n_2)/(n_1+n_2)$.

EXERCISE 7.1-1

Quarter-Wave Film as an Antireflection Coating. Specially designed thin dielectric films are often used to reduce or eliminate reflection at the boundary between two media of different refractive indices. Consider a thin film of refractive index n_2 and thickness d sandwiched between media of refractive indices n_1 and n_3. Derive an expression for the B element of the **M** matrix for this multilayer medium. Show that light incident from medium 1 has zero reflectance if $d = \lambda/4$ and $n_2 = \sqrt{n_1 n_3}$, where $\lambda = \lambda_o/n_2$.

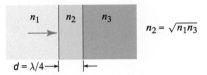

Figure 7.1-3 Antireflection coating.

Off-Axis Waves in Layered Media

When an oblique wave is incident on a layered medium, the transmitted and reflected waves, along with their reflections and transmissions in turn, bounce back and forth between the layers, as illustrated by its real part as shown in Fig. 7.1-4(a). The laws of reflection and refraction ensure that, within the same layer, all of the forward waves are parallel, and all of the backward waves are parallel. Moreover, within any given layer the forward and backward waves travel at the same angle, when measured from the $+z$ and $-z$ directions, respectively.

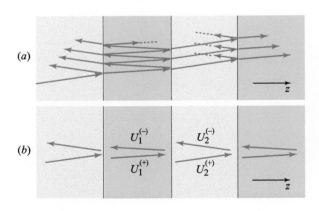

Figure 7.1-4 (a) Reflections of a single incident oblique wave at the boundaries of a multilayered medium. (b) In each layer, the forward waves are lumped into a collected forward wave while the backward waves are lumped into a collected backward wave.

The "macro" approach that was used earlier for normally incident waves is similarly applicable for oblique waves. The distinction is that the Fresnel transmittances and reflectances at a boundary, t_{12}, r_{21}, r_{12}, and t_{21}, are angle-dependent as well as polarization-dependent (see Sec. 6.2).

The simplest example is propagation a distance d through a homogeneous medium of refractive index n, at an angle θ measured from the z axis. The wave-transfer matrix **M** is then given by (7.1-4), where the phase is now $\varphi = nk_o d \cos\theta$. Two other examples are presented below.

EXAMPLE 7.1-4. *Single Boundary: Oblique TE Wave.* A wave transmitted through a planar boundary between media of refractive indices n_1 and n_2 at angles θ_1 and θ_2, satisfying Snell's law ($n_1 \sin \theta_1 = n_2 \sin \theta_2$), is described by an **S** matrix determined from the Fresnel equations (6.2-8) and (6.2-9), and its corresponding **M** matrix:

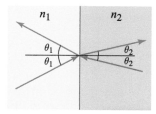

$$\mathbf{S} = \begin{bmatrix} t_{12} & r_{21} \\ r_{12} & t_{21} \end{bmatrix} = \frac{1}{\tilde{n}_1 + \tilde{n}_2} \begin{bmatrix} 2a_{12}\tilde{n}_1 & \tilde{n}_2 - \tilde{n}_1 \\ \tilde{n}_1 - \tilde{n}_2 & 2a_{21}\tilde{n}_2 \end{bmatrix}, \qquad (7.1\text{-}23)$$

$$\mathbf{M} = \begin{bmatrix} A & B \\ C & D \end{bmatrix} = \frac{1}{2a_{21}\tilde{n}_2} \begin{bmatrix} \tilde{n}_1 + \tilde{n}_2 & \tilde{n}_2 - \tilde{n}_1 \\ \tilde{n}_2 - \tilde{n}_1 & \tilde{n}_1 + \tilde{n}_2 \end{bmatrix}. \qquad (7.1\text{-}24)$$

These expressions are applicable for both TE and TM polarized waves with the following definitions:

TE: $\quad \tilde{n}_1 = n_1 \cos \theta_1, \quad \tilde{n}_2 = n_2 \cos \theta_2, \quad a_{12} = a_{21} = 1,$

TM: $\quad \tilde{n}_1 = n_1 \sec \theta_1, \quad \tilde{n}_2 = n_2 \sec \theta_2, \quad a_{12} = \cos \theta_1 / \cos \theta_2 = 1/a_{21}.$

EXAMPLE 7.1-5. *Dielectric Slab: Off-Axis Wave.* We now consider an oblique wave traveling through the system described in Example 7.1-3: a slab of thickness d and refractive index n_2 in a medium of refractive index n_1. The wave-transfer matrix for an oblique wave is a generalization of the on-axis result:

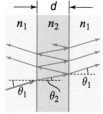

$$\mathbf{M} = \frac{1}{4\tilde{n}_1\tilde{n}_2} \begin{bmatrix} \tilde{n}_1 + \tilde{n}_2 & \tilde{n}_1 - \tilde{n}_2 \\ \tilde{n}_1 - \tilde{n}_2 & \tilde{n}_1 + \tilde{n}_2 \end{bmatrix} \begin{bmatrix} e^{-j\tilde{\varphi}} & 0 \\ 0 & e^{j\tilde{\varphi}} \end{bmatrix} \begin{bmatrix} \tilde{n}_2 + \tilde{n}_1 & \tilde{n}_2 - \tilde{n}_1 \\ \tilde{n}_2 - \tilde{n}_1 & \tilde{n}_2 + \tilde{n}_1 \end{bmatrix}, \qquad (7.1\text{-}25)$$

where $\tilde{\varphi} = n_2 k_o d \cos \theta_2$, and, as in Example 7.1-4, $\tilde{n}_1 = n_1 \cos \theta_1$ and $\tilde{n}_2 = n_2 \cos \theta_2$ for the TE polarization, and $\tilde{n}_1 = n_1 \sec \theta_1$ and $\tilde{n}_2 = n_2 \sec \theta_2$ for the TM polarization.

The expression for the matrix **M** in (7.1-25) is identical to that provided in (7.1-19), which describes the on-axis system, except that the parameters n_1, n_2, and φ are replaced by the angle- and polarization-dependent parameters $\tilde{n}_1$ and $\tilde{n}_2$, and by the angle-dependent parameter $\tilde{\varphi}$, respectively. Note that the factors a_{12} and a_{21}, which appear in (7.1-24) at each boundary, cancel out since $a_{12}a_{21} = 1$. With these substitutions, the expressions developed in (7.1-22) for the on-axis transmittance and reflectance in Example 7.1-3 generalize to the following off-axis, polarization-dependent formulas:

$$t = \frac{4\tilde{n}_1\tilde{n}_2 \exp(-j\tilde{\varphi})}{(\tilde{n}_1 + \tilde{n}_2)^2 - (\tilde{n}_1 - \tilde{n}_2)^2 \exp(-j2\tilde{\varphi})}, \qquad r = -j \left[\frac{\tilde{n}_2^2 - \tilde{n}_1^2}{4\tilde{n}_1\tilde{n}_2} \sin \tilde{\varphi} \right] t. \qquad (7.1\text{-}26)$$

B. Fabry–Perot Etalon

The Fabry–Perot etalon was introduced in Sec. 2.5B; it is an interferometer comprising two parallel and highly reflective mirrors that transmit light only at a set of specific, uniformly spaced frequencies, which depend on the optical pathlength between the mirrors. It is used both as a filter and as a spectrum analyzer, and is controlled by varying the pathlength, e.g., by moving one of the mirrors with respect to the other. It is also used as an optical resonator, as discussed in Sec. 11.1. In this section, we examine this multilayer device using the matrix methods developed in this chapter.

Mirror Fabry–Perot Etalon

Consider two lossless partially reflective mirrors with amplitude transmittances t_1 and t_2, and amplitude reflectances r_1 and r_2, separated by a distance d filled with a medium of refractive index n. The overall system is described by the matrix product

$$\mathbf{M} = \begin{bmatrix} 1/t_1^* & r_1/t_1 \\ r_1^*/t_1^* & 1/t_1 \end{bmatrix} \begin{bmatrix} \exp(-j\varphi) & 0 \\ 0 & \exp(j\varphi) \end{bmatrix} \begin{bmatrix} 1/t_2^* & r_2/t_2 \\ r_2^*/t_2^* & 1/t_2 \end{bmatrix}, \quad (7.1\text{-}27)$$

where $\varphi = nk_o d$. Since the system is lossless and symmetric, $\mathbf{M}$ takes the simplified form provided in (7.1-17) and the amplitude transmittance t is therefore the inverse of the D element of $\mathbf{M}$, so that

$$t = \frac{t_1 t_2 \exp(-j\varphi)}{1 - r_1 r_2 \exp(-j2\varphi)}. \quad (7.1\text{-}28)$$

This relation may also be derived by direct use of the Airy formula (7.1-8).

As a result, the intensity transmittance of the etalon is

$$\mathcal{T} = |t|^2 = \frac{|t_1 t_2|^2}{|1 - r_1 r_2 \exp(-j2\varphi)|^2}. \quad (7.1\text{-}29)$$

This expression is similar to (2.5-16) for the intensity of an infinite number of waves with equal phase differences, and with amplitudes that decrease at a geometric rate, as described in Sec. 2.5B. Assuming that $\arg\{r_1 r_2\} = 0$, (7.1-29) can be written in the form[†]

$$\mathcal{T} = \frac{\mathcal{T}_{\max}}{1 + (2\mathcal{F}/\pi)^2 \sin^2 \varphi}, \quad (7.1\text{-}30)$$

where

$$\mathcal{T}_{\max} = \frac{|t_1 t_2|^2}{(1 - |r_1 r_2|)^2} = \frac{(1 - |r_1|^2)(1 - |r_2|^2)}{(1 - |r_1 r_2|)^2} \quad (7.1\text{-}31)$$

and

$$\boxed{\mathcal{F} = \frac{\pi \sqrt{|r_1 r_2|}}{1 - |r_1 r_2|}.} \quad (7.1\text{-}32)$$
Finesse

The parameter $\mathcal{F}$, called the **finesse**, is a monotonic increasing function of the reflectance product $r_1 r_2$, and is a measure of the quality of the etalon. For example, if $r_1 r_2 = 0.99$, then $\mathcal{F} \approx 313$.

As described in Sec. 2.5B, the transmittance $\mathcal{T}$ is a periodic function of φ, now with period π. It reaches its maximum value of $\mathcal{T}_{\max}$, which equals unity if $|r_1| = |r_2|$, when φ is an integer multiple of π. When the finesse $\mathcal{F}$ is large (i.e., when $|r_1 r_2| \approx 1$), $\mathcal{T}$ becomes a sharply peaked function of φ of approximate width $\pi/\mathcal{F}$. Thus, the higher the finesse $\mathcal{F}$, the sharper the peaks of the transmittance as a function of the phase φ.

[†] Equation (7.1-30), in which $\varphi = nk_o d$, reproduces (2.5-18), in which $\varphi = 2nk_o d$.

The phase $\varphi = nk_o d = (\omega/c)d$ is proportional to the frequency, so that the condition $\varphi = \pi$ corresponds to $\omega = \omega_F$, or $\nu = \nu_F$, where

$$\nu_F = \frac{c}{2d}, \quad \omega_F = \frac{\pi c}{d} \qquad (7.1\text{-}33)$$
Free Spectral Range

is called the **free spectral range**. It follows that the transmittance as a function of frequency, $\mathcal{T}(\nu)$, is a periodic function of period ν_F,

$$\mathcal{T}(\nu) = \frac{\mathcal{T}_{\max}}{1 + (2\mathcal{F}/\pi)^2 \sin^2(\pi\nu/\nu_F)}, \qquad (7.1\text{-}34)$$
Transmittance
(Fabry–Perot Etalon)

as illustrated in Fig. 7.1-5. It reaches its peak value of $\mathcal{T}_{\max}$ at the resonance frequencies $\nu_q = q\nu_F$, where q is an integer. When the finesse $\mathcal{F} \gg 1$, $\mathcal{T}(\nu)$ drops sharply as the frequency deviates slightly from ν_q, so that $\mathcal{T}(\nu)$ takes the form of a comb-like function. The spectral width of each of these high-transmittance lines is

$$\delta\nu = \frac{\nu_F}{\mathcal{F}}, \qquad (7.1\text{-}35)$$

i.e., is a factor of $\mathcal{F}$ smaller than the spacing between the resonance frequencies.

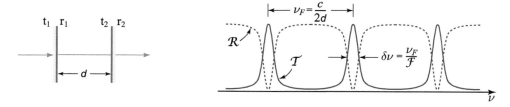

Figure 7.1-5 Intensity transmittance and reflectance, $\mathcal{T}$ and $\mathcal{R} = 1 - \mathcal{T}$, of the Fabry–Perot etalon as a function of frequency ν.

The Fabry–Perot etalon may be used as a sharply tuned optical filter or a spectrum analyzer. Because of the periodic nature of the spectral response, however, the spectral width of the measured light must be narrower than the free spectral range $\nu_F = c/2d$ in order to avoid ambiguity. The filter is tuned (i.e., the resonance frequencies are shifted) by adjusting the distance d between the mirrors. A slight change in mirror spacing Δd shifts the resonance frequency $\nu_q = qc/2d$ by a relatively large amount $\Delta\nu_q = -(qc/2d^2)\Delta d = -\nu_q \Delta d/d$. Although the frequency spacing ν_F also changes, it is by the far smaller amount $-\nu_F \Delta d/d$. As an example, a mirror separation of $d = 1.5$ cm leads to a free spectral range $\nu_F = 10$ GHz when $n = 1$. For a typical optical frequency of $\nu = 10^{14}$ Hz, corresponding to $q = 10^4$, a change of d by a factor of 10^{-4} ($\Delta d = 1.5$ μm) translates the peak frequency by $\Delta\nu_q = 10$ GHz, whereas the free spectral range is altered by only 1 MHz, becoming 9.999 GHz.

Applications of the Fabry–Perot etalon as a resonator are described in Sec. 11.1.

Off-Axis Transmittance of the Fabry–Perot Etalon

For an oblique wave traveling at an angle θ with the axis of a mirror etalon, the amplitude transmittance is given by (7.1-28) with the phase φ replaced by $\tilde{\varphi} = nk_o d \cos\theta$. It follows that the intensity transmittance in (7.1-34) is generalized to

$$\mathcal{T}(\nu) = \frac{\mathcal{T}_{\max}}{1 + (2\mathcal{F}/\pi)^2 \sin^2(\pi \cos\theta\, \nu/\nu_F)} \qquad (7.1\text{-}36)$$

in the off-axis case.

Maximum transmittance occurs at frequencies for which

$$\boxed{\nu = q\nu_F \sec\theta\,,} \quad q = 1, 2, \ldots, \quad \nu_F = c/2d. \qquad (7.1\text{-}37)$$
Resonance Condition

If the finesse of the etalon is large, transmission occurs at these frequencies and is almost completely blocked at all other frequencies. The plot of this relation provided in Fig. 7.1-6(c) shows that at each angle θ only a set of discrete frequencies are transmitted. Likewise, a wave at frequency ν is transmitted at only a set of angles, so that a cone of incident broad-spectrum (white) light creates a set of concentric rings spread like a rainbow, as illustrated in Fig. 7.1-6(b). For incident light with a spectral width smaller than the free spectral range ν_F, each frequency component corresponds to one and only one angle, so that the etalon can be used as a spectrum analyzer.

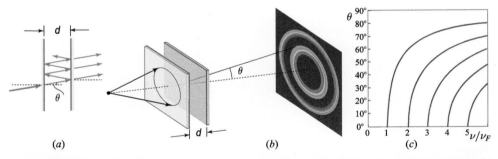

Figure 7.1-6 (a) An off-axis wave transmitted through a mirror Fabry–Perot etalon. (b) White light from a point source transmitted through the etalon creates a set of concentric rings of different frequencies (colors). (c) Frequencies and angles that satisfy the condition of peak transmittance, as set forth in (7.1-37).

EXAMPLE 7.1-6. *The Dielectric-Slab Beamsplitter.* The transmittance and reflectance of a dielectric slab of width d and refractive index n_2 surrounded by a medium of refractive index n_1 are given in (7.1-26). If the wave is incident at an angle θ_1, then for TE polarization we have $\tilde{n}_1 = n_1 \cos\theta_1$, $\tilde{n}_2 = n_2 \cos\theta_2$, and $\tilde{\varphi} = n_2 k_o d \cos\theta_2$, as in Example 7.1-4, with $\sin\theta_2 = (n_1/n_2)\sin\theta_1$. This expression reproduces (7.1-28) for the Fabry–Perot etalon if we substitute $t_1 t_2 = 4\tilde{n}_1\tilde{n}_2/(\tilde{n}_1 + \tilde{n}_2)$ and $r_1 r_2 = (\tilde{n}_1 - \tilde{n}_2)^2/(\tilde{n}_1 + \tilde{n}_2)^2$. It follows that the expressions for the intensity transmittance of the mirror etalon, (7.1-30) and (7.1-34), are also applicable for the dielectric slab. Using (7.1-32), the finesse of the slab is thus given by

$$\mathcal{F} = \frac{\pi}{4} \frac{|\tilde{n}_2^2 - \tilde{n}_1^2|}{\tilde{n}_1 \tilde{n}_2}. \qquad (7.1\text{-}38)$$

Large values of $\mathcal{F}$ are typically not obtained in slab etalons. As an example, for $n_1 = 1.5$ (the refractive index of SiO$_2$) and $n_2 = 3.5$ (the refractive index of Si), and an angle of incidence $\theta_1 = 45°$, the finesse $\mathcal{F} = 1.89$. As illustrated in Fig. 7.1-7, the dependence of $\mathcal{T}$ and $\mathcal{R}$ on the phase $\widetilde{\varphi}$, which is proportional to the frequency, does not exhibit the sharp peaks observed in etalons with highly reflective mirrors (see, e.g., the illustration in Fig. 7.1-5). Higher values of $\mathcal{F}$ are obtained by coating the surfaces of the slab to enhance internal reflection.

The dielectric slab may be used as a beamsplitter. If the slab width $d = 1$ mm, for example, the range between two consecutive transmittance peaks $\widetilde{\varphi} = n_2 k_o d \cos\theta_2 = \pi$ corresponds to a frequency of ≈ 45 GHz. As illustrated in Fig. 7.1-7, the transmittance and reflectance near the center of this interval are reasonably flat. Since beamsplitters are routinely used in interferometers, it is important to be cognizant of the relation between the phases of the reflected and transmitted waves. As illustrated in the figure, the relative phase between the reflected and transmitted waves is always $\pm\pi/2$, with the sign changing at points of peak transmittance.

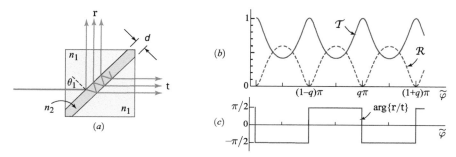

Figure 7.1-7 (a) A dielectric slab used as a beamsplitter. (b) Dependence of the intensity transmittance and reflectance ($\mathcal{T}$ and $\mathcal{R}$, respectively) on the phase $\widetilde{\varphi} = n_2 k_o d \cos\theta_2$ for $n_1 = 1.5$ (refractive index of SiO$_2$), $n_2 = 3.5$ (refractive index of Si), and an angle of incidence $\theta_1 = 45°$. Transmission peaks and reflectance minima occur at $\widetilde{\varphi} = q\pi$, where q is an integer. (c) The relative phase between the reflected and transmitted waves is $\pm\pi/2$; the sign switches at $\widetilde{\varphi} = q\pi$.

C. Bragg Grating

The Bragg grating was introduced in Exercise 2.5-3 as a set of uniformly spaced parallel partially reflective planar mirrors. Such a structure has angular and frequency selectivity that is useful in many applications. In this section, we generalize the definition of the Bragg grating to include a set of N uniformly spaced identical multilayer segments, and develop a theory for light reflection based on matrix wave optics. Devices fabricated according to this prescription include **distributed Bragg reflectors (DBRs)** and **fiber Bragg gratings (FBGs)**, which are often used in resonators and lasers.

Simplified Theory

The reflectance of the Bragg grating was determined in Exercise 2.5-3 under two assumptions: (1) the mirrors are weakly reflective so that the incident wave is not depleted as it propagates; and (2) secondary reflections (i.e., reflections of the reflected waves) are negligible. In this approximation, the reflectance $\mathcal{R}_N$ of an N-mirror grating is related to the reflectance $\mathcal{R}$ of a single mirror by the relation[†]

$$\mathcal{R}_N = \frac{\sin^2 N\varphi}{\sin^2 \varphi}\mathcal{R}. \qquad (7.1\text{-}39)$$

[†] In Exercise 2.5-3, the quantity φ denotes the phase between successive phasors whereas here that phase is denoted 2φ since it represents a round trip.

As described in Sec. 2.5B, the factor $\sin^2 N\varphi / \sin^2 \varphi$ represents the intensity of the sum of N phasors of unit amplitude and phase difference 2φ. This function has a peak value of N^2 when the Bragg condition is satisfied, i.e., when 2φ equals $q2\pi$, where $q = 0, 1, 2, \ldots$. It drops away from these values sharply, with a width that is inversely proportional to N. In this simplified model, the intensity of the total reflected wave is, at most, a factor of N^2 greater than the intensity of the wave reflected from a single segment.

For a Bragg grating comprising partially reflective mirrors separated from each other by a distance Λ and a round-trip phase $2\varphi = 2k\Lambda \cos\theta$, where θ is the angle of incidence. Therefore, maximum reflection occurs when $2k\Lambda \cos\theta = 2q\pi$ or

$$\cos\theta = q\frac{\lambda}{2\Lambda} = q\frac{\omega_B}{\omega} = q\frac{\nu_B}{\nu}, \qquad (7.1\text{-}40)$$
Bragg Condition

where

$$\nu_B = \frac{c}{2\Lambda}, \qquad \omega_B = \frac{\pi c}{\Lambda}, \qquad (7.1\text{-}41)$$
Bragg Frequency

is the Bragg frequency.

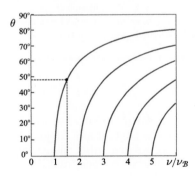

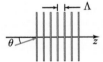

Figure 7.1-8 Locus of frequencies ν and angles θ at which the Bragg condition is satisfied. For example, if $\nu = 1.5\,\nu_B$ (dot-dash line), we have $\theta = 48.2°$. This corresponds to a Bragg angle $\theta_B = 41.8°$ (measured from the plane of the grating.)

At normal incidence ($\theta = 0°$), peak reflectance occurs at frequencies that are integer multiples of the Bragg frequency, i.e., $\nu = q\nu_B$. At frequencies such that $\nu < \nu_B$, the Bragg condition cannot be satisfied at any angle. At frequencies $\nu_B < \nu < 2\nu_B$, the Bragg condition is satisfied at one angle $\theta = \cos^{-1}(\lambda/2\Lambda) = \cos^{-1}(\nu_B/\nu)$. The complement of this angle, $\theta_B = \pi/2 - \theta$, is the Bragg angle (see (2.5-13) and Fig. 2.5-8),

$$\theta_B = \sin^{-1}(\lambda/2\Lambda). \qquad (7.1\text{-}42)$$
Bragg Angle

At frequencies $\nu \geq 2\nu_B$, the Bragg condition is satisfied at more than one angle. Figure 7.1-8 illustrates the spectral and angular dependence of reflections from a Bragg grating, based on the simplified theory.

7.1 OPTICS OF DIELECTRIC LAYERED MEDIA

Matrix Theory

We now use the matrix approach introduced in the previous section to develop an exact theory of Bragg reflection that includes multiple transmissions and reflections, as well as depletion of the incident wave. It turns out that the collaborative effects of the reflections, and the reflections of reflections, can lead to enhancement of the total reflected wave, and a phenomenon whereby total reflection occurs not only at single frequencies that are multiples of $\nu_B/\cos\theta$, but over extended spectral bands surrounding these frequencies!

Consider a grating comprising a stack of N identical generic segments (Fig. 7.1-9), each described by a unimodular wave-transfer matrix $\mathbf{M}_o$ satisfying the conservation relations for a lossless, symmetrical system, so that

$$\mathbf{M}_o = \begin{bmatrix} 1/t^* & r/t \\ r^*/t^* & 1/t \end{bmatrix}, \tag{7.1-43}$$

where t and r are complex amplitude transmittance and reflectance satisfying the conditions set forth in (7.1-15), and $\mathcal{T} = |t|^2$ and $\mathcal{R} = |r|^2$ are the corresponding intensity transmittance and reflectance.

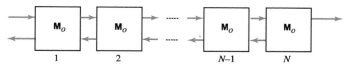

Figure 7.1-9 Bragg grating made of N segments, each of which is described by a matrix $\mathbf{M}_o$.

In accordance with (7.1-2), the wave-transfer matrix $\mathbf{M}$ for the N segments is simply the product $\mathbf{M} = \mathbf{M}_o^N$. Since $\mathbf{M}_o$ is unimodular, i.e., $\det \mathbf{M}_o = 1$, it satisfies the property

$$\mathbf{M}_o^N = \Psi_N \mathbf{M}_o - \Psi_{N-1} \mathbf{I}, \tag{7.1-44}$$

where

$$\boxed{\Psi_N = \frac{\sin N\Phi}{\sin \Phi}}, \tag{7.1-45}$$

$$\boxed{\cos \Phi = \mathrm{Re}\{1/t\},} \tag{7.1-46}$$

and $\mathbf{I}$ is the identity matrix. Equation (7.1-44) may be proved by induction (i.e., demonstrating that this relation is valid for N segments if it is valid for $N-1$ segments; this may be done by direct substitution with the help of trigonometric identities).

Since the N-segment system is also lossless and symmetric, its matrix may be written in the form

$$\mathbf{M}_o^N = \begin{bmatrix} 1/t_N^* & r_N/t_N \\ r_N^*/t_N^* & 1/t_N \end{bmatrix}, \tag{7.1-47}$$

where t_N and r_N are the N-segment amplitude transmittance and reflectance, respectively. Substituting from (7.1-43) and (7.1-47) into (7.1-44), and comparing the diagonal and off-diagonal elements of the matrices on both sides of the equation, leads to

$$\frac{1}{t_N} = \Psi_N \frac{1}{t} - \Psi_{N-1} \qquad (7.1\text{-}48)$$

$$\frac{r_N}{t_N} = \Psi_N \frac{r}{t}. \qquad (7.1\text{-}49)$$

These two equations define t_N and r_N in terms of t and r.

The intensity transmittance $\mathcal{T}_N = |t_N|^2$ is obtained by taking the absolute-squared value of (7.1-49) and using the relation $\mathcal{R} = 1 - \mathcal{T}$,

$$\mathcal{T}_N = \frac{\mathcal{T}}{\mathcal{T} + \Psi_N^2(1 - \mathcal{T})}. \qquad (7.1\text{-}50)$$

It follows that the power reflectance $\mathcal{R}_N = 1 - \mathcal{T}_N$ is given by

$$\boxed{\mathcal{R}_N = \frac{\Psi_N^2 \mathcal{R}}{1 - \mathcal{R} + \Psi_N^2 \mathcal{R}}.} \qquad (7.1\text{-}51)$$
Bragg-Grating Reflectance

Summary

The reflectance $\mathcal{R}_N$ of a medium comprising N identical segments is related to the single-segment reflectance $\mathcal{R}$ by the nonlinear relation (7.1-51), which contains a factor Ψ_N that results from the interference effects associated with collective reflections from the N segments of the grating. Defined by (7.1-45), Ψ_N depends on the number of segments N and on an additional parameter Φ that is related to the single-segment complex amplitude transmittance t via (7.1-46).

The dependence of $\mathcal{R}_N$ on $\mathcal{R}$, described by (7.1-51), takes simpler forms in certain limits. If the single-segment reflectance is very small, i.e., $\mathcal{R} \ll 1$, and if Ψ_N^2 is not too large so that $\Psi_N^2 \mathcal{R} \ll 1$, then (7.1-51) may be approximated by:

$$\mathcal{R}_N \approx \Psi_N^2 \mathcal{R} = \frac{\sin^2 N\Phi}{\sin^2 \Phi} \mathcal{R}. \qquad (7.1\text{-}52)$$

This relation is now similar in form to the approximate relation (7.1-39), with Φ playing the role of the phase φ.

In the opposite limit for which $\Psi_N^2 \gg 1$, the reflectance $\mathcal{R}_N \approx \Psi_N^2 \mathcal{R}/(1 + \Psi_N^2 \mathcal{R})$. This nonlinear relation between $\mathcal{R}_N$ and $\mathcal{R}$ exhibits saturation and is typical of systems with feedback, which in this case results from multiple internal reflections at the segment boundaries. Ultimately, if $\Psi_N^2 \mathcal{R} \gg 1$, then $\mathcal{R}_N$ approaches its maximum value of unity, so that the N-segment device acts as perfect mirror even though the single segment is only partially reflective. A large interference factor Ψ_N accelerates the rise of $\mathcal{R}_N$ to unity as $\mathcal{R}$ increases.

The interference factor Ψ_N, which depends on $\Phi = \cos^{-1}(\text{Re}\{1/t\})$ via (7.1-45), has two distinct regimes: (1) a normal regime for which Φ is real and the grating exhibits partial reflection/transmission (including zero reflection, or total transmission);

7.1 OPTICS OF DIELECTRIC LAYERED MEDIA 273

and (2) an anomalous regime for which Φ is complex and Ψ_N can be extremely large, corresponding to total reflection.

Partial- and Zero-Reflection Regime

This regime is defined by the condition $|\text{Re}\{1/t\}| \leq 1$, which ensures that $\Phi = \cos^{-1}(\text{Re}\{1/t\})$ is real. In this case, $\mathcal{R}_N$ depends on $\mathcal{R}$ and Ψ_N in accordance with (7.1-45) and (7.1-51). Maximum reflectance occurs when Ψ_N has its maximum value of N. In this case, $\mathcal{R}_N = N^2\mathcal{R}/(1 - \mathcal{R} + N^2\mathcal{R})$. Therefore, $\mathcal{R}_N$ cannot exactly equal unity unless $\mathcal{R} = 1$, exactly. For example, for $N = 10$, if $\mathcal{R} = 0.5$, then the maximum value of $\mathcal{R}_N \approx 0.99$.

Zero reflectance, or **total transmittance**, is possible, even if the reflectance R of the individual segment is substantial. This occurs when $\Psi_N = 0$, i.e., when $\sin N\Phi = 0$, or $\Phi = q\pi/N$ for $q = 0, 1, \ldots, N-1$. The N frequencies at which this complete transparency occurs are resonance frequencies of the grating. The phenomenon represents some form of tunneling through the individually reflective segments.

Total-Reflection Regime

In this regime, $|\text{Re}\{1/t\}| = |\cos \Phi| > 1$ so that Φ is a complex variable $\Phi = \Phi_R + j\Phi_I$. Using the identity $\cos(\Phi_R + j\Phi_I) = \cos \Phi_R \cosh \Phi_I - j \sin \Phi_R \sinh \Phi_I$, and equating the real and imaginary parts of both sides of (7.1-46), we obtain $\sin \Phi_R = 0$ so that $\Phi_R = m\pi$ and $\cos \Phi_R = +1$, or -1, when m is an even or odd integer, respectively, which results in

$$\cosh \Phi_I = |\text{Re}\{1/t\}|. \qquad (7.1\text{-}53)$$
Total-Reflection Regime

The factor $\Psi_N = \sin N\Phi / \sin \Phi$ then becomes

$$\Psi_N = \pm \frac{\sinh N\Phi_I}{\sinh \Phi_I}, \qquad (7.1\text{-}54)$$
Total-Reflection Regime

where the $\pm$ sign is the sign of the factor $\cos(Nm\pi)/\cos(m\pi)$. Since $\sinh(\cdot)$ increases exponentially with N for large N, $|\Psi_N|$ can be much greater than N. In this case, in accordance with (7.1-51), the reflectance $\mathcal{R}_N \approx 1$ and the grating acts as a total reflector. The forward waves become evanescent and do not penetrate the multisegment medium, much as occurs with total internal reflection.

Because Φ depends on t, which depends on the frequency ν, the two regimes correspond to distinct spectral bands, as illustrated in the following examples. The spectral bands associated with the total-reflection regime are called **stop bands** since they represent bands within which light transmission is almost completely blocked. The other regime corresponds to **passbands**. Total transmission (zero reflection) occurs at specific resonance frequencies within the passbands.

EXAMPLE 7.1-7. *Stack of Partially Reflective Mirrors.* Consider a grating made of a stack of N identical partially reflective mirrors (beamsplitters) that are mutually separated by a distance Λ and embedded in a homogeneous medium of refractive index n, as illustrated in Fig. 7.1-10(a). A single segment comprises a distance Λ in a homogeneous medium, followed by a partially reflective mirror of amplitude transmittance t and amplitude reflectance r.

The wave-transfer matrix $\mathbf{M}_o$ for this segment is determined by multiplying the matrix in (7.1-18) by the matrix in (7.1-4):

$$\mathbf{M}_o = \frac{1}{|\mathsf{t}|}\begin{bmatrix} e^{-j\varphi} & j|\mathsf{r}|e^{j\varphi} \\ -j|\mathsf{r}|e^{-j\varphi} & e^{j\varphi} \end{bmatrix}, \quad \varphi = nk_o\Lambda = \pi\nu/\nu_\mathcal{B}, \quad (7.1\text{-}55)$$

where $\nu_\mathcal{B} = c/2\Lambda$ is the Bragg frequency. This provides $\mathsf{t} = |\mathsf{t}|e^{j\varphi}$, and therefore Φ via

$$\cos\Phi = \frac{1}{|\mathsf{t}|}\cos\varphi \quad \text{for} \quad |\cos\varphi| \leq |\mathsf{t}|, \quad (7.1\text{-}56)$$

$$\cosh\Phi_I = \frac{1}{|\mathsf{t}|}|\cos\varphi| \quad \text{for} \quad |\cos\varphi| > |\mathsf{t}|. \quad (7.1\text{-}57)$$

The relationships between Φ and φ, and between Φ_I and φ, are nonlinear and unusual, as illustrated in Fig. 7.1-10(b). The corresponding dependence of the power reflectance $\mathcal{R}_N$ on φ is shown in Fig. 7.1-10(c). In the normal regime (indicated by the shaded regions), Φ is real and the reflectance exhibits multiple peaks with zeros between. None of the peaks approaches unity, despite the fact that Ψ_N reaches a maximum value of $N = 10$.

The situation is quite different in the total-reflection regime (unshaded regions), where Φ is complex. The factor Ψ_N reaches a value ≈ 3000 at the center of the band ($\varphi = \pi$) when $|\mathsf{t}|^2 = 0.5$. These regions represent ranges of φ where total reflection occurs ($\mathcal{R}_N \approx 1$). Since φ is proportional to the frequency ν, Fig. 7.1-10(c) is actually a display of the spectral reflectance, and the unshaded regions correspond to the stop bands.

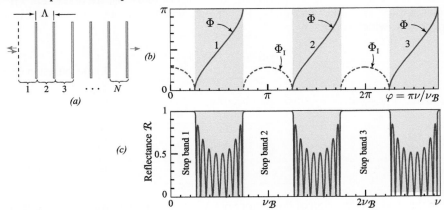

Figure 7.1-10 (a) Bragg grating comprising $N = 10$ identical mirrors, each with a power reflectance $|\mathsf{r}|^2 = 0.5$. (b) Dependence of Φ on the inter-mirror phase delay $\varphi = nk_o\Lambda$. Within the shaded regions, Φ is complex and its imaginary part Φ_I is represented by the dashed curves. (c) Reflectance $\mathcal{R}$ as a function of frequency (in units of the Bragg frequency $\nu_\mathcal{B} = c/2\Lambda$). Within the stop bands, the reflectance is approximately unity.

EXAMPLE 7.1-8. *Dielectric Bragg Grating.* A grating is made of N identical dielectric layers of refractive index n_2, each of width d_2, buried in a medium of refractive index n_1 and separated by a distance d_1, as illustrated in Fig. 7.1-11. This multisegment system is a stack of N identical double layers, each of the type described in Example 7.1-3. The $A = 1/\mathsf{t}^*$ element of the wave-transfer matrix $\mathbf{M}_o$ is given by (7.1-20), from which

$$\text{Re}\left\{\frac{1}{\mathsf{t}}\right\} = \frac{(n_1+n_2)^2}{4n_1n_2}\cos(\varphi_1+\varphi_2) - \frac{(n_2-n_1)^2}{4n_1n_2}\cos(\varphi_1-\varphi_2), \quad (7.1\text{-}58)$$

where $\varphi_1 = n_1k_od_1$ and $\varphi_2 = n_2k_od_2$ are the phases introduced by the two layers of a segment. This result can be used in conjunction with (7.1-45), (7.1-46), (7.1-51), (7.1-53), and (7.1-54) to determine the reflectance of the grating.

The spectral dependence of the reflectance can be computed as a function of ν by noting that $\varphi_1 + \varphi_2 = k_o(n_1d_1 + n_2d_2) = \pi\nu/\nu_\mathcal{B}$, where $\nu_\mathcal{B} = (c_o/\bar{n})/2\Lambda$, and $\bar{n} = (n_1d_1 + n_2d_2)/\Lambda$ is the average refractive index. The Bragg frequency ν_B is the frequency at which the single-segment round-trip phase $2k_o(n_1 d_1 + n_2 d_2) = 2\pi$. The phase difference $\varphi_1 - \varphi_2 = \zeta \pi \nu/\nu_B$, with $\zeta = (n_1 d_1 - n_2 d_2)/(n_1 d_1 + n_2 d_2)$, is also proportional to the frequency. Figure 7.1-11(b) provides an example of the spectral reflectance as a function of ν.

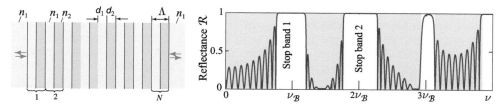

Figure 7.1-11 Power reflectance as a function of frequency for a dielectric Bragg grating comprising $N = 10$ segments, each of which has two layers of thickness $d_1 = d_2$ and refractive indices $n_1 = 1.5$ and $n_2 = 3.5$. The grating is placed in a medium with matching refractive index n_1. The reflectance is approximately unity within the stop bands centered about multiples of $\nu_B = c/2\Lambda$, where $c = c_o/\bar{n}$ and $\bar{n}$ is the mean refractive index.

EXAMPLE 7.1-9. *Dielectric Bragg Grating: Oblique Incidence.* The results in Example 7.1-8 may be generalized to oblique waves with angle of incidence θ_1 in medium 1, corresponding to angle θ_2 in layer 2, where $n_1 \sin \theta_1 = n_2 \sin \theta_2$. In this case, (7.1-58) becomes

$$\operatorname{Re}\left\{\frac{1}{t}\right\} = \frac{(\tilde{n}_1 + \tilde{n}_2)^2}{4\tilde{n}_1 \tilde{n}_2} \cos(\tilde{\varphi}_1 + \tilde{\varphi}_2) - \frac{(\tilde{n}_2 - \tilde{n}_1)^2}{4\tilde{n}_1 \tilde{n}_2} \cos(\tilde{\varphi}_1 - \tilde{\varphi}_2), \qquad (7.1\text{-}59)$$

where $\tilde{\varphi}_1 = n_1 k_o d_1 \cos \theta_1$ and $\tilde{\varphi}_2 = n_2 k_o d_2 \cos \theta_2$; $\tilde{n}_1 = n_1 \cos \theta_1$ and $\tilde{n}_2 = n_2 \cos \theta_2$ for TE polarization; and $\tilde{n}_1 = n_1 \sec \theta_1$ and $\tilde{n}_2 = n_2 \sec \theta_2$ for TM polarization. This relation may be used to compute the spectral reflectance at any angle of incidence. Figure 7.1-12 illustrates the dependence of the power reflectance R_N on frequency and the angle of incidence for both TE and TM polarization for a high-contrast grating. The range of angles over which unity reflectance obtains increases with increasing refractive-index contrast ratio n_2/n_1.

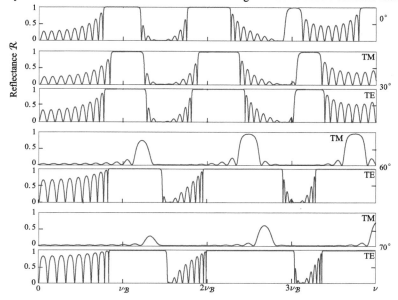

Figure 7.1-12 Spectral dependence of the reflectance $\mathcal{R}$ for the 10-segment dielectric Bragg grating displayed in Fig. 7.1-11, at various angles of incidence θ_1 and for TE and TM polarizations.

Bragg Grating in an Unmatched Medium

In the previous analysis, the Bragg grating was assumed to be made of N identical segments. If each segment is made of multiple dielectric layers, this requires that grating be placed in a matched medium, i.e., a medium with a refractive index equal to that of the front layer, so that the incident light undergoes no additional reflection at the front boundary, and reflects at the back boundary as if it were entering another layer of the grating. The device described in Example 7.1-8 meets this condition.

In most applications, the grating is placed in an unmatched medium, such as air, and boundary effects must be accounted for. This may be accomplished by writing the wave-transfer matrix **M** of the composite system, including all boundaries, and finding the corresponding scattering matrix **S** by use of the conversion relation. The reflectance of the composite system may be readily determined from **S**.

If $\mathbf{M}_o^N$ is the wave-transfer matrix of an N-segment grating in a medium matched to the front layer, then the overall wave transfer function takes the form

$$\mathbf{M} = \mathbf{M}_e \mathbf{M}_o^{N-1} \mathbf{M}_i, \qquad (7.1\text{-}60)$$

where $\mathbf{M}_i$ is the wave-transfer matrix of the entrance boundary, and $\mathbf{M}_e$ is the wave-transfer matrix of the Nth segment with a boundary into the unmatched medium.

EXAMPLE 7.1-10. *Reflectance of a Dielectric Bragg Grating in an Unmatched Medium.* An N-segment Bragg grating is made of alternating layers of refractive indices n_1 and n_2, and widths d_1 and d_2, placed in a medium of refractive index n_0. We wish to determine the reflectance for a wave incident at an angle θ_0 in the external medium, corresponding to angles θ_1 and θ_2 in the first and second layer of each segment, as determined by Snell's law ($n_1 \sin\theta_1 = n_2 \sin\theta_2$).

In this case, (7.1-60) may be used with the following wave-transfer matrices: (1) $\mathbf{M}_i$ represents a boundary between media of refractive indices n_0 and n_1, as described in Example 7.1-4; (2) $\mathbf{M}_o$ represents a single segment of the grating, as described in Example 7.1-5; (3) $\mathbf{M}_e$ represents propagation a distance d_1 in a medium with refractive index n_1 followed by a slab of width d_2 and refractive index n_2, with boundary into a medium of refractive index n_0. Once the **M** matrix is determined, we use the conversion relation (7.1-6) to determine the corresponding scattering matrix **S**. The overall reflectance is the element r_{12} in (7.1-4).

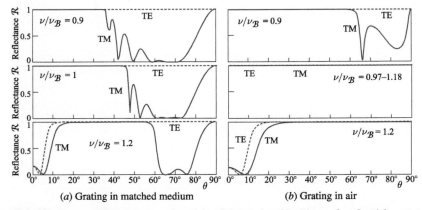

Figure 7.1-13 Power reflectance as a function of the angle of incidence θ at fixed frequencies for the grating described in Fig. 7.1-11. (*a*) Grating is placed in a matched medium ($n = n_1$). (*b*) Grating is placed in air ($n = 1$). In air, the grating has unity reflectance at all angles, for both TE and TM polarizations, at frequencies in the band $0.97\nu_\mathcal{B}$ to $1.18\nu_\mathcal{B}$.

7.2 ONE-DIMENSIONAL PHOTONIC CRYSTALS

One-dimensional (1D) photonic crystals are dielectric structures whose optical properties vary periodically in one direction, called the axis of periodicity, and are constant in the orthogonal directions. These structures exhibit unique optical properties, particularly when the period is of the same order of magnitude as the wavelength. If the axis of periodicity is taken to be the z direction, then optical parameters such as the permittivity $\epsilon(z)$ and the impermeability $\eta(z) = \epsilon_o/\epsilon(z)$ are periodic functions of z, satisfying

$$\eta(z + \Lambda) = \eta(z), \qquad (7.2\text{-}1)$$

for all z, where Λ is the period. Wave propagation in such periodic media may be studied by solving the generalized Helmholtz equations (7.0-2), for periodic $\eta(z)$.

For an on-axis wave traveling along the z axis and polarized in the x direction, the electric and the magnetic field components E_x and H_y are functions of z, independent of x and y, so that (7.0-2) becomes

$$-\frac{d}{dz}\left[\eta(z)\frac{d}{dz}\right]H_y = \frac{\omega^2}{c_o^2} H_y. \qquad (7.2\text{-}2)$$

For an off-axis wave, i.e., a wave traveling in an arbitrary direction in the x–z plane, the generalized Helmholtz equation has a more complex form. For example, for a TM-polarized off-axis wave, the magnetic field points in the y direction and (7.0-2) gives:

$$\left\{-\frac{\partial}{\partial z}\left[\eta(z)\frac{\partial}{\partial z}\right] + \eta(z)\frac{\partial^2}{\partial x^2}\right\}H_y = \frac{\omega^2}{c_o^2} H_y. \qquad (7.2\text{-}3)$$

Note that (7.2-2) and (7.2-3) are cast in the form of an eigenvalue problem from which the modes $H_y(x, z)$ can be determined.

Before embarking on finding solutions to these eigenvalue problems, we first examine the conditions imposed on the propagating modes by the translational symmetry associated with the periodicity.

A. Bloch Modes

Consider first a *homogeneous medium*, which is invariant to an arbitrary translation of the coordinate system. For this medium, an optical mode is a wave that is unaltered by such a translation; it changes only by a multiplicative constant of unity magnitude (a phase factor). The plane wave $\exp(-jkz)$ is such a mode since, upon translation by a distance d, it becomes $\exp[-jk(z+d)] = \exp(-jkd)\exp(-jkz)$. The phase factor $\exp(-jkd)$ is the eigenvalue of the translation operation, as discussed in Appendix C.

On-Axis Bloch Modes

Consider now a *1D periodic medium*, which is invariant to translation by the distance Λ along the axis of periodicity. Its optical modes are waves that maintain their form upon such translation, changing only by a phase factor. As explained in Appendix C, these modes must have the form

$$U(z) = p_K(z)\exp(-jKz), \qquad (7.2\text{-}4)$$
Bloch Mode

where U represents any of the field components E_x, E_y, H_x, or H_y; K is a constant, and $p_K(z)$ is a periodic function of period Λ. This form satisfies the condition that a translation Λ alters the wave by only a phase factor $\exp(-jK\Lambda)$ since the periodic function is unaltered by such translation. This optical wave is known as a **Bloch mode**, and the parameter K, which specifies the mode and its associated periodic function $p_K(z)$, is called the **Bloch wavenumber**.

The Bloch mode is thus a plane wave $\exp(-jKz)$ with propagation constant K, modulated by a periodic function $p_K(z)$, which has the character of a standing wave, as illustrated by its real part displayed in Fig. 7.2-1(a). Since a periodic function of period Λ can be expanded in a Fourier series as a superposition of harmonic functions of the form $\exp(-jmgz), m = 0, \pm 1, \pm 2, \ldots$, with

$$g = 2\pi/\Lambda, \qquad (7.2\text{-}5)$$

it follows that the Bloch wave is a superposition of plane waves of multiple spatial frequencies $K + mg$. The fundamental spatial frequency g of the periodic structure and its harmonics mg, added to the Bloch wavenumber K, constitute the spatial spectrum of the Bloch wave, as shown in Fig. 7.2-1(b). The spatial frequency shift introduced by the periodic medium is analogous to the temporal frequency (Doppler) shift introduced by reflection from a moving object.

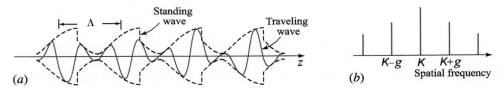

Figure 7.2-1 (a) The Bloch mode. (b) Spatial spectrum of the Bloch mode.

Two modes with Bloch wavenumbers K and $K' = K + g$ are equivalent since they correspond to the same phase factor, $\exp(-jK'\Lambda) = \exp(-jK\Lambda)\exp(-j2\pi) = \exp(-jK\Lambda)$. This is also evident since the factor $\exp(-jg\,z)$ is itself periodic and can be lumped with the periodic function $p_K(z)$. Therefore, for a complete specification of all modes, we need only consider values of K in a spatial-frequency interval of width $g = 2\pi/\Lambda$. The interval $[-g/2, g/2] = [-\pi/\Lambda, \pi/\Lambda]$, known as the first **Brillouin zone**, is a commonly used construct.

Off-Axis Bloch Modes

Off-axis optical modes traveling at some angles in the x–z plane assume the Bloch form

$$U(x, y, z) = p_K(z)\exp(-jKz)\exp(-jk_x x). \qquad (7.2\text{-}6)$$
Off-Axis Bloch Mode

The uniformity of the medium in the x direction constrains the x dependence of the optical mode to the harmonic form $\exp(-jk_x x)$, posing no other restriction on the transverse component k_x of the wavevector. At a location where the refractive index is n, $k_x = nk_o \sin\theta$, where θ is the inclination angle of the wave with respect to the z axis. As the wave travels through the various layers of the inhomogeneous medium, this angle changes, but in view of Snell's law, $n\sin\theta$ and k_x are unaltered.

Normal-to-Axis Bloch Modes

When the angle of incidence in the densest medium is greater than the critical angle, the modes do not travel along the axis of periodicity (the z direction). Rather, they are normal-to-axis modes traveling along the lateral x direction that take the Bloch form (7.2-6) with $K = 0$,

$$U(x,y,z) = p_0(z)\exp(-jk_x x), \qquad (7.2\text{-}7)$$

Normal-to-Axis Bloch Mode

where $p_0(z)$ is a periodic function representing a standing wave along the axis of periodicity.

Eigenvalue Problem, Dispersion Relation, and Photonic Bandgaps

Now that we have established the mathematical form of the modes, as imposed by the translational symmetry of the periodic medium, the next step is to solve the eigenvalue problem described by the generalized Helmholtz equation. For a mode with a Bloch wavenumber K, the eigenvalues ω^2/c_o^2 provide a discrete set of frequencies ω. These values are used to construct the ω-K dispersion relation. The eigenfunctions help us determine the Bloch periodic functions $p_K(z)$ for each of the values of ω associated with each K.

The ω-K relation is a periodic multivalued function of K with period g, the fundamental spatial frequency of the periodic structure; it is often plotted over the Brillouin zone $[-g/2 < k \leq g/2]$, as illustrated schematically in Fig. 7.2-2(a). When visualized as a monotonically increasing function of k, it appears as a continuous function with discrete jumps at values of K equal to integer multiples of $g/2$. These discontinuities correspond to the photonic bandgaps, which are spectral bands not crossed by the dispersion lines, so that no propagating modes exist.

The origin of the discontinuities in the dispersion relation lies in the special symmetry that emerges when $k = g/2$, i.e., when the period of the medium equals exactly half the period of the traveling wave. Consider the two modes with $k = \pm g/2$ and Bloch periodic functions $p_K(z) = p_{\pm g/2}(z)$. Since these modes travel with the same wavenumber, but in opposite directions, i.e. see inverted versions of the medium, $p_{-g/2}(z) = p_{g/2}(-z)$. But these two modes are in fact one and the same, because their Bloch wavenumbers differ by g. It therefore follows that at the edge of a Brillouin zone, there are two Bloch periodic functions that are inverted versions of one another. Since the medium is inhomogeneous or piecewise homogeneous within a unit cell, these two distinct functions interact with the medium differently, and therefore have two distinct eigenvalues, i.e., distinct values of ω. This explains the discontinuity that emerges as the continuous ω-K line crosses the boundary of the Brillouin zone. A similar argument explains the discontinuities that occur when K equals other integer multiples of $g/2$.

The variational principle (see Appendix C) is helpful in pointing out certain features of these eigenfunctions. Based on this principle, the eigenfunctions of a Hermitian operator are orthogonal distributions that minimize the variational energy. The variational energy associated with the linear operator $\mathcal{L}$ in the eigenvalue equation (7.0-2) is $E_v = \frac{1}{2}(\mathbf{H}, \mathcal{L}\mathbf{H})/(\mathbf{H}, \mathbf{H})$. By use of Maxwell's equations, it can be shown that $(\mathbf{H}, \mathcal{L}\mathbf{H}) = (\mathbf{H}, \nabla \times [\eta(\mathbf{r}) \nabla \times \mathbf{H}]) = \int |\mathbf{D}(\mathbf{r})|^2/\epsilon(\mathbf{r})d\mathbf{r}$, so that minimization of E_v is achieved by distributions for which higher displacement fields $\mathbf{D}(\mathbf{r})$ are located at positions of lower $1/\epsilon(\mathbf{r})$, i.e. higher refractive index. For example, if the periodic medium is made of two alternating dielectric layers, as illustrated in Fig. 7.2-2(b), then at a discontinuity the eigenfunction of the lower frequency concentrates its displacement field in the layer with the greater refractive index, whereas the eigenfunction of the higher frequency has an inverted distribution for which the displacement field is concentrated in the layer with the lower refractive index.

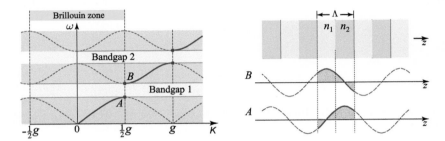

Figure 7.2-2 (a) The dispersion relation is a multivalued periodic function with period $g = 2\pi/\Lambda$ and discontinuities at k equals integer multiples of $g/2$. (b) Bloch functions at the points A and B at the edge of the Brillouin zone for an alternating dielectric-layer periodic medium with $n_2 > n_1$.

The challenging problem now is the solution of the eigenvalue problem associated with the Helmholtz equation. There are two approaches:

- The first approach is based on expanding the periodic function $\eta(z)$ of the medium and the periodic function $p_K(z)$ of the Bloch mode in Fourier series and converting the Helmholtz differential equation into a set of algebraic equation cast in the form of a matrix eigenvalue problem, which are solved numerically. This approach is called the **Fourier Optics** approach.
- The second approach is applicable to layered (piecewise homogeneous) media with planar boundaries. Instead of solving the Helmholtz equation, we make direct use of the laws of propagation and reflection/refraction at boundaries, which are known consequences of Maxwell's equations. We then use the matrix methods developed for multilayer media in Sec. 7.1A and applied to Bragg gratings in Sec. 7.1C. This **Matrix Optics** approach leads to a 2×2 matrix eigenvalue problem from which the dispersion relation and the Bloch modes are determined.

The matrix-optics approach is discussed next, and the Fourier-optics approach is examined in Sec. 7.2C.

B. Matrix Optics of Periodic Media

A one-dimensional periodic medium comprises identical segments, called **unit cells**, that are repeated periodically along one direction (the z axis) and separated by the period Λ (Fig. 7.2-3). Each unit cell contains a succession of lossless dielectric layers or partially reflective mirrors in some order, forming a symmetric system represented by a generic unimodular wave-transfer matrix

$$\mathbf{M}_o = \begin{bmatrix} 1/t^* & r/t \\ r^*/t^* & 1/t \end{bmatrix}, \qquad (7.2\text{-}8)$$

where t and r are complex amplitude transmittance and reflectance satisfying the conditions set forth in (7.1-17), and $\mathcal{T} = |t|^2$ and $\mathcal{R} = |r|^2$ are the corresponding intensity transmittance and reflectance. The medium is a Bragg grating, like that described in Sec. 7.1C, with an infinite number of segments. A wave traveling through the medium undergoes multiple transmissions and reflections that add up to one forward and one backward wave at every plane. We now use the matrix method developed in Sec. 7.1A to determine the Bloch modes.

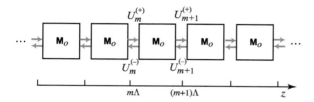

Figure 7.2-3 Wave-transfer matrix representation of a periodic medium.

Let $\{U_m^{(\pm)}\}$ be the complex amplitudes of the forward and backward waves at the initial position $z = m\Lambda$ of unit cell m. Knowing these amplitudes, the amplitudes elsewhere within the cell can be determined by straightforward application of the appropriate wave-transfer matrices, as described in Sec. 7.1. We therefore direct our attention to the dynamics of the amplitudes $\{U_m^{(\pm)}\}$ as they vary from one cell to the next. These dynamics are described by the recurrence relations

$$\begin{bmatrix} U_{m+1}^{(+)} \\ U_{m+1}^{(-)} \end{bmatrix} = \mathbf{M}_o \begin{bmatrix} U_m^{(+)} \\ U_m^{(-)} \end{bmatrix}, \qquad (7.2\text{-}9)$$

which can be used to determine the amplitudes at a particular cell if the amplitudes at the previous cell are known.

Eigenvalue Problem and Bloch Modes

By definition, the modes of the periodic medium are self-reproducing waves, for which

$$\begin{bmatrix} U_{m+1}^{(+)} \\ U_{m+1}^{(-)} \end{bmatrix} = e^{-j\Phi} \begin{bmatrix} U_m^{(+)} \\ U_m^{(-)} \end{bmatrix}, \qquad m = 1, 2, \ldots; \qquad (7.2\text{-}10)$$

after transmission through a distance Λ (in this case a unit cell), the magnitudes of the forward and backward waves remain unchanged and the phases are altered by a common shift Φ, called the **Bloch phase**. The corresponding Bloch wavenumber is $K = \Phi/\Lambda$, so that

$$\boxed{\Phi = K\Lambda.} \qquad (7.2\text{-}11)$$
Bloch Phase

Determination of the complex amplitudes $U_m^{(\pm)}$ and the phase $\Phi = K\Lambda$ that satisfy the self-reproduction condition (7.2-10) can be cast as an eigenvalue problem. This is accomplished by using (7.2-9) with $m = 0$ to write (7.2-10) in the form

$$\mathbf{M}_o \begin{bmatrix} U_0^{(+)} \\ U_0^{(-)} \end{bmatrix} = e^{-j\Phi} \begin{bmatrix} U_0^{(+)} \\ U_0^{(-)} \end{bmatrix}. \qquad (7.2\text{-}12)$$

This is an eigenvalue problem for the 2×2 unit-cell matrix $\mathbf{M}_o$. The factor $e^{-j\Phi}$ is the eigenvalue and the vector with components $U_0^{(+)}$ and $U_0^{(-)}$ is the eigenvector.

The eigenvalues are determined by equating the determinant of the matrix $\mathbf{M}_o - e^{-j\Phi}\mathbf{I}$ to zero. Noting that $|\mathrm{t}|^2+|\mathrm{r}|^2 = 1$, the solution to the ensuing quadratic equation yields $e^{-j\Phi} = \frac{1}{2}(1/\mathrm{t} + 1/\mathrm{t}^*) \pm j\{1 - [\frac{1}{2}(1/\mathrm{t} + 1/\mathrm{t}^*)]^2\}^{1/2}$, from which

$$\cos \Phi = \mathrm{Re}\left\{\frac{1}{\mathrm{t}}\right\}. \qquad (7.2\text{-}13)$$

Equation (7.2-13) is identical to (7.1-46) for the Bragg grating. This is gratifying inasmuch as the periodic medium at hand is an extended Bragg grating with an infinite number of segments.

Since $\mathbf{M}_o$ is a 2×2 matrix, it has two eigenvalues. Hence, only two of the multiple solutions of (7.2-13) are independent. Since the $\cos^{-1}(\cdot)$ function is even, the two solutions within the interval $[-\pi, \pi]$ have equal magnitudes and opposite signs. They correspond to Bloch modes traveling in the forward and backward directions. Other solutions obtained by adding multiples of 2π are not independent since they are irrelevant to the phase factor $e^{-j\Phi}$.

The associated eigenvectors of $\mathbf{M}_o$ are therefore

$$\begin{bmatrix} U_0^{(+)} \\ U_0^{(-)} \end{bmatrix} \propto \begin{bmatrix} \mathrm{r}/\mathrm{t} \\ e^{-j\Phi} - 1/\mathrm{t}^* \end{bmatrix}, \qquad (7.2\text{-}14)$$

as can be ascertained by operating on the right-hand side of (7.2-14) with the $\mathbf{M}_o$ matrix; the result is again the right-hand side of (7.2-14) to within a constant.

The periodic function $p_K(z)$ associated with the Bloch wave can be determined by propagating the amplitudes $U_0^{(+)}$ and $U_0^{(-)}$ through the unit cell. For example, if the initial layer in the unit cell is a homogeneous medium of refractive index n_1 and width d_1, then the wave at distance z into this layer is

$$p_K(z)\,e^{-jKz} = U_0^{(+)}e^{-jn_1 k_0 z} + U_0^{(-)}e^{jn_1 k_0 z}, \quad 0 < z < d_1. \qquad (7.2\text{-}15)$$

Using (7.2-14) and (7.2-11), (7.2-15) then provides

$$p_K(z) \propto [-\mathrm{r}e^{-jn_1 k_0 z} + (e^{-jK\Lambda} - 1)\,e^{jn_1 k_0 z}]\,e^{jKz}, \quad 0 < z < d_1. \qquad (7.2\text{-}16)$$

The waves in (7.2-16) may be propagated further into the subsequent layers within the cell by using the appropriate $\mathbf{M}$ matrices.

Dispersion Relation and Photonic Band Structure

The **dispersion relation** is an equation relating the Bloch wavenumber K and the angular frequency ω. Equation (7.2-13), which provides the eigenvalues $\exp(-j\Phi)$ of the unit-cell matrix, is the progenitor of the dispersion relation for the 1D periodic medium. The phase $\Phi = K\Lambda$ is proportional to K, and $\mathrm{t} = \mathrm{t}(\omega)$ is related to ω through the phase delay associated with propagation through the unit cell, so that (7.2-13), written in the form

$$\cos\left(2\pi \frac{K}{g}\right) = \mathrm{Re}\left\{\frac{1}{\mathrm{t}(\omega)}\right\}, \qquad (7.2\text{-}17)$$
Dispersion Relation

is the $\omega - K$ dispersion relation. Here, $g = 2\pi/\Lambda$ is the fundamental spatial frequency of the periodic medium.

The function $\cos(2\pi K/g)$ is a periodic function of K of period $g = 2\pi/\Lambda$, so that for a given ω, (7.2-17) has multiple solutions. However, solutions separated by the period g are not independent since they lead to identical Bloch waves. It is therefore common to limit the domain of the dispersion relation to a period with values of K in the interval $[-g/2, g/2]$ or $[-\pi/\Lambda, \pi/\Lambda]$, which is the Brillouin zone. This corresponds precisely to limiting the phase Φ to the interval $[-\pi, \pi]$. Also, since $\cos(2\pi K/g)$ is an even function of K, at each value ω there are two independent values of K of equal magnitudes and opposite signs within the Brillouin zone. They correspond to independent Bloch modes traveling in the forward and backward directions.

The dispersion relation exhibits multiple spectral bands classified into two regimes:
- **Propagation regime.** Spectral bands within which K is real correspond to propagating modes. Defined by the condition $|\text{Re}\{1/t(\omega)\}| \leq 1$, these bands are numbered, 1, 2, ..., starting with the lowest-frequency band.
- **Photonic-bandgap regime.** Spectral bands within which K is complex correspond to evanescent waves that are rapidly attenuated. Defined by the condition $|\text{Re}\{1/t(\omega)\}| > 1$, these bands correspond to the stop bands of the diffraction grating discussed in Sec. 7.1C. They are also called **photonic bandgaps** (**PBG**) or **forbidden gaps** since propagating modes do not exist.

The dispersion relation is often plotted with K measured in units of $g = 2\pi/\Lambda$, the fundamental spatial frequency of the periodic structure, whereas ω is measured in units of the Bragg frequency $\omega_\mathcal{B} = \pi c/\Lambda$, where $c = c_o/\bar{n}$ and $\bar{n}$ is the average refractive index of the periodic medium. The ratio $\omega_\mathcal{B}/(g/2) = c$, which is the slope of the dispersion relation $\omega = cK$ for propagation in a homogeneous medium with the average refractive index.

EXAMPLE 7.2-1. *Periodic Stack of Partially Reflective Mirrors.* The dispersion relation for a wave traveling along the axis of a periodic stack of identical partially reflective lossless mirrors with power reflectance $|\mathsf{r}|^2$ and intensity transmittance $|\mathsf{t}|^2 = 1 - |\mathsf{r}|^2$, separated by a distance Λ, is determined directly from Example 7.1-7. Using the results obtained there, namely $\mathsf{t} = |\mathsf{t}|e^{j\varphi}$ with $\varphi = nk_o\Lambda = (\omega/c)\Lambda$, in conjunction with (7.2-13), provides the dispersion relation

$$\cos\left(2\pi\frac{K}{g}\right) = \frac{1}{|\mathsf{t}|}\cos\left(\pi\frac{\omega}{\omega_\mathcal{B}}\right), \qquad (7.2\text{-}18)$$

where $g = 2\pi/\Lambda$, and $\omega_\mathcal{B} = c\pi/\Lambda$ is the Bragg frequency. This result is plotted in Fig. 7.2-4.

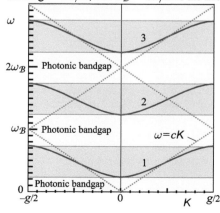

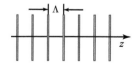

Figure 7.2-4 Dispersion diagram of a periodic set of mirrors, each with intensity transmittance $|\mathsf{t}|^2 = 0.5$, separated by a distance Λ. Here, $\omega_\mathcal{B} = \pi c/\Lambda$ and $g = 2\pi/\Lambda$. The dotted straight lines represent propagation in a homogeneous medium for which $\omega/K = \omega_\mathcal{B}\big/(g/2) = c$.

The photonic bandgaps, which correspond to frequency regions where (7.2-18) does not admit a real solution, are centered at ω_B, $2\omega_B$, ... These frequency regions do not permit propagating modes; rather, they correspond to the stop bands that exhibit unity reflectance in Fig. 7.1-10. In this system, the onset of the lowest photonic bandgap is at $\omega = 0$.

EXAMPLE 7.2-2. *Alternating Dielectric Layers.* A periodic medium comprises alternating dielectric layers of refractive indices n_1 and n_2, with corresponding widths d_1 and d_2, and period $\Lambda = d_1 + d_2$. This system is the dielectric Bragg grating described in Example 7.1-8 with $N = \infty$. For a wave traveling along the axis of periodicity, $\text{Re}\{1/t\} = \text{Re}\{A\}$ is given by (7.1-58). Using the relations $\varphi_1 + \varphi_2 = k_o(n_1 d_1 + n_2 d_2) = \pi\omega/\omega_B$ and $\varphi_1 - \varphi_2 = \zeta\pi\omega/\omega_B$, where $\omega_B = (c_o/\bar{n})(\pi/\Lambda)$ is the Bragg frequency, $\bar{n} = (n_1 d_1 + n_2 d_2)/\Lambda$ is the average refractive index and $\zeta = (n_1 d_1 - n_2 d_2)/(n_1 d_1 + n_2 d_2)$, (7.2-13) provides the dispersion relation

$$\cos\left(2\pi \frac{K}{g}\right) = \frac{1}{t_{12} t_{21}} \left[\cos\left(\pi \frac{\omega}{\omega_B}\right) - |\mathsf{r}_{12}|^2 \cos\left(\pi \zeta \frac{\omega}{\omega_B}\right)\right], \qquad (7.2\text{-}19)$$

where $t_{12} t_{21} = 4 n_1 n_2/(n_1 + n_2)^2$ and $|\mathsf{r}_{12}|^2 = (n_2 - n_1)^2/(n_1 + n_2)^2$.

An example of this dispersion relation is plotted in Fig. 7.2-5 for dielectric materials with $n_1 = 1.5$ and $n_2 = 3.5$, and $d_1 = d_2$. As with the periodic stack of partially reflective mirrors considered in Example 7.2-1, the photonic bandgaps are centered at the frequencies ω_B and its multiples, and occur at either the center of the Brillouin zone ($K = 0$) or at its edge ($K = g/2$). In this case, however, the frequency region surrounding $\omega = 0$ admits propagating modes instead of a forbidden gap. Dielectric materials with lower contrast have bandgaps of smaller width, but the bandgaps exist no matter how small the contrast.

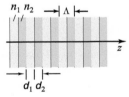

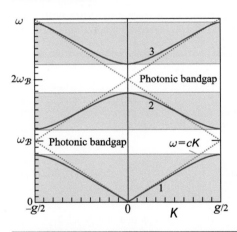

Figure 7.2-5 Dispersion diagram of an alternating-layer periodic dielectric medium with $n_1 = 1.5$ and $n_2 = 3.5$, and $d_1 = d_2$. Here, $\omega_B = \pi c_o/\Lambda\bar{n}$ and $g = 2\pi/\Lambda$. The dotted straight lines represent propagation in a homogeneous medium of mean refractive index $\bar{n}$, so that $\omega/K = \omega_B/(g/2) = c_o/\bar{n} = c$.

Phase and Group Velocities

The propagation constant K corresponds to a phase velocity ω/K and an effective refractive index $n_{\text{eff}} = c_o K/\omega$. The group velocity $v = d\omega/dK$, which governs pulse propagation in the medium, is associated with an effective group index $N_{\text{eff}} = c_o dK/d\omega$ (see Sec. 5.7). These indices can be determined at any point on the ω-K dispersion curve by finding the slope $d\omega/dK$, and the ratio ω/K, i.e., the slope of a line joining the point with the origin. Figure 7.2-6 is a schematic illustration of a dispersion relation of an alternating-layer dielectric periodic medium, together with the effective index and group index, at frequencies extending over two photonic bands with a photonic bandgap in-between.

At low frequencies within the first photonic band, n_{eff} is approximately equal to the average refractive index $\bar{n}$. This is expected since at long wavelengths the material

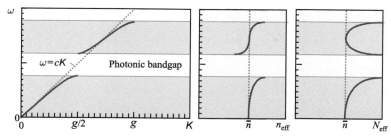

Figure 7.2-6 Frequency dependence of the effective refractive index n_{eff}, which determines the phase velocity, and the effective group index N_{eff}, which determines the group velocity.

behaves as a homogeneous medium with the average refractive index. With increase of frequency, n_{eff} increases above $\bar{n}$, reaching its highest value at the band edge. At the bottom of the second band, n_{eff} is smaller than $\bar{n}$ but increases at higher frequencies, approaching $\bar{n}$ in the middle of the band.

This drop of n_{eff} from a value above average just below the bandgap to a value below average just above the bandgap is attributed to the significantly different spatial distributions of the corresponding Bloch modes, which are orthogonal. The mode at the top of the lower band, has greater energy in the dielectric layers with the higher refractive index, so that its effective index is greater than the average. For the mode at the bottom of the upper band, greater energy is localized in the layers with the lower refractive index, and the effective index is therefore lower than the average.

The frequency dependence of the effective group index follows a different pattern, as shown in Fig. 7.2-6. As the edges of the bandgap are approached, from below or above, this index increases substantially, so that the group velocity is much smaller, i.e., optical pulses are very slow near the edges of the bandgap.

Off-Axis Dispersion Relation and Band Structure

The dispersion relation for off-axis waves may be determined by using the same equation, $\cos(K\Lambda) = \text{Re}\{1/t(\omega)\}$, where $\text{Re}\{1/t(\omega)\}$ now depends on the angles of incidence within the layers of each segment and on the state of polarization (TE or TM). For example, for a periodic medium made of alternating dielectric layers, $\text{Re}\{1/t(\omega)\}$ takes the more general form in (7.1-59).

Since the same transverse component k_x of the wavevector determines the angles of incidence at the two layers ($k_x = n_1 k_o \sin\theta_1 = n_2 k_o \sin\theta_2$), it is more convenient to express the dispersion relation as a function of k_x, in the form of a three-dimensional surface $\omega = \omega(K, k_x)$. Every value of k_x yields a dispersion diagram with bands and bandgaps similar to those of Fig. 7.2-5.

A simpler representation of the $\omega(K, k_x)$ 3D surface is the **projected dispersion diagram**, which displays in a two-dimensional plot of the edges of the bands and bandgaps at each value of k_x, for both TE and TM polarizations, as illustrated in Fig. 7.2-7. This figure is constructed by determining the ranges of angular frequencies over which photonic bands and bandgaps exist in the dispersion diagram for a particular value of k_x, and then projecting these onto corresponding vertical lines at that value of k_x in the projected dispersion diagram. The loci of all such vertical lines for the bands at different values of k_x correspond to the shaded (green) areas displayed in Fig. 7.2-7; the unshaded (white) areas represent the bandgaps.

In this diagram, each angle of incidence is represented by a straight line passing through the origin. For example, the incidence angle θ_1 in layer 1 corresponds to the line $k_x = (\omega/c_1)\sin\theta_1$, i.e., $\omega = (c_1/\sin\theta_1)k_x$, where $c_1 = c_o/n_1$. The line $\omega = c_1 k_x$, called the **light line** corresponds to $\theta_1 = 90°$. Similar lines may be drawn for

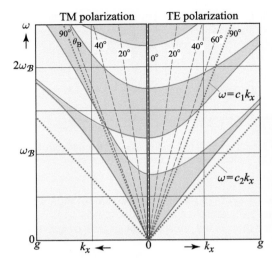
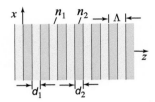

Figure 7.2-7 Projected dispersion diagram for an alternating-layer periodic dielectric medium with $n_1 = 1.5$, $n_2 = 3.5$, and $d_1 = d_2 = \Lambda/2$. Here, $\omega_{\mathcal{B}} = \pi c_o/\Lambda \bar{n}$ and $g = 2\pi/\Lambda$. Photonic bands are shaded (green). The dashed lines represent fixed angles of incidence θ_1 in layer 1, including the Brewster angle $\theta_B = 66.8°$. Points within the region bounded by the light lines $\omega = c_1 k_x$ and $\omega = c_2 k_x$ represent normal-to-axis waves.

the incidence angles in medium 2; Fig. 7.2-7 shows only the light line $\omega = c_2 k_x$, assuming that $n_2 > n_1$, i.e., $c_2 < c_1$. Points in the region bounded by the two light lines represent normal-to-axis modes, which travel in the lateral direction by undergoing total reflection in the denser medium (medium 2).

The question arises as to whether there exists a frequency range over which propagation is forbidden at all angles of incidence θ_1 and θ_2 and for both polarizations. This could occur if the forbidden gaps at all values of k_x between the lines $k_x = 0$ and $k_x = \omega/c_2$, and for both polarizations, were to align in such a way as to create a common or **complete photonic bandgap**. This is clearly not the case in the example in Fig. 7.2-7. It turns out that this is not possible; complete photonic bandgaps cannot exist within 1D periodic structures. However, they can occur in 2D and 3D periodic structures, as we shall see in Sec. 7.3.

Indeed, there is one special case in which a photonic bandgap cannot occur at all, and that is an oblique TM wave propagating at the Brewster angle $\theta_B = \tan^{-1}(n_2/n_1)$ in layer 1. As shown in Fig. 7.2-7, the line at the Brewster angle does not pass through a gap. This is not surprising since at this angle, the reflectance of a unit cell is zero, and the forward and backward waves are uncoupled so that the collective effect that leads to total reflection is absent.

C. Fourier Optics of Periodic Media

The matrix analysis of periodic media presented in the previous section is applicable to layered (i.e., piecewise homogeneous) media. A more general approach, applicable for arbitrary periodic media, including continuous media, is based on a Fourier-series representation of periodic functions and conversion of the Helmholtz equation into a set of algebraic equations whose solution provides the dispersion relation and the Bloch modes. This approach can also be generalized to 2D and 3D periodic media, as will be shown in Sec. 7.3.

A wave traveling along the axis of a 1D periodic medium (the z axis) and polarized in the x direction is described by the generalized Helmholtz equation (7.2-2). Since $\eta(z)$ is periodic with period Λ, it can be expanded in a Fourier series,

$$\eta(z) = \sum_{\ell=-\infty}^{\infty} \eta_\ell \exp(-j\ell g z), \qquad (7.2\text{-}20)$$

where $g = 2\pi/\Lambda$ is the spatial frequency (rad/mm) of the periodic structure and η_ℓ is the Fourier coefficient representing the ℓth harmonic. The impermeability $\eta(z)$ is real, so that $\eta_{-\ell} = \eta_\ell^*$.

The periodic portion of the Bloch wave $p_K(z)$ in (7.2-4) may also be expanded in a Fourier series,

$$p_K(z) = \sum_{m=-\infty}^{\infty} C_m \exp\left(-jmgz\right), \qquad (7.2\text{-}21)$$

whereupon the Bloch wave representation of the magnetic field may be written as

$$H_y(z) = \sum_{m=-\infty}^{\infty} C_m \exp\left[-j(K + mg)z\right]. \qquad (7.2\text{-}22)$$

For brevity, the dependence of the Fourier coefficients $\{C_m\}$ on the Bloch wavenumber K is suppressed. Substituting these expansions into the Helmholtz equation (7.2-2) and equating harmonic terms of the same spatial frequency, we obtain

$$\sum_{\ell=-\infty}^{\infty} F_{m\ell} C_\ell = \frac{\omega^2}{c_o^2} C_m, \quad F_{m\ell} = (K + mg)(K + \ell g)\eta_{m-\ell}, \qquad (7.2\text{-}23)$$

where $m = 0, \pm 1, \pm 2, \ldots$.

The differential equation (7.2-2) has now been converted into a set of linear equations (7.2-23) for the unknown Fourier coefficients $\{C_m\}$. These equations may be cast in the form of a matrix eigenvalue problem. For each K, the eigenvalues ω^2/c_o^2 correspond to multiple values of ω, from which the ω-K dispersion relation may be constructed. The eigenvectors are sets of $\{C_m\}$ coefficients, which determine the periodic function $p_K(z)$ of the Bloch mode for each K.

Posed as an eigenvalue problem for a matrix **F** with elements $F_{m\ell}$, this set of coupled equations may be solved using standard numerical techniques. Since $\eta_{m-\ell} = \eta_{\ell-m}^*$, the matrix **F** is Hermitian, i.e., $F_{m\ell} = F_{\ell m}^*$. Note that if we were to use the electric-field Helmholtz equation instead of the magnetic-field Helmholtz equation (7.2-2), we would obtain another matrix representation of the eigenvalue problem, but the matrix would be non-Hermitian, and therefore more difficult to solve. This is the rationale for working with the Helmholtz equation for the magnetic field.[†]

Approximate Solution of the Eigenvalue Problem

In (7.2-23), the harmonics of the optical wave are coupled via the harmonics of the periodic medium. An optical-wave harmonic of spatial frequency $K + \ell g$ mixes with a medium harmonic of spatial frequency $(m - \ell)g$ and contributes to the optical-wave harmonic of spatial frequency $(K + \ell g) + (m - \ell)g = K + mg$.

The conditions under which strong coupling emerges can be determined by separating out the mth term in (7.2-23), which leads to

$$C_m = \sum_{\ell \neq m} \frac{\eta_{m-\ell}}{\eta_o} \frac{(K + \ell g)(K + mg)}{(\bar{n}\omega/c_o)^2 - (K + mg)^2} C_\ell, \quad m = 0, \pm 1, \pm 2, \ldots, \qquad (7.2\text{-}24)$$

[†] It can be shown that the differential operator in the generalized magnetic-field Helmholtz equation (7.0-2) is Hermitian whereas that in the electric-field Helmholtz equation (7.0-2) is non-Hermitian.

where $\bar{n} = 1/\sqrt{\eta_0}$ is an average refractive index of the medium. Strong coupling between the mth harmonic of the wave and other harmonics exists if the denominator in (7.2-24) is small, i.e.,

$$\omega \bar{n}/c_o \approx |K + mg|. \quad (7.2\text{-}25)$$

This equation represents a resonance condition for interaction between the harmonics. It can also be regarded as a **phase-matching condition**.

Figure 7.2-8 is a plot of (7.2-25) as an equality. For each value of m, the ω-K relation is a V-shaped curve. The intersection points of these curves represent common values of ω and K at which (7.2-25) is simultaneously satisfied for two harmonics. The intersections between the $m = 0$ curve (dashed) and the curves for $m = -1, m = -2,$ ..., are marked by filled circles; they correspond to the lowest-order bandgaps 1, 2, ..., respectively. At each intersection point, K is an integer multiple of $\frac{1}{2}g$, and ω is an integer multiple of the Bragg frequency $\omega_B = (c_o/\bar{n})g/2$ or $\omega_B/2\pi = \nu_B = (c_o/\bar{n})/2\Lambda$. This corresponds to the Bragg wavelength $\lambda_B = 2\Lambda$ in the medium, and therefore to total reflection. Unmarked intersections in Fig. 7.2-8 are not independent since each of these has the same ω as a marked intersection, and a value of K differing by a reciprocal lattice constant g.

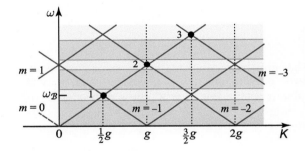

Figure 7.2-8 Plot of (7.2-25), as an equality, for various values of m. The $m = 0$ curve is indicated as dashed. Strong coupling between the harmonics of the optical wave and those of the medium occurs at the intersection points 1, 2,..., which correspond to the lowest-order bandgaps.

The lowest-order bandgap occurs at the intersection of the $m = 0$ and $m = -1$ curves (point 1 in Fig. 7.2-8). In this case, only the coefficients C_0 and C_{-1} are strongly coupled, so that (7.2-24) yields two coupled equations

$$C_0 = \frac{\eta_1}{\eta_0} \frac{(K-g)K}{\omega^2 \bar{n}^2/c_o^2 - K^2} C_{-1}, \quad (7.2\text{-}26)$$

$$C_{-1} = \frac{\eta_1^*}{\eta_0} \frac{K(K-g)}{\omega^2 \bar{n}^2/c_o^2 - (K-g)^2} C_0, \quad (7.2\text{-}27)$$

where $\eta_{-1} = \eta_1^*$. These equations are self-consistent if

$$\boxed{\frac{|\eta_1|^2}{\eta_0^2} K^2(K-g)^2 = \left[\omega^2 \frac{\bar{n}^2}{c_o^2} - K^2\right]\left[\omega^2 \frac{\bar{n}^2}{c_o^2} - (K-g)^2\right].} \quad \begin{array}{c}(7.2\text{-}28)\\ \text{Dispersion} \\ \text{Relation}\end{array}$$

A plot of this equation (Fig. 7.2-9) yields the ω-K dispersion relation near the edge of the bandgap, where the equation is valid. For $K = \frac{1}{2}g$, (7.2-28) yields two frequencies,

$$\omega_\pm = \omega_B \sqrt{1 \pm |\eta_1|/\eta_0}, \quad (7.2\text{-}29)$$

representing the edges of the first photonic bandgap. The center of the bandgap is at the Bragg frequency $\omega_B = (c_o/\bar{n})(g/2) = (\pi/\Lambda)(c_o/\bar{n})$. The ratio of the gap width to the midgap frequency, which is called the **gap-midgap ratio**, grows with increasing impermeability contrast ratio $|\eta_1|/\eta_0$.

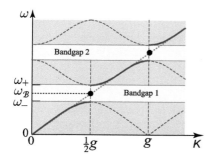

Figure 7.2-9 Dispersion relation in the vicinity of photonic bandgaps.

A similar procedure can be followed to determine the spectral width of higher-order bandgaps. The width of the mth bandgap is determined by a formula identical to (7.2-29), but the ratio $|\eta_m|/\eta_0$ replaces $|\eta_1|/\eta_0$, so that higher bandgaps are governed by higher spatial harmonics of the periodic function $\eta(z)$.

Off-Axis Waves

The dispersion relation for off-axis waves may be determined by use of the same Fourier expansion technique. For a TM-polarized off-axis wave traveling in an arbitrary direction in the x–z plane, the Helmholtz equation is given by (7.2-3). The Bloch wave is a generalization of (7.2-22) obtained via (7.2-6),

$$H_y(z) = \sum_{m=-\infty}^{\infty} C_m \exp\left[-j(K+mg)z\right] \exp(-jk_x x). \tag{7.2-30}$$

Carrying out calculations similar to the on-axis case, leads to the following set of algebraic equations for the C_m coefficients:

$$\sum_{\ell=-\infty}^{\infty} F_{m\ell} C_\ell = \frac{\omega^2}{c_o^2} C_m, \quad F_{m\ell} = [(K+\ell g)(K+mg) + k_x^2]\eta_{m-\ell}. \tag{7.2-31}$$

Equation (7.2-31) is a generalization of (7.2-23) for the off-axis wave. The dispersion relation may be determined by solving this matrix eigenvalue problem for the set of frequencies ω associated with each pair of values of K and k_x.

D. Boundaries Between Periodic and Homogeneous Media

The study of light wave propagation in periodic media has so far been limited to determining the dispersion relation and its associated band structure, as well as estimating the phase and group velocity of such waves. By definition, the periodic medium extend indefinitely in all directions. The next step is to examine reflection and transmission at boundaries between periodic and homogeneous media. We first examine reflection from a single boundary and subsequently consider a slab of periodic medium embedded in a homogeneous medium. Other configurations made of homogeneous structures such slabs or holes embedded in extended periodic media are described in Sec. 10.4 and Sec. 11.4D.

Omnidirectional Reflection at a Single Boundary

We examine the reflection and transmission of an optical wave at the boundary between a semi-infinite homogeneous medium and a semi-infinite one-dimensional periodic medium, as portrayed in Fig. 7.2-10. We demonstrate that, under certain conditions and within a specified range of angular frequencies, the periodic medium acts as a perfect mirror, totally reflecting waves incident from any direction and with any polarization!

Wave transmission and reflection at the boundary between two media is governed by the phase-matching condition. At the boundary between two homogeneous media, for example, the transverse components of the wavevector k_x must be the same on both sides of the boundary. Since $k_x = k\sin\theta = (\omega/c_o)n\sin\theta$, this condition means that the product $n\sin\theta$ is invariant. This is the origin of Snell's law of refraction, as explained in Sec. 2.4A.

Similarly, for a wave incident from a homogeneous medium into a one-dimensional periodic medium, k_x must remain the same. Thus, if the incident wave has angular frequency ω and angle of incidence θ, we have $k_x = (\omega/c_o)n\sin\theta$, where n is the refractive index of the homogeneous medium. Knowing k_x and ω, we can use the dispersion relation $\omega = \omega(K, k_x)$ for the periodic medium at the appropriate polarization to determine the Bloch wavenumber K. If the angular frequency ω lies within a forbidden gap at this value of k_x, the incident wave will not propagate into the periodic medium but will, instead, be totally reflected. This process is repeated for all frequencies, angles of incidence, and polarizations of the incident wave.

We now consider the possibility that the boundary acts as an omnidirectional reflector (a perfect mirror). For this purpose, we use the projected dispersion diagram, which displays the bandgaps for each value of k_x, as illustrated in the example provided in Fig. 7.2-10. On the same diagram, we delineate by a red dashed line the ω-k_x region that can be accessed by waves entering from the homogeneous medium. This region is defined by the equation $k_x = (\omega/c_o)n\sin\theta$, which dictates that $k_x < (\omega/c_o)n$ or $\omega > (c_o/n)k_x$; it is thus bounded by the line $\omega = (c_o/n)k_x$, or $\omega/\omega_B = (\bar{n}/n)[k_x/(g/2)]$, known as the light line. This line corresponds to an angle $\theta = 90°$ in the surrounding medium.

Figure 7.2-10 reproduces Fig. 7.2-7 with the light lines added, and the permissible ω-k_x regions within the light lines highlighted. Waves incident from the homogeneous medium at all angles, and both polarizations, are represented by points within this region; points outside this region are not accessible by waves incident from the homogeneous medium regardless of their angle of incidence or polarization. The spectral band bounded by the angular frequencies ω_1 and ω_2, as defined in Fig. 7.2-10, is of particular interest inasmuch as all ω-k_x points lying in this band are within a photonic bandgap. In this spectral band, therefore, no incident wave, regardless of its angle or polarization, can be matched with a propagating wave in the periodic medium — the boundary then acts as a perfect omnidirectional reflector. Also illustrated in Fig. 7.2-10 is a second spectral band, at higher angular frequencies, that behaves in the same way.

Slab of Periodic Material in a Homogeneous Medium

A slab of 1D periodic material embedded in a homogeneous medium is nothing but a 1D Bragg grating with a finite number of segments. Reflection and transmission from the Bragg grating has already been examined in Sec. 7.1C.

One would expect that a Bragg grating with a large, but finite, number of segments N captures the basic properties of a periodic medium made of the same unit cell. This is in fact the case since the passbands and stop bands of the grating are mathematically identical to the photonic bands and bandgaps of the extended periodic medium. However, the spectral transmittance and reflectance of the Bragg grating, which exhibit oscillatory properties sensitive to the size of the grating and the presence of its boundaries, do not have their counterparts in the extended periodic structure.

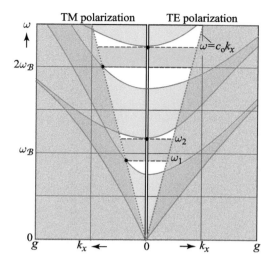
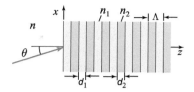

Figure 7.2-10 Projected dispersion diagram for an alternating-layer dielectric medium with $n_1 = 1.5$, $n_2 = 3.5$, and $d_1 = d_2 = \Lambda/2$. The dotted lines (red) are light lines for a homogeneous medium with refractive index $n = 1$. In the spectral band between ω_1 and ω_2, the medium acts as a perfect omnidirectional reflector for all polarizations. A similar band is shown at higher angular frequencies.

Likewise, the phase and group velocities and the associated effective refractive indices determined from the dispersion relation in the extended periodic medium do not have direct counterparts in the finite-size Bragg grating. Nevertheless, such parameters can be defined for a grating by determining the complex amplitude transmittance $t(\omega)$ and matching it with an effective homogeneous medium of the same total thickness d such that $\arg\{t_N\} = (\omega/c_o)n_{\text{eff}}d$. An effective group index $N_{\text{eff}} = n_{\text{eff}} + \omega dn_{\text{eff}}/d\omega$ is then determined [see (5.7-2)]. The dependence of these effective indices on frequency is different from that shown in Fig. 7.2-6 for the extended periodic medium in that it exhibits oscillations within the passbands. However, for sufficiently large N, say greater than 100, these oscillations are washed out and the effective indices become nearly the same as those of the extended periodic medium.

Another configuration of interest is a slab of homogeneous medium embedded in a periodic medium. In this configuration, the light may be trapped in the slab by omnidirectional reflection from the surrounding periodic medium, so that the slab becomes an optical waveguide. This configuration is discussed in Sec. 9.5.

7.3 TWO- AND THREE-DIMENSIONAL PHOTONIC CRYSTALS

The concepts introduced in Sec. 7.2 for the study of optical-wave propagation in 1D periodic media can be readily generalized to 2D and 3D structures. These include Bloch waves as the modes of the periodic medium and ω-K dispersion relations with photonic bands and bandgaps. In contrast to 1D structures, 2D photonic crystals have complete 2D photonic bandgaps, i.e., common bandgaps for waves of both polarization traveling in any direction in the plane of periodicity. However, complete 3D photonic bandgaps, i.e., common bandgaps for all directions and polarizations, can be achieved only in 3D photonic crystals. The mathematical treatment of 2D and 3D periodic media is more elaborate and the visualization of the dispersion diagrams is more difficult because of the additional degrees of freedom involved, but the concepts are essentially the same as those encountered for 1D periodic media. This section begins with a simple treatment of 2D structures followed by a more detailed 3D treatment.

A. Two-Dimensional Photonic Crystals

2D Periodic Structures

Consider a 2D periodic structure such as a set of identical parallel rods, tubes, or veins embedded in a homogeneous host medium [Fig. 7.3-1(a)] and organized at the points of a rectangular lattice, as illustrated in Fig. 7.3-1(b). The impermeability $\eta(x, y) = \epsilon_o/\epsilon(x,y)$ is periodic in the transverse directions, x and y, and uniform in the axial direction z. If a_1 and a_2 are the periods in the x and y directions, then $\eta(x,y)$ satisfies the translational symmetry relation

$$\eta(x + m_1 a_1, y + m_2 a_2) = \eta(x,y), \tag{7.3-1}$$

for all integers m_1 and m_2. This periodic function is represented as a 2D Fourier series,

$$\eta(x,y) = \sum_{\ell_1=-\infty}^{\infty} \sum_{\ell_2=-\infty}^{\infty} \eta_{\ell_1,\ell_2} \exp\left(-j\ell_1 g_1 x\right) \exp\left(-j\ell_2 g_2 x\right), \tag{7.3-2}$$

where $g_1 = 2\pi/a_1$ and $g_2 = 2\pi/a_2$ are fundamental spatial frequencies (radians/mm) in the x and y directions, and $\ell_1 g_1$ and $\ell_2 g_2$ are their harmonics. The coefficients η_{ℓ_1,ℓ_2} depend on the actual profile of the periodic function, e.g., the size of the rods.

The 2D Fourier transform of the periodic function is composed of points (delta functions) on a rectilinear lattice, as shown in Fig. 7.3-1(c). This Fourier-domain lattice is known to solid-state physicists as the **reciprocal lattice**.

What are the optical modes of a medium with such symmetry? The answer is a simple generalization of the 1D case given in (7.2-4). For waves traveling in a direction parallel to the x–y plane, the modes are 2D Bloch waves,

$$U(x,y) = p_{K_x,K_y}(x,y) \exp(-jK_x x) \exp(-jK_y y), \tag{7.3-3}$$

where $p_{K_x,K_y}(x,y)$ is a periodic function with the same periods as the medium. The wave is specified by a pair of Bloch wavenumbers (K_x, K_y). Another wave with Bloch wavenumbers $(K_x + g_1, K_y + g_2)$ is not a new mode. As shown in Fig. 7.3-1(c) a complete set of modes in the Fourier plane has Bloch wavenumbers located at points in a rectangle defined by $[-g_1/2 < K_x \leqslant g_1/2]$ and $[-g_2/2 < K_y \leqslant g_2/2]$, which is the first Brillouin zone.

Other symmetries may be used to reduce the set of independent Bloch wavevectors within the Brillouin zone. When all symmetries are included, the result is an area called the **irreducible Brillouin zone**. For example, the rotational symmetry inherent in the square lattice results in an irreducible Brillouin zone in the form of a triangle, as shown in Fig 7.3-1(d).

2D Skew-Periodic Structures

An example of another class of 2D periodic structures is a set of parallel cylindrical holes placed at the points of a triangular lattice, as illustrated in Fig. 7.3-2(a). Since the lattice points are skewed (not aligned with x and y axis), we use two primitive *vectors* $\mathbf{a}_1$ and $\mathbf{a}_2$ [Fig. 7.3-2(b)] to generate the lattice via the lattice vector $\mathbf{R} = m_1\mathbf{a}_1 + m_2\mathbf{a}_2$, where m_1 and m_2 are integers. We also define a position vector $\mathbf{r}_T = (x,y)$ so that the periodic function $\epsilon(\mathbf{r}_T) \equiv \epsilon(x,y)$ satisfies the translational symmetry relation $\epsilon(\mathbf{r}_T + \mathbf{R}) = \epsilon(\mathbf{r}_T)$ (the subscript "T" indicates transverse).

The 2D Fourier series of such a function is a set of points on a reciprocal lattice defined by the vectors $\mathbf{g}_1$ and $\mathbf{g}_2$, which are orthogonal to $\mathbf{a}_1$ and $\mathbf{a}_2$, respectively, and have magnitudes $g_1 = 2\pi/a_1 \sin\theta$ and $g_2 = 2\pi/a_2 \sin\theta$, where θ is the angle between $\mathbf{a}_1$ and $\mathbf{a}_2$. The 2D reciprocal lattice is also a triangular lattice generated by

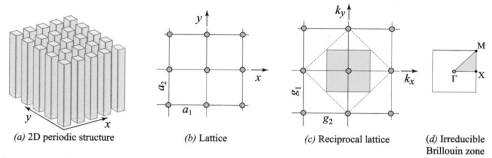

(a) 2D periodic structure   (b) Lattice   (c) Reciprocal lattice   (d) Irreducible Brillouin zone

Figure 7.3-1 (a) A 2D periodic structure comprising parallel rods. (b) The rectangular lattice at which the rods are placed. (c) The 2D Fourier transform of the lattice points is another set of points forming a reciprocal lattice with periods $g_1 = 2\pi/a_1$ and $g_2 = 2\pi/a_2$. The shaded (yellow) area is the Brillouin zone. (d) For a square lattice ($a_1 = a_2 = a$), the irreducible Brillouin zone is the triangle ΓMX.

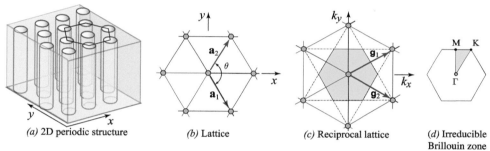

(a) 2D periodic structure   (b) Lattice   (c) Reciprocal lattice   (d) Irreducible Brillouin zone

Figure 7.3-2 (a) A 2D periodic structure comprising parallel cylindrical holes. (b) The triangular lattice at which the holes are placed. In this diagram the magnitudes $a_1 = a_2 = a$ and $\theta = 120°$. (c) Reciprocal lattice; the shaded (yellow) area is the Brillouin zone, a hexagon. (d) The irreducible Brillouin zone is the triangle ΓMK.

the vector $\mathbf{G} = \ell_1 \mathbf{g}_1 + \ell_2 \mathbf{g}_2$, where ℓ_1 and ℓ_2 are integers, as illustrated in Fig. 7.3-2(c).

For waves traveling in a direction parallel to the x–y plane, the Bloch modes are

$$U(\mathbf{r}_\mathrm{T}) = p_\mathbf{K}(\mathbf{r}_\mathrm{T}) \exp(-j\mathbf{K}_\mathrm{T} \cdot \mathbf{r}_\mathrm{T}), \qquad (7.3\text{-}4)$$

where $\mathbf{K}_\mathrm{T} = (K_x, K_y)$ is the Bloch wavevector and $p_{\mathbf{K}_\mathrm{T}}(\mathbf{r}_\mathrm{T})$ is a 2D periodic function on the same lattice. Two Bloch modes with Bloch wavevectors $\mathbf{K}_\mathrm{T}$ and $\mathbf{K}_\mathrm{T} + \mathbf{G}$ are equivalent. To cover a complete set of Bloch wavevectors, we therefore need only consider vectors within the Brillouin zone shown in Fig. 7.3-2(c).

The dispersion relation can be determined by ensuring that the Bloch wave in (7.3-3) or (7.3-4) satisfies the generalized Helmholtz equation. The calculations are facilitated by use of a Fourier series approach, as was done in the 1D case and as will be described (in a more general form) in the 3D case.

EXAMPLE 7.3-1. *Cylindrical Holes on a Triangular Lattice.* A 2D photonic crystal comprises a homogeneous medium ($n = 3.6$) with air-filled cylindrical holes of radius $0.48\,a$ organized at the points of a triangular lattice with lattice constant a. The calculated dispersion relation, shown in Fig. 7.3-3, for TE and TM waves traveling in the plane of periodicity ($k_z = 0$) exhibits a complete

2D photonic bandgap at frequencies near the angular frequency $\omega_0 = \pi c_o/a$.[†] As in the 1D case, the gap can be made wider by use of materials with greater refractive-index contrast. Indeed, most geometries exhibit photonic bandgaps if the constituent materials have sufficiently high contrast.

This photonic-crystal structure finds use as a "holey" optical fiber, which has a number of salutary properties (see Sec. 10.4).

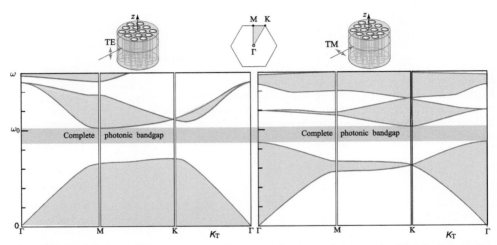

Figure 7.3-3 Calculated band structure of a 2D photonic crystal consisting of a homogeneous medium ($n = 3.6$) with air-filled cylindrical holes of radius $0.48\,a$ organized at the points of a triangular lattice with lattice constant a. The abscissa spans Bloch wavevectors defined by points on the periphery of the irreducible Brillouin zone, the ΓMK triangle. The ordinate is plotted in units of $\omega_0 = \pi c_o/a$. The wave travels in the plane of periodicity and has TE polarization (left) and TM polarization (right). The complete 2D photonic bandgap in the vicinity of ω_0 is highlighted.

For an oblique wave traveling at an angle with respect to the x–y plane, the Bloch wave in (7.3-4) becomes

$$U(\mathbf{r}_\text{T}) = p_\mathbf{K}(\mathbf{r}_\text{T}) \exp(-j\mathbf{K}_\text{T} \cdot \mathbf{r}_\text{T}) \exp(-jk_z z), \qquad (7.3\text{-}5)$$

where k_z is a constant. The band structure then takes the form of a set of surfaces of $\omega = \omega(\mathbf{K}_\text{T}, k_z)$.

A complete 3D photonic bandgap is a range of frequencies ω crossed by none of these surfaces, i.e., values of ω that are not obtained by any combination of real $\mathbf{K}_\text{T}$ and k_z. While a complete 2D photonic bandgap exists for $k_z = 0$, as illustrated by the example in Fig. 7.3-3, a photonic bandgap for all off-axis waves is not attainable in 2D periodic structures.

⋆B. Three-Dimensional Photonic Crystals

Crystal Structure

A 3D photonic crystal is generated by the placement of copies of a basic dielectric structure, such as a sphere or a cube, at points of a 3D lattice generated by the lattice vectors $\mathbf{R} = m_1\mathbf{a}_1 + m_2\mathbf{a}_2 + m_3\mathbf{a}_3$, where m_1, m_2, and m_3 are integers, and $\mathbf{a}_1$, $\mathbf{a}_2$, and $\mathbf{a}_3$ are primitive vectors defining the lattice unit cell. The overall structure is periodic

---

[†] See S. G. Johnson and J. D. Joannopoulos, Block-Iterative Frequency-Domain Methods for Maxwell's Equations in a Planewave Basis, *Optics Express*, vol. 8, pp. 173–190, 2001.

and its physical properties, such as the permittivity $\epsilon(\mathbf{r})$ and the impermeability $\eta(\mathbf{r}) = \epsilon_o/\epsilon(\mathbf{r})$, are invariant to translation by $\mathbf{R}$, so that

$$\eta(\mathbf{r} + \mathbf{R}) = \eta(\mathbf{r}) \qquad (7.3\text{-}6)$$

for all positions $\mathbf{r}$.

This periodic function may therefore be expanded in a 3D Fourier series,

$$\eta(\mathbf{r}) = \sum_{\mathbf{G}} \eta_{\mathbf{G}} \exp(-j\,\mathbf{G}\cdot\mathbf{r}), \qquad (7.3\text{-}7)$$

where $\mathbf{G} = \ell_1 \mathbf{g}_1 + \ell_2 \mathbf{g}_2 + \ell_3 \mathbf{g}_3$ is a vector defined by the primitive vectors $\mathbf{g}_1$, $\mathbf{g}_2$, and $\mathbf{g}_3$, of another lattice, the **reciprocal lattice**, and ℓ_1, ℓ_2, and ℓ_3, are integers. The $\mathbf{g}$ vectors are related to the $\mathbf{a}$ vectors via

$$\mathbf{g}_1 = 2\pi \frac{\mathbf{a}_2 \times \mathbf{a}_3}{\mathbf{a}_1 \cdot \mathbf{a}_2 \times \mathbf{a}_3}, \quad \mathbf{g}_2 = 2\pi \frac{\mathbf{a}_3 \times \mathbf{a}_1}{\mathbf{a}_1 \cdot \mathbf{a}_2 \times \mathbf{a}_3}, \quad \mathbf{g}_3 = 2\pi \frac{\mathbf{a}_1 \times \mathbf{a}_2}{\mathbf{a}_1 \cdot \mathbf{a}_2 \times \mathbf{a}_3}, \qquad (7.3\text{-}8)$$

so that $\mathbf{g}_1 \cdot \mathbf{a}_1 = 2\pi$, $\mathbf{g}_1 \cdot \mathbf{a}_2 = 0$, and $\mathbf{g}_1 \cdot \mathbf{a}_3 = 0$, i.e., $\mathbf{g}_1$ is orthogonal to $\mathbf{a}_2$ and $\mathbf{a}_3$ and its length is inversely proportional to $\mathbf{a}_1$. Similar properties apply to $\mathbf{g}_2$ and $\mathbf{g}_3$. It can also be shown that $\mathbf{G} \cdot \mathbf{R} = 2\pi$.

If $\mathbf{a}_1$, $\mathbf{a}_2$, and $\mathbf{a}_3$ are mutually orthogonal, then $\mathbf{g}_1$, $\mathbf{g}_2$, and $\mathbf{g}_3$ are also mutually orthogonal and the magnitudes $g_1 = 2\pi/a_1$, $g_2 = 2\pi/a_2$, and $g_3 = 2\pi/a_3$ are the spatial frequencies associated with the periodicities in the three directions, respectively. An example of a 3D crystal lattice and its corresponding reciprocal lattice is shown in Fig. 7.3-4.

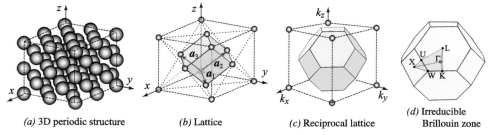

(a) 3D periodic structure (b) Lattice (c) Reciprocal lattice (d) Irreducible Brillouin zone

Figure 7.3-4 (a) A 3D periodic structure comprising dielectric spheres. (b) The spheres are placed at the points of a diamond (face-centered cubic) lattice for which $\mathbf{a}_1 = (a/\sqrt{2})(\hat{\mathbf{x}} + \hat{\mathbf{y}})$, $\mathbf{a}_2 = (a/\sqrt{2})(\hat{\mathbf{y}}+\hat{\mathbf{z}})$, and $\mathbf{a}_3 = (a/\sqrt{2})(\hat{\mathbf{x}}+\hat{\mathbf{z}})$, where a is the lattice constant. (c) The corresponding reciprocal lattice is a body-centered cubic lattice with a Brillouin zone indicated by the shaded volume, known as a Wigner–Seitz cell. (d) The irreducible Brillouin zone is the polyhedron whose corner points are marked by the crystallographic symbols ΓXULKW.

Bloch Modes

The modes of a 3D periodic medium are waves that maintain their shape upon translation by a lattice vector $\mathbf{R}$, changing only by a multiplicative constant of unity magnitude. These modes have the Bloch form $p_{\mathbf{K}}(\mathbf{r})\exp(-j\mathbf{K}\cdot\mathbf{r})$ where $p_{\mathbf{K}}(\mathbf{r})$ is a 3D periodic function, with the periodicity described by the same lattice vector $\mathbf{R}$; $\mathbf{K}$ is the **Bloch wavevector**; and $\hat{\mathbf{e}}$ is a unit vector in the direction of polarization. The Bloch mode is a traveling plane wave $\exp(-j\mathbf{K}\cdot\mathbf{r})$ modulated by a periodic function $p_{\mathbf{K}}(\mathbf{r})$.

Translation by **R** results in multiplication by a phase factor $\exp(-j\mathbf{K}\cdot\mathbf{R})$, which depends on **K**.

Two modes with Bloch wavevectors **K** and $\mathbf{K}' = \mathbf{K} + \mathbf{G}$ are equivalent since $\exp(-j\mathbf{K}'\cdot\mathbf{R}) = \exp(-j\mathbf{K}\cdot\mathbf{R})$, i.e., translation by **R** is equivalent to multiplication by the same phase factor. This is because $\exp(-j\mathbf{G}\cdot\mathbf{R}) = \exp(-j2\pi) = 1$. Therefore, for the complete specification of all modes, we need only consider values of **K** within a finite volume of the reciprocal lattice, the Brillouin zone. The Brillouin zone is the volume of points that are closer to one specific reciprocal lattice point (the origin of the zone, denoted Γ) than to any other lattice point. Other symmetries of the lattice permit further reduction of that volume to the irreducible Brillouin zone, as illustrated by the example in Fig. 7.3-4.

Photonic Band Structure

To determine the ω-**K** dispersion relation for a 3D periodic medium, we begin with the eigenvalue problem described by the generalized Helmholtz equation (7.0-2). One approach for solving this problem is to generalize the Fourier method that was introduced in Sec. 7.2C for 1D periodic structures. By expanding the periodic functions $\eta(\mathbf{r})$ and $p_\mathbf{K}(\mathbf{r})$ in Fourier series, the differential equation (7.0-2) is converted into a set of algebraic equations leading to a matrix eigenvalue problem that can be solved numerically using matrix methods. As discussed at the end of Sec. 7.2C, we work with the magnetic field to ensure Hermiticity of the matrix representation.

Expanding the periodic function $p_\mathbf{K}(\mathbf{r})$ in the Bloch wave into a 3D Fourier series

$$p_\mathbf{K}(\mathbf{r}) = \sum_\mathbf{G} C_\mathbf{G} \exp(-j\,\mathbf{G}\cdot\mathbf{r}), \qquad (7.3\text{-}9)$$

we write the magnetic-field vector in the Bloch form

$$\mathbf{H}(\mathbf{r}) = p_\mathbf{K}(\mathbf{r})\exp(-j\mathbf{K}\cdot\mathbf{r})\widehat{\mathbf{e}} = \sum_\mathbf{G} C_\mathbf{G}\exp[-j(\mathbf{K}+\mathbf{G})\cdot\mathbf{r}]\,\widehat{\mathbf{e}}. \qquad (7.3\text{-}10)$$

For notational simplicity, the dependence of the Fourier coefficients $C_\mathbf{G}$ on the Bloch wavevector **K** is not explicitly indicated. Substituting (7.3-7) and (7.3-10) into (7.0-2), using the relation $\nabla\times\exp(-j\mathbf{K}\cdot\mathbf{r})\,\widehat{\mathbf{e}} = -j(\mathbf{K}\times\widehat{\mathbf{e}})\exp(-j\mathbf{K}\cdot\mathbf{r})$, and equating harmonic terms of the same spatial frequency yields

$$-\sum_{\mathbf{G}'}(\mathbf{K}+\mathbf{G})\times\left[(\mathbf{K}+\mathbf{G}')\times\widehat{\mathbf{e}}\right]\eta_{\mathbf{G}-\mathbf{G}'}\,C_{\mathbf{G}'} = \frac{\omega^2}{c_o^2}\,C_\mathbf{G}\,\widehat{\mathbf{e}}. \qquad (7.3\text{-}11)$$

Forming a dot product with $\widehat{\mathbf{e}}$ on both sides, and using the vector identity $\mathbf{A}\cdot(\mathbf{B}\times\mathbf{C}) = -(\mathbf{B}\times\mathbf{A})\cdot\mathbf{C}$ leads to

$$\sum_{\mathbf{G}'} F_{\mathbf{GG}'}\,C_{\mathbf{G}'} = \frac{\omega^2}{c_o^2}\,C_\mathbf{G}, \quad F_{\mathbf{GG}'} = \left[(\mathbf{K}+\mathbf{G})\times\widehat{\mathbf{e}}\right]\cdot\left[(\mathbf{K}+\mathbf{G}')\times\widehat{\mathbf{e}}\right]\eta_{\mathbf{G}-\mathbf{G}'}. \quad (7.3\text{-}12)$$

The Helmholtz differential equation has now been converted into a set of linear equations for the Fourier coefficients $\{C_\mathbf{G}\}$. Since $\eta(z)$ is real, $\eta_{\mathbf{G}-\mathbf{G}'} = \eta^*_{\mathbf{G}'-\mathbf{G}}$, and the matrix $F_{\mathbf{GG}'}$ is Hermitian. Hence, (7.3-12) represents an eigenvalue problem for a Hermitian matrix. For each Bloch wavevector **K**, the eigenvalues ω^2/c_o^2 provide multiple values of ω, which are used to construct the ω-**K** diagram and the photonic band structure. The eigenvectors $\{C_\mathbf{G}\}$ determine the periodic function $p_\mathbf{K}(\mathbf{r})$ of the Bloch wave.

Examples of Structures

Spherical holes on a diamond lattice. An example of a 3D photonic crystal that has been shown to exhibit a complete 3D photonic bandgap comprises air spheres embedded in a high-index material at the points of a diamond (face-centered cubic) lattice (see Fig. 7.3-4). The radii of the air spheres are sufficiently large such that the spheres overlap, thereby creating intersecting veins. The calculated band structure, shown in Fig. 7.3-5, has a relatively wide complete 3D photonic bandgap between the two lowest bands. Silicon photonic crystals with an inverse-opal structure have been fabricated by inserting silicon into the voids of a self-assembled opal template comprising close-packed silica spheres; these are connected by small "necks" formed during sintering. As a final step the silica template is removed.[†]

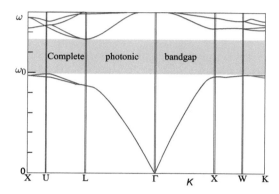

Figure 7.3-5 Calculated band structure of a 3D photonic crystal with a diamond (face-centered cubic) lattice of lattice constant a. The structure comprises air spheres of radius $0.325\,a$ embedded in a homogeneous material of refractive index $n = 3.6$. The complete photonic bandgap extends from approximately $\omega_0 = \pi c_o/a$ to $1.32\,\omega_0$ (see footnote on page 294).

Yablonovite. The first experimental observation of a complete 3D photonic bandgap was made by Eli Yablonovitch and colleagues in 1991 using a variant of the diamond lattice structure now known as **Yablonovite**.[‡] This slanted-pore structure is fabricated by drilling a periodic array of cylindrical holes at specified angles in a dielectric slab. Three holes are drilled at each point of a 2D triangular lattice at the surface of the slab; the directions of the holes are parallel to three of the axes of the diamond lattice, as shown in Fig. 7.3-6(a). This structure exhibits a complete 3D photonic bandgap with a gap-midgap ratio of 0.19 when the refractive index of the material is $n = 3.6$.

Woodpile. Another 3D photonic-crystal structure, which is simpler to fabricate, is made of a 1D periodic stack of alternating layers, each of which is itself a 2D photonic crystal. For example, the woodpile structure illustrated in Fig. 7.3-6(b) uses layers of parallel logs with a stacking sequence that repeats itself every four layers. The orientation of the logs in adjacent layers is rotated $90°$, and the logs are shifted by half the pitch every two layers. The resulting structure has a face-centered-tetragonal lattice symmetry. Fabricated using silicon micromachining, with a minimum feature size of 180 nm, this structure manifested a complete 3D photonic bandgap in the wavelength range $\lambda = 1.35$–$1.95\ \mu m$.[*]

[†] See A. Blanco, E. Chomski, S. Grabtchak, M. Ibisate, S. John, S. W. Leonard, C. Lopez, F. Meseguer, H. Miguez, J. P. Mondia, G. A. Ozin, O. Toader, and H. M. van Driel, Large-Scale Synthesis of a Silicon Photonic Crystal with a Complete Three-Dimensional Bandgap Near 1.5 Micrometres, *Nature*, vol. 405, pp. 437–440, 2000.

[‡] See E. Yablonovitch, T. J. Gmitter, and K. M. Leung, Photonic Band Structures: The Face-Centered Cubic Case Employing Non-Spherical Atoms, *Physical Review Letters*, vol. 67, pp. 2295–2298, 1991.

[*] See J. G. Fleming and S.-Y. Lin, Three-Dimensional Photonic Crystal with a Stop Band from 1.35 to 1.95 μm, *Optics Letters*, vol. 24, pp. 49–51, 1999.

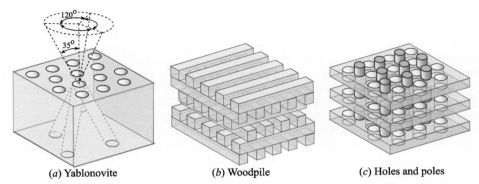

(a) Yablonovite (b) Woodpile (c) Holes and poles

Figure 7.3-6 (a) The Yablonovite photonic crystal is fabricated by drilling cylindrical holes through a dielectric slab. At each point of a 2D triangular lattice at the surface, three holes are drilled along directions that make an angle of 35° with the normal and are separated azimuthally by 120°. (b) The woodpile photonic crystal comprises alternating layers of parallel rods, with adjacent layers oriented at 90°. (c) The holes-and-poles structure consists of alternating layers of 2D periodic structures: a layer of parallel cylindrical holes on a hexagonal lattice, followed by a layer of parallel rods lined up to fit between the holes.

Holes and poles. Yet another example is the holes-and-poles structure illustrated in Fig. 7.3-6(c). Here, two complementary types of 2D-periodic photonic-crystal slabs are used: dielectric rods in air and air holes in a dielectric. Fabricated in silicon, this structure exhibited a stop-band for all tilt angles in the $\lambda = 1.15$–1.6 μm telecommunications band.[†]

Both the holes-and-poles and the woodpile structures offer the opportunity to introduce arbitrary **point defects**, such as a missing hole or rod, thereby providing means for fabricating devices such as photonic-crystal waveguides (Sec. 9.5), photonic-crystal resonators (Sec. 11.4D), and specially controlled light emitters[‡] (Sec. 18.5C). Indeed, the ability to insert a defect at will may well be the most valuable feature of 2D and 3D photonic structures since 1D periodic media serve admirably as omnidirectional reflectors.

Fabrication Methods

In the early 1990s, photonic crystals were fabricated using adaptations of conventional semiconductor nanofabrication techniques. A decade later, by the early 2000s, a host of new ("bottom-up" and "top-down") techniques for the fabrication of 3D photonic crystals had been developed and implemented with varying degrees of success. The most prominent of these are colloidal self-assembly, holographic lithography, and direct writing via 3D multiphoton microlithography.

Submicrometer colloidal spheres of uniform size tend to spontaneously assemble on a face-centered cubic lattice. The resulting material is a synthetic opal, which serves as a template that can be impregnated with a semiconductor material such as silicon. Subsequent removal of the template yields a 3D photonic crystal, known as inverse opal, that consists of the semiconductor material containing periodic spheres of air. A challenge in using this method is growing the initial opal without forming polycrystalline segments, which suffer from deleterious lattice defects.

[†] See M. Qi, E. Lidorikis, P. T. Rakich, S. G. Johnson, J. D. Joannopoulos, E. P. Ippen, and H. I. Smith, A Three-Dimensional Optical Photonic Crystal with Designed Point Defects, *Nature*, vol. 429, pp. 538–542, 2004.

[‡] See S. P. Ogawa, M. Imada, S. Yoshimoto, M. Okano, and S. Noda, Control of Light Emission by 3D Photonic Crystals, *Science*, vol. 305, pp. 227–229, 2004.

In the 3D holographic-lithography approach, multiple laser beams generate interference patterns that illuminate a film of photoresist. Highly exposed regions of the photoresist become insoluble and the unexposed regions are washed away. The result is a periodic structure of cross-linked polymer containing air-filled voids. The use of liquid-crystal spatial light modulators to govern the phase properties of the various beams offers (dynamically tunable) interference patterns of arbitrary form. Four beams, generated by a single laser, are sufficient to create any desired photonic band structure.

In 3D multiphoton microlithography, femtosecond laser pulses are delivered via a lens to a particular location in a specially designed transparent polymeric material. The laser power is set at a level that effects multiphoton polymerization only in the vicinity of the focal region of the lens, where the optical intensity is sufficiently high (see Sec. 14.5B). The light is able to reach that region without polymerizing the intervening material because its intensity lies below the polymerization threshold outside the focal region. The photonic-crystal structure is fabricated by moving the focal point of the lens to all desired locations, thereby writing the three-dimensional microstructure. The strong thresholding behavior of the polymerization nonlinearity serves to enhance the resolution of the process beyond the diffraction limit.

Novel variations accommodate the fabrication of electrically responsive and graded-index (GRIN) photonic crystals.

READING LIST

Layered and Periodic Media

O. Stenzel, *The Physics of Thin Film Optical Spectra: An Introduction*, Springer-Verlag, 2nd ed. 2015.

A. B. Shvartsburg and A. A. Maradudin, *Waves in Gradient Metamaterials*, World Scientific/Imperial College Press, 2013.

R. Kashyap, *Fiber Bragg Gratings*, Academic Press, 2nd ed. 2010.

Š. Višňovský, *Optics in Magnetic Multilayers and Nanostructures*, CRC Press, 2006.

P. Yeh, *Optical Waves in Layered Media*, Wiley, 2005.

M. Nevière and E. Popov, *Light Propagation in Periodic Media: Differential Theory and Design*, CRC Press, 2003.

M. Born and E. Wolf, *Principles of Optics*, Cambridge University Press, 7th expanded and corrected ed. 2002, Sec. 1.6.

W. C. Chew, *Waves and Fields in Inhomogeneous Media*, Van Nostrand Reinhold, 1990; IEEE Press, reprinted 1995.

A. Yariv and P. Yeh, *Optical Waves in Crystals: Propagation and Control of Laser Radiation*, Wiley, 1985, paperback reprint 2002.

L. Brillouin, *Wave Propagation in Periodic Structures: Electric Filters and Crystal Lattices*, Dover, 2nd ed. 1953, reprinted 2003.

Photonic Crystals

Q. Gong and X. Hu, *Photonic Crystals: Principles and Applications*, CRC Press/Taylor & Francis, 2013.

M. F. Limonov and R. M. De La Rue, *Optical Properties of Photonic Structures: Interplay of Order and Disorder*, CRC Press/Taylor & Francis, 2012.

D. W. Prather, S. Shi, A. Sharkawy, J. Murakowski, and G. J. Schneider, *Photonic Crystals: Theory, Applications, and Fabrication*, Wiley, 2009.

I. A. Sukhoivanov and I. V. Guryev, *Photonic Crystals: Physics and Practical Modeling*, Springer-Verlag, 2009.

M. Skorobogatiy and J. Yang, *Fundamentals of Photonic Crystal Guiding*, Cambridge University Press, 2009.

J. D. Joannopoulos, S. G. Johnson, J. N. Winn, and R. D. Meade, *Photonic Crystals: Molding the Flow of Light*, Princeton University Press, 1995, 2nd ed. 2008.

J.-M. Lourtioz, H. Benisty, V. Berger, J.-M. Gérard, D. Maystre, and A. Tchelnokov, *Photonic Crystals: Towards Nanoscale Photonic Devices*, Springer-Verlag, 2nd ed. 2008.

P. Markoš and C. M. Soukoulis, *Wave Propagation: From Electrons to Photonic Crystals and Left-Handed Materials*, Princeton University Press, 2008.

K. Sakoda, *Optical Properties of Photonic Crystals*, Springer-Verlag, 2nd ed. 2005.

K. Inoue and K. Ohtaka, eds., *Photonic Crystals: Physics, Fabrication and Applications*, Springer-Verlag, 2004.

S. Noda and T. Baba, eds., *Roadmap on Photonic Crystals*, Kluwer, 2003.

V. Kochergin, *Omnidirectional Optical Filters*, Kluwer, 2003.

Seminal and Review Articles

L. Nucara, F. Greco, and V. Mattoli, Electrically Responsive Photonic Crystals: A Review, *Journal of Materials Chemistry C*, vol. 3, pp. 8449–8467, 2015.

Q. Zhu, L. Jin, and Y. Fu, Graded Index Photonic Crystals: A Review, *Annalen der Physik*, vol. 527, pp. 205–218, 2015.

G. von Freymann, V. Kitaev, B. V. Lotsch, and G. A. Ozin, Bottom-Up Assembly Of Photonic Crystals, *Chemical Society Reviews*, vol. 42, pp. 2528–2554, 2013.

D. Xia, J. Zhang, X. He, and S. R. J. Brueck, Fabrication of Three-Dimensional Photonic Crystal Structures by Interferometric Lithography and Nanoparticle Self-Assembly, *Applied Physics Letters*, vol. 93, 071105, 2008.

T. Asano, B.-S. Song, Y. Akahane, and S. Noda, Ultrahigh-Q Nanocavities in Two-Dimensional Photonic Crystal Slabs, *IEEE Journal of Selected Topics in Quantum Electronics*, vol. 12, pp. 1123–1134, 2006.

Á. Blanco, P. D. García, D. Golmayo, B. H. Juárez, and C. López, Opals for Photonic Band-Gap Applications, *IEEE Journal of Selected Topics in Quantum Electronics*, vol. 12, pp. 1143–1150, 2006.

M. Qi, E. Lidorikis, P. T. Rakich, S. G. Johnson, J. D. Joannopoulos, E. P. Ippen, and H. I. Smith, A Three-Dimensional Optical Photonic Crystal with Designed Point Defects, *Nature*, vol. 429, pp. 538–542, 2004.

S. P. Ogawa, M. Imada, S. Yoshimoto, M. Okano, and S. Noda, Control of Light Emission by 3D Photonic Crystals, *Science*, vol. 305, pp. 227–229, 2004.

E. Yablonovitch, Photonic Crystals: Semiconductors of Light, *Scientific American*, vol. 285, no. 6, pp. 47–55, 2001.

Y. A. Vlasov, X. Z. Bo, J. C. Sturm, and D. J. Norris, On-Chip Natural Assembly of Silicon Photonic Bandgap Crystals, *Nature*, vol. 414, pp. 289–293, 2001.

S. G. Johnson and J. D. Joannopoulos, Block-Iterative Frequency-Domain Methods for Maxwell's Equations in a Planewave Basis, *Optics Express*, vol. 8, pp. 173–190, 2001.

A. Blanco, E. Chomski, S. Grabtchak, M. Ibisate, S. John, S. W. Leonard, C. Lopez, F. Meseguer, H. Miguez, J. P. Mondia, G. A. Ozin, O. Toader, and H. M. van Driel, Large-Scale Synthesis of a Silicon Photonic Crystal with a Complete Three-Dimensional Bandgap Near 1.5 Micrometres, *Nature*, vol. 405, pp. 437–440, 2000.

M. Campbell, D. N. Sharp, M. T. Harrison, R. G. Denning, and A. J. Turberfield, Fabrication of Photonic Crystals for the Visible Spectrum by Holographic Lithography, *Nature*, vol. 404, pp. 53–56, 2000.

H. B. Sun, S. Matsuo, and H. Misawa, Three-Dimensional Photonic Crystal Structures Achieved with Two-Photon-Absorption Photopolymerization of Resin, *Applied Physics Letters*, vol. 74, pp. 786–788, 1999.

J. G. Fleming and S.-Y. Lin, Three-Dimensional Photonic Crystal with a Stop Band from 1.35 to 1.95 μm, *Optics Letters*, vol. 24, pp. 49–51, 1999.

J. D. Joannopoulos, P. R. Villeneuve, and S. Fan, Photonic Crystals: Putting a New Twist on Light, *Nature*, vol. 386, pp. 143–149, 1997.

P. R. Villeneuve and M. Piché, Photonic Bandgaps in Periodic Dielectric Structures, *Progress in Quantum Electronics*, vol. 18, pp. 153–200, 1994.

S. John, Localization of Light, *Physics Today*, vol. 44, no. 5, pp. 32–40, 1991.

E. Yablonovitch, T. J. Gmitter, and K. M. Leung, Photonic Band Structures: The Face-Centered Cubic Case Employing Non-Spherical Atoms, *Physical Review Letters*, vol. 67, pp. 2295–2298, 1991.

S. John, Strong Localization of Photons in Certain Disordered Dielectric Superlattices, *Physical Review Letters*, vol. 58, pp. 2486–2489, 1987.

E. Yablonovitch, Inhibited Spontaneous Emission in Solid-State Physics and Electronics, *Physical Review Letters*, vol. 58, pp. 2059–2062, 1987.

PROBLEMS

7.1-2 **Beamsplitter Slab.** A lossless, dielectric slab of refractive index n and width d, oriented at $45°$ with respect to an incident beam, is used as a beamsplitter. Derive expressions for the transmittance and reflectance for TM polarization and sketch their spectral dependence. Compare the results with those expected for TE polarization, as provided in Example 7.1-6.

7.1-3 **Air Gap in Glass.** Determine the transmittance through a thin planar air gap of width $d = \lambda/2$ in glass of refractive index n. Assume (a) normal incidence, and (b) a TE wave incident at an angle greater than the critical angle. Can the wave penetrate (tunnel) through the gap?

7.1-4 **Multilayer Device in an Unmatched Medium.** The complex amplitude reflectance of a multilayer device is r_m when it is placed in a medium with refractive index n_1 matching its front layer. If the device is instead placed in a medium with refractive index n, show that the amplitude reflectance is $r = (r_b + r_m)/(1 + r_b r_m)$, where $r_b = (n - n_1)/(n + n_1)$ is the reflectance of the new boundary. Determine r in each of the following limiting cases: $r_b = 0$, $r_b = 1$, $r_m = 0$, and $r_m = 1$.

7.1-5 **Quarter-Wave Film: Angular Dependence of Reflectance.** Consider the quarter-wave antireflection coating described in Exercise 7.1-1. Derive an expression for the reflectance as a function of the angle of incidence.

7.1-6 **Transmittance of a Fabry–Perot Etalon.** The transmittance of a symmetric Fabry–Perot etalon was measured by using light from a tunable monochromatic light source. The transmittance versus frequency exhibits periodic peaks of period 150 MHz, each of width (FWHM) 5 MHz. Assuming that the medium within the resonator mirrors is a gas with $n = 1$, determine the length and finesse of the resonator. Assuming further that the only source of loss is associated with the mirrors, find their reflectances.

7.1-7 **Quarter-Wave and Half-Wave Stacks.** Derive expressions for the reflectance of a stack of N double layers of dielectric materials of equal optical thickness, $n_1 d_1 = n_2 d_2$, equal to $\lambda_o/4$ and $\lambda_o/2$.

7.1-8 **GaAs/AlAs Bragg Grating Reflector.** A Bragg grating reflector comprises N units of alternating layers of GaAs ($n_1 = 3.57$) and AlAs ($n_2 = 2.94$) of widths d_1 and d_2 equal to a quarter wavelength in each medium. The grating is placed in an extended GaAs medium. Calculate and plot the transmittance and reflectance of the grating as functions of N, for $N = 1, 2, \ldots, 10$, at a frequency equal to the Bragg frequency.

7.1-9 **Bragg Grating: Angular and Spectral Dependence of Reflectance.** Based on matrix algebra, determine the wave-transfer matrix and the reflectance of an N-layer alternating-layer dielectric Bragg grating. Use your program to verify the graphs presented in Fig. 7.1-12 and Fig. 7.1-13 for the spectral and angular dependence of the reflectance, respectively.

7.2-1 **Gap–Midgap Ratio.** Using a Fourier optics approach, determine the Bragg frequency and the gap–midgap ratio for the lowest bandgap of a 1D periodic structure comprising a stack of dielectric layers of equal optical thickness, with $n_1 = 1.5$ and $n_2 = 3.5$, and period $\Lambda = 2\ \mu\text{m}$. Assume that the wave travels along the axis of periodicity. Repeat the process for $n_1 = 3.4$ and $n_2 = 3.6$. Compare your results.

7.2-2 **Off-Axis Wave in 1D Periodic Medium.** Derive equations analogous to those provided in (7.2-24)–(7.2-28) for an off-axis wave traveling through a 1D periodic medium with a transverse wavevector k_x.

7.2-3 **Normal-to-Axis Wave in a 1D Periodic Medium.** Use the results of Prob. 7.2-2 to show that there are no bandgaps for a wave traveling along the lateral direction of a 1D periodic medium, i.e., for $K = 0$.

7.2-4 **Omnidirectional Reflector.** A periodic stack of double layers of dielectric materials with $n_1 d_1 = n_2 d_2$, $n_2 = 2n_1$ and $\Lambda = d_1 + d_2$ is to be used as an omnidirectional reflector in air. Plot the projected dispersion relation showing the light line for air (a diagram similar to Fig. 7.2-10). Determine the frequency range (in units of ω_B) for omnidirectional reflection.

CHAPTER 8

METAL AND METAMATERIAL OPTICS

8.1   SINGLE- AND DOUBLE-NEGATIVE MEDIA   306
     A.   Wave Propagation in SNG and DNG Media
     B.   Waves at Boundaries Between DPS, SNG, and DNG Media
    *C.   Hyperbolic Media

8.2   METAL OPTICS: PLASMONICS   320
     A.   Optical Properties of Metals
     B.   Metal–Dielectric Boundary: Surface Plasmon Polaritons
     C.   The Metallic Nanosphere: Localized Surface Plasmons
     D.   Optical Antennas

8.3   METAMATERIAL OPTICS   334
     A.   Metamaterials
     B.   Metasurfaces

*8.4   TRANSFORMATION OPTICS   343
     A.   Transformation Optics
     B.   Invisibility Cloaks

Paul Karl Ludwig Drude (1863–1906), a German physicist, worked toward integrating optics with Maxwell's electromagnetics. He forged a theory, commonly known as the Drude model, for describing the behavior of electrons in metals.

Viktor Georgievich Veselago (born 1929), a Russian physicist, established theoretically in the 1960s that materials whose electric permittivity and magnetic permeability were both negative would exhibit unexpected and unusual properties.

Sir John Pendry (born 1943), a British theoretical physicist, showed in 2000 that a slab of negative-index material acts as a lens with theoretically perfect focus. In 2006, he proposed the use of transformation optics for creating invisibility cloaks.

Fundamentals of Photonics, Third Edition. Bahaa E. A. Saleh and Malvin Carl Teich.
©2019 John Wiley & Sons, Inc. Published 2019 by John Wiley & Sons, Inc.

Visible light cannot propagate through highly conductive media such as metals. When a light beam crosses the boundary into such a medium, its intensity rapidly diminishes within a short distance known as the penetration depth, which can be substantially smaller than a wavelength. A metallic surface acts rather like a mirror from which light is fully reflected back into the contiguous dielectric medium whence it came.

In the previous chapters of this book, metallic components have indeed played the role of simple mirrors. It turns out, however, that metals *can* support light waves, provided that they travel along the boundaries of the metal, in regions confined to subwavelength dimensions. Such light waves travel along a metal surface in the form of a guided surface wave. It may propagate *on*, but not *in*, metal wires, and it may be guided by subwavelength metallic structures configured as integrated-photonic circuits. Such structures may also serve as resonators within which light may be confined, or from which light can scatter strongly at specific resonance frequencies.

Advances in nanotechnology permit such subwavelength metallic structures to be embedded and distributed within dielectric materials to form synthetic photonic materials known as **metamaterials**. Unavailable in nature, these materials have the merit that they may be endowed with highly useful optical properties. Metamaterials offer numerous novel applications and have come to play an important role in photonics.

Metals and metamaterials possess a number of unique optical characteristics:

- A metal–dielectric boundary can serve as a waveguide that supports an optical wave traveling along the boundary. Such a surface wave, known as a **surface plasmon polariton (SPP)**, is highly confined to the vicinity of the boundary and is accompanied by an electric-charge-density longitudinal wave (a plasma wave) of the same frequency that concomitantly travels along the metal surface. The tight confinement and short wavelength of SPPs can provide a significant increase in local field intensity. Biosensing applications are predicated on the sensitivity of the SPP to the properties of the surrounding dielectric media.

- Metallic structures of subwavelength dimensions (e.g., nanospheres) embedded in dielectric media support plasmonic oscillations at their boundaries. These oscillations, called **localized surface plasmons (LSPs)**, exhibit resonance when the excitation frequency matches the resonance frequency of the structure, which typically falls in the visible or ultraviolet region of the spectrum. As a result of resonantly enhanced absorption and scattering, such nanoparticles can exhibit intense colors in the visible, both in transmission and reflection. This metal-optics technology, known as **plasmonics**, finds use in applications ranging from fabricating stained glass to probing the dielectric properties of a host medium.

- Metallic nanostructures may be fabricated in the form of circuit elements and **nanoantennas** that are analogous to their radiowave and microwave counterparts, but that operate in the infrared and visible. This plasmonics technology seeks to couple the domains of highly integrated compact electronics (dimensions < 100 nm) and optical-frequency ultrafast photonics (bandwidths > 100 THz), and is expected to find use in applications such as intrachip interconnects.

- An array of metallic structures printed at the boundary between two dielectric media creates a **metasurface**, which exhibits unique optical properties that are dictated by the shape of the elements and the geometrical configuration of the array. Behaving much like arrays of optical antennas, metasurfaces bend light in unusual ways that do not obey the ordinary laws of reflection and refraction prevailing at dielectric boundaries.

- Materials with embedded metallic and dielectric structures of subwavelength dimensions, distributed at wavelength or subwavelength distances throughout the volume, exhibit novel electrical and magnetic properties that result from electric

charges and currents induced in the constituent conductive metal components. Such metamaterials may be engineered to exhibit remarkable optical properties such as **negative refractive index**, whereby a negative angle of refraction is imparted at a boundary with a conventional dielectric medium.
- Metamaterials may be designed with spatially varying (graded) properties that transform optical waves in unusual ways so that they can, for example, wrap around objects and render them invisible. A particularly intriguing application is **optical cloaking**.

The optical properties of metals and metamaterials are described by the electromagnetic theory of light introduced in Chapter 5, just as for dielectric media. The principal distinction is that the electric permittivity ϵ and the magnetic permeability μ may take on negative values for metals and metamaterials. Materials in which one of these parameters is negative are referred to as **single-negative (SNG) media**, while those in which both are negative are known as **double-negative (DNG) media**. Conventional lossless dielectric media are **double-positive (DPS) media**.

This Chapter

The chapter opens with an introductory section that examines some of the optical properties of SNG and DNG media (Sec. 8.1). The existence of surface guided waves at a DPS-SNG boundary is demonstrated, negative refraction at a DPS-DNG boundary is highlighted, and some of the potential applications of negative-refractive-index materials are described.

Section 8.2 provides an introduction to metal optics and plasmonics. It begins by presenting a physical model for the optics of metals, the **Drude model**, and shows that a metal can behave as a SNG material under certain conditions. SPP waves at a metal–dielectric boundary are examined. The Rayleigh scattering of light from a metal nanosphere is considered and contrasted with Rayleigh scattering from a dielectric sphere (as discussed in Sec. 5.6B). Optical antennas are briefly described.

Section 8.3 considers metamaterial optics. It deals with the various shapes and organizational patterns of metal objects that can be embedded in a dielectric medium, or on a dielectric surface, to endow it with particular macroscopic properties: $\epsilon(\omega)$ and $\mu(\omega)$ with real and imaginary components of desired signs. Of special interest is the design of negative-refractive-index media.

Section 8.4 introduces transformation optics, a mathematical tool that facilitates the design of special graded optical materials that guide light along desired trajectories. The goal is often to fabricate metamaterials that effect a desired result, such as creating trajectories that bypass an object so that it is rendered invisible (cloaked).

Since metamaterial optics, one of the principal topics of this chapter, involves the interaction of light with components of subwavelength (nanometer) scale, it lies in the domain of **nanophotonics**, also called **nano-optics**. Previous chapters, in contrast, have focused principally on the propagation of light through dielectric materials and components with dimensions substantially greater than the wavelength of light, which is the domain of **bulk optics**.

Upcoming chapters on guided-wave optics (Chapter 9), fiber optics (Chapter 10), and resonator optics (Chapter 11), have traditionally dealt with light propagating through structures whose dimensions are of the order of micrometers, a domain known as **micro-optics**. In recent years, however, the reach of nano-optics, via plasmonics, has penetrated these domains, as illustrated by the following examples: The ability of metal–dielectric structures to support surface plasmon polaritons at subwavelength spatial scales has led to the use of plasmonic waveguides, as reported in Chapter 9. Similarly, the ability of metallic structures of subwavelength dimensions to support localized surface plasmons at their boundaries with dielectrics has led to the development of plasmonic resonators and plasmonic nanolasers, considered in Chapters 11 and 18, respectively.

8.1 SINGLE- AND DOUBLE-NEGATIVE MEDIA

As described in Chapter 5, the propagation of an electromagnetic wave through a linear, isotropic medium is governed by the electric permittivity ϵ and magnetic permeability μ of the material. In general, these quantities are frequency-dependent and complex-valued. Wave properties, such as the propagation constant, velocity, attenuation coefficient, impedance, and dispersion relation, can be readily determined from ϵ and μ. The signs of the real and imaginary components of ϵ and μ at a given frequency govern the various regimes of wave propagation:

- For media in which μ is real and positive (indicating that there is neither magnetic absorption nor amplification), the wave-propagation characteristics depend on the signs of the real and imaginary parts of ϵ, as set forth in Chapter 5:
  - If ϵ is real, the medium exhibits neither dielectric absorption nor gain (it is lossless and passive), and waves propagate without attenuation, as described in Secs. 5.1–5.4.
  - In the presence of absorption (Sec. 5.5), ϵ is complex and $\mathrm{Im}\{\epsilon\} \equiv \epsilon''$ is negative, but $\mathrm{Re}\{\epsilon\} \equiv \epsilon'$ can be either positive or negative. For the resonant medium described in Sec. 5.5C, for example, the imaginary part of the electric susceptibility, $\mathrm{Im}\{\chi\} \equiv \chi''$, is negative (see Fig. 5.5-6) and therefore so too is $\epsilon'' = \epsilon_o \chi''$. The behavior of the real part of the susceptibility χ', also displayed in Fig. 5.5-6, is such that $\epsilon' = \epsilon_o(1 + \chi')$ is positive for frequencies below the resonance frequency but may be negative above the resonance frequency.
  - For an active medium that exhibits gain, such as a laser medium, χ'' is positive (gain represents negative absorption; see Sec. 15.1A). In this case, ϵ'' is positive, while ϵ' may be either positive or negative.
- Similarly, for media with real and positive ϵ, the magnetic properties, described by μ, dictate the nature of wave propagation. Magnetic media, including media with metal components that carry induced electric currents and generate magnetic fields, generally have complex values of μ, with real and imaginary parts that may be either positive or negative.
- In the most general case, the manner in which the signs of the real and imaginary components of ϵ and μ dictate the characteristics of wave propagation is more subtle.

In the exposition that follows, we confine ourselves principally to lossless and passive media, in which there is neither absorption nor gain, indicating that we are, for example, away from dielectric and magnetic resonances. Under these circumstances, both ϵ and μ are real, and their signs may be positive or negative at a given frequency. Four regimes ensue:

- **Double-positive (DPS)** materials (both ϵ and μ are positive). These materials are transparent and have positive refractive index. Ordinary dielectric media fall into this category.
- **Single-negative (SNG)** materials (either ϵ or μ is negative). These materials are opaque but they support optical surface waves at boundaries with DPS materials. As will become apparent in Sec. 8.2, metals such as gold and silver exhibit negative ϵ while maintaining positive μ in the infrared and visible spectral regions. Ferrites have positive ϵ and negative μ at microwave frequencies.

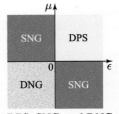

DPS, SNG, and DNG media are defined by the signs of ϵ and μ.

- **Double-negative (DNG)** materials (both ϵ and μ are negative). These materials, also called **left-handed media** for reasons that will become apparent in the sequel, are transparent and have negative refractive index, signifying that the application of Snell's law at a DPS-DNG boundary results in a negative angle of refraction. The ramifications of this property for optical components with multiple boundaries are most interesting. Metamaterials may be designed to exhibit such properties in specific frequency bands.

We begin by examining the optical properties of linear, lossless, passive media in the DPS, SNG, and DNG regimes. The ramifications of the presence of loss in a DNG medium are considered at the end of Sec. 8.1A. Initially we pay little heed to whether such media exist naturally or must be fabricated synthetically; this issue is addressed in Secs. 8.2 and 8.3.

A. Wave Propagation in SNG and DNG Media

The propagation of a monochromatic electromagnetic wave in a linear, homogeneous, and isotropic medium with electric permittivity ϵ and magnetic permeability μ is described in Secs. 5.3 and 5.4. Maxwell's equations (5.3-12)–(5.3-15), along with the Helmholtz equation (5.3-16), are applicable for arbitrary complex ϵ and μ, regardless of the signs of their real and imaginary parts.

For simplicity, we consider a monochromatic plane wave with electric and magnetic complex-amplitude vectors given by $\mathbf{E}(\mathbf{r}) = \mathbf{E}_0 \exp(-j\mathbf{k} \cdot \mathbf{r})$ and $\mathbf{H}(\mathbf{r}) = \mathbf{H}_0 \exp(-j\mathbf{k} \cdot \mathbf{r})$, respectively, and with wavevector $\mathbf{k}$. Maxwell's equations then require

$$\mathbf{k} \times \mathbf{H}_0 = -\omega\,\epsilon\,\mathbf{E}_0 \qquad (8.1\text{-}1)$$

$$\mathbf{k} \times \mathbf{E}_0 = \omega\,\mu\,\mathbf{H}_0, \qquad (8.1\text{-}2)$$

in accordance with (5.4-1)–(5.4-4). The associated wavenumber (magnitude of the vector $\mathbf{k}$) is

$$k = \omega\sqrt{\epsilon\mu}, \qquad (8.1\text{-}3)$$

and the impedance (ratio of the magnitudes of E_0 and H_0) is given by

$$\eta = \frac{\omega\mu}{k} = \sqrt{\frac{\mu}{\epsilon}}, \qquad (8.1\text{-}4)$$

as specified in (5.4-5). Equation (8.1-1) indicates that $\mathbf{E}_0$ is orthogonal to both $\mathbf{k}$ and $\mathbf{H}_0$, while (8.1-2) indicates that $\mathbf{H}_0$ is orthogonal to both $\mathbf{k}$ and $\mathbf{E}_0$, so that the three vectors form a mutually orthogonal set. For fixed orthogonal directions of the fields $\mathbf{E}_0$ and $\mathbf{H}_0$, the wavevector $\mathbf{k}$ is thus orthogonal to the plane defined by these field vectors, but its actual direction depends on the signs of ϵ and μ, as we shall see shortly.

Since k is in general complex, we write $k = \beta - j\gamma$, where β and γ are real, so that

$$\beta - j\gamma = \omega\sqrt{\epsilon\mu}. \qquad (8.1\text{-}5)$$

The propagation constant $\beta = \omega/c$ determines both the wave velocity $c = c_o/n$ and the refractive index n, whereas γ represents the field attenuation coefficient ($\gamma = \frac{1}{2}\alpha$, where α is the intensity attenuation coefficient; see Sec. 5.5A).

We now consider the implications of these equations for media in which ϵ and μ are real, where either or both may be negative.

DPS Medium

The double-positive (DPS) medium provides a simple and familiar benchmark. Both ϵ and μ are positive, so that k and η are real, whereupon

$$\gamma = 0, \qquad \beta = nk_o, \qquad n = \sqrt{\frac{\epsilon\,\mu}{\epsilon_o\,\mu_o}}, \qquad \eta = \sqrt{\frac{\mu}{\epsilon}}. \qquad (8.1\text{-}6)$$

As discussed in Sec. 5.4A, such media support transverse electromagnetic (TEM) waves for which the vectors $\mathbf{E}_0$, $\mathbf{H}_0$, and $\mathbf{k}$ are mutually orthogonal and form a right-handed system, as illustrated in Fig. 8.1-1(a). The Poynting vector $\mathbf{S} = \frac{1}{2}\mathbf{E}_0 \times \mathbf{H}_0^*$ points along the same direction as the wave vector $\mathbf{k}$, and the intensity of the wave (power flow per unit area) is given by $I = \text{Re}\{S\} = |E_0|^2/2\eta$.

SNG Medium

In a single-negative (SNG) medium, either ϵ or μ is negative so that k and η are both imaginary, whereupon (8.1-5) provides

$$\gamma = \omega\sqrt{|\epsilon||\mu|}, \qquad \beta = 0, \qquad \eta = j\sqrt{\frac{|\mu|}{|\epsilon|}}. \qquad (8.1\text{-}7)$$

These parameters correspond to an exponentially decaying field that behaves as $\exp(-\gamma z)$, where z is the propagation distance. Since $\beta = 0$, a SNG medium does not support propagating waves. The optical intensity is attenuated by the factor e^{-1} at a depth $d_p = 1/2\gamma = \lambda_o/4\pi\sqrt{|\epsilon/\epsilon_o|\,|\mu/\mu_o|}$. The quantity d_p is known as the **penetration depth** or **skin depth**.[†] The imaginary impedance η indicates that there is a $\pi/2$ phase shift between the electric and magnetic fields. Moreover, the Poynting vector $\mathbf{S} = \frac{1}{2}\mathbf{E}_0 \times \mathbf{H}_0^*$ is imaginary so that the intensity $I = \text{Re}\{S\} = 0$, indicating that no power is transported through such a medium.

DNG Medium

In a double-negative (DNG) medium, both ϵ and μ are negative so that (8.1-5) leads to $k = \omega\sqrt{|\epsilon||\mu|}$, which is real, whereupon

$$\gamma = 0, \qquad \beta = nk_o, \qquad n = -\sqrt{\frac{|\epsilon|\,|\mu|}{\epsilon_o\,\mu_o}}, \qquad \eta = \sqrt{\frac{|\mu|}{|\epsilon|}}, \qquad (8.1\text{-}8)$$

indicating that the refractive index is negative. Since $\gamma = 0$, the medium sustains wave propagation without attenuation. The choice of signs for the square roots in (8.1-8) is established by examining the directions of the vectors $\mathbf{E}_0$, $\mathbf{H}_0$, and $\mathbf{k}$, which may be determined directly from Maxwell's equations. For the DNG medium, (8.1-1) and (8.1-2) yield

$$\mathbf{k} \times \mathbf{H}_0 = \omega\,|\epsilon|\,\mathbf{E}_0 \qquad (8.1\text{-}9)$$

$$\mathbf{k} \times \mathbf{E}_0 = -\omega\,|\mu|\,\mathbf{H}_0. \qquad (8.1\text{-}10)$$

As with the DPS medium, the vectors $\mathbf{E}_0$, $\mathbf{H}_0$, and $\mathbf{k}$ are mutually orthogonal. However, the reversal of signs in (8.1-9) and (8.1-10), relative to those in (8.1-1) and (8.1-2)

[†] The penetration depth is sometimes defined as the distance over which the field, rather than the intensity, is attenuated by a factor e^{-1}, in which case $d_p' = 1/\gamma$.

for the DPS medium, is tantamount to exchanging the roles of the electric and magnetic fields.

Figure 8.1-1 illustrates the directions of flow of the wavefronts and power in DPS and DNG media. Comparing Figs. 8.1-1(a) and (c), it is apparent that $\mathbf{E}_0$, $\mathbf{H}_0$, and $\mathbf{k}$ in a DPS medium form the usual right-handed set of vectors, whereas in a DNG medium they form a left-handed set (the medium is then said to be **left-handed**). This has profound implications since it signifies that in a DNG medium the complex Poynting vector $\mathbf{S} = \frac{1}{2}\mathbf{E}_0 \times \mathbf{H}_0^*$ introduced in (5.3-10) is antiparallel to the wavevector $\mathbf{k}$. Since the impedance η positive, the wavenumber k is taken to be negative, and therefore so too is the refractive index n, as stipulated in (8.1-8). The DNG material is therefore a **negative-index material (NIM)**. The implications of negative refractive index will be considered in Sec. 8.1B.

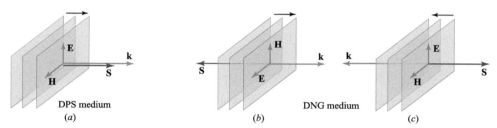

Figure 8.1-1 (a) Plane wave propagating in an ordinary double-positive (DPS) medium. The vectors $\mathbf{E}$, $\mathbf{H}$, and $\mathbf{k}$ form a right-handed set and the wavefronts travel in the same direction as the power flow. (b) Plane wave propagating in a double-negative (DNG) medium. The vectors $\mathbf{E}$, $\mathbf{H}$, and $\mathbf{k}$ form a left-handed set and the wavefronts travel in a direction opposite to that of the power flow. (c) Equivalent representation to that depicted in (b), obtained by effecting a 90° rotation about the horizontal ($\mathbf{S}$ and $\mathbf{k}$) axis that renders $\mathbf{E}$ vertical, followed by a 180° rotation about the new vertical ($\mathbf{E}$) axis.

Media with Complex ϵ and μ

The prospects of ϵ and μ both being *real* and *negative* at some frequency are remote. For example, if both parameters are described by a resonant-medium model, such as that considered in Sec. 5.5C, then throughout the frequency range over which the real part is negative, the imaginary part cannot be zero, by virtue of the Kramers–Kronig relations (see Sec. 5.5B and Sec. B.1 of Appendix B). It turns out, however, that left-handedness, and thus negative refractive index, can be exhibited in conjunction with absorption.

For absorptive media, $\epsilon = \epsilon' + j\epsilon''$ and $\mu = \mu' + j\mu''$ are complex with negative imaginary parts ϵ'' and μ'', respectively. We now demonstrate that if the real parts, ϵ' and μ', are both negative, the medium is indeed left-handed even for nonvanishing ϵ'' and μ''. As previously, consider a wave $\exp(-j\beta z)\exp(-\gamma z)$ with positive γ so that it decays along the $+z$ direction. The propagation constant β may be obtained by writing (8.1-5) in the form $(\beta - j\gamma)^2 = \omega^2(\epsilon' + j\epsilon'')(\mu' + j\mu'')$ and then matching the imaginary parts to obtain $2\gamma\beta = \omega^2(-\mu''\epsilon' - \epsilon''\mu')$. If both ϵ' and μ' are negative, then β is negative, and so too is the refractive index. The wavefront then travels in the $-z$ direction, opposite to the direction of decay. We now proceed to show that the power flow is always in the same direction as the power decay, so that the direction of the wavevector is opposite to that of the power flow and the medium is left-handed.

Power flow is determined by the Poynting vector $\text{Re}\{\frac{1}{2}\mathbf{E} \times \mathbf{H}^*\}$. Since $E_0 = \eta H_0$, where $\eta = \sqrt{\mu/\epsilon}$ is the characteristic impedance, and since $\text{Re}\{\eta\}$ is always positive for passive media, the power flow must be in the $+z$ direction for the field configuration shown in Fig. 8.1-1(c). It can be explicitly shown that $\text{Re}\{\eta\} > 0$ by writing $\arg\{\eta\} = \frac{1}{2}(\arg\{\mu\} - \arg\{\epsilon\})$; since ϵ'' and μ'' are both negative, we have $\pi < \arg\{\epsilon\} < 2\pi$ and $\pi < \arg\{\mu\} < 2\pi$ whereupon $-\frac{1}{2}\pi < \arg\{\eta\} < \frac{1}{2}\pi$ so that $\text{Re}\{\eta\} > 0$.

Although the condition that the real part of both ϵ and μ be negative is *sufficient* for achieving left handedness, it is not *necessary*. It is possible for left-handedness to be exhibited for an absorptive medium with only one of the two parameters, ϵ' or μ', being negative, indicating that the class of left-handed media transcends that of double-negative media. The definitive *necessary and sufficient* condition for left-handedness turns out to be[†]

$$\epsilon'/|\epsilon| + \mu'/|\mu| < 0. \tag{8.1-11}$$

Materials for which both ϵ and μ are real, but only one is negative, do not satisfy (8.1-11) and therefore cannot be left-handed. Nor do media for which one of the material parameters, ϵ or μ, is real and positive, whatever the real and imaginary values of the other. It is clear, then, that nonmagnetic materials cannot be left-handed.

B. Waves at Boundaries Between DPS, SNG, and DNG Media

We now consider waves at boundaries between DPS and SNG media, and between DPS and DNG media.

Reflection at a DPS-SNG Boundary

A wave from an ordinary DPS medium that impinges on its boundary with a SNG medium is fully reflected since it cannot propagate through the latter. Since the impedance of the DPS medium is real, and that of the SNG medium is imaginary, the magnitude of the reflection coefficient must be unity (see Sec. 6.2). Such reflection is accompanied by the creation of an evanescent field in the SNG medium; this bears a resemblance to total internal reflection at the boundary between two dielectric (DPS) media. However, total internal reflection occurs only for angles of incidence greater than the critical angle whereas reflection at the DPS–SNG boundary is total regardless of the angle of incidence, as illustrated in Fig. 8.1-2. Reflection at the DPS–SNG boundary is therefore more akin to that at a perfect mirror.

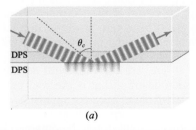
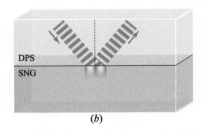

Figure 8.1-2 (a) Total internal reflection at a boundary between two DPS media takes place for angles of incidence greater than the critical angle θ_c. (b) Total reflection at a DPS–SNG boundary occurs for all angles of incidence. In both cases, an evanescent wave is created in the vicinity of the boundary.

[†] See R. A. Depine and A. Lakhtakia, A New Condition to Identify Isotropic Dielectric–Magnetic Materials Displaying Negative Phase Velocity, *Microwave and Optical Technology Letters*, vol. 41, pp. 315–316, 2004.

Surface Waves at a DPS-SNG Boundary

In the limit of grazing incidence (angle of incidence approaching 90°) at a DPS–SNG boundary, a **surface wave** may be created that propagates along the boundary but is evanescent on both sides of it. As a specific example, consider the boundary between a medium with positive electric permittivity ϵ_1 (medium 1) and another with negative permittivity ϵ_2 (medium 2). Both media are assumed to have the same (positive) magnetic permeability μ and both are lossless and passive so that ϵ_1, ϵ_2, and μ are real.

We proceed to demonstrate that such a DPS–SNG boundary can support a surface guided wave that travels along the boundary without changing its shape, as illustrated in Fig. 8.1-3(a). We assume that the wave is a TM wave, for which the magnetic field is parallel to the boundary and orthogonal to the direction of propagation. Each of the three field components H_x, E_y, and E_z varies as

$$\exp(-\gamma_1 y)\exp(-j\beta z), \quad y > 0 \text{ (medium 1)} \tag{8.1-12}$$

$$\exp(+\gamma_2 y)\exp(-j\beta z), \quad y < 0 \text{ (medium 2)}, \tag{8.1-13}$$

where β is a common propagation constant and γ_1 and γ_2 are positive field extinction coefficients. To satisfy the Helmholz equation (5.3-16) in each medium, we must have

$$-\gamma_1^2 + \beta^2 = \omega^2 \mu \epsilon_1, \qquad -\gamma_2^2 + \beta^2 = \omega^2 \mu \epsilon_2. \tag{8.1-14}$$

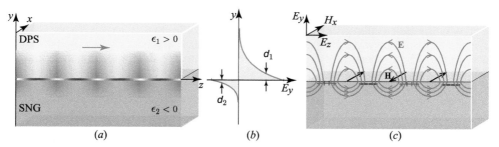

Figure 8.1-3 (a) Schematic representation of an optical surface wave traveling along the boundary between a medium of positive electric permittivity (DPS medium, such as a dielectric material) and another of negative permittivity (SNG medium, such as a metal below the plasma frequency). The associated longitudinal surface-charge wave is shown in dark blue; the width of each segment represents its wavelength, which is given by $2\pi/\beta$. (b) The electric field E_y as a function of the distance y from the boundary. (c) Electric and magnetic field lines and the associated charge. The plasmon wave penetrates to a depth d_1 and d_2 in the DPS and SNG media, respectively, and extends a distance d_b along the boundary.

The amplitudes of the three field components in each medium are related by Maxwell's equations, and the amplitudes in different media are related by the boundary conditions. Since the H_x component must be continuous, this component has a common amplitude, say H_0, in both media. Maxwell's equations (8.1-1) dictate that the amplitudes of the E_y components in the two media are $(-\beta/\omega\epsilon_1)H_0$ and $(-\beta/\omega\epsilon_2)H_0$. The condition that $D_y = \epsilon E_y$ is continuous at the boundary is therefore automatically satisfied. The amplitudes of the E_z components are $(-\gamma_1/j\omega\epsilon_1)H_0$ in medium 1 and $(\gamma_2/j\omega\epsilon_2)H_0$ in medium 2. Since the E_z components must be continuous at the boundary, it follows that

$$-\frac{\gamma_1}{\epsilon_1} = \frac{\gamma_2}{\epsilon_2}. \tag{8.1-15}$$

Because γ_1 and γ_2 are positive, this can only be attained if ϵ_1 and ϵ_2 have opposite signs. We therefore conclude that such a surface wave cannot exist at the boundary between two media with positive electric permittivities, but may exist if the media have electric permittivities of opposite sign.

Since ϵE_y is continuous and ϵ changes sign at the boundary, E_y must also reverse sign at the boundary, as shown in Fig. 8.1-3(b). This in turn requires the existence of surface electric charge in the form of a charge-density longitudinal wave that oscillates at the optical frequency ω. The field lines and the charge distribution are illustrated in Fig. 8.1-3(c). The combination of the charge-density wave and the optical wave is known as a **surface plasmon polariton (SPP)**.[†]

The properties of the SPP surface wave may be deduced from γ_1 and γ_2, in accordance with (8.1-14):

$$\beta = n_b k_o, \qquad n_b = \sqrt{\frac{\epsilon_b}{\epsilon_o}}, \qquad \epsilon_b = \frac{\epsilon_1 \epsilon_2}{\epsilon_1 + \epsilon_2}, \qquad (8.1\text{-}16)$$

$$\gamma_1 = \sqrt{\frac{-\epsilon_1^2}{\epsilon_o(\epsilon_1 + \epsilon_2)}}\, k_o, \qquad \gamma_2 = \sqrt{\frac{-\epsilon_2^2}{\epsilon_o(\epsilon_1 + \epsilon_2)}}\, k_o, \qquad \begin{array}{c}(8.1\text{-}17)\\ \text{SPP Wave}\\ \text{DPS-SNG Boundary}\end{array}$$

where $k_o = \omega\sqrt{\epsilon_o \mu}$ is the free-space wavenumber; and n_b and ϵ_b are, respectively, the refractive index and electric permittivity associated with the SPP (the subscript "b" signifies "boundary"). For the surface wave to be a traveling wave, β must be real, which requires that ϵ_b be positive. This is possible only if $|\epsilon_2| > \epsilon_1$, in which case the SPP surface wave exists. This same condition also renders γ_1 and γ_2 positive, as required. The properties of the surface wave, summarized below, are substantially influenced by the ratio $|\epsilon_2|/\epsilon_1$:

- The velocity is c_o/n_b, and the propagation wavelength, called the plasmon wavelength, is λ_o/n_b. If $|\epsilon_2| \approx \epsilon_1$, then n_b is large, the wave is slow, and the plasmon wavelength is much smaller than the free-space wavelength λ_o.

- The field extinction coefficient in the SNG medium, γ_2, is greater than that in the DNG medium, γ_1, by the factor $|\epsilon_2|/\epsilon_1$, which is greater than unity. The penetration depth in the SNG medium, $d_2 = 1/2\gamma_2$, is therefore always smaller than that in the DPS medium, $d_1 = 1/2\gamma_1$, by the same factor, as depicted in Fig. 8.1-3. If $|\epsilon_2| \approx \epsilon_1$, then $\epsilon_1 + \epsilon_2$ is small and both penetration depths are significantly smaller than the free-space wavelength λ_o so that the SPP is highly confined to the boundary. This is a remarkable property that can be exploited in many applications.

- The optical power flow in the two media may be determined from $\text{Re}\{\mathbf{S}\}$, where $\mathbf{S} = \frac{1}{2}\mathbf{E} \times \mathbf{H}^*$ is the complex Poynting vector [see (5.3-10)]. Since the components H_x and E_z are out of phase by $90°$, there is no power flow in the y direction. The power flows in the z and $-z$ directions in the DPS and SNG media, respectively, with intensities given by the magnitude of $\text{Re}\{\mathbf{S}\}$:

$$I_1(y) = \frac{\beta}{2\omega\epsilon_1}|H_0|^2 \exp(-2\gamma_1 y), \quad I_2(y) = \frac{\beta}{2\omega|\epsilon_2|}|H_0|^2 \exp(+2\gamma_2 y). \quad (8.1\text{-}18)$$

[†] A plasma is a collection of positive ions and free electrons whose overall net charge is approximately zero. A plasmon is a quasiparticle (a quantum) of the plasma oscillation, just as a photon is a quantum of the electromagnetic field. A polariton is a quasiparticle that results from the coupling of the electromagnetic field (photons) with the electric or magnetic excitation of a material. Surface plasmon polaritons result from the coupling of photons with surface plasmons.

The powers in the two media (areas under the $I_1(y)$ and $I_2(y)$ distributions) are:

$$P_1 = \frac{\beta}{4\omega\epsilon_1\gamma_1}|H_0|^2, \qquad P_2 = \frac{\beta}{4\omega|\epsilon_2|\gamma_2}|H_0|^2, \qquad (8.1\text{-}19)$$

so that the net power flow $P_1 - P_2$ is proportional to $[(\epsilon_1\gamma_1)^{-1} - (|\epsilon_2|\gamma_2)^{-1}]$, which, in view of (8.1-15), is proportional to $(\epsilon_2^2 - \epsilon_1^2)$. Therefore, in the limit of $|\epsilon_2| \approx \epsilon_1$, the net power flow approaches zero.

EXAMPLE 8.1-1. *SPP Wave.* The boundary between two media with equal (positive) magnetic permeabilities $\mu_1 = \mu_2$, and with permittivities $\epsilon_1 = 1.41\,\epsilon_o$ and $\epsilon_2 = -47\,\epsilon_o$, supports a SPP wave at a free-space wavelength $\lambda_o = 1000$ nm with the following characteristics:

$$n_b = 1.206, \quad \lambda_o/n_b = 829.4 \text{ nm}, \quad d_1 = 381.1 \text{ nm}, \quad d_2 = 11.43 \text{ nm}.$$

Summary: SPP Waves at a DPS-SNG Boundary

If the condition $-\epsilon_2 > \epsilon_1$ is satisfied, the boundary between a DPS medium and a SNG medium with negative permittivity supports a TM optical surface wave along with an associated longitudinal surface-charge wave; the combination is a surface plasmon polariton (SPP). Because the penetration depths of the SPP into the two media are substantially smaller than an optical wavelength, the SPP is tightly confined to the boundary, resulting in a significant increase in local field intensity. The SPP has the distinct merit that it can be manipulated at the nanometer spatial scale while oscillating at an optical frequency. Using a similar analysis, it can be shown that the boundary between a DPS medium and a SNG medium with negative permeability supports a TE surface wave if the condition $-\mu_2 > \mu_1$ is satisfied. It can also be shown, as illustrated diagrammatically in Fig. 8.1-4, that the boundary between two SNG media may support a surface wave if ϵ_1 and ϵ_2 are of opposite signs, μ_1 and μ_2 are also of opposite signs, and a number of other conditions on the magnitudes of these parameters are satisfied. Surface waves are not supported at the boundary between two DPS media nor at the boundary between two DNG media.

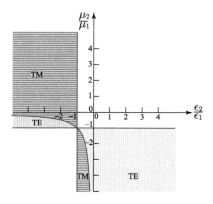

Figure 8.1-4 The boundary between two media with real permittivities and permeabilities can support TM or TE surface waves for values of the ratios ϵ_2/ϵ_1 and μ_2/μ_1 that fall in the shaded regions. Example 8.1-1 corresponds to $\mu_2/\mu_1 = 1$ and $\epsilon_2/\epsilon_1 < -1$.

SPP at boundary between DPS medium and lossy SNG medium. The result for a lossless SNG medium may be extended to a lossy medium by taking ϵ_2 to be complex ($\epsilon_2 = \epsilon_2' + j\epsilon_2''$), with negative real part ϵ_2', while maintaining $\mu_1 = \mu_2$ and ϵ_1 real and positive. Equations (8.1-16) and (8.1-17) for the parameters of the SPP wave in the lossless medium remain valid with slight modifications: the parameters β, ϵ_b, γ_1, and γ_2 become complex. Following the approach used in writing (5.5-5) and (8.1-5), we rewrite (8.1-16) as

$$n_b - j\frac{\gamma_b}{k_o} = \sqrt{\frac{\epsilon_b}{\epsilon_o}}, \qquad \epsilon_b = \frac{\epsilon_1 \epsilon_2}{\epsilon_1 + \epsilon_2}, \qquad (8.1\text{-}20)$$

where γ_b is the amplitude attenuation coefficient of the traveling SPP wave. It is clear that the SPP-wave refractive index then becomes $n_b = \text{Re}\{\sqrt{\epsilon_b/\epsilon_o}\}$. Equation (8.1-20) can be used to calculate the velocity of the SPP wave c_o/n_b, the plasmon wavelength λ_o/n_b, the intensity attenuation coefficient $\alpha_b = 2\gamma_b$, and the propagation length $d_b = 1/\alpha_b = 1/2\gamma_b$. Equation (8.1-17) can be used to calculate the complex parameters γ_1 and γ_2, from which the penetration depths on both sides of the boundary, $d_1 = 1/2\,\text{Re}\{\gamma_1\}$ and $d_2 = 1/2\,\text{Re}\{\gamma_2\}$, may be determined. The dimensions of the plasmon wave are d_1 and d_2 in the transverse direction, and d_b along the boundary.

EXAMPLE 8.1-2. *SPP at Gold-Si_3N_4 Interface.* At a free-space wavelength $\lambda_o = 1000$ nm, the permittivities of Si_3N_4 and gold are $\epsilon_1 = 1.41\,\epsilon_o$ and $\epsilon_2 = (-47 + j3.4)\,\epsilon_o$, respectively. These values are identical to those set forth in Example 8.1-1 except that we now accommodate a small imaginary component of ϵ_2. An SPP wave propagating at the gold-Si_3N_4 interface thus has approximately the same values of n_b, d_1, and d_2 as those cited in Example 8.1-1. The propagation length, on the other hand, is determined by the imaginary component of ϵ_2 via

$$d_b = 1/2\gamma_b = \lambda_o/4\pi\,\text{Im}\{\sqrt{\epsilon_b/\epsilon_o}\}. \qquad (8.1\text{-}21)$$

With the help of (8.1-20) we obtain

$$\epsilon_b/\epsilon_o = 1.453 + j\,0.003234 \quad \text{so that} \quad \sqrt{\epsilon_b/\epsilon_o} = 1.206 + j0.001341,$$

which leads to $d_b = 59.3\,\mu\text{m}$ for $\lambda_o = 1\,\mu\text{m}$.

Negative Refraction at a DPS-DNG Boundary

The refraction of light at the boundary between two ordinary dielectric (DPS) media obeys Snell's law, $n_1 \sin\theta_1 = n_2 \sin\theta_2$, which results from matching the components of the wavevectors $\mathbf{k}_1$ and $\mathbf{k}_2$ along the direction of the boundary [Figs. 8.1-5(a) and 2.4-2]. If one of the media, say medium 2, is instead a DNG medium with negative refractive index n_2 then we have $n_1 \sin\theta_1 = -|n_2|\sin\theta_2$, which reveals that *the angle of refraction θ_2 must be negative and the refracted and incident rays both lie on same side of the normal to the boundary.* This outcome also can be understood as arising from matching the components of the wavevectors $\mathbf{k}_1$ and $\mathbf{k}_2$ along the direction of the boundary [Fig. 8.1-5(b)].

It is thus clear that the optics of planar boundaries and lenses is altered significantly when a DPS medium is replaced by a DNG medium. Indeed, a convex lens of DNG material behaves like a concave lens of DPS material, and vice-versa. Also unexpected is the observation that a planar boundary between positive- and negative-index materials possesses focusing power, as illustrated in Fig. 8.1-6(a) for the special case $n_2 = -n_1$, which provides $\theta_2 = -\theta_1$. The planar boundary then acts on optical rays in the same way as does a convex spherical boundary between two DPS media, as depicted in Fig. 1.2-13: it forms an uninverted image with unity magnification. Moreover, for DPS and DNG media with permittivities and permeabilities that have the same magnitudes ($\epsilon_2 = -\epsilon_1$ and $\mu_2 = -\mu_1$), the impedances $\eta_1 = \sqrt{\mu_1/\epsilon_1}$ and

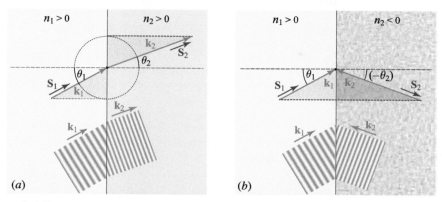

Figure 8.1-5 (a) Refraction at the boundary between two positive-index media. The directions of $\mathbf{S}_2$ and $\mathbf{k}_2$ are the same. (b) Refraction at the boundary between positive- and negative-index media. The directions of $\mathbf{S}_2$ and $\mathbf{k}_2$ are opposite. In both cases, the projections of the wavevectors $\mathbf{k}_1$ and $\mathbf{k}_2$ along the boundary are equal in magnitude and parallel in direction.

$\eta_2 = \sqrt{\mu_2/\epsilon_2}$ are of equal magnitude and sign, so that no reflection occurs at the boundary, at any inclination, regardless of the polarization.

NIM Slab as a Near-Field Imaging System

A slab of DNG material (NIM) with parameters $\epsilon = -\epsilon_o$, $\mu = -\mu_o$, $n = -1$, and $\eta = \eta_o$, located in free space, acts as a lens. As illustrated in Fig. 8.1-6(b), each of the two DPS-DNG boundaries has focusing power so that one image is formed inside the slab and another beyond it. For a slab of width d_0, the imaging equation is simply $d_1 + d_2 = d_0$. Since the impedances match at the boundaries, no reflection occurs and the NIM slab has unity transmittance.

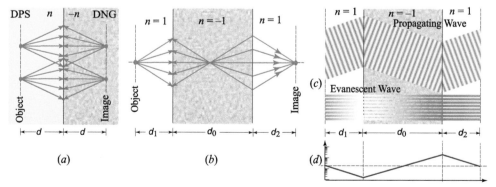

Figure 8.1-6 (a) Focusing of rays by a boundary between DPS and DNG media with refractive indices of equal magnitudes. The boundary forms an upright image inside the DNG medium. (b) Image formation using a slab with negative index $n = -1$ in free space. Each boundary has focusing power so that images are formed both inside and outside the slab. (c) Transmission of a propagating wave (spatial frequency smaller than the inverse wavelength), and an evanescent wave (spatial frequency greater than the inverse wavelength), through the slab. (d) The amplitude of an evanescent wave associated with a harmonic component of an object distribution with spatial frequency greater than the inverse wavelength is attenuated in free space but is amplified by the DNG medium so that its amplitude is restored in the image plane. A semilogarithmic sketch is provided.

As an imaging system, the NIM slab also has the remarkable property of forming an image of the object field distribution with subwavelength resolution, i.e., it offers imaging details finer than the wavelength (i.e., spatial frequencies greater than the inverse wavelength). As explained in Sec. 4.4D, imaging systems that make use of ordinary optics cannot transmit these high spatial frequencies since they correspond to attenuated evanescent waves. This remarkable property of the NIM slab means that, in principle, it is a "perfect lens" (also called a "superlens").[†]

The NIM slab offers resolution that exceeds the diffraction limit, as we now demonstrate with the help of the Fourier-optics approach described in Chapter 4. Referring to the illustration provided in Fig. 8.1-6(c), the transfer function for the propagation of a wave a distance d_1 through free space is

$$H_1(\nu_x, \nu_y) = \exp(-jk_z d_1), \quad k_z = \sqrt{k_o^2 - k_x^2 - k_y^2} = 2\pi\sqrt{\lambda^{-2} - \nu_x^2 - \nu_y^2},$$
(8.1-22)

where $(\nu_x, \nu_y) = (k_x/2\pi, k_y/2\pi)$ are the spatial frequencies [see (4.1-9)]. Similarly, propagation a distance d_0 through a DNG medium is described by the transfer function $H_0(\nu_x, \nu_y) = \exp(-jk'_z d_0)$, where $k'_z = -k_z$. The third segment of the imaging system portrayed in Fig. 8.1-6(c) is propagation a distance d_2 through free space, which has the transfer function $H_2(\nu_x, \nu_y) = \exp(-jk_z d_2)$. The overall transfer function of this imaging system is thus the product $H = H_1 H_0 H_2$. When the imaging equation $d_1 + d_2 = d_0$ is satisfied, H becomes unity, which reveals that the system is an all-pass spatial filter and hence provides "perfect" imaging.

However, these three transfer functions have distinctly different behaviors for spatial frequencies that lie below and above λ^{-1} (i.e., for spatial frequencies smaller and larger than those for which $k_x^2 + k_y^2 = k_o^2$; see Fig. 4.1-11). Below λ^{-1}, the quantities under the square-root signs in (8.1-22) are positive so that all transfer functions are phase factors. In this domain, the phase of the function H_0 has a sign opposite to those of the free-space functions H_1 and H_2 so that the NIM phase factor compensates the phase shifts introduced by the two stretches of free space. The net result is a propagating wave [Fig. 8.1-6(c)].

For spatial frequencies larger than λ^{-1}, in contrast, the wavevector components[‡] k_z and $k'_z = -k_z$ become imaginary, $k_z = -j\sqrt{k_x^2 + k_y^2 - k_o^2} = -j2\pi\sqrt{\nu_x^2 + \nu_y^2 - \lambda^{-2}}$, whereupon

$$H_1 = \exp(-\gamma d_1), \quad H_2 = \exp(-\gamma d_2), \quad H_0 = \exp(+\gamma d_0), \quad \gamma = 2\pi\sqrt{\nu_x^2 + \nu_y^2 - \lambda^{-2}},$$
(8.1-23)

where γ is real. The factors H_1 and H_2 then represent attenuated evanescent waves, whereas H_0 represents an amplified evanescent wave. Consequently, the high spatial frequencies that are severely attenuated by propagation through free space, both before and after the slab, are amplified by an equal factor in the DNG medium, and are therefore fully restored [Fig. 8.1-6(d)]. Amplification of the evanescent wave in the NIM is not inconsistent with conservation of energy.

While the perfect restoration of an evanescent field that has diminished significantly as a result of exponential decay is possible in principle, slight energy dissipation in the NIM slab (evidenced by a small imaginary part of ϵ or μ) may thwart the restoration process. The distance from the object to the slab, the thickness of the slab, and the overall distance between the object and image planes, must be small in comparison with the wavelength, particularly if the spatial frequencies to be recovered are much

---

[†] See J. B. Pendry, Negative Refraction Makes a Perfect Lens, *Physical Review Letters*, vol. 85, pp. 3966–3969, 2000.

[‡] The choice of the minus sign for k_z, which is allowed by virtue of (4.1-3), ensures consistency with the development provided in Sec. 4.1B.

higher than the inverse wavelength. This requirement means that the NIM slab is a near-field imaging system.

Far-field imaging with subwavelength resolution. However, evanescent waves restored in the near field of the DNG slab may be converted into propagating waves, which can then be used to produce an image in the far field. This conversion may be implemented by making use of a periodic element of high spatial frequency, such as a nanoscale corrugation on the exit surface of the slab, as illustrated in Fig. 8.1-7. When an evanescent wave $\exp(-jk_x x)$ of high spatial frequency k_x is modulated by a harmonic function $\exp(jqx)$ of high spatial frequency q, the result is a propagating wave $\exp[-j(k_x - q)x]$ of lower spatial frequency $|k_x - q|$, provided that $|k_x - q| < k_o$. The far-field image formed by the downconverted spatial frequencies may, in principle, be processed to obtain a perfect replica of the original spatial distribution.

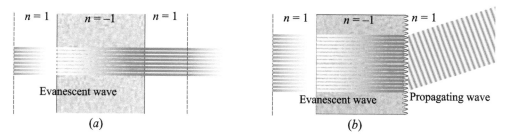

Figure 8.1-7 (a) An evanescent wave restored by a DNG slab attenuates in air beyond the near field. (b) An evanescent wave restored by a NIM slab can be converted into a propagating wave by making use of a periodic surface (grating) at the boundary.

*C. Hyperbolic Media

Anisotropic media may have permittivity and permeability tensors with positive or negative principal values. The designation of the medium as DPS, SNG, or DNG is then dependent on the direction of propagation of the wave as well as on its polarization. To demonstrate one facet of this rich behavior, we consider media endowed with isotropic magnetic properties and positive permeability μ, but with anisotropic dielectric properties.

Wave propagation in anisotropic media was described in Sec. 6.3 for media with positive principal values of the electric permittivity tensor $\boldsymbol{\epsilon}$, namely ϵ_1, ϵ_2 and ϵ_3. If these parameters take on mixed signs instead, wave propagation exhibits a number of unusual properties. For simplicity, we consider the special case of a uniaxial medium with $\epsilon_1 = \epsilon_2 > 0$, and compare wave propagation in media with positive and negative ϵ_3.

As shown in Sec. 6.3, the dispersion surfaces, also called the **k** surfaces or surfaces of constant $\omega(\mathbf{k})$, are, for the ordinary wave, a sphere of radius

$$k = n_o k_o, \qquad (8.1\text{-}24)$$

where $n_o = \sqrt{\epsilon_1/\epsilon_o}$, and, for the extraordinary wave, a quadric surface

$$\frac{k_1^2 + k_2^2}{\epsilon_3} + \frac{k_3^2}{\epsilon_1} = \frac{k_o^2}{\epsilon_o}, \qquad (8.1\text{-}25)$$

as provided in (6.3-21) and (6.3-22), respectively.

Considering the two cases at hand, we find:

- When ϵ_3 is positive, the extraordinary **k** surface is an ellipsoid of revolution, as displayed in Figs. 6.3-11(b) and 8.1-8(a). The ordinary and extraordinary waves have refractive indices n_o and $n(\theta)$, respectively, where $n(\theta)$, which is given by (6.3-15), varies between n_o and $n_e = \sqrt{\epsilon_3/\epsilon_o}$, as the angle θ varies between 0 and 90°.
- When ϵ_3 is negative, the extraordinary **k** surface is instead a hyperboloid of revolution (in two sheets), as portrayed in Fig. 8.1-8(b), and the material is known as a **hyperbolic medium**. The refractive index $n(\theta)$ for an extraordinary wave at an angle θ in the k_2–k_3 plane is then given by

$$\frac{1}{n^2(\theta)} = \frac{\cos^2\theta}{n_o^2} - \frac{\sin^2\theta}{n_e^2}, \qquad (8.1\text{-}26)$$

where $n_e = \sqrt{|\epsilon_3|/\epsilon_o}$. As θ increases from 0 to $\theta_{\max} = \tan^{-1}(n_e/n_o)$, the refractive index $n(\theta)$ increases from n_o to ∞, as can be discerned from Fig. 8.1-8(b), signifying that the wavelength in the medium becomes progressively smaller and the wave slows to a halt at $\theta = \theta_{\max}$.

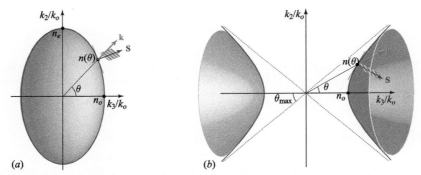

Figure 8.1-8 Contours of the **k** surfaces in the k_2–k_3 plane for a uniaxial anisotropic medium with principal values of the dielectric tensor $\epsilon_1 = \epsilon_2 > 0$ in two cases: (a) $\epsilon_3 > 0$, and (b) $\epsilon_3 < 0$. The contours are shown only for the extraordinary waves (for the ordinary waves, they are spheres in both cases). (a) The **k** contour is an ellipse and the medium is DPS for all directions of propagation. (b) The **k** contour is a hyperbola and the waves can propagate only in directions that lie within a cone of half-angle $\theta_{\max}$. Outside this cone, the medium acts like a SNG medium, which does not support propagating waves.

The Hyperbolic Slab as a Far-Field Imaging System with Subwavelength Resolution

An important property of a hyperbolic medium is that for a plane wave with wavevector components (k_1, k_2, k_3), no matter how large k_1 and k_2, there is a real value of k_3 that satisfies (8.1-25) when ϵ_3 is negative, indicating that the wave can propagate through the medium. This signifies that spatial frequencies greater than an inverse wavelength in any plane do not correspond to evanescent waves, as they do for ordinary media (see Sec. 4.1B), but rather can be transmitted over long distances. Although propagation is accompanied by Fresnel diffraction, the hyperbolic medium has a transfer function with no spatial frequency cutoff.

Moreover, Fresnel diffraction may be significantly reduced in a hyperbolic medium. If $n_0 \ll n_e$, then $\theta_{\max} = \tan^{-1}(n_e/n_o)$ approaches $\pi/2$, whereupon the hyperboloid of revolution in Fig. 8.1-8(b) flattens and becomes approximately planar, corresponding to a constant $k_3 = n_o k_o$, for all k_1 and k_2. The transfer function $\exp(-jk_3 z)$ associated with propagation in the slab is then independent of the spatial frequencies (k_1, k_2) of the input-field distribution. A point in the input plane is thus imaged to a point in the output plane, and propagation may be described by ray optics. The slab then acts as perfect near-field imaging system (i.e., one with subwavelength resolution), as illustrated in Fig. 8.1-9(a). Furthermore, the slab may be curved in such a way that the image is geometrically magnified, as shown in Fig. 8.1-9(b). If the spatial-frequency components of the magnified image become smaller than the inverse wavelength, they generate propagating waves that may be captured with the help of a conventional lens, thereby forming a far-field image. A cylindrical slab such as this is known as a **hyperlens**.

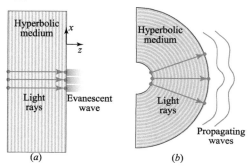

Figure 8.1-9 (a) A hyperbolic slab with $\epsilon_3 < 0$ and $0 < \epsilon_1 = \epsilon_2 \ll |\epsilon_3|$ has a planar dispersion relation ($k_3 =$ constant) so that propagation along the optic axis (z direction) is diffraction-free. (b) A hyperbolic slab curved to form an inhomogeneous cylindrical structure, with the local optic axis pointing in the radial direction, acts as a magnifier of subwavelength details. If the details of the magnified image are larger than the wavelength, it produces propagating waves in the outer medium.

Refraction at the Boundary of a Hyperbolic Medium

Refraction at a boundary between two media may be determined by drawing the **k** surfaces for the two media and matching the components of **k** along the boundary (see, e.g., the analysis of double refraction in Sec. 6.3E). The **k** surfaces for a boundary between an isotropic DPS medium and a hyperbolic medium are shown in Fig. 8.1-10(a), with only the extraordinary wave displayed for the hyperbolic medium.

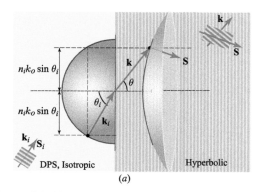 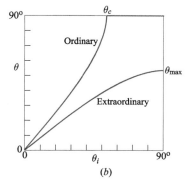

Figure 8.1-10 Refraction at a boundary between an isotropic DPS medium of refractive index n_i and a uniaxial hyperbolic medium with refractive indices n_o and n_e. (a) The **k** surface for the isotropic medium is a sphere, and that for the extraordinary wave of the hyperbolic medium is a hyperboloid of revolution. A form of negative refraction is evident since the Poynting vector **S** of the incident wave points upward whereas that of the refracted wave points downward. (b) Angle of refraction θ as a function of the angle of incidence θ_i for $n_o = 0.8\, n_i$ and $n_e = 1.25\, n_i$. For the ordinary wave, refraction occurs for angles of incidence θ_i smaller than the critical angle θ_c. For the extraordinary wave, refraction occurs for all angles θ_i, but the angle of refraction is limited to $\theta < \theta_{\max}$.

Because the **k** component along the boundary for the hyperbolic medium extends from 0 to ∞, it is always possible to find a match with that for any wave incident from the DPS medium, as may be confirmed by solving the matching equation, $n_i \sin\theta_i = n(\theta)\sin\theta$, together with (8.1-26). Total internal reflection therefore does not occur at this boundary. A plot of the angle of refraction θ as a function of the angle of incidence θ_i is displayed in Fig. 8.1-10(b). This type of **all-angle refraction** is consistent with the observation that the extraordinary wave in the hyperbolic medium cannot be evanescent.

8.2 METAL OPTICS: PLASMONICS

A. Optical Properties of Metals

Conductive Media

Conductive media such as metals, semiconductors, doped dielectrics, and ionized gases have free electric charges and an associated electric current density $\mathcal{J}$. In such materials, the first of the source-free Maxwell's equations, (5.1-7), must be modified by including the current density $\mathcal{J}$ along with the displacement current density $\partial\mathcal{D}/\partial t$, so that

$$\nabla \times \mathcal{H} = \frac{\partial \mathcal{D}}{\partial t} + \mathcal{J}. \tag{8.2-1}$$

The three other source-free Maxwell's equations, (5.1-8)–(5.1-10), remain unmodified, with $\mu = \mu_o$, since naturally occurring materials do not exhibit magnetic effects that vary at optical frequencies.

For a monochromatic wave of angular frequency ω, (8.2-1) takes the form

$$\nabla \times \mathbf{H} = j\omega \mathbf{D} + \mathbf{J}. \tag{8.2-2}$$

If the medium has linear dielectric properties, the electric flux density is given by $\mathbf{D} = \epsilon\mathbf{E} = \epsilon_o(1+\chi)\mathbf{E}$. Similarly, if the medium has linear conductive properties, and the **conductivity** is denoted σ, the electric current density is proportional to the electric field,

$$\mathbf{J} = \sigma\mathbf{E}, \tag{8.2-3}$$

which is a form of Ohm's law. Under these conditions, the right-hand side of (8.2-2) becomes $(j\omega\epsilon + \sigma)\mathbf{E} = j\omega(\epsilon + \sigma/j\omega)\mathbf{E}$, so that

$$\nabla \times \mathbf{H} = j\omega\epsilon_c \mathbf{E}, \tag{8.2-4}$$

where the effective electric permittivity ϵ_c is given by

$$\boxed{\epsilon_c = \epsilon\left(1 + \frac{\sigma}{j\omega\epsilon}\right).} \tag{8.2-5}$$
Effective Permittivity

The effective permittivity ϵ_c is a complex, frequency-dependent parameter that represents a combination of the dielectric and conductive properties of the medium. Since the second term in (8.2-5) varies inversely with frequency, the contribution of the conductive component diminishes with increasing frequency.

Since (8.2-4) takes the same form as the analogous equation for a dielectric medium, (5.3-12), the laws of wave propagation derived in Secs. 5.3–5.5 are applicable even in the presence of conductivity. Hence, the wavenumber in (5.5-2) and (5.5-3) takes the form $k = \beta - j\gamma = \omega\sqrt{\epsilon_c \mu_o}$, the impedance in (5.5-6) becomes $\eta = \sqrt{\mu_o/\epsilon_c}$, and the refractive index n and field attenuation coefficient $\gamma = \frac{1}{2}\alpha$ in (5.5-5) are determined from the complex equation

$$n - j\frac{\gamma}{k_o} = \sqrt{\frac{\epsilon_c}{\epsilon_o}} = \sqrt{\frac{\epsilon}{\epsilon_o}}\sqrt{1 + \frac{\sigma}{j\omega\epsilon}}, \qquad (8.2\text{-}6)$$

where $k_o = \omega/c_o = \omega\sqrt{\epsilon_o \mu_o}$.

The ratio $\sigma/\omega\epsilon$ in (8.2-6) has two limiting regimes. When $\sigma/\omega\epsilon \ll 1$, dielectric effects dominate and conductive effects constitute only a minor correction to the wavenumber. When $\sigma/\omega\epsilon \gg 1$, in contrast, conductive effects dominate and $\epsilon_c \approx \sigma/j\omega$. We can use the Taylor-series expansion $\sqrt{1+x} \approx 1 + x/2$ for $x \ll 1$, along with the identity $\sqrt{j} = (1+j)/\sqrt{2}$, to obtain approximate expressions for the wave parameters when σ is real and frequency-independent. The results are provided in Table 8.2-1 for both limiting regimes:

Table 8.2-1 Refractive index n, attenuation coefficient α, and impedance η for a medium with real-valued, frequency-independent conductivity σ, in the limits of small and large values of $\sigma/\omega\epsilon$. These equations are not applicable when the conductivity σ is complex and/or frequency-dependent, as discussed in the next section in connection with the Drude model.

$\sigma/\omega\epsilon \ll 1$	$\sigma/\omega\epsilon \gg 1$	
$n \approx \sqrt{\epsilon/\epsilon_o}$	$n \approx \sqrt{\sigma/2\omega\epsilon_o}$	(8.2-7)
$\alpha \approx \sigma\sqrt{\mu_o/\epsilon}$	$\alpha \approx \sqrt{2\omega\mu_o\sigma}$	(8.2-8)
$\eta \approx \sqrt{\mu_o/\epsilon}$	$\eta \approx (1+j)\sqrt{\omega\mu_o/2\sigma}$	(8.2-9)

It is apparent from Table 8.2-1 that at a fixed frequency ω, the attenuation coefficient α is proportional to the conductivity σ for small values of σ, but becomes proportional to $\sqrt{\sigma}$ for large values of σ. The impedance η, initially real and independent of σ, eventually acquires a 45° phase shift and becomes inversely proportional to $\sqrt{\sigma}$ as σ becomes large. In this limit, η itself becomes small so that the material becomes highly reflective at the boundary with a non-conductive medium.

For a fixed value of σ at low frequencies, the refractive index n is proportional to $1/\sqrt{\omega}$, whereas the attenuation coefficient α and the impedance η are directly proportional to $\sqrt{\omega}$. As ω increases to a value such that the ratio $\sigma/\omega\epsilon$ becomes very small, all three of the wave propagation parameters become frequency independent and the material behaves as a nondispersive dielectric medium with loss.

For a perfect conductor, $\sigma \to \infty$ so that $\alpha \to \infty$ and the penetration depth $d_p = 1/\alpha \to 0$. Also, $\eta \to 0$ so that at the boundary with a dielectric medium the power reflectance $\mathcal{R} \to 1$; the material then behaves as a **perfect mirror**.

As discussed in the next section, the conductivities of real metals at optical frequencies are often complex-valued and frequency-dependent, in which case the results provided in Table 8.2-1 are not applicable. Under those conditions, the frequency dependencies of the wave parameters can differ considerably from those presented in the table.

Metals: The Drude Model

When the relation between $\mathcal{J}$ and $\mathcal{E}$ is dynamic rather than static, the conductivity σ must be frequency-dependent with a finite bandwidth and a finite response time. Treating the conduction electrons as an ideal gas of independent particles that move freely between scattering events, the **Drude model** (also called the **Drude–Lorentz model**) prescribes a frequency-dependent complex conductivity of the form

$$\sigma = \frac{\sigma_0}{1 + j\omega\tau}, \qquad (8.2\text{-}10)$$

where σ_0 is the low-frequency conductivity and τ is a scattering time (or collision time). For sufficiently low frequencies such that $\omega \ll 1/\tau$, $\sigma \approx \sigma_0$ is real and frequency-independent, in which case the results provided in Table 8.2-1 apply.

If the medium has free-space-like dielectric properties with no other losses ($\epsilon = \epsilon_o$), inserting (8.2-10) into (8.2-5) leads to a relative effective permittivity given by

$$\frac{\epsilon_c}{\epsilon_o} = 1 + \frac{\omega_p^2}{-\omega^2 + j\omega/\tau} = 1 + \frac{\omega_p^2}{-\omega^2 + j\omega\zeta}, \qquad (8.2\text{-}11)$$

where $\zeta = 1/\tau$ is the scattering rate (collision frequency). The **plasma frequency** ω_p (strictly speaking, the plasma angular frequency) is defined as

$$\omega_p = \sqrt{\frac{\sigma_0}{\epsilon_o \tau}} \qquad (8.2\text{-}12)$$

and the free-space **plasma wavelength** λ_p is

$$\lambda_p = \frac{2\pi c_o}{\omega_p}. \qquad (8.2\text{-}13)$$

Some lossless media exhibit dielectric properties that differ from those of free space, as manifested by a residual frequency-independent relative permittivity $1 + \chi_m$ that persists to frequencies $\omega \gg \omega_p$. In this case, the relative effective permittivity in (8.2-11) becomes

$$\frac{\epsilon_c}{\epsilon_o} = 1 + \chi_m + \frac{\omega_p^2}{-\omega^2 + j\omega\zeta}. \qquad (8.2\text{-}14)$$

Observed values of ω_p, λ_p, τ, and ζ for a number of metals are provided in Table 8.2-2.

Table 8.2-2 Plasma frequency ω_p, free-space plasma wavelength λ_p, scattering time τ, and scattering rate $\zeta = 1/\tau$, for Al, Ag, Au, and Cu.[a]

	ω_p (rad/s)	λ_p (nm)	τ (fs)	ζ (s^{-1})
Al	1.83×10^{16}	103	5.10	1.96×10^{14}
Ag	1.37×10^{16}	138	31.3	0.32×10^{14}
Au	1.35×10^{16}	139	9.25	1.08×10^{14}
Cu	1.33×10^{16}	142	6.90	1.45×10^{14}

[a]See, e.g., E. J. Zeman and G. C. Schatz, An Accurate Electromagnetic Theory Study of Surface Enhancement Factors for Ag, Au, Cu, Li, Na, Al, Ga, In, Zn, and Cd, *Journal of Physical Chemistry*, vol. 91, pp. 634–643, 1987.

These results are related to those for the resonant medium described by the Lorentz oscillator model, which treats the motion of an electron bound to a nucleus as a harmonic oscillator (see Sec. 5.5C). However, here the electrons are free and there is no restoring force so that the elastic constant $\kappa = 0$ and the resonance frequency $\omega_0 = \sqrt{\kappa/m} = 0$. The Lorentz-model equation of motion (5.5-16) then becomes $d^2x/dt^2 + \zeta dx/dt = -e\mathcal{E}/m$, and the corresponding polarization density $\mathcal{P} = -Nex$ obeys the equation $d^2\mathcal{P}/dt^2 + \zeta d\mathcal{P}/dt = (Ne^2/m)\mathcal{E}$, where N is the electron density of the medium. For a field oscillating at an angular frequency ω, this gives rise to $-\omega^2 P + j\omega\zeta P = (Ne^2/m)E$, which leads to a susceptibility

$$\chi(\omega) = \frac{P}{\epsilon_o E} = \frac{\omega_p^2}{-\omega^2 + j\omega\zeta}, \quad (8.2\text{-}15)$$

where

$$\boxed{\omega_p = \sqrt{\frac{Ne^2}{\epsilon_o m}}.} \quad (8.2\text{-}16)$$
Plasma Frequency
Drude Model

Equation (8.2-15), together with the relation $\epsilon_c = \epsilon_o(1 + \chi)$ for a medium with free-space dielectric properties, leads to (8.2-11) and (8.2-12), provided that

$$\sigma_0 = \frac{Ne^2\tau}{m}. \quad (8.2\text{-}17)$$

Equation (8.2-15) is a special case of (5.5-19) for the Lorentz resonant medium, with $\omega_0 = 0$ and $\zeta = \Delta\omega = 2\pi\Delta\nu$, where $\Delta\nu$ is the spectral width. A comparison of the frequency dependence of the real and imaginary parts of the relative permittivity $\epsilon_r = \epsilon_c/\epsilon_o$ for a Lorentz dielectric resonant medium, based on (5.5-19), or (5.5-20) and (5.5-21), and for a Drude metal based on (8.2-11), is provided in Fig. 8.2-1.

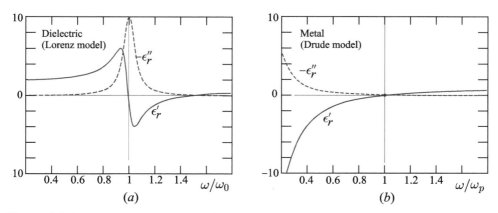

Figure 8.2-1 Frequency dependence of the real part (solid curve) and imaginary part (dashed curve) of the relative effective permittivity $\epsilon_r = \epsilon_c/\epsilon_o = \epsilon_r' + j\epsilon_r''$. (a) Dielectric resonant medium described by the Lorenz model, (5.5-20) and (5.5-21), with resonance frequency ω_0, $Q = \omega_0/\Delta\omega = 10$, and $\chi_o = 1$. (b) Metal described by the Drude model (8.2-11) with plasma frequency ω_p, $\omega_p\tau = 10$, and $\epsilon = \epsilon_o$. The real part of the permittivity is negative for $\omega < \omega_p\sqrt{1 - (\omega_p\tau)^{-2}} \approx \omega_p$.

EXAMPLE 8.2-1. *Permittivity and Reflectance of Silver.* As an example of the range of validity of the Drude model, Fig. 8.2-2(a) displays the measured relative effective permittivity $\epsilon_r = \epsilon_c/\epsilon_o$ for silver (dashed curves) along with the best-fitting functions (solid curves) consistent with the Drude model [see, e.g., Fig. 8.2-1(b)]. The experimental permittivities and reflectances for various metals, as functions of wavelength, are available from a number of sources.[†] It is apparent from Fig. 8.2-2(a) that the Drude-model relative permittivity provided in (8.2-14) fits the experimental data quite well over the 450–600-nm wavelength range, but is clearly inadequate at shorter wavelengths. The best-fitting model parameters over the wavelength range displayed (200–600 nm) are $\chi_m = 4.45$, $\omega_p = 1.47 \times 10^{16}$ rad/s ($\lambda_p = 128$ nm), and $\tau = 12$ fs ($\zeta = 0.84 \times 10^{14}$ s^{-1}). These values differ from those presented in Table 8.2-2 because of the constraints inherent in this analysis.

In Fig. 8.2-2(b), the measured power reflectance for light normally incident on the Ag-air boundary (dashed curve) is compared with the reflectance calculated on the basis of the Drude model (solid curve) using the relation $\mathcal{R} = |(\eta - \eta_o)/(\eta + \eta_o)|^2$, where $\eta = \sqrt{\epsilon_c/\mu_o}$ and $\eta_o = \sqrt{\epsilon_o/\mu_o}$ are the impedances of Ag and air, respectively [see (6.2-8)]. The fit is quite good over the 400–600-nm wavelength range, where Ag exhibits near-perfect reflectance (as it does at longer wavelengths). Other metals exhibit similar behavior. Gold and copper also have free-space plasma wavelengths that lie well into the ultraviolet (see Table 8.2-2) and those metals do not become good reflectors until the wavelength exceeds ~ 550 nm (thereby explaining their reddish color). The origin of this behavior is interband absorption, i.e., absorption by bound electrons, which is not accommodated in the Drude free-electron model (see Prob. 8.2-3). Aluminum, in contrast, hews quite closely to the predictions of the Drude model, exhibiting near-unity reflectance over a wavelength range that stretches from 200 nm to beyond 10 μm.

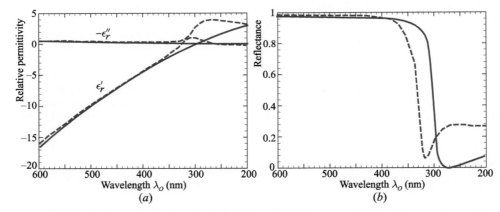

Figure 8.2-2 (a) Real and imaginary parts of the relative effective permittivity $\epsilon_r = \epsilon'_r + j\epsilon''_r$ as a function of wavelength for Ag. The dashed curves represent experimental values whereas the solid curves are fits to the Drude model. (b) Power reflectance $\mathcal{R}$ vs. wavelength at the Ag-air boundary. The dashed curve is the experimental reflectance while the solid curve is the calculated reflectance based on the Drude model. Note that the wavelength decreases from left to right (corresponding to increasing frequency); the free-space plasma wavelength λ_p lies beyond the right edges of the graphs.

Simplified Drude model. We now consider the behavior of the Drude model for angular frequencies sufficiently high such that $\omega \gg 1/\tau$ ($\omega \gg \zeta$), which is equivalent to neglecting damping ($\zeta \to 0$) in (5.5-16). Under these conditions, (8.2-10) becomes $\sigma \approx \sigma_0/j\omega\tau$, which is imaginary, and (8.2-14), with $\chi_m = 0$, reduces to the effective permittivity for the **simplified Drude model**, which is real:

[†] The experimental data presented in Fig. 8.2-2 were drawn from P. B. Johnson and R. W. Christy, Optical Constants of the Noble Metals, *Physical Review B*, vol. 6, pp. 4370–4379, 1972.

$$\epsilon_c \approx \epsilon_o \left(1 - \frac{\omega_p^2}{\omega^2}\right).$$

(8.2-18)
Effective Permittivity
Simplified Drude Model

It is clear from (8.2-18) that the presence of conductivity in the medium serves to suppress the permittivity to a value below ϵ_o and to impart to it a functional form that is inversely proportional to the square of the frequency.

The simplified Drude model is useful for describing the optical behavior of metals in the near-infrared and visible regions of the spectrum [see Fig. 2.0-1]. For wavelengths shorter than $\sim 1\,\mu$m, which corresponds to an angular frequency $\omega = 2\pi\nu = 1.9 \times 10^{15}$ rad/s, $\omega \gg \zeta$ for the metals listed in Table 8.2-2.

In summary, for frequencies that are sufficiently high such that $\omega \gg 1/\tau$, the simplified Drude model for a metal is a special case of the Lorentz model for a dielectric medium in which there is neither a restoring force ($\kappa = 0$) nor damping ($\zeta = 0$). The permittivity (8.2-18) is real and negative for frequencies below the plasma frequency ω_p, where the metal behaves as a SNG medium, and real and positive for frequencies above ω_p, where the metal behaves as a DPS medium.

Wave propagation in a medium described by the simplified Drude model is characterized by a propagation constant $\beta = nk_o = \sqrt{\epsilon_c/\epsilon_o}\,(\omega/c_o)$ and a dispersion relation

$$\beta = \frac{\omega}{c_o}\sqrt{1 - \frac{\omega_p^2}{\omega^2}},$$

(8.2-19)
Dispersion Relation
Simplified Drude Model

as illustrated in Fig. 8.2-3. The corresponding relative permittivity ϵ_c/ϵ_o, refractive index n, and attenuation coefficient α are also displayed in the figure.

It is evident from Fig. 8.2-3 that wave propagation in a metal described by the simplified Drude model exhibits distinctly different behavior below, at, and above the plasma frequency ω_p:

- At frequencies below the plasma frequency ($\omega < \omega_p$), the effective permittivity ϵ_c is negative. The metal then behaves as a SNG medium with imaginary wavenumber, $k = j\omega\sqrt{|\epsilon_c|\mu_o}$, which corresponds to attenuation without propagation. This spectral region may therefore be regarded as a **forbidden band**. The attenuation coefficient $\alpha = 2k_o\sqrt{\omega_p^2/\omega^2 - 1}$ decreases monotonically with increasing frequency and vanishes at the plasma frequency. The free electrons then undergo longitudinal collective oscillations that take the form of **plasmons**, the quanta of the plasma wave (much as photons are the quanta of the electromagnetic wave). The negative permittivity corresponds to an imaginary impedance, which indicates that total reflection occurs at the boundary with a DPS medium (see Sec. 6.2), provided that $\lambda_o > \lambda_p$. Doped semiconductors, in contrast, are not reflective in the visible region. This is because the plasma frequency of such materials lies in the infrared since the free-electron density N is far smaller than that in metals [see (8.2-16)]. Although the Drude model provides a good starting point for determining the optical properties of metals and doped semiconductors, it is by no means the final word, as was illustrated in Example 8.2-1.

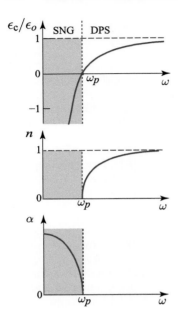
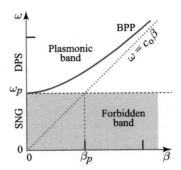

Figure 8.2-3 Left: Frequency dependence of the relative effective permittivity ϵ_c/ϵ_o, the refractive index n, and the attenuation coefficient α for a medium described by the simplified Drude model. Right: The dispersion relation $\omega = \sqrt{\omega_p^2 + c_o^2\beta^2}$. Metals support propagating optical waves, known as bulk plasmon polaritons (BPPs), but only when the frequency of the wave falls in the plasmonic band (which comprises frequencies greater than the plasma frequency ω_p).

- At frequencies above the plasma frequency ($\omega > \omega_p$), the effective permittivity is positive and real so that the medium behaves like a lossless dielectric material, albeit with unique dispersion characteristics. The propagation constant becomes $\beta = \sqrt{\omega^2 - \omega_p^2}/c_o$, while the refractive index $n = \sqrt{1 - \omega_p^2/\omega^2}$ lies below unity and is very small near the plasma frequency. This spectral band is referred to as the **plasmonic band** and waves that travel through the metal in this band are called **bulk plasmon polaritons (BPPs)**.

B. Metal–Dielectric Boundary: Surface Plasmon Polaritons

Since a metal behaves as a SNG medium at optical frequencies below the plasma frequency ($\omega < \omega_p$), it can support a surface plasmon polariton (SPP) wave at the boundary with an ordinary dielectric (DPS) medium. As indicated in Sec. 8.1B, a SPP is a guided optical surface wave accompanied by a longitudinal electron-density wave that oscillates at the optical frequency. Because the SPP hugs the metal–dielectric boundary over distances far smaller than an optical wavelength, it can be controlled and manipulated at nanometer spatial scales without sacrificing its optical temporal frequency. The tight binding of the wave to the boundary carries a concomitant enhancement of the local field intensity.

In this section we apply the analysis of SPP waves at a DPS-SNG boundary, developed in Sec. 8.1B, to a metal–dielectric boundary. Specifically, we combine the effective permittivity of a metal described by the simplified Drude model, set forth in (8.2-18), with the general wave-parameter results summarized in (8.1-16) and (8.1-17). Based on the analysis provided in Sec. 8.1B, a SPP wave can exist at the boundary provided that the magnitude of the permittivity of the SNG medium $|\epsilon_2|$ is greater than that of the DPS medium ϵ_1. If the metal is described by the simplified Drude model, (8.2-18) reveals that its effective permittivity is $\epsilon_2 = \epsilon_o(1 - \omega_p^2/\omega^2)$, from which it is clear that the material behaves as a SNG medium if $\omega < \omega_p$. As illustrated in Fig. 8.2-

4(a), the condition $|\epsilon_2| > \epsilon_1$ is equivalent to the condition $\omega < \omega_s$, where

$$\omega_s = \frac{\omega_p}{\sqrt{1+\epsilon_{r1}}}, \tag{8.2-20}$$

with $\epsilon_{r1} = \epsilon_1/\epsilon_o$ representing the relative permittivity of the dielectric medium. The frequency band over which a SPP wave can be supported, $\omega < \omega_s$, is smaller than that over which the simple Drude metal behaves as a SNG medium, $0 < \omega < \omega_p$.

After working through the algebra involved in combining (8.1-16) and (8.2-18), it turns out that β, n_b, and ϵ_b obey (8.2-21). Equation (8.2-22) is carried over intact from (8.1-17). Again, $k_o = \omega\sqrt{\epsilon_o \mu} = 2\pi/\lambda_o$ represents the free-space wavenumber, and the penetration depth is given by $d_p = 1/2\gamma$.

$$\beta = n_b k_o, \quad n_b = \sqrt{\frac{\epsilon_b}{\epsilon_o}}, \quad \epsilon_b = \epsilon_1 \frac{1-\omega^2/\omega_p^2}{1-\omega^2/\omega_s^2}, \tag{8.2-21}$$

$$\gamma_1 = \sqrt{\frac{-\epsilon_1^2}{\epsilon_o(\epsilon_1+\epsilon_2)}}\, k_o, \quad \gamma_2 = \sqrt{\frac{-\epsilon_2^2}{\epsilon_o(\epsilon_1+\epsilon_2)}}\, k_o. \tag{8.2-22}$$
SPP Wave
Metal–Dielectric

As illustrated in Fig. 8.2-4, the behavior of the parameters provided in (8.2-21) differs for the three salient frequency bands:

- For $\omega < \omega_s$, the simple Drude metal behaves as a SNG medium, and SPP waves may be guided along its boundary with a dielectric medium. The frequency band $\omega < \omega_s$ lies within the forbidden band of the *bulk* metal, $0 < \omega < \omega_p$, in which bulk propagating waves are not permitted. The properties of the SPP wave are dependent on the ratio $|\epsilon_2|/\epsilon_1$, which is a monotonically decreasing function of the ratio ω/ω_s, and approaches the critical value of unity when $\omega = \omega_s$. Forging a comparison with waves in the bulk dielectric medium, the SPP velocity c_o/n_b is smaller and the plasmon wavelength λ_o/n_b is shorter. This is also evident from the dispersion curve displayed in Fig. 8.2-4(c) since $\beta > \omega/c_1$ for the SPP wave. In accordance with (8.1-17), the penetration depth in the metal, $d_2 = 1/2\gamma_2$, is smaller than that in the dielectric, $d_1 = 1/2\gamma_1$, and both are smaller than the wavelength in the bulk dielectric medium. As ω/ω_s increases, the SPP slows and the associated plasmon wavelength decreases. As explained in Sec. 8.1B, the SPP also becomes more localized, exhibiting smaller penetration depths in both the metal and the dielectric. At $\omega/\omega_s = 1$, the permittivities of the metal and dielectric become equal in magnitude and opposite in sign ($\epsilon_1 + \epsilon_2 = 0$), whereupon the velocity of the SPP becomes zero.

- For $\omega_s < \omega < \omega_p$, the simple Drude metal is a SNG medium, but SPP waves are not permitted since $|\epsilon_2| < \epsilon_1$. Hence, bulk waves propagating in the dielectric medium are totally reflected from the DPS-SNG boundary at all angles of incidence.

- For $\omega > \omega_p$, the simple Drude metal acts as a DPS medium. DPS-DPS boundaries cannot support SPP waves. However, much like a dielectric medium, the bulk metal supports propagating waves, namely bulk plasmon polaritons (BPPs). At these high frequencies, the reflection and refraction of waves at the metal–dielectric boundary comply with the conventional laws obeyed by two dielectric media. Total internal reflection occurs, but only for angles of incidence greater than the critical angle.

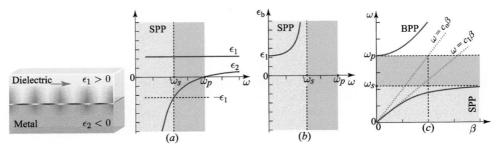

Figure 8.2-4 Surface plasmon polariton (SPP) wave at a metal–dielectric boundary, as depicted in Fig. 8.1-3. (a) Frequency dependence of the permittivities of the dielectric and metallic media, ϵ_1 and ϵ_2, respectively. The condition $|\epsilon_2| > \epsilon_1$, required for the existence of the SPP wave, is satisfied for $\omega < \omega_s$. (b) Frequency dependence of the effective permittivity ϵ_b of the SPP wave. The wave velocity is c_o/n_b, where $n_b = \sqrt{\epsilon_b/\epsilon_o}$. (c) Dispersion relations for the BPP (bulk plasmon polariton) and SPP waves. These plots were computed from (8.2-19) and (8.2-21), respectively, using $\epsilon_{r1} = 2.25$ (the relative permittivity of glass), so that $\omega_s = \omega_p/\sqrt{1+\epsilon_{r1}} = 0.55\,\omega_p$. Light lines in free space and in the bulk dielectric medium are shown as dotted red.

EXAMPLE 8.2-2. *SPP at Ag-Air and Ag-SiO$_2$ Boundaries.* SPP waves are supported at the boundaries of dielectrics with metals such as silver (Ag) and gold (Au). For an optical wave of free-space wavelength $\lambda_o = 800$ nm, the permittivity of Ag is $\epsilon_2 \approx -32.6\,\epsilon_o$.[†] The plasma frequency of Ag, $\omega_p = 1.37 \times 10^{16}$ s^{-1}, corresponds to a free-space plasma wavelength $\lambda_p = 2\pi c_o/\omega_p = 138$ nm (see Table 8.2-2). At a Ag-air boundary ($\epsilon_1 = \epsilon_o$), (8.2-20) provides that the free-space wavelength corresponding to ω_s is $\lambda_s = \sqrt{2}\,\lambda_p = 195$ nm. SPP waves must therefore have a free-space wavelength longer than 195 nm to be viable. In accordance with (8.2-21) and (8.2-22), a SPP wave of frequency corresponding to a free-space wavelength $\lambda_o = 800$ nm at this boundary has the following properties:

$$n_b = 1.016, \quad \lambda_o/n_b = 788 \text{ nm}, \quad d_1 = 358 \text{ nm}, \quad d_2 = 11 \text{ nm}.$$

At a Ag-SiO$_2$ boundary ($\epsilon_1 = 3.9\,\epsilon_o$ at $\lambda_o = 800$ nm), the free-space plasma wavelengths corresponding to ω_p and ω_s are $\lambda_p = 138$ nm and $\lambda_s = 305$ nm, respectively. At this boundary, a SPP wave of frequency corresponding to a free-space wavelength $\lambda_o = 800$ nm has the following properties:

$$n_b = 2.104, \quad \lambda_o/n_b = 380 \text{ nm}, \quad d_1 = 87 \text{ nm}, \quad d_2 = 10 \text{ nm}.$$

Comparing the results for both dielectric media, it is apparent that the higher-index material, SiO$_2$, results in a SPP wave with lower velocity, shorter wavelength, and shallower penetration into both the Ag and the dielectric medium. Moreover, $d_2 < d_1$ for both air and SiO$_2$, confirming the deeper penetration into the dielectric side of the boundary, as depicted in Figs. 8.1-3 and 8.2-4.

Generation and Detection of Surface Plasmon Polaritons

Since the propagation constant of a SPP wave traveling along a metal–dielectric boundary is greater than that of an ordinary optical wave of the same frequency propagating in the dielectric medium [see Fig. 8.2-4(c)], it is difficult to couple the two waves. One way in which this can be achieved, however, is to couple an evanescent wave, generated by total reflection at a metal–dielectric boundary, to the SPP wave at the

---

[†] The experimental value for the permittivity of Ag at a free-space wavelength of $\lambda_o = 800$ nm is $\epsilon_2 = (-32.6 + j\,0.5)\,\epsilon_o$. The calculated value of the effective permittivity at this wavelength, based on (8.2-18) for the simplified Drude model (no damping), provides $\epsilon_2' = -32.8\,\epsilon_o$, in close agreement with the experimental value.

opposite metal–dielectric boundary, as portrayed in the prism-coupler configuration shown in Fig. 8.2-5(a). A similar approach is used to couple light into a waveguide via a prism, as will become clear in Fig. 9.4-4.

The coupling takes place only when the two waves are phase matched, i.e., when their propagation constants are precisely equal. From the interior of a prism of refractive index n_p, this condition is satisfied at an angle of incidence $\theta_p = \theta_r$, where $n_p k_o \sin \theta_r = \beta$. The parameter β is the propagation constant of the SPP wave, given in (8.2-21) and (8.2-20), which is dependent on the refractive index $n_1 = \sqrt{\epsilon_1/\epsilon_o}$ of the dielectric medium adjacent to the remote side of the metallic film [which is represented in Fig. 8.2-5(a) as air]. It is straightforward to verify that this condition is met at an angle of incidence θ_r given by

$$\sin \theta_r = \frac{n_1}{n_p} \sqrt{\frac{1 - \omega^2/\omega_p^2}{1 - (1 + n_1^2)\,\omega^2/\omega_p^2}}\,. \qquad (8.2\text{-}23)$$

When the phase-matching condition is satisfied, optical power is transferred to launch the SPP wave via a form of **frustrated total internal reflection** (FTIR), so that the power of the reflected optical wave at the boundary of the prism diminishes significantly. The change in reflectance is manifested as a sharp, resonance-like function of the incidence angle, as portrayed in Fig. 8.2-5(b). Since the angle θ_r depends

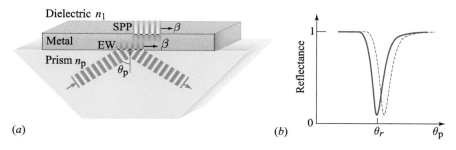

Figure 8.2-5 (a) Generation of a SPP wave by use of a prism coupler. An evanescent wave (EW) generated by total internal reflection of an optical wave at the prism–metal boundary excites a SPP wave at the opposite metal–dielectric boundary. (b) Dependence of the reflectance of the optical wave on the angle of incidence θ_p. When the phase-matching condition is satisfied, at $\theta_p = \theta_r$, a SPP wave is launched, which results in a decrease of the reflected-wave intensity. A slight change in the refractive index n_1 alters θ_r in accordance with (8.2-23), which gives rise to the dashed curve. The system thus serves as a precise sensor.

on n_1, it is sensitive to changes in the physical environment surrounding the metal film, which determine n_1. This measurement technique, known as **surface plasmon resonance (SPR) spectroscopy**, has found extensive use for chemical and biological sensing applications; examples are gas detection and the measurement of molecular adsorption. Another method of exciting an SPP wave is to scatter light from a periodic subwavelength structure (grating) deposited on the metallic surface, which contributes a spatial-frequency component that compensates for the propagation-constant mismatch.

The detection of an SPP wave may be achieved by converting it into a proportional optical wave, using a prism coupler or grating operated in a reverse conversion process.

C. The Metallic Nanosphere: Localized Surface Plasmons

A metallic structure of subwavelength dimensions supports plasmonic oscillations at its (external or internal) boundary with a dielectric medium. Examples of such structures are metallic nanospheres, nanodisks, and other nanoparticles. These oscillations are known as **localized surface plasmon polaritons**, or simply **localized surface plasmons (LSPs)**. When the excitation frequency matches the resonance frequency of the structure, the result is a **surface plasmon resonance (SPR)**. A LSP is to be distinguished from a long-range SPP, which is a SPP wave that propagates along an extended metal–dielectric boundary, as described in Sec. 8.2B. Similarly, the surface plasmon resonance frequency is to be distinguished from the plasma frequency of the metal, although they are related. Gold and silver nanoparticles have plasmon resonance frequencies that lie in the visible region of the spectrum, whereas the associated plasma frequencies of these metals lie well into the ultraviolet. By virtue of the curved surfaces of nanoparticles, SPRs may be excited by direct-light illumination. The resultant intense colors exhibited by such particles, both in transmission and in reflection, are attributable to resonantly enhanced scattering and absorption.

The Metallic Nanosphere

A metallic nanosphere embedded in a surrounding dielectric medium supports LSP oscillations. The distribution of the optical field is obtained by solving Maxwell's equations in the metal and dielectric, which have negative and positive permittivities, respectively, and accommodating surface charges and the appropriate boundary conditions. The field distribution of the lowest-order plasmonic mode is depicted in Fig. 8.2-6.

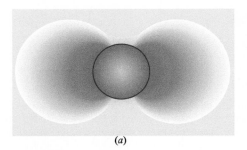
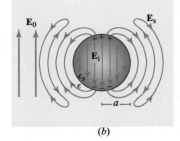

Figure 8.2-6 (a) Magnitude of the optical field outside a metallic nanosphere supporting the lowest-order plasmonic mode of this resonator. The internal field is not shown. (b) Field lines of a plasmonic mode excited by an incident plane wave.

The metallic nanosphere is a resonant scatterer. It is clear from the theory of Rayleigh scattering provided in Sec. 5.6B that a plane wave with electric field E_0 incident on a small sphere creates a parallel internal field E_i, which in turn generates an oscillating electric dipole that radiates a scattered dipole wave E_s (see Fig. 5.6-3). In accordance with (5.6-12), the total scattered optical power is $P_s = \sigma_s I_0$, where σ_s is the scattering cross section and I_0 is the intensity of the incident wave. Equations (5.6-16) and (5.6-17), which are applicable when the radius-to-wavelength ratio $a/\lambda \ll 1$, reveal that σ_s depends on the permittivities ϵ_s and ϵ of the nanosphere and the surrounding medium, respectively, as well as on a/λ:

$$\sigma_s = \pi a^2 Q_s, \qquad Q_s = \frac{8}{3}\left|\frac{\epsilon_s - \epsilon}{\epsilon_s + 2\epsilon}\right|^2 \left(2\pi\frac{a}{\lambda}\right)^4, \qquad (8.2\text{-}24)$$

$$E_{\rm i} = \frac{3\epsilon}{\epsilon_{\rm s} + 2\epsilon} E_0 . \qquad (8.2\text{-}25)$$

For a metal described by the simplified Drude model, the effective permittivity of the metallic medium is given by $\epsilon_{\rm s} = \epsilon_o(1 - \omega_p^2/\omega^2)$, as provided in (8.2-18); $\epsilon_{\rm s}$ is thus negative or positive, depending on whether ω lies below or above the plasma frequency ω_p, respectively. It follows that the denominator $(\epsilon_{\rm s} + 2\epsilon)$ of (8.2-24) and (8.2-25) can be either negative or positive, but vanishes when $\epsilon_{\rm s} = -2\epsilon$, where $\sigma_{\rm s}$ and $E_{\rm i}$ increase without limit. The LSP resonance frequency at which this occurs is established by setting $\epsilon_o(1 - \omega_p^2/\omega^2) = -2\epsilon$, which gives rise to

$$\omega_0 = \frac{\omega_p}{\sqrt{1 + 2\epsilon_r}} , \qquad (8.2\text{-}26)$$
LSP Resonance Frequency

with $\epsilon_r = \epsilon/\epsilon_o$. The surface plasmon resonance frequency ω_0 is to be distinguished from both the plasma frequency of the metal ω_p [see (8.2-16)], and the maximum frequency at which a SPP can exist ω_s [see (8.2-20)], although all three are closely related.

At frequencies near ω_0, the scattering cross section $\sigma_{\rm s}$ and the internal field $E_{\rm i}$ are substantially enhanced, as depicted in Fig. 8.2-7, which displays $\sigma_{\rm s}$ and $E_{\rm i}$ as functions of ω. Below resonance ($\omega < \omega_0$), the dipole created by the incident field points in the same direction as the incident field, while the internal field is in the opposite direction, i.e., out of phase since $E_{\rm i}/E_0$ is negative. The opposite situation prevails above resonance ($\omega > \omega_0$), as it does for the dielectric nanosphere (Fig. 5.6-3). In the

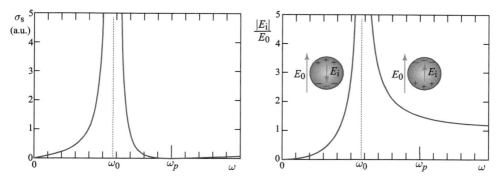

Figure 8.2-7 Resonance characteristics of the scattering cross section $\sigma_{\rm s}$ and the internal field $E_{\rm i}$ for a metallic nanosphere described by the simplified Drude model in air ($\epsilon_r = 1$). In accordance with (8.2-26), the resonance frequency is $\omega_0 = \omega_p/\sqrt{3}$, where ω_p is the plasma frequency of the metal. Below resonance, the internal field is opposite in direction to the incident external field (out of phase), whereas above resonance it is in the same direction (in phase).

vicinity of resonance, both the internal field $E_{\rm i}$ and the scattered field $E_{\rm s}$ in the near-field zone can be significantly enhanced with respect to the incident field E_0. This field enhancement is accompanied by spatial localization of electromechanical energy at a scale that corresponds to the size of the nanosphere.

The simplified Drude model used to generate the graphs presented in Fig. 8.2-7 assumes that the metal has no absorption; this idealization is the origin of the infinite cross section and infinite internal field at resonance. Of course, real metals have finite resistivity and finite absorption, which is mathematically accommodated by introducing an imaginary component into the permittivity. Denoting the complex permittivity as $\epsilon_s = \epsilon'_s + j\epsilon''_s$, resonance occurs when the real parts of the denominators in (8.2-24) and (8.2-25) vanish, namely when $\epsilon'_s = -2\epsilon$. Use of the complex permittivity thus results in residual denominators of $j\epsilon''_s$, rather than zero, which gives rise to a finite value of σ_s at resonance:

$$\sigma_s = \pi a^2 Q_s, \qquad Q_s = \frac{8}{3}\left(2\pi\frac{a}{\lambda}\right)^4 \frac{\epsilon''^2_s + 9\epsilon^2}{\epsilon''^2_s}. \qquad (8.2\text{-}27)$$

Aside from being a resonant scatterer, the metallic nanosphere is also a resonant absorber. As discussed in Sec. 5.6D, the Rayleigh scattering of an incident wave by a nanosphere of complex permittivity ϵ_s in a surrounding medium of real permittivity ϵ is accompanied by absorption. Both absorption and scattering contribute to the attenuation (extinction) of the incident wave. In accordance with (5.6-22), the absorption cross section of a small, spherical scatterer of radius a, $\sigma_a = \pi a^2 Q_a$, exhibits the same resonance condition as the scattering cross section σ_s set forth in (8.2-24), namely $\epsilon'_s = -2\epsilon$. Substituting $\epsilon_s = \epsilon'_s + j\epsilon''_s$ into (5.6-22) leads to an absorption efficiency Q_a given by

$$Q_a \approx -4\left(2\pi\frac{a}{\lambda}\right) \mathrm{Im}\left\{\frac{\epsilon_s - \epsilon}{\epsilon_s + 2\epsilon}\right\} = \left(2\pi\frac{a}{\lambda}\right)\frac{-12\epsilon\epsilon''_s}{(\epsilon'_s + 2\epsilon)^2 + \epsilon''^2_s}. \qquad (8.2\text{-}28)$$

At resonance, the denominator on the right becomes simply ϵ''^2_s, which leads to a peak value of the absorption coefficient given by $Q_a = -(2\pi a/\lambda)(12\,\epsilon/\epsilon''_s)$. The larger the resistivity of the metal, which is represented by ϵ''_s, the broader the resonance profile and the smaller the peak values of σ_a and σ_s.

Metallic nanospheres whose localized surface plasmon (LSP) resonance frequencies lie in the visible and ultraviolet bands are used in applications that exploit their wavelength-selective absorption and scattering resonances, along with the attendant field enhancement and localization. Nanospheres embedded in stained glass, for example, produce brilliant colors as a result of the extinction of specific wavelengths; an example is provided by the North Rose Window at Notre Dame Cathedral in Paris, pictured at right. As another example, the dependence of the LSP resonance frequency on the relative permittivity ϵ_r of the host medium gives rise to a sensor responsive to the dielectric properties of the surrounding medium; a host medium with increased permittivity results in a decreased resonance frequency and an increased resonance wavelength, as is understood from (8.2-26).

The North Rose Window at Notre Dame Cathedral in Paris dates from 1260. Metallic nanostructures embedded in the stained glass exhibit plasmonic resonances with dramatic optical characteristics. (Adapted from a photograph by Krzysztof Mizera, August 30, 2008, Wikimedia Commons.)

D. Optical Antennas

An antenna is an electrically conductive structure that converts an oscillating electric current into an electromagnetic field, and vice versa. It is a key component in transmitters and receivers of electromagnetic radiation. At radiowave and microwave frequencies, antennas take the form of metallic wires, poles, loops, and microstrips whose dimensions are of the order of the wavelength [Fig. 8.2-8(a)]. Antennas such as these are resonant structures. A monopole antenna comprising a metal pole of length L mounted on a conducting plate, for example, has a resonance frequency $c/4L$, corresponding to a wavelength $\lambda = 4L$. Equivalently, a dipole antenna comprising two poles separated by a small gap, each pole of length L, exhibits resonance when $L = \lambda/4$.

An antenna may also take the form of an electrically conductive structure that intercepts an electromagnetic wave and alters its angular distribution [Fig. 8.2-8(b)]. At microwave frequencies, these include the horn antenna (a metallic horn connected to the end of a waveguide) and the dish antenna (a paraboloidal metal surface with the end of a waveguide situated at its focus). These antennas are not necessarily resonant and their dimensions may be substantially greater than the wavelength of the radiation.

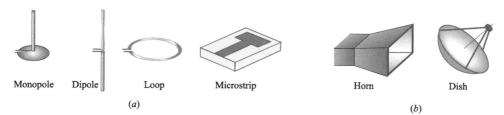

Figure 8.2-8 (a) Radiowave antennas. (b) Microwave antennas.

Figure 8.2-9 (a) Optical antennas made of metallic structures that exhibit resonance at optical frequencies. (b) Non-resonant optical antennas.

Resonant optical antennas may be constructed by fabricating metallic structures similar to those used for radiowave antennas, but with scaled-down dimensions [Fig. 8.2-9(a)]. However, since the length of an optical quarter-wave dipole antenna lies in the nanometer region, fabrication can be challenging. At optical frequencies, the optical field interacts with metallic antennas such as these via SPP waves, which have propagation wavelengths that are even smaller than the free-space optical wavelength. These **plasmonic antennas** operate as scatterers that convert the incoming light into localized SPP waves that in turn radiate light with a modified spatial distribution.

Optical antennas in the non-resonant category include the metal-coated tapered optical-fiber tip used in near-field microscopy and the paraboloidal mirror used in telescopes, as illustrated in Fig. 8.2-9(b). The dimensions of these optical antennas are typically far greater than the optical wavelength.

An example of an optical antenna that exhibits resonance is provided by an incoming planar optical wave illuminating a metallic nanosphere and exciting a localized SPP wave, which in turn radiates an optical dipole wave as discussed in Sec. 8.2C. At the resonance frequency, the field in the vicinity of the nanosphere is enhanced and localized, and the scattering cross-section increases sharply so that more of the incoming light is captured and scattered. The nanosphere thus functions as a resonant optical antenna. Other metallic structures with nanoscale dimensions, such as the split ring and the double split ring shown in Fig. 8.2-9(a), also exhibit resonances at optical frequencies; their resonance properties are shape- and material-dependent.

Resonant optical antennas may be used to localize and couple light into small absorbers, such as single molecules. In the near-field microscopy arrangement depicted in Fig. 8.2-10(a), for example, a metal rod on a conducting pedestal may be placed at the end of a tapered optical fiber to create a monopole antenna. A nanosphere at the end of a pointed glass tip, as illustrated in Fig. 8.2-10(b), carries out a similar function. In general, a resonant optical antenna placed between an emitter and an absorber can serve to enhance their interaction by facilitating the processes of radiation and detection.

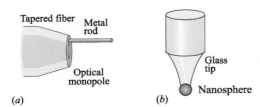

Figure 8.2-10 Optical antennas used to localize light in near-field microscopy. (a) Monopole antenna at the end of a tapered fiber. (b) Nanosphere antenna at the end of a glass tip.

The greatest challenge in modeling the interaction of an optical wave with a metallic structure arises for structures whose dimensions are comparable with the optical wavelength ($\sim\mu$m). This typically requires a complete analysis incorporating the electromagnetic fields and electric-charge distributions. Conventional bulk optical structures, which have far larger dimensions, are readily analyzed using the usual techniques of optics. At the opposite extreme, metallic nanostructures are handled quite well with effective-circuit models, as will become evident in Sec. 8.3A.

8.3 METAMATERIAL OPTICS

Optical metamaterials are synthetic composite materials constructed with carefully designed spatial patterns and with dimensions that are smaller than the optical wavelength. They owe their special optical properties to the atomic and molecular structures of the constituent materials as well as to the geometry and dimensions of the spatial patterns in relation to the wavelength. They can be engineered to have distinct and unusual optical properties that are not available in natural materials. Metamaterials form the basis for various exotic optical devices.

Photonic crystals, considered in Chapter 7, are a special class of optical metamaterials. These periodic *dielectric* structures exhibit photonic bandgaps similar to the electronic bandgaps observed in semiconductor materials. Another special class of metamaterials, described in this section, makes use of *metallic* elements of subwavelength dimensions, such as rods and rings, that are embedded in dielectric media and organized in periodic or random patterns at a subwavelength spatial scale. The shapes of these building blocks, and the patterns in which they are arranged, are designed so that the effective electric and magnetic material parameters, ϵ and μ respectively, are rendered either positive or negative. This in turn allows the synthesis of SNG and DNG materials, with their attendant special optical properties, as detailed in Sec. 8.1.

When the metallic elements exhibit resonance, the effective behavior of ϵ and μ can display the familiar frequency dependence of the susceptibility of a resonant medium, portrayed in Fig. 5.5-6. At frequencies above the resonance frequency, as shown in Fig. 8.3-1, the real part can be negative so that the medium can be SNG or DNG. Although the imaginary part, corresponding to attenuation, is significant near resonance, a narrow band with negative real part can persist at sufficiently high frequencies where the attenuation is minimal.

Since $\mu = \mu_o$ at optical frequencies for naturally occurring nonmagnetic materials, DNG (negative-index) media cannot be readily created without the use of metamaterials. The fabrication of optical DNG metamaterials is undeniably challenging. First, the metallic elements must be of subwavelength dimensions so that resonance frequencies lie in the optical band. This requires nanoscale fabrication technology. Second, the electric and magnetic resonance frequencies must be sufficiently close to each other so that the bands of negative ϵ and μ align, as depicted in Fig. 8.3-1. A great deal of effort has been directed toward addressing these challenges since the field of metamaterials has come to the fore. The principles underlying metamaterials also offer promise in the domains of acoustics, mechanics, and thermodynamics.

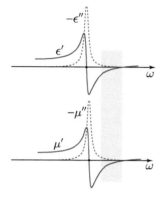

Figure 8.3-1 A medium with resonant permittivity and permeability can behave as a DNG medium in a frequency band above both resonance frequencies (shaded area).

We consider three-dimensional optical metamaterials, and metamaterials of reduced dimensionality, known as metasurfaces, in turn.

A. Metamaterials

The effective electromagnetic parameters ϵ and μ for metamaterials comprising three-dimensional distributions of metallic elements embedded in a dielectric host material may be determined by making use of approximate models, complex analytical techniques, or numerical methods. Approximate models include the effective-medium approach and the effective-circuit approach. The effective-medium approach based on the use of the Maxwell–Garnett formula described in Sec. 5.6D is, strictly speaking, applicable only for nanospheres. However, it is widely used in practice for metallic particles of other shapes as well.

The effective-circuit approach, on the other hand, is applicable for metallic elements of various shapes, provided they are sufficiently small. The dielectric/magnetic properties of a natural material are usually determined by summing the electric/magnetic dipoles of the constituent atoms induced by the applied electric/magnetic fields. This enables the polarization and magnetization densities to be determined, and thence the electric permittivity ϵ and magnetic permeability μ. A similar approach may be adopted for metamaterials, in which each constituent element, regarded as a Rayleigh scatterer, is modeled by electric and/or magnetic dipoles. To determine the electric and magnetic dipole moments, metallic structures of subwavelength dimensions are considered as electric circuit elements, an approach known as the **point-dipole approximation**.

Larger elements may be modeled using Mie scattering theory (see Sec. 5.6C), in which the dipole contributions become the leading terms of multipole series expansions. More complex circuit models can also be used. However, when metallic nanoparticles are juxtaposed in sufficiently close proximity, such that their localized plasmonic fields overlap, the foregoing approximations fail to account for element-to-element interactions and inter-element resonances. Such effects may be accommodated via the circuit approach by considering mutual couplings between neighboring elements, such as mutual inductances, and by treating the composite circuits as transmission lines or circuit networks. When approximations such as these fail, numerical methods offer a fallback position.

We proceed to examine several approximate models based on both the effective-medium and effective-circuit approaches. The models we consider describe metamaterials exhibiting negative permittivity, negative permeability, negative index, as well as hyperbolic properties.

Negative-Permittivity Metamaterial: Metallic Nanospheres in a Dielectric Medium

As described in Sec. 5.6D and illustrated in Fig. 8.3-2, a dielectric medium of electric permittivity ϵ that is uniformly filled with small nanospheres of complex permittivity ϵ_s yields an isotropic composite medium of effective permittivity ϵ_e that obeys the Maxwell-Garnett mixing rule (5.6-20),

$$\epsilon_e \approx \epsilon \frac{2(1-f)\epsilon + (1+2f)\epsilon_s}{(2+f)\epsilon + (1-f)\epsilon_s}, \qquad (8.3\text{-}1)$$

where f is the volume fraction of the inclusions (filling ratio).

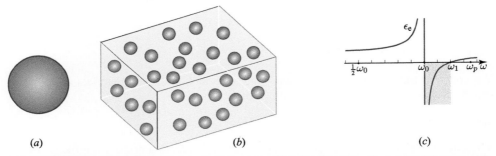

Figure 8.3-2 Negative-permittivity metamaterial. (*a*) Metallic nanosphere. (*b*) Metamaterial comprising uniformly distributed metallic nanospheres embedded in a dielectric medium. (*c*) The effective permittivity ϵ_e has a pole at the resonance frequency ω_0 and a zero at ω_1. Since ϵ_e is negative in the band lying between $\omega_0 < \omega < \omega_1$, this metamaterial is single-negative (SNG) if μ is positive.

For metallic elements described by the simplified Drude model, in accordance with (8.2-18) we have $\epsilon_s = \epsilon_o(1 - \omega_p^2/\omega^2)$, where ω_p is the plasma frequency, so that (8.3-1) yields

$$\epsilon_e = \epsilon_L \frac{1 - \omega^2/\omega_1^2}{1 - \omega^2/\omega_0^2}, \qquad (8.3\text{-}2)$$

where

$$\omega_0 = \frac{\omega_p}{\sqrt{1+\epsilon_{r0}}}, \qquad \omega_1 = \frac{\omega_p}{\sqrt{1+\epsilon_{r1}}}, \qquad (8.3\text{-}3)$$

$$\epsilon_L = \frac{1+2f}{1-f}\epsilon, \qquad \epsilon_{r0} = \frac{2+f}{1-f}\epsilon_r, \qquad \epsilon_{r1} = \frac{2(1-f)}{1+2f}\epsilon_r, \qquad (8.3\text{-}4)$$

and where $\epsilon_r = \epsilon/\epsilon_o$ is the relative permittivity of the host medium, which is assumed to be frequency-independent.

As shown in Fig. 8.3-2(*c*), the effective permittivity ϵ_e has a pole at ω_0 and a zero at ω_1. Since $\epsilon_{r0} > \epsilon_r$, the resonance frequency ω_0 falls below that of the isolated nanosphere, which is given by (8.2-26). Also, since $\epsilon_{r1} < \epsilon_{r0}$, we see that $\omega_1 > \omega_0$, so that ϵ_e is negative within the spectral band between ω_0 and ω_1, which lies below the plasma frequency ω_p of the metal. With μ positive, this metamaterial is therefore single-negative (SNG), much like a homogeneous metal below its plasma frequency [see Fig. 8.2-1(b)].

Negative-Permittivity Metamaterial: Thin Metallic Rods Isotropically Distributed in a Dielectric Medium

The inductance L of a cylindrical metallic rod of length a and radius w ($a \gg w$) [Fig. 8.3-3(a)] is given by $L \approx (\mu_o a/2\pi)\,[\ln(2a/w) - 3/4]$. The effective electric permittivity of a medium comprising parallel rods separated by a distance a, as depicted in Fig. 8.3-3(b), is determined by observing that an electric field E along a rod develops a voltage $V = aE$ between its two ends. This in turn generates an electric current $i = V/j\omega L$ in the inductor, which corresponds to a charge $q = i/j\omega$ and an electric dipole moment $\mathrm{p} = qa$. Since the number of rods per unit volume is $N = 1/a^3$, the polarization density is given by $P = N\mathrm{p} = \mathrm{p}/a^3$. The effective susceptibility of the medium is thus $\chi_e = P/\epsilon_o E$ and the effective permittivity is $\epsilon_e = \epsilon_o(1 + \chi_e)$.

Combining these equations leads to an expression for the effective permittivity that is identical in form to that of a simple Drude metal,

$$\epsilon_e = \epsilon_o\left(1 - \frac{\omega_p^2}{\omega^2}\right), \qquad \omega_p = \frac{1}{\sqrt{\epsilon_o a L}} = 2\pi\frac{c_o}{a}\frac{1}{\sqrt{2\pi\ln(2a/w) - \frac{3}{2}\pi}}, \qquad (8.3\text{-}5)$$

with a plasma frequency ω_p determined by the dimensions of the rod, a and w, via its inductance L, where we have assumed that the dielectric medium has the permittivity of free space. With μ positive, this metamaterial is thus single-negative (SNG). Though the rod is assumed to be a perfect conductor in the calculations set forth above, loss is readily accommodated by adding a resistance R to the impedance $j\omega L$ of the rod.

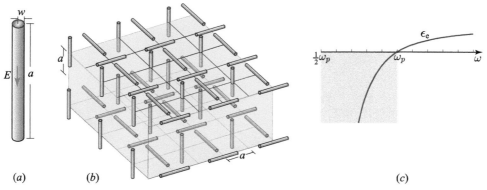

Figure 8.3-3 Negative-permittivity metamaterial. (a) Thin metallic rods of length a and radius w, oriented in (b) three orthogonal directions at every point of a cubic lattice of dimension a, create an isotropic metamaterial. (c) The effective electric permittivity ϵ_e has a frequency dependence identical to that of the simplified Drude model. This metamaterial is thus single-negative (SNG), provided that μ is positive.

Negative-Permeability Metamaterial: Split-Ring Metallic Elements in a Dielectric Medium

A metallic split ring [Fig. 8.3-4(a)], modeled as an inductor L in series with a capacitor C (the open section of the ring), forms a resonant circuit of resonance frequency $\omega_0 =$

$1/\sqrt{LC}$. When the size of the split ring is $\lesssim 100$ nm and the gap is $\lesssim 10$ nm, ω_0 lies in the optical region of the spectrum.

The effective magnetic permeability μ_e of a metamaterial consisting of a collection of such split rings, organized uniformly and in three directions at the vertices of a periodic lattice, as shown in Fig. 8.3-4(b), may be established by calculating the magnetic dipole moment m induced by a magnetic field H applied along an axis normal to the plane of the ring [Fig. 8.3-4(a)]. The voltage V induced in the loop is equal to the rate of change of the magnetic flux so that $V = -j\omega\mu_o AH$, where A is the area of the ring. This voltage generates an electric current $i = V/Z$, where the circuit impedance is $Z = j\omega L + 1/j\omega C$. This electric current in turn results in a magnetic dipole moment $\text{m} = Ai$.

A density of N split rings per unit volume gives rise to a magnetization density $M = N\text{m}$ so that the effective magnetic permeability $\mu_e = \mu_o(H + M)/H$ is given by

$$\mu_e = \mu_o \frac{1 - \omega^2/\omega_1^2}{1 - \omega^2/\omega_0^2}, \qquad \omega_0 = \frac{1}{\sqrt{LC}}, \qquad \omega_1 = \frac{\omega_0}{\sqrt{1 - \mu_o NA^2/L}}. \qquad (8.3\text{-}6)$$

The inductance of the ring is $L \approx \mu_o b \left[\ln(8b/a) - 7/4\right]$, where b and a are the ring and wire radii, respectively ($b \gg a$). As displayed in Fig. 8.3-4(c), μ_e exhibits a resonance at ω_0 and a zero at ω_1, and is negative in the intervening region. For a positive electric permittivity ϵ, this structure behaves as a single-negative (SNG) metamaterial. The frequency dependence of μ_e revealed in (8.3-6) is the same as that of the effective permittivity ϵ_e for metallic nanospheres set forth in (8.3-2).

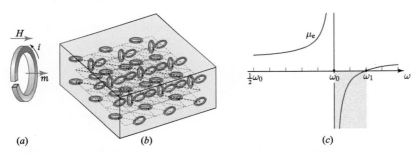

Figure 8.3-4 Negative-permeability metamaterial. (a) A metallic split ring excited by a magnetic field exhibits a magnetic dipole moment m. (b) An isotropic metamaterial fabricated by configuring such split rings in three directions at the vertices of a cubic lattice. (c) The frequency dependence of the effective magnetic permeability μ_e exhibits a pole at ω_0, a zero at ω_1, and is negative in the intervening range. When ϵ is positive, this structure serves as a single-negative (SNG) metamaterial.

Negative-Index Metamaterials

The negative electric permittivity of the metallic-rod metamaterial [Fig. 8.3-3(c)] may be combined with the negative magnetic permeability of the metallic split-ring metamaterial [Fig. 8.3-4(c)] to create a double-negative (DNG) metamaterial that serves as a negative-index material (NIM). Implementation is achieved by repeating the combined rod and double split-ring element, displayed in Fig. 8.3-5(a), in two directions, as illustrated in Fig. 8.3-5(b). This approach requires oblique incidence of the wave, as indicated, so that surface plasmon resonances (SPR) can be excited with out-of-plane magnetic fields. This design was first experimentally demonstrated in the microwave

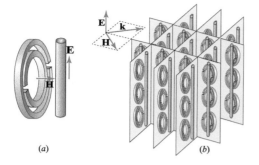

Figure 8.3-5 Negative-index metamaterial. (*a*) Combined rod and double split-ring element. (*b*) DNG metamaterial comprising an array of the elements shown in (*a*), oriented along two orthogonal directions. For waves traveling along directions in the horizontal plane, the medium is double-negative, and the refractive index is negative, at frequencies above the permittivity and permeability resonances, as schematized in Fig. 8.3-1.

region, and its dimensions were subsequently scaled down for operation at optical frequencies.

Alternative designs for optical NIMs that are easier to fabricate have subsequently been developed. One such design is a "fishnet" metal–dielectric multilayer structure, a simplified version of which is illustrated in Fig. 8.3-6. In this configuration, the optical wave is normally incident on the fishnet surface and the electric and magnetic fields are aligned with the metal strips, as shown. The strips aligned with the electric field are responsible for the negative permittivity. The strips aligned with the magnetic field support anti-symmetric resonant modes between pairs of coupled strips, which results in negative permeability above the resonance frequency. Fishnet nanostructures serve as NIMs that operate in the visible region.

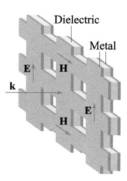

Figure 8.3-6 Simplified version of a "fishnet" metal–dielectric nanostructured composite metamaterial consisting of a stacked network of intersecting subwavelength plasmonic waveguides. The structure serves as a negative-index material (NIM) in the visible region of the spectrum.

*Hyperbolic Metamaterial: Parallel Metallic Rods in a Dielectric Medium

As described in Sec. 8.1C, an anisotropic medium is said to be hyperbolic if the effective principal values of its permittivity (or permeability) tensor have mixed signs. We demonstrate that an array of parallel metallic rods embedded in a dielectric medium of permittivity ϵ, such as those illustrated in Fig. 8.3-7, may exhibit hyperbolic behavior when the rods have extreme anisotropy by virtue of a large aspect ratio.

For this purpose, we use the effective-medium approach to write expressions for the principal values of the permittivity tensor along the x, y, z directions (denoted 1, 2, 3, respectively). An analysis similar to that carried out for the dielectric sphere, the result of which is provided in (5.6-17), reveals that the ratio of internal and external fields for a dielectric cylinder takes the form $E_i/E_0 = 2\epsilon/(\epsilon_s + \epsilon)$. Using the Maxwell–Garnett mixing rule then yields effective permittivities for fields in the plane of the cross section, and along the axial direction (where $E_i/E_0 = 1$), that are given by:

$$\epsilon_{e1} = \epsilon_{e2} = \epsilon \frac{2f\epsilon_s + (1-f)(\epsilon_s + \epsilon)}{\epsilon_s + \epsilon + f(\epsilon - \epsilon_s)}, \qquad \epsilon_{e3} = (1-f)\epsilon + f\epsilon_s. \qquad (8.3\text{-}7)$$

Making use of the simplified Drude model (8.2-18) for ϵ_s, combining the results for metallic nanospheres (8.3-2)–(8.3-4) and metallic rods (8.3-5), and using (8.2-20), we obtain the following expressions for the permittivity principal values:

$$\epsilon_{e1} = \epsilon_{e2} = \epsilon_L \frac{1 - \omega^2/\omega_1^2}{1 - \omega^2/\omega_0^2}, \qquad \epsilon_{e3} = \epsilon_H \left(1 - \frac{\omega_3^2}{\omega^2}\right), \qquad (8.3\text{-}8)$$

where

$$\omega_0 = \frac{\omega_p}{\sqrt{1 + \epsilon_{r0}}}, \qquad \omega_1 = \frac{\omega_p}{\sqrt{1 + \epsilon_{r1}}}, \qquad \omega_3 = \frac{\omega_p}{\sqrt{1 + \epsilon_{r3}}}, \qquad (8.3\text{-}9)$$

$$\epsilon_{r0} = \frac{1 + f}{1 - f}\epsilon_r, \qquad \epsilon_{r1} = \frac{1 - f}{1 + f}\epsilon_r, \qquad \epsilon_{r3} = \frac{1 - f}{f}\epsilon_r, \qquad (8.3\text{-}10)$$

$$\epsilon_L = \frac{1 + f}{1 - f}\epsilon, \qquad\qquad \epsilon_H = \epsilon_o\left[f + (1 - f)\epsilon_r\right], \qquad (8.3\text{-}11)$$

with $\epsilon_r = \epsilon/\epsilon_o$.

The effective ordinary permittivity $\epsilon_{e1} = \epsilon_{e2}$ exhibits a pole at the resonance frequency ω_0 and a zero at ω_1, while the effective extraordinary permittivity ϵ_{e3} varies from negative values at low frequencies to positive values at high frequencies, passing through zero at ω_3, much like a pure metal. As illustrated in Fig. 8.3-7, an extensive frequency band exists for which $\epsilon_{e1} = \epsilon_{e2}$ and ϵ_{e3} have opposite signs, so that the anisotropic medium is hyperbolic. Within this band, the material behaves like a dielectric in one direction and a metal in the orthogonal direction.

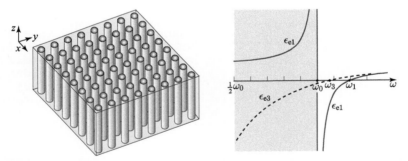

Figure 8.3-7 Hyperbolic metamaterial. Frequency dependence of the principal components of the effective electric permittivity tensor for a metamaterial comprising parallel metallic rods described by the simplified Drude model embedded in a host medium of permittivity ϵ. The ordinary effective permittivities $\epsilon_{e1} = \epsilon_{e2}$ have resonance frequencies ω_0 and exhibit positive values below resonance. The extraordinary effective permittivity ϵ_{e3} is negative in the frequency range $\omega < \omega_3$. The spectral band over which the effective medium is hyperbolic is delineated by the shaded region, where $\epsilon_{e1} = \epsilon_{e2}$ and ϵ_{e3} have opposite signs.

B. Metasurfaces

Metasurfaces are metamaterials whose dimensionality is reduced from three to two. Examples are ultrathin arrays of subwavelength-scale metallic elements, deposited in periodic, aperiodic, or random patterns, on the surface of a dielectric substrate. **Complementary metasurfaces** comprise arrays of subwavelength-scale dielectric elements, such as holes and their separations, arrayed on an ultrathin metallic surface. The shapes of the individual elements, and the geometry of their layout on the surface, endow metasurfaces with distinctive optical properties that are a consequence of the coupling of light and SPP waves generated at the metal–dielectric boundary.

Metasurface as a Phase Modulator

As explained in Sec. 2.4, a wave traveling in the z direction, on transmission through a dielectric plate of fixed thickness d and graded refractive index $n(x, y)$ in the x–y plane, undergoes a spatially varying phase shift $\varphi(x, y) = n(x, y)k_o d$, which modifies its wavefront [see (2.4-14)]. Achieving a phase shift of 2π requires a local thickness equal to the wavelength of light in the medium. A planar metasurface has the merit that it can introduce a phase shift of similar magnitude with far less thickness. The metallic elements of the metasurface function much like optical antennas that modify the optical wavefront. A resonant antenna acts as a scatterer that introduces a frequency-dependent phase shift that ranges from $-\pi/2$ to $\pi/2$ for frequencies below to above resonance, respectively.

A spatially varying phase shift $\varphi(x, y)$ may be implemented by making use of a metasurface comprising elements of spatially graded sizes and geometries that correspond to spatially varying resonance frequencies. An incoming wave of fixed frequency is then subjected to a spatially varying phase shift so that the metasurface acts as a phase modulator. An example is provided in Fig. 8.3-8(a). Since the metasurface is ultra thin, it may be modeled mathematically as an optical component that introduces a spatially varying phase discontinuity (i.e., a phase shift that takes place over a distance $d \to 0$).

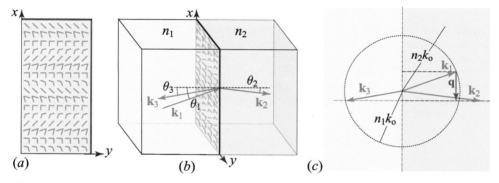

Figure 8.3-8 (a) A metasurface using an array of metallic elements whose shapes and resonance frequencies vary in the x direction. The shapes of the elements are engineered such that the phase shift they introduce is a linear function $\varphi = qx$ for one of the polarization components. (b) Negative reflection and negative refraction at a boundary between two media of refractive indices n_1 and n_2 by virtue of the presence of the metasurface portrayed in (a) between the two media. (c) Phase-matching condition for the incident and refracted waves, and for the incident and reflected waves, at the metasurface boundary.

The phase modulation introduced by such a metasurface may modify an incoming optical wave in any of the many ways described in Sec. 2.4. A salutary feature of this approach is that the wave undergoes minimal spatial spread (diffraction) as it crosses the infinitesimally thin metasurface. Consider, for example, a phase $\varphi(x, y)$ that varies linearly along the metasurface at a rate q, so that $\varphi = qx$. The complex amplitude of the incoming wave is then modulated by the factor $\exp(-jqx)$, which is a periodic function of the spatial frequency $\nu_x = q/2\pi$, as explained in Sec. 4.1A. An incoming plane wave of wavevector $\mathbf{k}_1$ will then generate refracted and reflected plane waves with wavevectors $\mathbf{k}_2$ and $\mathbf{k}_3$, respectively.

To ensure phase matching at both sides of the surface, as depicted in Fig. 8.3-8(c), the component of the vector $\mathbf{k}_2$ parallel to the surface must match that of $\mathbf{k}_1 + \mathbf{q}$, where $\mathbf{q}$ is a vector of magnitude q pointing in the x direction. Likewise, for the reflected wave, the component of the vector $\mathbf{k}_3$ along the surface must match that of $\mathbf{k}_1 + \mathbf{q}$. Hence, if the metasurface lies at the boundary between two ordinary media of refractive indices n_1 and n_2, its presence causes the conventional Snell's law of refraction and reflection to assume the following modified form:

$$n_2 k_o \sin\theta_2 = n_1 k_o \sin\theta_1 + q, \qquad (8.3\text{-}12)$$
Metasurface Refraction

$$n_1 k_o \sin\theta_3 = n_1 k_o \sin\theta_1 + q, \qquad (8.3\text{-}13)$$
Metasurface Reflection

where θ_1, θ_2, and θ_3 are the angles of incidence, refraction, and reflection, respectively.

With appropriate choice of the magnitude and sign of q, the presence of the metasurface can result in *negative reflection* and *negative refraction* at the boundary, as illustrated in Fig. 8.3-8(*b*). Equations (8.3-12) and (8.3-13) properly reduce to Snell's law when $q = 0$.

For a phase discontinuity $\varphi(x)$ that varies slowly with the position x, the derivative $q = d\varphi/dx$ may be regarded as the local spatial frequency at x. This quantity determines the local tilt imparted to an incoming wavefront, and thus the angles of reflection and refraction as a function of x. This approach can clearly be generalized to metasurfaces that introduce a two-dimensional phase discontinuity $\varphi(x, y)$. In that case, the vector $\mathbf{q} = \nabla\varphi$ represents the magnitude and direction of the local spatial frequency of the phase modulation. The metasurface can therefore be designed to introduced desired local tilts of the wavefront in both the x–z and y–z planes, much like an antenna array or an optical phase plate. The metasurface can also be engineered to introduce position-dependent amplitude modulation, imparted by the shape of the local elements. The combination of phase and amplitude modulation can serve as a hologram with complex transmittance that is designed to simulate the wavefront of light generated by an object.

Extraordinary Optical Transmission Through Subwavelength Holes in a Metallic Film

A metasurface consisting of a periodic array of subwavelength holes and separations perforated in a planar metallic film may exhibit extraordinarily high optical-power transmittance at certain wavelengths and angles of incidence. The power transmittance $\mathfrak{T}(\lambda, \theta)$, as a function of the wavelength λ and the angle of incidence θ, is found to exhibit sharp peaks with values that significantly exceed those predicated on the basis of conventional diffraction theory. Indeed, if $\mathfrak{T}_h$ is the total hole area per unit area of the film, the peak values of the transmittance $\mathfrak{T}(\lambda, 0)$ for a plane wave traveling in a direction orthogonal to the film may exceed $\mathfrak{T}_h$ by orders of magnitude.

This phenomenon is attributable to the generation of SPP waves through the holes and the attendant radiation of light by the oscillating charges created in the metallic film. The array of subwavelength holes in the metal film should thus be considered as an active radiating element, rather than as a passive geometrical aperture through which the light is transmitted.

Maximum transmission occurs at frequencies for which the incident optical wave and the excited SPP wave are phase matched. For a periodic array of holes in the form of a square lattice with period a_0, the phase-matching condition is

$$\beta = k_\perp \pm m_x g_x \pm m_y g_y, \qquad (8.3\text{-}14)$$

where $\beta = (\omega/c_o)\sqrt{\epsilon_b/\epsilon_o}$ is the propagation constant of the SPP wave [see (8.2-21)]; $k_\perp = (2\pi/\lambda)\sin\theta$ is the component of the wavevector of the incident light in the plane of the array; $g_x = g_y = 2\pi/a_0$ are the fundamental spatial frequencies of the periodic array; and m_x and m_y are integers representing associated spatial harmonics (the scattering order). The transmittance $\mathcal{T}(\lambda, \theta)$ as a function of the angle of incidence θ also exhibits photonic band gaps,[†] much like those observed in photonic crystals, which are three-dimensional periodic dielectric structures of subwavelength dimensions (see Sec. 7.3).

Similar extraordinary optical transmission is expected for holes in a perfect conductor, rather than a real metal. The medium is then characterized by an effective dielectric function that has a plasmon form, with a plasma frequency dictated by the geometry of the holes.[‡]

*8.4 TRANSFORMATION OPTICS

In other chapters of this book, graded-index optics is considered from an *analysis* perspective, with a mandate to *determine how light propagates in a medium endowed with particular dielectric and magnetic properties*. Examples are provided in the context of ray optics, wave optics, and electromagnetic optics:

- Graded-index (GRIN) materials allow optical rays to follow curved trajectories governed by the profile of the refractive index $n(\mathbf{r})$ (Sec. 1.3).
- The trajectories of scalar waves in GRIN materials can be described in terms of the Eikonal equation (Secs. 2.3 and 10.2C).
- For isotropic materials with graded electric permittivity $\epsilon(\mathbf{r})$ and magnetic permeability $\mu(\mathbf{r})$, Maxwell's equations give rise to the generalized Helmholtz equations (5.2-16) and (5.2-17); these equations were solved in Chapter 7 for layered and periodic structures such as photonic crystals.
- The optics of anisotropic graded media are described by Maxwell's equations with position-dependent tensors $\boldsymbol{\epsilon}(\mathbf{r})$ and $\boldsymbol{\mu}(\mathbf{r})$; the solution requires full vector analysis.
- GRIN optics is useful for fabricating a variety of optical components, including GRIN lenses (Exercise 1.3-1) and GRIN optical fibers (Sec. 10.1B).

In this section, in contrast, we consider graded-index optics from a *synthesis* or *design* perspective, where the goal is to *determine the dielectric and magnetic properties of a medium that realize a desired pattern of light propagation*. The synthesis problem is more challenging than the analysis problem in two respects: (1) the required mathematical tools are more advanced; and (2) physical implementation of the graded medium often requires the use of metamaterials constructed from components that are available, configured in a particular spatial arrangement, and amenable to being fabricated with current technology.

A. Transformation Optics

Transformation optics is a mathematical tool that facilitates the design of optical materials that guide light along desired trajectories. The underlying concept relies on a geometrical transformation that converts simple trajectories into desired ones. In order that Maxwell's equations remain valid, the optical parameters associated with

---

[†] T. W. Ebbesen, H. J. Lezec, H. F. Ghaemi, T. Thio, and P. A. Wolff, Extraordinary Optical Transmission Through Sub-Wavelength Hole Arrays, *Nature*, vol. 391, pp. 667–669, 1998.

[‡] F. J. Garcia-Vidal, L. Martín-Moreno, and J. B. Pendry, Surfaces with Holes in Them: New Plasmonic Metamaterials, *Journal of Optics A: Pure and Applied Optics*, vol. 7, pp. S97–S101, 2005.

the transformed equivalent system must also be modified, and this establishes the character of the required optical material. As a simple example of such equivalence, a local compression of the coordinate system by a scaling factor is equivalent to a local increase of the refractive index by the same factor, so that the optical pathlength (product of length and refractive index) remains unchanged.

A three-step design procedure provides a guide:

- Begin with a *pilot* physical system for which the optical trajectories are known, such as a homogeneous and isotropic material.
- Find a coordinate transformation that converts these trajectories to the desired ones.
- Determine the transformed physical parameters of the equivalent material. The new material will implement the desired optical trajectories in the original coordinate system.

Since geometrical transformations generally involve changes of directions and introduce direction-dependent scaling, the transformed parameters are generally both anisotropic and spatially varying.

Transformation Principle

Let $\{\epsilon_{ij}\}$ and $\{\mu_{ij}\}$ be the elements of the permittivity and permeability tensors of the original material in the original coordinate system (x_1, x_2, x_3). The elements of the permittivity and permeability tensors of the equivalent material (denoted by the superscript "e") in the transformed coordinate system (u_1, u_2, u_3) are then related to the original elements by the matrix equations[†]

$$\boldsymbol{\epsilon}^e = |\det \mathbf{A}|^{-1} \mathbf{A}^\mathsf{T} \boldsymbol{\epsilon} \mathbf{A}, \quad \boldsymbol{\mu}^e = |\det \mathbf{A}|^{-1} \mathbf{A}^\mathsf{T} \boldsymbol{\mu} \mathbf{A}. \qquad (8.4\text{-}1)$$

Here $\mathbf{A}$ is the 3×3 Jacobian transformation matrix, whose elements are the partial derivatives

$$A_{ij} = \frac{\partial u_i}{\partial x_j}, \quad i,j = 1,2,3. \qquad (8.4\text{-}2)$$

The quantity $\mathbf{A}^\mathsf{T}$ is the transpose of $\mathbf{A}$, and $\boldsymbol{\epsilon}$ and $\boldsymbol{\mu}$ are 3×3 matrices whose elements are $\{\epsilon_{ij}\}$ and $\{\mu_{ij}\}$, respectively. Since $\mathbf{A}$ is generally dependent on (x_1, x_2, x_3), the equivalent material is generally inhomogeneous, even if the original material is homogeneous.

In the special case for which the original material is both homogeneous and isotropic, say free space, then $\boldsymbol{\epsilon}$ and $\boldsymbol{\mu}$ are diagonal with equal diagonal elements ϵ_o and μ_o, respectively, whereupon

$$\epsilon_o^{-1} \boldsymbol{\epsilon}^e = \mu_o^{-1} \boldsymbol{\mu}^e = |\det \mathbf{A}|^{-1} \mathbf{A}^\mathsf{T} \mathbf{A}. \qquad (8.4\text{-}3)$$

The tensors $\boldsymbol{\epsilon}^e$ and $\boldsymbol{\mu}^e$ are then identical except for a scaling factor. Under these conditions the impedance, which depends on their ratio, remains unchanged for all polarizations, which in turn implies that the equivalent medium introduces no reflection at any boundary with free space.

We provide a number of examples to illustrate the transformation principle:

[†] See, e.g., J. B. Pendry, Y. Luo, and R. Zhao, Transforming the Optical Landscape, *Science*, vol. 348, pp. 521–524, 2015.

EXAMPLE 8.4-1. *Refraction Without Reflection.* In this example, we design an optical material implementing ray trajectories that refract without reflection at a planar surface, as shown in Fig. 8.4-1(*a*).

We begin with an initial homogeneous medium, say free space, with rays that follow parallel straight trajectories at an angle θ_1, as shown in Fig. 8.4-1(*b*). We now apply a geometrical transformation that stretches the coordinate system by a scale factor s along the x_3 direction in the region $x_3 > 0$. The desired refraction is achieved by choosing s as the ratio of the initial and desired slopes, $s = \tan\theta_1 / \tan\theta_2$ [Fig. 8.4-1(*c*)]. This transformation is implemented by the relations

$$u_1 = x_1, \quad u_2 = x_2, \quad u_3 = s^{-1}x_3. \tag{8.4-4}$$

This type of scaling of the Cartesian coordinate system, in which the directions of the axes do not change, converts a cube into a cuboid. Based on (8.4-2), the Jacobian matrix $\mathbf{A}$ is diagonal with diagonal elements $(1, 1, s^{-1})$ and determinant $\det \mathbf{A} = s^{-1}$, so that (8.4-3) provides

$$\epsilon_o^{-1}\boldsymbol{\epsilon}^e = \mu_o^{-1}\boldsymbol{\mu}^e = \begin{bmatrix} s & 0 & 0 \\ 0 & s & 0 \\ 0 & 0 & s^{-1} \end{bmatrix}. \tag{8.4-5}$$

Since the matrices $\boldsymbol{\epsilon}^e$ and $\boldsymbol{\mu}^e$ are diagonal, the anisotropic material has principal axes pointing along the axes of the coordinate system. The principal values are: $\epsilon_1 = s\epsilon_o$, $\epsilon_2 = s\epsilon_o$, and $\epsilon_3 = s^{-1}\epsilon_o$ together with $\mu_1 = s\mu_o$, $\mu_2 = s\mu_o$, and $\mu_3 = s^{-1}\mu_o$.

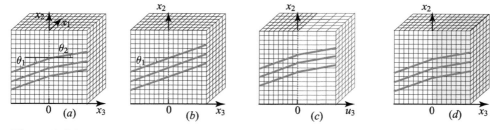

Figure 8.4-1 Geometrical transformation implementing refraction without reflection. (*a*) Desired optical trajectories. (*b*) Free space with straight-line optical trajectories. (*c*) Stretching the coordinate system by the factor s for $x_3 > 0$ causes the rays to change slope and follow the desired trajectories. (*d*) Equivalent anisotropic, homogeneous material that causes the rays to change slope in an identical way.

The parameters of the equivalent material may also be obtained by matching the phase shift encountered when a plane wave crosses the stretched free-space segment with that encountered when the wave is transmitted through an unstretched segment filled with the new material. To determine the parameters, we consider three waves in turn, each with the electric field along one of the coordinates:
– Wave 1 is a plane wave traveling along the x_3 direction with electric and magnetic fields in the x_1 and x_2 directions, respectively. The appropriate permittivity and permeability are thus ϵ_1 and μ_2 so that $k = \omega\sqrt{\epsilon_1\mu_2} = \omega\sqrt{s\epsilon_o s\mu_o} = sk_o$, corresponding to a refractive index $n_1 = s$. The phase shift accumulated over the distance d is therefore $sk_o d$, as expected, and the impedance is $\eta_1 = \sqrt{\mu_2/\epsilon_1} = \eta_o$.
– Wave 2 is also taken to travel along the x_3 direction but the electric and magnetic fields are now in the x_2 and $-x_1$ directions, respectively. This wave also travels with a refractive index $n_2 = s$ and has an impedance $\eta_2 = \eta_o$.
– Wave 3 travels along the x_2 direction with electric and magnetic fields in the x_3 and x_1 directions, respectively. The appropriate permittivity and permeability are ϵ_3 and μ_1 so that $k = \omega\sqrt{\epsilon_3\mu_1} = \omega\sqrt{s^{-1}\epsilon_o s\mu_o} = k_o$, corresponding to a refractive index $n_3 = 1$. The phase shift is $k_o d$, as expected, since there is no stretching in the x_2 direction. The impedance is $\eta_3 = \sqrt{\mu_1/\epsilon_3} = s\eta_o$.

Using these results, we conclude that the final design is a piecewise homogeneous medium with free space in the left half plane and an anisotropic uniaxial material in the right half plane [Fig. 8.4-1(d)]. The anisotropic material is birefringent with $n_1 = s$, $n_2 = s$, and $n_3 = 1$, but it introduces no reflection at the boundary with free space since the impedances are the same as that of free space: $\eta_1 = \eta_2 = \eta_o$.

Two factors distinguish refraction at the boundary of the synthesized anisotropic medium from conventional refraction at the boundary of a homogeneous and isotropic medium: (1) the refraction is not accompanied by reflection; and (2) the relationship between the angle of refraction and the angle of incidence, $s \tan \theta_2 = \tan \theta_1$, differs from Snell's law.

EXAMPLE 8.4-2. *Refraction at Normal Incidence.* We next consider the design of an optical material that implements refraction by an angle θ at a normal planar surface, as depicted in Fig. 8.4-2(a). This type of refraction cannot occur at the boundary between two isotropic dielectric materials, but can occur at the boundary between an isotropic and an anisotropic material, as described in Sec. 6.3E.

We begin with a pilot system of free space with ray trajectories along horizontal parallel straight lines [Fig. 8.4-2(b)], and implement the coordinate transformation

$$u_1 = x_1, \quad u_2 = x_2 + sx_3, \quad u_3 = x_3 \tag{8.4-6}$$

for $x_3 > 0$, with $s = \tan \theta$. This deflects the trajectories, as desired, by shearing along the x_2 direction [Fig. 8.4-1(c)]. The permittivity and permeability tensors of the equivalent anisotropic material corresponding to this coordinate transformation, as determined by use of (8.4-2) and (8.4-3), are:

$$\epsilon_o^{-1}\boldsymbol{\epsilon}^e = \mu_o^{-1}\boldsymbol{\mu}^e = \begin{bmatrix} 1 & 0 & 0 \\ 0 & 1 & s \\ 0 & s & 1+s^2 \end{bmatrix}. \tag{8.4-7}$$

This represents a homogeneous, but anisotropic, medium. When placed in the $x_3 > 0$ region, it introduces the desired refraction at normal incidence [Fig. 8.4-2(d)].

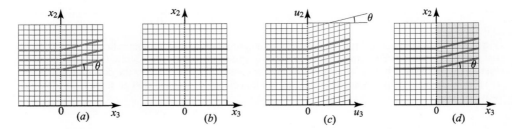

Figure 8.4-2 Geometrical transformation implementing refraction at normal incidence. (a) Desired optical trajectories. (b) Free space with horizontal straight-line optical trajectories. (c) Shearing the coordinate system along the x_2 direction for $x_3 > 0$ refracts the trajectories as desired. (d) Equivalent anisotropic material that exhibits identical refraction.

EXAMPLE 8.4-3. *Cylindrical Focusing.* In this case, parallel straight-line trajectories are to be refracted at a planar boundary such that they all meet at a common focal point at a distance f from the boundary, as shown in Fig. 8.4-3(a).

We begin with the straight trajectories shown in Fig. 8.4-3(b) in a Cartesian coordinate system and apply the coordinate transformation

$$u_1 = x_1, \quad u_2 = (f - x_3)\sin(x_2/f), \quad u_3 = f - (f - x_3)\cos(x_2/f), \tag{8.4-8}$$

for $x_3 > 0$. The result is a cylindrical coordinate system centered at $(u_2 = 0, u_3 = f)$, as shown in Fig. 8.4-3(c). This transformation converts a line $x_2 = a$ in the plane $x_1 = 0$ in the original coordinate system into a line $u_2 = (f - u_3)\tan(a/f)$ in the new coordinate system. It also converts a line $x_3 = b$ in the plane $x_1 = 0$ into a circle $u_2^2 + (f - u_3)^2 = (f - b)^2$ of radius $(f - b)$ centered at the point $(u_2, u_3) = (0, f)$. Based on (8.4-1) and (8.4-2), the transformation yields the diagonal matrix

$$\epsilon_o^{-1} \boldsymbol{\epsilon}^e = \mu_o^{-1} \boldsymbol{\mu}^e = \begin{bmatrix} s & 0 & 0 \\ 0 & s^{-1} & 0 \\ 0 & 0 & s \end{bmatrix}, \qquad s = \frac{f}{|x_3 - f|}. \qquad (8.4\text{-}9)$$

The permittivity and permeability tensors of the equivalent medium therefore have principal axes along the (x_1, x_2, x_3) axes, with principal values that are dependent on the position x_3, i.e., the equivalent material is graded along the x_3 direction with larger anisotropy near the focal line [Fig. 8.4-3(d)].

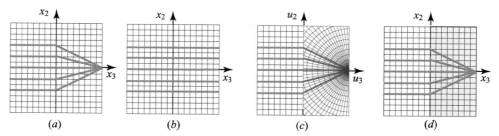

Figure 8.4-3 Geometrical transformation implementing cylindrical focusing. (*a*) Desired optical trajectories. (*b*) Free space with Cartesian coordinate system and parallel straight-line optical trajectories. (*c*) Conversion to a cylindrical coordinate system centered at $x_3 = f$ for $x_3 > 0$ produces the desired trajectories. (*d*) Equivalent anisotropic material with identical trajectories.

B. Invisibility Cloaks

An invisibility cloak is a device that guides light around an object such that the object appears transparent, and therefore invisible. For example, the trajectories shown in Fig. 8.4-4(*a*) avoid a sphere of radius *a*, emerging as if they had followed straight lines and passed right through it.

Following the prescribed design steps for transformation optics, we begin with the straight-line trajectories shown in Fig. 8.4-4(*b*) in a Cartesian-coordinate system (x_1, x_2, x_3). We next convert to a coordinate system (u_1, u_2, u_3) such that points of a sphere with radius $r = \sqrt{x_1^2 + x_2^2 + x_3^2}$ are mapped to points of a sphere of radius $u = \sqrt{u_1^2 + u_2^2 + u_3^2} > a$, thereby avoiding the sphere of radius *a*, as desired. This is accomplished for all points $r < b$, where $b > a$, via the linear relation

$$u = a + \frac{b-a}{b} r. \qquad (8.4\text{-}10)$$

As r varies from 0 to b, u varies from a to b so that points of the sphere $0 < r < b$ in the original coordinate system are mapped into points in a spherical shell $a < u < b$ in the new coordinate system. This mapping may also be written as $u = s(r)r$, where

$$s(r) = \frac{a}{r} + \frac{b-a}{b} \qquad (8.4\text{-}11)$$

is a position-dependent scaling factor.

When applied isotropically, this scaling produces the coordinate transformation

$$u_1 = s(r)\, x_1, \quad u_2 = s(r)\, x_2, \quad u_3 = s(r)\, x_3. \qquad (8.4\text{-}12)$$

As can be shown by simple substitution, points on the straight line $x_2 = f$ in the plane $x_1 = 0$ are mapped into the curved trajectory

$$u_2^2 + u_3^2 = a^2 \left[\frac{u_2}{u_2 - (b-a)f/b} \right]^2 \qquad (8.4\text{-}13)$$

in the u_2–u_3 plane. The red curved trajectories shown in Fig. 8.4-4 are computed from (8.4-13) for four values of f. The grid shown in Fig. 8.4-4(c) within the shell $a < u < b$ is computed by use of (8.4-13) and a similar equation determined by mapping the straight lines $x_3 = f$ in the plane $x_1 = 0$ into the u_2–u_3 plane. The parameters of the

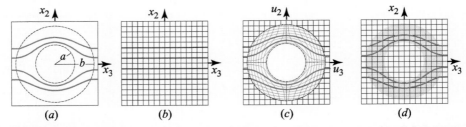

Figure 8.4-4 Geometrical transformation implementing cloaking of a sphere. (*a*) Desired optical trajectories. (*b*) Free space with Cartesian coordinate system and parallel straight-line optical trajectories. (*c*) Coordinate system transformation mapping points inside the sphere to points within a spherical shell outside the sphere, thereby producing the desired trajectories. (*d*) Equivalent anisotropic material with identical trajectories.

equivalent material to be placed in the $a < r < b$ spherical shell shown in Fig. 8.4-4(*d*) may be determined by use of (8.4-1) and (8.4-2) together with (8.4-12). The result is

$$\epsilon_o^{-1}\boldsymbol{\epsilon}^e = \mu_o^{-1}\boldsymbol{\mu}^e = \frac{b}{b-a}\frac{1}{v^2}\begin{bmatrix} v^2 - u_1^2 & -u_1 u_2 & -u_1 u_3 \\ -u_2 u_1 & v^2 - u_2^2 & -u_2 u_3 \\ -u_3 u_1 & -u_3 u_2 & v^2 - u_3^2 \end{bmatrix}, \quad v^2 = \frac{u^4}{2au - a^2}. \qquad (8.4\text{-}14)$$

Clearly, the dielectric and magnetic properties of the material in the spherical shell are both inhomogeneous and anisotropic. For example, at points on the x_1 axis $(u,0,0)$, we have

$$\epsilon_o^{-1}\boldsymbol{\epsilon}^e = \mu_o^{-1}\boldsymbol{\mu}^e = \frac{b}{b-a}\begin{bmatrix} 1 - u^2/v^2 & 0 & 0 \\ 0 & 1 & 0 \\ 0 & 0 & 1 \end{bmatrix}. \qquad (8.4\text{-}15)$$

At these points, the principal axes are aligned with the Cartesian coordinates (x_1, x_2, x_3). The principal value ϵ_1 varies from 0 to $\epsilon_o(b-a)/b$ as u varies from a to b, while the principal values ϵ_2 and ϵ_3 are fixed at the value $\epsilon_o b/(b-a)$. Similar results apply for μ.

At optical wavelengths, the implementation of invisibility cloaks via metamaterials requires the use of advanced nanofabrication technologies such as electron-beam or focused ion-beam lithography. The constituent dielectric and magnetic elements, which have various shapes and dimensions, must be intricately designed and precisely laid out. Since such elements are highly resonant, the electromagnetic properties of the metamaterial depend strongly on wavelength so that such devices typically operate only over narrow spectral bandwidths.

READING LIST

Single- and Double-Negative Media

S. A. Ramakrishna and T. M. Grzegorczyk, *Physics and Applications of Negative Refractive Index Materials*, CRC Press/Taylor & Francis, 2008.

J. B. Pendry and D. R. Smith, Reversing Light with Negative Refraction, *Physics Today*, vol. 57, no. 6, pp. 37–43, 2004.

R. A. Depine and A. Lakhtakia, A New Condition to Identify Isotropic Dielectric–Magnetic Materials Displaying Negative Phase Velocity, *Microwave and Optical Technology Letters*, vol. 41, pp. 315–316, 2004.

S. A. Darmanyan, M. Nevière, and A. A. Zakhidov, Surface Modes at the Interface of Conventional and Left-Handed Media, *Optics Communications*, vol. 225, pp. 233–240, 2003.

W. S. Weiglhofer and A. Lakhtakia, eds., *Introduction to Complex Mediums for Electromagnetics and Optics*, SPIE Optical Engineering Press, 2003.

M. W. McCall, A. Lakhtakia, and W. S. Weiglhofer, The Negative Index of Refraction Demystified, *European Journal of Physics*, vol. 23, pp. 353–359, 2002.

R. A. Shelby, D. R. Smith, and S. Schultz, Experimental Verification of a Negative Index of Refraction, *Science*, vol. 292, pp. 77–99, 2001.

J. B. Pendry, Negative Refraction Makes a Perfect Lens, *Physical Review Letters*, vol. 85, pp. 3966–3969, 2000.

V. G. Veselago, The Electrodynamics of Substances with Simultaneously Negative Values of ϵ and μ, *Soviet Physics–Uspekhi*, vol. 10, pp. 509–514, 1968 [*Uspekhi Fizicheskikh Nauk*, vol. 92, pp. 517–526, 1964].

Plasmonics

M. I. Stockman *et al.*, Roadmap on Plasmonics, *Journal of Optics*, vol. 20, 043001, 2018.

A. Trügler, *Optical Properties of Metallic Nanoparticles: Basic Principles and Simulation*, Springer-Verlag, 2016.

I. D. Mayergoyz, *Plasmon Resonances in Nanoparticles*, World Scientific, 2013.

M. Pelton and G. Bryant, *Introduction to Metal–Nanoparticle Plasmonics*, Wiley, 2013.

A. V. Zayats and S. A. Maier, eds., *Active Plasmonics and Tuneable Plasmonic Metamaterials*, Wiley–Science Wise, 2013.

Z. Han and S. I. Bozhevolnyi, Radiation Guiding with Surface Plasmon Polaritons, *Reports on Progress in Physics*, vol. 76, 016402, 2013.

S. Enoch and N. Bonod, eds., *Plasmonics: From Basics to Advanced Topics*, Springer-Verlag, 2012.

S. Adachi, *The Handbook on Optical Constants of Metals: In Tables and Figures*, World Scientific, 2012.

S. Roh, T. Chung, and B. Lee, Overview of the Characteristics of Micro- and Nano-Structured Surface Plasmon Resonance Sensors, *Sensors*, vol. 11, pp. 1565–1588, 2011.

D. Sarid and W. Challener, *Modern Introduction to Surface Plasmons*, Cambridge, 2010.

P. Berini, Long-Range Surface Plasmon Polaritons, *Advances in Optics and Photonics*, vol. 1, pp. 484–588, 2009.

S. A. Maier, *Plasmonics: Fundamentals and Applications*, Springer-Verlag, 2007.

M. L. Brongersma and P. G. Kik, eds., *Surface Plasmon Nanophotonics*, Springer-Verlag, 2007.

H. A. Atwater, The Promise of Plasmonics, *Scientific American*, vol. 296, no. 4, pp. 56–63, 2007.

W. L. Barnes, A. Dereux, and T. W. Ebbesen, Surface Plasmon Subwavelength Optics, *Nature*, vol. 424, pp. 824–830, 2003.

E. J. Zeman and G. C. Schatz, An Accurate Electromagnetic Theory Study of Surface Enhancement Factors for Ag, Au, Cu, Li, Na, Al, Ga, In, Zn, and Cd, *Journal of Physical Chemistry*, vol. 91, pp. 634–643, 1987.

H. J. Hagemann, W. Gudat, and C. Kunz, Optical Constants from the Far Infrared to the X-Ray Region: Mg, Al, Cu, Ag, Au, Bi, C, and Al_2O_3, *Deutsches Elektronen-Synchrotron Report DESY-SR74-7*, May 1974.

P. B. Johnson and R. W. Christy, Optical Constants of the Noble Metals, *Physical Review B*, vol. 6, pp. 4370–4379, 1972.

Near-Field Optics and Nanophotonics

M. Bertolotti, C. Sibilia, and A. M. Guzmán, *Evanescent Waves in Optics: An Introduction to Plasmonics*, Springer-Verlag, 2017.

N. Rotenberg and L. Kuipers, Mapping Nanoscale Light Fields, *Nature Photonics*, vol. 8, pp. 919–926, 2014.

L. Novotny and B. Hecht, *Principles of Nano-Optics*, Cambridge University Press, 2nd ed. 2012.

Z. Zalevsky and I. Abdulhalim, *Integrated Nanophotonic Devices*, Elsevier, 2nd ed. 2014.

L. Novotny, From Near-Field Optics to Optical Antennas, *Physics Today*, vol. 64, no. 7, pp. 47–52, 2011.

S. V. Gaponenko, *Introduction to Nanophotonics*, Cambridge University Press, 2010.

M. Ohtsu, K. Kobayashi, T. Kawazoe, T. Yatsui, and M. Naruse, *Principles of Nanophotonics*, CRC Press/Taylor & Francis, 2008.

D. Courjon, *Near-Field Microscopy and Near-Field Optics*, World Scientific, 2003.

S. Jutamulia, ed., *Selected Papers on Near-Field Optics*, SPIE Optical Engineering Press (Milestone Series Volume 172), 2002.

S. Kawata, M. Ohtsu, and M. Irie, eds., *Near-Field Optics and Surface Plasmon Polaritons*, Springer-Verlag, 2001.

Metamaterials

I. Liberal, A. M. Mahmoud, Y. Li, B. Edwards, and N. Engheta, Photonic Doping of Epsilon-Near-Zero Media, *Science*, vol. 355, pp. 1058–1062, 2017.

I. Liberal and N. Engheta, Zero-Index Platforms: Where Light Defies Geometry, *Optics & Photonics News*, vol. 27, no. 7/8, pp. 26–33, 2016.

N. Engheta, 150 Years of Maxwell's Equations, *Science*, vol. 349, pp. 136–137, 2015.

N. M. Litchinitser and V. M. Shalaev, Metamaterials: State-of-the Art and Future Directions, in D. L. Andrews, ed., *Photonics: Scientific Foundations, Technology and Applications*, Volume II, *Nanophotonic Structures and Materials*, Wiley–Science Wise, 2015.

A. Poddubny, I. Iorsh, P. Belov, and Y. Kivshar, Hyperbolic Metamaterials, *Nature Photonics*, vol. 7, pp. 958–967, 2013.

R. Marqués, F. Martín, and M. Sorolla, *Metamaterials with Negative Parameters: Theory, Design and Microwave Applications*, Wiley, 2013.

O. Hess, J. B. Pendry, S. A. Maier, R. F. Oulton, J. M. Hamm, and K. L. Tsakmakidis, Active Nanoplasmonic Metamaterials, *Nature Materials*, vol. 11, pp. 573–584, 2012.

N. I. Zheludev and Y. S. Kivshar, From Metamaterials to Metadevices, *Nature Materials*, vol. 11, pp. 917–924, 2012.

Y. Liu and X. Zhang, Metamaterials: A New Frontier of Science and Technology, *Chemical Society Reviews*, vol. 40, pp. 2494–2507, 2011.

M. Wegener and S. Linden, Shaping Optical Space with Metamaterials, *Physics Today*, vol. 63, no. 10, pp. 32–36, 2010.

T. J. Cui, D. R. Smith, and R. Liu, eds., *Metamaterials: Theory, Design, and Applications*, Springer-Verlag, 2010.

W. Cai and V. Shalaev, *Optical Metamaterials: Fundamentals and Applications*, Springer-Verlag, 2009.

A. K. Sarychev and V. M. Shalaev, *Electrodynamics of Metamaterials*, World Scientific, 2007.

N. Engheta, Circuits with Light at Nanoscales: Optical Nanocircuits Inspired by Metamaterials, *Science*, vol. 317, pp. 1698–1702, 2007.

N. Engheta and R. W. Ziolkowski, eds., *Electromagnetic Metamaterials: Physics and Engineering Explorations*, Wiley–IEEE, 2006.

S. Linden, C. Enkrich, G. Dolling, M. W. Klein, J. Zhou, T. Koschny, C. M. Soukoulis, S. Burger, F. Schmidt, and M. Wegener, Photonic Metamaterials: Magnetism at Optical Frequencies, *IEEE Journal of Selected Topics in Quantum Electronics*, vol. 12, pp. 1097–1105, 2006.

G. V. Eleftheriades and K. G. Balmain, eds., *Negative-Refraction Metamaterials: Fundamental Principles and Applications*, Wiley–IEEE, 2005.

D. R. Smith, J. B. Pendry, and M. C. K. Wiltshire, Metamaterials and Negative Refractive Index, *Science*, vol. 305, 788–792, 2004.

Optical Antennas and Metasurfaces

N. Meinzer, W. L. Barnes, and I. R. Hooper, Plasmonic Meta-Atoms and Metasurfaces, *Nature Photonics*, vol. 8, pp. 889–898, 2014.

N. Yu and F. Capasso, Flat Optics with Designer Metasurfaces, *Nature Materials*, vol. 13, pp. 139–150, 2014.

M. Agio and A. Alù, eds., *Optical Antennas*, Cambridge University Press, 2013.

A. V. Kildishev, A. Boltasseva, and V. M. Shalaev, Planar Photonics with Metasurfaces, *Science*, vol. 339, 1232009, 2013.

A. A. Maradudin, *Structured Surfaces as Optical Metamaterials*, Cambridge University Press, 2011.

N. Yu, P. Genevet, M. A. Kats, F. Aieta, J.-P. Tetienne, F. Capasso, and Z. Gaburro, Light Propagation with Phase Discontinuities: Generalized Laws of Reflection and Refraction, *Science*, vol. 334, pp. 333–337, 2011.

R. Gordon, A. G. Brolo, D. Sinton, and K. L. Kavanagh, Resonant Optical Transmission Through Hole-Arrays in Metal Films: Physics and Applications, *Laser & Photonics Reviews*, vol. 4, pp. 311–335, 2010.

P. Bharadwaj, B. Deutsch, and L. Novotny, Optical Antennas, *Advances in Optics and Photonics*, vol. 1, pp. 438–483, 2009.

J. Bravo-Abad, L. Martín-Moreno, and F. J. Garcia-Vidal, Resonant Transmission of Light Through Subwavelength Holes in Thick Metal Films, *IEEE Journal of Selected Topics in Quantum Electronics*, vol. 12, pp. 1221–1227, 2006.

F. J. Garcia-Vidal, L. Martín-Moreno, and J. B. Pendry, Surfaces with Holes in Them: New Plasmonic Metamaterials, *Journal of Optics A: Pure and Applied Optics*, vol. 7, pp. S97–S101, 2005.

T. W. Ebbesen, H. J. Lezec, H. F. Ghaemi, T. Thio, and P. A. Wolff, Extraordinary Optical Transmission Through Sub-Wavelength Hole Arrays, *Nature*, vol. 391, pp. 667–669, 1998.

Transformation Optics

J. B. Pendry, Y. Luo, and R. Zhao, Transforming the Optical Landscape, *Science*, vol. 348, pp. 521–524, 2015.

D. H. Werner and D.-H. Kwon, *Transformation Electromagnetics and Metamaterials: Fundamental Principles and Applications*, Springer-Verlag, 2014.

A. B. Shvartsburg and A. A. Maradudin, *Waves in Gradient Metamaterials*, World Scientific/Imperial College Press, 2013.

M. McCall, Transformation Optics and Cloaking, *Contemporary Physics*, vol. 54, pp. 273–286, 2013.

J. B. Pendry, A. Aubry, D. R. Smith, and S. A. Maier, Transformation Optics and Subwavelength Control of Light, *Science*, vol. 227, pp. 549–552, 2012.

U. Leonhardt and T. G. Philbin, Transformation Optics and the Geometry of Light, in E. Wolf, ed., *Progress in Optics*, Elsevier, 2009, vol. 53, pp. 69–152.

D. Schurig, J. B. Pendry, and D. R. Smith, Calculation of Material Properties and Ray Tracing in Transformation Media, *Optics Express*, vol. 14, pp. 9794–9804, 2006.

J. B. Pendry, D. Schurig, and D. R. Smith, Controlling Electromagnetic Fields, *Science*, vol. 312, pp. 1780–1782, 2006.

A. Alù and N. Engheta, Achieving Transparency with Plasmonic and Metamaterial Coatings, *Physical Review E*, vol. 72, 016623, 2005.

C. Gomez-Reino, M. V. Perez, and C. Bao, *Gradient-Index Optics: Fundamentals and Applications*, Springer-Verlag, 2002, paperback ed. 2010.

PROBLEMS

8.1-1 SPP at Boundary Between DPS Medium and Lossy SNG Medium. Consider a DPS-SNG boundary with $\mu_1 = \mu_2 = \mu_o$ and ϵ_1 real and positive, and with $\epsilon_2 = \epsilon_2' + j\epsilon_2''$ complex with ϵ_2' real and negative. Given that $|\epsilon_2''| \ll \epsilon_2'$, demonstrate that the plasmon wavelength λ_o/n_b and the propagation length d_b may be calculated with the help of the following approximate formulas:

$$n_b \approx \sqrt{\frac{\epsilon_b}{\epsilon_o}}, \qquad \epsilon_b \approx \frac{\epsilon_1 \epsilon_2'}{\epsilon_1 + \epsilon_2'}, \qquad d_b \approx \frac{\lambda_o}{2\pi} \frac{1}{n_b^3} \frac{\epsilon_2'^2}{\epsilon_o \epsilon_2''}.$$

8.1-2 **NIM Slab as a Near-Field Imaging System.** Demonstrate the near-field imaging capability of a slab of refractive index $n = -2$ in air ($n = 1$) by constructing a graph similar to that displayed in Fig. 8.1-6(b), and by determining the imaging equation. Find the reflection and transmission coefficients for a wave normally incident on the boundaries of the slab. Sketch the amplitude of an evanescent wave transmitted through the slab by constructing a profile similar to that portrayed in Fig. 8.1-6(d).

8.1-3 **Subwavelength-Resolution Near-Field Imaging with a Lossy NIM Slab.** Consider a slab of refractive index $n = -2$ in air ($n = 1$). An evanescent wave entering the slab from air is amplified by the slab material, as demonstrated in Prob. 8.1-2. If the material is lossy, there will also be attenuation. If the attenuation coefficient $\gamma = 0.1 k_o$, where $k_o = 2\pi/\lambda_o$ is the wavenumber in free space, determine the spatial angular frequency k_x (in units of k_o) at which the amplification will be smaller than the attenuation? What is the corresponding resolution in units of wavelength? Assume that $k_y = 0$.

*8.1-4 **Type-II Hyperbolic Medium.** For the hyperbolic medium described in Sec. 8.1C, ϵ_1 and ϵ_2 are positive, while ϵ_3 is negative. This is called a *Type-I hyperbolic medium*. Find the **k** surface for a *Type-II hyperbolic medium*, in which ϵ_1 and ϵ_2 are negative, while ϵ_3 is positive. Show that a Type-II hyperbolic medium also supports propagating waves with very short wavelengths, but can be highly reflective.

8.2-1 **Group Velocity in a Metal.** For a medium described by the simplified Drude model with effective permittivity (8.2-18), show that the product of the phase velocity and the group velocity is c_o^2.

8.2-2 **Prism Coupler for Exciting a SPP Wave.** A SPP wave is to be generated at the interface between air and a 60-nm-thick layer of Ag by making use of a SiN prism (refractive index $n = 2.0$) that abuts the opposite face of the Ag layer (see Fig. 8.2-5). At a free-space wavelength $\lambda_o = 700$ nm, the relative permittivity of Ag is $-20 + j\,1.3$. Calculate the angle of incidence of the light inside the prism required to couple to the Ag-air SPP. Assume that the dispersion relation of the SPP wave at the Ag-air interface is undisturbed by the presence of the prism.

8.2-3 **Silver Nanosphere in Glass.** Consider a silver nanosphere of radius $a = 10$ nm embedded in glass ($n = 1.45$). Calculate and plot the scattering efficiency Q_s, the absorption efficiency Q_a, and the normalized internal field E_i/E_0 as functions of the free-space wavelength λ_o over the 350–1000-nm wavelength range. Identify the resonance frequency and determine the scattering and absorption coefficients, α_s and α_a, respectively, at resonance. Use the fitted modified-Drude relative-permittivity function for bulk Ag:

$$\frac{\epsilon_s}{\epsilon_o} = 1 + \frac{\omega_p^2}{-\omega^2 + j\omega\zeta} + \frac{2.2\,\omega_L^2}{\omega_L^2 - \omega^2 + j\omega\zeta_L},$$

where $\zeta = 0.00229\,\omega_p$, $\omega_L = 0.575\,\omega_p$, and $\zeta_L = 0.124\,\omega_p$. The quantity ω_p corresponds to a free-space plasma wavelength $\lambda_p = 135.2$ nm. This model incorporates an interband absorption (bound-electron) contribution to the complex permittivity.

8.3-1 **Negative-Permittivity Metamaterial: Silver Nanospheres in Water.** Use the Maxwell-Garnett mixing rule (5.6-20) to calculate and plot the real and imaginary parts of the effective permittivity ϵ_e of water with inclusions of silver nanospheres [see Fig. 8.3-2(b)], as a function of the free-space wavelength λ_o, over the wavelength range from 250 to 1000 nm. Use the relative-permittivity function of Ag defined in Prob. 8.2-3 and take the refractive index of water to be constant at $n = 1.33$. Assume that the volume fraction f $= 3\%$. Identify wavelength ranges within which the real part of ϵ_e is negative. Investigate the effects of changing both f and n.

*8.3-2 **Layered Metal–Dielectric Hyperbolic Metamaterial.** Planar layers of metal are alternately stacked with layers of dielectric material. The anisotropic medium has an effective dielectric tensor with components $\epsilon_1 = \epsilon_2$ in the plane of the layers and ϵ_3 in the orthogonal direction. Use an effective-medium approximation,

$$\epsilon_1 = \epsilon_2 = \frac{\epsilon_m d_m + \epsilon_d d_d}{d_m + d_d}, \qquad 1/\epsilon_3 = \frac{d_m/\epsilon_m + d_d/\epsilon_d}{d_m + d_d},$$

where ϵ_m and ϵ_d are the permittivities of the metal and dielectric layers, respectively, and d_m and d_d are their thicknesses. With the help of the simplified Drude-model expression for ϵ_m, show that this structure can behave as a hyperbolic material. Identify which of the components, ϵ_1 or ϵ_3, is negative. Sketch the **k** surface.

CHAPTER

9

GUIDED-WAVE OPTICS

9.1	PLANAR-MIRROR WAVEGUIDES	355
9.2	PLANAR DIELECTRIC WAVEGUIDES	363
	A. Waveguide Modes	
	B. Field Distributions	
	C. Dispersion Relation and Group Velocities	
9.3	TWO-DIMENSIONAL WAVEGUIDES	372
9.4	OPTICAL COUPLING IN WAVEGUIDES	376
	A. Input Couplers	
	B. Coupled Waveguides	
	*C. Waveguide Arrays	
9.5	PHOTONIC-CRYSTAL WAVEGUIDES	385
9.6	PLASMONIC WAVEGUIDES	386

Jean-Daniel Colladon
(1802–1893)

John Tyndall
(1820–1893)

Total internal reflection, the basis of guided-wave optics, was first demonstrated in water jets in the mid-1800s by the Swiss physicist Jean-Daniel Colladon and by the Irish-born physicist John Tyndall.

Fundamentals of Photonics, Third Edition. Bahaa E. A. Saleh and Malvin Carl Teich.
©2019 John Wiley & Sons, Inc. Published 2019 by John Wiley & Sons, Inc.

Traditional optical instruments and systems make use of bulk optics, in which light is transmitted between different locations in the form of beams that are collimated, relayed, focused, and scanned by mirrors, lenses, and prisms. The beams diffract and broaden as they propagate although they can be refocused by the use of lenses and mirrors. However, the bulk optical components that comprise such systems are often large and unwieldy, and objects in the paths of the beams can obstruct or scatter them.

In many circumstances it is advantageous to transmit optical beams through dielectric conduits rather than through free space. The technology for achieving this is known as **guided-wave optics**. It was initially developed to provide long-distance light transmission without the necessity of using relay lenses. This technology now has many important applications. A few examples are: carrying light over long distances for optical fiber communications, biomedical imaging where light must reach awkward locations, and connecting components within miniaturized optical and optoelectronic devices and systems.

The underlying principle of optical confinement is straightforward. A medium of refractive index n_1, embedded in a medium of lower refractive index $n_2 < n_1$, acts as a light "trap" within which optical rays remain confined by multiple total internal reflections at the boundaries. Because this effect facilitates the confinement of light generated inside a medium of high refractive index [see Exercise (1.2-6)], it can be exploited in making light conduits — guides that transport light from one location to another. An **optical waveguide** is a light conduit consisting of a slab, strip, or cylinder of dielectric material embedded in another dielectric material of lower refractive index (Fig. 9.0-1). The light is transported through the inner medium without radiating into the surrounding medium. The most widely used of these waveguides is the optical fiber, comprising two concentric cylinders of low-loss dielectric material such as glass (see Chapter 10). Other forms of optical waveguides make use of photonic crystals (Chapter 7) and metal–dielectric structures (Chapter 8).

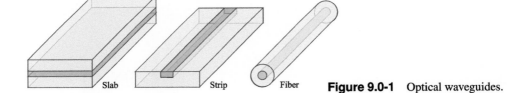

Figure 9.0-1 Optical waveguides.

Integrated photonics, also known as **integrated optics**, is the technology of combining, on a single substrate ("**chip**"), various optical devices and components useful for generating, focusing, splitting, combining, isolating, polarizing, coupling, switching, modulating, and detecting light. Optical waveguides provide the links among these components. Such chips, called **photonic integrated circuits (PICs)**, are optical versions of electronic integrated circuits. An example of a transceiver (transmitter/receiver) chip is schematized in Fig. 9.0-2. Integrated photonics serves to miniaturize optics in much the same way that silicon integrated circuits have miniaturized electronics.

This Chapter

The basic theory of optical waveguides is presented in this and the following chapter. In this chapter, we consider rectangular waveguides, which are used extensively in integrated photonics. Chapter 10 deals with cylindrical waveguides, i.e., optical fibers.

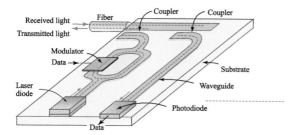

Figure 9.0-2 Schematic of a photonic integrated circuit that serves as an elementary optical transceiver (transmitter/receiver). Received light enters via a waveguide and is directed to a photodiode where it is detected. Light from a laser diode is guided, modulated, and coupled into a fiber for transmission.

If reflectors are placed at the two ends of a short waveguide, the result is a structure that traps and stores light — an optical resonator. These devices, which are essential to the operation of lasers, are described in Chapter 11. Various integrated-photonic components and devices (such as laser diodes, detectors, modulators, interconnects, and switches) are considered in the chapters that deal specifically with those components and devices. Photonic integrated circuits based on silicon photonics are addressed in Sec. 25.1E. Optical fiber communication systems are discussed in detail in Chapter 25.

9.1 PLANAR-MIRROR WAVEGUIDES

We begin by examining wave propagation in a waveguide comprising two parallel infinite planar *mirrors* separated by a distance d (Fig. 9.1-1). The mirrors are assumed to reflect light without loss. A ray of light, say in the y–z plane, making an angle θ with the mirrors reflects and bounces between them without loss of energy. The ray is thus guided along the z direction.

This waveguide appears to provide a perfect conduit for light rays. It is *not* used in practical applications, however, principally because of the difficulty and cost of fabricating low-loss mirrors. Nevertheless, we study this simple example in detail because it provides a valuable pedagogical introduction to the dielectric waveguide, which we examine in Sec. 9.2, and to the optical resonator, which is the subject of Chapter 11.

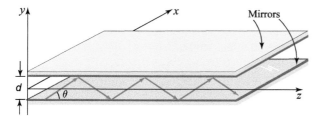

Figure 9.1-1 Planar-mirror waveguide.

Waveguide Modes

The ray-optics picture of light being guided by multiple reflections cannot explain a number of important effects that require the use of electromagnetic theory. A simple approach for carrying out an electromagnetic analysis is to associate with each optical ray a transverse electromagnetic (TEM) plane wave. The total electromagnetic field is then the sum of these plane waves.

Consider a monochromatic TEM plane wave of wavelength $\lambda = \lambda_o/n$, wavenumber $k = nk_o$, and phase velocity $c = c_o/n$, where n is the refractive index of the medium between the mirrors. The wave is polarized in the x direction and its wavevector lies in the y–z plane at an angle θ with the z axis (Fig. 9.1-1). Like the optical ray, the wave reflects from the upper mirror, travels at an angle $-\theta$, reflects from the lower mirror, and travels once more at an angle θ, and so on. Since the electric-field vector is parallel to the mirror, each reflection is accompanied by a phase shift π for a perfect mirror, but the amplitude and polarization are not changed. The π phase shift ensures that the sum of each wave and its own reflection vanishes so that the total field is zero at the mirrors. At each point within the waveguide we have TEM waves traveling upward at an angle θ and others traveling downward at an angle $-\theta$; all waves are polarized in the x direction.

We now impose a self-consistency condition by requiring that as the wave reflects twice, it reproduces itself [see Fig. 9.1-2(a)], so that we have only two distinct plane waves. Fields that satisfy this condition are called the modes (or eigenfunctions) of the waveguide (see Appendix C). *Modes are fields that maintain the same transverse distribution and polarization at all locations along the waveguide axis.* We shall see that self-consistency guarantees this shape invariance. In connection with Fig. 9.1-2, the phase shift $\Delta\varphi$ encountered by the original wave in traveling from A to B must be equal to, or differ by an integer multiple of 2π, from that encountered when the wave reflects, travels from A to C, and reflects once more. Accounting for a phase shift of π at each reflection, we have $\Delta\varphi = 2\pi\overline{AC}/\lambda - 2\pi - 2\pi\overline{AB}/\lambda = 2\pi q$, where $q = 0, 1, 2, \ldots$, so that $2\pi(\overline{AC} - \overline{AB})/\lambda = 2\pi(q+1)$. The geometry portrayed in Fig. 9.1-2(a), together with the identity $\cos(2x) = 1 - 2\sin^2 x$, provides $\overline{AC} - \overline{AB} = 2d\sin\theta$, where d is the distance between the mirrors. Thus, $2\pi(2d\sin\theta)/\lambda = 2\pi(q+1)$ so that

$$\frac{2\pi}{\lambda} 2d \sin\theta = 2\pi m, \qquad m = 1, 2, \ldots . \tag{9.1-1}$$

where $m = q + 1$. The self-consistency condition is therefore satisfied only for certain bounce angles $\theta = \theta_m$ satisfying

$$\boxed{\sin\theta_m = m\frac{\lambda}{2d}, \qquad m = 1, 2, \ldots .} \tag{9.1-2}$$
Bounce Angles

Each integer m corresponds to a bounce angle θ_m, and the corresponding field is called the mth mode. The $m = 1$ mode has the smallest angle, $\theta_1 = \sin^{-1}(\lambda/2d)$; modes with larger m are composed of more oblique plane-wave components.

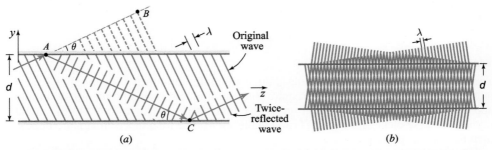

Figure 9.1-2 (a) Condition of self-consistency: as a wave reflects twice it duplicates itself. (b) At angles for which self-consistency is satisfied, the two waves interfere and create a pattern that does not change with z.

When the self-consistency condition is satisfied, the phases of the upward and downward plane waves at points on the z axis differ by half the round-trip phase shift $q\pi$, $q = 0, 1, \ldots$, or $(m - 1)\pi$, $m = 1, 2, \ldots$, so that they add for odd m and subtract for even m.

Since the y component of the propagation constant is given by $k_y = nk_o \sin \theta$, it is quantized to the values $k_{ym} = nk_o \sin \theta_m = (2\pi/\lambda) \sin \theta_m$. Using (9.1-2), we obtain

$$k_{ym} = m\frac{\pi}{d}, \qquad m = 1, 2, 3 \ldots, \qquad (9.1\text{-}3)$$

Wavevector Transverse Component

so that the k_{ym} are spaced by π/d. Equation (9.1-3) states that the phase shift encountered when a wave travels a distance $2d$ (one round trip) in the y direction, with propagation constant k_{ym}, must be a multiple of 2π.

Propagation Constants

A guided wave is composed of two distinct plane waves traveling in the y–z plane at angles $\pm\theta$ with the z axis. Their wavevectors have components $(0, k_y, k_z)$ and $(0, -k_y, k_z)$. Their sum or difference therefore varies with z as $\exp(-jk_z z)$, so that the propagation constant of the guided wave is $\beta \equiv k_z = k \cos \theta$. Thus, β is quantized to the values $\beta_m = k \cos \theta_m$, from which $\beta_m^2 = k^2(1 - \sin^2 \theta_m)$. Using (9.1-2), we obtain

$$\beta_m^2 = k^2 - \frac{m^2 \pi^2}{d^2}. \qquad (9.1\text{-}4)$$

Propagation Constants

Higher-order (more oblique) modes travel with smaller propagation constants. The values of θ_m, k_{ym}, and β_m for the different modes are illustrated in Fig. 9.1-3.

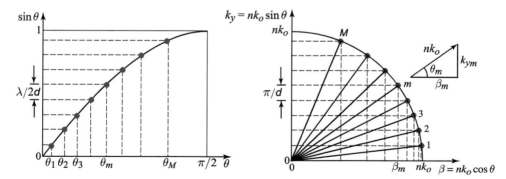

Figure 9.1-3 The bounce angles θ_m and the wavevector components of the modes of a planar-mirror waveguide (indicated by dots). The transverse components $k_{ym} = k \sin \theta_m$ are spaced uniformly at multiples of π/d, but the bounce angles θ_m and the propagation constants β_m are not equally spaced. Mode $m = 1$ has the smallest bounce angle and the largest propagation constant.

Field Distributions

The complex amplitude of the total field in the waveguide is the superposition of the two bouncing TEM plane waves. If $A_m \exp(-jk_{ym}y - j\beta_m z)$ is the upward wave, then $e^{j(m-1)\pi} A_m \exp(+jk_{ym}y - j\beta_m z)$ must be the downward wave [at $y = 0$, the two waves differ by a phase shift $(m-1)\pi$]. There are therefore symmetric modes, for which the two plane-wave components are added, and anti-symmetric modes, for which they are subtracted. The total field turns out to be $E_x(y,z) = 2A_m \cos(k_{ym}y) \exp(-j\beta_m z)$ for odd modes and $2jA_m \sin(k_{ym}y) \exp(-j\beta_m z)$ for even modes.

Using (9.1-3) we write the complex amplitude of the electric field in the form

$$E_x(y,z) = a_m u_m(y) \exp(-j\beta_m z), \qquad (9.1\text{-}5)$$

where

$$u_m(y) = \begin{cases} \sqrt{\dfrac{2}{d}} \cos\left(m\pi \dfrac{y}{d}\right), & m = 1, 3, 5, \ldots \\ \sqrt{\dfrac{2}{d}} \sin\left(m\pi \dfrac{y}{d}\right), & m = 2, 4, 6, \ldots, \end{cases} \qquad (9.1\text{-}6)$$

with $a_m = \sqrt{2d} A_m$ and $j\sqrt{2d} A_m$, for odd m and even m, respectively. The functions $u_m(y)$ have been normalized to satisfy

$$\int_{-d/2}^{d/2} u_m^2(y)\, dy = 1. \qquad (9.1\text{-}7)$$

Thus, a_m is the amplitude of mode m. It can be shown that the functions $u_m(y)$ also satisfy

$$\int_{-d/2}^{d/2} u_m(y)\, u_l(y)\, dy = 0, \qquad l \neq m, \qquad (9.1\text{-}8)$$

i.e., they are orthogonal in the $[-d/2, d/2]$ interval.

The transverse distributions $u_m(y)$ are plotted in Fig. 9.1-4. Each mode can be viewed as a standing wave in the y direction, traveling in the z direction. Modes of large m vary in the transverse plane at a greater rate k_y and travel with a smaller propagation constant β. The field vanishes at $y = \pm d/2$ for all modes, so that the boundary conditions at the surface of the mirrors are always satisfied.

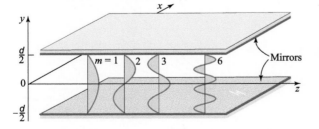

Figure 9.1-4 Field distributions of the modes of a planar-mirror waveguide.

Since we assumed at the outset that the bouncing TEM plane wave is polarized in the x direction, the total electric field is also in the x direction and the guided wave is a transverse-electric (TE) wave. Transverse magnetic (TM) waves are analyzed in a similar fashion as will be seen subsequently.

EXERCISE 9.1-1

Optical Power. Show that the optical power flow in the z direction associated with the TE mode $E_x(y,z) = a_m u_m(y) \exp(-j\beta_m z)$ is $(|a_m|^2/2\eta) \cos\theta_m$, where $\eta = \eta_o/n$ and $\eta_o = \sqrt{\mu_o/\epsilon_o}$ is the impedance of free space.

Number of Modes

Since $\sin\theta_m = m\lambda/2d$, $m = 1, 2, \ldots$, and taking $\sin\theta_m < 1$, the maximum allowed value of m is the greatest integer smaller than $1/(\lambda/2d)$,

$$M \doteq \frac{2d}{\lambda}. \qquad (9.1\text{-}9)$$

Number of Modes

The symbol $\doteq$ denotes that $2d/\lambda$ is reduced to the nearest integer. As examples, when $2d/\lambda = 0.9$, 1, and 1.1, we have $M = 0$, 0, and 1, respectively. Thus, M is the number of modes of the waveguide. Light can be transmitted through the waveguide in one, two, or many modes. The actual number of modes that carry optical power depends on the source of excitation, but the maximum number is M.

The number of modes increases with increasing ratio of the mirror separation to the wavelength. Under conditions such that $2d/\lambda \leq 1$, corresponding to $d \leq \lambda/2$, M is seen to be 0, which indicates that the self-consistency condition cannot be met and the waveguide cannot support any modes. The wavelength $\lambda_c = 2d$ is called the **cutoff wavelength** of the waveguide. It is the longest wavelength that can be guided by the structure. It corresponds to the **cutoff frequency**

$$\nu_c = \frac{c}{2d}, \qquad (9.1\text{-}10)$$

Cutoff Frequency

or the cutoff angular frequency $\omega_c = 2\pi\nu_c = \pi c/d$, the lowest frequency of light that can be guided by the waveguide. If $1 < 2d/\lambda \leq 2$ (i.e., $d \leq \lambda < 2d$ or $\nu_c \leq \nu < 2\nu_c$), only one mode is allowed. The structure is then said to be a **single-mode waveguide**. If $d = 5$ μm, for example, the waveguide has a cutoff wavelength $\lambda_c = 10$ μm; it supports a single mode for 5 μm $\leq \lambda < 10$ μm, and more modes for $\lambda < 5$ μm. Equation (9.1-9) can also be written in terms of the frequency ν, $M \doteq \nu/\nu_c = \omega/\omega_c$, so that the number of modes increases by unity when the angular frequency ω is incremented by ω_c, as illustrated in Fig. 9.1-5(a).

Dispersion Relation

The relation between the propagation constant β and the angular frequency ω is an important characteristic of the waveguide, known as the **dispersion relation**. For a homogeneous medium, the dispersion relation is simply $\omega = c\beta$. For mode m of a planar-mirror waveguide, β_m and ω are related by (9.1-4) so that

$$\beta_m^2 = (\omega/c)^2 - m^2\pi^2/d^2. \qquad (9.1\text{-}11)$$

This relation may be written in terms of the cutoff angular frequency $\omega_c = 2\pi\nu_c = \pi c/d$ as

$$\beta_m = \frac{\omega}{c}\sqrt{1 - m^2 \frac{\omega_c^2}{\omega^2}}.$$

(9.1-12)
Dispersion Relation

As shown in Fig. 9.1-5(b) for $m = 1, 2, \ldots$, the propagation constant β for mode m is zero at angular frequency $\omega = m\omega_c$, increases monotonically with angular frequency, and ultimately approaches the linear relation $\beta = \omega/c$ for sufficiently large values of β.

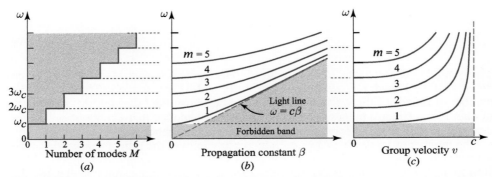

Figure 9.1-5 (a) Number of modes M as a function of angular frequency ω. Modes are not permitted for angular frequencies below the cutoff, $\omega_c = \pi c/d$. M increments by unity as ω increases by ω_c. (b) Dispersion relation. A forbidden band exists for angular frequencies below ω_c. (c) Group velocities of the modes as a function of angular frequency.

Group Velocities

In a medium with a given ω-β dispersion relation, a pulse of light (wavepacket) that has an angular frequency centered at ω travels with a velocity $v = d\omega/d\beta$, known as the group velocity (see Sec. 5.7). Taking the derivative of (9.1-12) and assuming that c is independent of ω (i.e., ignoring dispersion in the waveguide material), we obtain $2\beta_m\, d\beta_m/d\omega = 2\omega/c^2$, so that $d\omega/d\beta_m = c^2\beta_m/\omega = c^2 k \cos\theta_m/\omega = c\cos\theta_m$, from which the group velocity of mode m is

$$v_m = c\cos\theta_m = c\sqrt{1 - m^2\frac{\omega_c^2}{\omega^2}}.$$

(9.1-13)
Group Velocity

It follows that more oblique modes travel with smaller group velocities since they are delayed by the longer paths of the zigzagging process. The dependence of the group velocity on angular frequency is illustrated in Fig. 9.1-5(c), which shows that for each mode, the group velocity increases monotonically from 0 to c as the angular frequency increases above the mode cutoff frequency.

Equation (9.1-13) may also be obtained geometrically by examining the plane wave as it bounces between the mirrors and determining the distance advanced in the z

direction and the time taken by the zigzagging process. For the trip from the bottom mirror to the top mirror (Fig. 9.1-6) we have

$$v = \frac{\text{distance}}{\text{time}} = \frac{d \cot \theta}{d \csc \theta / c} = c \cos \theta. \quad (9.1\text{-}14)$$

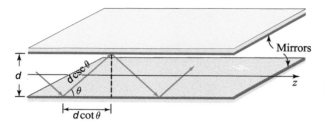

Figure 9.1-6 A plane wave bouncing at an angle θ advances in the z direction by a distance $d \cot \theta$ in a time $d \csc \theta / c$. The velocity is $c \cos \theta$.

TM Modes

Only TE modes (electric field in the x direction) have been considered to this point. TM modes (magnetic field in the x direction) can also be supported by the mirror waveguide. They can be studied by means of a TEM plane wave with the magnetic field in the x direction, traveling at an angle θ and reflecting from the two mirrors (Fig. 9.1-7). The electric-field complex amplitude then has components in the y and z directions. Since the z component is parallel to the mirror, it behaves precisely like the x component of the TE mode (i.e., it undergoes a phase shift π at each reflection and vanishes at the mirrors). When the self-consistency condition is applied to this component the result is mathematically identical to that of the TE case. The angles θ, the transverse wavevector components k_y, and the propagation constants β of the TM modes associated with this component are identical to those of the TE modes. There are $M = 2d/\lambda$ TM modes (and a total of $2M$ modes) supported by the waveguide.

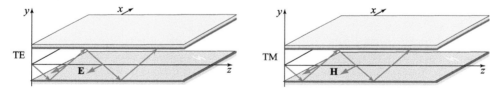

Figure 9.1-7 TE and TM polarized guided waves.

The z component of the electric-field complex amplitude of mode m, as previously, is the sum of an upward plane wave $A_m \exp(-jk_{ym}y)\exp(-j\beta_m z)$ and a downward plane wave $e^{j(m-1)\pi} A_m \exp(jk_{ym}y)\exp(-j\beta_m z)$, with equal amplitudes and phase shift $(m-1)\pi$, so that

$$E_z(y,z) = \begin{cases} a_m \sqrt{\dfrac{2}{d}} \cos\left(m\pi \dfrac{y}{d}\right) \exp(-j\beta_m z), & m = 1, 3, 5, \ldots \\ a_m \sqrt{\dfrac{2}{d}} \sin\left(m\pi \dfrac{y}{d}\right) \exp(-j\beta_m z), & m = 2, 4, 6, \ldots, \end{cases} \quad (9.1\text{-}15)$$

where $a_m = \sqrt{2d} A_m$ and $j\sqrt{2d} A_m$ for odd and even m, respectively. Since the electric-field vector of a TEM plane wave is normal to its direction of propagation, it makes an angle $\pi/2 + \theta_m$ with the z axis for the upward wave, and $\pi/2 - \theta_m$ for the downward wave.

The y components of the electric field of these waves are

$$A_m \cot\theta_m \exp(-jk_{ym}y)\exp(-j\beta_m z) \quad \text{and} \quad e^{jm\pi}A_m \cot\theta_m \exp(jk_{ym}y)\exp(-j\beta_m z), \tag{9.1-16}$$

so that

$$E_y(y,z) = \begin{cases} a_m\sqrt{\dfrac{2}{d}} \cot\theta_m \cos\left(m\pi\dfrac{y}{d}\right) \exp(-j\beta_m z), & m = 1, 3, 5, \ldots \\ a_m\sqrt{\dfrac{2}{d}} \cot\theta_m \sin\left(m\pi\dfrac{y}{d}\right) \exp(-j\beta_m z), & m = 2, 4, 6, \ldots \end{cases} \tag{9.1-17}$$

Satisfaction of the boundary conditions is assured because $E_z(y,z)$ vanishes at the mirrors. The magnetic field component $H_x(y,z)$ may be similarly determined by noting that the ratio of the electric to the magnetic fields of a TEM wave is the impedance of the medium η. The resultant fields $E_y(y,z)$, $E_z(y,z)$, and $H_x(y,z)$ do, of course, satisfy Maxwell's equations.

Multimode Fields

For light to be guided by the mirrors, it is not necessary that it have the distribution of one of the modes. In fact, a field satisfying the boundary conditions (vanishing at the mirrors) but otherwise having an arbitrary distribution in the transverse plane *can* be guided by the waveguide. The optical power, however, is then divided among the modes. Since different modes travel with different propagation constants and different group velocities, the transverse distribution of the field will alter as it travels through the waveguide. Fig. 9.1-8 illustrates how the transverse distribution of a single mode is invariant to propagation, whereas the multimode distribution varies with z (the illustration is for the *intensity* distribution).

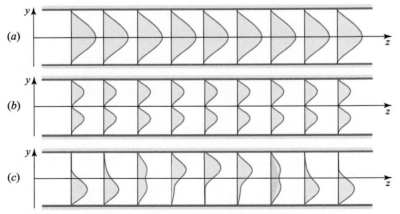

Figure 9.1-8 Variation of the intensity distribution in the transverse direction y at different axial distances z. (*a*) The electric-field complex amplitude in mode 1 is $E(y,z) = u_1(y)\exp(-j\beta_1 z)$, where $u_1(y) = \sqrt{2/d}\cos(\pi y/d)$. The intensity does not vary with z. (*b*) The complex amplitude in mode 2 is $E(y,z) = u_2(y)\exp(-j\beta_2 z)$, where $u_2(y) = \sqrt{2/d}\sin(2\pi y/d)$. The intensity does not vary with z. (*c*) The complex amplitude in a mixture of modes 1 and 2, $E(y,z) = u_1(y)\exp(-j\beta_1 z) + u_2(y)\exp(-j\beta_2 z)$. Since $\beta_1 \neq \beta_2$, the intensity distribution changes with z.

An arbitrary field polarized in the x direction and satisfying the boundary conditions can be written as a weighted superposition of the TE modes,

$$E_x(y,z) = \sum_{m=0}^{M} a_m u_m(y) \exp(-j\beta_m z), \qquad (9.1\text{-}18)$$

where a_m, the superposition weights, are the amplitudes of the different modes.

EXERCISE 9.1-2

Optical Power in a Multimode Field. Show that the optical power flow in the z direction associated with the multimode field in (9.1-18) is the sum of the powers $(|a_m|^2/2\eta)\cos\theta_m$ carried by each of the modes.

9.2 PLANAR DIELECTRIC WAVEGUIDES

A planar dielectric waveguide is a slab of dielectric material surrounded by media of lower refractive indices. The light is guided inside the slab by total internal reflection. In thin-film devices the slab is called the "film" and the upper and lower media are called the "cover" and the "substrate," respectively. The inner medium and outer media may also be called the "core" and the "cladding" of the waveguide, respectively. In this section we study the propagation of light in a symmetric planar dielectric waveguide made of a slab of width d and refractive index n_1 surrounded by a cladding of smaller refractive index n_2, as illustrated in Fig. 9.2-1. All materials are assumed to be lossless.

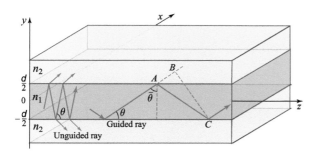

Figure 9.2-1 Planar dielectric (slab) waveguide. Rays making an angle $\theta < \bar{\theta}_c = \cos^{-1}(n_2/n_1)$ are guided by total internal reflection.

Light rays making angles θ with the z axis, in the y–z plane, undergo multiple total internal reflections at the slab boundaries, provided that θ is smaller than the complement of the critical angle $\bar{\theta}_c = \pi/2 - \sin^{-1}(n_2/n_1) = \cos^{-1}(n_2/n_1)$ [see (1.2-5) and Figs. 6.2-3 and 6.2-5]. They travel in the z direction by bouncing between the slab surfaces without loss of power. Rays making larger angles refract, losing a portion of their power at each reflection, and eventually vanish.

To determine the waveguide modes, a formal approach may be pursued by developing solutions to Maxwell's equations in the inner and outer media with the appropriate boundary conditions imposed (see Prob. 9.2-6). We shall instead write the solution in terms of TEM plane waves bouncing between the surfaces of the slab. By imposing the self-consistency condition, we determine the bounce angles of the waveguide modes from which the propagation constants, field distributions, and group velocities are determined. The analysis is analogous to that used in the previous section for the planar-mirror waveguide.

A. Waveguide Modes

Assume that the field in the slab is in the form of a monochromatic TEM plane wave of wavelength $\lambda = \lambda_o/n_1$ bouncing back and forth at an angle θ smaller than the complementary critical angle $\overline{\theta}_c$. The wave travels with a phase velocity $c_1 = c_o/n_1$, has a wavenumber $n_1 k_o$, and has wavevector components $k_x = 0$, $k_y = n_1 k_o \sin\theta$, and $k_z = n_1 k_o \cos\theta$. To determine the modes we impose the self-consistency condition that a wave reproduces itself after each round trip.

In one round trip, the twice-reflected wave lags behind the original wave by a distance $\overline{AC} - \overline{AB} = 2d\sin\theta$, as in Fig. 9.1-2. There is also a phase φ_r introduced by each internal reflection at the dielectric boundary (see Sec. 6.2). For self-consistency, the phase shift between the two waves must be zero or a multiple of 2π,

$$\frac{2\pi}{\lambda} 2d\sin\theta - 2\varphi_r = 2\pi m, \qquad m = 0, 1, 2, \ldots \tag{9.2-1}$$

or

$$2k_y d - 2\varphi_r = 2\pi m. \tag{9.2-2}$$

The only difference between this condition and the corresponding condition in the mirror waveguide, (9.1-1) and (9.1-3), is that the phase shift π introduced by the mirror is replaced here by the phase shift φ_r introduced at the dielectric boundary.

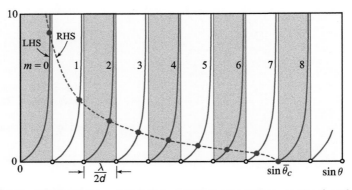

Figure 9.2-2 Graphical solution of (9.2-4) to determine the bounce angles θ_m of the modes of a planar dielectric waveguide. The right-hand side (RHS) and left-hand side (LHS) of (9.2-4) are plotted versus $\sin\theta$. The intersection points, marked by filled circles, determine $\sin\theta_m$. Each branch of the tan or cot function on the left-hand side corresponds to a mode. In this plot $\sin\overline{\theta}_c = 8(\lambda/2d)$ and the number of modes is $M = 9$. The open circles mark $\sin\theta_m = m\lambda/2d$, which provide the bounce angles of the modes of a planar-mirror waveguide of the same dimensions.

The reflection phase shift φ_r is a function of the angle θ. It also depends on the polarization of the incident wave, TE or TM. In the TE case (the electric field is in the

9.2 PLANAR DIELECTRIC WAVEGUIDES

x direction), substituting $\theta_1 = \pi/2 - \theta$ and $\theta_c = \pi/2 - \bar{\theta}_c$ in (6.2-11) gives

$$\tan \frac{\varphi_r}{2} = \sqrt{\frac{\sin^2 \bar{\theta}_c}{\sin^2 \theta} - 1}, \qquad (9.2\text{-}3)$$

so that φ_r varies from π to 0 as θ varies from 0 to $\bar{\theta}_c$. Rewriting (9.2-1) in the form $\tan(\pi d \sin \theta / \lambda - m\pi/2) = \tan(\varphi_r/2)$ and using (9.2-3), we obtain

$$\boxed{\tan\left(\pi \frac{d}{\lambda} \sin \theta - m \frac{\pi}{2}\right) = \sqrt{\frac{\sin^2 \bar{\theta}_c}{\sin^2 \theta} - 1}.} \qquad (9.2\text{-}4)$$
Self-Consistency Condition (TE Modes)

This is a transcendental equation in one variable, $\sin \theta$. Its solutions yield the bounce angles θ_m of the modes. A graphic solution is instructive. The right- and left-hand sides of (9.2-4) are plotted in Fig. 9.2-2 as functions of $\sin \theta$. Solutions are given by the intersection points. The right-hand side, $\tan(\varphi_r/2)$, is a monotonic decreasing function of $\sin \theta$ that reaches 0 when $\sin \theta = \sin \bar{\theta}_c$. The left-hand side generates two families of curves, $\tan[(\pi d/\lambda) \sin \theta]$ and $\cot[(\pi d/\lambda) \sin \theta]$, when m is even and odd, respectively. The intersection points determine the angles θ_m of the modes. The bounce angles of the modes of a *mirror* waveguide of mirror separation d may be obtained from this diagram by using $\varphi_r = \pi$ or, equivalently, $\tan(\varphi_r/2) = \infty$. For comparison, these angles are marked by open circles.

The angles θ_m lie between 0 and $\bar{\theta}_c$. They correspond to wavevectors with components $(0, n_1 k_o \sin \theta_m, n_1 k_o \cos \theta_m)$. The z components are the propagation constants

$$\boxed{\beta_m = n_1 k_o \cos \theta_m.} \qquad (9.2\text{-}5)$$
Propagation Constants

Since $\cos \theta_m$ lies between 1 and $\cos \bar{\theta}_c = n_2/n_1$, β_m lies between $n_2 k_o$ and $n_1 k_o$, as illustrated in Fig. 9.2-3.

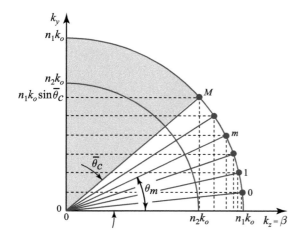

Figure 9.2-3 The bounce angles θ_m and the corresponding components k_z and k_y of the wavevector of the waveguide modes are indicated by dots. The angles θ_m lie between 0 and $\bar{\theta}_c$, and the propagation constants β_m lie between $n_2 k_o$ and $n_1 k_o$. These results should be compared with those shown in Fig. 9.1-3 for the planar-mirror waveguide.

The bounce angles θ_m and the propagation constants β_m of TM modes can be found by using the same equation (9.2-1), but with the phase shift φ_r given by (6.2-13). Similar results are obtained.

Number of Modes

To determine the number of TE modes supported by the dielectric waveguide we examine the diagram in Fig. 9.2-2. The abscissa is divided into equal intervals of width $\lambda/2d$, each of which contains a mode marked by a filled circle. This extends over angles for which $\sin\theta \leq \sin\bar{\theta}_c$. The number of TE modes is therefore the smallest integer greater than $\sin\bar{\theta}_c/(\lambda/2d)$, so that

$$M \doteq \frac{\sin\bar{\theta}_c}{\lambda/2d}. \tag{9.2-6}$$

The symbol $\doteq$ denotes that $\sin\bar{\theta}_c/(\lambda/2d)$ is increased to the nearest integer. For example, if $\sin\bar{\theta}_c/(\lambda/2d) = 0.9$, 1, or 1.1, then $M = 1$, 2, and 2, respectively. Substituting $\cos\bar{\theta}_c = n_2/n_1$ into (9.2-6), we obtain

$$\boxed{M \doteq \frac{2d}{\lambda_o} \text{NA},} \tag{9.2-7}$$
Number of TE Modes

where

$$\boxed{\text{NA} = \sqrt{n_1^2 - n_2^2}} \tag{9.2-8}$$
Numerical Aperture

is the numerical aperture of the waveguide (the NA is the sine of the angle of acceptance of rays from air into the slab; see Exercise 1.2-5). If $d/\lambda_o = 10$, $n_1 = 1.47$, and $n_2 = 1.46$, for example, then $\bar{\theta}_c = 6.7°$, NA = 0.171, and $M = 4$ TE modes. A similar expression can be obtained for the TM modes.

When $\lambda/2d > \sin\bar{\theta}_c$ or $(2d/\lambda_o)\text{NA} < 1$, only one mode is allowed. The waveguide is then a **single-mode waveguide**. This occurs when the slab is sufficiently thin or the wavelength is sufficiently long. Unlike the mirror waveguide, the dielectric waveguide has no absolute cutoff wavelength (or cutoff frequency). In a dielectric waveguide there is at least one TE mode, since the fundamental mode $m = 0$ is always allowed. Each of the modes $m = 1, 2, \ldots$ has its own cutoff wavelength, however.

Stated in terms of frequency, the condition for single-mode operation is that $\nu < \nu_c$, or $\omega < \omega_c$, where the mode cutoff frequency is

$$\boxed{\nu_c = \omega_c/2\pi = \frac{1}{\text{NA}} \frac{c_o}{2d}.} \tag{9.2-9}$$
Mode Cutoff Frequency

The number of modes is then $M \doteq \nu/\nu_c = \omega/\omega_c$, which is the relation illustrated in Fig. 9.2-4. M is incremented by unity as ω increases by ω_c. Identical expressions for the number of TM modes are obtained via a similar derivation.

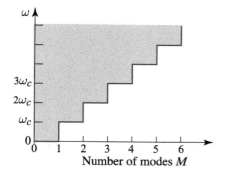

Figure 9.2-4 Number of TE modes as a function of frequency. Compare with Fig. 9.1-5(a) for the planar-mirror waveguide. There is no forbidden band in the case at hand.

EXAMPLE 9.2-1. *Number of Modes in an AlGaAs Waveguide.* Consider a waveguide fabricated by sandwiching a layer of $Al_xGa_{1-x}As$ between two layers of $Al_yGa_{1-y}As$. The refractive index of this ternary semiconductor depends on the relative proportions of Al and Ga. Assume that x and y are chosen such that $n_1 = 3.50$ and $n_1 - n_2 = 0.05$ at an operating wavelength of $\lambda_o = 0.9$ μm. If the core has width $d = 10$ μm, in accordance with (9.2-7) and (9.2-8) there will be $M = 14$ supported TE modes. Only a single mode is allowed when $d < 0.76$ μm.

B. Field Distributions

We now determine the field distributions of the TE modes.

Internal Field

The field inside the slab is composed of two TEM plane waves traveling at angles θ_m and $-\theta_m$ with the z axis with wavevector components $(0, \pm n_1 k_o \sin\theta_m, n_1 k_o \cos\theta_m)$. They have the same amplitude and phase shift $m\pi$ (half that of a round trip) at the center of the slab. The electric-field complex amplitude is therefore $E_x(y, z) = a_m u_m(y)\exp(-j\beta_m z)$, where $\beta_m = n_1 k_o \cos\theta_m$ is the propagation constant, a_m is a constant,

$$u_m(y) \propto \begin{cases} \cos\left(2\pi\dfrac{\sin\theta_m}{\lambda}y\right), & m = 0, 2, 4, \ldots \\ \sin\left(2\pi\dfrac{\sin\theta_m}{\lambda}y\right), & m = 1, 3, 5, \ldots, \end{cases} \quad -\dfrac{d}{2} \le y \le \dfrac{d}{2}, \quad (9.2\text{-}10)$$

and $\lambda = \lambda_o/n_1$. Note that although the field is harmonic, it does not vanish at the slab boundary. As m increases, $\sin\theta_m$ increases, so that higher-order modes vary more rapidly with y.

External Field

The external field must match the internal field at all boundary points $y = \pm d/2$. It is therefore clear that it must vary with z as $\exp(-j\beta_m z)$. Substituting the field $E_x(y,z) = a_m u_m(y)\exp(-j\beta_m z)$ into the Helmholtz equation $(\nabla^2 + n_2^2 k_o^2)E_x(y,z) = 0$ leads to

$$\frac{d^2 u_m}{dy^2} - \gamma_m^2 u_m = 0, \qquad (9.2\text{-}11)$$

where

$$\gamma_m^2 = \beta_m^2 - n_2^2 k_o^2. \qquad (9.2\text{-}12)$$

Since $\beta_m > n_2 k_o$ for guided modes (See Fig. 9.2-3), $\gamma_m^2 > 0$, so that (9.2-11) is satisfied by the exponential functions $\exp(-\gamma_m y)$ and $\exp(\gamma_m y)$. Since the field must decay away from the slab, we choose $\exp(-\gamma_m y)$ in the upper medium and $\exp(\gamma_m y)$ in the lower medium

$$u_m(y) \propto \begin{cases} \exp(-\gamma_m y), & y > d/2 \\ \exp(\gamma_m y), & y < -d/2. \end{cases} \qquad (9.2\text{-}13)$$

The decay rate γ_m is the field extinction coefficient. The wave is said to be an **evanescent wave**. Substituting $\beta_m = n_1 k_o \cos\theta_m$ and $\cos\bar{\theta}_c = n_2/n_1$ into (9.2-12), we obtain

$$\boxed{\gamma_m = n_2 k_o \sqrt{\frac{\cos^2\theta_m}{\cos^2\bar{\theta}_c} - 1}\,.} \qquad \begin{array}{c}(9.2\text{-}14)\\ \text{Extinction Coefficient}\end{array}$$

As the mode number m increases, θ_m increases, and γ_m decreases. Higher-order modes therefore penetrate deeper into the cover and substrate.

To determine the proportionality constants in (9.2-10) and (9.2-13), we match the internal and external fields at $y = d/2$ and use the normalization

$$\int_{-\infty}^{\infty} u_m^2(y)\, dy = 1. \qquad (9.2\text{-}15)$$

This gives an expression for $u_m(y)$ valid for all y. These functions are illustrated in Fig. 9.2-5. As in the mirror waveguide, all of the $u_m(y)$ are orthogonal, i.e.,

$$\int_{-\infty}^{\infty} u_m(y)\, u_l(y)\, dy = 0, \qquad l \neq m. \qquad (9.2\text{-}16)$$

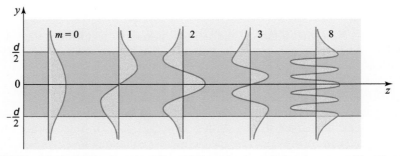

Figure 9.2-5 Field distributions for TE guided modes in a dielectric waveguide. These results should be compared with those shown in Fig. 9.1-4 for the planar-mirror waveguide.

An arbitrary TE field in the dielectric waveguide can be written as a superposition of these modes:

$$E_x(y,z) = \sum_m a_m u_m(y) \exp(-j\beta_m z), \qquad (9.2\text{-}17)$$

where a_m is the amplitude of mode m.

EXERCISE 9.2-1

Confinement Factor. The power confinement factor is the ratio of power in the slab to the total power

$$\Gamma_m = \frac{\int_0^{d/2} u_m^2(y)\,dy}{\int_0^{\infty} u_m^2(y)\,dy}. \qquad (9.2\text{-}18)$$

Derive an expression for Γ_m as a function of the angle θ_m and the ratio d/λ. Demonstrate that the lowest-order mode (smallest θ_m) has the highest power confinement factor.

The field distributions of the TM modes may be similarly determined (Fig. 9.2-6). Since it is parallel to the slab boundary, the z component of the electric field behaves similarly to the x component of the TE electric field. The analysis may start by determining $E_z(y,z)$. Using the properties of the constituent TEM waves, the other components $E_y(y,z)$ and $H_x(y,z)$ may readily be determined, as was done for mirror waveguides. Alternatively, Maxwell's equations may be used to determine these fields.

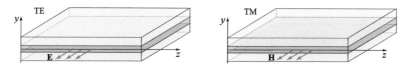

Figure 9.2-6 TE and TM modes in a planar dielectric waveguide.

The field distribution of the lowest-order TE mode ($m = 0$) is similar in shape to that of the Gaussian beam (see Chapter 3). However, unlike the Gaussian beam, guided light does not spread in the transverse direction as it propagates in the axial direction (see Fig. 9.2-7). In a waveguide, the tendency of light to diffract is compensated by the guiding action of the medium.

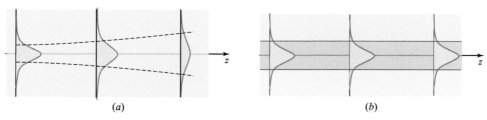

Figure 9.2-7 (a) Gaussian beam in a homogeneous medium. (b) Guided mode in a dielectric waveguide.

C. Dispersion Relation and Group Velocities

The dispersion relation (ω versus β) is obtained by writing the self-consistency equation (9.2-2) in terms of β and ω. Since $k_y^2 = (\omega/c_1)^2 - \beta^2$, (9.2-2) gives

$$2d\sqrt{\frac{\omega^2}{c_1^2} - \beta^2} = 2\varphi_r + 2\pi m. \qquad (9.2\text{-}19)$$

Since $\cos\theta = \beta/(\omega/c_1)$ and $\cos\bar{\theta}_c = n_2/n_1 = c_1/c_2$, (9.2-3) becomes

$$\tan^2\frac{\varphi_r}{2} = \frac{\beta^2 - \omega^2/c_2^2}{\omega^2/c_1^2 - \beta^2}. \qquad (9.2\text{-}20)$$

Substituting (9.2-20) into (9.2-19) we obtain

$$\boxed{\tan^2\left(\frac{d}{2}\sqrt{\frac{\omega^2}{c_1^2} - \beta^2} - m\frac{\pi}{2}\right) = \frac{\beta^2 - \omega^2/c_2^2}{\omega^2/c_1^2 - \beta^2}.} \qquad (9.2\text{-}21)$$
Dispersion Relation
(TE Modes)

This relation may be plotted by rewriting it in parametric form,

$$\frac{\omega}{\omega_c} = \frac{\sqrt{n_1^2 - n_2^2}}{\sqrt{n_1^2 - n^2}}\left(m + \frac{2}{\pi}\tan^{-1}\sqrt{\frac{n^2 - n_2^2}{n_1^2 - n^2}}\right), \quad \beta = n\omega/c_o, \qquad (9.2\text{-}22)$$

in terms of the **effective refractive index** n defined in (9.2-22), where $\omega_c/2\pi = c_o/2d\text{NA}$ is the mode-cutoff angular frequency. As shown in the schematic plot in Fig. 9.2-8(a), the dispersion relations for the different modes lie between the lines $\omega = c_2\beta$ and $\omega = c_1\beta$, the **light lines** representing propagation in homogeneous media with the refractive indices of the surrounding medium and the slab, respectively. As the frequency increases above the mode cutoff frequency, the dispersion relation moves from the light line of the surrounding medium toward the light line of the slab, i.e., the effective refractive index n increases from n_2 to n_1. This effect is indicative of a stronger confinement of waves of shorter wavelength in the medium of higher refractive index.

The group velocity is obtained from the dispersion relation by determining the slope $v = d\omega/d\beta$ for each of the guided modes. The dependence of the group velocity on the angular frequency is illustrated schematically in Fig. 9.2-8(b). As the angular frequency increases above the mode cutoff frequency for each mode, the group velocity decreases from its maximum value c_2, reaches a minimum value slightly below c_1, and then asymptotically returns back toward c_1. The group velocities of the allowed modes thus range from c_2 to a value slightly below c_1.

In propagating through a multimode waveguide, optical pulses spread in time since the modes have different velocities, an effect called **modal dispersion**. In a single-mode waveguide, an optical pulse spreads as a result of the dependence of the group velocity on frequency. This effect is called **group velocity dispersion** (GVD). As shown in Sec. 5.7, GVD occurs in homogeneous materials by virtue of the frequency dependence of the refractive index of the material. Moreover, GVD occurs in waveguides even in the absence of material dispersion. It is then a consequence of the frequency dependence of the propagation coefficients, which are determined by the dependence of wave confinement on wavelength. As illustrated in Fig. 9.2-8(b), each

9.2 PLANAR DIELECTRIC WAVEGUIDES

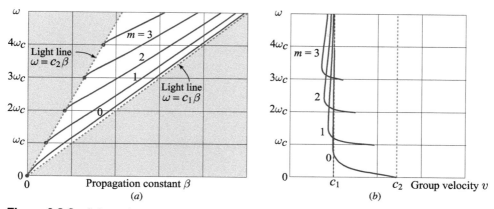

Figure 9.2-8 Schematic representations of (a) the dispersion relation for the different TE modes, $m = 0, 1, 2, \ldots$; and (b) the frequency dependence of the group velocity, which is the derivative of the dispersion relation, $v = d\omega/d\beta$.

mode has a particular angular frequency at which the group velocity changes slowly with frequency (the point at which v reaches its minimum value so that its derivative with respect to ω is zero). At this frequency, the GVD coefficient is zero and pulse spreading is negligible.

An approximate expression for the group velocity may be obtained by taking the total derivative of (9.2-19) with respect to β,

$$\frac{2d}{2k_y}\left(\frac{2\omega}{c_1^2}\frac{d\omega}{d\beta} - 2\beta\right) = 2\frac{\partial \varphi_r}{\partial \beta} + 2\frac{\partial \varphi_r}{\partial \omega}\frac{d\omega}{d\beta}. \qquad (9.2\text{-}23)$$

Substituting $d\omega/d\beta = v$, $k_y/(\omega/c_1) = \sin\theta$, and $k_y/\beta = \tan\theta$, and introducing the new parameters

$$\Delta z = \frac{\partial \varphi_r}{\partial \beta}, \qquad \Delta \tau = -\frac{\partial \varphi_r}{\partial \omega}, \qquad (9.2\text{-}24)$$

we obtain

$$v = \frac{d\cot\theta + \Delta z}{d\csc\theta/c_1 + \Delta \tau}. \qquad (9.2\text{-}25)$$

As we recall from (9.1-14) and Fig. 9.1-6 for the planar-mirror waveguide, $d\cot\theta$ is the distance traveled in the z direction as a ray travels once between the two boundaries. This takes a time $d\csc\theta/c_1$. The ratio $d\cot\theta/(d\csc\theta)/c_1 = c_1\cos\theta$ yields the group velocity for the mirror waveguide. The expression (9.2-25) for the group velocity in a dielectric waveguide indicates that the ray travels an additional distance $\Delta z = \partial\varphi_r/\partial\beta$, a trip that lasts a time $\Delta\tau = -\partial\varphi_r/\partial\omega$. We can think of this as an effective penetration of the ray into the cladding, or as an effective lateral shift of the ray, as shown in Fig. 9.2-9. The penetration of a ray undergoing total internal reflection is known as the **Goos–Hänchen effect** (see Prob. 6.2-6). Using (9.2-24) it can be shown that $\Delta z/\Delta\tau = \omega/\beta = c_1/\cos\theta$.

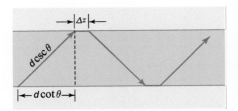

Figure 9.2-9 A ray model that replaces the reflection phase shift with an additional distance Δz traversed at velocity $c_1/\cos\theta$.

EXERCISE 9.2-2

The Asymmetric Planar Waveguide. Examine the TE field in an asymmetric planar waveguide consisting of a dielectric slab of width d and refractive index n_1 placed on a substrate of lower refractive index n_2 and covered with a medium of refractive index $n_3 < n_2 < n_1$, as illustrated in Fig. 9.2-10.

(a) Determine an expression for the maximum inclination angle θ of plane waves undergoing total internal reflection, and the corresponding numerical aperture NA of the waveguide.

(b) Write an expression for the self-consistency condition, similar to (9.2-4).

(c) Determine an approximate expression for the number of modes M (valid when M is very large).

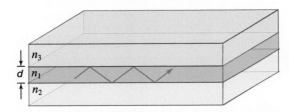

Figure 9.2-10 Asymmetric planar waveguide.

9.3 TWO-DIMENSIONAL WAVEGUIDES

The planar-mirror waveguide and the planar dielectric waveguide studied in the preceding two sections confine light in one transverse direction (the y direction) while guiding it along the z direction. Two-dimensional waveguides confine light in the two transverse directions (the x and y directions). The principle of operation and the underlying modal structure of two-dimensional waveguides is basically the same as planar waveguides; only the mathematical description is lengthier. This section is a brief description of the nature of modes in two-dimensional waveguides. Details can be found in specialized books. Chapter 10 is devoted to an important example of two-dimensional waveguides, the cylindrical dielectric waveguide used in optical fibers.

Rectangular Mirror Waveguide

The simplest generalization of the planar waveguide is the rectangular waveguide (Fig. 9.3-1). If the walls of the waveguide are mirrors, then, as in the planar case, light is guided by multiple reflections at all angles. For simplicity, we assume that the cross section of the waveguide is a square of width d. If a plane wave of wavevector (k_x, k_y, k_z) and its multiple reflections are to exist self-consistently inside the waveguide, it must satisfy the conditions:

$$2k_x d = 2\pi m_x, \quad m_x = 1, 2, \ldots$$
$$2k_y d = 2\pi m_y, \quad m_y = 1, 2, \ldots, \qquad (9.3\text{-}1)$$

which are obvious generalizations of (9.1-3).

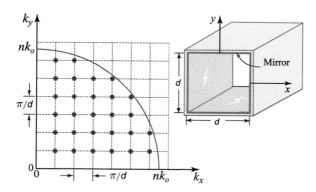

Figure 9.3-1 Modes of a rectangular mirror waveguide are characterized by a finite number of discrete values of k_x and k_y represented by dots.

The propagation constant $\beta = k_z$ can be determined from k_x and k_y by using the relation $k_x^2 + k_y^2 + \beta^2 = n^2 k_o^2$. The three components of the wavevector therefore have discrete values, yielding a finite number of modes. Each mode is identified by two indices m_x and m_y (instead of one index m). All positive integer values of m_x and m_y are allowed as long as $k_x^2 + k_y^2 \leq n^2 k_o^2$, as illustrated in Fig. 9.3-1.

The number of modes M can be easily determined by counting the number of dots within a quarter circle of radius nk_o in the k_y versus k_x diagram (Fig. 9.3-1). If this number is large, it may be approximated by the ratio of the area $\pi(nk_o)^2/4$ to the area of a unit cell $(\pi/d)^2$:

$$M \approx \frac{\pi}{4}\left(\frac{2d}{\lambda}\right)^2. \qquad (9.3\text{-}2)$$

Since there are two polarizations per mode, the total number of modes is actually $2M$. Comparing this to the number of modes in a one-dimensional mirror waveguide, $M \approx 2d/\lambda$, we see that increase of the dimensionality yields approximately the square of the number of modes. The number of modes is a measure of the degrees of freedom. When we add a second dimension we simply multiply the number of degrees of freedom.

The field distributions associated with these modes are generalizations of those in the planar case. Patterns such as those in Fig. 9.1-4 are obtained in each of the x and y directions depending on the mode indices m_x and m_y.

Rectangular Dielectric Waveguide

A dielectric cylinder of refractive index n_1 with square cross section of width d is embedded in a medium of slightly lower refractive index n_2. The waveguide nodes can be determined using a similar theory. Components of the wavevector (k_x, k_y, k_z) must satisfy the condition $k_x^2 + k_y^2 \leq n_1^2 k_o^2 \sin^2 \bar{\theta}_c$, where $\bar{\theta}_c = \cos^{-1}(n_2/n_1)$, so that k_x and k_y lie in the area shown in Fig. 9.3-2. The values of k_x and k_y for the different modes can be obtained from a self-consistency condition in which the phase shifts at the dielectric boundary are included, as was done in the planar case.

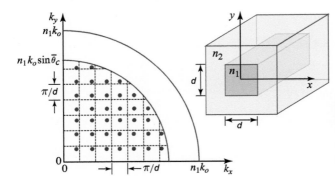

Figure 9.3-2 Geometry of a rectangular dielectric waveguide. The values of k_x and k_y for the waveguide modes are marked by dots.

Unlike the mirror waveguide, k_x and k_y of the modes are not uniformly spaced. However, two consecutive values of k_x (or k_y) are separated by an *average* value of π/d (the same as for the mirror waveguide). The number of modes can therefore be approximated by counting the number of dots in the inner circle in the k_y versus k_x diagram of Fig. 9.3-2, assuming an average spacing of π/d. The result is $M \approx (\pi/4)(n_1 k_o \sin \overline{\theta}_c)^2/(\pi/d)^2$, from which

$$M \approx \frac{\pi}{4}\left(\frac{2d}{\lambda_o}\right)^2 (\text{NA})^2, \qquad (9.3\text{-}3)$$
Number of TE Modes

where $\text{NA} = \sqrt{n_1^2 - n_2^2}$ is the numerical aperture. The approximation is satisfactory when M is large. There is also an identical number M of TM modes. The number of modes is roughly the square of that for the planar dielectric waveguide (9.2-7).

Geometries for Channel Waveguides

As illustrated in Fig. 9.3-3, channel waveguides can take many forms. Representative examples include immersed-strip (or buried channel), embedded-strip, ridge, rib, and strip-loaded geometries. Exact analysis for many of these geometries can be rather complex, but approximations serve well. The reader is referred to specialized texts for details pertaining to this topic.

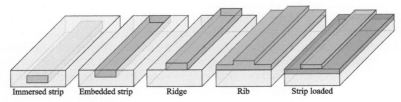

Figure 9.3-3 Various waveguide geometries. The darker the shading, the higher the refractive index.

Waveguides may also be fabricated in a variety of configurations, as illustrated in Fig. 9.3-4 for the embedded-strip geometry. S-bends are used to offset the propagation axis. The Y-branch plays the role of a beamsplitter or beam combiner. A pair of Y-branches may be used to construct a Mach–Zehnder interferometer. Two waveguides in close proximity, or intersecting with each other, can exchange power and be used as directional couplers, as will become apparent in Sec. 9.4B.

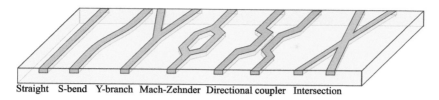

Straight   S-bend   Y-branch   Mach-Zehnder   Directional coupler   Intersection

Figure 9.3-4 Different waveguide configurations, in this case for the embedded-strip geometry.

Materials

The earliest optical waveguides were fabricated from electro-optic materials, principally lithium niobate (**LiNbO$_3$**). As shown in Fig. 9.3-5(a), an embedded-strip waveguide using this material may be fabricated by indiffusing titanium (Ti) in a lithium niobate substrate to increase its refractive index in the region of the strip.

Semiconductors are also commonly used. A **GaAs** rib waveguide may be fabricated by using layers of GaAs and AlGaAs, which has lower refractive index [Fig. 9.3-5(b)]. Another semiconductor material of substantial importance in optical waveguides is **InP**. Its refractive index may be controlled by making use of n-type and p-type dopants, or by using the quaternary semiconductor InGaAsP with various mixing ratios. The ridge waveguide illustrated in Fig. 9.3-5(c) offers strong optical confinement because it is surrounded on three sides by lower index materials — air on two sides and InGaAsP of a different composition on the third.

Waveguides may also be fabricated from **silicon-on-insulator (SOI)**, usually **silica-on-silicon** (SiO$_2$/Si), by making use of standard silicon and oxide etching tools. Since the refractive index of Si is $\approx$ 3.5, and that of silica is $<$ 1.5, this combination of materials exhibits a large refractive-index difference Δn. A typical SOI structure takes the form of a Si rib waveguide atop a layer of silica, which serves as a lower cladding, supported by a silicon substrate [Fig. 9.3-5(d)]. Silicon processing and fabrication has been extraordinarily well developed by the microelectronics industry, and compatibility with **complementary metal-oxide-semiconductor (CMOS)** fabrication technology offers an important advantage. This approach lies in the domain of **silicon photonics** (Sec. 25.1E).

Glass waveguides fabricated by ion exchange, as well as polymer waveguides, are also emerging as viable technologies.

The ability to modulate the refractive index is an important requirement for materials used in integrated-photonic devices such as light modulators and switches, as will become evident in Chapters 21 and 24.

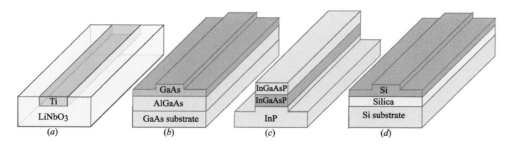

Figure 9.3-5 (a) Ti:LiNbO$_3$ embedded-strip waveguide. (b) Rib waveguide with GaAs core, AlGaAs lower cladding, and GaAs substrate. (c) InGaAsP ridge waveguide with air and lower-index InGaAsP cladding. (d) SOI rib waveguide with Si core, silica lower cladding, and Si substrate compatible with CMOS electronics technology.

9.4 OPTICAL COUPLING IN WAVEGUIDES

A. Input Couplers

Mode Excitation

As indicated in previous sections, light propagates in a waveguide in the form of modes. The complex amplitude of the optical field is generally a superposition of these modes,

$$E(y, z) = \sum_m a_m u_m(y) \exp(-j\beta_m z), \quad (9.4\text{-}1)$$

where a_m is the amplitude, $u_m(y)$ is the transverse distribution (assumed to be real), and β_m is the propagation constant of mode m.

The amplitudes of the different modes depend on the nature of the light source used to excite the waveguide. If the source has a distribution that is a perfect match to a specific mode, only that mode will be excited. In general, a source of arbitrary distribution $s(y)$ excites different modes at different levels. The fraction of power transferred from the source to mode m depends on the degree of similarity between $s(y)$ and $u_m(y)$. To establish this, we write $s(y)$ as an expansion (a weighted superposition) of the orthogonal functions $u_m(y)$,

$$s(y) = \sum_m a_m u_m(y), \quad (9.4\text{-}2)$$

where the coefficient a_l, which represents the amplitude of the excited mode l, is

$$a_l = \int_{-\infty}^{\infty} s(y) u_l(y) \, dy. \quad (9.4\text{-}3)$$

This expression can be derived by multiplying both sides of (9.4-2) by $u_l(y)$, integrating with respect to y, and using the orthogonality relation $\int_{-\infty}^{\infty} u_l(y) u_m(y) \, dy = 0$ for $l \neq m$ along with the normalization condition. The coefficient a_l represents the degree of similarity (or correlation) between the source distribution $s(y)$ and the mode distribution $u_l(y)$.

Input Couplers

Light may be coupled into a waveguide by directly focusing it at one end (Fig. 9.4-1). To excite a given mode, the transverse distribution of the incident light $s(y)$ should match that of the mode. The polarization of the incident light must also match that of the desired mode. Because of the small dimensions of the waveguide slab, focusing and alignment are usually difficult and coupling using this method is inefficient.

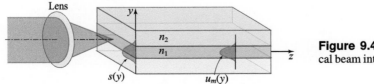

Figure 9.4-1 Coupling an optical beam into an optical waveguide.

In a multimode waveguide, the amount of coupling can be assessed by using a ray-optics approach (Fig. 9.4-2). The guided rays within the waveguide are confined to an angle $\bar{\theta}_c = \cos^{-1}(n_2/n_1)$. Because of refraction at the input to the waveguide, this corresponds to an external angle θ_a satisfying $\mathrm{NA} = \sin\theta_a = n_1 \sin\bar{\theta}_c = n_1\sqrt{1-(n_2/n_1)^2} = \sqrt{n_1^2 - n_2^2}$, where NA is the numerical aperture of the waveguide (see Exercise 1.2-5). For maximum coupling efficiency the incident light should be focused within the angle θ_a.

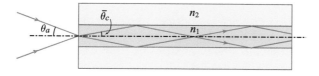

Figure 9.4-2 Focusing rays into a multimode waveguide.

Light may also be coupled from a semiconductor source (a light-emitting diode or a laser diode) into a waveguide by simply aligning the ends of the source and the waveguide, leaving a small space that is selected for maximum coupling (Fig. 9.4-3). In light-emitting diodes, light originates from a semiconductor junction region and is emitted in all directions. In a laser diode, the emitted light is confined in a waveguide of its own (light-emitting diodes and laser diodes are described in Chapter 18). Other methods of coupling light into waveguides include the use of prisms, diffraction gratings, and other waveguides, as discussed below.

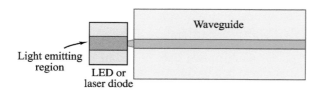

Figure 9.4-3 End butt coupling from a light-emitting diode or laser diode into a waveguide.

Prism and Grating Side Couplers

Can optical power be coupled into a guided mode of a waveguide by use of a source wave entering from the side at some angle θ_i in the cladding, as shown in Fig. 9.4-4(a)? The condition for such coupling is that the axial component of the wavevector of the incident wave, $n_2 k_o \cos\theta_i$, equals the propagation constant β_m of the guided mode. Since $\beta_m > n_2 k_o$ (see Fig. 9.4-4), it is not possible to achieve the required phase-matching condition $\beta_m = n_2 k_o \cos\theta_i$. The axial component of the wavevector of the incident wave is simply too small. However, the problem may be alleviated by use of a prism or a grating.

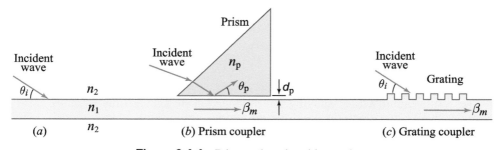

Figure 9.4-4 Prism and grating side couplers.

As illustrated in Fig. 9.4-4(b), a prism of refractive index $n_p > n_2$ is placed at a small distance d_p from the waveguide slab. The incident wave is refracted into the prism where it undergoes total internal reflection at an angle θ_p. The incident and reflected waves form a wave traveling in the z direction with propagation constant $\beta_p = n_p k_o \cos\theta_p$. The transverse field distribution extends into the space separating the prism and the slab as an exponentially decaying evanescent wave. If the distance d_p is sufficiently small, the wave couples to a mode of the slab waveguide with a matching propagation constant $\beta_m \approx \beta_p = n_p k_o \cos\theta_p$. Since $n_p > n_2$, phase matching is possible, and if an appropriate interaction distance is selected, frustrated total internal reflection ensues and significant power can be coupled into the waveguide. The operation may also be reversed to make an output coupler, extracting light from the slab waveguide into free space. This is the same approach as that used to excite a surface plasmon polariton wave at a metal–dielectric boundary, as illustrated in Fig. 8.2-5.

The grating [Fig. 9.4-4(c)] addresses the phase-matching problem by modifying the wavevector of the incoming wave. A grating with period Λ modulates the incoming wave by phase factors $2\pi q/\Lambda z$, where $q = \pm 1, \pm 2, \ldots$. These are equivalent to changes of the axial component of the wavevector by factors $2\pi q/\Lambda$. The phase-matching condition can now be satisfied if $n_2 k_o \cos\theta_i + 2\pi q/\Lambda = \beta_m$, with $q = 1$, for example. The grating may even be designed to enhance the $q = 1$ component.

B. Coupled Waveguides

If two waveguides are sufficiently close such that their fields overlap, light can be coupled from one into the other. Optical power can then be transferred between the waveguides, an effect that can be used to make optical couplers and switches. The basic principle of waveguide coupling is presented here; couplers and switches are discussed in Chapters 24 and 25.

Consider two parallel planar waveguides made of two slabs of widths d, separation $2a$, and refractive indices n_1 and n_2, embedded in a medium of refraction index n that is slightly smaller than n_1 and n_2, as illustrated in Fig. 9.4-5. Each of the waveguides is assumed to be single-mode. The separation between the waveguides is such that the optical field outside the slab of one waveguide (in the absence of the other) overlaps slightly with the slab of the other waveguide.

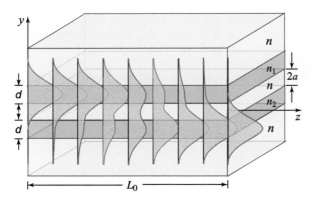

Figure 9.4-5 Coupling between two parallel planar waveguides. At $z = 0$ the light is located principally in the upper waveguide (waveguide 1); at $z = L_0/2$ the light is divided equally between the two waveguides; and at $z = L_0$ the light is located principally in the lower waveguide (waveguide 2). The distance L_0 at which the power is completely transferred from one waveguide to the other is called the **coupling length** or **transfer distance**.

The formal approach to studying the propagation of light in this structure is to write Maxwell's equations for the different regions and use the boundary conditions to determine the modes of the overall system. These modes are different from those of each of the waveguides in isolation. An exact analysis is not easy and is beyond the scope of this book. For weak coupling, however, a simplified approximate theory, known as coupled-mode theory, is often satisfactory.

9.4 OPTICAL COUPLING IN WAVEGUIDES

Coupled-mode theory assumes that the mode of each waveguide is determined as if the other waveguide were absent. In the presence of both waveguides, the modes are taken to remain approximately unchanged, say $u_1(y)\exp(-j\beta_1 z)$ and $u_2(y)\exp(-j\beta_2 z)$. Coupling is assumed to modify only the *amplitudes* of these modes without affecting either their transverse spatial distributions or their propagation constants. The amplitudes of the modes of waveguides 1 and 2 are therefore functions of z, $a_1(z)$, and $a_2(z)$. The theory is directed toward determining $a_1(z)$ and $a_2(z)$ under appropriate boundary conditions.

Coupling can be regarded as a scattering effect. The field of waveguide 1 is scattered from waveguide 2, creating a source of light that changes the amplitude of the field in waveguide 2. The field of waveguide 2 has a similar effect on waveguide 1. An analysis of this mutual interaction leads to two coupled differential equations that govern the variation of the amplitudes $a_1(z)$ and $a_2(z)$.

It can be shown (see the derivation at the end of this section) that the amplitudes $a_1(z)$ and $a_2(z)$ are governed by two coupled first-order differential equations

$$\frac{da_1}{dz} = -j\mathcal{C}_{21}\exp(j\Delta\beta\, z)\, a_2(z) \tag{9.4-4a}$$

$$\frac{da_2}{dz} = -j\mathcal{C}_{12}\exp(-j\Delta\beta\, z)\, a_1(z), \tag{9.4-4b}$$

Coupled-Mode Equations

where

$$\Delta\beta = \beta_1 - \beta_2 \tag{9.4-5}$$

is the phase mismatch per unit length and

$$\mathcal{C}_{21} = \tfrac{1}{2}(n_2^2 - n^2)\frac{k_o^2}{\beta_1}\int_a^{a+d} u_1(y)\,u_2(y)\,dy,$$
$$\mathcal{C}_{12} = \tfrac{1}{2}(n_1^2 - n^2)\frac{k_o^2}{\beta_2}\int_{-a-d}^{-a} u_2(y)\,u_1(y)\,dy \tag{9.4-6}$$

are coupling coefficients. We see from (9.4-4) that the rate of variation of a_1 is proportional to a_2, and *vice versa*. The coefficient of proportionality is the product of the coupling coefficient and the phase mismatch factor $\exp(j\Delta\beta\, z)$.

The coupled-mode equations in (9.4-4) may be solved by beginning with harmonic trial solutions of the form $a_1(z) = b_1\exp(j\gamma z)\exp(j\Delta\beta\, z/2)$ and $a_2(z) = b_2\exp(j\gamma z)\exp(-j\Delta\beta\, z/2)$, where b_1 and b_2 are constants. These solution satisfy (9.4-4) provided that the following condition is satisfied:

$$\gamma = \pm\sqrt{\left(\frac{\Delta\beta}{2}\right)^2 + \mathcal{C}^2},\quad \mathcal{C} = \sqrt{\mathcal{C}_{12}\mathcal{C}_{21}}. \tag{9.4-7}$$

Since γ has two possible values, we modify the trial solutions to be superpositions of $\exp(j\gamma z)$ and $\exp(-j\gamma z)$, or of $\sin(\gamma z)$ and $\cos(\gamma z)$, where γ is the positive value of the square root in (9.4-7). The weights of the superposition are established from the boundary values $a_1(0)$ and $a_2(0)$. The final outcome is

$$a_1(z) = A(z)a_1(0) + B(z)a_2(0) \tag{9.4-8a}$$
$$a_2(z) = C(z)a_1(0) + D(z)a_2(0), \tag{9.4-8b}$$

where
$$A(z) = D^*(z) = \exp\left(\frac{j\,\Delta\beta\,z}{2}\right)\left(\cos\gamma z - j\frac{\Delta\beta}{2\gamma}\sin\gamma z\right) \quad (9.4\text{-}9a)$$

$$B(z) = \frac{\mathcal{C}_{21}}{j\gamma}\exp\left(j\frac{\Delta\beta\,z}{2}\right)\sin\gamma z \quad (9.4\text{-}9b)$$

$$C(z) = \frac{\mathcal{C}_{12}}{j\gamma}\exp\left(-j\frac{\Delta\beta\,z}{2}\right)\sin\gamma z \quad (9.4\text{-}9c)$$

are elements of a transmission matrix **T** that relates the output and input fields.

If we assume that no light enters waveguide 2 so that $a_2(0) = 0$, then the optical powers $P_1(z) \propto |a_1(z)|^2$ and $P_2(z) \propto |a_2(z)|^2$ are

$$P_1(z) = P_1(0)\left[\cos^2\gamma z + \left(\frac{\Delta\beta}{2\gamma}\right)^2\sin^2\gamma z\right] \quad (9.4\text{-}10a)$$

$$P_2(z) = P_1(0)\frac{|\mathcal{C}_{21}|^2}{\gamma^2}\sin^2\gamma z. \quad (9.4\text{-}10b)$$

Thus, power is exchanged periodically between the two waveguides, as illustrated in Fig. 9.4-6(a). The period is π/γ.

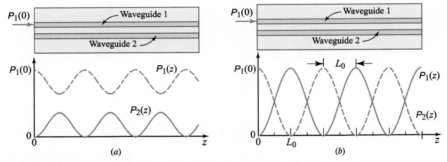

Figure 9.4-6 Periodic exchange of power between waveguides 1 and 2: (a) Phase-mismatched case; (b) phase-matched case.

When the waveguides are identical, i.e., $n_1 = n_2$, $\beta_1 = \beta_2$, and $\Delta\beta = 0$, the two guided waves are said to be phase matched. In this case, $\gamma = \mathcal{C}$, $\mathcal{C}_{12} = \mathcal{C}_{21} = \mathcal{C}$, and the transmission matrix takes the simpler form

$$\mathbf{T} = \begin{bmatrix} A(z) & B(z) \\ C(z) & D(z) \end{bmatrix} = \begin{bmatrix} \cos\mathcal{C}z & -j\sin\mathcal{C}z \\ -j\sin\mathcal{C}z & \cos\mathcal{C}z \end{bmatrix}. \quad (9.4\text{-}11)$$

Equations (9.4-10) then simplify to

$$P_1(z) = P_1(0)\cos^2\mathcal{C}z \quad (9.4\text{-}12a)$$

$$P_2(z) = P_1(0)\sin^2\mathcal{C}z. \quad (9.4\text{-}12b)$$

The exchange of power between the waveguides can then be complete, as illustrated in Fig. 9.4-6(b).

We thus have a device capable of coupling any desired fraction of optical power from one waveguide into another. At a distance $z = L_0 = \pi/2\mathcal{C}$, called the **coupling length** or the **transfer distance**, the power is transferred completely from waveguide 1 into waveguide 2 [Fig. 9.4-7(a)]. At a distance $L_0/2$, half the power is transferred, so that the device acts as a 3-dB coupler, i.e., a 50/50 beamsplitter [Fig. 9.4-7(b)].

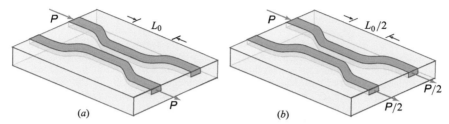

Figure 9.4-7 Optical couplers: (a) switching power from one waveguide to another; (b) a 3-dB coupler.

Switching by Control of Phase Mismatch

A waveguide coupler of fixed length, $L_0 = \pi/2\mathcal{C}$ for example, changes its power-transfer ratio if a small phase mismatch $\Delta\beta$ is introduced. Using (9.4-10b) and (9.4-7), the power-transfer ratio $\mathcal{T} = P_2(L_0)/P_1(0)$ may be written as a function of $\Delta\beta$,

$$\mathcal{T} = \frac{\pi^2}{4} \operatorname{sinc}^2 \left[\frac{1}{2}\sqrt{1 + \left(\frac{\Delta\beta L_0}{\pi}\right)^2} \right], \quad (9.4\text{-}13)$$

Power-Transfer Ratio

where $\operatorname{sinc}(x) = \sin(\pi x)/(\pi x)$. Figure 9.4-8 illustrates the dependence of the power-transfer ratio $\mathcal{T}$ on the mismatch parameter $\Delta\beta L_0$. The ratio achieves a maximum value of unity at $\Delta\beta L_0 = 0$, decreases with increasing $\Delta\beta L_0$, and then vanishes when $\Delta\beta L_0 = \sqrt{3}\,\pi$.

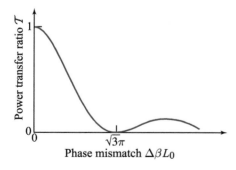

Figure 9.4-8 Dependence of the power transfer ratio $\mathcal{T} = P_2(L_0)/P_1(0)$ on the phase-mismatch parameter $\Delta\beta L_0$. The waveguide length is chosen such that for $\Delta\beta = 0$ (the phase-matched case), maximum power is transferred to waveguide 2, i.e., $\mathcal{T} = 1$.

The dependence of the transferred power on the phase mismatch can be utilized in making electrically activated directional couplers. If the mismatch $\Delta\beta L_0$ is switched between 0 and $\sqrt{3}\,\pi$, the light is switched from waveguide 2 to waveguide 1. Electrical control of $\Delta\beta$ can be achieved if the material of the waveguides is electro-optic (i.e., if its refractive index can be altered by applying an electric field). Such devices will be examined in Secs. 21.1D and 24.3B in connection with electro-optic switches.

382   CHAPTER 9   GUIDED-WAVE OPTICS

☐ ***Derivation of the Coupled Wave Equations.*** We proceed to derive the differential equations (9.4-4) that govern the amplitudes $a_1(z)$ and $a_2(z)$ of the coupled modes. When the two waveguides are not interacting they carry optical fields whose complex amplitudes are of the form

$$E_1(y,z) = a_1 u_1(y) \exp(-j\beta_1 z) \tag{9.4-14a}$$

$$E_2(y,z) = a_2 u_2(y) \exp(-j\beta_2 z). \tag{9.4-14b}$$

The amplitudes a_1 and a_2 are then constant. In the presence of coupling, we assume that the amplitudes a_1 and a_2 become functions of z but the transverse functions $u_1(y)$ and $u_2(y)$, and the propagation constants β_1 and β_2, are not altered. The amplitudes a_1 and a_2 are assumed to be slowly varying functions of z in comparison with the distance β^{-1} (the inverse of the propagation constant, β_1 or β_2), which is of the order of magnitude of the wavelength of light.

The presence of waveguide 2 is regarded as a perturbation of the medium outside waveguide 1 in the form of a slab of refractive index $n_2 - n$ and width d at a distance $2a$. The excess refractive index $(n_2 - n)$ and the field E_2 correspond to an excess polarization density $P = (\epsilon_2 - \epsilon)E_2 = \epsilon_o(n_2^2 - n^2)E_2$, which creates a source of optical radiation into waveguide 1 [see (5.2-25)] $S_1 = -\mu_o \partial^2 P/\partial t^2$ with complex amplitude

$$S_1 = \mu_o \omega^2 P = \mu_o \omega^2 \epsilon_o \left(n_2^2 - n^2\right) E_2 = \left(n_2^2 - n^2\right) k_o^2 E_2$$

$$= \left(k_2^2 - k^2\right) E_2. \tag{9.4-15}$$

Here ϵ_2 and ϵ are the electric permittivities associated with the refractive indices n_2 and n, respectively, and $k_2 = n_2 k_o$. This source is present only in the slab of waveguide 2.

To determine the effect of such a source on the field in waveguide 1, we write the Helmholtz equation in the presence of a source as

$$\nabla^2 E_1 + k_1^2 E_1 = -S_1 = -\left(k_2^2 - k^2\right) E_2. \tag{9.4-16a}$$

We similarly write the Helmholtz equation for the wave in waveguide 2 with a source generated as a result of the field in waveguide 1,

$$\nabla^2 E_2 + k_2^2 E_2 = -S_2 = -\left(k_1^2 - k^2\right) E_1, \tag{9.4-16b}$$

where $k_1 = n_1 k_o$. Equations (9.4-16) are two coupled partial different equations that we solve to determine E_1 and E_2. This type of perturbation analysis is valid only for weakly coupled waveguides.

We now write $E_1(y,z) = a_1(z) e_1(y,z)$ and $E_2(y,z) = a_2(z) e_2(y,z)$, where $e_1(y,z) = u_1(y) \exp(-j\beta_1 z)$ and $e_2(y,z) = u_2(y) \exp(-j\beta_2 z)$ and note that e_1 and e_2 must satisfy the Helmholtz equations,

$$\nabla^2 e_1 + k_1^2 e_1 = 0 \tag{9.4-17a}$$

$$\nabla^2 e_2 + k_2^2 e_2 = 0, \tag{9.4-17b}$$

where $k_1 = n_1 k_o$ and $k_2 = n_2 k_o$ for points inside the slabs of waveguides 1 and 2, respectively, and $k_1 = k_2 = n k_o$ elsewhere. Substituting $E_1 = a_1 e_1$ into (9.4-16a), we obtain

$$\frac{d^2 a_1}{dz^2} e_1 + 2\frac{da_1}{dz}\frac{de_1}{dz} = -\left(k_2^2 - k^2\right) a_2 e_2. \tag{9.4-18}$$

Noting that a_1 varies slowly, whereas e_1 varies rapidly with z, we neglect the first term of (9.4-18) in comparison with the second. The ratio between these terms is $[(d\Psi/dz)e_1]/[2\Psi de_1/dz] = [(d\Psi/dz)e_1]/[2\Psi(-j\beta_1 e_1)] = j(d\Psi/\Psi)/2\beta_1 dz$ where $\Psi = da_1/dz$. The approximation is valid if $d\Psi/\Psi \ll \beta_1 dz$, i.e., if the variation in $a_1(z)$ is slow in comparison with the length β_1^{-1}.

We proceed by substituting $e_1 = u_1 \exp(-j\beta_1 z)$ and $e_2 = u_2 \exp(-j\beta_2 z)$ into (9.4-18). Neglecting the first term leads to

$$2\frac{da_1}{dz}(-j\beta_1) u_1(y) e^{-j\beta_1 z} = -\left(k_2^2 - k^2\right) a_2 u_2(y) e^{-j\beta_2 z}. \tag{9.4-19}$$

Multiplying both sides of (9.4-19) by $u_1(y)$, integrating with respect to y, and using the fact that $u_1^2(y)$ is normalized so that its integral is unity, we finally obtain

$$\frac{da_1}{dz} e^{-j\beta_1 z} = -j\mathcal{C}_{21} a_2(z) e^{-j\beta_2 z}, \qquad (9.4\text{-}20)$$

where $\mathcal{C}_{21}$ is given by (9.4-6). A similar equation is obtained by repeating the procedure for waveguide 2. These equations yield the coupled differential equations (9.4-4). ∎

---

*C. Waveguide Arrays

The foregoing analysis of light propagation in a pair of weakly coupled waveguides, as presented in Sec. 9.4B, may be generalized to light propagation in waveguide arrays. Consider an array of N identical parallel slab waveguides separated by equal distances, under the assumption that the coupling is sufficiently weak so that only next-neighbor coupling is significant.

If $a_n(z)$ represents the complex amplitude of light in the nth waveguide, then the set of N coupled-mode equations can be written as

$$\frac{da_n}{dz} = -j\mathcal{C}(a_{n+1} + a_{n-1}), \quad n = 1, \ldots, N, \qquad (9.4\text{-}21)$$

where $\mathcal{C}$ is the coupling coefficient and $a_0 = a_{N+1} = 0$. If the amplitudes are represented by a vector $\mathbf{a}$ of dimension N, comprising the elements $\{a_n\}$, then (9.4-21) may be expressed in matrix form as $d\mathbf{a}/dz = -j\mathbf{H}\mathbf{a}$, where $\mathbf{H}$ is an $N \times N$ matrix whose elements are $H_{nm} = \mathcal{C}$ for $m = n \pm 1$, and zero otherwise. The solution to this equation is $\mathbf{a}(z) = \mathbf{T}\mathbf{a}(0)$, where $\mathbf{T} = \exp(-jz\mathbf{H})$ is the transmission matrix.

The transmission of light through such an array is best described in terms of modes (see Appendix C). The modes of the waveguide array, known as **supermodes**, are to be distinguished from the modes of individual isolated waveguides. The supermodes are determined by diagonalizing the matrix $\mathbf{H}$. This $N \times N$ matrix has N eigenvalues λ_r, and corresponding eigenvectors $\mathbf{b}_r$ comprising the elements $\{b_{rn}\}$, given by

$$\lambda_r = 2\mathcal{C}\cos\left(\frac{r\pi}{N+1}\right), \quad b_{rn} = \sqrt{\frac{2}{N+1}}\sin\left(\frac{r\pi n}{N+1}\right), \quad r = 1, 2, \ldots, N.$$
$$(9.4\text{-}22)$$

The related transmission matrix $\mathbf{T} = \exp(-jz\mathbf{H})$ has eigenvalues $\exp(-j\lambda_r z)$ and the same eigenvectors $\mathbf{b}_r$ corresponding to N modes. If the initial amplitudes $\{a_n(0)\}$ are equal to the amplitudes $\{b_{rn}\}$ of the rth mode, they evolve in accordance with the simple relation $a_n(z) = a_n(0)e^{-j\lambda_r z}$, independent of the other modes. The associated optical fields then propagate with a single propagation constant $\beta_r = \beta_0 + \lambda_r$, where β_0 is the propagation constant in an isolated waveguide. Since $-2\mathcal{C} \leq \lambda_r \leq 2\mathcal{C}$, the propagation constants of the modes lie in the range $\beta_0 - 2\mathcal{C} \leq \beta_r \leq \beta_0 + 2\mathcal{C}$.

An arbitrary input distribution $\{a_n(0)\}$ to the waveguide array can be expressed as a superpositions of the modes, $a_n(0) = \sum_{r=1}^{N} w_r b_{rn}$, where $w_r = \sum_{n=1}^{N} a_n(0) b_{rn}$ are the superposition weights. The amplitudes at a distance z are then given by

$$a_n(z) = \sum_{r=1}^{N} w_r b_{rn} e^{-j\lambda_r z}. \qquad (9.4\text{-}23)$$

The modal analysis enables us to determine $a_n(z)$ for arbitrary $a_n(0)$.

EXAMPLE 9.4-1. *Supermodes of Two Coupled Waveguides.* For an array of two coupled waveguides ($N = 2$), $\mathbf{H} = \begin{bmatrix} 0 & \mathcal{C} \\ \mathcal{C} & 0 \end{bmatrix}$ and the transmission matrix $\mathbf{T} = \exp(-jz\mathbf{H}) = \begin{bmatrix} \cos \mathcal{C}z & -j\sin \mathcal{C}z \\ -j\sin \mathcal{C}z & \cos \mathcal{C}z \end{bmatrix}$, which reproduces (9.4-11). The two modes have eigenvalues $\lambda_r = \pm\mathcal{C}$ and corresponding propagation constants $\beta_0 \pm \mathcal{C}$. The eigenvectors are $\frac{1}{\sqrt{2}}\begin{bmatrix} 1 \\ 1 \end{bmatrix}$ and $\frac{1}{\sqrt{2}}\begin{bmatrix} 1 \\ -1 \end{bmatrix}$, corresponding to equal or opposite excitation of the two waveguides. An input field at a single waveguide, say $\mathbf{a}(0) = \begin{bmatrix} 1 \\ 0 \end{bmatrix} = \frac{1}{2}\begin{bmatrix} 1 \\ 1 \end{bmatrix} + \frac{1}{2}\begin{bmatrix} 1 \\ -1 \end{bmatrix}$, excites both supermodes, which have different propagation constants, and the result is an exchange of power between the two waveguides.

Periodic Waveguides

In the limit of large N, the waveguide array may be regarded as a periodic medium and the theory presented in Sec. 7.2 may be readily applied. In particular, it is instructive to compare the dispersion diagram for light propagation in an isolated slab dielectric waveguide to that for light propagation in an array with a finite number of parallel slab waveguides, and to a periodic dielectric medium comprising an infinite set of such waveguides. These diagrams are presented in Fig. 9.4-9. In the single-slab waveguide [Fig. 9.4-9(a)], light travels in modes, each with a dispersion curve lying in the region between the light lines $\omega = c_1\beta$ and $\omega = c_2\beta$. At any frequency, there is at least one mode. In an array of N waveguides [Fig. 9.4-9(b)], each dispersion curve splits into N curves, representing the supermodes. The shapes of these curves are dependent on the coupling coefficient $\mathcal{C}$, which is frequency dependent in accordance with (9.4-6). In the periodic waveguide [Fig. 9.4-9(c)], the dispersion curves broaden into bands that lie between the light lines, and the bands are separated by photonic bandgaps.

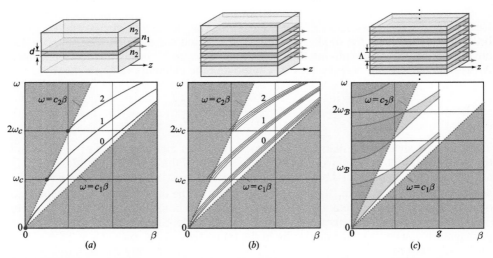

Figure 9.4-9 (a) Dispersion diagram of a slab waveguide with cutoff angular frequency $\omega_c = (\pi/d)(c_o/\mathrm{NA})$, as displayed in Fig. 9.2-8(a). (b) Dispersion diagram of the supermodes of a waveguide array. (c) Dispersion diagram of a periodic waveguide with period Λ, spatial frequency $g = 2\pi/\Lambda$, and Bragg angular frequency $\omega_\mathcal{B} = (\pi/\Lambda)(c_o/\bar{n})$, as shown in Fig. 7.2-7. Here, the waves travel in the z direction, which is parallel to the layers, and the higher-index medium is denoted n_1. In Fig. 7.2-7, in contrast, the direction parallel to the layers is the x direction and the higher-index medium is denoted n_2.

9.5 PHOTONIC-CRYSTAL WAVEGUIDES

Bragg-Grating Waveguide

We have seen earlier in this chapter that light may be guided by bouncing between two parallel reflectors — e.g., planar mirrors as described in Sec. 9.1; or planar dielectric boundaries at which the light undergoes total internal reflection, as described in Sec. 9.2. Alternatively, Bragg grating reflectors (see Sec. 7.1C) may be used to guide light, as illustrated in Fig. 9.5-1. The Bragg grating reflector (BGR) is a stack of alternating dielectric layers that has special angle- and frequency-dependent reflectance. For a given angle, the reflectance is close to unity at frequencies within a stop band. Similarly, at a given frequency, the reflectance is close to unity within a range of angles, but omnidirectional reflection is also possible. Thus, a wave with a given frequency can be guided through the waveguide by repeated reflections within a range of bounce angles. Within this angular range, the self-consistency condition is satisfied at a discrete set of angles, each corresponding to a propagating mode. The field distribution of a propagating mode is confined principally to the slab; decaying (evanescent) tails reach into the adjacent grating layers, as illustrated in Fig. 9.5-1.

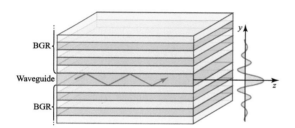

Figure 9.5-1 Planar waveguide comprising a dielectric slab sandwiched between two Bragg-grating reflectors (BGRs).

Bragg-Grating Waveguide as a Photonic Crystal with a Defect Layer

If the upper and lower gratings of a Bragg-grating waveguide are identical, and the slab thickness is comparable to the thickness of the periodic layers constituting the gratings, then the entire medium may be regarded as a 1D periodic structure, i.e., a 1D photonic crystal, but with a **defect**. For example, the device shown in Fig. 9.5-1 is periodic everywhere except for the slab, which is a layer of different thickness and/or different refractive index; the slab may therefore be viewed as a "defective" layer. As described in Sec. 7.2, a perfect photonic crystal has a dispersion relation, or energy-band diagram, containing bandgaps within which no propagating modes exist. In the presence of the "defective" layer, however, a mode whose frequency lies within the bandgap may exist, but it is confined primarily within that layer. Such a mode corresponds to a frequency in the dispersion diagram that lies within the photonic bandgap, as illustrated in Fig. 9.5-2. Such a frequency is the analog of a defect energy level that lies within the bandgap of a semiconductor crystal.

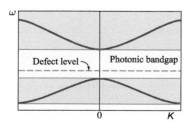

Figure 9.5-2 Dispersion diagram of a photonic crystal with a defect layer.

2D Photonic-Crystal Waveguides

Waveguides may also be created by introducing a path of defects in a 2D photonic crystal. In the example illustrated in Fig. 9.5-3(a), a 2D photonic crystal comprising a set of parallel cylindrical holes, placed in a dielectric material at the points of a periodic triangular lattice, exhibits a complete photonic bandgap for waves traveling along directions parallel to the plane of periodicity (normal to the cylindrical holes). The defect waveguide may take the form of a line of absent holes. A wave entering the waveguide at frequencies within the photonic bandgap does not leak into the surrounding periodic media so that the light is guided through the waveguide. A schematic profile of the propagating mode is illustrated in Fig. 9.5-3(a).

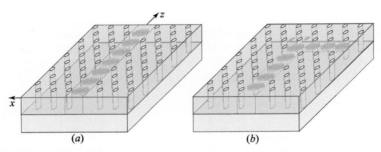

Figure 9.5-3 (a) Propagating mode in a photonic-crystal waveguide. (b) An L-shaped photonic-crystal waveguide.

Moreover, because of the omnidirectional nature of the photonic bandgap, light may be guided through photonic-crystal waveguides with sharp bends and corners without losing energy into the surrounding medium, as illustrated by the L-shaped waveguide configuration shown in Fig. 9.5-3(b). Such behavior is not possible with conventional dielectric waveguides based on total internal reflection.

9.6 PLASMONIC WAVEGUIDES

As demonstrated earlier in this chapter, it is difficult to confine a propagating wave to dimensions much smaller than its wavelength (see also Sec. 4.4D). For the perfect-mirror waveguide described in Sec. 9.1, it is demonstrated in Fig. 9.6-1(a) that a wave of wavelength λ can be guided if the mirror separation $d > \lambda/2$, but it cannot be guided if d is smaller. For the dielectric waveguide described in Sec. 9.2, it was shown that if the slab width d is reduced below $\lambda/2$, only a single guided mode is supported, and as d is further reduced, the guided wave spreads substantially outside the slab and into the surrounding dielectric medium (see Fig. 9.2-5).

Light can, however, be confined and guided at subwavelength spatial scales by making use of metallic structures. As shown in Sec. 8.2B, a metal–dielectric boundary supports a *surface* wave, called a surface plasmon polariton (SPP), that is tightly confined to the boundary, with penetration depths on both sides of the boundary that are much smaller than the wavelength [see Fig. 9.6-1(b)]. It will become apparent below that an optical wave may be guided within an ultrathin dielectric slab embedded in metal cladding if its width is much smaller than the wavelength, as illustrated in Fig. 9.6-1(c). Under these conditions, the SPP waves at the two boundaries couple with each other and combine into modes that extend through the dielectric medium. Similarly, an ultrathin metallic film can guide an optical wave of subwavelength scale [Fig. 9.6-1(d)].

Other metal–dielectric structures with more complex configurations may also be constructed to guide light through various optical circuits. As discussed in Sec. 8.2, this branch of integrated photonics is called **plasmonics**. The propagation lengths of plasmonic waveguides are limited by metallic losses.

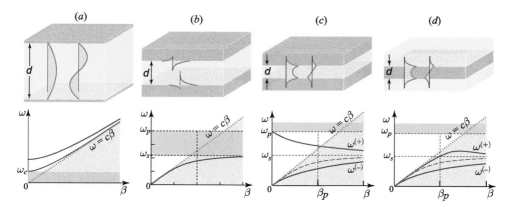

Figure 9.6-1 Configurations and dispersion relations for various optical and plasmonic waveguides. (a) A perfect-mirror waveguide supports optical guided modes if its width $d > \lambda/2$, i.e., if $\omega > \omega_c$ where $\omega_c = \pi c/d$ is the cutoff frequency. (b) A metal–dielectric–metal waveguide of width $d < \lambda/2$ does not support optical guided modes, but does support independent surface plasmon polariton (SPP) waves at the two boundaries if $\omega < \omega_s$, where $\omega_s = \omega_p/\sqrt{1 + \epsilon_{r1}}$, ω_p is the plasma frequency of the metal, and $\epsilon_{r1} = \epsilon_1/\epsilon_o$ is the relative permittivity of the dielectric. (c) A metal–insulator–metal (MIM) waveguide of width $d \ll \lambda$ supports a symmetric guided mode $\omega^{(+)}$ for $\omega_s < \omega < \omega_p$, and an anti-symmetric mode $\omega^{(-)}$ for $\omega < \omega_s$. (d) A thin-metal slab of width $d \ll \lambda$, called a metal-slab waveguide, supports two guided modes, one below ω_s and the other that extends slightly above ω_s. For the plots in (c) and (d), $d = \lambda_p/10$, where $\lambda_p = 2\pi c_o/\omega_p$, and $\epsilon_{r1} = 2.25$, so that $\omega_s = 0.55\,\omega_p$. The dashed blue curves represent the dispersion relation for the single-boundary SPP wave. In (a)–(d) the dotted red lines are the dielectric-medium light lines $\omega = c\beta$.

Metal–Insulator–Metal Waveguide

A dielectric slab surrounded by metal claddings forms a metal–dielectric–metal, or **metal–insulator–metal (MIM)** waveguide. If the slab thickness is greater than twice the penetration depth of the SPP waves at the boundaries [see (8.2-22)], the structure supports two independent SPP waves [Fig. 9.6-1(b)]. For a thinner dielectric slab, these surface waves overlap and couple, splitting the SPP dispersion curve into two branches and thereby creating two distinct guided modes, labeled $\omega^{(-)}$ and $\omega^{(+)}$ in Fig. 9.6-1(c). These modes correspond to anti-symmetric and symmetric field distributions, respectively.

The dispersion relations for these two modes may be derived by matching the boundary conditions at the metal–dielectric interfaces, much as was done for the dielectric waveguide (Sec. 9.2). The result for the TM wave is similar to (9.2-4) for the dielectric waveguide:

$$\tanh\left(\frac{d}{2}\gamma_1\right) = -\frac{\epsilon_1}{\epsilon_2}\frac{\gamma_2}{\gamma_1}, \qquad \coth\left(\frac{d}{2}\gamma_1\right) = -\frac{\epsilon_1}{\epsilon_2}\frac{\gamma_2}{\gamma_1}, \qquad (9.6\text{-}1)$$

with

$$\gamma_1 = \sqrt{\beta^2 - \omega^2\mu\epsilon_1}, \qquad \gamma_2 = \sqrt{\beta^2 - \omega^2\mu\epsilon_2}, \qquad (9.6\text{-}2)$$

where ϵ_1 and ϵ_2 are the permittivities of the dielectric and metal materials, respectively. This dispersion relation is plotted in Fig. 9.6-1(c) for a metal described by the Drude model, $\epsilon_2 = \epsilon_o(1 - \omega_p^2/\omega^2)$, where ω_p is the bulk-metal plasma frequency. Note that the upper branch of the dispersion curve crosses the light line, indicating that the wave travels at a phase velocity greater than the velocity of light c in the dielectric medium.

Since the two branches of the dispersion relation extend over the frequency range $0 < \omega < \omega_p$ a wave at any frequency $\omega < \omega_p$ can be guided in dielectric slabs that are significantly smaller than the wavelength. Modes at near-infrared wavelengths, for example, can be localized at the nanometer scale.

Metal-Slab Waveguide

Similarly, a thin metallic film of width $d \ll \lambda$ embedded in a dielectric medium can serve as a plasmonic waveguide [Fig. 9.6-1(d)]. If the film thickness is smaller than the penetration depth of the SPP waves from the boundaries into the metal, then these waves overlap and coalesce into two distinct waveguide modes. Again, the dispersion relation may be obtained by matching the boundary conditions, yielding

$$\tanh\left(\frac{d}{2}\gamma_2\right) = -\frac{\epsilon_1}{\epsilon_2}\frac{\gamma_2}{\gamma_1}, \qquad \coth\left(\frac{d}{2}\gamma_2\right) = -\frac{\epsilon_1}{\epsilon_2}\frac{\gamma_2}{\gamma_1}, \qquad (9.6\text{-}3)$$

where γ_1 and γ_2 are given by (9.6-2) and, as before, ϵ_1 and ϵ_2 are the permittivities of the dielectric and metal materials, respectively. The Drude model for the metal is again characterized by $\epsilon_2 = \epsilon_o(1-\omega_p^2/\omega^2)$. Note that (9.6-3) is the same as (9.6-1) except that the subscripts 1 and 2 are interchanged on the left-hand sides of both equations. This dispersion relation is plotted in Fig. 9.6-1(d). As with the MIM waveguide considered above, the dispersion relation for the SPP at a single metal–dielectric boundary splits into two branches, corresponding to symmetric and anti-symmetric modes; in this case both lie below the light line for the bulk dielectric material $\omega = c\beta$.

*Periodic Metal–Dielectric Arrays

A periodic structure comprising an array of metal–dielectric slabs functions as a photonic crystal. In analogy with the all-dielectric arrays discussed in Chapter 7, and illustrated in Figs. 9.4-9(a) and (c), the dispersion curve of the single metal–dielectric boundary depicted as the dashed blue curve in Fig. 9.6-2(b) splits to create bands, as shown. The frequency ω of a mode with propagation constant β may only lie within two separated spectral bands, both of which are within the range $\omega < \omega_p$.

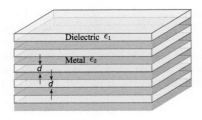
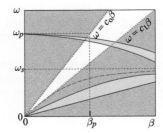

Figure 9.6-2 Metal–insulator periodic structure and its dispersion relation for light traveling in the direction of the layers. Two bands lie within the $\omega < \omega_p$ frequency range. The dashed blue curve in (b) is the dispersion relation for the single-boundary SPP wave and the dotted red lines are the free-space and dielectric-medium light lines.

READING LIST

Books and Seminal Articles

See also the reading list on layered and periodic media in Chapter 7, the reading list in Chapter 10, and the reading list on photonic integrated circuits in Chapter 25.

S. Bhadra and A. Ghatak, eds., *Guided Wave Optics and Photonic Devices*, CRC Press/Taylor & Francis, 2013.

L. N. Binh, *Guided Wave Photonics: Fundamentals and Applications with* MATLAB, CRC Press/Taylor & Francis, 2012.

D. Dai, J. Bauters, and J. E. Bowers, Passive Technologies for Future Large-Scale Photonic Integrated Circuits on Silicon: Polarization Handling, Light Non-Reciprocity and Loss Reduction, *Light: Science & Applications* (2012) **1**, e1; doi:10.1038/lsa.2012.1

R. G. Hunsperger, *Integrated Optics: Theory and Technology*, Springer-Verlag, 1982, 6th ed. 2010.

J. Bures, *Guided Optics*, Wiley–VCH, 2009.

C.-L. Chen, *Foundations for Guided Wave Optics*, Wiley, 2007.

K. Iga and Y. Kokobun, eds., *Encyclopedic Handbook of Integrated Optics*, CRC Press/Taylor & Francis, 2006.

K. Okamoto, *Fundamentals of Optical Waveguides*, Elsevier, 2nd ed. 2005.

J.-M. Liu, *Photonic Devices*, Cambridge University Press, 2005, paperback ed. 2009.

G. Lifante, *Integrated Photonics: Fundamentals*, Wiley, 2003.

C. Pollock and M. Lipson, *Integrated Photonics*, Kluwer, 2003.

A. A. Barybin and V. A. Dmitriev, *Modern Electrodynamics and Coupled-Mode Theory: Application to Guided-Wave Optics*, Rinton Press, 2002.

K. Iizuka, *Elements of Photonics, Volume 2: For Fiber and Integrated Optics*, Wiley, 2002.

M. Young, *Optics and Lasers: Including Fibers and Optical Waveguides*, Springer-Verlag, 5th ed. 2000.

A. R. Mickelson, *Guided Wave Optics*, Springer-Verlag, 1993, paperback ed. 2012.

K. J. Ebeling, *Integrated Optoelectronics: Waveguide Optics, Photonics, Semiconductors*, Springer-Verlag, 1993, paperback ed. 2011.

D. G. Hall, ed., *Selected Papers on Coupled Mode Theory in Guided-Wave Optics*, SPIE Optical Engineering Press (Milestone Series Volume 84), 1993.

D. Marcuse, *Theory of Dielectric Optical Waveguides*, Academic Press, 1974, 2nd ed. 1991.

PROBLEMS

9.1-3 **Field Distribution.**
  (a) Demonstrate that a single TEM plane wave $E_x(y,z) = A\exp(-jk_y y)\exp(-j\beta z)$ cannot satisfy the boundary conditions, $E_x(\pm d/2, z) = 0$ at all z, for the mirror waveguide illustrated in Fig. 9.1-1.
  (b) Show that the sum of two TEM plane waves written as $E_x(y,z) = A_1\exp(-jk_{y1}y)\exp(-j\beta_1 z) + A_2\exp(-jk_{y2}y)\exp(-j\beta_2 z)$ does not satisfy the boundary conditions if $A_1 = \pm A_2$, $\beta_1 = \beta_2$, and $k_{y1} = -k_{y2} = m\pi/d$ where $m = 1, 2, \ldots$.

9.1-4 **Modal Dispersion.** Light of wavelength $\lambda_o = 0.633$ μm is transmitted through a mirror waveguide of mirror separation $d = 10$ μm and $n = 1$. Determine the number of TE and TM modes. Determine the group velocities of the fastest and the slowest modes. If a narrow pulse of light is carried by all modes for a distance 1 m in the waveguide, how much does the pulse spread as a result of the differences of the group velocities?

9.2-3 **Parameters of a Dielectric Waveguide.** Light of free-space wavelength $\lambda_o = 0.87$ μm is guided by a thin planar film of width $d = 2$ μm and refractive index $n_1 = 1.6$ surrounded by a medium of refractive index $n_2 = 1.4$.
  (a) Determine the critical angle θ_c and its complement $\bar{\theta}_c$, the numerical aperture NA, and the maximum acceptance angle for light originating in air ($n = 1$).

(b) Determine the number of TE modes.

(c) Determine the bounce angle θ and the group velocity v of the $m = 0$ TE mode.

9.2-4 **Effect of Cladding.** Redo Prob. 9.2-3 under the proviso that the thin film is suspended in air ($n_2 = 1$). Compare the results.

9.2-5 **Field Distribution.** The transverse distribution $u_m(y)$ of the electric-field complex amplitude of a TE mode in a slab waveguide is given by (9.2-10) and (9.2-13). Derive an expression for the ratio of the proportionality constants. Plot the distribution of the $m = 0$ TE mode for a slab waveguide with parameters $n_1 = 1.48$, $n_2 = 1.46$, $d = 0.5$ μm, and $\lambda_o = 0.85$ μm, and determine its confinement factor (percentage of power in the slab).

9.2-6 **Derivation of the Field Distributions Using Maxwell's Equations.** Assuming that the electric field in a symmetric dielectric waveguide is harmonic within the slab and exponential outside the slab and has a propagation constant β in both media, we may write $E_x(y,z) = u(y)\,e^{-j\beta z}$, where

$$u(y) = \begin{cases} A\cos(k_y y + \varphi), & -d/2 \leq y \leq d/2 \\ B\exp(-\gamma y), & y > d/2 \\ B\exp(\gamma y), & y < d/2. \end{cases}$$

Satisfying the Helmholtz equation requires $k_y^2 + \beta^2 = n_1^2 k_o^2$ and $-\gamma^2 + \beta^2 = n_2^2 k_o^2$. Use Maxwell's equations to derive expressions for $H_y(y,z)$ and $H_z(y,z)$. Show that the boundary conditions are satisfied if β, γ, and k_y take the values β_m, γ_m, and k_{ym} derived in the text and verify the self-consistency condition (9.2-4).

9.2-7 **Single-Mode Waveguide.** What is the largest thickness d of a planar symmetric dielectric waveguide with refractive indices $n_1 = 1.50$ and $n_2 = 1.46$ for which there is only one TE mode at $\lambda_o = 1.3$ μm? What is the number of modes if a waveguide with this thickness is used at $\lambda_o = 0.85$ μm instead?

9.2-8 **Mode Cutoff.** Show that the cutoff condition for TE mode $m > 0$ in a symmetric slab waveguide with $n_1 \approx n_2$ is approximately $\lambda_o^2 \approx 8 n_1 \Delta n\, d^2/m^2$, where $\Delta n = n_1 - n_2$.

9.2-9 **TM Mode Bounce Angles.** Derive an expression for the bounce angles of the TM modes similar to (9.2-4). Generate a plot similar to Fig. 9.2-2 for TM modes in a waveguide with $\sin\overline{\theta}_c = 0.3$ and $\lambda/2d = 0.1$. What is the number of TM modes?

9.3-1 **Modes of a Rectangular Dielectric Waveguide.** A rectangular dielectric waveguide has a square cross section of area 10^{-2} mm^2 and numerical aperture NA $= 0.1$. Use (9.3-3) to plot the number of TE modes as a function of frequency ν. Compare your results with Fig. 9.2-4.

9.4-1 **Coupling Coefficient Between Two Slabs.**

(a) Use (9.4-6) to determine the coupling coefficient between two *identical* slab waveguides of width $d = 0.5$ μm, spacing $2a = 1.0$ μm, and refractive indices $n_1 = n_2 = 1.48$, in a medium of refractive index $n = 1.46$, at $\lambda_o = 0.85$ μm. Assume that both waveguides are operating in the $m = 0$ TE mode and use the results of Prob. 9.2-5 to determine the transverse distributions.

(b) Determine the length of the waveguides that makes the device act as a 3-dB coupler.

*9.6-1 **Silver-Slab Waveguide.** Calculate and plot the dispersion relation for the symmetric and anti-symmetric modes of a silver-slab waveguide of thickness $d = 20$ nm in a host medium of frequency-independent refractive index $n_2 = 2$. Assume that the permittivity of silver is described by the Drude model (8.2-18), with a plasma frequency ω_p corresponding to a free-space wavelength of 138 nm. Determine the properties (velocity, propagation wavelength, and penetration depths) of the symmetric mode at a free-space wavelength of 400 nm. Compare these properties with those of a SPP propagating on a single interface between silver and the same host medium.

CHAPTER

10

FIBER OPTICS

10.1	GUIDED RAYS	393
	A. Step-Index Fibers	
	B. Graded-Index Fibers	
10.2	GUIDED WAVES	397
	A. Step-Index Fibers	
	B. Single-Mode Fibers	
*C.	Quasi-Plane Waves in Step-Index and Graded-Index Fibers	
	D. Multicore Fibers and Fiber Couplers	
10.3	ATTENUATION AND DISPERSION	415
	A. Attenuation	
	B. Dispersion	
10.4	HOLEY AND PHOTONIC-CRYSTAL FIBERS	426
10.5	FIBER MATERIALS	429

Working at Corning in the early 1970s, **Peter C. Schultz (born 1942)**, left, **Donald B. Keck (born 1941)**, center, and **Robert D. Maurer (born 1924)**, right, developed ultra-low-loss silica-glass optical fibers that permitted light to propagate over exceptionally long distances, thereby paving the way for worldwide optical fiber communications. Billions of kilometers of optical fiber span the globe.

Fundamentals of Photonics, Third Edition. Bahaa E. A. Saleh and Malvin Carl Teich.
©2019 John Wiley & Sons, Inc. Published 2019 by John Wiley & Sons, Inc.

Fiber optics is an enabling technology for telecommunications, data transmission, and information science. The availability of ultra-low-loss optical fibers is, in large part, responsible for the commercial viability of optical fiber communications.

An optical fiber is a cylindrical dielectric waveguide fabricated from a low-loss material such as silica glass. It has a central **core** in which the light is guided, embedded in an outer **cladding** of slightly lower refractive index (Fig. 10.0-1). Light rays in the core incident on the core–cladding boundary at angles greater than the critical angle undergo total internal reflection and are thereby guided through the core without refraction into the cladding and without loss. Rays at greater inclination to the fiber axis lose a portion of their power into the cladding at each reflection and are not guided.

Figure 10.0-1 An optical fiber is a cylindrical dielectric waveguide with an inner core and an outer cladding of refractive index lower than that of the core.

Technological advances in the fabrication of optical fibers over the past several decades allow light to be guided through 1 km of silica-glass fiber with a loss as low as ≈ 0.15 dB ($\approx 3.4\%$) at the wavelength of maximum transparency. Because of this low loss, silica-glass optical fibers long ago replaced copper coaxial cables as the preferred transmission medium for terrestrial and sub-oceanic voice as well as for data communications. In recent years, optical fibers have transcended their monolithic silica-glass origins and have come to play a central role in the arenas of sensing, security, transportation, defense, and biomedicine. This has been facilitated by the development of photonic-crystal, specialty, multimaterial, and multifunctional optical fibers.

In this chapter we introduce the principles of light transmission in optical fibers. These principles are essentially the same as those applicable to planar dielectric waveguides (Chapter 9); the most notable distinction is that optical fibers have cylindrical geometry. In both types of waveguide, light propagates in the form of modes. Each mode travels along the axis of the waveguide with a distinct propagation constant and group velocity, maintaining its transverse spatial distribution and polarization. When the core diameter is small, only a single mode is supported and the optical fiber is said to be a **single-mode fiber**.

Optical fibers with large core diameters are **multimode fibers**. One of the difficulties associated with the propagation of light in multimode fibers arises from the differences among the group velocities of the modes. This results in a spread of travel times and leads to the broadening of a light pulse as it travels through the fiber. This effect, called **modal dispersion**, limits the rate at which adjacent pulses can be launched without resulting in pulse overlap at the far end of the fiber. Modal dispersion therefore limits the speed at which multimode optical fiber communication systems can operate.

Modal dispersion can be reduced by grading the refractive index of the fiber core from a maximum value at its center to a minimum value at the core–cladding boundary. The fiber is then called a **graded-index fiber**, or GRIN fiber, whereas conventional fibers with constant refractive indices in the core and the cladding are known as **step-index fibers**. In a graded-index fiber the travel velocity increases with radial distance from the core axis (since the refractive index decreases). Although rays of greater inclination to the fiber axis must travel farther, they travel faster. This permits the travel times of the different modes to be equalized.

Optical fibers are thus classified as step-index or graded-index, and multimode or single-mode, as illustrated in Fig. 10.0-2.

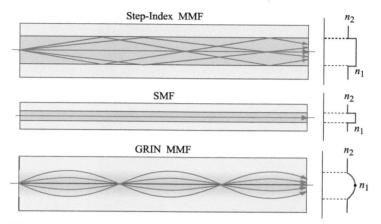

Figure 10.0-2 Geometry, refractive-index profile, and typical rays in a step-index multimode fiber (MMF), a single-mode fiber (SMF), and a graded-index multimode fiber (GRIN MMF).

This Chapter

The chapter begins with ray-optics descriptions of step-index and graded-index fibers (Sec. 10.1). An electromagnetic-optics approach, which highlights the nature of optical modes and single-mode propagation, follows in Sec. 10.2. In the simplified approximate approach set forth in Sec. 10.2C, the field is treated as a quasi-plane wave in analogy with the bouncing plane-wave construct for planar dielectric waveguides considered in Sec. 9.2. The optical properties of the fiber material (often fused silica), including attenuation and material dispersion as well as modal, waveguide, polarization-mode, and nonlinear dispersion, are presented in Sec. 10.3. Holey and photonic-crystal fibers, which have more complex refractive-index profiles, along with unusual dispersion characteristics, are introduced in Sec. 10.4. Finally, multimaterial and multifunctional fibers, including mid-infrared and specialty fibers, are considered in Sec. 10.5. We return to a discussion of fiber optics in Chapters 23 and 25, which are devoted to ultrafast optics and optical fiber communications, respectively.

10.1 GUIDED RAYS

A. Step-Index Fibers

A step-index fiber is a cylindrical dielectric waveguide specified by the refractive indices of its core and cladding, n_1 and n_2, respectively, and their radii a and b (see Fig. 10.0-1). Examples of standard core-to-cladding diameter ratios (in units of $\mu\text{m}/\mu\text{m}$) are $2a/2b = 8/125, 50/125, 62.5/125, 85/125$, and $100/140$. The refractive indices of the core and cladding differ only slightly, so that the fractional refractive-index change is small:

$$\Delta \equiv \frac{n_1^2 - n_2^2}{2n_1^2} \approx \frac{n_1 - n_2}{n_1} \ll 1. \qquad (10.1\text{-}1)$$

Most fibers used in currently implemented optical fiber communication systems are made of fused silica glass (SiO_2) of high chemical purity. Slight changes in the refractive index are effected by adding low concentrations of doping materials (e.g., titanium, germanium, boron). The refractive index n_1 ranges from 1.44 to 1.46, depending on the wavelength, and Δ typically lies between 0.001 and 0.02.

An optical ray in a step-index fiber is guided by total internal reflections within the fiber core if its angle of incidence at the core–cladding boundary is greater than the critical angle $\theta_c = \sin^{-1}(n_2/n_1)$, and remains so as the ray bounces.

Meridional Rays

Meridional rays, which are rays confined to planes that pass through the fiber axis, have a particularly simple guiding condition, as illustrated in Fig. 10.1-1. These rays intersect the fiber axis and reflect in the same plane without changing their angle of incidence, behaving as if they were in a planar waveguide. Meridional rays are guided if the angle θ they make with the fiber axis is smaller than the complement of the critical angle, i.e., if $\theta < \overline{\theta}_c = \pi/2 - \theta_c = \cos^{-1}(n_2/n_1)$. Since $n_1 \approx n_2$, $\overline{\theta}_c$ is usually small and the guided rays are approximately paraxial.

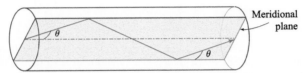

Figure 10.1-1 The trajectory of a meridional ray lies in a plane that passes through the fiber axis. The ray is guided if $\theta < \overline{\theta}_c = \cos^{-1}(n_2/n_1)$.

Skewed Rays

An arbitrary ray is identified by its plane of incidence, which is a plane parallel to the fiber axis through which the ray passes, and by the angle with that axis, as illustrated in Fig. 10.1-2. The plane of incidence intersects the core–cladding cylindrical boundary at an angle ϕ with respect to the normal to the boundary and lies at a distance R from the fiber axis. The ray is identified by its angle θ with the fiber axis and by the angle ϕ of its plane. When $\phi \neq 0$ ($R \neq 0$) the ray is said to be skewed. For meridional rays $\phi = 0$ and $R = 0$.

A skewed ray reflects repeatedly into planes that make the same angle ϕ with the core–cladding boundary; it follows a helical trajectory confined within a cylindrical shell of inner and outer radii R and a, respectively, as illustrated in Fig. 10.1-2. The projection of the trajectory onto the transverse (x–y) plane is a regular polygon that is not necessarily closed. The condition for a skewed ray to always undergo total internal reflection is that its angle with the z axis be smaller than the complementary critical angle, i.e., $\theta < \overline{\theta}_c$.

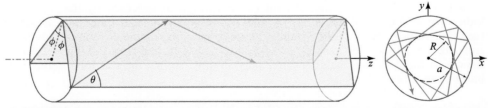

Figure 10.1-2 A skewed ray lies in a plane offset from the fiber axis by a distance R. The ray is identified by the angles θ and ϕ. It follows a helical trajectory confined within a cylindrical shell with inner and outer radii R and a, respectively. The projection of the ray on the transverse plane is a regular polygon that is not necessarily closed.

Numerical Aperture

A ray incident from air into the fiber becomes a guided ray if, upon refraction into the core, it makes an angle θ with the fiber axis that is smaller than $\bar{\theta}_c$. As shown in Fig. 10.1-3(a), if Snell's law is applied at the air–core boundary, the angle θ_a in air corresponding to the angle $\bar{\theta}_c$ in the core is obtained from $1 \cdot \sin \theta_a = n_1 \sin \bar{\theta}_c$, which leads to $\sin \theta_a = n_1 \sqrt{1 - \cos^2 \bar{\theta}_c} = n_1 \sqrt{1 - (n_2/n_1)^2} = \sqrt{n_1^2 - n_2^2}$ (see Exercise 1.2-5). The **acceptance angle** of the fiber is therefore

$$\theta_a = \sin^{-1} \text{NA}, \qquad (10.1\text{-}2)$$

where the numerical aperture (NA) of the fiber is given by

$$\boxed{\text{NA} = \sqrt{n_1^2 - n_2^2} \approx n_1 \sqrt{2\Delta}} \qquad (10.1\text{-}3)$$
Numerical Aperture

since $n_1 - n_2 = n_1 \Delta$ and $n_1 + n_2 \approx 2n_1$.

The acceptance angle θ_a of the fiber determines the cone of external rays that are guided by the fiber. Rays incident at angles greater than θ_a are refracted into the fiber but are guided only for a short distance since they do not undergo total internal reflection. The numerical aperture therefore describes the light-gathering capacity of the fiber, as illustrated in Fig. 10.1-3(b).

When the guided rays arrive at the terminus of the fiber, they are refracted back into a cone of angle θ_a. The acceptance angle is thus a crucial design parameter for coupling light into and out of a fiber.

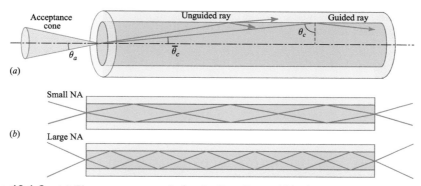

Figure 10.1-3 (a) The acceptance angle θ_a of a fiber. Rays within the acceptance cone are guided by total internal reflection. The numerical aperture $\text{NA} = \sin \theta_a$. The angles θ_a and $\bar{\theta}_c$ are typically quite small; they are exaggerated here for clarity. (b) The light-gathering capacity of a large NA fiber is greater than that of a small NA fiber.

EXAMPLE 10.1-1. *Cladded and Uncladded Fibers.* In a silica-glass fiber with $n_1 = 1.46$ and $\Delta = (n_1 - n_2)/n_1 = 0.01$, the complementary critical angle $\bar{\theta}_c = \cos^{-1}(n_2/n_1) = 8.1°$, and the acceptance angle $\theta_a = 11.9°$, corresponding to a numerical aperture $\text{NA} = 0.206$. By comparison, a fiber with silica-glass core ($n_1 = 1.46$) and a cladding with a much smaller refractive index $n_2 = 1.064$ has $\bar{\theta}_c = 43.2°$, $\theta_a = 90°$, and $\text{NA} = 1$. Rays incident from *all* directions are guided since they reflect within a cone of angle $\bar{\theta}_c = 43.2°$ inside the core. Likewise, for an uncladded fiber ($n_2 = 1$), $\bar{\theta}_c = 46.8°$, and rays incident from air at any angle are also refracted into guided rays. Although its light-gathering capacity is high, the uncladded fiber is generally not suitable for use as an optical waveguide because of the large number of modes it supports, as will be explained subsequently.

B. Graded-Index Fibers

Index grading is an ingenious method for reducing the pulse spreading caused by differences in the group velocities of the modes in a multimode fiber. The core of a graded-index (GRIN) fiber has a refractive index that varies; it is highest in the center of the fiber and decreases gradually to its lowest value where the core meets the cladding. The phase velocity of light is therefore minimum at the center and increases gradually with radial distance. Rays of the most axial mode thus travel the shortest distance, but they do so at the smallest phase velocity. Rays of the most oblique mode zigzag at a greater angle and travel a longer distance, but mostly in a medium where the phase velocity is high. The disparities in distances are thus compensated by opposite disparities in the phase velocities. As a consequence, the differences in the travel times associated with a light pulse are reduced. In this section we examine the propagation of light in GRIN fibers.

The core refractive index of a GRIN fiber is a function $n(r)$ of the radial position r. As illustrated in Fig. 10.1-4, the largest value of $n(r)$ is at the core center, $n(0) = n_1$, while the smallest value occurs at the core radius, $n(a) = n_2$. The cladding refractive index is maintained constant at n_2.

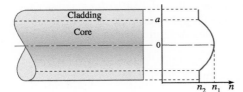

Figure 10.1-4 Geometry and refractive-index profile of a graded-index optical fiber.

A versatile refractive-index profile that exhibits this generic behavior is described by the power-law function

$$n^2(r) = n_1^2 \left[1 - 2\left(\frac{r}{a}\right)^p \Delta\right], \qquad r \leq a, \qquad (10.1\text{-}4)$$

where

$$\Delta = \frac{n_1^2 - n_2^2}{2n_1^2} \approx \frac{n_1 - n_2}{n_1}. \qquad (10.1\text{-}5)$$

The **grade profile parameter** p determines the steepness of the profile. As illustrated in Fig. 10.1-5, $n^2(r)$ is a linear function of r for $p = 1$ and a quadratic function for $p = 2$. The quantity $n^2(r)$ becomes increasingly steep as p becomes larger, and ultimately approaches a step function for $p \to \infty$. The step-index fiber is thus a special case of the GRIN fiber.

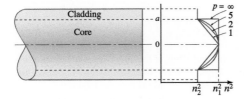

Figure 10.1-5 Power-law refractive-index profile $n^2(r)$ for various values of p.

The transmission of light rays through a GRIN medium with parabolic-index profile was discussed in Sec. 1.3. Rays in meridional planes follow oscillatory planar trajectories, whereas skewed rays follow helical trajectories. For an arbitrary refractive-index profile, the turning points form cylindrical caustic surfaces, as illustrated in Fig. 10.1-6. Guided rays are confined within the core and do not reach the cladding.

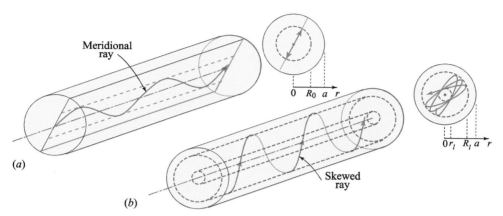

Figure 10.1-6 Guided rays in the core of a GRIN fiber. (a) A meridional ray confined to a meridional plane inside a cylinder of radius R_0. (b) A skewed ray follows a helical trajectory confined within two cylindrical shells of radii r_l and R_l. For a parabolic-index profile, the trajectory projects to a stationary ellipse, as in Fig. 1.3-7.

The numerical aperture of a GRIN optical fiber may be determined by identifying the largest angle of the incident ray that is guided within the GRIN core without reaching the cladding. For meridional rays in a GRIN fiber with parabolic profile, the numerical aperture is given by (10.1-3) (see Exercise 1.3-2).

10.2 GUIDED WAVES

We now proceed to develop an electromagnetic-optics theory of light propagation in fibers. We seek to determine the electric and magnetic fields of guided waves by using Maxwell's equations and the boundary conditions imposed by the cylindrical dielectric core and cladding. As with all waveguides, there are certain special solutions, known as modes (see Appendix C), each of which has a distinct propagation constant, a characteristic field distribution in the transverse plane, and two independent polarization states. Since an exact solution is rather difficult, a number of approximations will be used.

Helmholtz Equation

The optical fiber is a dielectric medium with refractive index $n(r)$. In a step-index fiber, $n(r) = n_1$ in the core $(r < a)$ and $n(r) = n_2$ in the cladding $(r > a)$. In a GRIN fiber, $n(r)$ is a continuous function in the core and has a constant value $n(r) = n_2$ in the cladding. In either case, we assume that the outer radius b of the cladding is sufficiently large so that it can be taken to be infinite when considering guided light in the core and near the core–cladding boundary.

Each of the components of the monochromatic electric and magnetic fields obeys the Helmholtz equation, $\nabla^2 U + n^2(r) k_o^2 U = 0$, where $k_o = 2\pi/\lambda_o$. This equation is obeyed exactly in each of the two regions of the step-index fiber, and is obeyed approximately within the core of the GRIN fiber if $n(r)$ varies slowly within a wavelength (see

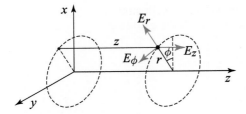

Figure 10.2-1 Cylindrical fiber coordinate system.

Sec. 5.3). In a cylindrical coordinate system (see Fig. 10.2-1) the Helmholtz equation is written as

$$\frac{\partial^2 U}{\partial r^2} + \frac{1}{r}\frac{\partial U}{\partial r} + \frac{1}{r^2}\frac{\partial^2 U}{\partial \phi^2} + \frac{\partial^2 U}{\partial z^2} + n^2 k_o^2 U = 0, \qquad (10.2\text{-}1)$$

where $U = U(r, \phi, z)$. The guided modes are waves traveling in the z direction with propagation constant β, so that the z dependence of U is of the form $e^{-j\beta z}$. They are periodic in the angle ϕ with period 2π, so that they take the harmonic form $e^{-jl\phi}$, where l is an integer. Substituting

$$U(r, \phi, z) = u(r) e^{-jl\phi} e^{-j\beta z}, \qquad l = 0, \pm 1, \pm 2, \ldots \qquad (10.2\text{-}2)$$

into (10.2-1) leads to an ordinary differential equation for the radial profile $u(r)$:

$$\frac{d^2 u}{dr^2} + \frac{1}{r}\frac{du}{dr} + \left(n^2(r)k_o^2 - \beta^2 - \frac{l^2}{r^2}\right) u = 0. \qquad (10.2\text{-}3)$$

A. Step-Index Fibers

As we discovered in Sec. 9.2B, the wave is guided (or bound) if the propagation constant is smaller than the wavenumber in the core ($\beta < n_1 k_o$) and greater than the wavenumber in the cladding ($\beta > n_2 k_o$). It is therefore convenient to define the quantities

$$k_T^2 = n_1^2 k_o^2 - \beta^2 \qquad (10.2\text{-}4a)$$

and

$$\gamma^2 = \beta^2 - n_2^2 k_o^2, \qquad (10.2\text{-}4b)$$

so that, for guided waves, k_T^2 and γ^2 are positive and k_T and γ are real. Equation (10.2-3) may then be written in the core and cladding separately:

$$\frac{d^2 u}{dr^2} + \frac{1}{r}\frac{du}{dr} + \left(k_T^2 - \frac{l^2}{r^2}\right) u = 0, \qquad r < a \quad \text{(core)}, \qquad (10.2\text{-}5a)$$

$$\frac{d^2 u}{dr^2} + \frac{1}{r}\frac{du}{dr} - \left(\gamma^2 + \frac{l^2}{r^2}\right) u = 0, \qquad r > a \quad \text{(cladding)}. \qquad (10.2\text{-}5b)$$

Equations (10.2-5) are well-known differential equations whose solutions comprise the family of Bessel functions. Excluding functions that approach ∞ at $r = 0$ in the core, or at $r \to \infty$ in the cladding, we obtain the bounded solutions:

$$u(r) \propto \begin{cases} J_l(k_T r), & r < a \quad \text{(core)} \\ K_l(\gamma r), & r > a \quad \text{(cladding)}, \end{cases} \quad (10.2\text{-}6)$$

where $J_l(x)$ is the Bessel function of the first kind and order l, and $K_l(x)$ is the modified Bessel function of the second kind and order l. The function $J_l(x)$ oscillates like the sine or cosine function but with a decaying amplitude. The function $K_l(x)$ decays exponentially at large x. Two representative examples of the radial distribution $u(r)$ are displayed in Fig. 10.2-2.

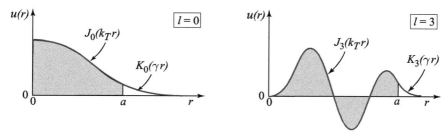

Figure 10.2-2 Examples of the radial distribution $u(r)$ provided in (10.2-6) for $l = 0$ and $l = 3$. The shaded and unshaded areas represent the fiber core and cladding, respectively. The parameters k_T and γ, and the two proportionality constants in (10.2-6), have been selected such that $u(r)$ is continuous and has a continuous derivative at $r = a$. Larger values of k_T and γ lead to a greater number of oscillations in $u(r)$.

The parameters k_T and γ determine the rate of change of $u(r)$ in the core and in the cladding, respectively. A large value of k_T means more oscillation of the radial distribution in the core. A large value of γ means more rapid decay and therefore smaller penetration of the wave into the cladding. As can be seen from (10.2-4), the sum of the squares of k_T and γ is a constant:

$$k_T^2 + \gamma^2 = \left(n_1^2 - n_2^2\right) k_o^2 = (\text{NA})^2 \cdot k_o^2, \quad (10.2\text{-}7)$$

so that as k_T increases, γ decreases and the field penetrates more deeply into the cladding. For those values of k_T that exceed $\text{NA} \cdot k_o$, the quantity γ becomes imaginary and the wave ceases to be bound to the core.

Fiber V Parameter

It is convenient to normalize k_T and γ by defining the quantities

$$X = k_T a, \qquad Y = \gamma a. \quad (10.2\text{-}8)$$

In view of (10.2-7), we have

$$X^2 + Y^2 = V^2, \quad (10.2\text{-}9)$$

where $V = \text{NA} \cdot k_o a$, from which

$$\boxed{V = 2\pi \frac{a}{\lambda_o} \text{NA}.} \quad (10.2\text{-}10)$$
V Parameter

It is important to recall that for the wave to be guided, X must be smaller than V.

As we shall see shortly, V is an important parameter that governs the number of modes of the fiber and their propagation constants. It is called the **fiber parameter** or the **V parameter**. It is directly proportional to the radius-to-wavelength ratio a/λ_o, and to the numerical aperture NA. Equation (10.2-10) is not unlike (9.2-7) for the number of TE modes in a planar dielectric waveguide.

Modes

We now consider the boundary conditions. We begin by writing the axial components of the electric- and magnetic-field complex amplitudes, E_z and H_z, in the form of (10.2-2). The condition that these components must be continuous at the core–cladding boundary $r = a$ establishes a relation between the coefficients of proportionality in (10.2-6), so that we have only one unknown for E_z and one unknown for H_z. With the help of Maxwell's equations, $j\omega\epsilon_o n^2 \mathbf{E} = \nabla \times \mathbf{H}$ and $-j\omega\mu_o \mathbf{H} = \nabla \times \mathbf{E}$ [see (5.3-12) and (5.3-13), respectively], the remaining four components, E_ϕ, H_ϕ, E_r, and H_r, are determined in terms of E_z and H_z. Continuity of E_ϕ and H_ϕ at $r = a$ yields two additional equations. One equation relates the two unknown coefficients of proportionality in E_z and H_z; the other provides a condition that the propagation constant β must satisfy. This condition, called the **characteristic equation** or **dispersion relation**, is an equation for β with the ratio a/λ_o and the fiber indices n_1, n_2 as known parameters.

For each azimuthal index l, the characteristic equation has multiple solutions yielding discrete propagation constants β_{lm}, $m = 1, 2, \ldots$, each solution representing a mode. The corresponding values of k_T and γ, which govern the spatial distributions in the core and in the cladding, respectively, are determined by using (10.2-4) and are denoted k_{Tlm} and γ_{lm}. A mode is therefore described by the indices l and m, characterizing its azimuthal and radial distributions, respectively. The function $u(r)$ depends on both l and m; $l = 0$ corresponds to meridional rays. Moreover, there are two independent configurations of the $\mathbf{E}$ and $\mathbf{H}$ vectors for each mode, corresponding to the two states of polarization. The classification and labeling of these configurations are generally quite involved (details are provided in specialized books in the reading list).

Characteristic Equation (Weakly Guiding Fiber)

Most fibers are weakly guiding (i.e., $n_1 \approx n_2$ or $\Delta \ll 1$) so that the guided rays are paraxial, i.e., approximately parallel to the fiber axis. The longitudinal components of the electric and magnetic fields are then far weaker than the transverse components and the guided waves are approximately transverse electromagnetic (TEM) in nature. The linear polarization in the x and y directions then form orthogonal states of polarization. The linearly polarized (LP) mode with indices (l, m) is usually labeled as the LP$_{lm}$ mode. The two polarizations of mode (l, m) travel with the same propagation constant and have the same spatial distribution.

For weakly guiding fibers the characteristic equation obtained using the procedure outlined earlier turns out to be approximately equivalent to the conditions that the scalar function $u(r)$ in (10.2-6) is continuous and has a continuous derivative at $r = a$. These two conditions are satisfied if

$$\frac{(k_T a) J_l'(k_T a)}{J_l(k_T a)} = \frac{(\gamma a) K_l'(\gamma a)}{K_l(\gamma a)}. \tag{10.2-11}$$

The derivatives J_l' and K_l' of the Bessel functions satisfy the identities

$$J_l'(x) = \pm J_{l\mp 1}(x) \mp l \frac{J_l(x)}{x} \tag{10.2-12}$$

$$K'_l(x) = -K_{l\mp1}(x) \mp l\frac{K_l(x)}{x}. \qquad (10.2\text{-}13)$$

Substituting these identities into (10.2-11) and using the normalized parameters $X = k_T a$ and $Y = \gamma a$ leads to the characteristic equation

$$\boxed{X\frac{J_{l\pm1}(X)}{J_l(X)} = \pm Y\frac{K_{l\pm1}(Y)}{K_l(Y)}, \quad Y = \sqrt{V^2 - X^2}.} \qquad (10.2\text{-}14)$$

Characteristic Equation

Given V and l, the characteristic equation contains a single unknown variable X. Note that $J_{-l}(x) = (-1)^l J_l(x)$ and $K_{-l}(x) = K_l(x)$, so that the equation remains unchanged if l is replaced by $-l$.

The characteristic equation may be solved graphically by plotting its right- and left-hand sides (RHS and LHS, respectively) versus X and finding the intersections. As illustrated in Fig. 10.2-3 for $l = 0$, the LHS has multiple branches whereas the right-hand side decreases monotonically with increasing X until it vanishes at $X = V$ ($Y = 0$). There are therefore multiple intersections in the interval $0 < X \le V$. Each intersection point corresponds to a fiber mode with a distinct value of X. These values are denoted X_{lm}, $m = 1, 2, \ldots, M_l$ in order of increasing X. Once the X_{lm} are found, (10.2-8), (10.2-4), and (10.2-6) allow us to determine the corresponding transverse propagation constants k_{Tlm}, the decay parameters γ_{lm}, the propagation constants β_{lm}, and the radial distribution functions $u_{lm}(r)$. The graph in Fig. 10.2-3 is similar in character to that in Fig. 9.2-2, which governs the modes of a planar dielectric waveguide.

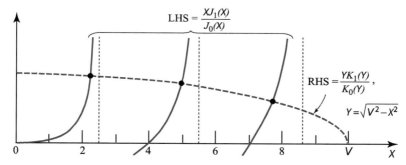

Figure 10.2-3 Graphical construction for solving the characteristic equation (10.2-14). The left- and right-hand sides are plotted as functions of X. The intersection points are the solutions. The left-hand side (LHS) has multiple branches intersecting the abscissa at the roots of $J_{l\pm1}(X)$. The right-hand side (RHS) intersects each branch once and meets the abscissa at $X = V$. The number of modes therefore equals the number of roots of $J_{l\pm1}(X)$ that are smaller than V. In this plot $l = 0$, $V = 10$, and either the $-$ or $+$ signs in (10.2-14) may be used.

Each mode has a distinct radial distribution. As examples, the two radial distributions $u(r)$ illustrated in Fig. 10.2-2 correspond to the LP$_{01}$ mode ($l = 0$, $m = 1$) in a fiber with $V = 5$, and the LP$_{34}$ mode ($l = 3$, $m = 4$) in a fiber with $V = 25$, respectively. Modes with $l > 0$ exist in pairs with azimuthal dependencies given by $\exp(\pm jl\phi)$, in analogy with the Laguerre–Gaussian optical beam discussed in Sec. 3.4. Since the (l,m) and $(-l,m)$ modes have the same propagation constants, the azimuthal behavior of the modes is revealed by examining the transverse intensity distribution of their equal-weight superpositions, as is understood from the commentary associated with Fig. 3.4-2. Specifically, the complex amplitude of the sum is proportional to $u_{lm}(r)\cos l\phi \exp(-j\beta_{lm}z)$; the intensity, which is proportional to $u_{lm}^2(r)\cos^2 l\phi$, is illustrated in Fig. 10.2-4 for several LP$_{lm}$ modes.

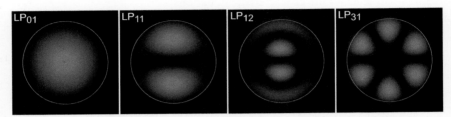

Figure 10.2-4 Intensity distributions in the transverse plane for several LP$_{lm}$ modes for a step-index fiber with fiber parameter $V = 10$. The white circles depict core–cladding boundaries. The intensity distribution of the fundamental LP$_{01}$ mode resembles that of the Gaussian beam displayed in Fig. 3.1-1. Each panel represents a superposition of a pair of modes with values of l that are identical but opposite in sign. The intensity distributions are proportional to $\cos^2 l\phi$ and thus display $2l$ azimuthal interference fringes.

Mode Cutoff

It is evident from the graphical construction in Fig. 10.2-3 that as V increases, the number of intersections (modes) increases since the left-hand side of the characteristic equation (10.2-14) is independent of V, whereas the right-hand side moves rightward as V increases. Considering the minus signs in the characteristic equation, branches of the left-hand side intersect the abscissa when $J_{l-1}(X) = 0$. These roots are denoted x_{lm}, $m = 1, 2, \ldots$. The number of modes M_l is therefore equal to the number of roots of $J_{l-1}(X)$ that are smaller than V. The (l, m) mode is allowed if $V > x_{lm}$. The mode reaches its cutoff point when $V = x_{lm}$. As V decreases, the $(l, m-1)$ mode also reaches its cutoff point whereupon a new root is reached, and so on. The smallest root of $J_{l-1}(X)$ is $x_{01} = 0$ for $l = 0$ and the next smallest is $x_{11} = 2.405$ for $l = 1$. The numerical values of some of these roots are provided in Table 10.2-1.

Table 10.2-1 Cutoff V parameter for low-order LP$_{lm}$ modes.[a]

l	$m = 1$	2	3	4	5
0	0	3.832	7.016	10.174	13.324
1	2.405	5.520	8.654	11.792	14.931
2	3.832	7.016	10.174	13.324	16.471
3	5.136	8.417	11.620	14.796	17.960
4	6.380	9.761	13.015	16.224	19.409
5	7.588	11.065	14.373	17.616	20.218
6	8.772	12.339	15.700	18.980	22.218

[a]The cutoffs of the $l = 0$ modes occur at the roots of $J_{-1}(X) = -J_1(X)$. The $l = 1$ modes are cut off at the roots of $J_0(X)$, and so on.

When $V < 2.405$ all modes, with the exception of the fundamental LP$_{01}$ mode, are cut off. The fiber then operates as a single-mode waveguide. The condition for single-mode operation is therefore

$$V < 2.405. \qquad (10.2\text{-}15)$$

Single-Mode Condition

Since V is proportional to the optical frequency [see (10.2-10)], the cutoff condition for the fundamental mode provided in (10.2-15) yields a corresponding cutoff frequency:

$$\boxed{\nu_c = \omega_c/2\pi = \frac{1}{\text{NA}} \frac{c_o}{2.61a}.}$$ (10.2-16)
Cutoff Frequency

By comparison, in accordance with (9.2-9), the cutoff frequency of the lowest-order mode in a dielectric slab waveguide of width d is $\nu_c = (1/\text{NA})(c_o/2d)$.

Number of Modes

A plot of the number of modes M_l as a function of V therefore takes the form of a staircase function that increases by unity at each of the roots x_{lm} of the Bessel function $J_{l-1}(X)$. A composite count of the total number of modes M (for all values of l), as a function of V, is provided in Fig. 10.2-5. Each root must be counted twice since, for each mode of azimuthal index $l > 0$, there is a corresponding mode $-l$ that is identical except for opposite polarity of the angle ϕ (corresponding to rays with helical trajectories of opposite senses), as can be seen by using the plus signs in the characteristic equation. Moreover, each mode has two states of polarization and must therefore be counted twice.

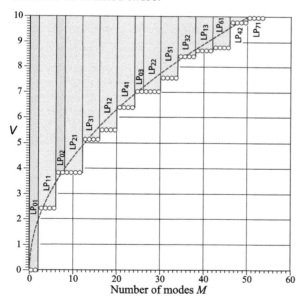

Figure 10.2-5 Total number of modes M versus fiber parameter $V = 2\pi(a/\lambda_o)\text{NA}$. Included in the count are two helical polarities for each mode with $l > 0$ as well as two polarizations per mode. For $V < 2.405$, there is only a single mode, the fundamental LP_{01} mode with two polarizations. For $V = 4$, modes LP_{11} and LP_{21} come into play, each with two helicities, along with LP_{01} and LP_{02} with zero helicities; accommodating the two polarizations in each mode leads to a total of 12 modes. The dotted curve is the relation $M = \frac{1}{2}V^2$ set forth in (10.2-18), which provides an approximate result for the number of modes when $V \gg 1$.

Though there are no explicit exact formulas for the roots of $J_l(X)$, for $X \gg l^2$ we can write $J_l(X) \approx (2/\pi X)^{1/2} \cos[X - (l + \frac{1}{2})\frac{\pi}{2}]$ in which case the roots are approximately given by $x_{lm} = (l + \frac{1}{2})\frac{\pi}{2} + (2m - 1)\frac{\pi}{2} = (l + 2m - \frac{1}{2})\frac{\pi}{2}$. This relation may be used to estimate the number of modes for fibers with large V parameter and consequently a large number of modes. When m is large the cutoff points of modes (l, m), which are the roots of $J_{l\pm1}(X)$, are

$$x_{lm} \approx (l + 2m - \frac{1}{2} \pm 1)\frac{\pi}{2} \approx (l + 2m)\frac{\pi}{2}, \quad l = 0, 1, \ldots; \quad m \gg 1. \quad (10.2\text{-}17)$$

For fixed l, these roots are uniformly spaced at a distance π, in which case the number of roots M_l satisfies $l\frac{\pi}{2} + m\pi = V$, from which $M_l \approx V/\pi - l/2$. M_l then decreases linearly with increasing l, beginning with $M_l \approx V/\pi$ for $l = 0$ and ending at $M_l = 0$ when $l = l_{\max}$, where $l_{\max} = 2V/\pi$. Accommodating the two degrees of freedom associated with positive and negative l for $l > 0$, and the two polarizations for each index (l, m), leads to a total number of modes given by $M = 2M_0 + 4 \sum_{l=1}^{l_{\max}} M_l$. With the help of the relation $\sum_{l=1}^{L} l = \frac{1}{2} L(L+1)$, we obtain the approximate expression $M \approx (4/\pi^2) V^2$.

However, this expression underestimates the actual number of modes by virtue of the fact that M_l includes lower-order modes for which the separation distances are less than π, as evinced in Table 10.2-1. As illustrated in Fig. 10.2-5, a good fit to the exact number of modes is provided by the approximation

$$M \approx \tfrac{1}{2} V^2, \qquad (10.2\text{-}18)$$

Number of Modes ($V \gg 1$)

an expression that obtains in the quasi-plane-wave approach, as shown in Sec. 10.2C [see (10.2-35)].

Based on (10.2-18), when V is large the approximate number of modes in the circular waveguide is given by $M \approx \tfrac{1}{2} V^2 = 2\pi(\pi a^2/\lambda_o^2)(\text{NA})^2$. This expression is analogous to that for the number of modes in a square dielectric waveguide of cross sectional area d^2, in which case $M \approx 2\pi(d^2/\lambda_o^2)(\text{NA})^2$ when both TE and TM polarizations are accommodated [see (9.3-3)].

EXAMPLE 10.2-1. *Number of Modes.* A silica-glass fiber with $n_1 = 1.452$ and $\Delta = 0.01$ has a numerical aperture $\text{NA} = \sqrt{n_1^2 - n_2^2} \approx n_1 \sqrt{2\Delta} \approx 0.205$. If $\lambda_o = 1.55~\mu\text{m}$ and the core radius $a = 20~\mu\text{m}$, then $V = 2\pi(a/\lambda_o)\text{NA} \approx 16.6$. There are therefore approximately $M \approx \tfrac{1}{2} V^2 \approx 138$ modes. If the cladding is stripped away so that the core is in direct contact with air, $n_2 = 1$ and $\text{NA} = 1$, whereupon $V \approx 81.1$ and approximately 3,286 modes are allowed.

Propagation Constants and Group Velocities

As indicated earlier, the propagation constants can be determined by solving the characteristic equation (10.2-14) for the X_{lm} and using (10.2-4a) and (10.2-8) to obtain $\beta_{lm} = (n_1^2 k_o^2 - X_{lm}^2/a^2)^{1/2}$. Since $V = 2\pi(a/\lambda_o)\text{NA}$ and $\text{NA} = n_1\sqrt{2\Delta}$, we have $\beta_{lm} = n_1 k_o (1 - 2\Delta \cdot X_{lm}^2/V^2)^{1/2}$. Because $\Delta \ll 1$ and $X_{lm} < V$, we use the expansion $(1 + \delta)^{1/2} \approx 1 + \tfrac{1}{2}\delta$ for $|\delta| \ll 1$ to obtain

$$\beta_{lm} \approx n_1 k_o \left[1 - \frac{X_{lm}^2}{V^2} \Delta \right]. \qquad (10.2\text{-}19)$$

Propagation Constants
$0 < X_{lm} < V$

For fibers with large V, corresponding to a large number of modes, X_{lm} spans the range $0 < X_{lm} < V$ so that β_{lm} varies approximately between $n_1 k_o$ and $n_1 k_o (1 - \Delta) \approx n_2 k_o$, as predicted by ray optics.

To determine the group velocity of the (l, m) mode, $v_{lm} = d\omega/d\beta_{lm}$, we express β_{lm} as an explicit function of ω by substituting $n_1 k_o = \omega/c_1$ and $V = a(\omega/c_o)\text{NA}$ into (10.2-19), and then calculating $(d\beta_{lm}/d\omega)^{-1}$. Assuming that c_1 and Δ are independent of ω, and using the approximation $(1 + \delta)^{-1} \approx 1 - \delta$ for $|\delta| \ll 1$, the group velocity becomes

$$\boxed{v_{lm} \approx c_1 \left[1 - \frac{X_{lm}^2}{V^2} \Delta \right].}$$

(10.2-20)
Group Velocities
$0 < X_{lm} < V$

For $V \gg 1$, the group velocity varies approximately between c_1 and $c_1(1 - \Delta) = c_1(n_2/n_1)$. In this case, the group velocities of the low-order modes are approximately equal to the phase velocity of the core material, whereas those of the high-order modes are smaller.

The fractional group-velocity change between the fastest and the slowest mode is roughly equal to Δ, the fractional refractive index change of the fiber. Fibers with large Δ, although endowed with a large NA and therefore large light-gathering capacity, also have a large number of modes, large modal dispersion, and consequently high pulse-spreading rates. These effects are particularly severe if the cladding is removed altogether.

B. Single-Mode Fibers

As discussed earlier, a fiber with core radius a and numerical aperture NA operates as a single-mode fiber in the fundamental LP_{01} mode if $V = 2\pi(a/\lambda_o)NA < 2.405$. Single-mode operation is therefore achieved via a small core diameter and small numerical aperture (indicating that n_2 is close to n_1), or by operating at a sufficiently low optical frequency [below the cutoff frequency $\nu_c = (1/NA)(c_o/2.61a)$].

The fundamental LP_{01} mode has a bell-shaped spatial distribution (Figs. 10.2-2 and 10.2-4 for $l = 0$) similar to that of the simple Gaussian beam (Fig. 3.1-1). It provides the greatest confinement of light power within the core.

EXAMPLE 10.2-2. *Single-Mode Operation.* A silica-glass fiber with $n_1 = 1.447$ and $\Delta = 0.01$ (NA $= 0.205$) operates at $\lambda_o = 1.3$ μm as a single-mode fiber if $V = 2\pi(a/\lambda_o)NA < 2.405$, i.e., if the core diameter $2a < 4.86$ μm. If Δ is reduced to 0.0025, single-mode operation is maintained for a diameter $2a < 9.72$ μm.

The dependence of the effective refractive index $n = \beta/k_o$ on the V parameter for the fundamental mode is displayed in Fig. 10.2-6(a), and the corresponding dispersion relation (ω versus β) is illustrated in Fig. 10.2-6(b). As the V parameter increases, i.e., as the frequency increases or the fiber diameter increases, the effective refractive index n increases from n_2 to n_1. This is expected since the mode is more confined in the core at shorter wavelengths.

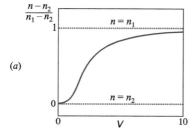
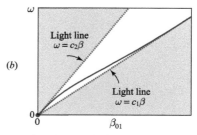

Figure 10.2-6 Schematic illustrations of the propagation characteristics of the fundamental LP_{01} mode. (a) Effective refractive index $n = \beta/k_o$ as a function of the V parameter. (b) Dispersion relation (ω versus β_{01}).

There are numerous advantages of using single-mode fibers in optical fiber communication systems. As explained earlier, the modes of a multimode fiber travel at different group velocities so that a short-duration pulse of multimode light suffers a range of delays and therefore spreads in time. Quantitative measures of modal dispersion are examined in Sec. 10.3B. In a single-mode fiber, on the other hand, there is only one mode with a single group velocity, so that a short pulse of light arrives without delay distortion. As explained in Sec. 10.3B, pulse spreading in single-mode fibers does nevertheless result from other dispersive mechanisms, but these are significantly smaller than modal dispersion.

Moreover, as shown in Sec. 10.3A, the rate of power attenuation is lower in a single-mode fiber than in a multimode fiber. This, together with the smaller rate of pulse spreading, permits substantially higher data rates to be transmitted over single-mode fibers than over multimode fibers. This topic is addressed further in Chapters 23 and 25.

Another difficulty with the use of multimode fibers stems from the random interference of the modes. As a result of uncontrollable imperfections, strains, and temperature fluctuations, each mode undergoes a random phase shift so that the sum of the complex amplitudes of the modes exhibits an intensity that is random in time and space. This randomness is known as **modal noise** or **speckle**. This effect is similar to the fading of radio signals resulting from multiple-path transmission. In a single-mode fiber there is only one path and therefore no modal noise.

Polarization-Maintaining Fibers

In a fiber with circular cross section, each mode has two independent states of polarization with the same propagation constant. Thus, the fundamental LP_{01} mode in a single-mode weakly guiding fiber may be polarized in the x or y direction; the two orthogonal polarizations have the same propagation constant and the same group velocity.

In principle, there should be no exchange of power between the two polarization components. If the power of the light source is delivered exclusively into one polarization, the power should remain in that polarization. In practice, however, slight random imperfections and uncontrollable strains in the fiber result in random power transfer between the two polarizations, as illustrated in Fig. 10.2-7.

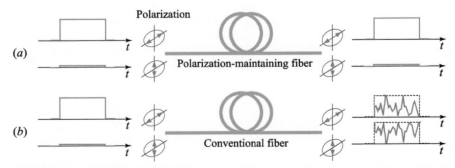

Figure 10.2-7 (*a*) Ideal polarization-maintaining fiber. (*b*) Random transfer of power between two polarizations in a conventional fiber.

Such coupling is facilitated because the two polarizations have the same propagation constant and their phases are therefore matched. Thus, linearly polarized light at the fiber input is generally transformed into elliptically polarized light at the fiber output.

In spite of the fact that the total optical power remains fixed (see Fig. 10.2-7), the ellipticity of the received light fluctuates randomly with time as a result of fluctuations in the material strain and temperature, and of the source wavelength. The randomization of the power division between the two polarization components poses no difficulty if the object is solely to transmit light power, provided that the total power is collected.

However, in many areas where fiber optics is used, e.g., in integrated-photonic devices, optical sensors based on interferometric techniques, and coherent optical communications, the fiber must transmit the complex amplitude (magnitude and phase) of a specific polarization. Polarization-maintaining fibers are required for such applications. To construct a polarization-maintaining fiber, the circular symmetry of the conventional fiber must be abandoned, for example by using fibers with elliptical cross section or stress-induced anisotropy of the refractive index. This eliminates the polarization degeneracy, thereby making the propagation constants of the two polarizations different. The introduction of such phase mismatch serves to reduce the coupling efficiency.

*C. Quasi-Plane Waves in Step-Index and Graded-Index Fibers

As shown in Sec. 10.1A, the modes of a step-index fiber are determined by writing the Helmholtz equation (10.2-1) with $n = n(r)$, solving for the spatial distributions of the field components, and using Maxwell's equations and the boundary conditions to obtain the characteristic equation. However, carrying out this procedure for a graded-index fiber is generally a difficult proposition.

In this section we rely instead on an approximate approach based on picturing the field distribution as a quasi-plane wave traveling within the core, approximately along the trajectory of an optical ray. A quasi-plane wave is a wave that is locally identical to a plane wave, but slowly changes its direction and amplitude as it travels. This approach permits us to maintain the simplicity of ray optics while at the same time retaining the phase associated with the wave, so that the self-consistency condition for determining the propagation constants of the guided modes can be used (as was done for the planar dielectric waveguide in Sec. 9.2). This approximate technique, which makes use of the WKB (Wentzel–Kramers–Brillouin) method, is applicable only for fibers with a large number of modes (large V parameter). This approach also allows us to conveniently compare the behavior of step-index and graded-index fibers.

Quasi-Plane Waves

Consider a solution of the Helmholtz equation (10.2-1) that takes the form of a quasi-plane wave (see Sec. 2.3)

$$U(\mathbf{r}) = a(\mathbf{r}) \exp\left[-jk_o S(\mathbf{r})\right], \tag{10.2-21}$$

where $a(\mathbf{r})$ and $S(\mathbf{r})$ are real functions of position that are slowly varying in comparison with the wavelength $\lambda_o = 2\pi/k_o$. It is known from (2.3-4) that $S(\mathbf{r})$ approximately satisfies the eikonal equation $|\nabla S|^2 \approx n^2$, and that the rays travel in the direction of the gradient ∇S. If we take $k_o S(\mathbf{r}) = k_o s(r) + l\phi + \beta z$, where $s(r)$ is a slowly varying function of r, the eikonal equation yields

$$\left(k_o \frac{ds}{dr}\right)^2 + \beta^2 + \frac{l^2}{r^2} = n^2(r)\, k_o^2. \tag{10.2-22}$$

The local spatial frequency of the wave in the radial direction is the partial derivative of the phase $k_o S(\mathbf{r})$ with respect to r,

$$k_r = k_o \frac{ds}{dr}, \tag{10.2-23}$$

so that (10.2-21) becomes

$$U(r) = \mathrm{a}(r) \exp\left(-j \int_0^r k_r \, dr\right) e^{-jl\phi} e^{-j\beta z} \qquad (10.2\text{-}24)$$
Quasi-Plane Wave

and (10.2-22) provides

$$k_r^2 = n^2(r) k_o^2 - \beta^2 - l^2/r^2. \qquad (10.2\text{-}25)$$

Defining $k_\phi = l/r$ so that $\exp(-jl\phi) = \exp(-jk_\phi r\phi)$, and $k_z = \beta$, (10.2-25) yields $k_r^2 + k_\phi^2 + k_z^2 = n^2(r) k_o^2$. The quasi-plane wave therefore has a local wavevector $\mathbf{k}$ with magnitude $n(r)k_o$ and cylindrical-coordinate components (k_r, k_ϕ, k_z). Since $n(r)$ and k_ϕ are functions of r, k_r is also generally position dependent. The direction of $\mathbf{k}$ changes slowly with r (see Fig. 10.2-8), and follows a helical trajectory similar to that of the skewed ray shown earlier in Fig. 10.1-6(b).

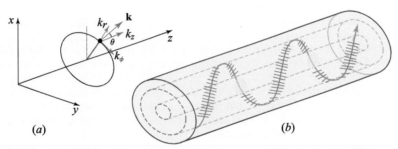

Figure 10.2-8 (a) The wavevector $\mathbf{k} = (k_r, k_\phi, k_z)$ in a cylindrical coordinate system. (b) Quasi-plane wave following the direction of a ray.

To establish the region of the core within which the wave is bound, we determine the values of r for which k_r is real, or $k_r^2 > 0$. For given values of l and β we plot $k_r^2 = [n^2(r) k_o^2 - l^2/r^2 - \beta^2]$ as a function of r. The term $n^2(r) k_o^2$ is first plotted as a function of r [thick solid curve in Fig. 10.2-9(a)]. The term l^2/r^2 is then subtracted, yielding the dashed curve. The value of β^2 is marked by the thin solid vertical line. It follows that k_r^2 is represented by the difference between the dashed curve and the thin solid line, i.e., by the shaded area. Regions where k_r^2 is positive and negative are indicated by $+$ and $-$ signs, respectively.

For the step-index fiber, we have $n(r) = n_1$ for $r < a$, and $n(r) = n_2$ for $r > a$. In this case the quasi-plane wave is guided in the core by reflecting from the core–cladding boundary at $r = a$. As illustrated in Fig. 10.2-9(a), the region of confinement is then $r_l < r < a$, where

$$n_1^2 k_o^2 - l^2/r_l^2 - \beta^2 = 0. \qquad (10.2\text{-}26)$$

The wave bounces back and forth helically like the skewed ray illustrated in Fig. 10.1-2. In the cladding ($r > a$), and near the center of the core ($r < r_l$), k_r^2 is negative so that k_r is imaginary; the wave therefore decays exponentially in these regions. Note that r_l depends on β. For large β (or large l), r_l is large so that the wave is confined to a thin cylindrical shell near the boundary of the core.

For the graded-index fiber illustrated in Fig. 10.2-9(b), k_r is real in the region $r_l < r < R_l$, where r_l and R_l are the roots of the equation

$$n^2(r)\, k_o^2 - l^2/r^2 - \beta^2 = 0. \tag{10.2-27}$$

It follows that the wave is essentially confined within a cylindrical shell of radii r_l and R_l, just as for the helical ray trajectory shown in Fig. 10.1-6(b).

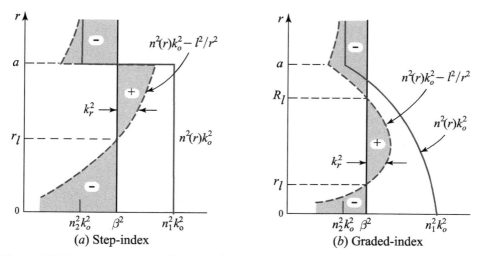

Figure 10.2-9 Dependence of $n^2(r)\, k_o^2$, $n^2(r)\, k_o^2 - l^2/r^2$, and $k_r^2 = n^2(r)\, k_o^2 - l^2/r^2 - \beta^2$ on the position r. At any r, k_r^2 is the width of the shaded area with the $+$ and $-$ signs denoting positive and negative values of k_r^2, respectively. (a) Step-index fiber: k_r^2 is positive in the region $r_l < r < a$. (b) Graded-index fiber: k_r^2 is positive in the region $r_l < r < R_l$.

Modes

The modes of the fiber are determined by imposing the self-consistency condition that the wave reproduce itself after one helical period of travel between r_l and R_l and back. The azimuthal pathlength corresponding to an angle 2π must correspond to a multiple of 2π phase shift, i.e., $k_\phi 2\pi r = 2\pi l$; $l = 0, \pm 1, \pm 2, \ldots$. This condition is evidently satisfied since $k_\phi = l/r$. Furthermore, since the component k_r vanishes at r_l and R_l, the phase shift encountered in traveling between these turning points must be a multiple of π, much like the case of a standing wave between two mirrors. Thus,

$$\int_{r_l}^{R_l} k_r\, dr = \pi m, \qquad m = 1, 2, \ldots, M_l, \tag{10.2-28}$$

where $R_l = a$ for the step-index fiber. This condition provides the characteristic equation from which the propagation constants β_{lm} of the modes are determined. These values are represented schematically in Fig. 10.2-10; the mode $m = 1$ has the largest value of β (approximately $n_1 k_o$) whereas the mode $m = M_l$ has the smallest value (approximately $n_2 k_o$). The WKB method, which is applicable for oscillatory solutions between turning points, yields more accurate results: the quantities l^2 and m above are replaced by $l^2 + 1/4$ and $m + 1/4$, corrections that are particularly important for small values of l and m.

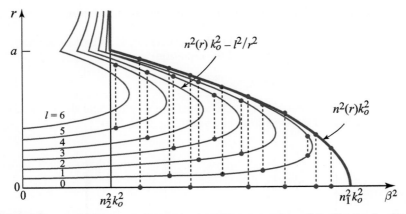

Figure 10.2-10 The propagation constants and confinement regions of the fiber modes. Each curve corresponds to an index l, which stretches from 0 to 6 in this plot. Each mode (corresponding to a certain value of m) is schematically indicated by two dots connected by a dashed vertical line. The ordinates of the dots denote the radii r_l and R_l of the cylindrical shell within which the mode is confined. Values on the abscissa are the squared propagation constants of the modes, β^2.

Number of Modes

The total number of modes can be determined by adding the number of modes M_l for $l = 0, 1, \ldots, l_{\max}$. We approach this computation using a different procedure, however. We first determine the number q_β of modes with propagation constants greater than a given value β. For each l, the number of modes $M_l(\beta)$ with propagation constant greater than β is the number of multiples of 2π the integral in (10.2-28) yields, i.e.,

$$M_l(\beta) = \frac{1}{\pi} \int_{r_l}^{R_l} k_r \, dr = \frac{1}{\pi} \int_{r_l}^{R_l} \sqrt{n^2(r) k_o^2 - l^2/r^2 - \beta^2} \, dr, \qquad (10.2\text{-}29)$$

where r_l and R_l are the radii of confinement corresponding to the propagation constant β, as provided in (10.2-27). Clearly, r_l and R_l depend on β, and $R_l = a$ for the step-index fiber.

The total number of modes with propagation constant greater than β is therefore

$$q_\beta = 4 \sum_{l=0}^{l_{\max}(\beta)} M_l(\beta), \qquad (10.2\text{-}30)$$

where $l_{\max}(\beta)$ is the maximum value of l that yields a bound mode with propagation constants greater than β, i.e., for which the peak value of the function $n^2(r) k_o^2 - l^2/r^2$ is greater than β^2. The grand-total mode count M is q_β for $\beta = n_2 k_o$. The factor of 4 in (10.2-30) accommodates the two possible polarities of the angle ϕ, corresponding to positive and negative helical trajectories for each (l, m), and the two possible polarizations. If the number of modes is sufficiently large, we can replace the summation in (10.2-30) by an integration, whereupon

$$q_\beta \approx 4 \int_0^{l_{\max}(\beta)} M_l(\beta) \, dl. \qquad (10.2\text{-}31)$$

For fibers with power-law refractive-index profiles, we insert (10.1-4) into (10.2-

29), and thence into (10.2-31). Evaluation of the integral then yields

$$q_\beta \approx M \left[\frac{1 - (\beta/n_1 k_o)^2}{2\Delta}\right]^{\frac{p+2}{p}} \tag{10.2-32}$$

with

$$M \approx \frac{p}{p+2} n_1^2 k_o^2 a^2 \Delta = \frac{p}{p+2} \frac{V^2}{2}, \tag{10.2-33}$$

where $\Delta = (n_1 - n_2)/n_1$ and $V = 2\pi(a/\lambda_o)\mathrm{NA}$ is the fiber V parameter. Since $q_\beta \approx M$ at $\beta = n_2 k_o$, M is indeed the total number of modes.

For step-index fibers ($p \to \infty$), (10.2-32) and (10.2-33) become

$$q_\beta \approx M \left[\frac{1 - (\beta/n_1 k_o)^2}{2\Delta}\right] \tag{10.2-34}$$

and

$$\boxed{M \approx \tfrac{1}{2} V^2,} \tag{10.2-35}$$

Number of Modes
(Step-Index)

respectively. This expression for M is the same as that set forth in (10.2-18), which was found to be a good fit to the exact number as a function of V, as shown in Fig. 10.2-5.

Propagation Constants

The propagation constant β_q for mode q is obtained by inverting (10.2-32):

$$\beta_q \approx n_1 k_o \sqrt{1 - 2\left(\frac{q}{M}\right)^{p/(p+2)} \Delta}, \qquad q = 1, 2, \ldots, M, \tag{10.2-36}$$

where the index q_β has been replaced by q, and β has been replaced by β_q. Since $\Delta \ll 1$, the approximation $\sqrt{1+\delta} \approx 1 + \tfrac{1}{2}\delta$ (applicable for $|\delta| \ll 1$) can be applied to (10.2-36), yielding

$$\boxed{\beta_q \approx n_1 k_o \left[1 - \left(\frac{q}{M}\right)^{p/(p+2)} \Delta\right].} \tag{10.2-37}$$

Propagation Constants

The propagation constant β_q therefore decreases from $\approx n_1 k_o$ (for $q = 1$) to $n_2 k_o$ (for $q = M$), as illustrated in Fig. 10.2-11. For the step-index fiber ($p \to \infty$), (10.2-36) reduces to

$$\boxed{\beta_q \approx n_1 k_o \left(1 - \frac{q}{M}\Delta\right),} \tag{10.2-38}$$

Propagation Constants
(Step-Index Fiber)

which mimics the behavior of (10.2-19).

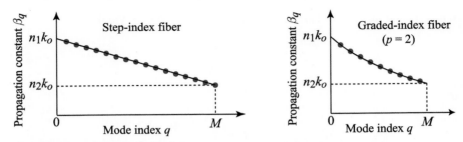

Figure 10.2-11 Dependence of the propagation constants β_q on the mode index $q = 1, 2, \ldots, M$ for (a) a step-index fiber ($p \to \infty$) and (b) an optimal graded-index fiber ($p = 2$).

Group Velocities

To determine the group velocity $v_q = d\omega/d\beta_q$, we write β_q as a function of ω by substituting (10.2-33) into (10.2-37), substituting $n_1 k_o = \omega/c_1$ into the result, and evaluating $v_q = (d\beta_q/d\omega)^{-1}$. With the help of the approximation $(1+\delta)^{-1} \approx 1 - \delta$ (valid for $|\delta| \ll 1$), and assuming that c_1 and Δ are independent of ω (i.e., ignoring material dispersion), we obtain

$$v_q \approx c_1 \left[1 - \frac{p-2}{p+2} \left(\frac{q}{M}\right)^{p/(p+2)} \Delta \right]. \qquad (10.2\text{-}39)$$

Group Velocities

For the step-index fiber ($p \to \infty$), (10.2-39) yields

$$v_q \approx c_1 \left(1 - \frac{q}{M} \Delta \right). \qquad (10.2\text{-}40)$$

The group velocity thus varies from approximately c_1 to $c_1(1 - \Delta)$, as illustrated in Fig. 10.2-12(a). The form of this equation harks back to (10.2-20).

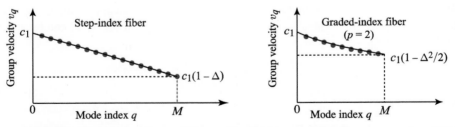

Figure 10.2-12 Group velocities v_q of the modes of (a) a step-index fiber ($p \to \infty$) and (b) an optimal graded-index fiber ($p = 2$).

Optimal Index Profile

Equation (10.2-39) indicates that the grade profile parameter $p = 2$ yields a group velocity $v_q \approx c_1$ for all q, so that all modes travel at approximately the same velocity c_1. This highlights the advantage of the graded-index fiber for multimode transmission.

To determine the group velocity with better accuracy, we return to the derivation of v_q from (10.2-36) for $p = 2$. Carrying the Taylor-series expansion to three terms instead of two, i.e., $\sqrt{1+\delta} \approx 1 + \frac{1}{2}\delta - \frac{1}{8}\delta^2$, gives rise to

$$v_q \approx c_1 \left(1 - \frac{q}{M}\frac{\Delta^2}{2}\right).$$

(10.2-41)
Group Velocities
(Graded-Index, $p=2$)

Thus, the group velocities vary from approximately c_1 at $q = 1$ to approximately $c_1(1 - \Delta^2/2)$ at $q = M$. Comparison with the results for the step-index fiber is provided in Fig. 10.2-12. The group-velocity difference for the parabolically graded fiber is $\Delta^2/2$, which is substantially smaller than the group-velocity difference Δ for the step-index fiber. Under ideal conditions, the graded-index fiber therefore reduces the group-velocity difference by a factor $\Delta/2$, thus realizing its intended purpose of equalizing the modal velocities. However, since the analysis leading to (10.2-41) is based on a number of approximations, this improvement factor is only a rough estimate — indeed it is not fully attained in practice.

The number of modes M in a graded-index fiber with grade profile parameter p is specified by (10.2-33). For $p = 2$, this becomes

$$M \approx \tfrac{1}{4}V^2.$$

(10.2-42)
Number of Modes
(Graded-Index, $p=2$)

Comparing this with the result for the step-index fiber provided in (10.2-35), $M \approx \tfrac{1}{2}V^2$, reveals that the number of modes in an optimal graded-index fiber is roughly half that in a step-index fiber with the same parameters n_1, n_2, and a.

D. Multicore Fibers and Fiber Couplers

Multicore Fibers

A **multicore fiber** (MCF) is a fiber with multiple cores embedded in a common cladding (Fig. 10.2-13). The application in which such a fiber is to be used determines the number of cores, the manner in which the cores are arranged, their diameters, and their separations (pitch). Cores with small (large) diameter function as single-mode (multimode) waveguides, respectively. Cores that are sufficiently well separated exhibit minimal inter-core crosstalk and hence behave as independent waveguides. MCFs find use in high-capacity optical networks, in inter-chip communications in datacenters, and as sensors since they respond differentially to mechanical changes (e.g., bending).

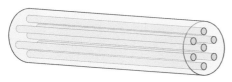

Figure 10.2-13 A multicore fiber (MCF) with seven cores.

Multicore Couplers

Optical fiber couplers are important components of optical fiber systems. **Multicore couplers** are used to connect one or more fibers at the input of the device to one or more fibers at its output. Several examples implementing various coupling configurations are illustrated in Fig. 10.2-14 (additional examples are provided in Sec. 24.1B). It should be pointed out, however, that other technologies are also useful for achieving coupling, as discussed in Sec. 24.1A.

A key consideration in the design of an optical fiber coupler is the optical loss that it introduces; the reduction of insertion loss is particularly challenging for single-mode fibers. In certain applications, it is desired to render the distribution of optical power at the output fibers of the coupler insensitive to particular wave properties, such as wavelength or polarization. In other applications, the coupler may be deliberately designed so that coupling is governed by one of these properties, such as different wavelengths or polarizations being directed to different output fibers, as considered in Sec. 24.1.

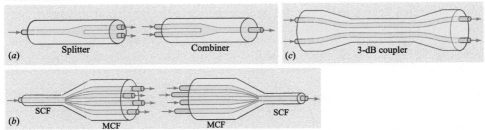

Figure 10.2-14 Multicore fiber-optic couplers. (*a*) Dual-core fiber used as a splitter or a combiner. (*b*) Fan-out and fan-in couplers using a conical fused taper to connect a single-core fiber (SCF) to a multicore fiber (MCF). (*c*) A 3-dB coupler using a dual-core fiber incorporating a biconical fused taper. Adjacent fibers are weakly coupled, which allows light to gradually migrate from one fiber to the other, much as in the integrated-photonic coupler displayed in Fig. 9.4-7(*b*).

Photonic Lantern

A **photonic lantern** is a fiber-optic coupler that connects a single multimode fiber (MMF) to multiple single-mode fibers (SMFs). As illustrated in Fig. 10.2-15, working our way from right to left, a collection of SMFs are fused into a single glass body, thereby forming a multicore fiber (MCF). This body in turn is tapered in cross section and connected to a single-core MMF at the narrow end of the taper. The device is reciprocal and can be used in either a fan-out or a fan-in configuration.

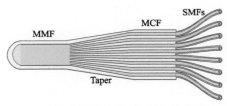

Figure 10.2-15 A photonic lantern provides an interface between the modes of one multimode fiber (MMF) and many single-mode fibers (SMFs). It can be operated in a fan-out configuration, in which light travels from left to right, or in a fan-in configuration, in which light travels from right to left.

When operated in a fan-out configuration, incoming light excites the modes of the MMF, and each mode in turn excites several, or all, of the multiple cores in the tapered region. Within the narrow portion of the tapered region, the cores are weakly coupled and thus behave as an array of coupled single-mode waveguides, wherein the light in one waveguide can migrate into neighboring waveguides (see Sec. 9.4C). In the wider portion of the tapered region, the cores are sufficiently separated so that they form a MCF comprising independent single-mode waveguides that serve to feed the outgoing SMFs.

While this structure may be used as a simple fan-out coupler that distributes the power entering the MMF among the outgoing SMFs, much like the fan-out configuration depicted in Fig. 10.2-14(*b*), it has additional capabilities. Ideally, each of the MMF modes would couple to one, and only one, of the SMFs, whereupon the device would serve as a demultiplexer separating the information carried by each of the incoming modes into distinct SMF channels (or, in the case of the reciprocal fan-in configuration, as a multiplexer). Unfortunately, however, at the entrance to the taper, and throughout its narrow region where the constituent fibers are not well separated, each of the MMF modes couples to more than one of the SMFs, thereby obviating such a one-to-one correspondence.

Nevertheless, the resultant mixing of information can be undone using computational methods. The relation between the complex amplitudes of the output waves in the SMFs and those of the modes in the input MMF is mathematically represented by a matrix with appropriate weights for such a linear superposition. This matrix is established by the geometry of the taper and the transmission through the array of coupled waveguides, as described in Sec. 9.4C for waveguide arrays. If this matrix is able to be determined, then matrix inversion can be used to computationally extract the information carried by the MMF modes from the complex amplitudes of the outgoing light in the SMFs.

Photonic lanterns find use in optical fiber communication systems in which MMF modes serve as individual communication channels (spatial-mode multiplexing is considered in Sec. 25.3B). Fan-out lanterns may also be used to distribute light carried by one MMF to several SMFs for processing by optical components designed for single-mode operation, such as single-mode fiber Bragg grating (FBG) spectral filters or optical amplifiers, before being recombined and directed to another MMF via a fan-in lantern. An example of the use of a photonic lantern in astronomical instrumentation relies on directing the multimode light delivered by a telescope to a set of single-mode fiber filters for spectral filtering.

10.3 ATTENUATION AND DISPERSION

Attenuation and dispersion limit the performance of the optical-fiber medium as a data-transmission channel. Attenuation, associated with losses of various kinds, limits the magnitude of the optical power transmitted. Dispersion, which is responsible for the temporal spread of optical pulses, limits the rate at which data-carrying pulses may be transmitted.

A. Attenuation

Attenuation Coefficient

The power of a light beam traveling through an optical fiber decreases exponentially with distance as a result of absorption and scattering. The associated **attenuation coefficient** is conventionally defined in units of decibels per kilometer (dB/km) and is denoted by the symbol α,

$$\alpha = \frac{1}{L} 10 \log_{10} \frac{1}{\mathcal{T}}, \qquad (10.3\text{-}1)$$

where $\mathcal{T} = P(L)/P(0)$ is the **power transmission ratio** (the ratio of transmitted to incident power) for a fiber of length L km. The conversion of a ratio to dB units is illustrated in Fig. 10.3-1. An attenuation of 3 dB/km, for example, corresponds to a power transmission of $\mathcal{T} = 0.5$ through a fiber of length $L = 1$ km.

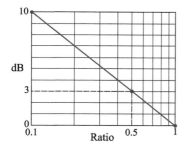

Figure 10.3-1 The dB value of a ratio. For example, 3 dB is equivalent to a ratio of 0.5; 10 dB corresponds to $\mathcal{T} = 0.1$; and 20 dB corresponds to $\mathcal{T} = 0.01$.

For light traveling through a cascade of lossy systems, the overall transmission ratio is the *product* of the constituent transmission ratios. By virtue of the logarithmic relation in (10.3-1), the overall loss in dB thus becomes the *sum* of the constituent dB losses. For a propagation distance of z km, the loss is αz dB. The associated power transmission ratio, which is obtained by inverting (10.3-1), is then

$$P(z)/P(0) = 10^{-\alpha z/10} \approx e^{-0.23\,\alpha z}, \qquad \alpha \text{ in dB/km}. \qquad (10.3\text{-}2)$$

Equation (10.3-2) applies when the quantity α is specified in units of dB/km. If, instead, the attenuation coefficient α is specified in units of km^{-1}, we have

$$P(z)/P(0) = e^{-\alpha z}, \qquad \alpha \text{ in km}^{-1}, \qquad (10.3\text{-}3)$$

where $\alpha \approx 0.23\,\alpha$. For components other than optical fibers, the attenuation coefficient α is usually specified in units of cm^{-1} in which case the power attenuation is described by (10.3-3) with z specified in cm.

Absorption

The attenuation coefficient α of fused silica (SiO$_2$) is strongly dependent on wavelength, as illustrated in Fig. 10.3-2. This material has two strong absorption bands: a mid-infrared absorption band resulting from vibrational transitions and an ultraviolet absorption band arising from electronic and molecular transitions. The tails of these bands form a window in the near infrared region of the spectrum in which there is little intrinsic absorption.

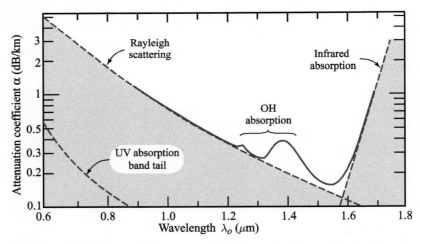

Figure 10.3-2 Attenuation coefficient α of silica glass versus free-space wavelength λ_o. There is a local minimum at 1.3 μm ($\alpha \approx 0.3$ dB/km) and an absolute minimum at 1.55 μm ($\alpha \approx 0.15$ dB/km).

Scattering

Rayleigh scattering is another intrinsic effect that contributes to the attenuation of light in glass. The random localized variations of the molecular positions in the glass itself create random inhomogeneities in the refractive index that act as tiny scattering centers. The amplitude of the scattered field is proportional to ω^2, where ω is the angular frequency of the light.[†] As discussed in Sec. 5.6B, the scattered intensity is

---

[†] The scattering medium creates a polarization density $\mathcal{P}$, which corresponds to a source of radiation proportional to $d^2\mathcal{P}/dt^2 = -\omega^2 \mathcal{P}$; see (5.2-25).

therefore proportional to ω^4, or to $1/\lambda_o^4$, so that short wavelengths are scattered more than long wavelengths. Light in the visible region of the spectrum is therefore scattered more than light in the infrared.

The functional form of Rayleigh scattering, which decreases with wavelength as $1/\lambda_o^4$, is known as the **Rayleigh inverse fourth-power law**. In the visible region of the spectrum, Rayleigh scattering is a more significant source of loss than is the tail of the ultraviolet absorption band, as shown in Fig. 10.3-2. However, Rayleigh loss becomes negligible in comparison with infrared absorption in silica glass for wavelengths greater than $\approx 1.6~\mu$m.

We conclude that the transparent window in silica glass is bounded by Rayleigh scattering on the short-wavelength side and by infrared absorption on the long-wavelength side (indicated by the dashed curves in Fig. 10.3-2). Near-infrared communication systems using silica-glass fibers are deliberately designed to operate in this window.

Extrinsic Effects

Aside from these intrinsic effects there are extrinsic absorption bands that result from the presence of impurities in silica glass, principally metallic ions (e.g., Fe, Cu, Cr, and Ni) and OH radicals associated with water vapor dissolved in the glass. Most metal impurities can be readily removed. Initially, OH impurities proved to be more difficult to eliminate but methods were ultimately developed to do so. Wavelengths at which glass fibers are used for optical fiber communications were traditionally selected to avoid the OH absorption bands.

Light-scattering losses may also be accentuated when dopants are added, as they often are for purposes of index grading. The attenuation coefficient for guided light in glass fibers depends on the absorption and scattering in the core and cladding materials. Each mode has a different penetration depth into the cladding, causing the rays to travel different effective distances and rendering the attenuation coefficient mode-dependent. It is generally higher for higher-order modes. Single-mode fibers therefore typically have smaller attenuation coefficients than multimode fibers (Fig. 10.3-3). Losses are also introduced by small random variations in the geometry of the fiber and by bends.

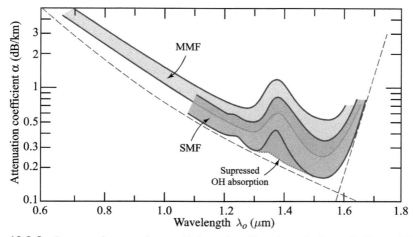

Figure 10.3-3 Ranges of attenuation coefficients for silica-glass single-mode fibers (SMF) and multimode fibers (MMF).

B. Dispersion

When a short pulse of light travels through an optical fiber, its power is "dispersed" in time so that the pulse spreads into a wider time interval. There are five principal sources of dispersion in optical fibers:
- Modal dispersion
- Material dispersion
- Waveguide dispersion
- Polarization mode dispersion
- Nonlinear dispersion

The combined contributions of these effects to the spread of pulses in time are not necessarily additive, as will be understood in the sequel.

Modal Dispersion

Modal dispersion occurs in multimode fibers as a result of the differences in the group velocities of the various modes. A single impulse of light entering an M-mode fiber at $z = 0$ disperses into M pulses whose differential delay increases as a function of z. For a fiber of length L, the time delays engendered by the different velocities are $\tau_q = L/v_q$, $q = 1, \ldots, M$, where v_q is the group velocity of mode q. If $v_{\min}$ and $v_{\max}$ are the smallest and largest group velocities, respectively, the received pulse spreads over a time interval $L/v_{\min} - L/v_{\max}$. Since the modes are usually not excited equally, the overall shape of the received pulse generally has a smooth envelope, as illustrated in Fig. 10.3-4. An estimate of the overall pulse duration (assuming a triangular envelope and using the FWHM definition of the width) is $\sigma_\tau = \frac{1}{2}(L/v_{\min} - L/v_{\max})$, which represents the modal-dispersion response time of the fiber.

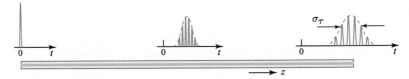

Figure 10.3-4 Pulse spreading caused by modal dispersion.

In a step-index fiber with a large number of modes, $v_{\min} \approx c_1(1-\Delta)$ and $v_{\max} \approx c_1$ [see Sec. 10.2C and Fig. 10.2-12(a)]. Since $(1-\Delta)^{-1} \approx 1+\Delta$ for $\Delta \ll 1$, the response time turns out to be a fraction $\Delta/2$ of the delay time L/c_1:

$$\sigma_\tau \approx \frac{L}{c_1} \cdot \frac{\Delta}{2}.$$

(10.3-4)
Response Time
(Multimode Step-Index)

Modal dispersion is far smaller in graded-index (GRIN) fibers than in step-index fibers since the group velocities are equalized and the differences between the delay times of the modes, $\tau_q = L/v_q$, are reduced. It was shown in Sec. 10.2C and in Fig. 10.2-12(b) that a graded-index fiber with an optimal index profile and a large number of modes has $v_{\max} \approx c_1$ and $v_{\min} \approx c_1(1 - \Delta^2/2)$. The response time in this case is therefore a factor of $\Delta/2$ smaller than that in a step-index fiber:

$$\sigma_\tau \approx \frac{L}{c_1} \cdot \frac{\Delta^2}{4}.$$

(10.3-5)
Response Time
(Graded-Index)

EXAMPLE 10.3-1. *Multimode Pulse Broadening Rate.* In a step-index fiber with $\Delta = 0.01$ and $n = 1.46$, pulses spread at a rate of approximately $\sigma_\tau/L = \Delta/2c_1 = n_1\Delta/2c_o \approx 24$ ns/km. In a 100-km fiber, therefore, an impulse spreads to a width of ≈ 2.4 μs. If the same fiber is optimally index-graded, the pulse broadening rate is approximately $n_1\Delta^2/4c_o \approx 122$ ps/km, a substantial reduction.

The pulse broadening arising from modal dispersion is proportional to the fiber length L in both step-index and GRIN fibers. Because of mode coupling, however, this dependence does not necessarily apply for fibers longer than a certain critical length. Coupling occurs among modes that have approximately the same propagation constants as a result of small imperfections in the fiber, such as random irregularities at its surface or inhomogeneities in its refractive index. This permits optical power to be exchanged between the modes. Under certain conditions, the response time σ_τ of mode-coupled fibers is proportional to L for small fiber lengths and to $\sqrt{L}$ when a critical length is exceeded, whereupon the pulses are broadened at a reduced rate.[†]

Material Dispersion

Glass is a dispersive medium, i.e., its refractive index is a function of wavelength. As discussed in Sec. 5.7, an optical pulse travels in a dispersive medium of refractive index n with a group velocity $v = c_o/N$, where $N = n - \lambda_o\, dn/d\lambda_o$. Since the pulse is a wavepacket comprising a collection of components of different wavelengths, each traveling at a different group velocity, its width spreads. The temporal duration of an optical impulse of spectral width σ_λ (nm), after traveling a distance L through a dispersive material, is $\sigma_\tau = |(d/d\lambda_o)(L/v)|\sigma_\lambda = |(d/d\lambda_o)(LN/c_o)|\sigma_\lambda$. This leads to a response time [see (5.7-2), (5.7-7), and (5.7-8)]

$$\sigma_\tau = |D_\lambda|\sigma_\lambda L, \qquad (10.3\text{-}6)$$

Response Time
(Material Dispersion)

where the material dispersion coefficient D_λ is

$$D_\lambda = -\frac{\lambda_o}{c_o}\frac{d^2 n}{d\lambda_o^2}. \qquad (10.3\text{-}7)$$

The response time increases linearly with the distance L. Usually, L is measured in km, σ_τ in ps, and σ_λ in nm, so that D_λ has units of ps/km-nm. This type of dispersion is called **material dispersion**.

The wavelength dependence of the dispersion coefficient D_λ for a silica-glass fiber is displayed in Fig. 10.3-5. At wavelengths shorter than 1.3 μm the dispersion coefficient is negative, so that wavepackets of long wavelength travel faster than those of short wavelength. At a wavelength $\lambda_o = 0.87$ μm, for example, the dispersion coefficient D_λ is approximately -80 ps/km-nm. At $\lambda_o = 1.55$ μm, on the other hand, $D_\lambda \approx +17$ ps/km-nm. At $\lambda_o \approx 1.312$ μm the dispersion coefficient vanishes, so that σ_τ in (10.3-6) vanishes. A more precise expression for σ_τ that incorporates the spread of the spectral width σ_λ about $\lambda_o = 1.312$ μm yields a very small, but nonzero, width.

[†] See, e.g., J. E. Midwinter, *Optical Fibers for Transmission*, Wiley, 1979; Krieger, reissued 1992.

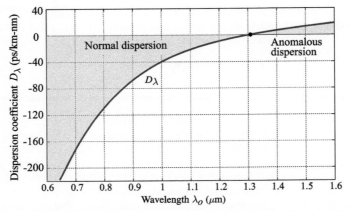

Figure 10.3-5 Observed dispersion coefficient D_λ for silica-glass fiber as a function of the wavelength λ_o. The result differs slightly from that calculated for bulk fused silica (see Fig. 5.7-5).

EXAMPLE 10.3-2. *Pulse Broadening Associated with Material Dispersion.* The dispersion coefficient D_λ for a silica-glass fiber is approximately -80 ps/km-nm at $\lambda_o = 0.87$ µm. For a source of spectral linewidth $\sigma_\lambda = 50$ nm (generated by an LED, for example) the pulse-spread rate in a single-mode fiber with no other sources of dispersion is $|D_\lambda|\sigma_\lambda = 4$ ns/km. An impulse of light traveling a distance $L = 100$ km in the fiber is therefore broadened to a width $\sigma_\tau = |D_\lambda|\sigma_\lambda L = 0.4$ µs. The response time of the fiber is thus $\sigma_\tau = 0.4$ µs. As another example, an impulse with narrower spectral linewidth $\sigma_\lambda = 2$ nm (generated by a laser diode, for example), operating near 1.3 µm where the dispersion coefficient is 1 ps/km-nm, spreads at a rate of only 2 ps/km. In this case, therefore, a 100-km fiber has a substantially shorter response time, $\sigma_\tau = 0.2$ ns.

Combined Material and Modal Dispersion

The effect of material dispersion on pulse broadening in multimode fibers may be determined by returning to the original equations for the propagation constants β_q of the modes and determining the group velocities $v_q = (d\beta_q/d\omega)^{-1}$ with n_1 and n_2 provided as functions of ω. Consider, for example, the propagation constants of a graded-index fiber with a large number of modes, which are given by (10.2-37) and (10.2-33). Although n_1 and n_2 are dependent on ω, it is reasonable to assume that the ratio $\Delta = (n_1 - n_2)/n_1$ is approximately independent of ω. Using this approximation and evaluating $v_q = (d\beta_q/d\omega)^{-1}$, we obtain

$$v_q \approx \frac{c_o}{N_1}\left[1 - \frac{p-2}{p+2}\left(\frac{q}{M}\right)^{p/(p+2)}\Delta\right], \qquad (10.3\text{-}8)$$

where $N_1 = (d/d\omega)(\omega n_1) = n_1 - \lambda_o(dn_1/d\lambda_o)$ is the group index of the core material. Under this approximation, the earlier expression (10.2-39) for v_q remains intact, except that the refractive index n_1 is replaced with the group index N_1. For a step-index fiber ($p \to \infty$), the group velocities of the modes vary from c_o/N_1 to $(c_o/N_1)(1-\Delta)$, so that the response time is

$$\boxed{\sigma_\tau \approx \frac{L}{(c_o/N_1)} \cdot \frac{\Delta}{2}.} \qquad (10.3\text{-}9)$$

Response Time
(Multimode Step-Index,
Material Dispersion)

This expression should be compared with (10.3-4), which is applicable in the absence of material dispersion.

EXERCISE 10.3-1

Optimal Grade Profile Parameter. Use (10.2-37) and (10.2-33) to derive the following expression for the group velocity v_q when both n_1 and Δ are wavelength dependent:

$$v_q \approx \frac{c_o}{N_1}\left[1 - \frac{p-2-p_s}{p+2}\left(\frac{q}{M}\right)^{p/(p+2)}\Delta\right], \quad q=1,2,\ldots,M \quad (10.3\text{-}10)$$

with $p_s = 2(n_1/N_1)(\omega/\Delta)\,d\Delta/d\omega$. What is the optimal value of the grade profile parameter p for minimizing modal dispersion?

Waveguide Dispersion

The group velocities of the modes in a waveguide are dependent on the wavelength even if material dispersion is negligible. This dependence, known as **waveguide dispersion**, results from the dependence of the field distribution in the fiber on the ratio of the core radius to the wavelength (a/λ_o). The relative portions of optical power in the core and cladding thus depend on λ_o. Since the phase velocities in the core and cladding differ, the group velocity of the mode is altered. Waveguide dispersion is particularly important in single-mode fibers where modal dispersion is not present, and at wavelengths for which material dispersion is small (near $\lambda_o = 1.3$ μm in silica glass), since it then dominates.

As discussed in Sec. 10.2A, the group velocity $v = (d\beta/d\omega)^{-1}$ and the propagation constant β are determined from the characteristic equation, which is governed by the fiber V parameter, $V = 2\pi(a/\lambda_o)\text{NA} = (a\cdot\text{NA}/c_o)\omega$. In the absence of material dispersion (i.e., when NA is independent of ω), V is directly proportional to ω, so that

$$\frac{1}{v} = \frac{d\beta}{d\omega} = \frac{d\beta}{dV}\frac{dV}{d\omega} = \frac{a\cdot\text{NA}}{c_o}\frac{d\beta}{dV}. \quad (10.3\text{-}11)$$

The pulse broadening associated with a source of spectral width σ_λ is related to the time delay L/v by $\sigma_\tau = |(d/d\lambda_o)(L/v)|\sigma_\lambda$. Thus,

$$\sigma_\tau = |D_w|\sigma_\lambda L, \quad (10.3\text{-}12)$$

where the waveguide dispersion coefficient D_w is given by

$$D_w = \frac{d}{d\lambda_o}\left(\frac{1}{v}\right) = -\frac{\omega}{\lambda_o}\frac{d}{d\omega}\left(\frac{1}{v}\right). \quad (10.3\text{-}13)$$

Substituting (10.3-11) into (10.3-13) leads to

$$D_w = -\left(\frac{1}{2\pi c_o}\right)V^2\frac{d^2\beta}{dV^2}. \quad (10.3\text{-}14)$$

Thus, the group velocity is inversely proportional to $d\beta/dV$ and the waveguide dispersion coefficient is proportional to $V^2 d^2\beta/dV^2$. The dependence of β on V is displayed in Fig. 10.2-6(a) for the fundamental LP_{01} mode. Since β varies nonlinearly with V, the waveguide dispersion coefficient D_w is itself a function of V and is therefore also a function of the wavelength.[†] The dependence of D_w on λ_o may be controlled by altering the radius of the core or, for graded-index fibers, the index grading profile.

Combined Material and Waveguide Dispersion (Chromatic Dispersion)

The combined effects of material dispersion and waveguide dispersion (which we refer to as **chromatic dispersion**) may be determined by including the wavelength dependence of the refractive indices, n_1 and n_2 and therefore NA, when determining $d\beta/d\omega$ from the characteristic equation. Although generally smaller than material dispersion, waveguide dispersion does shift the wavelength at which the total chromatic dispersion is minimum.

Since chromatic dispersion limits the performance of single-mode fibers, more advanced fiber designs aim at reducing this effect by using graded-index cores with refractive-index profiles selected such that the wavelength at which waveguide dispersion compensates material dispersion is shifted to the wavelength at which the fiber is to be used (Fig. 10.3-6).

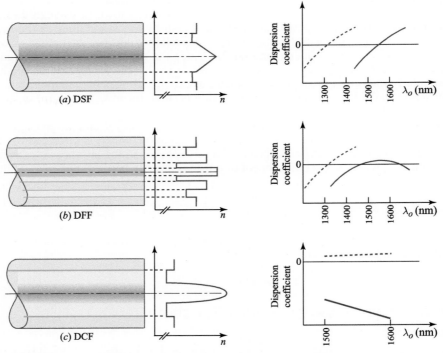

Figure 10.3-6 Refractive-index profiles with schematic wavelength dependences of the silica-glass material dispersion coefficient (dashed curves) and the combined material and waveguide dispersion coefficients (solid curves) for (a) dispersion-shifted fiber (DSF); (b) dispersion-flattened fiber (DFF); and (c) dispersion-compensating fiber (DCF).

Dispersion-shifted fibers have been successfully fabricated by using a linearly tapered core refractive index and a reduced core radius, as illustrated in Fig. 10.3-6(a). This technique can be used to shift the zero-chromatic-dispersion wavelength

---

[†] For further details on this topic, see the reading list, particularly the seminal articles by D. Gloge.

from 1.3 μm to 1.55 μm, where silica-glass fiber has its lowest attenuation. Other grading profiles have been developed for which the chromatic dispersion vanishes at two wavelengths and is reduced for intermediate wavelengths. These fibers, called **dispersion-flattened**, have been implemented by using a quadruple-clad layered grading, as illustrated in Fig. 10.3-6(b). Note, however, that the process of index grading itself introduces losses since dopants are used.

Fibers with other refractive index profiles may be engineered such that the combined material and waveguide dispersion coefficient is proportional to that of a conventional step-index fiber but has the opposite sign. This can be achieved over an extended wavelength band, as illustrated in Fig. 10.3-6(c). The pulse spread introduced by a conventional fiber can then be reversed by concatenating the two types of fiber. A fiber with a reversed dispersion coefficient is known as a **dispersion-compensating fiber (DCF)**. A short segment of the DCF may be used to compensate the dispersion introduced by a long segment of conventional fiber.

Polarization Mode Dispersion (PMD)

As indicated earlier, the fundamental spatial mode (LP_{01}) of an optical fiber has two polarization modes, say linearly polarized in the x and y directions. If the fiber has perfect circular symmetry about its axis, and its material is perfectly isotropic, then the two polarization modes are degenerate, i.e., they travel with the same velocity. However, fibers exposed to real environmental conditions exhibit a small birefringence that varies randomly along their length. This is caused by slight variations in the refractive indices and fiber cross-section ellipticity. Although the effects of such inhomogeneities and anisotropies on the polarization modes, and on the dispersion of optical pulses, are generally difficult to assess, we consider these effects in terms of simple models.

Consider first a fiber modeled as a homogeneous anisotropic medium with principal axes in the x and y directions and principal refractive indices n_x and n_y. The third principal axis lies along the fiber axis (the z direction). The fiber material is assumed to be dispersive so that n_x and n_y are frequency dependent, but the principal axes are taken to be frequency independent within the spectral band of interest. If the input pulse is linearly polarized in the x direction, over a length of fiber L it will undergo a group delay $\tau_x = N_x L/c_o$; if it is linearly polarized in the y direction, the group delay will be $\tau_y = N_y L/c_o$. Here, N_x and N_y are the group indices associated with n_x and n_y (see Sec. 5.7). A pulse in a polarization state that includes both linear polarizations will undergo a **differential group delay (DGD)** $\delta\tau = |\tau_y - \tau_x|$ given by

$$\delta\tau = \Delta N L/c_o, \qquad (10.3\text{-}15)$$
Differential Group Delay

where $\Delta N = |N_y - N_x|$. Upon propagation, therefore, the pulse will split into two orthogonally polarized components whose centers will separate in time as the pulses travel (see Fig. 10.3-7). The DGD corresponds to **polarization mode dispersion (PMD)** that increases linearly with the fiber length at the rate $\Delta N/c_o$, which is usually expressed in units of ps/km.

Figure 10.3-7 Differential group delay (DGD) associated with polarization mode dispersion (PMD).

Since a long fiber is typically exposed to environmental and structural factors that vary along its axis, the simple model considered above is often inadequate, however. Under these conditions, a more realistic model comprises a sequence of short homogeneous fiber segments, each with its own principal axes and principal indices. The principal axes are taken to be slightly misaligned (rotated) from one segment to the next. Such a cascaded system is generally described by a 2×2 Jones matrix $\mathbf{T}$, which is a product of the Jones matrices of the individual segments (see Sec. 6.1B). The polarization modes of the combined system are the eigenvectors of $\mathbf{T}$ and are not necessarily linearly polarized modes. If the fiber is taken to be lossless, the matrix $\mathbf{T}$ is unitary. Its eigenvalues are then phase factors $\exp(j\varphi_1)$ and $\exp(j\varphi_2)$, which may be written in the form $\exp(jn_1 k_o L)$ and $\exp(jn_2 k_o L)$, where n_1 and n_2 are the effective refractive indices of the two polarization modes and L is the fiber length. The propagation of light through such a length of fiber may then be determined by analyzing the input wave into components along the two polarization modes; these components travel with effective refractive indices n_1 and n_2.

Since the fiber is dispersive, $\mathbf{T}$ is frequency dependent and so too are the indices n_1 and n_2 of the modes, as well as their corresponding group indices N_1 and N_2. An input pulse with a polarization state that is the same as that of the fiber's first polarization mode travels with an effective group index N_1. Similarly, if the pulse is in the second polarization mode, it travels with an effective group index N_2. However, an input pulse with a component in each of the fiber's polarization modes suffers DGD, as provided in (10.3-15), with $\Delta N = |N_1 - N_2|$.

A statistical model describing the random variations in the magnitude and orientation of birefringence along the length of the fiber leads to an expression of the RMS value of the pulse broadening associated with DGD. This turns out to be proportional to $\sqrt{L}$ instead of L,

$$\sigma_{\text{PMD}} = D_{\text{PMD}}\sqrt{L}, \qquad (10.3\text{-}16)$$
Polarization Mode Dispersion

where D_{PMD} is a dispersion parameter typically ranging from 0.1 to 1 ps/$\sqrt{\text{km}}$.

Aside from DGD, higher-order dispersion effects are also present. Each of the polarization modes is spread by group velocity dispersion (GVD) with dispersion coefficients proportional to the second derivative of its refractive index (see Sec. 5.7).

Another higher-order effect relates to the coupled nature of the spectral and polarization properties of the system. Since the matrix $\mathbf{T}$ is frequency dependent, not only are the eigenvalues (i.e., the principal indices n_1 and n_2) frequency dependent, but so too are the eigenvectors (i.e., the polarization modes). If the pulse spectral width is sufficiently narrow (i.e., the pulse is not too short), we may approximately use the polarization modes at the central frequency. For ultrashort pulses, however, a more detailed analysis that includes a combined polarization and spectral description of the system is required. Polarization states may be found such that the group delays are frequency insensitive so that their associated GVD is minimal. However, these are not eigenvectors of the Jones matrix so that the input and output polarization states are not the same.[†]

[†] For further details on this topic, see C. D. Poole and R. E. Wagner, Phenomenological Approach to Polarization Dispersion in Long Single-Mode Fibers, *Electronics Letters*, vol. 22, pp. 1029-1030, 1986.

EXERCISE 10.3-2

Differential Group Delay in a Two-Segment Fiber. Consider the propagation of an optical pulse through a fiber of length 1 km comprising two segments of equal length. Each segment is a single-mode anisotropic fiber with principal group indices $N_x = 1.462$ and $N_y = 1.463$. The corresponding group velocity dispersion coefficients are $D_x = D_y = 20$ ps/km-nm. The principal axes of one segment are at an angle of $45°$ with respect to the other, as illustrated in Fig. 10.3-8.

(a) If the input pulse has a width of 100 ps and is linearly polarized at $45°$ with respect to the fiber x and y directions, sketch the temporal profile of the pulse at the output end of the fiber. Assume that the pulse source has a spectral linewidth of 50 nm.

(b) Determine the polarization modes of the full fiber and determine the temporal profile of the output pulse if the input pulse is in one of the polarization modes.

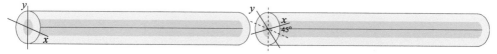

Figure 10.3-8 Two-segment birefringent fiber.

Nonlinear Dispersion

Yet another dispersion effect occurs when the intensity of light in the core of the fiber is sufficiently high, since the refractive index then becomes intensity dependent and the material exhibits nonlinear behavior. Since the phase is proportional to the refractive index, the high-intensity portions of an optical pulse undergo phase shifts different from the low-intensity portions, resulting in instantaneous frequencies shifted by different amounts. This nonlinear effect, called **self-phase modulation** (SPM), contributes to pulse dispersion. Under certain conditions, SPM can compensate the group velocity dispersion (GVD) associated with material dispersion, and the pulse can travel without altering its temporal profile. Such a guided wave is known as a soliton. Nonlinear optics is introduced in Chapter 22 and optical solitons are discussed in Chapter 23.

Summary

The propagation of pulses in optical fibers is governed by attenuation and several types of dispersion. Figure 10.3-9 provides a schematic illustration in which the profiles of pulses traveling through different types of fibers are compared.

- In a multimode fiber (MMF), modal dispersion dominates the width of the pulse received at the terminus of the fiber. It is governed by the disparity in the group delays of the individual modes.
- In a single-mode fiber (SMF), there is no modal dispersion and the transmission of optical pulses is limited by combined material and waveguide dispersion (called chromatic dispersion). The width of the output pulse is governed by group velocity dispersion (GVD).
- Material dispersion is usually much stronger than waveguide dispersion. However, at wavelengths where material dispersion is small, waveguide dispersion becomes important. Fibers with special index profiles may then be used to alter the chromatic-dispersion characteristics, creating dispersion-flattened, dispersion-shifted, and dispersion-compensating fibers.

- Pulse propagation in long single-mode fibers for which chromatic dispersion is negligible is dominated by polarization mode dispersion (PMD). Small anisotropic changes in the fiber, caused, for example, by environmental conditions, alter the polarization modes so that the input pulse travels in two polarization modes with different group indices. This differential group delay (DGD) results in a small pulse spread.

- Under certain conditions an intense pulse, called an optical soliton, can render a fiber nonlinear and travel through it without broadening. This results from a balance between material dispersion and self-phase modulation (the dependence of the refractive index on the light intensity), as discussed in Chapter 23.

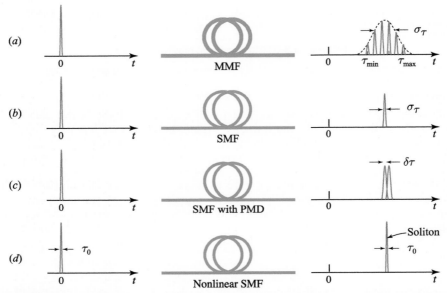

Figure 10.3-9 Broadening of a short optical pulse after transmission through different types of optical fibers. (*a*) Modal dispersion in a multimode fiber (MMF). (*b*) Material and waveguide dispersion (chromatic dispersion) in a single-mode fiber (SMF). (*c*) Polarization mode dispersion (PMD) in a SMF. (*d*) Soliton transmission in a nonlinear SMF.

10.4 HOLEY AND PHOTONIC-CRYSTAL FIBERS

A **holey fiber** is a fiber that contains multiple cylindrical air holes parallel to, and along the length of, its axis. Usually fabricated from pure silica glass, its holes are organized in a regular periodic pattern. As illustrated in Fig. 10.4-1, the core of the fiber is defined by a **defect**, or fault, in the periodic structure, such as a missing hole, a hole of a different size, or an extra hole. The holes are characterized by the spacing between their centers Λ, and their diameters d. The quantity Λ, which is also called the **pitch**, is typically in the range 1–10 μm. It is not necessary to include dopants in the glass. Holey fibers guide optical waves via one of two mechanisms: **effective-index guidance** or **photonic-bandgap guidance**, which we consider in turn below.

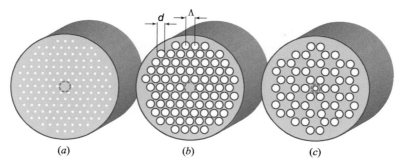

Figure 10.4-1 Various forms of holey fibers. (*a*) Solid core (dotted circle) surrounded by a cladding of the same material but suffused with an array of cylindrical air holes with diameters much smaller than a wavelength. The average refractive index of the cladding is lower than that of the core. (*b*) Photonic-crystal holey fiber with cladding that contains a periodic array of large air holes and a solid core (dotted circle). (*c*) Photonic-crystal holey fiber with cladding that contains a periodic array of large air holes and a core that is an air hole of a different size (dotted circle).

Effective-Index Guidance

If the hole diameter is much smaller than the wavelength of light ($d \ll \lambda$), then the periodic cladding behaves approximately as a homogeneous medium whose **effective refractive index** n_2 is equal to the average refractive index of the holey material [see Fig. 10.4-1(*a*)]. Waveguiding is then achieved by making use of a solid core with index $n_1 > n_2$, so that the light is guided by total internal reflection as with conventional optical fibers. In this configuration, the holes serve merely as distributed "negative dopants" that reduce the refractive index of the cladding below that of the solid core. The holes can therefore be randomly, rather than periodically, arrayed and they need not be axially continuous.

If the size of the holes is not much smaller than the wavelength, then the holey cladding must be treated as a two-dimensional periodic medium [Fig. 10.4-1(*b*)]. The effective refractive index n_2 is then given by the average refractive index, weighted by the optical intensity distribution in the medium, and is therefore strongly dependent on the wavelength, as well as on the size and the geometry of the holes. Since waves of shorter wavelength are more confined in the medium of higher refractive index, the effective refractive index of the cladding $n_2(\lambda)$ is a decreasing function of the wavelength. A similar effect occurs in a 1D photonic crystal, for which the effective refractive index is an increasing function of frequency at frequencies in the lowest photonic band (see Fig. 7.2-6). The holey fiber is therefore endowed with strong waveguide dispersion, which can be a useful feature.

One consequence of the waveguide dispersion is that the holey fiber may operate as a single-mode structure over a broad range of wavelengths, possibly stretching from the infrared to the ultraviolet.[†] This property, called **endless single-mode guidance**, results when the *V* parameter of the fiber, $V = (2\pi a/\lambda)\sqrt{n_1^2 - n_2^2(\lambda)}$, is approximately independent of λ. This condition arises when the effective index $n_2(\lambda)$ decreases with increasing λ in such a way that $\sqrt{n_1^2 - n_2^2(\lambda)} \propto \lambda$. For a conventional optical fiber, in contrast, *V* is inversely proportional to λ so that single-mode behavior at a particular wavelength ($V < 2.405$) morphs into multimode behavior for wavelengths that are sufficiently short such that *V* exceeds 2.405.

Another interesting feature is the feasibility of achieving **large mode-area** (LMA) single-mode operation. Optical fibers with large mode areas are useful for applications

---

[†] See T. A. Birks, J. C. Knight, and P. St J. Russell, Endlessly Single-Mode Photonic Crystal Fibre, *Optics Letters*, vol. 22, pp. 961–963, 1997.

requiring the delivery of high optical powers. In a conventional optical fiber the condition for single-mode operation $[V = 2\pi(a/\lambda_o)\mathrm{NA} < 2.405]$ can be met for a large core diameter $2a$ by making use of a small numerical aperture. Similarly, in holey fibers the guided-mode size can be increased by increasing the hole-to-hole spacing Λ (thereby increasing the core diameter) and concomitantly using holes of smaller diameter d (thereby reducing the numerical aperture and allowing the field to penetrate farther into the cladding). Dramatic increases in mode area for relatively small changes in hole size have been observed and mode areas several order of magnitudes greater than those in conventional optical fibers have been reported.

Photonic-Bandgap Guidance

The cladding of a holey fiber may be regarded as a two-dimensional **photonic crystal**. The triangular-hole microstructure shown in Fig. 10.4-1(b), for example, has a dispersion diagram endowed with photonic bandgaps, as shown in Fig. 7.3-3 and discussed in Sec. 7.3A. If the optical frequency lies within the photonic bandgap, propagation through the cladding is prohibited and the fiber serves as a photonic-crystal waveguide (see Sec. 9.5).

A **photonic-crystal fiber** (PCF) may have a solid or hollow core, as illustrated in Figs. 10.4-1(b) and (c), respectively. Fibers with hollow cores cannot operate by means of effective-index guidance, i.e., guidance cannot be based on total internal reflection. In any case, there are a number of merits to using hollow-core PCFs: (1) A guided wave traveling in an air-core PCF suffers lower losses as well as reduced nonlinear effects and can thus carry greater optical power; (2) light in the mid-ultraviolet region that would cause damage and degradation in solid-core fibers can be guided and transmitted; (3) light can be guided at wavelengths where transparent materials are not available; and (4) atoms, molecules, gases, and other subwavelength structures can be placed within the hollow core.

Applications

Photonic-crystal fibers offer many unique design possibilities and applications. As an example, dispersion flattening over broad wavelength ranges can be achieved, and the dispersion can be shifted to wavelengths below that at which the material dispersion is zero. PCFs support the propagation of high-intensity femtosecond pulses and their compression to sub-carrier-cycle durations. Long interaction lengths and tight confinement permit nonlinear optical effects such as low-threshold stimulated Raman scattering, harmonic generation, and electromagnetic-induced transparency to be investigated. The birefringence properties of such fibers can be tailored in interesting and useful ways. Acousto-optic interactions can be fostered and examined. PCFs can be used to construct broadband (mid infrared to mid ultraviolet) supercontinuum sources, mode-locked fiber soliton lasers, and powerful fiber lasers operating over a broad range of wavelengths.

Photonic-crystal fibers can be readily used as dynamic sensors for strain, temperature, and electric field. Pressure and heat serve to modify the sizes of the air holes, which in turn modifies their optical properties. Hollow-core fibers can be used to inspect, characterize, manipulate, trap, and accelerate objects such as living cells, colloids, clusters, and nm-size particles. Gases, liquids, or metabolites can be diffused into a hollow-core PCF, where light is trapped, thereby allowing ultrasensitive optical measurements to be carried out. Moreover, the hollow channels within the PCFs can be filled with materials such as metals, semiconductors, or soft glasses. This allows nanometer-scale features to be incorporated in the fibers, thereby leading to devices with both fiber-optic and plasmonic features.

10.5 FIBER MATERIALS

As discussed throughout this chapter, a principal use of optical fibers is for near-infrared optical fiber communications and data-transmission systems. The preeminent material for fabricating these fibers is silica glass since this medium exhibits low loss in the 1.3–1.6-μm telecommunications band, as described in Sec. 10.3A. **Specialty fibers**, which are optical fibers endowed with at least one special property that distinguishes them from standard silica-glass optical fibers, are also used in a many applications. Such fibers may, for example, be double-clad, polarization-maintaining, radiation-resistant, or doped with laser-active rare-earth ions.

Mid-Infrared Fibers

Interest in fiber optics extends beyond the near-infrared region. The mid infrared, for example, offers a substantial number of material choices, including fluoride, germanate, tellurite, and chalcogenide glasses. Of these, the chalcogenides (which include sulfides, selenides, and tellurides) exhibit the broadest transparency windows. The melting/processing temperatures of these so called *soft glasses* are substantially lower than that of silica glass so they can be conveniently used to fabricate infrared fibers via thermal fiber drawing; indeed, fibers fabricated from these three glasses are commercially available.

Interest in these materials is significant since the Rayleigh inverse fourth-power law (see Secs. 5.6B and 10.3A) predicts a reduction of Rayleigh scattering, and absorption further into the infrared, than that in silica glass. The attenuation arising from Rayleigh scattering in fluoride-glass infrared fibers, for example, is expected to be approximately ten times smaller than that for silica-glass optical fibers, reaching a minimum of ≈ 0.01 dB/km at $\lambda_o \approx 2.5$ μm. This issue is, of course, significant only when extrinsic loss mechanisms do not dominate transmission loss.

In spite of the extensive choice of optical fiber materials in the mid infrared, in the current state of technology the optical and mechanical properties of these fibers are substantially inferior to those of conventional silica optical fibers. With the notable exception of fluoride-glass fibers, most mid-infrared fibers exhibit transmission losses in the dB/m range, whereas silica-glass fibers are far more transparent with losses in the dB/km range. The use of most mid-infrared fibers is thus currently limited to applications that involve fiber lengths of the order of meters rather than kilometers, i.e., to short-haul applications such as sensing, metrology, biomedicine, and the delivery of infrared-laser power. These limitations are ultimately expected to be overcome, however.

Multimaterial and Multifunctional Fibers

Hybrid mid-infrared fibers. The ability to co-draw fibers comprising heterogeneous materials has led to the development of hybrid fibers that offer promise for ameliorating some of the optical and mechanical shortcomings of conventional mid-infrared fibers discussed above. Several examples of hybrid mid-infrared fibers are:[†] (1) Step-index fibers comprising a chalcogenide glass core with claddings of polymer (as a protective jacket), silica-glass, or tellurite-glass; (2) Step-index fibers comprising a semiconductor crystalline core such as InSb, ZnSe, Si, or Ge, with borosilicate-glass or silica-glass cladding. In connection with photonic-crystal fibers (PCFs), as discussed in Sec. 10.4, examples are: (3) Chalcogenide or fluoride glass solid-core PCFs; and (4) Silica or chalcogenide hollow-core PCFs.

[†] See G. Tao, H. Ebendorff-Heidepriem, A. M. Stolyarov, S. Danto, J. V. Badding, Y. Fink, J. Ballato, and A. F. Abouraddy, Infrared Fibers, *Advances in Optics and Photonics*, vol. 7, pp. 379–458, 2015.

Multimaterial fibers. Multimaterial fibers comprise combinations of materials. As alluded to in Sec. 10.4, optical fibers have been developed that incorporate not only glasses, but also conductors, semiconductors, insulators, and gases, along with more specialized materials such as liquid crystals and piezoelectric media. Nowadays, thermally drawn fibers that are kilometers long can be internally structured with elaborate macroscopic device architectures that incorporate these materials. Moreover, the ultimate composition of a fiber can differ from that of the initial materials by virtue of chemical reactions initiated by the heating and drawing processes (e.g., initial ingredients of aluminum metal and silica glass can result in a fiber whose core is crystalline silicon). Multimaterial fibers can be configured to sense light, sound, and/or heat impinging on their lateral surfaces. They can also be operated in reverse, i.e., to emit light and sound.

Multifunctional fibers and fiber assemblies. Multifunctional fibers, on the other hand, accommodate multitudinal functionalities. Beyond integrating multiple functional components into individual fibers, large-scale fiber arrays and fiber textiles can be fabricated. Multimaterial and multifunctional fibers and fiber assemblies offer a panoply of functionalities: photonic, electronic, mechanical, thermal, acoustic, chemical, and biomedical. Moreover, subwavelength internal structures of nanometer length scales can be organized within fibers of kilometer length scales, resulting in devices that merge fiber-optic and plasmonic features.

Examples. Some examples of multimaterial and multifunctional fibers are:[†]
- An axially pumped photonic-bandgap fiber laser, comprising a core of organic dye dissolved in a solid host as the gain medium, that emits light *radially*.
- Fibers engineered to detect sound by incorporating a piezoelectric plastic component along the length of the fiber (a pressure wave induces a charge in the piezoelectric material).
- Fibers designed to detect heat by incorporating a semiconductor component along the length of the fiber (temperature modifies the conductivity).
- A flexible nonimaging textile camera woven from fibers that incorporate a photoconductive-semiconductor or photoconductive-glass component along the length of the fiber (light modifies the conductivity).
- A textile display woven from fibers that incorporate liquid-crystal channels (an applied field serves to block or transmit light).
- A multifunctional polymer-fiber probe that allows concurrent optical, electrical, and chemical interactions with a cell in a neural circuit.[‡]

READING LIST

Books
See also the reading list on photonic crystals in Chapter 7 and the reading list in Chapter 9.
F. Mitschke, *Fiber Optics: Physics and Technology*, Springer-Verlag, 2nd ed. 2016.

[†] See A. F. Abouraddy, M. Bayindir, G. Benoit, S. D. Hart, K. Kuriki, N. Orf, O. Shapira, F. Sorin, B. Temelkuran, and Y. Fink, Towards Multimaterial Multifunctional Fibres that See, Hear, Sense and Communicate, *Nature Materials*, vol. 6, pp. 336–347, 2007; G. Tao, A. F. Abouraddy, and A. M. Stolyarov, Multimaterial Fibers, *International Journal of Applied Glass Science*, vol. 3, pp. 349–368, 2012.

[‡] See A. Canales, X. Jia, U. P. Froriep, R. A. Koppes, C. M. Tringides, J. Selvidge, C. Lu, C. Hou, L. Wei, Y. Fink, and P. Anikeeva, Multifunctional Fibers for Simultaneous Optical, Electrical and Chemical Interrogation of Neural Circuits *In Vivo*, *Nature Biotechnology*, vol. 33, pp. 277–284, 2015.

Y. Koike, *Fundamentals of Plastic Optical Fibers*, Wiley–VCH, 2015.

J. Hecht, *Understanding Fiber Optics*, Laser Light Press, 5th ed. 2015.

F. Zolla, G. Renversez, A. Nicolet, B. Kuhlmey, S. Guenneau, D. Felbacq, A. Argyros, and S. Leon-Saval, *Foundations of Photonic Crystal Fibres*, Imperial College Press (London), 2nd ed. 2012.

A. Kumar and A. Ghatak, *Polarization of Light with Applications in Optical Fibers*, SPIE Optical Engineering Press, 2011.

R. Paschotta, *Field Guide to Optical Fiber Technology*, SPIE Optical Engineering Press, 2010.

F. Poli, A. Cucinotta, and S. Selleri, *Photonic Crystal Fibers: Properties and Applications*, Springer-Verlag, 2007, paperback ed. 2010.

M. G. Kuzyk, *Polymer Fiber Optics: Materials, Physics, and Applications*, CRC Press/Taylor & Francis, 2007.

A. Méndez and T. F. Morse, eds., *Specialty Optical Fibers Handbook*, Academic Press, 2007.

C. DeCusatis and C. J. Sher DeCusatis, *Fiber Optic Essentials*, Academic Press/Elsevier, 2005.

A. Galtarossa and C. R. Menyuk, eds., *Polarization Mode Dispersion*, Springer-Verlag, 2005.

J. A. Harrington, *Infrared Fibers and Their Applications*, SPIE Optical Engineering Press, 2004.

A. Bjarklev, J. Broeng, and A. S. Bjarklev, *Photonic Crystal Fibers*, Springer-Verlag, 2003, paperback ed. 2012.

J. Hecht, *City of Light: The Story of Fiber Optics*, Oxford University Press, 1999, paperback ed. 2004.

C. K. Kao, *Optical Fiber Systems: Technology, Design, and Applications*, McGraw–Hill, 1982.

D. Marcuse, *Light Transmission Optics*, Van Nostrand Reinhold, 1972, 2nd ed. 1982; Krieger, reissued 1989.

D. Marcuse, *Principles of Optical Fiber Measurements*, Academic Press, 1981.

Review Articles on Photonic-Crystal, Multicore, Multimaterial, and Infrared Fibers

G. Tao, H. Ebendorff-Heidepriem, A. M. Stolyarov, S. Danto, J. V. Badding, Y. Fink, J. Ballato, and A. F. Abouraddy, Infrared Fibers, *Advances in Optics and Photonics*, vol. 7, pp. 379–458, 2015.

T. Hayashi, Multi-Core Optical Fibers, in I. P. Kaminow, T. Li, and A. E. Willner, eds., *Optical Fiber Telecommunications VI-A: Components and Subsystems*, Academic Press/Elsevier, 6th ed. 2013.

G. Tao, A. F. Abouraddy, and A. M. Stolyarov, Multimaterial Fibers, *International Journal of Applied Glass Science*, vol. 3, pp. 349–368, 2012.

A. F. Abouraddy, M. Bayindir, G. Benoit, S. D. Hart, K. Kuriki, N. Orf, O. Shapira, F. Sorin, B. Temelkuran, and Y. Fink, Towards Multimaterial Multifunctional Fibres that See, Hear, Sense and Communicate, *Nature Materials*, vol. 6, pp. 336–347, 2007.

P. Russell, Photonic Crystal Fibers, *Science*, vol. 299, pp. 358–362, 2003.

J. C. Knight, J. Broeng, T. A. Birks, and P. St. J. Russell, Photonic Band Gap Guidance in Optical Fibers, *Science*, vol. 282, pp. 1476–1478, 1998.

Seminal Articles

P. J. Winzer, C. J. Chang-Hasnain, A. E. Willner, R. C. Alferness, R. W. Tkach, and T. G. Giallorenzi, eds., *A Third of a Century of Lightwave Technology: January 1983–April 2016*, IEEE–OSA, 2016.

C. K. Kao, Sand from Centuries Past: Send Future Voices Fast (Nobel Lecture in Physics, 2009), in L. Brink, ed., *Nobel Lectures in Physics 2006–2010*, World Scientific, 2014, pp. 253–264.

D. B. Keck, ed., *Selected Papers on Optical Fiber Technology*, SPIE Optical Engineering Press (Milestone Series Volume 38), 1992.

J. P. Pocholle, L. Jeunhomme, and L. D'Auria, Caractéristiques de la propagation guidée dans les fibres multimodes à gradient d'indice (FMGI), *Revue technique Thomson-CSF*, vol. 13, pp. 727–807, 1981.

P. J. B. Clarricoats, Optical Fibre Waveguides – A Review, in E. Wolf, ed., *Progress in Optics*, North-Holland, 1976, vol. 14, pp. 327–402.

D. B. Keck, R. D. Maurer, and P. C. Schultz, On the Ultimate Lower Limit of Attenuation in Glass Optical Waveguides, *Applied Physics Letters*, vol. 22, pp. 307–309, 1973.

D. Gloge and E. A. J. Marcatili, Multimode Theory of Graded-Core Fibers, *Bell System Technical Journal*, vol. 52, pp. 1563–1578, 1973.

D. Gloge, Weakly Guiding Fibers, *Applied Optics*, vol. 10, pp. 2252–2258, 1971.

D. Gloge, Dispersion in Weakly Guiding Fibers, *Applied Optics*, vol. 10, pp. 2442–2445, 1971.

PROBLEMS

10.1-1 **Coupling Efficiency.**
(a) A source emits light with optical power P_0 and a distribution $I(\theta) = (1/\pi)P_0 \cos\theta$, where $I(\theta)$ is the power per unit solid angle in the direction making an angle θ with the axis of a fiber. Show that the power collected by the fiber is $P = (\text{NA})^2 P_0$, so that the coupling efficiency is $(\text{NA})^2$, where NA is the numerical aperture of the fiber.
(b) If the source is a planar light-emitting diode of refractive index n_s bonded to the fiber, and the fiber cross-sectional area is larger than the LED emitting area, calculate the numerical aperture of the fiber and the coupling efficiency when $n_1 = 1.46$, $n_2 = 1.455$, and $n_s = 3.5$.

10.1-2 **Numerical Aperture of a Graded-Index Fiber.** Compare the numerical apertures of a step-index fiber with $n_1 = 1.45$ and $\Delta = 0.01$ and a graded-index fiber with $n_1 = 1.45$, $\Delta = 0.01$, and a parabolic refractive-index profile ($p = 2$). (See Exercise 1.3-2.)

10.2-1 **Modes.** A step-index fiber has radius $a = 5$ μm, core refractive index $n_1 = 1.45$, and fractional refractive-index change $\Delta = 0.002$. Determine the shortest wavelength λ_c for which the fiber is a single-mode waveguide. If the wavelength is changed to $\lambda_c/2$, identify the indices (l, m) of all the guided modes.

10.2-2 **Modal Dispersion.** A step-index fiber of numerical aperture $\text{NA} = 0.16$, core radius $a = 45$ μm, and core refractive index $n_1 = 1.45$ is used at $\lambda_o = 1.3$ μm, where material dispersion is negligible. If a light pulse of very short duration enters the fiber at $t = 0$ and travels a distance 1 km, sketch the shape of the received pulse:
(a) using ray optics and assuming that only meridional rays are allowed;
(b) using wave optics and assuming that only meridional ($l = 0$) modes are allowed.

10.2-3 **Propagation Constants and Group Velocities.** A step-index fiber with refractive indices $n_1 = 1.444$ and $n_2 = 1.443$ operates at $\lambda_o = 1.55$ μm. Determine the core radius at which the fiber V parameter is 10. Use Fig. 10.2-3 to estimate the propagation constants of all the guided modes with $l = 0$. If the core radius is now changed so that $V = 4$, use Fig. 10.2-6(a) to determine the phase velocity, the propagation constant, and the group velocity of the LP_{01} mode. Ignore the effect of material dispersion.

*10.2-4 **Propagation Constants and Wavevector (Step-Index Fiber).** A step-index fiber of radius $a = 20$ μm and refractive indices $n_1 = 1.47$ and $n_2 = 1.46$ operates at $\lambda_o = 1.55$ μm. Using the quasi-plane wave theory and considering only guided modes with $l = 1$:
(a) determine the smallest and largest propagation constants;
(b) for the mode with the smallest propagation constant, determine the radii of the cylindrical shell within which the wave is confined, and determine the components of the wavevector **k** at $r = 5$ μm.

*10.2-5 **Propagation Constants and Wavevector (Graded-Index Fiber).** Carry out the same calculations as in Prob. 10.2-4, but for a graded-index fiber with parabolic profile ($p = 2$).

10.3-3 **Scattering Loss.** At a wavelength of $\lambda_o = 820$ nm, the absorption loss of a fiber is 0.25 dB/km and the scattering loss is 2.25 dB/km. If the fiber is instead used at $\lambda_o = 600$ nm, and calorimetric measurements of the heat generated by light absorption reveal a loss of 2 dB/km, estimate the total attenuation at $\lambda_o = 600$ nm.

10.3-4 **Modal Dispersion in Step-Index Fibers.** Determine the core radius of a multimode step-index fiber with a numerical aperture $\text{NA} = 0.1$ if the number of modes $M = 5000$ when the wavelength is 0.87 μm. If the core refractive index $n_1 = 1.445$, the group index $N_1 = 1.456$, and Δ are approximately independent of wavelength, determine the modal-dispersion response time σ_τ for a 2-km-long fiber.

10.3-5 **Modal Dispersion in Graded-Index Fibers.** Consider a graded-index fiber with $a/\lambda_o = 10$, $n_1 = 1.45$, $\Delta = 0.01$, and power-law profile with index p. Determine the number of modes M, and the modal-dispersion pulse-broadening rate σ_τ/L, for $p = 1.9, 2, 2.1, \infty$.

10.3-6 **Pulse Propagation.** Consider a pulse of initial temporal width τ_0 transmitted through a graded-index fiber of length L km and power-law refractive-index profile with index p. The peak refractive index n_1 is wavelength-dependent with $D_\lambda = -(\lambda_o/c_o)\, d^2 n_1/d\lambda_o^2$, Δ is approximately independent of wavelength, σ_λ is the spectral width of the source, and λ_o is the operating wavelength. Discuss the effect of increasing each of the following parameters on the temporal width of the received pulse: L, τ_0, p, $|D_\lambda|$, σ_λ, and λ_o.

CHAPTER

11

RESONATOR OPTICS

11.1	PLANAR-MIRROR RESONATORS	436
	A. Resonator Modes	
	B. Off-Axis Resonator Modes	
11.2	SPHERICAL-MIRROR RESONATORS	447
	A. Ray Confinement	
	B. Gaussian Modes	
	C. Resonance Frequencies	
	D. Hermite–Gaussian Modes	
	*E. Finite Apertures and Diffraction Loss	
11.3	TWO- AND THREE-DIMENSIONAL RESONATORS	459
	A. Two-Dimensional Rectangular Resonators	
	B. Circular Resonators and Whispering-Gallery Modes	
	C. Three-Dimensional Rectangular Resonators	
11.4	MICRORESONATORS AND NANORESONATORS	463
	A. Rectangular Microresonators	
	B. Micropillars, Microdisks, and Microtoroids	
	C. Microspheres	
	D. Photonic-Crystal Microcavities	
	E. Plasmonic Resonators: Metallic Nanodisks and Nanospheres	

Charles Fabry
(1867–1945)

Alfred Perot
(1863–1925)

Working together, the French physicists Charles Fabry and Alfred Perot constructed an optical resonator for use as an interferometer. Now known as the Fabry–Perot etalon, it is used extensively in lasers.

Fundamentals of Photonics, Third Edition. Bahaa E. A. Saleh and Malvin Carl Teich.
©2019 John Wiley & Sons, Inc. Published 2019 by John Wiley & Sons, Inc.

An optical resonator is the optical counterpart of an electronic resonant circuit. It confines and stores light at resonance frequencies determined by its configuration. It is conveniently viewed as an optical transmission system that incorporates feedback: light is repeatedly reflected, or circulates, within its boundaries. Various optical-resonator configurations are depicted in Fig. 11.0-1. The simplest of these, the **Fabry–Perot resonator**, comprises two parallel planar mirrors; light is repeatedly reflected between them while experiencing little loss. Other mirror configurations include spherical mirrors, ring arrangements, and rectangular two- and three-dimensional resonators.

Fiber-ring resonators and **integrated-optic-ring resonators** are widely used. Light can also be trapped in defects within dielectric photonic-bandgap periodic structures, forming **photonic-crystal resonators**. **Guided-wave Fabry–Perot resonators** make use of Fresnel reflection at the boundaries between semiconductors and air. Periodic dielectric structures such as distributed Bragg reflectors (DBRs) can serve in lieu of conventional mirrors in Fabry–Perot resonators, providing feedback in structures such as **micropillar resonators**. **Dielectric resonators** make use of total internal reflection, in place of conventional reflection, at the boundary between low-loss dielectric materials. **Microdisks**, **microtoroids**, and **microspheres** support light that circulates via reflection at near-grazing incidence, in what are known as whispering-gallery modes. The confined rays skim around the inside rim of the resonator with an angle of incidence that is always greater than the critical angle so they do not refract out of the resonator. **Plasmonic resonators** are metallic structures of subwavelength dimensions, such as **nanodisks** and **nanospheres**, that support surface plasmon polariton waves and localized surface plasmon oscillations.

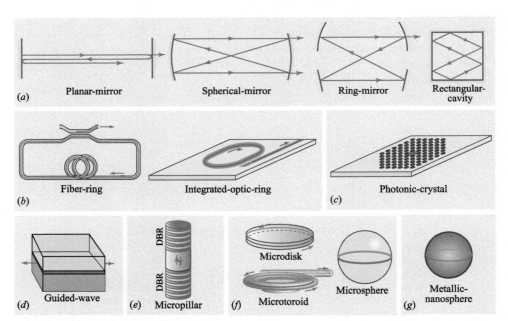

Figure 11.0-1 Storage of light in optical resonators via: (*a*) multiple reflections from mirrors; (*b*) propagation through closed-loop optical fibers and integrated-photonic waveguides; (*c*) trapping of light within defects in photonic crystals; (*d*) multiple Fresnel reflections at semiconductor–air boundaries; (*e*) reflections from periodic structures such as distributed Bragg reflectors (DBRs); (*f*) whispering-gallery mode reflections near the surfaces of dielectric microresonators such as disks, toroids, and spheres; and (*g*) localized surface plasmon oscillations in metallic nanospheres.

The size of an optical resonator can be of the same order of magnitude as its resonance wavelength, as with microresonators. But optical resonators can also be orders of magnitude larger than the resonance wavelength, as with bulk mirror resonators; or orders of magnitude smaller, as with metallic nanospheres. Figure 11.0-2 displays the relationship between resonator size and resonance wavelength for a number of electromagnetic resonators. Optical resonators are frequently characterized by two

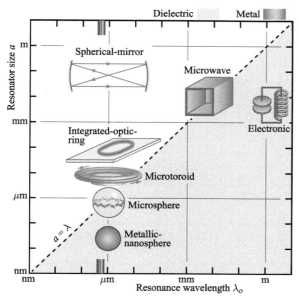

Figure 11.0-2 Resonator size a vs. resonance wavelength λ_o for various dielectric and metallic electromagnetic resonators. The size-to-wavelength ratio belongs to one of three regimes: $a/\lambda_o > 1$ (unshaded region), $a/\lambda_o \approx 1$ (dotted diagonal), and $a/\lambda_o < 1$ (shaded region). Metallic-nanosphere and electronic resonators lie well within the shaded region.

key parameters, representing the degrees of temporal and spatial light confinement, respectively:

1. The quality factor Q, which is proportional to the storage time of the resonator in units of optical period; a large value of Q indicates strong temporal confinement;
2. The modal volume V, which is the volume occupied by the confined optical mode; this measure is of particular interest for microresonators in which a small value of V indicates tight spatial confinement.

Because of their frequency selectivity, optical resonators can also serve as optical filters or spectrum analyzers (Sec. 7.1B). Their most important use, however, is as a "container" within which laser light can be generated and built up. The laser medium (Sec. 16.1A) amplifies light inside the optical resonator while the resonator determines, in part, the frequency and spatial distribution of the laser beam that is generated. Because resonators have the capacity to store energy, they can also be used to produce pulses of laser energy (Sec. 16.4A). Lasers are discussed in Chapters 16 and 18, and the material presented in this chapter is essential to their understanding.

This Chapter

The theoretical approaches considered in previous chapters are useful for describing the operation of optical resonators:

1. The simplest approach is based on *Ray Optics* (Chapter 1); optical rays are traced as they repeatedly reflect within the resonator and geometrical conditions are established that assure that the rays are confined.

2. *Wave Optics* (Chapter 2) is used to determine the modes of a resonator, i.e., the resonance frequencies and wavefunctions of the optical waves that are permitted to exist self-consistently within the resonator.
3. The study of *Beam Optics* (Chapter 3) is useful for understanding the behavior of spherical-mirror resonators; the modes of a resonator with spherical mirrors give rise to Gaussian and Laguerre–Gaussian optical beams.
4. *Fourier Optics* and the theory of light propagation and diffraction (Chapter 4) determine how the finite sizes of resonator mirrors affect resonator loss and the spatial characteristics of the ensuing modes.
5. *Electromagnetic Optics* (Chapter 5) provides a treatment of scattering that is fundamental to understanding the resonant feedback in coherent random lasers.
6. *Polarization Optics* (Chapter 6) sets forth the Fresnel relations, which establish the reflectances at the boundaries between semiconductors and air in guided-wave Fabry–Perot resonators.
7. *Photonic-Crystal Optics*, including the optics of multilayer media (Chapter 7), is important for describing optical resonators that make use of multiple dielectric layers and periodic media (e.g., distributed Bragg reflectors and photonic crystals) in lieu of conventional mirrors.
8. The analysis of oscillating electric charges (localized surface plasmons) in metals, considered in the *Optics of Metals and Metamaterials* (Chapter 8), is crucial for understanding the behavior of nanoresonators such as the metallic nanosphere.
9. The methods used in *Guided-Wave Optics* (Chapter 9) for determining the modes of planar-mirror and planar dielectric waveguides are similar to those required for the analysis of resonator modes.
10. *Fiber Optics* (Chapter 10) provides an analysis of the modes supported by optical fibers; the fiber laser resonator is an optical fiber bounded by end reflectors that result in the repeated reflection and confinement of the propagating light.

The optical resonator evidently provides a comprehensive venue for applying the theories of light discussed in earlier chapters. We begin with a study of planar-mirror resonators in Sec. 11.1 and spherical-mirror resonators in Sec. 11.2. We then introduce two- and three-dimensional resonators in Sec. 11.3 and finally consider microresonators and nanoresonators in Sec. 11.4.

11.1 PLANAR-MIRROR RESONATORS

A. Resonator Modes

In this section we examine the modes of an optical resonator constructed from two parallel, highly reflective, flat mirrors separated by a distance d (Fig. 11.1-1). This simple one-dimensional resonator is known as a **Fabry–Perot resonator**. We first consider an idealized version of this resonator in which the mirrors are lossless; the effect of losses is included subsequently.

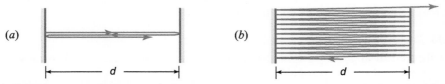

Figure 11.1-1 Two-mirror planar resonator (Fabry–Perot resonator). (*a*) Light rays perpendicular to the mirrors reflect back and forth without escaping. (*b*) Rays that are only slightly inclined eventually escape. Rays also escape if the mirrors are not perfectly parallel.

Resonator Modes as Standing Waves

As discussed in Secs. 2.2, 5.3, and 5.4, a monochromatic wave of frequency ν has a wave function

$$u(\mathbf{r}, t) = \text{Re}\left\{U(\mathbf{r}) \exp(j2\pi\nu t)\right\}, \tag{11.1-1}$$

representing a transverse component of the electric field. The complex amplitude $U(\mathbf{r})$ satisfies the Helmholtz equation, $\nabla^2 U(\mathbf{r}) + k^2 U(\mathbf{r}) = 0$, where $k = 2\pi\nu/c$ is the wavenumber and c is the speed of light in the medium. The resonator modes are the solutions to the Helmholtz equation under the appropriate boundary conditions. For the lossless planar-mirror resonator, the transverse components of the electric field vanish at the mirror surfaces (see Sec. 5.1), so that $U(\mathbf{r}) = 0$ at the planes $z = 0$ and $z = d$ in Fig. 11.1-2.

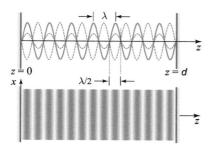

Figure 11.1-2 (a) Wave function $u(\mathbf{r},t)$ for an ideal planar-mirror mode as a function of z (for $x = y = 0$), portrayed at several different times. In this illustration, 14 half-wavelengths match the length of the resonator so that the mode number $q = d/(\lambda/2) = 14$. (b) Spatial distribution of the magnitude $|u(\mathbf{r},t)|$ as a function of x and z (for $y = 0$) at a particular time, represented on a color scale where red represents a large magnitude and white represents zero.

The standing wave $U(\mathbf{r}) = A \sin kz$, where A is a constant, satisfies the Helmholtz equation and vanishes at $z = 0$ and $z = d$ if k satisfies the condition $kd = q\pi$, where q is an integer. This restricts k to the values

$$k_q = q\frac{\pi}{d}, \qquad q = 1, 2, \ldots, \tag{11.1-2}$$

so that the modes have complex amplitudes

$$U(\mathbf{r}) = A_q \sin k_q z, \tag{11.1-3}$$

where the A_q are constants. Negative values of q do not constitute independent modes since $\sin k_{-q} z = -\sin k_q z$. Furthermore, the value $q = 0$ is associated with a mode that carries no energy since $k_0 = 0$ and $\sin k_0 z = 0$. The modes of the resonator are therefore the standing waves $A_q \sin k_q z$, where the positive integer $q = 1, 2, \ldots$ is called the **mode number**. An arbitrary wave inside the resonator can be written in terms of a superposition of the resonator modes:

$$U(\mathbf{r}) = \sum_q A_q \sin k_q z. \tag{11.1-4}$$

It follows from (11.1-2) that the associated frequencies $\nu = ck/2\pi$ are restricted to the discrete values

$$\nu_q = q\frac{c}{2d}, \qquad q = 1, 2, \ldots, \tag{11.1-5}$$

which are the resonance frequencies. As illustrated in Fig. 11.1-3, adjacent resonance frequencies are separated by a constant frequency difference, known as the **free spectral range**:

$$\nu_F = \frac{c}{2d} = \frac{c_o}{2nd}.$$
(11.1-6)
Frequency Spacing of Resonator Modes

The associated resonance wavelengths are $\lambda_q = c/\nu_q = 2d/q$. The round-trip distance traversed at resonance must therefore precisely equal an integer number of wavelengths:

$$2d = q\lambda_q, \qquad q = 1, 2, \ldots.$$
(11.1-7)

It is important to keep in mind that $c = c_o/n$ is the speed of light in the medium between the two mirrors, and that the λ_q represent wavelengths within that medium.

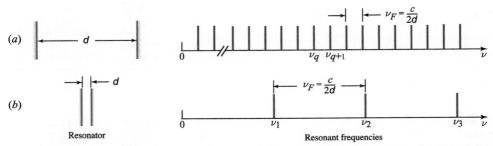

Figure 11.1-3 The adjacent resonance frequencies of a planar-mirror resonator are separated by $\nu_F = c/2d = c_o/2nd$, as illustrated by two examples: (a) A 30-cm long resonator ($d = 30$ cm) with air between the mirrors ($n = 1$) has a frequency spacing between modes given by $\nu_F = 500$ MHz. (b) A much shorter resonator with $d = 3$ μm has $\nu_F = 50$ THz, so that the first mode has a frequency corresponding to a wavelength of 6 μm and there are only two modes within the 700–900-nm optical band, which occupies a frequency range of 95 THz.

Resonator Modes as Traveling Waves

Alternatively, the resonator modes can be determined by following a wave as it travels back and forth between the two mirrors [Fig. 11.1-4(a)]. A mode is a wave that reproduces itself after a single round trip (see Appendix C). The phase shift imparted by a single round trip of propagation (a distance $2d$), $\varphi = k2d = 4\pi\nu d/c$, must therefore be a multiple of 2π:

$$\varphi = k2d = q2\pi, \qquad q = 1, 2, \ldots.$$
(11.1-8)

This result is not altered by an additional phase shift of 2π, which can be imparted by reflections at the two mirrors (see Sec. 6.2). As expected, we therefore obtain $kd = q\pi$, as in (11.1-2), and the same resonance frequencies as set forth in (11.1-5). Equation (11.1-8) may be viewed as a condition of positive feedback in the system displayed in Fig. 11.1-4(b); this requires that the output of the system be fed back *in phase* with the input.

We now demonstrate that only self-reproducing waves, or combinations thereof, can exist within the resonator under steady-state conditions. Consider a monochromatic

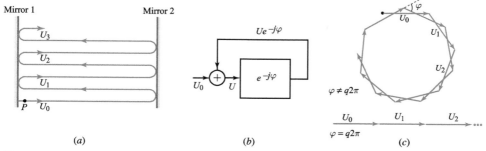

Figure 11.1-4 (a) A wave reflects back and forth between the resonator mirrors, suffering a phase shift φ on each round trip. (b) Block diagram of an optical feedback system with a phase delay φ. (c) Phasor diagram representing the sum $U = U_0 + U_1 + \cdots$ for $\varphi \neq q2\pi$ and for $\varphi = q2\pi$.

plane wave of complex amplitude U_0 at point P traveling to the right along the axis of the resonator [see Fig. 11.1-4(a)]. The wave is reflected from mirror 2 and propagates back to mirror 1 where it is again reflected. Its amplitude then becomes U_1. Yet another round trip results in a wave of complex amplitude U_2, and so on *ad infinitum*. Because the original wave U_0 is monochromatic, it is "eternal." Indeed, all of the partial waves, $U_0, U_1, U_2, \ldots$ are monochromatic and perpetually coexist. Moreover, their magnitudes are identical because it has been assumed that there is no loss associated with reflection and propagation. The total wave U is therefore represented by the sum of an infinite number of phasors of equal magnitude,

$$U = U_0 + U_1 + U_2 + \cdots, \tag{11.1-9}$$

as shown in Figs. 11.1-4(b) and (c).

The phase difference of two consecutive phasors imparted by a single round trip of propagation is $\varphi = k2d$. If the magnitude of the initial phasor is infinitesimal, the magnitude of each of these phasors must also be infinitesimal. The magnitude of the sum of this infinite number of infinitesimal phasors is itself infinitesimal unless they are aligned, i.e., unless $\varphi = q2\pi$, as illustrated at the bottom of Fig. 11.1-4(c). Thus, an infinitesimal initial wave can result in the buildup of finite power in the resonator, but only if $\varphi = q2\pi$.

Traveling-Wave Resonators

In a traveling-wave resonator, an optical mode travels in one direction along a closed path representing a round trip and retraces itself without reversing direction. Examples are the **ring resonator** and the **bow-tie resonator** illustrated in Fig. 11.1-5. The resonance frequencies of the modes may be obtained by equating the round-trip phase shift to 2π. Each of the set of modes traveling in the clockwise direction has a corresponding mode of the same resonance frequency traveling in the counterclockwise direction, and the matching modes are said to be degenerate.

EXERCISE 11.1-1

Resonance Frequencies of a Traveling-Wave Resonator. Derive expressions for the resonance frequencies ν_q and their frequency spacing ν_F for the three-mirror ring and the four-mirror bow-tie resonator shown in Fig. 11.1-5. Assume that each mirror reflection introduces a phase shift of π.

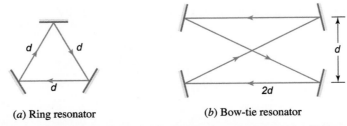

(a) Ring resonator   (b) Bow-tie resonator

Figure 11.1-5 Traveling-wave resonators. (a) Three-mirror ring resonator. (b) Four-mirror bow-tie resonator.

Density of Modes

The number of modes per unit frequency is the inverse of the frequency spacing between modes, i.e., $1/\nu_F = 2d/c$ in each of the two orthogonal polarizations. The density of modes $M(\nu)$, which is the number of modes per unit frequency per unit length of the resonator, is therefore

$$M(\nu) = \frac{4}{c}. \qquad (11.1\text{-}10)$$

Density of Modes
(1D Resonator)

The number of modes in a resonator of length d, in the frequency interval $\Delta\nu$, is thus $(4/c)d\Delta\nu$. This represents the number of degrees of freedom for the optical waves existing in the resonator, i.e., the number of independent ways in which these waves may be arranged.

Losses and Resonance Spectral Width

The strict condition on the frequencies of optical waves that are permitted to exist inside a resonator is relaxed when the resonator has losses. Consider again Fig. 11.1-4(a) and follow the initial wave inside the resonator, U_0, in its excursions between the two mirrors. As discussed above, the result is the infinite sum of phasors shown in Fig. 11.1-4(c) and the phase difference imparted by propagation through a single round trip is

$$\varphi = 2kd = 4\pi\nu d/c. \qquad (11.1\text{-}11)$$

Reflection at the two mirrors can impart an additional phase shift, usually 2π.

However, in the presence of loss the phasors are *not* all of equal magnitude. Two successive phasors are related by a complex round-trip amplitude attenuation factor $h = |\mathrm{r}|e^{-j\varphi}$ resulting from losses associated with the *two* mirror reflections and the absorption in the medium (the corresponding intensity attenuation factor for a round trip is $|\mathrm{r}|^2$ with $|\mathrm{r}| < 1$). Thus, $U_1 = hU_0$ and, in fact, U_2 is related to U_1 by this same complex factor h, as are all consecutive phasor pairs. The net result is the superposition of an infinite number of waves, each distinguished from the previous one by a constant phase shift and an amplitude that is geometrically reduced. It is readily seen that $U = U_0 + U_1 + U_2 + \cdots = U_0 + hU_0 + h^2 U_0 + \cdots = U_0(1 + h + h^2 + \cdots) = U_0/(1-h)$. The net result, $U = U_0/(1-h)$, is easily understood in terms of the simple feedback configuration pictured in Fig. 11.1-4(b).

The intensity of the light in the resonator is therefore given by

$$I = |U|^2 = \frac{|U_0|^2}{|1 - |r|e^{-j\varphi}|^2} = \frac{I_0}{1 + |r|^2 - 2|r|\cos\varphi}, \quad (11.1\text{-}12)$$

which can be written as

$$I = \frac{I_{\max}}{1 + (2\mathcal{F}/\pi)^2 \sin^2(\varphi/2)}, \quad I_{\max} = \frac{I_0}{(1 - |r|)^2}. \quad (11.1\text{-}13)$$

Here, $I_0 = |U_0|^2$ is the intensity of the initial wave, and the **finesse** of the resonator is

$$\boxed{\mathcal{F} = \frac{\pi\sqrt{|r|}}{1 - |r|}.} \quad (11.1\text{-}14)$$
Finesse

Again, $|r|$ is the magnitude of the *round-trip* attenuation factor.

The treatment offered above is nearly identical to that provided earlier in Sec. 2.5B, where the complex round-trip amplitude attenuation factor was chosen to be $h = |h|e^{+j\varphi}$. In the current context we instead select this factor to be $h = |r|e^{-jk2d} = |r|e^{-j\varphi}$ by dint of the fact that successive phasors arise from the *delay* of the wave as it bounces between the mirrors. This distinction is superficial, however, and has no bearing on the results.

Indeed, (11.1-13) is identical to (2.5-18), which is plotted in Fig. 2.5-10(b). The intensity $I(\varphi)$ is a periodic function of φ with period 2π. For large $\mathcal{F}$, $I(\varphi)$ has sharp peaks centered about the values $\varphi = q2\pi$, which correspond to the alignment of all phasors. The peaks have a full-width at half-maximum (FWHM) described by $\Delta\varphi \approx 2\pi/\mathcal{F}$, in accordance with (2.5-21).

The internal resonator intensity $I(\varphi)$ in (11.1-13) can alternately be expressed as a function of the optical frequency of an internal monochromatic wave, $I(\nu)$, by virtue of (11.1-11), which shows that $\varphi = 4\pi\nu d/c$. This function then takes the form

$$\boxed{I = \frac{I_{\max}}{1 + (2\mathcal{F}/\pi)^2 \sin^2(\pi\nu/\nu_F)}, \quad I_{\max} = \frac{I_0}{(1 - |r|)^2},} \quad (11.1\text{-}15)$$

with $\nu_F = c/2d$. This result is displayed in Fig. 11.1-6 and indeed it mirrors that depicted in Figs. 2.5-10 and 7.1-5. The maximum internal intensity $I = I_{\max}$ is attained when the second term in the denominator is zero, i.e., at the resonance frequencies

$$\nu = \nu_q = q\nu_F, \quad q = 1, 2, \ldots. \quad (11.1\text{-}16)$$

The minimum intensity, attained at the midpoints between the resonances, is

$$I_{\min} = \frac{I_{\max}}{1 + (2\mathcal{F}/\pi)^2}. \quad (11.1\text{-}17)$$

When the finesse is large ($\mathcal{F} \gg 1$), it is clear that the spectral response of the resonator is sharply peaked about the resonance frequencies and $I_{\min}/I_{\max}$ is small. In

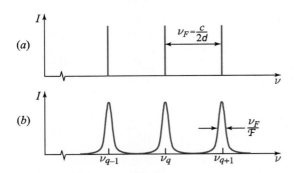

Figure 11.1-6 (a) In the steady state, a lossless resonator ($\mathcal{F} = \infty$) sustains light waves only at the precise resonance frequencies ν_q. (b) A lossy resonator best sustains waves in the immediate vicinity of the resonance frequencies, but it can sustain waves at other frequencies as well.

that case, the FWHM of the resonance peaks is $\delta\nu \approx \nu_F/\mathcal{F}$ since $\delta\nu = (c/4\pi d)\Delta\varphi$ and $\Delta\varphi \approx 2\pi/\mathcal{F}$ in accordance with (2.5-21). This simple result provides the rationale for the definition of the finesse given in (11.1-14).

In short, the spectral response of the Fabry–Perot optical resonator is characterized by two parameters:

- The frequency spacing ν_F between adjacent resonator modes:

$$\nu_F = \frac{c}{2d}.$$
(11.1-18)
Frequency Spacing

- The spectral width $\delta\nu$ of the individual resonator modes:

$$\delta\nu \approx \frac{\nu_F}{\mathcal{F}}.$$
(11.1-19)
Spectral Width

Equation (11.1-19) is valid in the usual case when $\mathcal{F} \gg 1$. The spectral width $\delta\nu$ is inversely proportional to the finesse $\mathcal{F}$. As the loss increases, $\mathcal{F}$ decreases and $\delta\nu$ therefore increases.

Sources of Resonator Loss

The two principal sources of loss in optical resonators are:

- Losses arising from imperfect reflection at the mirrors. There are two underlying sources of reduced reflection: (1) a partially transmitting mirror is often deliberately used in a resonator to permit laser light generated in the resonator to escape through it; and (2) the finite size of the mirrors causes a fraction of the light to leak around them and thereby to be lost. This latter effect also modifies the spatial distribution of the reflected wave by truncating it to the size of the mirror. The reflected light produces a diffraction pattern at the opposite mirror that is again truncated. Such diffraction loss may be regarded as an effective reduction of the mirror reflectance. Further details regarding diffraction loss are provided in Sec. 11.2E.

- Losses attributable to absorption and scattering that occurs in the medium between the mirrors. The round-trip power attenuation factor associated with these effects is $\exp(-2\alpha_s d)$, where α_s is the loss coefficient of the medium associated with absorption and scattering.

For mirrors of reflectances $\mathcal{R}_1 = |r_1|^2$ and $\mathcal{R}_2 = |r_2|^2$, the wave intensity decreases by the factor $\mathcal{R}_1 \mathcal{R}_2$ as a result of the two reflections associated with a single round trip. These are referred to as "lumped losses" since they occur only at the discrete locations where the mirrors are located. Accounting also for the "distributed losses" that take place within the intervening medium yields a round-trip intensity attenuation factor

$$|r|^2 = \mathcal{R}_1 \mathcal{R}_2 \exp(-2\alpha_s d), \qquad (11.1\text{-}20)$$

which is usually written in the form

$$|r|^2 = \exp(-2\alpha_r d), \qquad (11.1\text{-}21)$$

where α_r is an effective overall distributed-loss coefficient. Equating (11.1-20) and (11.1-21), and taking the natural logarithm of both sides, allows α_r to be written in terms of the distributed and lumped loss parameters, α_s and $\mathcal{R}_1 \mathcal{R}_2$, respectively:

$$\boxed{\alpha_r = \alpha_s + \frac{1}{2d} \ln \frac{1}{\mathcal{R}_1 \mathcal{R}_2}.} \qquad (11.1\text{-}22)$$
Loss Coefficient

This can also be written as

$$\alpha_r = \alpha_s + \alpha_{m1} + \alpha_{m2}, \qquad (11.1\text{-}23)$$

where the quantities

$$\alpha_{m1} = \frac{1}{2d} \ln \frac{1}{\mathcal{R}_1}, \qquad \alpha_{m2} = \frac{1}{2d} \ln \frac{1}{\mathcal{R}_2} \qquad (11.1\text{-}24)$$

represent the effective distributed-loss coefficients associated with mirrors 1 and 2, respectively.

These loss coefficients can be cast in a simpler form for mirrors of high reflectance. If $\mathcal{R}_1 \approx 1$, then $\ln(1/\mathcal{R}_1) = -\ln(\mathcal{R}_1) = -\ln[1 - (1 - \mathcal{R}_1)] \approx 1 - \mathcal{R}_1$, where we have used the Taylor-series approximation $\ln(1 - \Delta) \approx -\Delta$, which is valid for $|\Delta| \ll 1$. This allows us to write

$$\alpha_{m1} \approx \frac{1 - \mathcal{R}_1}{2d}. \qquad (11.1\text{-}25)$$

Similarly, if $\mathcal{R}_2 \approx 1$, we have $\alpha_{m2} \approx (1 - \mathcal{R}_2)/2d$. If, furthermore, $\mathcal{R}_1 = \mathcal{R}_2 = \mathcal{R} \approx 1$, then

$$\alpha_r \approx \alpha_s + \frac{1 - \mathcal{R}}{d}. \qquad (11.1\text{-}26)$$

The finesse $\mathcal{F}$ can be expressed as a function of the effective loss coefficient α_r by substituting (11.1-21) in (11.1-14). The result is

$$\mathcal{F} = \frac{\pi \exp(-\alpha_r d/2)}{1 - \exp(-\alpha_r d)}, \qquad (11.1\text{-}27)$$

which is plotted in Fig. 11.1-7. It is clear that the finesse decreases as the loss increases. If the loss factor $\alpha_r d \ll 1$, then $\exp(-\alpha_r d) \approx 1 - \alpha_r d$, whereupon

$$\boxed{\mathcal{F} \approx \frac{\pi}{\alpha_r d}.} \qquad (11.1\text{-}28)$$

Finesse and Loss Factor

The finesse is thus inversely proportional to the loss factor $\alpha_r d$ in this limit.

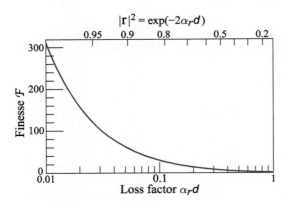

Figure 11.1-7 Finesse of an optical resonator versus the loss factor $\alpha_r d$, where α_r is the effective overall distributed-loss coefficient. The round-trip intensity attenuation factor $|\mathbf{r}|^2 = \exp(-2\alpha_r d)$ is shown on the upper abscissa.

EXERCISE 11.1-2

Resonator Modes and Spectral Width. Determine the frequency spacing, and spectral width, of the modes of a Fabry–Perot resonator whose mirrors have power reflectances of 0.98 and 0.99 and are separated by a distance $d = 100$ cm. Assume that the medium has refractive index $n = 1$ and negligible losses. Is the approximation used to derive (11.1-28) appropriate in this case?

Photon Lifetime

The relationship between the resonance linewidth and resonator loss may be viewed as a manifestation of the time–frequency uncertainty relation, as we now demonstrate. Substituting (11.1-18) and (11.1-28) in (11.1-19), we obtain

$$\delta\nu \approx \frac{c/2d}{\pi/\alpha_r d} = \frac{c\alpha_r}{2\pi}. \qquad (11.1\text{-}29)$$

Because α_r is the loss per unit length, $c\alpha_r$ represents the loss per unit time. Defining the characteristic decay time

$$\tau_p = \frac{1}{c\alpha_r} \qquad (11.1\text{-}30)$$

as the **resonator lifetime** or **photon lifetime**, we obtain

$$\delta\nu \approx \frac{1}{2\pi\tau_p}. \qquad (11.1\text{-}31)$$

The time–frequency uncertainty product is thus $\delta\nu \cdot \tau_p = 1/2\pi$. Resonance-line broadening may therefore be considered to be a consequence of optical-energy decay arising from resonator losses. An electric field that decays as $\exp(-t/2\tau_p)$, corresponding to an energy that decays as $\exp(-t/\tau_p)$, has a Fourier transform that is proportional to $1/(1 + j4\pi\nu\tau_p)$, which has a (FWHM) spectral width $\delta\nu = 1/2\pi\tau_p$ (see Sec. 14.3D).

Quality Factor Q

The **quality factor** Q is often used to characterize electrical resonance circuits and microwave resonators. This parameter is defined as

$$Q = 2\pi \frac{\text{stored energy}}{\text{energy loss per cycle}}. \qquad (11.1\text{-}32)$$

Large values of Q are associated with low-loss resonators. A series RLC circuit has resonance frequency $\nu_0 \approx 1/2\pi\sqrt{LC}$ and quality factor $Q = 2\pi\nu_0 L/R$, where R, L, and C are the resistance, inductance, and capacitance of the resonance circuit, respectively.

The quality factor of an optical resonator is determined by observing that the stored energy E is lost at the rate $c\alpha_r E$ (per unit time), which is equivalent to the rate $c\alpha_r E/\nu_0$ (per cycle of the optical field), so that

$$Q = \frac{2\pi\nu_0}{c\alpha_r}. \qquad (11.1\text{-}33)$$

Since $\delta\nu \approx c\alpha_r/2\pi$ in accordance with (11.1-29), we have

$$Q \approx \frac{\nu_0}{\delta\nu}. \qquad (11.1\text{-}34)$$

By virtue of (11.1-33), the quality factor is related to the resonator lifetime (photon lifetime) $\tau_p = 1/c\alpha_r$ via

$$Q = 2\pi\nu_0\tau_p. \qquad (11.1\text{-}35)$$

The quality factor Q is thus understood to be the storage time of the resonator in units of the optical period $T = 1/\nu_0$.

Finally, combining (11.1-19) and (11.1-34) leads to a relationship between Q and the finesse $\mathcal{F}$ of the resonator:

$$Q \approx \frac{\nu_0}{\nu_F}\mathcal{F}. \qquad (11.1\text{-}36)$$

Since optical resonator frequencies ν_0 are typically much greater than the mode spacing ν_F, we have $Q \gg \mathcal{F}$. Moreover, the quality factor of an optical resonator is typically far greater than that of a resonator at microwave frequencies.

Summary

- Two parameters are convenient for characterizing the losses in an optical resonator: the loss coefficient α_r (cm^{-1}) and the photon lifetime $\tau_p = 1/c\alpha_r$ (s).
- Two dimensionless parameters characterize the quality of an optical resonator of length d operated at frequency ν_0: the finesse $\mathcal{F} \approx \pi/\alpha_r d$ and the quality factor $Q = 2\pi\nu_0\tau_p$.
- Two frequencies describe the spectral characteristics of an optical resonator: the frequency spacing between the modes $\nu_F = c/2d$, known as the free spectral range, and the spectral width $\delta\nu \approx \nu_F/\mathcal{F}$.

B. Off-Axis Resonator Modes

An optical resonator with perfectly parallel planar mirrors of infinite dimensions can also support oblique, or off-axis, modes. A plane wave traveling at an angle θ with respect to the axis of the resonator (the z direction) bounces back and forth between lossless mirrors [see Fig. 11.1-8(a)] as a guided wave traveling in the transverse direction (the x direction). Such guided waves were described in Sec. 9.1.

The boundary conditions at the mirrors dictate that the axial component of the propagation constant, $k_z = k\cos\theta$, is an integer multiple of π/d. However, no such condition is imposed on the transverse component k_x since the resonator is open in the x direction. Since $k = 2\pi\nu/c$, the condition $k\cos\theta = q\pi/d$, where q is an integer, can be written in the form

$$\nu = q\nu_F \sec\theta, \qquad q = 1, 2, \ldots, \qquad (11.1\text{-}37)$$

where $\nu_F = c/2d$. This relation, which is plotted in Fig 11.1-8(b), is equivalent to the self-consistency condition for guided modes in planar-mirror waveguides (see Sec. 9.1). It is also identical to the condition (7.1-37) for the peak transmittance of an oblique wave through a Fabry–Perot etalon. As illustrated in Fig 11.1-8(c), at a given frequency ν, there are modes at a discrete set of angles θ_q that satisfy the condition $\cos\theta_q = q\nu_F/\nu$. These are the complements of the bounce angles of the guided modes of a waveguide. Also, at any fixed angle θ, the modal frequencies are $\nu_q = q\nu_F \sec\theta$, as illustrated in Fig 11.1-8(d). The larger the inclination angle, the greater the spacing between the modal frequencies.

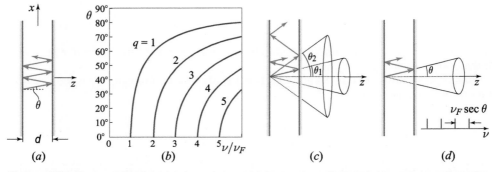

Figure 11.1-8 (a) Off-axis mode in a planar-mirror resonator. (b) Relation between mode angles and resonance frequencies. (c) Off-axis modes at a fixed frequency $\nu > \nu_F$. (d) Resonance frequencies of an off-axis mode at a prescribed angle θ.

11.2 SPHERICAL-MIRROR RESONATORS

The planar-mirror resonator configuration discussed in the preceding section is highly sensitive to misalignment. If the mirrors are not perfectly parallel, or the rays are not perfectly normal to the mirror surfaces, they undergo a sequence of lateral displacements that eventually causes them to wander out of the resonator [see Fig. 11.1-1(b)]. Spherical-mirror resonators, in contrast, provide a more stable configuration for the confinement of light that renders them less sensitive to misalignment under appropriate geometrical conditions.

A spherical-mirror resonator is constructed from two spherical mirrors of radii R_1 and R_2, separated by a distance d (Fig. 11.2-1). A line connecting the centers of the mirrors defines the optical axis (z axis), about which the system exhibits circular symmetry. Each of the mirrors can be concave ($R < 0$) or convex ($R > 0$). The planar-mirror resonator is a special case for which $R_1 = R_2 = \infty$. We first make use of the results set forth in Sec. 1.4D and examine the conditions required for ray confinement. Then, using the results derived in Chapter 3, we determine the resonator modes and resonance frequencies. Finally, we briefly discuss the implications of finite mirror size.

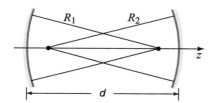

Figure 11.2-1 Geometry of a spherical-mirror resonator. In this illustration both mirrors are concave (their radii of curvature are negative).

A. Ray Confinement

We begin with ray optics to determine the conditions of confinement for light rays in a spherical-mirror resonator. We consider only meridional rays (rays lying in a plane that passes through the optical axis) and limit our consideration to paraxial rays (rays that make small angles with the optical axis). The matrix-optics methods introduced in Sec. 1.4, which are valid only for meridional and paraxial rays in a circularly symmetric system, are thus suitable for studying the trajectories of these rays as they travel inside the resonator.

A resonator is a periodic optical system, since a ray travels through the same system after a round trip of two reflections. We may therefore make use of the analysis of periodic optical systems presented in Sec. 1.4D. Let y_m and θ_m be the position and inclination of an optical ray after m round trips, as illustrated in Fig. 11.2-2. Given y_m and θ_m, we determine y_{m+1} and θ_{m+1} by tracing the ray through the system.

For paraxial rays, where all angles are small, the relation between (y_{m+1}, θ_{m+1}) and (y_m, θ_m) is linear and can be written in matrix form as [see (1.4-3)]

$$\begin{bmatrix} y_{m+1} \\ \theta_{m+1} \end{bmatrix} = \begin{bmatrix} A & B \\ C & D \end{bmatrix} \begin{bmatrix} y_m \\ \theta_m \end{bmatrix}. \tag{11.2-1}$$

Beginning at the left of Fig. 11.2-2 with y_0 and θ_0, the round-trip ray-transfer matrix for the ray pattern shown is

$$\begin{bmatrix} A & B \\ C & D \end{bmatrix} = \begin{bmatrix} 1 & 0 \\ \frac{2}{R_1} & 1 \end{bmatrix} \begin{bmatrix} 1 & d \\ 0 & 1 \end{bmatrix} \begin{bmatrix} 1 & 0 \\ \frac{2}{R_2} & 1 \end{bmatrix} \begin{bmatrix} 1 & d \\ 0 & 1 \end{bmatrix}. \tag{11.2-2}$$

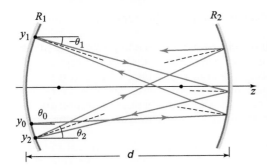

Figure 11.2-2 The position and inclination of a ray after m round trips are represented by y_m and θ_m, respectively, where $m = 0, 1, 2, \ldots$. In this diagram, $\theta_1 < 0$ since the ray is directed downward. Angles are exaggerated for the purposes of illustration; all rays are paraxial so that $\sin\theta \approx \tan\theta \approx \theta$ and the propagation distance of all rays between the mirrors is $\approx d$.

This cascade of ray-transfer matrices represents, from right to left [see (1.4-4) and (1.4-9)]:

- Propagation a distance d through free space
- Reflection from a mirror of radius R_2
- Propagation a distance d through free space
- Reflection from a mirror of radius R_1

As shown in Sec. 1.4D, the solution of the difference equation (11.2-1) is $y_m = y_{\max} F^m \sin(m\varphi + \varphi_0)$, where $F^2 = AD - BC$, $\varphi = \cos^{-1}(b/F)$, $b = (A + D)/2$, and $y_{\max}$ and φ_0 are constants determined from the initial position and inclination of the ray. For the case at hand $F = 1$, so that

$$y_m = y_{\max}\sin(m\varphi + \varphi_0), \qquad (11.2\text{-}3)$$

$$\varphi = \cos^{-1} b, \qquad b = 2\left(1 + \frac{d}{R_1}\right)\left(1 + \frac{d}{R_2}\right) - 1. \qquad (11.2\text{-}4)$$

The solution (11.2-3) is harmonic, and therefore bounded, provided $\varphi = \cos^{-1} b$ is real. This is ensured if $|b| \leq 1$, i.e., if $-1 \leq b \leq 1$, so that

$$0 \leq \left(1 + \frac{d}{R_1}\right)\left(1 + \frac{d}{R_2}\right) \leq 1. \qquad (11.2\text{-}5)$$

It is convenient to write this **confinement condition** in terms of the quantities $g_1 = 1 + d/R_1$ and $g_2 = 1 + d/R_2$, which are known as the g-**parameters**:

$$\boxed{0 \leq g_1 g_2 \leq 1.} \qquad (11.2\text{-}6)$$
Confinement Condition

The resonator is said to be **stable** when this condition is satisfied. This result also emerges from wave optics, as will be demonstrated subsequently [see (11.2-17)].

When the confinement condition (11.2-6) is not satisfied, φ is imaginary so that y_m in (11.2-3) becomes a hyperbolic sine function of m and increases without bound. The resonator is then said to be **unstable**. At the boundary of the confinement condition (when the inequalities are equalities), the resonator is said to be **conditionally stable**.

A useful graphical representation of the confinement condition (Fig. 11.2-3) identifies each combination (g_1, g_2) of the two g-parameters of a resonator as a point in a g_2 versus g_1 diagram. The left inequality in (11.2-6) is equivalent to $\{g_1 \geq 0$ and $g_2 \geq 0$; or $g_1 \leq 0$ and $g_2 \leq 0\}$ so that all stable points (g_1, g_2) must lie in the first or third quadrants. The right inequality in (11.2-6) signifies that stable points (g_1, g_2) must lie in a region bounded by the hyperbola $g_1 g_2 = 1$. The unshaded area in Fig. 11.2-3 represents the region for which both inequalities are satisfied, indicating that the resonator is stable.

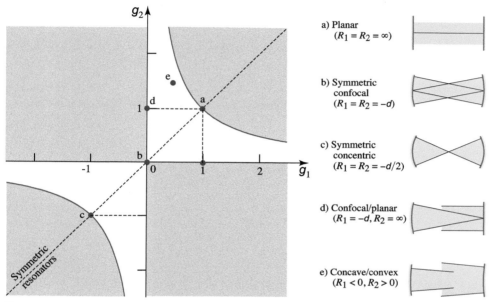

Figure 11.2-3 Resonator stability diagram. A spherical-mirror resonator is stable if the parameters $g_1 = 1 + d/R_1$ and $g_2 = 1 + d/R_2$ lie in the unshaded regions, which are bounded by the lines $g_1 = 0$ and $g_2 = 0$, and the hyperbola $g_2 = 1/g_1$. R is negative for a concave mirror and positive for a convex mirror. Commonly used resonator configurations are indicated by letters and are sketched at the right; shaded areas represent collections of rays perpendicular to the mirrors. All symmetric resonators lie along the line $g_2 = g_1$.

Symmetric resonators, by definition, have identical mirrors ($R_1 = R_2 = R$) so that $g_1 = g_2 = g$. Resonators in this class are thus represented in Fig. 11.2-3 by points lying along the line $g_2 = g_1$. The condition of stability then becomes $g^2 \leq 1$, or $-1 \leq 1 + d/R \leq 1$, which implies

$$0 \leq \frac{d}{(-R)} \leq 2. \qquad (11.2\text{-}7)$$

Confinement Condition
(Symmetric Resonator)

To satisfy (11.2-7) a stable symmetric resonator must use concave mirrors ($R < 0$) whose radii are greater than half the resonator length. Three examples within this class are of special interest: $d/(-R) = 0$, 1, and 2, corresponding to **planar**, **confocal**, and **concentric resonators**, respectively.

In the **symmetric confocal resonator**, $(-R) = d$ so that the center of curvature of each mirror lies on the other. Thus, $b = -1$ and $\varphi = \pi$ so that the ray position in (11.2-3) is prescribed to be $y_m = y_{\max} \sin(m\pi + \varphi_0)$, i.e., $y_m = (-1)^m y_0$. Rays initiated at position y_0, at any inclination, are thus imaged to position $y_1 = -y_0$, and then reimaged again to position $y_2 = y_0$, and so on, repeatedly. Each ray thus retraces itself after two round trips (Fig. 11.2-4). All paraxial rays are therefore confined, whatever their original position and inclination. This is a substantial improvement in comparison with the planar-mirror resonator, for which only rays of zero inclination retrace themselves as schematized in Fig. 11.1-1.

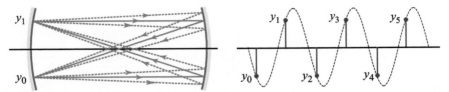

Figure 11.2-4 All paraxial rays in a symmetric confocal resonator retrace themselves after two round trips, whatever their original position and inclination. Angles are exaggerated in this drawing for purposes of illustration.

Summary

The confinement condition for paraxial rays in a spherical-mirror resonator, comprising mirrors of radii R_1 and R_2 separated by a distance d, is $0 \leq g_1 g_2 \leq 1$, where $g_1 = 1 + d/R_1$ and $g_2 = 1 + d/R_2$. The confinement condition for symmetric resonators is $0 \leq d/(-R) \leq 2$; this condition governs planar, symmetric confocal, and symmetric concentric mirror configurations.

EXERCISE 11.2-1

Maximum Resonator Length for Confined Rays. A resonator is constructed using concave mirrors of radii 50 cm and 100 cm. Determine the maximum resonator length for which rays satisfy the confinement condition.

B. Gaussian Modes

Although the ray-optics approach considered in the preceding section is useful for determining the geometrical conditions under which rays are confined, it cannot provide information about the resonance frequencies and spatial intensity distributions of the resonator modes. For those quantities we must appeal to wave optics. We now proceed to show that Gaussian beams are solutions of the paraxial Helmholtz equation for the boundary conditions imposed by a pair of spherical mirrors in a resonator configuration. More generally, we demonstrate that Hermite–Gaussian beams are modes of the spherical-mirror resonator. In the course of our analysis, we obtain expressions for the resonance frequencies and spatial intensity distributions of the resonator modes.

Gaussian Beams

As discussed in Chapter 3, the Gaussian beam is a circularly symmetric wave whose energy is confined about its axis (the z axis) and whose wavefront normals are paraxial rays (Fig. 11.2-5). In accordance with (3.1-12), at an axial distance z from the beam waist, the beam intensity I varies in the transverse x–y plane as the Gaussian distribution $I = I_0[W_0/W(z)]^2 \exp[-2(x^2+y^2)/W^2(z)]$. Its width is given by (3.1-8):

$$W(z) = W_0 \sqrt{1 + \left(\frac{z}{z_0}\right)^2}, \qquad (11.2\text{-}8)$$

where z_0 is the distance, known as the Rayleigh range, at which the beam wavefronts are most curved. The beam width (radius) $W(z)$ increases in both directions from its minimum value W_0 at the beam waist ($z = 0$). The radius of curvature of the wavefronts, given by (3.1-9),

$$R(z) = z \left[1 + \left(\frac{z_0}{z}\right)^2\right] \qquad (11.2\text{-}9)$$

decreases from ∞ at $z = 0$, to a minimum value at $z = z_0$, and thereafter grows linearly with z for large z. For $z > 0$, the wave diverges and $R(z) > 0$; for $z < 0$, the wave converges and $R(z) < 0$. The Rayleigh range z_0 is related to the beam waist radius W_0 by (3.1-11):

$$z_0 = \frac{\pi W_0^2}{\lambda}. \qquad (11.2\text{-}10)$$

The depth of focus is $2z_0$, i.e., twice the Rayleigh range.

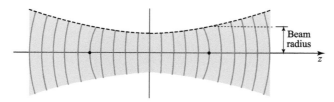

Figure 11.2-5 Gaussian beam wavefronts (solid curves) and beam width (dashed curve).

The Gaussian Beam is a Mode of the Spherical-Mirror Resonator

A Gaussian beam reflected from a spherical mirror will retrace the incident beam if the radius of curvature of its wavefront is the same as that of the mirror radius (see Sec. 3.2C). Hence, if the radii of curvature of the wavefronts of a Gaussian beam, at planes separated by a distance d, match the radii of two mirrors separated by the same distance d, a beam incident on the first mirror will reflect and retrace itself to the second mirror, where it once again will reflect and retrace itself back to the first mirror, and so on. The beam can then exist self-consistently within that spherical-mirror resonator, satisfying the Helmholtz equation and the boundary conditions imposed by the mirrors. Provided that the phase also retraces itself, as discussed in Sec. 11.2C, the Gaussian beam is then said to be a mode of the spherical-mirror resonator.

We now proceed to determine the Gaussian beam that matches a spherical-mirror resonator, whose mirrors have radii of curvature R_1 and R_2 and are separated by the

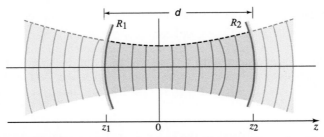

Figure 11.2-6 Fitting a Gaussian beam to two mirrors separated by a distance d. Their radii of curvature are R_1 and R_2. Both mirrors are taken to be concave so that R_1 and R_2 are negative, as is z_1.

distance d. The task is illustrated in Fig. 11.2-6 for the special case when both mirrors are concave ($R_1 < 0$ and $R_2 < 0$).

The z axis is defined by the centers of the mirrors. The center of the beam, which is yet to be determined, is assumed to be located at the origin $z = 0$; mirrors R_1 and R_2 are located at positions z_1 and

$$z_2 = z_1 + d, \qquad (11.2\text{-}11)$$

respectively. A negative value for z_1 indicates that the center of the beam lies to the right of mirror 1; a positive value indicates that it lies to the left. The values of z_1 and z_2 are determined by matching the radius of curvature of the beam, $R(z) = z + z_0^2/z$, to the radii R_1 at z_1 and R_2 at z_2. Careful attention must be paid to the signs. If both mirrors are concave, they have negative radii. But the beam radius of curvature was defined to be positive for $z > 0$ (at mirror 2) and negative for $z < 0$ (at mirror 1). We therefore equate $R_1 = R(z_1)$, but $-R_2 = R(z_2)$, to obtain

$$R_1 = z_1 + z_0^2/z_1 \qquad (11.2\text{-}12)$$

$$-R_2 = z_2 + z_0^2/z_2. \qquad (11.2\text{-}13)$$

Solving (11.2-11), (11.2-12), and (11.2-13) for z_1, z_2, and z_0 leads to

$$z_1 = \frac{-d(R_2 + d)}{R_2 + R_1 + 2d}, \qquad z_2 = z_1 + d, \qquad (11.2\text{-}14)$$

$$z_0^2 = \frac{-d(R_1 + d)(R_2 + d)(R_2 + R_1 + d)}{(R_2 + R_1 + 2d)^2}, \qquad (11.2\text{-}15)$$

which accord with (3.1-27) and (3.1-28) (if R_2 is replaced with $-R_2$).

Having determined the location of the beam center and the depth of focus $2z_0$, everything about the beam is known (see Sec. 3.1B). The waist radius is $W_0 = \sqrt{\lambda z_0/\pi}$, and the beam radii at the mirrors are

$$W_i = W_0 \sqrt{1 + \left(\frac{z_i}{z_0}\right)^2}, \qquad i = 1, 2. \qquad (11.2\text{-}16)$$

In order that the solution (11.2-14)–(11.2-15) indeed represents a Gaussian beam, z_0 must be real. An imaginary value of z_0 would signify that the Gaussian beam is a paraboloidal wave, which is an unconfined solution of the paraxial Helmholtz equation (see Sec. 3.1A). Using (11.2-15), it is not difficult to show that the condition $z_0^2 > 0$ is equivalent to

$$0 \leq \left(1 + \frac{d}{R_1}\right)\left(1 + \frac{d}{R_2}\right) \leq 1. \qquad (11.2\text{-}17)$$

This is precisely the confinement condition derived from ray optics as set forth in (11.2-5).

EXERCISE 11.2-2

A Plano-Concave Resonator. If mirror 1 is planar ($R_1 = \infty$), determine the confinement condition and the depth of focus, as well as the beam width at the waist and at each of the mirrors, as a function of $d/|R_2|$.

Gaussian Mode of a Symmetric Spherical-Mirror Resonator

The results provided in (11.2-11)–(11.2-15) simplify considerably for symmetric resonators with concave mirrors. Substituting $R_1 = R_2 = -|R|$ into (11.2-14) provides $z_1 = -d/2$ and $z_2 = d/2$. The beam center thus lies at the center of the resonator, and

$$z_0 = \frac{d}{2}\sqrt{2\frac{|R|}{d} - 1}, \qquad (11.2\text{-}18)$$

$$W_0^2 = \frac{\lambda d}{2\pi}\sqrt{2\frac{|R|}{d} - 1}, \qquad (11.2\text{-}19)$$

$$W_1^2 = W_2^2 = \frac{\lambda d/\pi}{\sqrt{(d/|R|)\,[2 - (d/|R|)]}}. \qquad (11.2\text{-}20)$$

The confinement condition (11.2-17) becomes

$$0 \leq \frac{d}{|R|} \leq 2. \qquad (11.2\text{-}21)$$

Given a resonator of fixed mirror separation d, we now examine the effect of increasing mirror curvature on the beam radius at the waist W_0, and at the mirrors $W_1 = W_2$. (Increasing curvature corresponds to increasing $d/|R|$ since the *radius of curvature* diminishes as the *curvature* increases.) The results are illustrated in Fig. 11.2-7. For a planar-mirror resonator, $d/|R| = 0$, so that W_0 and W_1 are infinite, corresponding to a plane wave rather than a Gaussian beam. As $d/|R|$ increases, W_0 decreases until it vanishes for the concentric resonator ($d/|R| = 2$); at this point $W_1 = W_2 = \infty$ and $W_0 = 0$. In this limit, the resonator supports a spherical wave instead of a Gaussian beam.

The width of the beam at the mirrors attains its minimum value, $W_1 = W_2 = \sqrt{\lambda d/\pi}$, when $d/|R| = 1$, i.e., for the symmetric confocal resonator. In this case

$$z_0 = d/2, \qquad (11.2\text{-}22)$$

$$W_0 = \sqrt{\lambda d/2\pi}, \qquad (11.2\text{-}23)$$

$$W_1 = W_2 = \sqrt{2}\,W_0. \qquad (11.2\text{-}24)$$

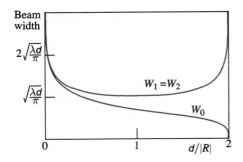

Figure 11.2-7 The beam width at the waist, W_0, and at the mirrors, $W_1 = W_2$, for a symmetric spherical-mirror resonator with concave mirrors, as a function of the ratio $d/|R|$. The planar-mirror resonator corresponds to $d/|R| = 0$. Symmetric confocal and concentric resonators correspond to $d/|R| = 1$ and $d/|R| = 2$, respectively.

The depth of focus $2z_0$ is then equal to the length of the resonator d, as shown in Fig. 11.2-8. This explains why the parameter $2z_0$ is sometimes called the confocal parameter. A long resonator has a long depth of focus. The waist radius is proportional to the square root of the mirror spacing. A Gaussian beam at $\lambda_0 = 633$ nm (a He–Ne laser wavelength) in a resonator with $d = 100$ cm, for example, has a waist radius $W_0 = \sqrt{\lambda d/2\pi} = 0.32$ mm, whereas a 25-cm-long resonator supports a Gaussian beam with a waist radius that is half as big at the same wavelength: 0.16 mm. The width of the beam at each of the mirrors is greater than it is at the waist by a factor of $\sqrt{2}$.

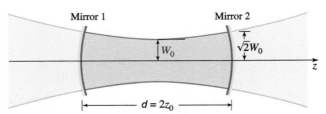

Figure 11.2-8 Gaussian beam in a symmetric confocal resonator with concave mirrors. The depth of focus $2z_0$ equals the length of the resonator d. The beam width at the mirrors is a factor of $\sqrt{2}$ greater than that at the waist.

C. Resonance Frequencies

As indicated in Sec. 11.2B, a Gaussian beam is a mode of the spherical-mirror resonator provided that the wavefront normals reflect back onto themselves, always retracing the same path, and that the phase retraces itself as well.

The phase of a Gaussian beam, in accordance with (3.1-23), is

$$\varphi(\rho, z) = kz - \zeta(z) + \frac{k\rho^2}{2R(z)}, \tag{11.2-25}$$

where $\zeta(z) = \tan^{-1}(z/z_0)$ and $\rho^2 = x^2 + y^2$. At points on the optical axis ($\rho = 0$), $\varphi(0, z) = kz - \zeta(z)$, so that the phase retardation relative to a plane wave is $\zeta(z)$. At the locations of the mirrors, z_1 and z_2, we therefore have

$$\varphi(0, z_1) = kz_1 - \zeta(z_1), \tag{11.2-26}$$

$$\varphi(0, z_2) = kz_2 - \zeta(z_2). \tag{11.2-27}$$

Because the mirror surface coincides with the wavefronts, all points on each mirror share the same phase. As the beam propagates from mirror 1 to mirror 2, its phase changes by

$$\varphi(0, z_2) - \varphi(0, z_1) = k(z_2 - z_1) - [\zeta(z_2) - \zeta(z_1)]$$
$$= kd - \Delta\zeta, \qquad (11.2\text{-}28)$$

where

$$\Delta\zeta = \zeta(z_2) - \zeta(z_1). \qquad (11.2\text{-}29)$$

As the traveling wave completes a round trip between the two mirrors, therefore, its phase changes by $2kd - 2\Delta\zeta$.

In order that the beam truly retrace itself, the round-trip phase change must be zero or a multiple of $\pm 2\pi$, i.e., $2kd - 2\Delta\zeta = 2\pi q$, $q = 0, \pm 1, \pm 2, \ldots$. Using the substitutions $k = 2\pi\nu/c$ and $\nu_F = c/2d$, the frequencies ν_q that satisfy this condition are

$$\boxed{\nu_q = q\nu_F + \frac{\Delta\zeta}{\pi}\nu_F.} \qquad (11.2\text{-}30)$$
Resonance Frequencies
Gaussian Modes

The frequency spacing of adjacent modes is therefore $\nu_F = c/2d$, which is identical to the result obtained in Sec. 11.1A for the planar-mirror resonator. For spherical-mirror resonators, this frequency spacing is evidently independent of the curvatures of the mirrors. The second term in (11.2-30), which does depend on the mirror curvatures, simply represents a displacement of all resonance frequencies.

EXERCISE 11.2-3

Resonance Frequencies of a Confocal Resonator. A symmetric confocal resonator has a length $d = 30$ cm, and the medium has refractive index $n = 1$. Determine the frequency spacing ν_F and the displacement frequency $(\Delta\zeta/\pi)\nu_F$. Determine all resonance frequencies that lie within the band $5 \times 10^{14} \pm 2 \times 10^9$ Hz.

D. Hermite–Gaussian Modes

In Sec. 3.3 it was shown that the Gaussian beam is not the only beam-like solution of the paraxial Helmholtz equation. The family of Hermite–Gaussian beams also provides solutions. Although a Hermite–Gaussian beam of order (l, m) has an amplitude distribution that differs from that of the Gaussian beam, their wavefronts are identical. As a result, the design of a resonator that "matches" a given beam (or the design of a beam that "fits" a given resonator) is the same as for the Gaussian beam, whatever the values of (l, m). It follows that all members of the family of Hermite–Gaussian beams represent modes of the spherical-mirror resonator.

The resonance frequencies of the (l, m) mode do, however, depend on the indices (l, m). This is because of the dependence of the Gouy phase shift on l and m. As is evident from (3.3-10), the phase of the (l, m) mode on the beam axis is

$$\varphi(0, z) = kz - (l + m + 1)\zeta(z). \qquad (11.2\text{-}31)$$

Again, the phase shift encountered by a traveling wave undergoing a single round trip through a resonator of length d must be set equal to zero or an integer multiple of $\pm 2\pi$ in order that the beam retrace itself. Thus,

$$2kd - 2(l+m+1)\Delta\zeta = 2\pi q, \qquad q = 0, \pm 1, \pm 2, \ldots, \qquad (11.2\text{-}32)$$

where, as previously, $\Delta\zeta = \zeta(z_2) - \zeta(z_1)$ and z_1, z_2 represent the positions of the two mirrors. With $k = 2\pi\nu/c$ and $\nu_F = c/2d$, this yields the resonance frequencies

$$\boxed{\nu_{l,m,q} = q\nu_F + (l+m+1)\frac{\Delta\zeta}{\pi}\nu_F.} \qquad (11.2\text{-}33)$$

Resonance Frequencies
Hermite–Gaussian Modes

Modes of different q, but the same (l, m), have identical intensity distributions [see (3.3-12)]. They are known as **longitudinal** or **axial modes**. The indices (l, m) label different spatial dependencies on the transverse coordinates x, y; these therefore represent different **transverse modes**, as illustrated in Fig. 3.3-2.

Equation (11.2-33) dictates that the resonance frequencies of the Hermite–Gaussian modes satisfy the following properties:

- Longitudinal modes corresponding to a given transverse mode have resonance frequencies spaced by $\nu_F = c/2d$ since $\nu_{l,m,q+1} - \nu_{l,m,q} = \nu_F$. This result is the same as that obtained for the (0,0) Gaussian mode and for the planar-mirror resonator.
- All transverse modes, for which the sum of the indices $l + m$ is the same, have the same resonance frequencies.
- Two transverse modes (l, m), (l', m') corresponding to the same longitudinal mode q have resonance frequencies spaced by

$$\nu_{l,m,q} - \nu_{l',m',q} = [(l+m) - (l'+m')]\frac{\Delta\zeta}{\pi}\nu_F. \qquad (11.2\text{-}34)$$

This expression determines the frequency shift between the sets of longitudinal modes of indices (l, m) and (l', m').

EXERCISE 11.2-4

Resonance Frequencies of the Symmetric Confocal Resonator. Show that for a symmetric confocal resonator, the longitudinal modes associated with different transverse modes are either the same, or are displaced by $\nu_F/2$, as illustrated in Fig. 11.2-9.

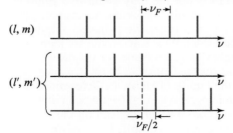

Figure 11.2-9 In a symmetric confocal resonator, the longitudinal modes associated with two transverse modes of indices (l, m) and (l', m') are either aligned or displaced by half a longitudinal mode spacing.

*E. Finite Apertures and Diffraction Loss

Since Gaussian and Hermite–Gaussian beams have infinite transverse extent whereas the resonator mirrors are of finite extent, a portion of the optical power leaks around the mirrors and escapes from the resonator on each pass. An estimate of the power loss may be obtained by calculating the fractional power of the beam that is not intercepted by the mirror. If the beam is Gaussian with width W and the mirror is circular with radius $a = 2W$, for example, a small fraction, $\exp(-2a^2/W^2) \approx 3.35 \times 10^{-4}$, of the beam power escapes on each pass [see (3.1-17)], the remainder being reflected (or transmitted through the mirror). Higher-order transverse modes suffer greater losses since they have greater spatial extent in the transverse plane.

When the mirror radius a is smaller than $2W$, the losses are greater. The resonator modes are then less well described by Gaussian and Hermite–Gaussian beams. The problem of determining the modes of a spherical-mirror resonator with finite-size mirrors is difficult. A wave is a mode if it retraces its amplitude (to within a multiplicative constant) and reproduces its phase (to within an integer multiple of 2π) after completing a round trip through the resonator. One oft-used method for determining the modes involves following a wave repeatedly as it bounces through the resonator, thereby determining its amplitude and phase, much as we determined the position and inclination of a ray bouncing within a resonator. After many round trips this process converges to one of the modes.

The configuration for implementing this approach in a spherical-mirror resonator is schematized in Fig. 11.2-10.

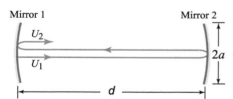

Figure 11.2-10 Propagation of a wave through a spherical-mirror resonator. The complex amplitude $U_1(x, y)$ corresponds to a mode if it reproduces itself after a round trip, i.e., if $U_2(x, y) = \mu U_1(x, y)$ and $\arg\{\mu\} = q2\pi$.

If $U_1(x, y)$ is the complex amplitude of a wave immediately to the right of mirror 1 in Fig. 11.2-10, and if $U_2(x, y)$ is the complex amplitude after one round trip of travel through the resonator, then $U_1(x, y)$ is a mode provided that $U_2(x, y) = \mu U_1(x, y)$ and provided that $\arg\{\mu\}$ is an integer multiple of 2π (i.e., μ is real and positive). After a single round trip, the mode intensity is attenuated by the factor μ^2, and the phase is reproduced. The methods of Fourier optics (Chapter 4) may be used to determine $U_2(x, y)$ from $U_1(x, y)$. These quantities may be regarded as the output and input, respectively, of a linear system (see Appendix B) characterized by an impulse response function $h(x, y; x', y')$, so that

$$U_2(x, y) = \iint_{-\infty}^{\infty} h(x, y; x', y')\, U_1(x', y')\, dx'\, dy'. \tag{11.2-35}$$

If the impulse response function h is known, the modes can be determined by solving the eigenvalue problem described by the integral equation (see Appendix C)

$$\iint_{-\infty}^{\infty} h(x, y; x', y')\, U(x', y')\, dx'\, dy' = \mu U(x, y). \tag{11.2-36}$$

The solutions determine the eigenfunctions $U_{l,m}(x,y)$, and the eigenvalues $\mu_{l,m}$, labeled by the indices (l,m). The eigenfunctions are the modes and the eigenvalues are the round-trip multiplicative factors. The squared magnitude $|\mu_{l,m}|^2$ is the round-trip intensity reduction factor for the (l,m) mode. Clearly, when the mirrors are infinite in size and the paraxial approximation is satisfied, the modes reduce to the family of Hermite–Gaussian beams discussed earlier.

It remains to determine $h(x,y;x',y')$ and to solve the integral equation (11.2-36). A single pass inside the resonator involves traveling a distance d, truncation by the mirror aperture, and reflection by the mirror. The remaining pass, needed to comprise a single round trip, is similar. The impulse response function $h(x,y;x',y')$ can then be determined by applying the theory of Fresnel diffraction (Sec. 4.3B). In general, however, the modes and their associated losses can be determined only by numerically solving the integral equation (11.2-36). An iterative numerical solution begins with an initial guess U_1, from which U_2 is computed and passed through the system one more round trip, and so on until the process converges.

This technique has been used to determine the losses associated with the various modes of a spherical-mirror resonator with circular mirror apertures of radius a. The results are illustrated in Fig. 11.2-11 for a symmetric confocal resonator. The loss is governed by a single parameter, the Fresnel number $N_F = a^2/\lambda d$. This is because the Fresnel number governs Fresnel diffraction between the two mirrors, as discussed in Sec. 4.3B. For the symmetric confocal resonator described by (11.2-23) and (11.2-24), the beam width at the mirrors is $W = \sqrt{\lambda d/\pi}$, so that $\lambda d = \pi W^2$, from which the Fresnel number is readily determined to be $N_F = a^2/\pi W^2$. N_F is therefore proportional to the ratio a^2/W^2; a higher Fresnel number corresponds to a smaller loss. From Fig. 11.2-11 we find that the loss per pass of the lowest-order symmetric-confocal-resonator mode $(l,m) = (0,0)$ is about 0.1% when $N_F \approx 0.94$. This Fresnel number corresponds to $a/W = 1.72$. If the beam were Gaussian with width W, the percentage of power contained outside a circle of radius $a = 1.72\,W$ would be $\exp(-2a^2/W^2) \approx 0.27\%$. This is larger than the 0.1% loss per pass for the actual resonator mode. Higher-order modes suffer from greater losses because of their greater spatial extent.

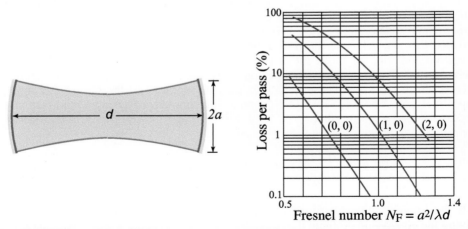

Figure 11.2-11 Percent diffraction loss per pass (half a round trip) as a function of the Fresnel number $N_F = a^2/\lambda d$ for the $(0,0)$, $(1,0)$, and $(2,0)$ modes in a symmetric confocal resonator. (Adapted from A. E. Siegman, *Lasers*, University Science, 1986, Fig. 19.19 left.)

11.3 TWO- AND THREE-DIMENSIONAL RESONATORS

A. Two-Dimensional Rectangular Resonators

A two-dimensional (2D) planar-mirror resonator is constructed from two orthogonal pairs of parallel mirrors, e.g., a pair normal to the z axis and another pair normal to the y axis. Light is confined in the z–y plane by a sequence of ray reflections, as illustrated in Fig. 11.3-1(a).

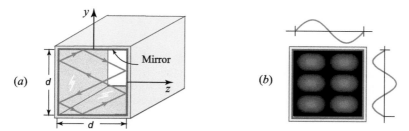

Figure 11.3-1 A two-dimensional planar-mirror resonator: (a) ray pattern; (b) standing-wave pattern with mode numbers $q_y = 3$ and $q_z = 2$. The curves represent the modal amplitudes while the brightness pattern depicts intensity.

The boundary conditions establish the resonator modes, much as for the one-dimensional Fabry–Perot resonator. If the mirror spacing is d, then for standing waves the components of the wavevector $\mathbf{k} = (k_y, k_z)$ are restricted to the values

$$k_y = q_y \frac{\pi}{d}, \quad k_z = q_z \frac{\pi}{d}, \quad q_y = 1, 2, \ldots, \quad q_z = 1, 2, \ldots, \quad (11.3\text{-}1)$$

where q_y and q_z are mode numbers for the y and z directions, respectively. These conditions are a generalization of (11.1-2). Each pair of integers (q_y, q_z) represents a resonator mode $U(\mathbf{r}) \propto \sin(q_y \pi y / d) \sin(q_z \pi z / d)$, as illustrated in Fig. 11.3-1(b). The lowest-order mode is $(1, 1)$ since the modes $(q_y, 0)$ and $(0, q_z)$ have zero amplitude, i.e., $U(\mathbf{r}) = 0$. Modes are conveniently represented by dots that indicate their values of k_y and k_z on a periodic lattice of spacing π/d (Fig. 11.3-2).

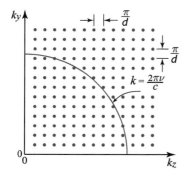

Figure 11.3-2 Dots denote the endpoints of the wavevectors $\mathbf{k} = (k_y, k_z)$ for modes in a two-dimensional resonator.

The wavenumber k of a mode is the distance of the dot from the origin. The associated frequency of the mode is $\nu = ck/2\pi$. The frequencies of the resonator modes are thus determined from

$$k^2 = k_y^2 + k_z^2 = \left(\frac{2\pi\nu}{c}\right)^2, \tag{11.3-2}$$

so that

$$\boxed{\nu_{\mathbf{q}} = \nu_F\sqrt{q_y^2 + q_z^2}, \quad q_y, q_z = 1, 2, \ldots,} \quad \nu_F = \frac{c}{2d}, \tag{11.3-3}$$
Resonance Frequencies

where $\mathbf{q} = (q_y, q_z)$.

The number of modes in a given frequency band, $\nu_1 < \nu < \nu_2$, is established by drawing two circles, of radii $k_1 = 2\pi\nu_1/c$ and $k_2 = 2\pi\nu_2/c$ in the k diagram of Fig. 11.3-2, and counting the number of dots that lie within the annulus. This procedure converts the allowed values of the vector $\mathbf{k}$ into allowed values of the frequency ν.

EXERCISE 11.3-1

Density of Modes in a Two-Dimensional Resonator.

(a) Determine an approximate expression for the number of modes in a two-dimensional resonator with frequencies lying between 0 and ν, assuming that $2\pi\nu/c \gg \pi/d$, i.e., $d \gg \lambda/2$, and allowing for two orthogonal polarizations per mode.

(b) Show that the number of modes per unit area lying within the frequency interval between ν and $\nu + d\nu$ is $M(\nu)d\nu$, where the density of modes $M(\nu)$ (modes per unit area per unit frequency) at frequency ν is given by

$$\boxed{M(\nu) = \frac{4\pi\nu}{c^2}.} \tag{11.3-4}$$
Density of Modes (2D Resonator)

The resonator modes described thus far in this section are in-plane modes, traveling in the plane of the 2D resonator (the y–z plane). Off-plane modes have a propagation constant with a component in the orthogonal direction (the x direction). These are guided modes traveling along the axis of a 2D waveguide such as that described in Sec. 9.3. Whereas the k_y and k_z components of the wavevector take discrete values dictated by the boundary conditions, the k_x component takes continuous values since the 2D resonator is open in the x direction.

B. Circular Resonators and Whispering-Gallery Modes

Light may be confined in a two-dimensional circular resonator by repeated reflections from the circular boundary. As illustrated in Fig. 11.3-3, a ray that self-reproduces after N reflections traces a path with round-trip pathlength Nd, where $d = 2a\sin(\pi/N)$ and a is the radius. For a traveling-wave mode, the resonance frequencies are determined by equating the round-trip pathlength to an integer number of wavelengths, as in (11.1-7). Ignoring the phase shift associated with each reflection, this leads to $Nd = q\lambda = qc/\nu$, i.e., to resonant frequencies $\nu_q = qc/Nd$, where $q = 1, 2, \ldots$. The spacing between these frequencies is therefore $\nu_F = c/Nd$.

For $N = 2$, we have $\nu_F = c/2d = c/4a$, which is identical to (11.1-5). Similarly, $N = 3$ yields $\nu_F = c/3d = c/3\sqrt{3}\,a$, which coincides with the result for the three-mirror resonator (Exercise 11.1-1). In the limit $N \to \infty$, the pathlength Nd approaches the cylindrical circumference $2\pi a$ and the corresponding spacing of the resonance frequencies becomes

$$\nu_F = \frac{c}{2\pi a}. \qquad (11.3\text{-}5)$$

Spacing of Resonance Frequencies

The rays then hug the interior boundary of the resonator, reflecting at near-grazing incidence, as illustrated in Fig. 11.3-3. Such optical modes are known as **whispering-gallery modes** (WGM). The optical modes then behave similarly to acoustic modes in the familiar acoustical whispering gallery, so-named because of the ease with which an acoustic whisper can bounce along the convex surface of a church dome or gallery.

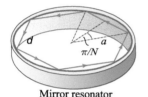
Mirror resonator

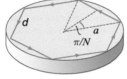
Dielectric resonator

Figure 11.3-3 Reflections in a circular resonator.

Two-dimensional resonators with other cross sections are also used. For example, the circular cross section can be squeezed into a stadium-shaped structure. This oblong configuration supports bow-tie modes [see Fig. 11.1-5(*b*)] in which the ray executes a round-trip path comprising localized reflections from the four locations on the perimeter of the resonator that match the curvature of a conventional spherical-mirror confocal resonator (see Sec. 11.2A).

C. Three-Dimensional Rectangular Resonators

A three-dimensional (3D) planar-mirror resonator is constructed from three pairs of parallel mirrors forming the walls of a closed rectangular box of dimensions d_x, d_y, and d_z. The structure is a three-dimensional resonator, as depicted in Fig. 11.3-4(*a*). Standing-wave solutions within the resonator require that the components of the wavevector $\mathbf{k} = (k_x, k_y, k_z)$ are discretized to obey

$$k_x = q_x \frac{\pi}{d_x}, \quad k_y = q_y \frac{\pi}{d_y}, \quad k_z = q_z \frac{\pi}{d_z}, \qquad q_x, q_y, q_z = 1, 2, \ldots, \qquad (11.3\text{-}6)$$

where q_x, q_y, and q_z are positive integers representing the respective mode numbers. Each mode **q**, which is characterized by the three integers (q_x, q_y, q_z), is represented by a dot in (k_x, k_y, k_z)-space. The spacing between these dots in a given direction is inversely proportional to the width of the resonator along that direction. Figure 11.3-4(*b*) illustrates the concept of the k-space for a cubic resonator with $d_x = d_y = d_z = d$.

The values of the wavenumbers k, and the corresponding resonance frequencies ν, satisfy

$$k^2 = k_x^2 + k_y^2 + k_z^2 = \left(\frac{2\pi\nu}{c}\right)^2. \qquad (11.3\text{-}7)$$

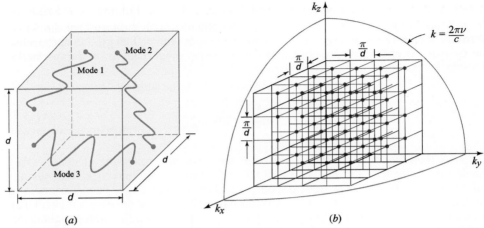

Figure 11.3-4 (a) Waves in a three-dimensional cubic resonator ($d_x = d_y = d_z = d$). (b) The endpoints of the wavevectors (k_x, k_y, k_z) of the modes in a three-dimensional resonator are marked by dots. The wavenumber k of a mode is the distance from the origin to the dot. Each point in k-space occupies a volume $(\pi/d)^3$. All modes of frequency smaller than ν lie inside the positive octant of a sphere of radius $k = 2\pi\nu/c$.

The surface of constant frequency ν is a sphere of radius $k = 2\pi\nu/c$. The resonance frequencies are determined from (11.3-6) and (11.3-7):

$$\nu_{\mathbf{q}} = \sqrt{q_x^2 \nu_{Fx}^2 + q_y^2 \nu_{Fy}^2 + q_z^2 \nu_{Fz}^2}\,, \quad q_x, q_y, q_z = 1, 2, \ldots, \tag{11.3-8}$$

Resonance Frequencies

where

$$\nu_{Fx} = \frac{c}{2d_x}, \quad \nu_{Fy} = \frac{c}{2d_y}, \quad \nu_{Fz} = \frac{c}{2d_z} \tag{11.3-9}$$

are frequency spacings that are inversely proportional to the resonator widths in the x, y, and z direction, respectively. For resonators whose dimensions are much greater than a wavelength, the frequency spacing is much smaller than the optical frequency. For example, for $d = 1$ cm and $n = 1$, $\nu_F = 15$ GHz. This is not so for microresonators, however, as will be discussed in Sec. 11.4.

Density of Modes

When all dimensions of the resonator are much greater than a wavelength, the frequency spacing $\nu_F = c/2d$ is small, and it is analytically difficult to enumerate the modes. In this case, it is useful to resort to a continuous approximation and introduce the concept of density of modes, the validity of which depends on the relative values of the bandwidth of interest and the frequency interval between successive modes.

The number of modes lying in the frequency interval between 0 and ν corresponds to the number of points lying in the volume of the positive octant of a sphere of radius k in the k diagram [Fig. 11.3-4(b)]. The number of modes in the positive octant of a sphere of radius k is $2(\frac{1}{8})(\frac{4}{3}\pi k^3)/(\pi/d)^3 = (k^3/3\pi^2)d^3$. The initial factor of 2 accounts for the two possible polarizations of each mode, whereas the denominator $(\pi/d)^3$ represents the volume in k-space per point. It follows that the number of modes with wavenumbers between k and $k + \Delta k$, per unit volume, is $\varrho(k)\Delta k = [(d/dk)(k^3/3\pi^2)]\Delta k = (k^2/\pi^2)\Delta k$, so that the density of modes in k-space is $\varrho(k) = k^2/\pi^2$. It is worthy of mention that this derivation is identical to that used for determining the density of allowed quantum states for electron waves confined within perfectly reflecting walls in a bulk semiconductor [see Sec. 17.1C and (17.1-6)].

Since $k = 2\pi\nu/c$, the number of modes lying between 0 and ν is $[(2\pi\nu/c)^3/3\pi^2]d^3 = (8\pi\nu^3/3c^3)d^3$. The number of modes in the incremental frequency interval lying between ν and $\nu + \Delta\nu$ is therefore given by $(d/d\nu)[(8\pi\nu^3/3c^3)d^3]\Delta\nu = (8\pi\nu^2/c^3)d^3\Delta\nu$. The density of modes $M(\nu)$, i.e., the number of modes per unit volume of the resonator, per unit bandwidth surrounding the frequency ν, is thus

$$M(\nu) = \frac{8\pi\nu^2}{c^3}.$$

(11.3-10)
Density of Modes
(3D Resonator)

This formula was used by Rayleigh and Jeans in connection with the spectrum of blackbody radiation (see Sec. 14.4B). The quantity $M(\nu)$ is a quadratically increasing function of frequency so that the number of modes within a fixed bandwidth $\Delta\nu$ increases with the frequency ν in the manner indicated in Fig. 11.3-5. As an example, at $\nu = 3 \times 10^{14}$ ($\lambda_o = 1\ \mu$m), $M(\nu) = 0.08$ modes/cm^3-Hz. Within a frequency band of width 1 GHz, there are therefore $\approx 8 \times 10^7$ modes/cm^3. The number of modes per unit volume within an arbitrary frequency interval $\nu_1 < \nu < \nu_2$ is simply the integral $\int_{\nu_1}^{\nu_2} M(\nu)\,d\nu$.

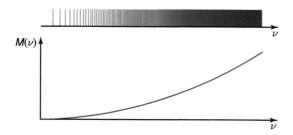

Figure 11.3-5 (a) The frequency spacing between adjacent modes decreases as the frequency increases. (b) The density of modes $M(\nu)$ for a three-dimensional optical resonator is a quadratically increasing function of frequency.

The density of modes in two and three dimensions were derived on the basis of square and cubic geometry, respectively. Nevertheless, the results are applicable for arbitrary geometries, provided that the resonator dimensions are large in comparison with the wavelength.

11.4 MICRORESONATORS AND NANORESONATORS

Microresonators are resonators in which one or more of the spatial dimensions assumes the size of a few wavelengths of light or smaller. The term **microcavity resonator**, or **microcavity** for short, is usually reserved for a microresonator that has small dimensions in all spatial directions, so that the modes exhibit large spacings in all directions of k-space and the resonance frequencies are sparse. However, these terms are often used interchangeably.

The absence of resonance modes in extended spectral bands can inhibit the emission of light from sources placed within a microcavity. At the same time, the emission of light into particular modes of a high-Q, small-volume microcavity can be enhanced relative to emission into ordinary optical modes, as described in Sec. 14.3E. These effects can be important in the operation of resonant-cavity light-emitting diodes (RCLEDs) and microcavity lasers (see Secs. 18.1B and 18.5, respectively).

Microresonators can be fabricated using dielectric materials configured in various geometries, such as (1) micropillars with Bragg-grating reflectors; (2) microdisks and microspheres in which light reflects near the surface in whispering-gallery modes; (3) microtoroids, which resemble small fiber rings; and (4) 2D photonic crystals containing light-trapping defects that function as microcavities. These technologies embrace two principal design objectives:

- The reduction of the modal volume V, the spatial integral of the optical energy density $\frac{1}{2}\epsilon\mathcal{E}^2$ of the mode, normalized to the maximum energy density.
- The enhancement of the quality factor Q.

Spatial confinement is improved by fabricating microresonators with special geometries, while enhancement of temporal confinement is realized by making use of low-loss materials and low-leakage configurations. Typical modal volumes and quality factors for these structures are summarized in Table 11.4-1.

Table 11.4-1 Normalized modal volume V/λ^3 and quality factor Q for various microresonators.

	Micropillar	Microdisk	Microtoroid	Microsphere	Photonic-Crystal
V/λ^3	5	5	10^3	10^3	1
Q	10^3	10^4	10^8	10^{10}	10^4

An exact analysis of the resonator modes of dielectric microresonators requires the full electromagnetic theory. The Helmholtz equation is solved in a coordinate system suitable for the geometry of the structure, and appropriate boundary conditions are applied to the electric and magnetic fields at the planar, cylindrical, or spherical boundaries. The solution yields the resonance frequencies of the modes and their spatial distributions, which may be used to determine the modal volume for each mode. Since the analysis is complex for all practical geometries, numerical solutions are often necessary.

In the next section, we describe some of the properties of a simple rectangular (box) microresonator whose walls are made of perfect mirrors. A simple analysis of the modes of such a structure provides the resonance frequencies and the spatial distributions of the modes. High-Q microresonators do not make use of mirrors because of their relatively high losses, and the box structure is also not among the geometries typically used in practical microresonators. Nevertheless, the analysis is useful for elucidating the relation between the resonance frequencies and the dimensions of the resonator, and for illustrating the frequency dependence of the density of modes for boxes with different aspect ratios.

A. Rectangular Microresonators

The simplest microresonator structure is a rectangular (box) resonator made of planar parallel mirrors. The modes are then sinusoidal standing waves in all three directions and the resonance frequencies are given by (11.3-8). When the dimensions of the box are small, only the lowest order modes lie within the optical band. For a cubic resonator, the resonance frequencies are provided in Table 11.4-2 in units of $\nu_F = c/2d$. As an example, if $d = 1$ μm and the medium has refractive index $n = 1.5$, we obtain $\nu_F = 100$ THz. The frequencies of the lowest-order modes then correspond to the free-space wavelengths $\lambda_o = 2.13, 1.73, 1.34, 1.22, 1.06, 1.00,$ and 0.87μm, which are widely spaced.

Table 11.4-2 Resonance frequencies for the lowest-order modes of a cubic microcavity resonator.

Mode $(q_x\, q_y\, q_z)^{(a)}$	$(011)^{(3)}$	$(111)^{(1)}$	$(012)^{(6)}$	$(112)^{(3)}$	$(022)^{(3)}$	$(122)^{(3)}$	$(222)^{(1)}$
Frequency (units of ν_F)	1.41	1.73	2.24	2.45	2.83	3	3.46

[a]Superscripts in parentheses indicate the modal degeneracy, i.e., the number of modes of the same resonance frequency. As an example, three modes have the same resonance frequency $1.41\,\nu_F$: (011), (101), and (110).

If the resonator has a mixture of dimensions both small and large, as with a box of large aspect ratio, the modes are placed at the points of an anisotropic grid in k-space [see Fig. 11.3-4(b)]. The grid is finely divided along the directions of the large dimensions and coarsely divided along the directions of the small dimensions. Mode counting may then be implemented by use of a continuous approximation only in those directions for which the grid is fine. The resultant modal density is displayed in Fig. 11.4-1 for various cases.

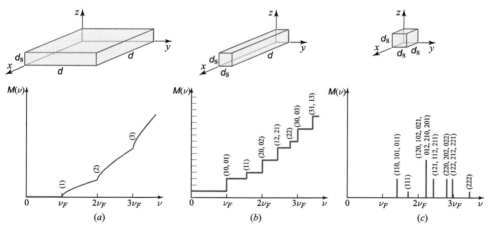

Figure 11.4-1 Modal density $M(\nu)$ for rectangular microresonators with (a) one; (b) two; and (c) three sides of small dimension $d_S \ll d$. The frequency spacing associated with the small dimension is $\nu_F = c/2d_S$. When all dimensions are small, as in (c), the resonance frequencies are discrete and their values are those provided in Table 11.4-2 for the cubic microcavity resonator. The result shown in (b) represents a combination of discrete modes associated with a 2D microresonator and continuous modes associated with a 1D large resonator, which has a uniform modal density [see (11.1-10)]. The result provided in (a) illustrates a combination of discrete modes associated with a 1D microresonator and a continuum of modes associated with a 2D large resonator, which has a modal density that is linearly proportional to frequency [see (11.3-4)].

B. Micropillars, Microdisks, and Microtoroids

Dielectric microresonators have been fabricated in a number of configurations, including micropillars, microdisks, and microtoroids, as illustrated in Fig. 11.4-2. Light is confined in these structures by total internal reflection (see Fig. 11.3-3).

The **micropillar**, or **micropost**, resonator is a cylinder of high-refractive-index material sandwiched between dielectric layers comprising distributed Bragg-grating reflectors, as illustrated in Fig. 11.4-2(a). Light is confined in the axial direction by reflection from the DBRs, as in a Fabry–Perot resonator; light is confined in the lateral direction by total internal reflection from the walls of the cylinder. Micropillars are typically fabricated from compound semiconductors via conventional lithographic and etching processes; DBR layers are often made of AlAs/GaAs or AlGaAs/GaAs. The pillar itself can contain an active region such as a multiquantum-well structure that provides optical gain when pumped (Sec. 18.5A).

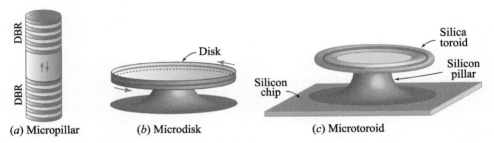

(a) Micropillar   (b) Microdisk   (c) Microtoroid

Figure 11.4-2 Micropillar, microdisk, and microtoroid resonators.

The **microdisk** resonator displayed in Fig. 11.4-2(b) is a circular resonator in which light travels at near-grazing incidence in whispering-gallery modes and is confined by total internal reflection from the circular boundary (see Sec. 11.3B). Micropillar and microdisk diameters usually range from 1 μm to tens of μm and their quality factors Q are substantially larger than those of mirror resonators since their losses are significantly lower (Table 11.4-1). Microdisk resonators fabricated from semiconductor materials are widely used as microdisk lasers because of their many salutary features (Sec. 18.5B).

The **microtoroid** dielectric microresonator illustrated in Fig. 11.4-2(c) is much like a fiber-ring resonator, in which the resonator modes are circulating guided waves. These microresonators are often fabricated from silica and are supported on a silicon chip by a silicon pillar. The toroid is formed by surface tension while the material is in a molten state; the outer boundary thus assumes a near atomic-scale surface finish and has significantly lower scattering losses than the microdisk resonator. Silica toroidal microresonators-on-a-chip exhibit exceptionally high values of the quality factor, $Q > 10^8$ (Table 11.4-1).

C. Microspheres

Dielectric **microspheres** are used as three-dimensional optical microresonators. Certain modes are guided along trajectories (orbits) that are tightly confined near a great circle of the sphere, resulting in whispering-gallery modes.

The modes of a dielectric sphere may be determined by solving the Helmholtz equation (5.3-16) for the electric and magnetic-field vectors, together with the appropriate boundary conditions. These modes are similar to the wavefunctions of an electron in a hydrogen atom (see Sec. 14.1A) because of the spherical symmetry of both problems, but there are also differences in view of the vector nature of the electromagnetic field.

The electric and magnetic vector fields are directly related to a scalar potential function U that satisfies the Helmholtz equation.[†] For a sphere of radius a and refractive

---

[†] For a detailed mathematical description, see, for example, A. N. Oraevsky, Whispering-Gallery Waves, *Kvantovaya Elektronika (Quantum Electronics)*, vol. 32, pp. 377–400, 2002.

index n, located in air, the separation-of-variables method in a spherical coordinate system (r, θ, ϕ) results in a solution of the form

$$U(r,\theta,\phi) \propto \sqrt{r}\, J_{\ell+1/2}(nk_o r)\, P_m^\ell(\cos\theta)\, \exp(\pm jm\phi), \quad r \leq a, \quad (11.4\text{-}1)$$

$$\propto \sqrt{r}\, H^{(1)}_{\ell+1/2}(nk_o r)\, P_m^\ell(\cos\theta)\, \exp(\pm jm\phi), \quad r > a, \quad (11.4\text{-}2)$$

where $J_\ell(\cdot)$ is the Bessel function of the first kind of order ℓ, $H^{(1)}_\ell(\cdot)$ is the Hankel function of the first kind of order ℓ, $P_m^\ell(\cdot)$ is the adjoint Legendre function, and m and ℓ are nonnegative integers. The boundary conditions at $r = a$ yield a characteristic equation that provides a discrete set of values for k_o, corresponding to the resonance frequencies. These are indexed by a third integer n. In addition, there are two polarization modes — an E mode for which $H_r = 0$ and an H mode for which $E_r = 0$.

The modes are generally oscillatory functions of r, θ, and ϕ characterized by the radial, polar, and azimuthal mode numbers n, ℓ, and m, respectively. There are n maxima in the radial direction within the sphere. The number of field maxima in the azimuthal direction is 2ℓ, while the number of field maxima in the polar direction (between the two poles) is $\ell - m + 1$.

The fundamental mode (n $= 1$, m $= \ell$) has a single peak in the radial direction within the sphere, and a single peak in the polar direction at $\theta = \pi/2$. For large m $= \ell$, the modes are highly confined near the equator. This is because $P_\ell^\ell(\cos\theta) \approx \sin^\ell \theta$ vanishes rapidly at angles slightly different from $\theta = \pi/2$, and $J_\ell(nk_o r)$ is small everywhere within the sphere except for a sharp peak near $r = a$. The mode therefore represents an optical beam traveling along the equator, as shown in Fig. 11.4-3(a), much like the whispering-gallery modes of the disk resonator displayed in Fig. 11.3-3. For sufficiently large $\ell = $ m, the resonance frequencies of these modes are approximately equal to $\nu_\ell \approx \ell c/2\pi a$. This is to be expected since the angular mode number ℓ is close to the number of wavelengths that comprise the optical length of the equator.

The whispering-gallery mode may be viewed from a ray-optics perspective in terms of quasi-plane waves with wavevectors parallel to the local rays (see Sec. 2.3 and Fig. 10.2-8) that zigzag near the equator, as shown in Fig. 11.4-3(b). The wavevector **k** has magnitude $k = \sqrt{\ell(\ell+1)}/a$ and azimuthal component $k_\phi = $ m$/a$. The inclination angle of the zigzagging rays is smallest ($\approx 1/\sqrt{\ell}$) for the fundamental mode m $= \ell$, while the m $= 0$ mode has a $90°$ inclination.

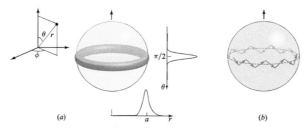

Figure 11.4-3 (a) Whispering-gallery mode in a microsphere resonator. (b) Ray model of the whispering-gallery mode.

Microspheres fabricated from low-loss fused silica have been used as optical resonators with ultrahigh values of Q. Like the toroidal resonator depicted in Fig. 11.4-2(c), the shape and surface finish of the sphere are determined by the surface tension in the molten state during fabrication; the result is near atomic perfection in the surface finish. The reduced surface scattering losses lead to remarkably high quality factors,

$Q > 10^{10}$ (see Table 11.4-1). Optical power may be coupled into the sphere via an optical fiber that is locally stripped of its cladding, as illustrated in Fig. 11.4-4.

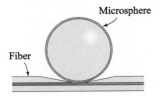

Figure 11.4-4 Coupling optical power from an optical fiber into a microsphere resonator.

D. Photonic-Crystal Microcavities

As described in Chapter 7, photonic crystals are periodic dielectric structures exhibiting photonic bandgaps, i.e., spectral bands within which light cannot propagate. The Bragg grating reflector (BGR) is an example of a 1D photonic crystal that serves as a reflector for frequencies within a photonic bandgap. The micropillar resonator shown in Fig. 11.4-2(a), for example, uses BGRs in lieu of mirrors. If the height of the microresonator equals one or just a few periods of the BGR, as illustrated in Fig. 11.4-5(a), the structure may also be regarded as an extended photonic crystal with the cavity acting as a defect in the crystal structure. The resonator is then called a **photonic-crystal resonator**.

This concept is also applicable to 2D photonic crystals. As schematized in Fig. 11.4-5(b), a *defect* in the 2D periodic crystal structure is a local alteration such as a missing hole in a periodic array of air holes drilled in a slab. For wavelengths that fall within the photonic-crystal bandgap, the periodic structure surrounding the defect does not support light propagation, so that light is trapped within the defect, much like electrons or holes are trapped by a defect in a semiconductor crystal. The defect then serves as a microcavity resonator. Stated differently, the defect produces new resonance frequencies that lie within the bandgap and correspond to optical modes that have spatial distributions centered within the microcavity and that decay rapidly in the surrounding photonic crystal.

Two-dimensional photonic crystals can be fabricated by using electron-beam lithography and reactive ion etching in semiconductor materials. Microcavities of dimensions close to a period of the photonic crystal, which can be of the order of a wavelength of light, can support modal volumes as small as λ^3. Hence, photonic-crystal microcavities have the smallest normalized modal volumes of the family of microresonators (see Table 11.4-1). The quality factors Q can be as high as 10^4. Because of these features, photonic-crystal resonators are often co-opted for use in photonic-crystal microcavity lasers (Sec. 18.5C).

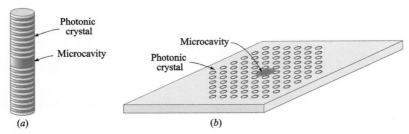

Figure 11.4-5 Photonic-crystal microresonators. (a) The micropillar resonator as a 1D photonic crystal in which the microcavity acts as a defect. (b) A 2D photonic-crystal resonator may be fabricated by drilling holes in a dielectric slab at the points of a planar hexagonal lattice; a missing hole serves as the microcavity.

E. Plasmonic Resonators: Metallic Nanodisks and Nanospheres

Metallic electromagnetic resonators are routinely used at microwave and radio frequencies. As illustrated in Fig. 11.0-2, microwave cavity resonators have dimensions that are close to the resonance wavelength (cm), whereas radio-frequency resonant circuits, comprising metallic capacitors and inductors, have dimensions (cm) much smaller than the resonance wavelength (m).

Plasmonic resonators contain metallic structures of subwavelength dimensions that operate at optical frequencies by supporting surface plasmon polariton (SPP) waves or localized surface plasmon (LSP) oscillations at their boundaries with dielectric media (Secs. 8.2B and 8.2C, respectively). They can take the form of nanodisks, nanospheres, or other nanoparticles. Their dimensions (~ 10 nm) can be much smaller than the resonance wavelength ($\sim \mu$m), while maintaining a size-to-wavelength ratio ($\sim 10^{-2}$) not unlike that of the radio-frequency electronic resonator portrayed in Fig. 11.0-2.

The mathematical modeling of structures whose dimensions are much larger than the resonance wavelength, such as resonators with large mirrors, is readily achieved by making use of electromagnetic optics. The electric and magnetic fields are then of paramount importance, and these are the quantities that we customarily encounter in the optics literature. At the opposite extreme, metallic structures whose dimensions are much smaller than the resonance wavelength can be well-modeled in terms of electrical voltages and currents. The key elements in this domain are lumped electrical components such as inductors and capacitors, as encountered in the electrical-engineering literature — as a simple example, the resonance frequency of the electronic resonator portrayed in Fig. 11.0-2 is $\omega_0 = 1/\sqrt{LC}$. Metallic structures whose dimensions are comparable to the resonance wavelength pose the greatest challenge in terms of modeling because the analysis must then generally be carried out in terms of voltages and currents as well as fields.

Metallic Nanodisk

A metallic cylinder with an interior dielectric material supports whispering-gallery *photonic* modes of light that reflect at the metallic boundary, as displayed in Fig. 11.4-6(a). This disk resonator also supports *plasmonic* modes in the form of SPP waves (Sec. 8.2B) at the interior boundary, as illustrated in Fig. 11.4-6(b). The internal optical field of the plasmonic mode is more tightly confined near the boundary than is the field of the photonic mode. Moreover, since the plasmonic wave penetrates more deeply into the metal than does the photonic mode, it suffers greater losses and the resonator exhibits a smaller quality factor Q. By comparison, a dielectric disk resonator (Sec. 11.3B) supports whispering-gallery modes with evanescent optical fields that extend into the exterior dielectric medium, as shown in Fig. 11.4-6(c).

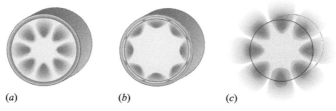

Figure 11.4-6 Schematic of optical-field distributions in disk resonators. (a) Photonic mode in a metal–dielectric disk: light is confined to the interior by multiple reflections at the metallic boundary. (b) Plasmonic mode in a metal–dielectric disk: a SPP wave travels along the interior boundary. (c) Photonic mode in a dielectric–dielectric disk: light is confined by total internal reflection at the boundary.

Nominal values for the normalized modal volume and quality factor of a plasmonic mode in a metal–dielectric disk resonator with a diameter of 100 nm are $V/\lambda^3 \sim 10^{-4}$ and $Q \sim 10$, respectively.[†] These values are substantially below those for photonic-crystal microcavities (see Table 11.4-1). Nevertheless, structures of this kind find use in surface plasmon polariton disk and ring nanolasers (Sec. 18.6).

Metallic Nanosphere

As discussed in Sec. 8.2C, a metallic nanosphere embedded in a dielectric medium supports resonant localized surface plasmon (LSP) oscillations. When the nanosphere is illuminated by an optical wave, the scattered field, as well as the internal field, are substantially enhanced at frequencies near resonance. This field enhancement is accompanied by spatial localization of energy at the nanoscale, so that the nanosphere serves as a nanoresonator. Since metals are relatively lossy, however, the resonator quality factor Q is far lower than that of dielectric resonators. Nevertheless, a metal nanosphere embedded in a specialized dielectric medium can serve as a localized surface plasmon nanolaser (Sec. 18.6).

READING LIST

Books

See also the reading list on lasers in Chapter 16.

A. H. W. Choi, ed., *Handbook of Optical Microcavities*, CRC Press/Taylor & Francis, 2015.

J. Heebner, R. Grover, and T. A. Ibrahim, *Optical Microresonators: Theory, Fabrication, and Applications*, Springer-Verlag, 2008, paperback ed. 2010.

A. V. Kavokin, J. J. Baumberg, G. Malpuech, and F. P Laussy, *Microcavities*, Oxford University Press, 2007.

D. G. Rabus, *Integrated Ring Resonators: The Compendium*, Springer-Verlag, 2007.

N. Hodgson and H. Weber, *Laser Resonators and Beam Propagation: Fundamentals, Advanced Concepts and Applications*, Springer-Verlag, 2nd ed. 2005.

K. J. Vahala, ed., *Optical Microcavities*, World Scientific, 2004.

K. Staliūnas and V. J. Sánchez-Morcillo, *Transverse Patterns in Nonlinear Optical Resonators*, Springer-Verlag, 2003.

A. N. Oraevsky, *Gaussian Beams and Optical Resonators*, Nova, 1996.

Yu. A. Anan'ev, *Laser Resonators and the Beam Divergence Problem*, CRC/Taylor & Francis, 1992.

J. M. Vaughan, *The Fabry–Perot Interferometer*, CRC Press, 1989.

G. Hernandez, *Fabry–Perot Interferometers*, Cambridge University Press, 1986, paperback ed. 1988.

A. E. Siegman, *Lasers*, University Science, 1986.

L. A. Vaynshteyn, *Open Resonators and Open Waveguides*, Golem Press, 1969.

Seminal Articles

K. J. Vahala, Optical Microcavities, *Nature*, vol. 424, pp. 839–846, 2003.

J. U. Nöckel and A. D. Stone, Ray and Wave Chaos in Asymmetric Resonant Optical Cavities, *Nature*, vol. 385, pp. 45–47, 1997.

H. Kogelnik and T. Li, Laser Beams and Resonators, *Applied Optics*, vol. 5, pp. 1550–1567, 1966 (published simultaneously in *Proceedings of the IEEE*, vol. 54, pp. 1312–1329, 1966).

A. E. Siegman, Unstable Optical Resonators for Laser Applications, *Proceedings of the IEEE*, vol. 53, pp. 277–287, 1965.

A. G. Fox and T. Li, Resonant Modes in a Maser Interferometer, *Bell System Technical Journal*, vol. 40, pp. 453–488, 1961.

G. D. Boyd and J. P. Gordon, Confocal Multimode Resonator for Millimeter Through Optical Wavelength Masers, *Bell System Technical Journal*, vol. 40, pp. 489–508, 1961.

[†] See, e.g., M. Kuttge, F. Javier García de Abajo, and A. Polman, Ultrasmall Mode Volume Plasmonic Nanodisk Resonators, *Nano Letters*, vol. 10, pp. 1537–1541, 2010.

PROBLEMS

11.1-3 **Resonance Frequencies of a Resonator with an Etalon.**

(a) Determine the spacing between adjacent resonance frequencies in a resonator constructed of two parallel planar mirrors separated by a distance $d = 15$ cm in air ($n = 1$).

(b) A transparent plate of thickness $d_1 = 2.5$ cm and refractive index $n = 1.5$ is placed inside the resonator and is tilted slightly to prevent light reflected from the plate from reaching the mirrors. Determine the spacing between the resonance frequencies of the resonator.

11.1-4 **Mirrorless Resonators.** Laser diodes are often fabricated from crystals whose surfaces are cleaved along crystal planes. These surfaces act as reflectors by virtue of Fresnel reflection, and therefore serve as the mirrors of a Fabry–Perot resonator. An expression for the power reflectance is provided in (6.2-15). Consider a crystal placed in air ($n = 1$) whose refractive index $n = 3.6$ and loss coefficient $\alpha_s = 1$ cm^{-1}. The light reflects between two parallel surfaces separated by a distance $d = 0.2$ mm. Determine the spacing between resonance frequencies ν_F, the overall distributed loss coefficient α_r, the finesse $\mathcal{F}$, the spectral width of a mode $\delta\nu$, and the quality factor Q. If the free-space wavelength of the generated light is $\lambda_o = 1.55$ μm, estimate the longitudinal mode number q.

11.1-5 **Fabry–Perot Etalon with Bragg Grating Reflectors.** A Fabry–Perot etalon is made by sandwiching a layer of GaAs between two of the GaAs/AlAs Bragg grating reflectors described in Prob. 7.1-8. Determine the finesse $\mathcal{F}$ of the resonator and quality factor Q. Determine the transmittance of a Bragg grating reflector comprised of $N = 10$ alternating layers of GaAs ($n_1 = 3.6$) and AlAs ($n_2 = 3.2$) of widths d_1 and d_2 equal to a quarter wavelength in each medium. Assume that the light is incident from an extended GaAs medium.

11.1-6 **Optical Energy Decay Time.** How much time does it take for the optical energy stored in a resonator of finesse $\mathcal{F} = 100$, length $d = 50$ cm, and refractive index $n = 1$, to decay to one-half of its initial value?

11.2-5 **Stability of Spherical-Mirror Resonators.**

(a) Can a resonator with two convex mirrors ever be stable?

(b) Can a resonator with one convex and one concave mirror ever be stable?

11.2-6 **A Planar-Mirror Resonator Containing a Lens.** A lens of focal length f is placed inside a planar-mirror resonator constructed of two flat mirrors separated by a distance d. The lens is located at a distance $d/2$ from each of the mirrors.

(a) Determine the ray-transfer matrix for a ray that begins at one of the mirrors and travels a round trip inside the resonator.

(b) Determine the condition of stability of the resonator.

(c) Under stable conditions sketch the Gaussian beam that fits this resonator.

11.2-7 **Self-Reproducing Rays in a Symmetric Resonator.** Consider a symmetric resonator using two concave mirrors of radii R separated by a distance $d = 3|R|/2$. After how many round trips through the resonator will a ray retrace its path?

11.2-8 **Ray Position in Unstable Resonators.** Show that for an unstable resonator the ray position after m round trips is given by $y_m = \alpha_1 h_1^m + \alpha_2 h_2^m$, where α_1 and α_2 are constants. Here $h_1 = b + \sqrt{b^2 - 1}$, $h_2 = b - \sqrt{b^2 - 1}$, and $b = 2(1 + d/R_1)(1 + d/R_2) - 1$. *Hint:* Use the results in Sec. 1.4D.

11.2-9 **Ray Position in Unstable Symmetric Resonators.** Verify that a symmetric resonator using two concave mirrors of radii $R = -30$ cm separated by a distance $d = 65$ cm is unstable. Find the position y_1 of a ray that begins at one of the mirrors, at position $y_0 = 0$ with an angle $\theta_0 = 0.1°$, and undergoes one round trip. If the mirrors have 5-cm-diameter apertures, after how many round trips does the ray leave the resonator? Plot y_m, $m = 2, 3, \ldots$, for $d = 50$ cm and $d = 65$ cm. You may use the results of Prob. 11.2-8.

11.2-10 **Gaussian-Beam Standing Waves.** Consider a wave formed by the sum of two identical Gaussian beams propagating in the $+z$ and $-z$ directions. Show that the result is a standing wave. Using the boundary conditions at two ideal mirrors placed such that they coincide with the wavefronts, derive the resonance frequencies (11.2-30).

11.2-11 **Gaussian Beam in a Symmetric Confocal Resonator.** A symmetric confocal resonator with mirror spacing $d = 16$ cm, mirror reflectances 0.995, and $n = 1$ is used in a laser operating at $\lambda_o = 1$ μm.
(a) Find the radii of curvature of the mirrors.
(b) Find the waist of the $(0,0)$ (Gaussian) mode.
(c) Sketch the intensity distribution of the $(1,0)$ modes at one of the mirrors and determine the distance between its two peaks.
(d) Determine the resonance frequencies of the $(0,0)$ and $(1,0)$ modes.
(e) Assuming that losses arise only from imperfect mirror reflectances, determine the distributed resonator loss coefficient α_r.

*11.2-12 **Diffraction Loss in a Symmetric Confocal Resonator.** The percent diffraction loss per pass for the different low-order modes of a symmetric confocal resonator is given in Fig. 11.2-11, as a function of the Fresnel number $N_F = a^2/\lambda d$ (where d is the mirror spacing and a is the radius of the mirror aperture). Using the parameters provided in Prob. 11.2-11, determine the mirror radius for which the loss per pass of the $(1,0)$ mode is 1%.

11.3-2 **Number of Modes in Resonators of Different Dimensions.** Consider light of wavelength $\lambda_o = 1.06$ μm and spectral width $\Delta\nu = 120$ GHz. How many modes have frequencies within this linewidth in the following resonators ($n = 1$):
(a) A one-dimensional resonator of length $d = 10$ cm?
(b) A 10 cm × 10 cm two-dimensional resonator?
(c) A 10 cm × 10 cm × 10 cm three-dimensional resonator?

CHAPTER

12

STATISTICAL OPTICS

12.1	STATISTICAL PROPERTIES OF RANDOM LIGHT	475
	A. Optical Intensity	
	B. Temporal Coherence and Spectrum	
	C. Spatial Coherence	
	D. Longitudinal Coherence	
12.2	INTERFERENCE OF PARTIALLY COHERENT LIGHT	489
	A. Interference of Two Partially Coherent Waves	
	B. Interferometry and Temporal Coherence	
	C. Interferometry and Spatial Coherence	
*12.3	TRANSMISSION OF PARTIALLY COHERENT LIGHT	497
	A. Propagation of Partially Coherent Light	
	B. Image Formation with Incoherent Light	
	C. Gain of Spatial Coherence by Propagation	
12.4	PARTIAL POLARIZATION	506

Max Born (1882–1970)   **Emil Wolf (1922–2018)**

The book *Principles of Optics*, first published in 1959 by Max Born and Emil Wolf, drew attention to the importance of coherence in optics. Emil Wolf is responsible for many advances in the theory of optical coherence.

Statistical optics is the study of the properties of random light. Randomness in light arises because of unpredictable fluctuations in the source of light itself or in the medium through which it propagates. Natural light, such as that radiated by a hot object (e.g., the sun) is random in time because it comprises a superposition of emissions from a very large number of atoms that radiate independently, and with different frequencies and phases. Randomness in light can also arise as a result of scattering from a rough surface, or transmission through a ground-glass diffuser or a turbulent fluid, which impart random spatial variations to the optical wavefront. The study of the random fluctuations of light is also known as the **theory of optical coherence**.

In earlier chapters it was assumed that light is deterministic or "coherent." As discussed in Sec. 2.2, an example of coherent light is the monochromatic wavefunction $u(\mathbf{r}, t) = \text{Re}\{U(\mathbf{r}) \exp(j2\pi\nu t)\}$, for which the complex amplitude $U(\mathbf{r})$ is a deterministic complex function, e.g., $U(\mathbf{r}) = (A_0/r) \exp(-jkr)$ for a spherical wave. The dependence of the wavefunction on time and position is then perfectly periodic and predictable [Fig. 12.0-1(a)]. For random light, in contrast, the dependence of the wavefunction on time and position is not totally predictable and generally requires statistical methods for its characterization [Fig. 12.0-1(b)].

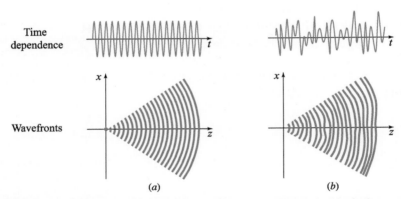

Figure 12.0-1 Time dependence and wavefronts of (a) a monochromatic spherical wave, which is an example of coherent light; (b) random light.

How can we go about extracting meaningful measures from the fluctuations of a random optical wave that will permit us to characterize it and distinguish it from other random waves? Examine, for instance, the three random optical waves displayed in Fig. 12.0-2, whose wavefunctions at some position vary with time as shown. A good beginning is to observe that wave (b) is more "intense" than wave (a), and that the envelope of wave (c) fluctuates "faster" than the envelopes of the other two waves.

Figure 12.0-2 Time dependence of the wavefunctions of three random waves.

To translate these casual qualitative observations into quantitative measures, we use the concept of statistical averaging to define a number of (nonrandom) measures of a wave. Because the random function $u(\mathbf{r}, t)$ satisfies certain laws (the wave equation and boundary conditions) its statistical averages must also satisfy certain laws. The theory of optical coherence is concerned with the definitions of these statistical averages, the laws that govern them, and the measures by means of which light is classified as **coherent, incoherent**, or, in general, **partially coherent**.

This Chapter

This chapter serves as an introduction to the theory of optical coherence. Familiarity with the theory of random fields (random functions of many variables — space and time) is useful for fully understanding the theory of partial coherence. However, the notions presented in this chapter are limited in scope, so that knowledge of the concept of statistical averaging suffices.

In Sec. 12.1 we define two statistical averages used to describe random light: the optical intensity and the mutual coherence function. Temporal and spatial coherence are delineated, and the connection between temporal coherence and monochromaticity is established. The examples of partially coherent light provided in Sec. 12.1 reveal that spatially coherent light need not be temporally coherent, and that monochromatic light need not be spatially coherent. One of the basic manifestations of the coherence of light is its ability to produce visible interference fringes. Section 12.2 is devoted to the laws of interference obeyed by random light. The propagation of partially coherent light in free space, and its transmission through various optical systems (including image-formation systems) is the subject of Sec. 12.3. The theory of polarization of random light (partial polarization), which relies on statistical averaging of the components of the optical field vector, is introduced in Sec. 12.4. With the exception of this latter section, our exposition is framed in the context of scalar wave optics.

12.1 STATISTICAL PROPERTIES OF RANDOM LIGHT

An arbitrary optical wave is described by a wavefunction $u(\mathbf{r}, t) = \text{Re}\{U(\mathbf{r}, t)\}$, where $U(\mathbf{r}, t)$ is the complex wavefunction. For example, $U(\mathbf{r}, t)$ may take the form $U(\mathbf{r}) \exp(j2\pi\nu t)$ for monochromatic light, or it may comprise a sum of such functions with many different values of ν for polychromatic light (see Sec. 2.6A for a discussion of the complex wavefunction). For random light, both functions, $u(\mathbf{r}, t)$ and $U(\mathbf{r}, t)$, are random and are characterized by a number of statistical averages that are introduced in this section.

A. Optical Intensity

As discussed in Secs. 2.2A and 2.6A, the intensity $I(\mathbf{r}, t)$ of coherent (deterministic) light is related to the absolute square of the complex wavefunction $U(\mathbf{r}, t)$ via

$$I(\mathbf{r}, t) = |U(\mathbf{r}, t)|^2. \qquad (12.1\text{-}1)$$

For pulsed light, the intensity is time varying whereas for monochromatic deterministic light it is independent of time.

For random light, $U(\mathbf{r}, t)$ is a random function of time and position. The intensity $|U(\mathbf{r}, t)|^2$ is therefore also random. The **average intensity** is then defined as

$$\boxed{I(\mathbf{r}, t) = \langle |U(\mathbf{r}, t)|^2 \rangle,} \qquad (12.1\text{-}2)$$
Average Intensity

where the symbol $\langle \cdot \rangle$ denotes an ensemble average over many realizations of the random function. Thus, although a random wave that is repeatedly reproduced under the same conditions leads to a different wavefunction on each trial, the average intensity at each time and position is determined by making use of (12.1-2). We denote $I(\mathbf{r}, t)$ the *intensity* of the light (with the modifier *average* implied), when there is no ambiguity in meaning. The unaveraged quantity $|U(\mathbf{r}, t)|^2$, in contrast, is called the **random intensity** or **instantaneous intensity**. For deterministic light, the averaging operation is superfluous since all trials produce exactly the same wavefunction, whereupon (12.1-2) is equivalent to (12.1-1).

The average intensity may be time independent or it may be a function of time, as illustrated in Figs. 12.1-1(a) and (b), respectively. The former case applies when the optical wave is statistically **stationary**, i.e., when its statistical averages are invariant to time. The instantaneous intensity $|U(\mathbf{r}, t)|^2$ then fluctuates randomly with time, but its average is constant. We denote the average intensity in this case by $I(\mathbf{r})$. It is clear that stationarity does not necessarily mean constancy; rather it signifies constancy of the average properties. An example of stationary random light is that emitted by an ordinary incandescent lamp whose filament is heated by a constant electric current. The average intensity $I(\mathbf{r})$ is then a function of distance from the lamp, but it does not vary with time. On the other hand, the random intensity $|U(\mathbf{r}, t)|^2$ fluctuates with both position and time, as illustrated in Fig. 12.1-1(a).

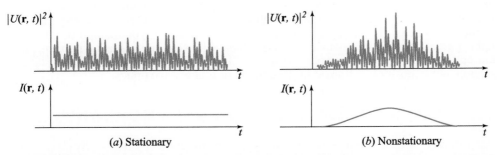

Figure 12.1-1 (a) A statistically stationary wave has an average intensity $I(\mathbf{r})$ that does not vary with time. (b) A statistically nonstationary wave has an average intensity $I(\mathbf{r}, t)$ that varies with time. These plots represent, for example, the intensity of light produced by an incandescent lamp driven by (a) a constant electric current, and (b) a pulse of electric current.

When the light is stationary, the statistical averaging operation over many realizations of the wave, as prescribed by (12.1-2), is usually equivalent to time averaging over a long time duration, which is written as

$$I(\mathbf{r}) = \lim_{T \to \infty} \frac{1}{2T} \int_{-T}^{T} |U(\mathbf{r}, t)|^2 \, dt. \qquad (12.1\text{-}3)$$

B. Temporal Coherence and Spectrum

Consider the fluctuations of *stationary* light at a fixed position $\mathbf{r}$, as a function of time. The stationary random function $U(\mathbf{r}, t)$ has a constant intensity $I(\mathbf{r}) = \langle |U(\mathbf{r}, t)|^2 \rangle$. For brevity, we need not explicitly indicate the $\mathbf{r}$ dependence since $\mathbf{r}$ is fixed, so that we may write $U(\mathbf{r}, t) = U(t)$ and $I(\mathbf{r}) = I$.

The random fluctuations of $U(t)$ are characterized by a time scale representing the "memory" of the random function. After this time, the process "forgets" itself, so that fluctuations at points separated by a time interval longer than this memory time are independent. The function appears to be relatively smooth within its memory time, but "rough" or "erratic" when examined over longer time scales as portrayed in Fig. 12.0-2. A quantitative measure of this temporal behavior is established by defining a statistical *average* called the autocorrelation function. This function describes the extent to which the wavefunction fluctuates in unison at two instants of time separated by a given time delay, and thus serves to establish the time scale of the process that characterizes the wavefunction.

Temporal Coherence Function

The **autocorrelation function** of a stationary complex random function $U(t)$ is defined as the average of the product of $U^*(t)$ and $U(t+\tau)$, as a function of the time delay τ,

$$G(\tau) = \langle U^*(t)\, U(t+\tau) \rangle \qquad (12.1\text{-}4)$$

Temporal Coherence Function

or

$$G(\tau) = \lim_{T\to\infty} \frac{1}{2T} \int_{-T}^{T} U^*(t)\, U(t+\tau)\, dt \qquad (12.1\text{-}5)$$

(see Sec. A.1 in Appendix A).

To understand the significance of the definition in (12.1-4), consider the case in which the average value of the complex wavefunction $\langle U(t) \rangle = 0$. This is applicable when the phase of the phasor $U(t)$ is equally likely to have any value between 0 and 2π, as illustrated in Fig. 12.1-2. The phase of the product $U^*(t)U(t+\tau)$ is the angle between phasors $U(t)$ and $U(t+\tau)$. If these two quantities are uncorrelated, the angle between their phasors varies randomly between 0 and 2π. The phasor $U^*(t)U(t+\tau)$ then has a totally uncertain angle, so that it is equally likely to take any direction, making its average, the autocorrelation function $G(\tau)$, vanish. On the other hand if, for a given value of τ, $U(t)$ and $U(t+\tau)$ are correlated, their phasors will maintain some relationship. Their fluctuations are then linked together so that the product phasor $U^*(t)U(t+\tau)$ will have a preferred direction and its average $G(\tau)$ will not vanish.

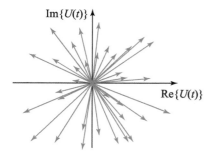

Figure 12.1-2 Variation of the phasor $U(t)$ with time when its argument is uniformly distributed between 0 and 2π. The average values of its real and imaginary parts are zero, so that $\langle U(t) \rangle = 0$.

In the language of optical coherence theory, the autocorrelation function $G(\tau)$ is known as the **temporal coherence function**. It is easy to show that $G(\tau)$ is a function with Hermitian symmetry, $G(-\tau) = G^*(\tau)$, and that the intensity I, defined by (12.1-2), is equal to $G(\tau)$ when $\tau = 0$:

$$I = G(0). \qquad (12.1\text{-}6)$$

Degree of Temporal Coherence

The temporal coherence function $G(\tau)$ carries information about both the intensity $I = G(0)$ and the degree of correlation (coherence) of stationary light. A measure of coherence that is insensitive to the intensity is provided by the normalized autocorrelation function,

$$g(\tau) = \frac{G(\tau)}{G(0)} = \frac{\langle U^*(t)U(t+\tau)\rangle}{\langle U^*(t)U(t)\rangle}, \qquad (12.1\text{-}7)$$

Complex Degree of Temporal Coherence

which is called the **complex degree of temporal coherence**. Its absolute value cannot exceed unity,

$$0 \leq |g(\tau)| \leq 1. \qquad (12.1\text{-}8)$$

The value of $|g(\tau)|$ is a measure of the degree of correlation between $U(t)$ and $U(t+\tau)$. When the light is deterministic and monochromatic, i.e., when $U(t) = A_0 \exp(j2\pi\nu_0 t)$ where A_0 is a constant, (12.1-7) yields

$$g(\tau) = \exp(j2\pi\nu_0\tau), \qquad (12.1\text{-}9)$$

so that $|g(\tau)| = 1$ for all τ. The variables $U(t)$ and $U(t+\tau)$ are then completely correlated for all time delays τ. For most sources of light, $|g(\tau)|$ decreases from its maximum value $|g(0)| = 1$ as τ increases, and the fluctuations become uncorrelated for sufficiently large τ.

Coherence Time

If $|g(\tau)|$ decreases monotonically with time delay, the value τ_c at which it drops to a prescribed value ($1/2$ or $1/e$, for example) serves as a measure of the memory time of the fluctuations. The quantity τ_c is known as the **coherence time** (see Fig. 12.1-3).

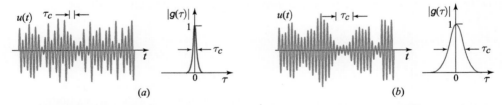

Figure 12.1-3 Illustrations of the wavefunction, magnitude of the complex degree of temporal coherence $|g(\tau)|$, and coherence time τ_c for an optical field with (a) short coherence time and (b) long coherence time. The amplitudes and phases of the wavefunctions vary randomly with time constants approximately equal to τ_c (which is greater than the duration of an optical cycle). Within the coherence time, the wave is rather predictable and can be approximated by a sinusoid. However, given the amplitude and phase of the wave at a particular time, the amplitude and phase at times beyond the coherence time cannot be predicted.

For $\tau < \tau_c$ the fluctuations are "strongly" correlated whereas for $\tau > \tau_c$ they are "weakly" correlated. In general, τ_c is the width of the function $|g(\tau)|$. Although the definition of the width of a function can take many forms (see Sec. A.2 of Appendix A), the power-equivalent width is commonly used as the definition of coherence time:

$$\tau_c = \int_{-\infty}^{\infty} |g(\tau)|^2 \, d\tau \qquad (12.1\text{-}10)$$
Coherence Time

[see (A.2-8) and note that $g(0) = 1$].

EXERCISE 12.1-1

Coherence Time. Verify that the following expressions for the complex degree of temporal coherence are consistent with the definition of τ_c given in (12.1-10):

$$g(\tau) = \begin{cases} \exp\left(-\dfrac{|\tau|}{\tau_c}\right) & \text{(exponential)} \\ \exp\left(-\dfrac{\pi\tau^2}{2\tau_c^2}\right). & \text{(Gaussian)} \end{cases} \qquad (12.1\text{-}11)$$

By what factor does $|g(\tau)|$ drop as τ increases from 0 to τ_c in each case?

The coherence time of monochromatic light is infinite since $|g(\tau)| = 1$ everywhere. Light for which the coherence time τ_c is much longer than differences of the time delays encountered in an optical system of interest is effectively completely coherent. Thus, light is effectively coherent if the distance $c\tau_c$ is much greater than all optical pathlength differences encountered. This distance is known as the **coherence length**:

$$l_c = c\tau_c \qquad (12.1\text{-}12)$$
Coherence Length

Power Spectral Density

To determine the *average* spectrum of random light, we carry out a Fourier decomposition of the random function $U(t)$. The amplitude of the component with frequency ν is the Fourier transform (see Appendix A)

$$V(\nu) = \int_{-\infty}^{\infty} U(t) \exp(-j2\pi\nu t) \, dt. \qquad (12.1\text{-}13)$$

The average energy per unit area of those components with frequencies in the interval between ν and $\nu + d\nu$ is $\langle |V(\nu)|^2 \rangle \, d\nu$, so that $\langle |V(\nu)|^2 \rangle$ represents the energy spectral density of the light (energy per unit area per unit frequency). Note that the complex wave function $U(t)$ has been defined so that $V(\nu) = 0$ for negative ν (see Sec. 2.6A).

Since a truly stationary function $U(t)$ is eternal and carries infinite energy, we consider instead the *power* spectral density. We first determine the energy spectral density of the function $U(t)$ observed over a window of time width T by finding the truncated Fourier transform

$$V_T(\nu) = \int_{-T/2}^{T/2} U(t) \exp(-j2\pi\nu t) \, dt \qquad (12.1\text{-}14)$$

and we then determine the energy spectral density $\langle |V_T(\nu)|^2 \rangle$. The power spectral density is the energy per unit time $(1/T)\langle |V_T(\nu)|^2 \rangle$. We can now extend the time window to infinity by taking the limit $T \to \infty$, which yields the **power spectral density**:

$$S(\nu) = \lim_{T \to \infty} \frac{1}{T} \langle |V_T(\nu)|^2 \rangle; \qquad (12.1\text{-}15)$$

$S(\nu)$ is nonzero only for positive frequencies. Because $U(t)$ was defined such that $|U(t)|^2$ represents power per unit area, or intensity (W/cm^2), $S(\nu)\,d\nu$ represents the average power per unit area carried by frequencies between ν and $\nu + d\nu$, so that $S(\nu)$ actually represents the **intensity spectral density** (W/cm^2-Hz). It is often referred to simply as the **spectral density** or the **spectrum**. The total average intensity is the integral

$$I = \int_0^\infty S(\nu)\,d\nu. \qquad (12.1\text{-}16)$$

The autocorrelation function $G(\tau)$, defined by (12.1-4), and the spectral density $S(\nu)$ defined by (12.1-15) can be shown to form a Fourier transform pair (see Prob. 12.1-5),

$$\boxed{S(\nu) = \int_{-\infty}^\infty G(\tau) \exp(-j2\pi\nu\tau)\,d\tau.} \qquad (12.1\text{-}17)$$
Power Spectral Density

This relation is known as the **Wiener–Khinchin theorem**.

An optical wave representing a color image, such as that illustrated in Fig. 12.1-4, has a spectrum that varies with position **r**; each spectral profile shown corresponds to a perceived color.

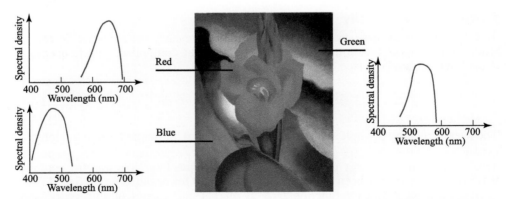

Figure 12.1-4 Spectral densities, plotted as a function of wavelength, at three locations in a color image (Georgia O'Keeffe, *Red Canna*, 1919, High Museum of Art, Atlanta).

Spectral Width

The spectrum of light is often confined to a narrow band centered about a central frequency ν_0. The **spectral width**, or **linewidth**, of light is the width $\Delta\nu$ of the spectral density $S(\nu)$. Because of the Fourier-transform relation between $S(\nu)$ and $G(\tau)$, their widths are inversely related. As illustrated in Fig. 12.1-5, a light source of broad spectral width has a short coherence time, whereas a light source of narrow spectral width has a long coherence time. In the limiting case of monochromatic light, $G(\tau) = I\exp(j2\pi\nu_0\tau)$, so that the corresponding intensity spectral density $S(\nu) = I\delta(\nu-\nu_0)$ contains only a single frequency component ν_0, in which case $\tau_c = \infty$ and $\Delta\nu = 0$. The coherence time of a source of light can be increased by passing it through an optical filter to reduce its spectral width. The resultant gain of coherence comes at the expense of a reduction of its intensity.

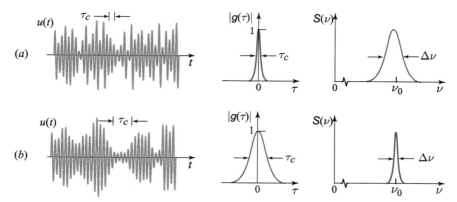

Figure 12.1-5 Two random waves together with the magnitudes of their complex degree of temporal coherence and their spectral densities. The widths of $S(\nu)$ and $|g(\tau)|$ are inversely related.

There are several definitions for spectral width (see Appendix A, Sec. A.2). The most common is the full-width at half-maximum (FWHM) of the function $S(\nu)$, which we denote $\Delta\nu_{\mathrm{FWHM}} \equiv \Delta\nu$. The relation between $\Delta\nu_{\mathrm{FWHM}}$ and the coherence time τ_c depends on the spectral profile of the source, as indicated in Table 12.1-1.

Table 12.1-1 Relation between spectral width $\Delta\nu_{\mathrm{FWHM}}$ and coherence time τ_c for light with several different spectral profiles.

Spectral Profile	Rectangular	Lorentzian	Gaussian
Spectral Width $\Delta\nu_{\mathrm{FWHM}}$	$\dfrac{1}{\tau_c}$	$\dfrac{1}{\pi\tau_c} \approx \dfrac{0.32}{\tau_c}$	$\dfrac{\sqrt{2\ln 2/\pi}}{\tau_c} \approx \dfrac{0.66}{\tau_c}$

An alternative convenient definition of the spectral width is

$$\Delta\nu_c = \frac{\left(\int_0^\infty S(\nu)\,d\nu\right)^2}{\int_0^\infty S^2(\nu)\,d\nu}. \tag{12.1-18}$$

Using the definition provided in (12.1-18) it can be shown that

$$\Delta\nu_c = \frac{1}{\tau_c},$$

(12.1-19)
Spectral Width

regardless of the spectral profile (see Exercise 12.1-2). As an example, if $S(\nu)$ is a rectangular function extending over a frequency interval from $\nu_0 - B/2$ to $\nu_0 + B/2$, then (12.1-18) yields $\Delta\nu_c = B$. For this profile, the coherence time $\tau_c = 1/B$, so that (12.1-19) is obeyed. The two definitions of bandwidth, $\Delta\nu_c$ and $\Delta\nu_{\text{FWHM}}$, differ by a factor that ranges from 0.32 to 1 for the spectral profiles presented in Table 12.1-1.

EXERCISE 12.1-2

Relation Between Spectral Width and Coherence Time. Show that the coherence time τ_c defined in (12.1-10) is related to the spectral width $\Delta\nu_c$ defined in (12.1-18) by the simple inverse relation $\tau_c = 1/\Delta\nu_c$. *Hint:* Use the definitions of $\Delta\nu_c$ and τ_c, the Fourier-transform relation between $S(\nu)$ and $G(\tau)$, and Parseval's theorem provided in (A.1-7) [Appendix A].

Representative spectral widths for several different sources of light, along with their associated coherence times and coherence lengths, are provided in Table 12.1-2.

Table 12.1-2 Spectral widths $\Delta\nu_c$ for various sources of light together with their coherence times τ_c and coherence lengths in free space $l_c = c_o \tau_c$.

Source	$\Delta\nu_c$ (Hz)	$\tau_c = 1/\Delta\nu_c$	$l_c = c_o\tau_c$
Filtered sunlight ($\lambda_o = 0.4$–0.8 μm)	3.74×10^{14}	2.67 fs	800 nm
Light-emitting diode ($\lambda_o = 1$ μm, $\Delta\lambda_o = 50$ nm)	1.5×10^{13}	67 fs	20 μm
Low-pressure sodium lamp	5×10^{11}	2 ps	600 μm
Multimode He–Ne laser ($\lambda_o = 633$ nm)	1.5×10^{9}	0.67 ns	20 cm
Single-mode He–Ne laser ($\lambda_o = 633$ nm)	1×10^{6}	1 μs	300 m

EXAMPLE 12.1-1. *A Wave Comprising a Random Sequence of Wavepackets.* Light emitted from an incoherent source may be modeled as a sequence of wavepackets emitted at random times (Fig. 12.1-6).[†] Each wavepacket is taken to have a random phase since it is emitted by a different atom.

Figure 12.1-6 Light comprising wavepackets emitted at random times has a coherence time equal to the duration of a wavepacket.

[†] See B. E. A. Saleh, D. Stoler, and M. C. Teich, Coherence and Photon Statistics for Optical Fields Generated by Poisson Random Emissions, *Physical Review A*, vol. 27, pp. 360–374, 1983.

The individual wavepackets may be sinusoidal with an exponentially decaying envelope, for example, so that at a given position a wavepacket emitted at $t = 0$ has a complex wavefunction given by

$$U_p(t) = \begin{cases} A_p \exp\left(-\dfrac{t}{\tau_c}\right) \exp(j2\pi\nu_0 t), & t \geq 0 \\ 0, & t < 0. \end{cases} \quad (12.1\text{-}20)$$

The emission times are totally random, and the random independent phases of the different emissions are included in A_p. The statistical properties of the total field may be determined by performing the necessary averaging operations using the rules of mathematical statistics. The result is a complex degree of coherence given by $g(\tau) = \exp(-|\tau|/\tau_c)\exp(j2\pi\nu_0\tau)$, whose magnitude is a double-sided exponential function. The corresponding power spectral density is Lorentzian, $S(\nu) = (\Delta\nu/2\pi)/[(\nu - \nu_0)^2 + (\Delta\nu/2)^2]$, where $\Delta\nu = 1/\pi\tau_c$ (see Table A.1-1 in Appendix A). The coherence time τ_c in this case turns out to be exactly the width of a wavepacket. The statement that this light is correlated within the coherence time therefore signifies that it is correlated within the duration of an individual wavepacket.

C. Spatial Coherence

Mutual Coherence Function

An important descriptor of the spatial and temporal fluctuations of the random function $U(\mathbf{r}, t)$ is the cross-correlation function of $U(\mathbf{r}_1, t)$ and $U(\mathbf{r}_2, t)$ at pairs of positions $\mathbf{r}_1$ and $\mathbf{r}_2$:

$$\boxed{G(\mathbf{r}_1, \mathbf{r}_2, \tau) = \langle U^*(\mathbf{r}_1, t)\, U(\mathbf{r}_2, t + \tau)\rangle.} \quad (12.1\text{-}21)$$
Mutual Coherence Function

This function of the two positions and the time delay τ is known as the **mutual coherence function**. Its normalized form is known as the **complex degree of coherence**:

$$\boxed{g(\mathbf{r}_1, \mathbf{r}_2, \tau) = \dfrac{G(\mathbf{r}_1, \mathbf{r}_2, \tau)}{\sqrt{I(\mathbf{r}_1)I(\mathbf{r}_2)}}.} \quad (12.1\text{-}22)$$
Complex Degree of Coherence

When the two points coincide so that $\mathbf{r}_1 = \mathbf{r}_2 = \mathbf{r}$, (12.1-21) and (12.1-22) reproduce the temporal coherence function and the complex degree of temporal coherence at the position $\mathbf{r}$, as provided in (12.1-4) and (12.1-7), respectively. When $\tau = 0$ as well, we recover the intensity $I(\mathbf{r}) = G(\mathbf{r}, \mathbf{r}, 0)$ at the position $\mathbf{r}$. The analogous cross-correlation functions in the *quantum* theory of optical coherence are defined in terms of operators rather than fields and the averages are carried out with respect to the quantum state of the light, in accordance with the precepts of quantum optics (Sec. 13.3).

The complex degree of coherence $g(\mathbf{r}_1, \mathbf{r}_2, \tau)$ is the cross-correlation coefficient of the random variables $U^*(\mathbf{r}_1, t)$ and $U(\mathbf{r}_2, t + \tau)$. As with the complex degree of temporal coherence [see (12.1-8)], its absolute value is bounded between zero and unity:

$$0 \leq |g(\mathbf{r}_1, \mathbf{r}_2, \tau)| \leq 1. \quad (12.1\text{-}23)$$

This quantity is therefore considered a measure of the degree of correlation between the fluctuations at $\mathbf{r}_1$ and those at $\mathbf{r}_2$ at a time τ later.

When the two phasors $U(\mathbf{r}_1, t)$ and $U(\mathbf{r}_2, t)$ fluctuate independently and their phases are totally random (each having a phase that is equally probable between 0 and 2π), $|g(\mathbf{r}_1, \mathbf{r}_2, \tau)| = 0$ since the average of the product $U^*(\mathbf{r}_1, t) U(\mathbf{r}_2, t + \tau)$ vanishes. The light fluctuations at the two points are then uncorrelated. The other limit, $|g(\mathbf{r}_1, \mathbf{r}_2, \tau)| = 1$, obtains when the light fluctuations at $\mathbf{r}_1$, and at $\mathbf{r}_2$ a time τ later, are fully correlated. Note that $|g(\mathbf{r}_1, \mathbf{r}_2, 0)|$ is not necessarily unity; however, by definition $|g(\mathbf{r}, \mathbf{r}, 0)| = 1$.

The dependence of $g(\mathbf{r}_1, \mathbf{r}_2, \tau)$ on the positions and on the time delay characterizes the spatial and temporal coherence of light. Two examples of the dependence of $|g(\mathbf{r}_1, \mathbf{r}_2, \tau)|$ on the distance $|\mathbf{r}_1 - \mathbf{r}_2|$ and on the time delay τ are illustrated in Fig. 12.1-7. The temporal and spatial fluctuations of light are interrelated since light propagates in waves and the complex wavefunction $U(\mathbf{r}, t)$ must satisfy the wave equation. This imposes certain conditions on the mutual coherence function (see Exercise 12.1-3). To illustrate this point, consider, for example, a plane wave of random light traveling in the z direction at velocity c in a homogeneous and nondispersive medium. Fluctuations at the points $\mathbf{r}_1 = (0, 0, z_1)$ and $\mathbf{r}_2 = (0, 0, z_2)$ are completely correlated when the time delay is $\tau = \tau_0 \equiv |z_2 - z_1|/c$, whereupon $|g(\mathbf{r}_1, \mathbf{r}_2, \tau_0)| = 1$. Considered as a function of τ, $|g(\mathbf{r}_1, \mathbf{r}_2, \tau)|$ then has its maximum value at $\tau = \tau_0$, as illustrated in Fig. 12.1-7(b). This example will be revisited in Sec. 12.1D.

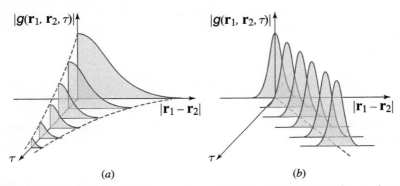

Figure 12.1-7 Two examples of $|g(\mathbf{r}_1, \mathbf{r}_2, \tau)|$ as a function of the separation $|\mathbf{r}_1 - \mathbf{r}_2|$ and the time delay τ. In (a) the maximum correlation for a given $|\mathbf{r}_1 - \mathbf{r}_2|$ occurs at $\tau = 0$, whereas in (b) the maximum correlation occurs at $|\mathbf{r}_1 - \mathbf{r}_2| = c\tau$.

EXERCISE 12.1-3

Differential Equations Governing the Mutual Coherence Function. In free space, $U(\mathbf{r}, t)$ must satisfy the wave equation, $\nabla^2 U - (1/c^2)\partial^2 U/\partial t^2 = 0$. Use the definition (12.1-21) to show that the mutual coherence function $G(\mathbf{r}_1, \mathbf{r}_2, \tau)$ satisfies a pair of partial differential equations known as the **Wolf equations**,

$$\nabla_1^2 G - \frac{1}{c^2} \frac{\partial^2 G}{\partial \tau^2} = 0 \qquad (12.1\text{-}24a)$$

$$\nabla_2^2 G - \frac{1}{c^2} \frac{\partial^2 G}{\partial \tau^2} = 0, \qquad (12.1\text{-}24b)$$

where ∇_1^2 and ∇_2^2 are the Laplacian operators with respect to $\mathbf{r}_1$ and $\mathbf{r}_2$, respectively.

Mutual Intensity

The spatial correlation of light may be assessed by examining the dependence of the mutual coherence function on position at a specified fixed time delay τ. In many situations the point $\tau = 0$ is the most appropriate to consider, as in the example illustrated in Fig. 12.1-7(a). The mutual coherence function at $\tau = 0$ is known as the **mutual intensity**,

$$G(\mathbf{r}_1, \mathbf{r}_2, 0) = \langle U^*(\mathbf{r}_1, t) U(\mathbf{r}_2, t) \rangle, \qquad (12.1\text{-}25)$$

and is usually denoted by $G(\mathbf{r}_1, \mathbf{r}_2)$ for simplicity. The diagonal values of the mutual intensity ($\mathbf{r}_1 = \mathbf{r}_2 = \mathbf{r}$) yield the intensity $I(\mathbf{r}) = G(\mathbf{r}, \mathbf{r})$. It is sometimes appropriate to use values of τ other than $\tau = 0$, however, as the example in Fig. 12.1-7(b) illustrates.

When the optical pathlength differences encountered in an optical system are much shorter than the coherence length $l_c = c\tau_c$, the light effectively possesses complete temporal coherence, in which case the mutual coherence function is a harmonic function of time [see (12.1-10)],

$$G(\mathbf{r}_1, \mathbf{r}_2, \tau) = G(\mathbf{r}_1, \mathbf{r}_2) \exp(j\omega_0 \tau), \qquad (12.1\text{-}26)$$

where $\nu_0 = \omega_0 / 2\pi$ is the central frequency. The light is then referred to as **quasi-monochromatic** and the mutual intensity $G(\mathbf{r}_1, \mathbf{r}_2)$ completely describes the spatial coherence.

At $\tau = 0$, the complex degree of coherence $g(\mathbf{r}_1, \mathbf{r}_2, 0)$ is called the **normalized mutual intensity** and is denoted $g(\mathbf{r}_1, \mathbf{r}_2)$:

$$\boxed{g(\mathbf{r}_1, \mathbf{r}_2) = \frac{G(\mathbf{r}_1, \mathbf{r}_2)}{\sqrt{I(\mathbf{r}_1) I(\mathbf{r}_2)}}.} \qquad (12.1\text{-}27)$$

Normalized Mutual Intensity

The magnitude $|g(\mathbf{r}_1, \mathbf{r}_2)|$ is bounded between zero and unity and is regarded as a measure of the degree of spatial coherence when the time delay $\tau = 0$. If the complex wavefunction $U(\mathbf{r}, t)$ is deterministic, $|g(\mathbf{r}_1, \mathbf{r}_2)| = 1$ for all $\mathbf{r}_1$ and $\mathbf{r}_2$, and the light is completely correlated everywhere.

Coherence Area

In a given plane, in the vicinity of a given position $\mathbf{r}_2$, the spatial coherence of quasi-monochromatic light is described by $|g(\mathbf{r}_1, \mathbf{r}_2)|$ as a function of the distance $|\mathbf{r}_1 - \mathbf{r}_2|$. This function is unity when $\mathbf{r}_1 = \mathbf{r}_2$ and decreases (but not necessarily monotonically) as $|\mathbf{r}_1 - \mathbf{r}_2|$ increases. The area scanned by the point $\mathbf{r}_1$ within which the function $|g(\mathbf{r}_1, \mathbf{r}_2)|$ is greater than some prescribed value ($1/2$ or $1/e$, for example) is called the **coherence area** A_c. It represents the spatial extent of $|g(\mathbf{r}_1, \mathbf{r}_2)|$ as a function of $\mathbf{r}_1$ for a fixed value of $\mathbf{r}_2$, as illustrated in Fig. 12.1-8. In the ideal limit of coherent light, the coherence area is infinite.

The coherence area is an important parameter for characterizing random light, but it must be viewed in relation to other pertinent dimensions of the optical system under consideration. For example, if the coherence area is greater than the size of an aperture through which the light is transmitted, we have $|g(\mathbf{r}_1, \mathbf{r}_2)| \approx 1$ at all points of interest, so that the light may be regarded as coherent (just as if A_c were infinite). Similarly, if the coherence area is smaller than the spatial resolution of the optical system, it can be regarded as infinitesimal, i.e., $g(\mathbf{r}_1, \mathbf{r}_2) \approx 0$ for practically all $\mathbf{r}_1 \neq \mathbf{r}_2$. In this limit, the light is said to be **incoherent**.

Light emitted from an extended radiating hot surface has a coherence area of the order of λ^2, where λ is the central wavelength, so that in most practical cases such light may be regarded as incoherent. Complete coherence and incoherence are therefore seen to be idealizations that represent the two limits of partial coherence.

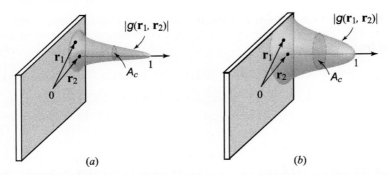

Figure 12.1-8 Two illustrative examples of the magnitude of the normalized mutual intensity as a function of $\mathbf{r}_1$ in the vicinity of a fixed point $\mathbf{r}_2$. The coherence area in (a) is smaller than that in (b).

Cross-Spectral Density

The mutual coherence function $G(\mathbf{r}_1, \mathbf{r}_2, \tau)$ describes the spatial correlation at each time delay τ. The time delay $\tau = 0$ is selected to define the mutual intensity $G(\mathbf{r}_1, \mathbf{r}_2) = G(\mathbf{r}_1, \mathbf{r}_2, 0)$, which is suitable for describing the spatial coherence of quasi-monochromatic light. A useful alternative is to describe coherence in the frequency domain by examining the spatial correlation at a fixed frequency. The **cross-spectral density** (or the cross-power spectrum) is defined as the Fourier transform of $G(\mathbf{r}_1, \mathbf{r}_2, \tau)$ with respect to τ:

$$S(\mathbf{r}_1, \mathbf{r}_2, \nu) = \int_{-\infty}^{\infty} G(\mathbf{r}_1, \mathbf{r}_2, \tau) \exp(-j2\pi\nu\tau) \, d\tau. \qquad (12.1\text{-}28)$$

Cross-Spectral Density

When $\mathbf{r}_1 = \mathbf{r}_2 = \mathbf{r}$, the cross-spectral density becomes the power-spectral density $S(\nu)$ at position $\mathbf{r}$, as defined in (12.1-17).

The **normalized cross-spectral density** is defined by

$$s(\mathbf{r}_1, \mathbf{r}_2, \nu) = \frac{S(\mathbf{r}_1, \mathbf{r}_2, \nu)}{\sqrt{S(\mathbf{r}_1, \mathbf{r}_1, \nu)\, S(\mathbf{r}_2, \mathbf{r}_2, \nu)}}. \qquad (12.1\text{-}29)$$

Its magnitude can be shown to be bounded between zero and unity, so that it serves as a measure of the degree of spatial coherence at the frequency ν. It represents the degree of correlation of the fluctuation components of frequency ν at positions $\mathbf{r}_1$ and $\mathbf{r}_2$.

In certain cases, the cross-spectral density factors into a product of a function of position and another of frequency, $S(\mathbf{r}_1, \mathbf{r}_2, \nu) = G(\mathbf{r}_1, \mathbf{r}_2)s(\nu)$, so that the spatial and spectral properties are separable. The light is then said to be **cross-spectrally pure**. The mutual coherence function must then also factor into a product of a function of position and another of time, $G(\mathbf{r}_1, \mathbf{r}_2, \tau) = G(\mathbf{r}_1, \mathbf{r}_2)g(\tau)$, where $g(\tau)$ is the inverse Fourier transform of $s(\nu)$. If the factorization parts are selected such that $\int s(\nu)\, d\nu = 1$, then $G(\mathbf{r}_1, \mathbf{r}_2) = G(\mathbf{r}_1, \mathbf{r}_2, 0)$, so that $G(\mathbf{r}_1, \mathbf{r}_2)$ is nothing but the mutual intensity. Cross-spectrally pure light has two important properties:

1. At a single position $\mathbf{r}$, $S(\mathbf{r}, \mathbf{r}, \nu) = G(\mathbf{r}, \mathbf{r})s(\nu) = I(\mathbf{r})s(\nu)$. The spectrum has the same profile at all positions. If the light represents a visible image, it would appear to have the same color everywhere but the brightness would vary.

2. The normalized cross-spectral density

$$s(\mathbf{r}_1, \mathbf{r}_2, \nu) = G(\mathbf{r}_1, \mathbf{r}_2)/\sqrt{G(\mathbf{r}_1, \mathbf{r}_1)\, G(\mathbf{r}_2, \mathbf{r}_2)} = g(\mathbf{r}_1, \mathbf{r}_2) \qquad (12.1\text{-}30)$$

is independent of frequency. In this case the normalized mutual intensity $g(\mathbf{r}_1, \mathbf{r}_2)$ describes the spatial coherence at all frequencies.

D. Longitudinal Coherence

In this section the concept of longitudinal coherence is introduced by taking examples of random waves with fixed wavefronts, such as plane and spherical waves.

Partially Coherent Plane Wave

Consider a plane wave

$$U(\mathbf{r}, t) = \mathfrak{a}\left(t - \frac{z}{c}\right) \exp\left[j\omega_0 \left(t - \frac{z}{c}\right)\right] \qquad (12.1\text{-}31)$$

traveling in the z direction in a homogeneous medium with velocity c, as considered in Sec. 2.6A. The complex wavefunction $U(\mathbf{r}, t)$ satisfies the wave equation for an arbitrary function $\mathfrak{a}(t)$. If $\mathfrak{a}(t)$ is a random function, $U(\mathbf{r}, t)$ represents partially coherent light. The mutual coherence function defined in (12.1-21) is then

$$G(\mathbf{r}_1, \mathbf{r}_2, \tau) = G_\mathfrak{a}\left(\tau - \frac{z_2 - z_1}{c}\right) \exp\left[j\omega_0 \left(\tau - \frac{z_2 - z_1}{c}\right)\right], \qquad (12.1\text{-}32)$$

where z_1 and z_2 are the z components of $\mathbf{r}_1$ and $\mathbf{r}_2$ and $G_\mathfrak{a}(\tau) = \langle \mathfrak{a}^*(t)\mathfrak{a}(t+\tau)\rangle$ is the autocorrelation function of $\mathfrak{a}(t)$, which is assumed to be independent of t [see Fig. 12.1-7(b)].

The intensity $I(\mathbf{r}) = G(\mathbf{r}, \mathbf{r}, 0) = G_\mathfrak{a}(0)$ is constant everywhere in space. Temporal coherence is characterized by the time function $G(\mathbf{r}, \mathbf{r}, \tau) = G_\mathfrak{a}(\tau)\exp(j\omega_0\tau)$, which is independent of position. The complex degree of coherence is $g(\mathbf{r}, \mathbf{r}, \tau) = g_\mathfrak{a}(\tau)\exp(j\omega_0\tau)$, where $g_\mathfrak{a}(\tau) = G_\mathfrak{a}(\tau)/G_\mathfrak{a}(0)$. The width of $|g_\mathfrak{a}(\tau)| = |g(\mathbf{r}, \mathbf{r}, \tau)|$, defined by an expression similar to (12.1-10), is the coherence time τ_c. It is the same at all positions.

The power spectral density is the Fourier transform of $G(\mathbf{r}, \mathbf{r}, \tau)$ with respect to τ. From (12.1-32), $S(\nu)$ is seen to be equal to the Fourier transform of $G_\mathfrak{a}(\tau)$ shifted by a frequency ν_0 (in accordance with the frequency shift property of the Fourier transform defined in Appendix A, Sec. A.1). The wave therefore has the same power spectral density everywhere in space.

The spatial coherence properties are described by

$$G(\mathbf{r}_1, \mathbf{r}_2, 0) = G_\mathfrak{a}\left(\frac{z_1 - z_2}{c}\right) \exp\left[j\omega_0 \frac{z_1 - z_2}{c}\right], \qquad (12.1\text{-}33)$$

and its normalized version

$$g(\mathbf{r}_1, \mathbf{r}_2, 0) = g_\mathfrak{a}\left(\frac{z_1 - z_2}{c}\right) \exp\left[j\omega_0 \frac{z_1 - z_2}{c}\right]. \qquad (12.1\text{-}34)$$

If the two points $\mathbf{r}_1$ and $\mathbf{r}_2$ lie in the same transverse plane, i.e., $z_1 = z_2$, then $|g(\mathbf{r}_1, \mathbf{r}_2, 0)| = |g_\mathfrak{a}(0)| = 1$. This means that fluctuations at points on a wavefront (a plane normal to the z axis) are completely correlated; the coherence area in any transverse plane is infinite (see Fig. 12.1-9). On the other hand, fluctuations at two points separated by an axial distance $z_2 - z_1$ such that $|z_2 - z_1|/c > \tau_c$, or $|z_2 - z_1| > l_c$ where $l_c = c\tau_c$ is the coherence length, are approximately uncorrelated.

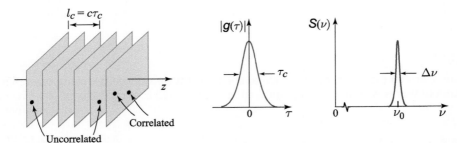

Figure 12.1-9 The fluctuations of a partially coherent plane wave at points on any wavefront (transverse plane) are completely correlated, whereas those at points on wavefronts separated by an axial distance greater than the coherence length $l_c = c\tau_c$ are approximately uncorrelated.

We conclude that the partially coherent plane wave is spatially coherent across each transverse plane, but only partially coherent in the axial direction. The axial (longitudinal) spatial coherence of the wave has a one-to-one correspondence with the temporal coherence. The relationship of the coherence length $l_c = c\tau_c$ to the maximum optical path difference in the system $l_{\max}$ governs the role played by coherence. If $l_c \gg l_{\max}$, the wave is effectively completely coherent. The coherence lengths of various light sources were provided in Table 12.1-2.

Partially Coherent Spherical Wave

A partially coherent spherical wave is described by the complex wavefunction (see Sec. 2.2B and Sec. 2.6A)

$$U(\mathbf{r}, t) = \frac{1}{r} \mathsf{a}\left(t - \frac{r}{c}\right) \exp\left[j\omega_0 \left(t - \frac{r}{c}\right)\right], \quad (12.1\text{-}35)$$

where $\mathsf{a}(t)$ is a random function. The corresponding mutual coherence function is

$$G(\mathbf{r}_1, \mathbf{r}_2, \tau) = \frac{1}{r_1 r_2} G_\mathsf{a}\left(\tau - \frac{r_2 - r_1}{c}\right) \exp\left[j\omega_0 \left(\tau - \frac{r_2 - r_1}{c}\right)\right], \quad (12.1\text{-}36)$$

with $G_\mathsf{a}(\tau) = \langle \mathsf{a}^*(t)\,\mathsf{a}(t + \tau)\rangle$.

The intensity $I(\mathbf{r}) = G_\mathsf{a}(0)/r^2$ varies in accordance with an inverse-square law. The coherence time τ_c is the width of the function $|g_\mathsf{a}(\tau)| = |G_\mathsf{a}(\tau)/G_\mathsf{a}(0)|$. It is the same everywhere in space, as is the power spectral density. For $\tau = 0$, fluctuations at all points on a spherical wavefront are completely correlated, whereas fluctuations at points on two wavefronts separated by the radial distance $|r_2 - r_1| \gg l_c = c\tau_c$ are uncorrelated (see Fig. 12.1-10).

An arbitrary partially coherent wave transmitted through a pinhole generates a partially coherent spherical wave. This process therefore imparts spatial coherence to the incident wave (points on any sphere centered about the pinhole become completely correlated). However, the wave remains temporally partially coherent. Points at different distances from the pinhole are only partially correlated. The pinhole imparts spatial coherence, but not temporal coherence, to the wave.

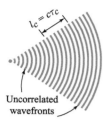

Figure 12.1-10 A partially coherent spherical wave exhibits complete spatial coherence at all points on a wavefront, while points on wavefronts separated by a distance greater than the coherence length $l_c = c\tau_c$ are approximately uncorrelated.

Suppose now that an optical filter of very narrow spectral width is placed at the pinhole, causing the transmitted wave to become approximately monochromatic. The wave will then have complete temporal as well as spatial coherence. Spatial coherence is imparted by the pinhole, which acts as a spatial filter, while temporal coherence is introduced by the narrowband filter. The price paid for obtaining such an ideally coherent wave is, of course, the loss of optical energy associated with the temporal and spatial filtering processes.

12.2 INTERFERENCE OF PARTIALLY COHERENT LIGHT

The interference of *coherent* light was discussed in Sec. 2.5. This section is devoted to the interference of *partially coherent* light.

A. Interference of Two Partially Coherent Waves

The statistical properties of two partially coherent waves U_1 and U_2 are characterized not only by their own mutual coherence functions but also by a measure of the degree to which their fluctuations are correlated. At a given position $\mathbf{r}$ and time t, the intensities of the two waves are $I_1 = \langle |U_1|^2 \rangle$ and $I_2 = \langle |U_2|^2 \rangle$, whereas their cross-correlation is described by the statistical average $G_{12} = \langle U_1^* U_2 \rangle$, along with its normalized version

$$g_{12} = \frac{\langle U_1^* U_2 \rangle}{\sqrt{I_1 I_2}}. \tag{12.2-1}$$

When the two waves are superposed, the average intensity of their sum is

$$\begin{aligned} I &= \langle |U_1 + U_2|^2 \rangle = \langle |U_1|^2 \rangle + \langle |U_2|^2 \rangle + \langle U_1^* U_2 \rangle + \langle U_1 U_2^* \rangle \\ &= I_1 + I_2 + G_{12} + G_{12}^* = I_1 + I_2 + 2\operatorname{Re}\{G_{12}\} \\ &= I_1 + I_2 + 2\sqrt{I_1 I_2}\,\operatorname{Re}\{g_{12}\}. \end{aligned} \tag{12.2-2}$$

We thus obtain

$$\boxed{I = I_1 + I_2 + 2\sqrt{I_1 I_2}\,|g_{12}|\cos\varphi,} \tag{12.2-3}$$
Interference Equation

where $\varphi = \arg\{g_{12}\}$ is the phase of g_{12}. The third term on the right-hand side of (12.2-3) represents optical interference.

It is useful to consider two limits of this equation:

1. For two *completely correlated* waves with $g_{12} = \exp(j\varphi)$ and $|g_{12}| = 1$, we recover the interference equation (2.5-4) for two coherent waves of phase difference φ.

2. For two *uncorrelated* waves with $g_{12} = 0$, the result is $I = I_1 + I_2$ and there is no interference.

In the general case, the normalized intensity versus the phase φ assumes the form of a sinusoidal pattern, as shown in Fig. 12.2-1. The strength of the interference is measured by the **visibility** $\mathcal{V}$ (also called the *modulation depth* or the *contrast* of the interference pattern):

$$\mathcal{V} = \frac{I_{\max} - I_{\min}}{I_{\max} + I_{\min}}, \qquad (12.2\text{-}4)$$

where $I_{\max}$ and $I_{\min}$ are, respectively, the maximum and minimum values that I takes as φ is varied. Since $\cos\varphi$ stretches between 1 and -1, inserting (12.2-3) into (12.2-4) yields

$$\mathcal{V} = \frac{2\sqrt{I_1 I_2}}{I_1 + I_2}\,|g_{12}|. \qquad (12.2\text{-}5)$$

The visibility is therefore proportional to the absolute value of the normalized cross-correlation $|g_{12}|$. In the special case when $I_1 = I_2$, this simplifies to

$$\boxed{\mathcal{V} = |g_{12}|.} \qquad (12.2\text{-}6)$$
Visibility

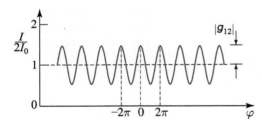

Figure 12.2-1 Normalized intensity $I/2I_0$ of the sum of two partially coherent waves of equal intensities ($I_1 = I_2 = I_0$), as a function of the phase φ of their normalized cross-correlation g_{12}. This sinusoidal pattern has visibility $\mathcal{V} = |g_{12}|$.

The interference equation (12.2-3) will now be considered in a number of specific contexts to highlight the effects that temporal and spatial coherence have on the interference of partially coherent light.

B. Interferometry and Temporal Coherence

Consider a partially coherent wave $U(t)$ with intensity I_0 and complex degree of temporal coherence $g(\tau) = \langle U^*(t)U(t+\tau)\rangle/I_0$. If $U(t)$ is simply added to a replica of itself that is delayed by the time τ, $U(t+\tau)$, what is the intensity I of the superposition?

Using the interference formula (12.2-2) with $U_1 = U(t)$, $U_2 = U(t+\tau)$, $I_1 = I_2 = I_0$, and $g_{12} = \langle U_1^* U_2\rangle/I_0 = \langle U^*(t)U(t+\tau)\rangle/I_0 = g(\tau)$, we obtain

$$I = 2I_0\left[1 + \text{Re}\left\{g(\tau)\right\}\right] = 2I_0\left[1 + |g(\tau)|\cos\varphi(\tau)\right], \qquad (12.2\text{-}7)$$

where $\varphi(\tau) = \arg\{g(\tau)\}$. It is thus apparent that the ability of a wave to interfere with a time delayed replica of itself is governed by its complex degree of temporal coherence at that time delay.

12.2 INTERFERENCE OF PARTIALLY COHERENT LIGHT

Implementing the addition of a wave with a time-delayed replica of itself may be achieved by using a beamsplitter to generate two identical waves, one of which is made to traverse a longer optical path than the other, and then recombining them at another (or the same) beamsplitter. This can be effected, for example, with the help of a Mach–Zehnder or a Michelson interferometer (see Fig. 2.5-3).

Consider, as an example, the partially coherent plane wave introduced in Sec. 12.1D [see (12.1-31)], whose complex degree of temporal coherence is $g(\tau) = g_a(\tau) \exp(j\omega_0\tau)$. The spectral width of the wave is $\Delta\nu_c = 1/\tau_c$, where τ_c (the width of $|g_a(\tau)|$) is the coherence time. Substituting this into (12.2-7), we obtain

$$I = 2I_0 \left\{ 1 + |g_a(\tau)| \cos\left[\omega_0\tau + \varphi_a(\tau)\right] \right\}, \qquad (12.2\text{-}8)$$

where $\varphi_a(\tau) = \arg\{g_a(\tau)\}$.

This relation between I and τ, which is known as an **interferogram**, is illustrated in Fig. 12.2-2. Assuming that $\Delta\nu_c = 1/\tau_c \ll \nu_0$, the functions $|g_a(\tau)|$ and $\varphi_a(\tau)$ vary slowly in comparison with the period $1/\nu_0$. The visibility of this interferogram in the vicinity of a particular time delay τ is $\mathcal{V} = |g(\tau)| = |g_a(\tau)|$. It has a peak value of unity near $\tau = 0$ and vanishes for $\tau \gg \tau_c$, i.e., when the optical path difference is much greater than the coherence length $l_c = c\tau_c$. For the Michelson interferometer illustrated in Fig. 12.2-2, $\tau = 2(d_2 - d_1)/c$. Interference occurs only when the optical path difference is smaller than the coherence length.

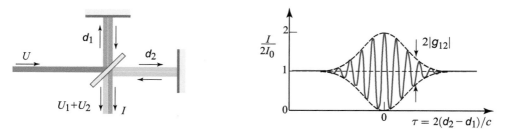

Figure 12.2-2 The normalized intensity $I/2I_0$, as a function of the time delay τ, when a partially coherent plane wave is introduced into a Michelson interferometer. The visibility is a measure of the magnitude of the complex degree of temporal coherence.

The magnitude of the complex degree of temporal coherence of a wave, $|g(\tau)|$, may therefore be measured by monitoring the visibility of the interference pattern as a function of time delay. The phase of $g(\tau)$ may be measured by observing the locations of the peaks of the pattern.

Fourier-Transform Spectroscopy

It is revealing to write (12.2-7) in terms of the power spectral density of the wave $S(\nu)$. Using the Fourier-transform relation between $G(\tau)$ and $S(\nu)$,

$$G(\tau) = I_0 g(\tau) = \int_0^\infty S(\nu) \exp(j2\pi\nu\tau)\, d\nu, \qquad (12.2\text{-}9)$$

substituting into (12.2-7), and noting that $S(\nu)$ is real and that $\int_0^\infty S(\nu)\, d\nu = I_0$, we obtain

$$I = 2\int_0^\infty S(\nu)\left[1 + \cos(2\pi\nu\tau)\right] d\nu. \qquad (12.2\text{-}10)$$

This equation can be interpreted as representing a weighted superposition of interferograms produced by each of the monochromatic components of the wave. Each component ν produces an interferogram with period $1/\nu$ and unity visibility, but the composite interferogram exhibits reduced visibility by virtue of the different periods.

Equation (12.2-10) suggests that the spectral density $S(\nu)$ of a light source can be determined by measuring the interferogram I versus τ and then inverting the result by means of Fourier-transform methods. This technique is known as **Fourier-transform spectroscopy**.

Optical Coherence Tomography

Optical coherence tomography (OCT) is an interferometric technique for profiling a multilayered medium, i.e., for measuring the reflectance and depth of each of its boundaries. In its simplest form, known as **time-domain OCT**, it makes use of a partially coherent light source of short coherence length and a Michelson interferometer. As illustrated in Fig. 12.2-3, a replica of the original wave, delayed by a movable mirror, is superposed with a collection of waves reflected from the multiple boundaries of the sample. Information about the sample profile is carried by the interferogram, which is the intensity measured at the detector as the movable mirror is translated. By virtue of the short coherence length of the source, the interferogram comprises sets of fringes centered at path delays of the movable mirror that match those of the reflecting boundaries.

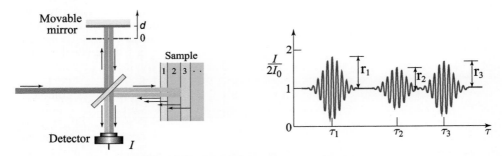

Figure 12.2-3 Optical coherence tomography.

Let $U(t - \tau)$ be the wave reflected from the movable mirror, with its associated time delay $\tau = d/c_o$, and let $r_i U(t - \tau_i)$, $i = 1, 2, \ldots$, be the waves reflected from the boundaries of the sample, where r_i represents the amplitude reflectance at the ith boundary; the associated time delays are designated τ_i. For a symmetric beamsplitter, the average intensity is then $I(\tau) = \langle |U(t - \tau) + \sum_i r_i U(t - \tau_i)|^2 \rangle$, which may be written in normalized form as

$$I/2I_0 = 1 + \sum_i r_i \operatorname{Re}\{g(\tau - \tau_i)\} + \sum_{ij} r_i r_j^* \operatorname{Re}\{g(\tau_j - \tau_i)\}, \qquad (12.2\text{-}11)$$

since the complex degree of temporal coherence of the source is characterized by $g(\tau) = \langle U^*(t)\, U(t + \tau) \rangle / \langle U^*(t) U(t) \rangle$.

The left-most summation on the right-hand side of (12.2-11) is of paramount importance since it represents interference between the reference wave from the movable mirror and each of the waves reflected from the sample boundaries. The right-most summation represents interference terms associated with pairs of reflections from the sample; since these terms are independent of the path delay of the movable mirror, $\tau = d/c$, they may be regarded as background contributions and ignored.

For a light source of central frequency ν_0, we have $g(\tau) = g_a(\tau)\exp(j\omega_0\tau)$, where the width of $g_a(\tau)$ is the coherence time τ_c. Equation (12.2-11) then becomes

$$I/2I_0 \approx 1 + \sum_i r_i |g_a(\tau - \tau_i)| \cos\left[\omega_0(\tau - \tau_i) + \varphi_a(\tau - \tau_i)\right], \qquad (12.2\text{-}12)$$

where $\varphi_a(\tau) = \arg\{g_a(\tau)\}$. If the source is of short coherence length, the function $g_a(\tau)$ is narrow. As illustrated in Fig. 12.2-3, the reflection from each sample boundary then generates a distinct set of interference fringes of brief duration τ_c, centered about its corresponding time delay. Measurement of the OCT interferogram therefore permits the reflectance at each boundary, as well as the width of each of the sample layers, to be determined.

Optical coherence tomography has proven to be an effective imaging technique in clinical medicine as well as in engineering. It can also be carried out in a **frequency-domain OCT** configuration in the spirit of Fourier-transform spectroscopy. A particularly useful configuration makes use of a narrowband optical source whose frequency is swept in time (e.g., a wavelength-swept laser); this approach offers improved detection sensitivity and data-acquisition rates.

C. Interferometry and Spatial Coherence

The effect of spatial coherence on interference is demonstrated by considering the Young's double-pinhole interference experiment discussed in Exercise 2.5-2 for coherent light. A partially coherent optical wave $U(\mathbf{r}, t)$ illuminates an opaque screen with two pinholes located at positions $\mathbf{r}_1$ and $\mathbf{r}_2$. The wave has mutual coherence function $G(\mathbf{r}_1, \mathbf{r}_2, \tau) = \langle U^*(\mathbf{r}_1, t) U(\mathbf{r}_2, t + \tau) \rangle$ and complex degree of coherence $g(\mathbf{r}_1, \mathbf{r}_2, \tau)$. The intensities at the pinholes are assumed to be equal.

Light is diffracted in the form of two spherical waves centered at the pinholes. The two waves interfere, and the intensity I of their sum is observed at a point $\mathbf{r}$ in the observation plane, at a distance d from the screen that is sufficiently large so that the paraboloidal approximation is applicable. In Cartesian coordinates, as shown in Fig. 12.2-4, $\mathbf{r}_1 = (-a, 0, 0)$, $\mathbf{r}_2 = (a, 0, 0)$, and $\mathbf{r} = (x, 0, d)$. The intensity is observed as a function of x. An important geometrical parameter is the angle $\theta \approx 2a/d$ subtended by the two pinholes.

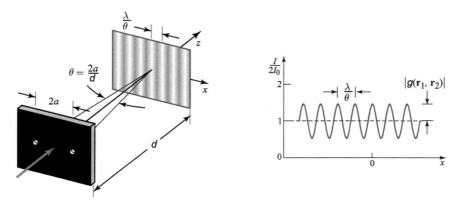

Figure 12.2-4 Young's double-pinhole interferometer illuminated by partially coherent light. The incident wave is quasi-monochromatic and has a normalized mutual intensity at the pinholes described by $g(\mathbf{r}_1, \mathbf{r}_2)$. The normalized intensity $I/2I_0$ in the observation plane, at a large distance from the pinhole plane, is a sinusoidal function of x with period λ/θ and visibility $\mathcal{V} = |g(\mathbf{r}_1, \mathbf{r}_2)|$.

In the paraboloidal (Fresnel) approximation [see (2.2-17)], the two diffracted spherical waves are approximately related to $U(\mathbf{r}, t)$ by

$$U_1(\mathbf{r}, t) \propto U\left(\mathbf{r}_1, t - \frac{|\mathbf{r} - \mathbf{r}_1|}{c}\right) \approx U\left(\mathbf{r}_1, t - \frac{d + (x + a)^2/2d}{c}\right) \quad (12.2\text{-}13\text{a})$$

$$U_2(\mathbf{r}, t) \propto U\left(\mathbf{r}_2, t - \frac{|\mathbf{r} - \mathbf{r}_2|}{c}\right) \approx U\left(\mathbf{r}_2, t - \frac{d + (x - a)^2/2d}{c}\right), \quad (12.2\text{-}13\text{b})$$

and have approximately equal intensities, $I_1 = I_2 = I_0$. The normalized cross-correlation between the two waves at $\mathbf{r}$ is

$$g_{12} = \frac{\langle U_1^*(\mathbf{r}, t) U_2(\mathbf{r}, t) \rangle}{I_0} = g(\mathbf{r}_1, \mathbf{r}_2, \tau_x), \quad (12.2\text{-}14)$$

where the difference in the time delays encountered by the two waves is given by

$$\tau_x = \frac{|\mathbf{r} - \mathbf{r}_1| - |\mathbf{r} - \mathbf{r}_2|}{c} = \frac{(x + a)^2 - (x - a)^2}{2dc} = \frac{2ax}{dc} = \frac{\theta}{c} x. \quad (12.2\text{-}15)$$

Substituting (12.2-14) into the interference formula (12.2-3) gives rise to an observed intensity $I \equiv I(x)$:

$$\boxed{I(x) = 2I_0 \left[1 + |g(\mathbf{r}_1, \mathbf{r}_2, \tau_x)| \cos \varphi_x\right],} \quad (12.2\text{-}16)$$

where $\varphi_x = \arg\{g(\mathbf{r}_1, \mathbf{r}_2, \tau_x)\}$. This equation describes the pattern of intensity observed as a function of position x in the observation plane, in terms of the magnitude and phase of the complex degree of coherence at the pinholes at time delay $\tau_x = \theta x/c$.

Quasi-Monochromatic Light

Moreover, if the light is quasi-monochromatic with central frequency $\nu_0 = \omega_0/2\pi$, i.e., if $g(\mathbf{r}_1, \mathbf{r}_2, \tau) \approx g(\mathbf{r}_1, \mathbf{r}_2) \exp(j\omega_0 \tau)$, then (12.2-16) provides

$$\boxed{I(x) = 2I_0 \left[1 + \mathcal{V} \cos\left(\frac{2\pi\theta}{\lambda} x + \varphi\right)\right],} \quad (12.2\text{-}17)$$

where $\lambda = c/\nu_0$, $\mathcal{V} = |g(\mathbf{r}_1, \mathbf{r}_2)|$, $\tau_x = \theta x/c$, and $\varphi = \arg\{g(\mathbf{r}_1, \mathbf{r}_2)\}$. The interference-fringe pattern is then sinusoidal with spatial period λ/θ and visibility $\mathcal{V}$. In analogy with the temporal case, the visibility of the interference pattern is equal to the magnitude of the complex degree of spatial coherence at the two pinholes (Fig. 12.2-4). The locations of the peaks depend on the phase φ.

If the incident wave in a Young's interferometer is a coherent plane wave traveling in the z direction, $U(\mathbf{r}, t) = \exp(-jkz)\exp(j\omega_0 t)$, then $g(\mathbf{r}_1, \mathbf{r}_2) = 1$, whereupon $|g(\mathbf{r}_1, \mathbf{r}_2)| = 1$ and $\arg\{g(\mathbf{r}_1, \mathbf{r}_2)\} = 0$. The interference pattern then has unity visibility and a peak at $x = 0$. However, if the illumination is, instead, a tilted plane wave arriving from a direction in the x–z plane that makes a small angle θ_x with respect to the z axis, i.e., $U(\mathbf{r}, t) \approx \exp[-j(kz + k\theta_x x)]\exp(j\omega_0 t)$, then $g(\mathbf{r}_1, \mathbf{r}_2) = \exp(-jk\theta_x 2a)$. The visibility remains at $\mathcal{V} = 1$, but the tilt results in a phase shift $\varphi = -k\theta_x 2a = -2\pi\theta_x 2a/\lambda$, so that the interference pattern is shifted laterally by a fraction $(2a\theta_x/\lambda)$ of a period. When $\varphi = 2\pi$, the pattern is shifted by one period.

Interference with Light from an Extended Source

Suppose now that the incident light is a collection of independent plane waves arriving from a source that subtends an angle θ_s at the plane of the pinhole (Fig. 12.2-5). The phase shift φ then takes values in the range $\pm 2\pi(\theta_s/2)2a/\lambda = \pm 2\pi\theta_s a/\lambda$ and the fringe pattern is a superposition of displaced sinusoids. If $\theta_s = \lambda/2a$, then φ takes on values in the range $\pm\pi$, which is sufficient to wash out the interference pattern and thereby reduce its visibility to zero.

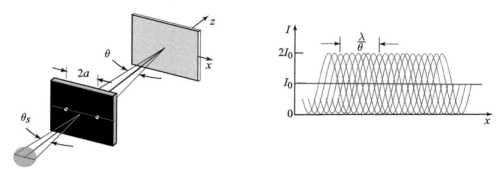

Figure 12.2-5 Young's interference fringes are washed out if the illumination emanates from a source of angular diameter $\theta_s > \lambda/2a$. If the distance $2a$ is smaller than λ/θ_s, the fringes become visible.

We conclude that the degree of spatial coherence at the two pinholes is very small when the angle subtended by the source is $\theta_s \geq \lambda/2a$. Consequently, a measure of the **coherence distance** in the plane of the screen is

$$\rho_c \approx \frac{\lambda}{\theta_s}, \qquad (12.2\text{-}18)$$
Coherence Distance

and a measure of the coherence area of light emitted from a source subtending an angle θ_s is described by

$$A_c \approx \left(\frac{\lambda}{\theta_s}\right)^2. \qquad (12.2\text{-}19)$$

The angle subtended by the sun, for example, is $0.5°$, so that the coherence distance for filtered sunlight of wavelength λ is $\rho_c \approx \lambda/\theta_s \approx 115\lambda$. At $\lambda = 0.5\ \mu\text{m}$, $\rho_c \approx 57.5\ \mu\text{m}$.

A more rigorous analysis (see Sec. 12.3C) reveals that the transverse coherence distance ρ_c for a circular incoherent light source of uniform intensity is

$$\rho_c = 1.22\frac{\lambda}{\theta_s}. \qquad (12.2\text{-}20)$$

Effect of Spectral Width on Interference

Finally, we examine the effect of the spectral width of the light on interference in the Young's double-pinhole interferometer. The power spectral density of the incident wave is assumed to be a narrow function of width $\Delta\nu_c$ centered about ν_0, and $\Delta\nu_c \ll \nu_0$. The complex degree of coherence then takes the form

$$g(\mathbf{r}_1, \mathbf{r}_2, \tau) = g_a(\mathbf{r}_1, \mathbf{r}_2, \tau) \exp(j\omega_0\tau), \tag{12.2-21}$$

where $g_a(\mathbf{r}_1, \mathbf{r}_2, \tau)$ is a slowly varying function of τ (in comparison with the period $1/\nu_0$). Substituting (12.2-21) into (12.2-16), we obtain

$$\boxed{I(x) = 2I_0 \left[1 + \mathcal{V}_x \cos\left(\frac{2\pi\theta}{\overline{\lambda}}x + \varphi_x\right)\right],} \tag{12.2-22}$$

where $\mathcal{V}_x = |g_a(\mathbf{r}_1, \mathbf{r}_2, \tau_x)|$, $\varphi_x = \arg\{g_a(\mathbf{r}_1, \mathbf{r}_2, \tau_x)\}$, $\tau_x = \theta x/c$, and $\overline{\lambda} = c/\nu_0$.

The resulting interference pattern is sinusoidal with period $\overline{\lambda}/\theta$ but with varying visibility $\mathcal{V}_x$ and varying phase φ_x, specified by the magnitude and phase of the complex degree of coherence at the two pinholes, respectively, evaluated at the time delay $\tau_x = \theta x/c$. If $|g_a(\mathbf{r}_1, \mathbf{r}_2, \tau)| = 1$ at $\tau = 0$ and decreases with increasing τ, vanishing for $\tau \gg \tau_c$, then the visibility $\mathcal{V}_x = 1$ at $x = 0$ and decreases with increasing x, vanishing for $x \gg x_c = c\tau_c/\theta$. The interference pattern is then visible over a distance

$$x_c = \frac{l_c}{\theta}, \tag{12.2-23}$$

where $l_c = c\tau_c$ is the coherence length and θ is the angle subtended by the two pinholes (Fig. 12.2-6).

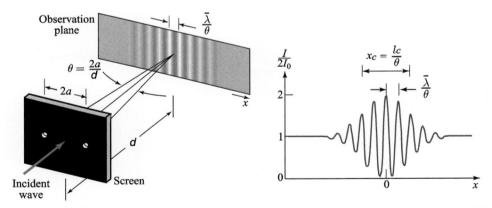

Figure 12.2-6 The visibility of Young's interference fringes at position x is the magnitude of the complex degree of coherence at the pinholes, at a time delay $\tau_x = \theta x/c$. For spatially coherent light the number of observable fringes is the ratio of the coherence length to the central wavelength, or, equivalently, the ratio of the central frequency to the spectral linewidth.

The number of observable fringes is thus $x_c/(\overline{\lambda}/\theta) = l_c/\overline{\lambda} = c\tau_c/\overline{\lambda} = \nu_0/\Delta\nu_c$. It is thus equal to the ratio of the coherence length to the central wavelength, $l_c/\overline{\lambda}$, or the ratio of the central frequency to the linewidth, $\nu_0/\Delta\nu_c$. Clearly, if $|g(\mathbf{r}_1, \mathbf{r}_2, 0)| < 1$, i.e., if the source is not spatially coherent, the visibility will be further reduced and even fewer fringes will be observable.

*12.3 TRANSMISSION OF PARTIALLY COHERENT LIGHT

The transmission of coherent light through thin optical components, apertures, and free space was discussed in Chapters 2 and 4. In this section we revisit this discussion for quasi-monochromatic partially coherent light. We assume that the spectral width of the light is sufficiently small so that the coherence length $l_c = c\tau_c = c/\Delta\nu_c$ is much greater than the differences of optical pathlengths in the system. The mutual coherence function may then be approximated by $G(\mathbf{r}_1, \mathbf{r}_2, \tau) \approx G(\mathbf{r}_1, \mathbf{r}_2)\exp(j2\pi\nu_0\tau)$, where $G(\mathbf{r}_1, \mathbf{r}_2)$ is the mutual intensity and ν_0 is the central frequency.

We observe at the outset that the laws of transmission applicable to the deterministic function $U(\mathbf{r})$ representing coherent light are also applicable to the random function $U(\mathbf{r})$ representing partially coherent light. However, for partially coherent light our interest is in the laws that govern statistical averages: the intensity $I(\mathbf{r})$ and the mutual intensity $G(\mathbf{r}_1, \mathbf{r}_2)$.

A. Propagation of Partially Coherent Light

Transmission Through Thin Optical Components

When a partially coherent wave is transmitted through a thin optical component characterized by an amplitude transmittance $t(x, y)$, the incident and transmitted waves are related by $U_2(\mathbf{r}) = t(\mathbf{r})U_1(\mathbf{r})$ where $\mathbf{r} = (x, y)$ is the position in the plane of the component (see Fig. 12.3-1). Using the definition of the mutual intensity, $G(\mathbf{r}_1, \mathbf{r}_2) = \langle U^*(\mathbf{r}_1)U(\mathbf{r}_2)\rangle$, we obtain

$$G_2(\mathbf{r}_1, \mathbf{r}_2) = t^*(\mathbf{r}_1)\,t(\mathbf{r}_2)\,G_1(\mathbf{r}_1, \mathbf{r}_2), \tag{12.3-1}$$

where $G_1(\mathbf{r}_1, \mathbf{r}_2)$ and $G_2(\mathbf{r}_1, \mathbf{r}_2)$ are the mutual intensities of the incident and transmitted light, respectively.

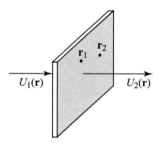

Figure 12.3-1 The absolute value of the degree of spatial coherence is not altered by transmission through a thin optical component.

Since the intensity at position $\mathbf{r}$ is equal to the mutual intensity at $\mathbf{r}_1 = \mathbf{r}_2 = \mathbf{r}$, we have

$$I_2(\mathbf{r}) = |t(\mathbf{r})|^2 I_1(\mathbf{r}). \tag{12.3-2}$$

The normalized mutual intensities defined by (12.1-27) therefore satisfy

$$|g_2(\mathbf{r}_1, \mathbf{r}_2)| = |g_1(\mathbf{r}_1, \mathbf{r}_2)|. \tag{12.3-3}$$

Although transmission through a thin optical component may change the intensity of partially coherent light, it does not alter the magnitude of its degree of spatial coherence. Of course, if the complex amplitude transmittance of the component itself were random, the coherence of the transmitted light would be altered accordingly.

Transmission Through an Arbitrary Optical System

We next consider transmission through an arbitrary optical system — one that includes propagation in free space and thick optical components. It was shown in Chapter 4 that the complex amplitude $U_2(\mathbf{r})$ at a point $\mathbf{r} = (x, y)$ in the output plane of such a system is generally a weighted superposition integral comprising contributions from the complex amplitudes $U_1(\mathbf{r})$ at points $\mathbf{r}' = (x', y')$ in the input plane,

$$U_2(\mathbf{r}) = \int h(\mathbf{r}; \mathbf{r}')\, U_1(\mathbf{r}')\, d\mathbf{r}', \qquad (12.3\text{-}4)$$

where $h(\mathbf{r}; \mathbf{r}')$ is the impulse response function of the system (see Fig. 12.3-2). Equation (12.3-4) is a double integral with respect to $\mathbf{r}' = (x', y')$ that extends over the entire input plane.

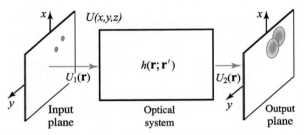

Figure 12.3-2 An optical system is characterized by its impulse response function $h(\mathbf{r}; \mathbf{r}')$.

To translate this relation between the random complex-amplitude functions $U_2(\mathbf{r})$ and $U_1(\mathbf{r})$ into a relation between their mutual intensities, we substitute (12.3-4) into the definition $G_2(\mathbf{r}_1, \mathbf{r}_2) = \langle U_2^*(\mathbf{r}_1)\, U_2(\mathbf{r}_2) \rangle$ and use the definition $G_1(\mathbf{r}_1, \mathbf{r}_2) = \langle U_1^*(\mathbf{r}_1)\, U_1(\mathbf{r}_2) \rangle$ to obtain

$$G_2(\mathbf{r}_1, \mathbf{r}_2) = \iint h^*(\mathbf{r}_1; \mathbf{r}_1')\, h(\mathbf{r}_2; \mathbf{r}_2')\, G_1(\mathbf{r}_1', \mathbf{r}_2')\, d\mathbf{r}_1'\, d\mathbf{r}_2'. \qquad (12.3\text{-}5)$$
Mutual Intensity

If the mutual intensity $G_1(\mathbf{r}_1, \mathbf{r}_2)$ of the input light, and the impulse response function $h(\mathbf{r}; \mathbf{r}')$ of the system are known, the mutual intensity of the output light $G_2(\mathbf{r}_1, \mathbf{r}_2)$ is readily determined by computing the integrals in (12.3-5).

The intensity of the output light is obtained by using the definition $I_2(\mathbf{r}) = G_2(\mathbf{r}, \mathbf{r})$, which, with the help of (12.3-5), reduces to

$$I_2(\mathbf{r}) = \iint h^*(\mathbf{r}; \mathbf{r}_1')\, h(\mathbf{r}; \mathbf{r}_2')\, G_1(\mathbf{r}_1', \mathbf{r}_2')\, d\mathbf{r}_1'\, d\mathbf{r}_2'. \qquad (12.3\text{-}6)$$
Image Intensity

Thus, to determine the intensity of the output light, we must know the mutual intensity of the input light. Knowledge of the input intensity $I_1(\mathbf{r})$ by itself is generally not sufficient to determine the output intensity $I_2(\mathbf{r})$.

B. Image Formation with Incoherent Light

We now consider the special case when the input light is incoherent. The mutual intensity $G_1(\mathbf{r}_1, \mathbf{r}_2)$ then vanishes when $\mathbf{r}_2$ is only slightly separated from $\mathbf{r}_1$, so that the coherence distance is much smaller than other pertinent dimensions in the system (for example, the resolution distance of an imaging system). According to (12.1-27), the mutual intensity may then be written in the form $G_1(\mathbf{r}_1, \mathbf{r}_2) = \sqrt{I_1(\mathbf{r}_1) I_1(\mathbf{r}_2)} \, g(\mathbf{r}_1 - \mathbf{r}_2)$, where $g(\mathbf{r}_1 - \mathbf{r}_2)$ is a very narrow function. When using this expression for $G_1(\mathbf{r}_1, \mathbf{r}_2)$ in the integrals in (12.3-5) and (12.3-6), it is convenient to approximate $g(\mathbf{r}_1 - \mathbf{r}_2)$ by a delta function, $g(\mathbf{r}_1 - \mathbf{r}_2) = \sigma \delta(\mathbf{r}_1 - \mathbf{r}_2)$, where $\sigma = \int g(\mathbf{r}) \, d\mathbf{r}$ is the area under $g(\mathbf{r})$, whereupon

$$G_1(\mathbf{r}_1, \mathbf{r}_2) \approx \sigma \sqrt{I_1(\mathbf{r}_1) I_1(\mathbf{r}_2)} \, \delta(\mathbf{r}_1 - \mathbf{r}_2). \tag{12.3-7}$$

Since the mutual intensity must remain finite and $\delta(0) \to \infty$, this equation cannot be used in general. However, it is suitable for evaluating integrals such as that in (12.3-6). Substituting (12.3-7) into (12.3-6), the delta function reduces the double integral to a single integral, which provides

$$\boxed{I_2(\mathbf{r}) \approx \int I_1(\mathbf{r}') \, h_i(\mathbf{r}; \mathbf{r}') \, d\mathbf{r}',} \tag{12.3-8}$$
Imaging Equation
(Incoherent Illumination)

where

$$\boxed{h_i(\mathbf{r}; \mathbf{r}') = \sigma |h(\mathbf{r}; \mathbf{r}')|^2.} \tag{12.3-9}$$
Impulse Response Function
(Incoherent Illumination)

Under these conditions, the relation between the intensities at the input and output planes describes a linear system of impulse response function $h_i(\mathbf{r}; \mathbf{r}')$, which is also known as the **point-spread function**. When the input light is completely incoherent, therefore, the intensity of the light at each point $\mathbf{r}$ on the output plane is a weighted superposition of contributions from the intensities at many points $\mathbf{r}'$ on the input plane; there is no interference and the intensities simply add (Fig. 12.3-3). This is to be contrasted with the completely coherent system, in which the complex amplitudes, rather than intensities, are related by a superposition integral, as evidenced in (12.3-4).

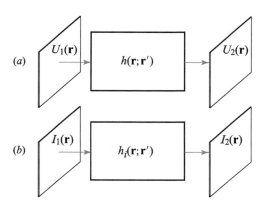

Figure 12.3-3 (a) The *complex amplitudes* of light at the input and output planes of an optical system illuminated by *coherent light* are related by a linear system with impulse response function $h(\mathbf{r}; \mathbf{r}')$. (b) The *intensities* of light at the input and output planes of an optical system illuminated by *incoherent light* are related by a linear system with impulse response function $h_i(\mathbf{r}; \mathbf{r}') = \sigma |h(\mathbf{r}; \mathbf{r}')|^2$.

In many optical systems, the impulse response function $h(\mathbf{r}; \mathbf{r}')$ is a function of $\mathbf{r} - \mathbf{r}'$, say $h(\mathbf{r} - \mathbf{r}')$. The system is then said to be **shift invariant** or **isoplanatic** (see Appendix B, Sec. B.2). In this case $h_i(\mathbf{r}; \mathbf{r}') = h_i(\mathbf{r} - \mathbf{r}')$. The integrals in (12.3-4) and (12.3-8) then represent two-dimensional convolutions and the systems can be described by transfer functions $H(\nu_x, \nu_y)$ and $H_i(\nu_x, \nu_y)$, which are the Fourier transforms of $h(\mathbf{r}) = h(x, y)$ and $h_i(\mathbf{r}) = h_i(x, y)$, respectively.

As an example, we apply the relations set forth above to an imaging system. The impulse response function of the single-lens focused imaging system illustrated in Fig. 12.3-4 was considered in Sec. 4.4C with coherent illumination. It was shown that, in the Fresnel approximation, it can be expressed as [see (4.4-12)]

$$h(\mathbf{r}) \propto P\left(\frac{x}{\lambda d_2}, \frac{y}{\lambda d_2}\right), \qquad (12.3\text{-}10)$$

where $P(\nu_x, \nu_y)$ is the Fourier transform of the pupil function $p(x, y)$ and d_2 is the distance from the lens to the image plane. The pupil function is unity within the aperture and zero elsewhere.

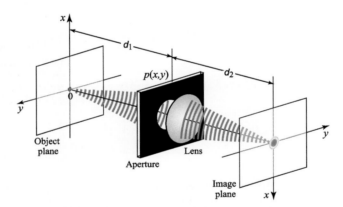

Figure 12.3-4 Single-lens imaging system.

When the illumination is quasi-monochromatic and spatially incoherent, the intensities of the light at the object and image planes are linearly related by a system with impulse response function

$$h_i(\mathbf{r}) = \sigma |h(\mathbf{r})|^2 \propto \left| P\left(\frac{x}{\lambda d_2}, \frac{y}{\lambda d_2}\right) \right|^2, \qquad (12.3\text{-}11)$$

where λ is the wavelength corresponding to the central frequency ν_0.

EXAMPLE 12.3-1. *Coherent and Incoherent Imaging Systems with Circular Apertures.* If the aperture in Fig. 12.3-4 is a circle of radius a (diameter $D = 2a$), the pupil function $p(x, y) = 1$ for x, y inside the circle, and 0 elsewhere. Its Fourier transform is (see Appendix A, Sec. A.3)

$$P(\nu_x, \nu_y) = \frac{a J_1(2\pi \nu_\rho a)}{\nu_\rho}, \qquad \nu_\rho = \sqrt{\nu_x^2 + \nu_y^2}, \qquad (12.3\text{-}12)$$

where $J_1(\cdot)$ is the Bessel function of order 1. The impulse response function of the coherent system is obtained by substituting into (12.3-10),

$$h(x,y) \propto \left[\frac{J_1(2\pi\nu_s\rho)}{\pi\nu_s\rho}\right], \qquad \rho = \sqrt{x^2+y^2}, \qquad (12.3\text{-}13)$$

where

$$\nu_s = \frac{\theta}{2\lambda}, \qquad \theta = \frac{2a}{d_2}, \qquad (12.3\text{-}14)$$

which accords with the result reported in Example 4.4-1 [see (4.4-13)].

For incoherent illumination, the impulse response function is therefore

$$h_i(x,y) \propto \left[\frac{J_1(2\pi\nu_s\rho)}{\pi\nu_s\rho}\right]^2. \qquad (12.3\text{-}15)$$

The impulse response functions $h(x,y)$ and $h_i(x,y)$ are illustrated in the top row of Fig. 12.3-5. Both functions reach their first zero when $2\pi\nu_s\rho = 3.832$, or $\rho = \rho_s \approx 3.832/2\pi\nu_s = 3.832\,\lambda/\pi\theta$, from which we obtain the two-point resolution

$$\boxed{\rho_s \approx 1.22\frac{\lambda}{\theta}.} \qquad (12.3\text{-}16)$$
Two-Point Resolution

Thus, the image of a point (impulse) at the input plane is a patch of intensity $h_i(x,y)$ and radius ρ_s. When the input distribution comprises two points (impulses) separated by a distance ρ_s, the image of one point vanishes at the center of the image of the other point. The distance ρ_s is therefore a measure of the resolution of the imaging system.

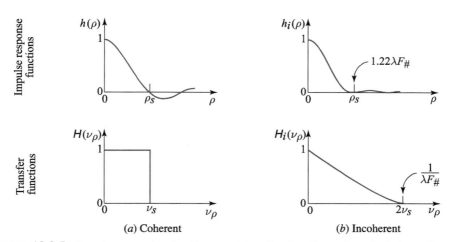

Figure 12.3-5 Impulse response functions and transfer functions of a single-lens, focused, diffraction-limited imaging system with a circular aperture and lens $F_\#$ under (a) coherent, and (b) incoherent illumination.

The transfer functions of linear systems with the impulse response functions $h(x,y)$ and $h_i(x,y)$ are, respectively, the Fourier transforms (see Appendix A, Sec. A.3),

$$H(\nu_x,\nu_y) = \begin{cases} 1, & \nu_\rho < \nu_s \\ 0, & \text{otherwise,} \end{cases} \qquad (12.3\text{-}17)$$

and

$$H_i(\nu_x, \nu_y) = \begin{cases} \dfrac{2}{\pi}\left[\left(\cos^{-1}\dfrac{\nu_\rho}{2\nu_s}\right) - \dfrac{\nu_\rho}{2\nu_s}\sqrt{1-\left(\dfrac{\nu_\rho}{2\nu_s}\right)^2}\right], & \nu_\rho < 2\nu_s \\ 0, & \text{otherwise,} \end{cases} \quad (12.3\text{-}18)$$

where $\nu_\rho = \sqrt{\nu_x^2 + \nu_y^2}$. Both functions have been normalized such that their values are unity at $\nu_\rho = 0$. These functions are illustrated in the bottom row of Fig. 12.3-5. For coherent illumination, the transfer function is flat and has a cutoff frequency $\nu_s = \theta/2\lambda$ lines/mm. For incoherent illumination, the transfer function decreases approximately linearly with the spatial frequency and has a cutoff frequency $2\nu_s = \theta/\lambda$ lines/mm.

If the object is placed at infinity so that $d_1 = \infty$, we have $d_2 = f$, where f is the focal length of the lens. The angle $\theta = 2a/f$ is then the inverse of the lens F-number since $F_\# = f/2a$. The cutoff frequencies ν_s and $2\nu_s$ are thus related to the lens F-number by [see (4.4-19)]

$$\text{Cutoff frequency} \atop \text{(lines/mm)} = \begin{cases} \dfrac{1}{2\lambda F_\#} & \text{(coherent illumination)} \\ \dfrac{1}{\lambda F_\#} & \text{(incoherent illumination).} \end{cases} \quad (12.3\text{-}19)$$

It should not be concluded, however, that incoherent illumination is superior to coherent illumination because it has twice the spatial bandwidth. The transfer functions for the two systems should not be compared directly since one describes imaging of the complex amplitude while the other describes imaging of the intensity.

C. Gain of Spatial Coherence by Propagation

Equation (12.3-5) describes the change of the mutual intensity when the light propagates through an optical system of impulse response function $h(\mathbf{r}; \mathbf{r}')$. When the input light is incoherent, the mutual intensity $G_1(\mathbf{r}_1, \mathbf{r}_2)$ may be approximated by (12.3-7), whereupon the double integral in (12.3-5) reduces to

$$G_2(\mathbf{r}_1, \mathbf{r}_2) = \sigma \int h^*(\mathbf{r}_1; \mathbf{r}) h(\mathbf{r}_2; \mathbf{r}) I_1(\mathbf{r}) \, d\mathbf{r}. \quad (12.3\text{-}20)$$
Mutual Intensity

It is evident that the transmitted light is no longer incoherent. In general, *light gains spatial coherence by the mere act of propagation*. This can be understood as follows. Although light fluctuations at different points of the input plane are uncorrelated, the radiation from each point spreads and overlaps with that from neighboring points. The light reaching two points at the output plane originates from many points at the input plane, some of which are common (see Fig. 12.3-6). These common contributions engender partial correlation between fluctuations at the output points.

This phenomenon is not unlike the transmission of an uncorrelated time signal (white noise) through a low-pass filter. The filter smoothes the function and reduces its spectral bandwidth, so that its coherence time increases and it is no longer uncorrelated. The propagation of light through an optical system is a form of spatial filtering that reduces the spatial bandwidth and therefore increases the coherence area.

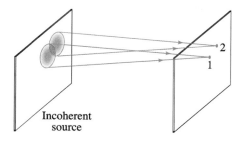

Figure 12.3-6 Gain of coherence by propagation is a result of the spreading of light. Although the light is completely uncorrelated at the source, the light fluctuations at points 1 and 2 share a common origin, the dark shaded area, and are therefore partially correlated.

van Cittert–Zernike Theorem

There is a mathematical identity between the expressions for the gain of *coherence of initially incoherent light* propagating through an optical system, and the change of the *amplitude of coherent light* traveling through the same system. With respect to (12.3-20), if the observation point $\mathbf{r}_1$ is fixed at the origin $\mathbf{0}$, for example, and the mutual intensity $G_2(\mathbf{0}, \mathbf{r}_2)$ is examined as a function of $\mathbf{r}_2$, then

$$G_2(\mathbf{0}, \mathbf{r}_2) = \sigma \int h^*(\mathbf{0}; \mathbf{r})\, h(\mathbf{r}_2; \mathbf{r})\, I_1(\mathbf{r})\, d\mathbf{r}. \qquad (12.3\text{-}21)$$

Defining $U_2(\mathbf{r}_2) = G_2(\mathbf{0}, \mathbf{r}_2)$ and $U_1(\mathbf{r}) = \sigma h^*(\mathbf{0}; \mathbf{r}) I_1(\mathbf{r})$, (12.3-21) may be written in the familiar form

$$U_2(\mathbf{r}_2) = \int h(\mathbf{r}_2; \mathbf{r})\, U_1(\mathbf{r})\, d\mathbf{r}, \qquad (12.3\text{-}22)$$

which is nothing other than the integral (12.3-4) that governs the propagation of coherent light. Thus, the observed mutual intensity $G(\mathbf{0}, \mathbf{r}_2)$ at the output of an optical system whose input is incoherent is mathematically identical to the observed complex amplitude of a coherent wave $U_1(\mathbf{r}) = \sigma h^*(\mathbf{0}; \mathbf{r}) I_1(\mathbf{r})$ presented at the input of the same system.

As an example, assume that the incoherent input wave has uniform intensity and extends over an aperture $p(\mathbf{r})$ [$p(\mathbf{r}) = 1$ within the aperture and zero elsewhere] so that $I_1(\mathbf{r}) = p(\mathbf{r})$; and assume also that the optical system is free space so that $h(\mathbf{r}'; \mathbf{r}) = \exp(-jk|\mathbf{r}' - \mathbf{r}|)/|\mathbf{r}' - \mathbf{r}|$. The mutual intensity $G_2(\mathbf{0}, \mathbf{r}_2)$ is then identical to the amplitude $U_2(\mathbf{r}_2)$ obtained when a coherent wave with input amplitude $U_1(\mathbf{r}) = \sigma h^*(\mathbf{0}; \mathbf{r}) p(\mathbf{r}) = \sigma p(\mathbf{r}) \exp(jkr)/r$ is transmitted through the same system. This is a spherical wave converging to the point $\mathbf{0}$ at the output plane and transmitted through the aperture.

This connection between the gain of spatial coherence of incoherent light, and the diffraction of coherent light, traveling through the same system is known as the **Van Cittert–Zernike theorem**.

Gain of Coherence in Free Space

Consider an optical system comprising free-space propagation between two parallel planes separated by a distance d (Fig. 12.3-7). Light at the input plane is quasi-monochromatic and spatially incoherent, and has intensity $I(x, y)$ extending over a finite area. The impulse response function is then described by the Fresnel diffraction formula [see (4.1-18)]:

$$h(\mathbf{r}; \mathbf{r}') = h_0 \exp\left[-j\pi \frac{(x - x')^2 + (y - y')^2}{\lambda d}\right], \qquad (12.3\text{-}23)$$

where $\mathbf{r} = (x, y, d)$ and $\mathbf{r}' = (x', y', 0)$ are the coordinates of points at the output and input planes, respectively, and $h_0 = (j/\lambda d) \exp(-j2\pi d/\lambda)$ is a constant.

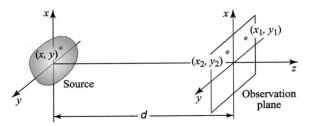

Figure 12.3-7 Radiation from an incoherent source in free space.

We determine the mutual coherence function $G(x_1, y_1, x_2, y_2)$ at two points (x_1, y_1) and (x_2, y_2) in the output plane, by substituting (12.3-23) into (12.3-20) to obtain

$$|G(x_1, y_1, x_2, y_2)| = \sigma_1 \left| \iint_{-\infty}^{\infty} \exp\left\{ j\frac{2\pi}{\lambda d}[(x_2 - x_1)x + (y_2 - y_1)y] \right\} I(x, y)\, dx\, dy \right|, \tag{12.3-24}$$

where $\sigma_1 = \sigma|h_0|^2 = \sigma/\lambda^2 d^2$ is another constant. Given $I(x, y)$, one can easily determine $|G(x_1, y_1, x_2, y_2)|$ in terms of the two-dimensional Fourier transform of $I(x, y)$,

$$\mathcal{I}(\nu_x, \nu_y) = \iint_{-\infty}^{\infty} \exp[j2\pi(\nu_x x + \nu_y y)] I(x, y)\, dx\, dy, \tag{12.3-25}$$

evaluated at $\nu_x = (x_2 - x_1)/\lambda d$ and $\nu_y = (y_2 - y_1)/\lambda d$. The magnitude of the corresponding normalized mutual intensity is then

$$\boxed{|g(x_1, y_1, x_2, y_2)| = \left| \mathcal{I}\left(\frac{x_2 - x_1}{\lambda d}, \frac{y_2 - y_1}{\lambda d} \right) \right| \Big/ \mathcal{I}(0, 0).} \tag{12.3-26}$$

This Fourier transform-relation between the intensity profile of an incoherent source and the degree of spatial coherence of its far field is similar to the Fourier-transform relation between the amplitude of coherent light at the input and output planes (see Sec. 4.2A). The similarity is expected in view of the Van Cittert–Zernike theorem.

The implications of (12.3-26) are profound. If the area of the source, i.e., the spatial extent of $I(x, y)$, is small, its Fourier transform $\mathcal{I}(\nu_x, \nu_y)$ is wide, so that the mutual intensity at the output plane extends over a wide area and the area of coherence at the output plane is large. In the extreme limit in which light at the input plane originates from a point, the area of coherence is infinite and the radiated field is spatially completely coherent. This confirms our earlier discussions in Sec. 12.1D with respect to the coherence of spherical waves. On the other hand, if the input light originates from a large extended source, the propagated light has a small area of coherence.

EXAMPLE 12.3-2. *Radiation from an Incoherent Circular Source.* For input light with uniform intensity $I(x,y) = I_0$ confined to a circular aperture of radius a, (12.3-26) yields

$$|g(x_1, y_1, x_2, y_2)| = \left| \frac{2J_1(\pi \rho \theta_s / \lambda)}{\pi \rho \theta_s / \lambda} \right|, \quad (12.3\text{-}27)$$

where $\rho = \sqrt{(x_2 - x_1)^2 + (y_2 - y_1)^2}$ is the distance between the two points, $\theta_s = 2a/d$ is the angle subtended by the source, and $J_1(\cdot)$ is the Bessel function. This relation is plotted in Fig. 12.3-8. The Bessel function reaches its first zero when its argument is 3.832. We can therefore define the area of coherence as a circle of radius $\rho_c = 3.832(\lambda/\pi\theta_s)$, so that

$$\boxed{\rho_c = 1.22 \frac{\lambda}{\theta_s}.} \quad (12.3\text{-}28)$$
Coherence Distance

A similar result, (12.2-18), was obtained using a less rigorous analysis. The area of coherence is inversely proportional to θ_s^2. An incoherent light source of wavelength $\lambda = 0.6$ μm and radius 1 cm observed at a distance $d = 100$ m, for example, has a coherence distance $\rho_c \approx 3.7$ mm.

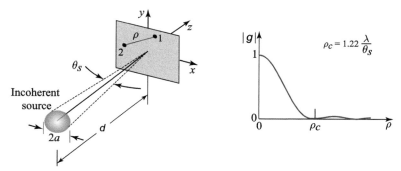

Figure 12.3-8 The magnitude of the degree of spatial coherence of light radiated from an incoherent circular light source subtending an angle θ_s, as a function of the separation ρ.

Measurement of the Angular Diameter of Stars: The Michelson Stellar Interferometer

Equation (12.3-28) is the basis of a method for measuring the angular diameters of stars. If the star is regarded as an incoherent disk of diameter $2a$ with uniform brilliance, then at an observation plane a distance d away from the star, the coherence function decreases to 0 when the separation between the two observation points reaches $\rho_c = 1.22\lambda/\theta_s$. Measuring ρ_c at a given value of λ permits us to determine the angular diameter $\theta_s = 2a/d$.

As an example, taking the angular diameter of the sun to be 0.5°, $\theta_s = 8.7 \times 10^{-3}$ radians, and assuming that the intensity is uniform, we obtain $\rho_c = 140\lambda$. For $\lambda = 0.5$ μm, we have $\rho_c = 70$ μm. To observe interference fringes in a Young double-slit apparatus, the slits would have to be separated by a distance smaller than 70 μm. Stars of smaller angular diameter have correspondingly larger areas of coherence. For example, the first star whose angular diameter was measured using this technique (α-Orion) has an angular diameter $\theta_s = 22.6 \times 10^{-8}$, so that at $\lambda = 0.57$ μm, we have $\rho_c = 3.1$ m. A Young interferometer can be modified to accommodate such large slit separations by using movable mirrors, as shown in Fig. 12.3-9.

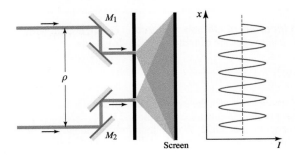

Figure 12.3-9 The Michelson stellar interferometer. The angular diameter of a star is estimated by measuring the mutual intensity at two points with variable separation ρ using a Young double-slit interferometer. The distance ρ between mirrors M_1 and M_2 is varied and the visibility of the interference fringes is measured. When $\rho = \rho_c = 1.22\lambda/\theta_s$, the visibility is 0.

12.4 PARTIAL POLARIZATION

As we have seen in Chapter 6, the scalar wave theory of light is often inadequate and a vector theory that includes the polarization of light is required. This section provides a brief discussion of the statistical theory of *random* light, including the effects of polarization. The **theory of partial polarization** is based on characterizing the components of the optical field vector by correlations and cross-correlations similar to those defined earlier in this chapter.

To simplify the presentation, we shall not be concerned with spatial effects. We therefore limit ourselves to light described by a transverse electromagnetic (TEM) plane wave traveling in the z direction. The electric-field vector has two components, in the x and y directions, with complex wavefunctions $U_x(t)$ and $U_y(t)$ that are generally random. Each function is characterized by its autocorrelation function (the temporal coherence function),

$$G_{xx}(\tau) = \langle U_x^*(t)U_x(t+\tau)\rangle \tag{12.4-1}$$

$$G_{yy}(\tau) = \langle U_y^*(t)U_y(t+\tau)\rangle. \tag{12.4-2}$$

An additional descriptor of the wave is the cross-correlation function of $U_x(t)$ and $U_y(t)$,

$$G_{xy}(\tau) = \langle U_x^*(t)U_y(t+\tau)\rangle. \tag{12.4-3}$$

The normalized function

$$g_{xy}(\tau) = \frac{G_{xy}(\tau)}{\sqrt{G_{xx}(0)G_{yy}(0)}} \tag{12.4-4}$$

is the cross-correlation coefficient of $U_x(t)$ and $U_y(t+\tau)$. It satisfies the inequality $0 \leq |g_{xy}(\tau)| \leq 1$. When the two components are uncorrelated at all times, $|g_{xy}(\tau)| = 0$; when they are completely correlated at all times, $|g_{xy}(\tau)| = 1$.

The spectral properties are, in general, tied to the polarization properties and the autocorrelation and cross-correlation functions can have different dependencies on τ. However, for quasi-monochromatic light, all dependencies on τ in (12.4-1)–(12.4-4) are approximately of the form $\exp(j2\pi\nu_0\tau)$, so that the polarization properties are described by the values at $\tau = 0$. The three numbers $G_{xx}(0)$, $G_{yy}(0)$, and $G_{xy}(0)$, hereafter denoted G_{xx}, G_{yy}, and G_{xy}, are then used to describe the polarization of the wave. Note that $G_{xx} = I_x$ and $G_{yy} = I_y$ are real numbers that represent the intensities of the x and y components, but G_{xy} is complex and $G_{yx} = G_{xy}^*$, as can easily be verified from the definition.

Coherency Matrix

It is convenient to write the four variables G_{xx}, G_{xy}, G_{yx}, and G_{yy} in the form of a 2×2 Hermitian matrix

$$\mathbf{G} = \begin{bmatrix} G_{xx} & G_{xy} \\ G_{yx} & G_{yy} \end{bmatrix} \qquad (12.4\text{-}5)$$

called the **coherency matrix**. The diagonal elements are the intensities I_x and I_y, whereas the off-diagonal elements are the cross-correlations. The trace of the matrix, given by $\text{Tr}\,\mathbf{G} = I_x + I_y \equiv \bar{I}$, is the total intensity.

The coherency matrix may also be written in terms of the Jones vector, $\mathbf{J} = \begin{bmatrix} U_x \\ U_y \end{bmatrix}$ defined in terms of the complex wavefunctions and complex amplitudes (instead of in terms of the complex envelopes as in Sec. 6.1B),

$$\langle \mathbf{J}^* \mathbf{J}^{\mathsf{T}} \rangle = \left\langle \begin{bmatrix} U_x^* \\ U_y^* \end{bmatrix} [U_x \ U_y] \right\rangle = \begin{bmatrix} \langle U_x^* U_x \rangle & \langle U_x^* U_y \rangle \\ \langle U_y^* U_x \rangle & \langle U_y^* U_y \rangle \end{bmatrix} = \mathbf{G}, \qquad (12.4\text{-}6)$$

where the superscript T denotes the transpose of a matrix, and U_x and U_y denote $U_x(t)$ and $U_y(t)$, respectively.

The Jones vector is transformed by polarization devices, such as polarizers and retarders, in accordance with the rule $\mathbf{J}' = \mathbf{TJ}$ [see (6.1-17)], where $\mathbf{T}$ is the Jones matrix representing the device [see (6.1-18) to (6.1-25)]. The coherency matrix is therefore transformed in accordance with $\mathbf{G}' = \langle \mathbf{T}^* \mathbf{J}^* (\mathbf{TJ})^{\mathsf{T}} \rangle = \langle \mathbf{T}^* \mathbf{J}^* \mathbf{J}^{\mathsf{T}} \mathbf{T}^{\mathsf{T}} \rangle = \mathbf{T}^* \langle \mathbf{J}^* \mathbf{J}^{\mathsf{T}} \rangle \mathbf{T}^{\mathsf{T}}$, so that

$$\boxed{\mathbf{G}' = \mathbf{T}^* \mathbf{G} \mathbf{T}^{\mathsf{T}}.} \qquad (12.4\text{-}7)$$

We thus have a formalism for determining the effect of polarization devices on the coherency matrix of partially polarized light.

Stokes Parameters and Poincaré-Sphere Representation

The Stokes parameters were defined in Sec. 6.1A for coherent light as a set of four real parameters related to the products of the x and y components of the complex envelope [see (6.1-9)]. This definition is readily generalized to partially coherent light as an average of these products:

$$\begin{aligned}
\mathsf{S}_0 &= \langle |U_x|^2 \rangle + \langle |U_y|^2 \rangle = G_{xx} + G_{yy} & (12.4\text{-}8\text{a}) \\
\mathsf{S}_1 &= \langle |U_x|^2 \rangle - \langle |U_y|^2 \rangle = G_{xx} - G_{yy} & (12.4\text{-}8\text{b}) \\
\mathsf{S}_2 &= 2\,\text{Re}\{\langle U_x^* U_y \rangle\} = 2\,\text{Re}\{G_{xy}\} & (12.4\text{-}8\text{c}) \\
\mathsf{S}_3 &= 2\,\text{Im}\{\langle U_x^* U_y \rangle\} = 2\,\text{Im}\{G_{xy}\}. & (12.4\text{-}8\text{d})
\end{aligned}$$

Stokes Parameters

Thus, the Stokes parameters are directly related to elements of the coherency matrix $\mathbf{G}$. The first parameter, S_0, is simply the sum of the diagonal elements, which is the total intensity $\bar{I}$. The second, S_1, is the difference of the diagonal elements, i.e., the difference between the intensities of the two polarization components. The third and fourth, S_2 and S_3, respectively, are proportional to the real and imaginary parts of the off-diagonal element, i.e., the cross-correlation function. Using these relations, it can be readily shown that the inequality $|G_{xy}|^2 \leq G_{xx}G_{yy}$ leads to the condition $S_1^2 + S_2^2 + S_3^2 \leq S_0^2$. For coherent light, these inequalities become equalities.

The state of polarization of partially polarized light may be represented geometrically on the Poincaré sphere (Fig. 6.1-5) as a point whose Cartesian coordinates are $(S_1/S_0, S_2/S_0, S_3/S_0)$. Since $S_1^2 + S_2^2 + S_3^2 \leq S_0^2$, such a point lies inside, or on, the surface of the sphere.

To understand the significance of the coherency matrix and the Stokes parameters, we next examine two limiting cases.

Unpolarized Light

Light of intensity $\bar{I}$ is said to be **unpolarized** if its two components have the same intensity and are uncorrelated, i.e., $I_x = I_y \equiv \frac{1}{2}\bar{I}$ and $G_{xy} = 0$. The coherency matrix is then

$$\mathbf{G} = \tfrac{1}{2}\bar{I} \begin{bmatrix} 1 & 0 \\ 0 & 1 \end{bmatrix}. \quad (12.4\text{-}9)$$

Unpolarized Light

By use of (12.4-7) and (6.1-22), it can be shown that (12.4-9) is invariant to rotation of the coordinate system, so that the two components always have equal intensities and are uncorrelated. Unpolarized light therefore has an electric-field vector that is statistically isotropic; it is equally likely to have any direction in the x–y plane, as illustrated in Fig. 12.4-1(a). Analogous results for partially polarized and right-circularly polarized (RCP) light are displayed in Figs. 12.4-1(b) and 12.4-1(c), respectively.

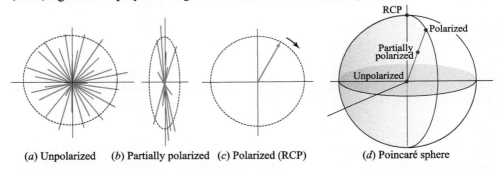

(a) Unpolarized (b) Partially polarized (c) Polarized (RCP) (d) Poincaré sphere

Figure 12.4-1 Fluctuations of the electric-field vector for (a) unpolarized light, (b) partially polarized light, and (c) polarized light with circular polarization. (d) Poincaré-sphere representation for unpolarized light (at the origin), partially polarized light (in the interior), and elliptically polarized light (on the surface).

When passed through a polarizer, unpolarized light becomes linearly polarized, but it remains random with an average intensity $\frac{1}{2}\bar{I}$. A wave retarder has no effect on unpolarized light since it only introduces a phase shift between two components that have a totally random phase to begin with. Similarly, unpolarized light transmitted through a polarization rotator remains unpolarized. These effects may be formally derived by use of (12.4-7) and (12.4-9) together with (6.1-18), (6.1-19), and (6.1-20), respectively.

The Stokes parameters describing unpolarized light are $(S_0, S_1, S_2, S_3) = (\bar{I}, 0, 0, 0)$ as can be readily shown by use of (12.4-8) and (12.4-9). The corresponding representation on the Poincaré sphere is a point with Cartesian coordinates $(S_1/S_0, S_2/S_0, S_3/S_0)$ = $(0, 0, 0)$ so that the point is located at the origin of the sphere [Fig. 12.4-1(d)].

Polarized Light

If the cross-correlation coefficient $g_{xy} = G_{xy}/\sqrt{I_x I_y}$ has unity magnitude, $|g_{xy}| = 1$, the two components of the optical field are perfectly correlated and the light is said to be completely polarized (or simply **polarized**). The coherency matrix then takes the form

$$\mathbf{G} = \begin{bmatrix} I_x & \sqrt{I_x I_y}\, e^{j\varphi} \\ \sqrt{I_x I_y}\, e^{-j\varphi} & I_y \end{bmatrix}, \qquad (12.4\text{-}10)$$

where φ is the argument of g_{xy}. Defining $U_x = \sqrt{I_x}$ and $U_y = \sqrt{I_y}\, e^{j\varphi}$, we have

$$\mathbf{G} = \begin{bmatrix} U_x^* U_x & U_x^* U_y \\ U_y^* U_x & U_y^* U_y \end{bmatrix} = \mathbf{J}^* \mathbf{J}^\mathsf{T}, \qquad (12.4\text{-}11)$$

where $\mathbf{J}$ is a Jones matrix with components U_x and U_y. Thus, $\mathbf{G}$ has the same form as the coherency matrix of a coherent wave. Using the Jones vectors provided in Table 6.1-1, we can determine the coherency matrices for different states of polarization. Two examples are:

Linearly polarized in the x direction: $\quad \mathbf{G} = \bar{I} \begin{bmatrix} 1 & 0 \\ 0 & 0 \end{bmatrix} \qquad$ Right-circularly polarized: $\quad \mathbf{G} = \tfrac{1}{2}\bar{I} \begin{bmatrix} 1 & j \\ -j & 1 \end{bmatrix}$

The Stokes parameters corresponding to (12.4-11) satisfy the relation $\mathsf{S}_1^2 + \mathsf{S}_2^2 + \mathsf{S}_3^2 = \mathsf{S}_0^2$, so that polarized light is represented by a point on the surface of, rather than inside, the Poincaré sphere [Fig. 12.4-1(d)].

It is instructive to examine the distinction between unpolarized light and circularly polarized light. In both cases the intensities of the x and y components are equal ($I_x = I_y$). For circularly polarized light the two components are completely correlated, but for unpolarized light they are uncorrelated. Circularly polarized light may be transformed into linearly polarized light by the use of a wave retarder, but unpolarized light remains unpolarized upon passage through such a device. Circularly polarized light is represented by a point at the north or south pole of the Poincaré sphere, while unpolarized light is represented by a point at the origin [Fig. 12.4-1(d)].

Degree of Polarization

Partial polarization is a general state of random polarization that lies between the two ideal limits of unpolarized and polarized light. One measure of the **degree of polarization** $\mathbb{P}$ is defined in terms of the determinant and the trace of the coherency matrix:

$$\mathbb{P} = \sqrt{1 - \frac{4\det \mathbf{G}}{(\operatorname{Tr}\mathbf{G})^2}} \qquad (12.4\text{-}12)$$

$$= \sqrt{1 - 4\left[\frac{I_x I_y}{(I_x + I_y)^2}\right](1 - |g_{xy}|^2)}. \qquad (12.4\text{-}13)$$

This measure is meaningful because of the following considerations:

- It satisfies the inequality $0 \leq \mathbb{P} \leq 1$.

- For polarized light, $\mathbb{P}$ has its highest value of 1, as can readily be seen by substituting $|g_{xy}| = 1$ into (12.4-13). For unpolarized light it has its lowest value $\mathbb{P} = 0$, since $I_x = I_y$ and $g_{xy} = 0$.

- It is invariant to rotation of the coordinate system (since the determinant and the trace of a matrix are invariant to unitary transformations).

- The degree of polarization in (12.4-13) may also be expressed in terms of the Stokes parameters as:

$$\mathbb{P} = \frac{\sqrt{S_1^2 + S_2^2 + S_3^2}}{S_0}, \qquad (12.4\text{-}14)$$

so that in the Poincaré-sphere representation [Fig. 12.4-1(d)], it is equal to the distance from the origin of the sphere.

- It can be shown (Exercise 12.4-1) that a partially polarized wave can always be regarded as a mixture of two uncorrelated waves: a completely polarized wave and an unpolarized wave, with the ratio of the intensity of the polarized component to the total intensity equal to the degree of polarization $\mathbb{P}$.

EXERCISE 12.4-1

Partially Polarized Light. Demonstrate that the superposition of unpolarized light of intensity $(I_x + I_y)(1 - \mathbb{P})$, and linearly polarized light with intensity $(I_x + I_y)\mathbb{P}$, where $\mathbb{P}$ is given by (12.4-13), yields light whose x and y components have intensities I_x and I_y and normalized cross-correlation $|g_{xy}|$.

READING LIST

Statistical Optics, Coherence, and Partial Polarization

J. J. Gil Pérez and R. Ossikovski, *Polarized Light and the Mueller Matrix Approach*, CRC Press/Taylor & Francis, 2016.

J. W. Goodman, *Statistical Optics*, Wiley, 2nd ed. 2015.

E. Wolf, *Introduction to the Theory of Coherence and Polarization of Light*, Cambridge University Press, 2007.

O. Marchenko, S. Kazantsev, and L. Windholz, *Demonstrational Optics: Part 2, Coherent and Statistical Optics*, Springer-Verlag, 2007.

M. Born and E. Wolf, *Principles of Optics*, Cambridge University Press, 7th expanded and corrected ed. 2002, Chapter 10.

B. R. Frieden, *Probability, Statistical Optics, and Data Testing: A Problem Solving Approach*, Springer-Verlag, 1983, 3rd ed. 2001.

C. Brosseau, *Fundamentals of Polarized Light: A Statistical Optics Approach*, Wiley, 1998.

L. Mandel and E. Wolf, *Optical Coherence and Quantum Optics*, Cambridge University Press, 1995.

L. Mandel and E. Wolf, eds., *Selected Papers on Coherence and Fluctuations of Light (1850–1966)*, SPIE Optical Engineering Press (Milestone Series Volume 19), 1990.

M. C. Teich and B. E. A. Saleh, Photon Bunching and Antibunching, in E. Wolf, ed., *Progress in Optics*, North-Holland, 1988, vol. 26, pp. 1–104.

J. Peřina, *Coherence of Light*, Reidel, 1971, 2nd ed. 1985.

B. E. A. Saleh, *Photoelectron Statistics with Applications to Spectroscopy and Optical Communication*, Springer-Verlag, 1978.

B. Crosignani, P. Di Porto, and M. Bertolotti, *Statistical Properties of Scattered Light*, Academic Press, 1975.

R. Hanbury-Brown, *The Intensity Interferometer: Its Application to Astronomy*, Taylor & Francis, 1974.

L. Mandel and E. Wolf, eds., *Selected Papers on Coherence and Fluctuations of Light*, Volumes 1 and 2, Dover, 1970.

M. J. Beran and G. B. Parrent, Jr., *Theory of Partial Coherence*, Prentice Hall, 1964; SPIE Optical Engineering Press, reissued 1974.

E. L. O'Neill, *Introduction to Statistical Optics*, Addison–Wesley, 1963; Dover, reissued 2003.

Imaging

D. F. Buscher, *Practical Optical Interferometry: Imaging at Visible and Infrared Wavelengths*, Cambridge University Press, 2015.

A. Donges and R. Noll, *Laser Measurement Technology: Fundamentals and Applications*, Springer-Verlag, 2015.

W. Drexler and J. G. Fujimoto, eds., *Optical Coherence Tomography: Technology and Applications*, Volumes. 1–3, Springer-Verlag, 2nd ed. 2015.

M. C. Teich, B. E. A. Saleh, F. N. C. Wong, and J. H. Shapiro, Variations on the Theme of Quantum Optical Coherence Tomography: A Review, *Quantum Information Processing*, vol. 11, pp. 903–923, 2012.

B. E. A. Saleh, *Introduction to Subsurface Imaging*, Cambridge University Press, 2011.

D. J. Brady, *Optical Imaging and Spectroscopy*, Wiley, 2009.

M. E. Brezinski, *Optical Coherence Tomography: Principles and Applications*, Academic Press/Elsevier, 2006.

H. H. Barrett and K. J. Myers, *Foundations of Image Science*, Wiley, 2004.

Random Processes, Random Media, and Speckle

B. Hajek, *Random Processes for Engineers*, Cambridge University Press, 2015.

V. Krishnan and K. Chandra, *Probability and Random Processes*, Wiley, 2nd ed. 2015.

R. G. Gallager, *Stochastic Processes: Theory for Applications*, Cambridge University Press, 2014.

H. Pishro-Nik, *Introduction to Probability, Statistics, and Random Processes*, Kappa Research, 2014.

O. Korotkova, *Random Light Beams: Theory and Applications*, CRC Press/Taylor & Francis, 2014.

N. Pinel and C. Bourlier, *Electromagnetic Wave Scattering from Random Rough Surfaces: Asymptotic Models*, Wiley–ISTE, 2013.

P. Bajorski, *Statistics for Imaging, Optics, and Photonics*, Wiley, 2012.

J. W. Goodman, *Speckle Phenomena in Optics: Theory and Applications*, Roberts, 2007, paperback ed. 2010.

L. C. Andrews and R. L. Phillips, *Laser Beam Propagation through Random Media*, SPIE Optical Engineering Press, 2nd ed. 2005.

A. A. Kokhanovsky, *Polarization Optics of Random Media*, Springer-Verlag/Praxis, 2003.

A. Papoulis and S. U. Pillai, *Probability, Random Variables, and Stochastic Processes*, McGraw–Hill, 1965, 4th ed. 2002.

B. R. Frieden, *Probability, Statistical Optics, and Data Testing: A Problem Solving Approach*, Springer-Verlag, 1983, 3rd ed. 2001.

H. E. Rowe, *Electromagnetic Propagation in Multi-Mode Random Media*, Wiley, 1999.

P. Meinlschmidt, K. D. Hinsch, and R. S. Sirohi, eds., *Selected Papers on Electronic Speckle Pattern Interferometry: Principles and Practice*, SPIE Optical Engineering Press (Milestone Series Volume 132), 1996.

R. S. Sirohi, ed., *Selected Papers on Speckle Metrology*, SPIE Optical Engineering Press (Milestone Series Volume 35), 1991.

M. Françon, *Laser Speckle and Application in Optics*, Academic Press, 1979.

A. Ishimaru, *Wave Propagation and Scattering in Random Media*, Volume 1 and Volume 2, Academic Press, 1978; IEEE Press/Oxford University Press, reissued 1997.

PROBLEMS

12.1-4 **Lorentzian Spectrum.** A light-emitting diode (LED) emits light of Lorentzian spectrum with a linewidth $\Delta \nu$ (FWHM) $= 10^{13}$ Hz centered about a frequency corresponding to a wavelength $\lambda_o = 0.7$ μm. Determine the linewidth $\Delta \lambda_o$ (in units of nm), the coherence time τ_c, and the coherence length l_c. What is the maximum time delay within which the magnitude of the complex degree of temporal coherence $|g(\tau)|$ is greater than 0.5?

12.1-5 **Proof of the Wiener–Khinchin Theorem.** Use the definitions in (12.1-4), (12.1-14), and (12.1-15) to prove that the spectral density $S(\nu)$ is the Fourier transform of the autocorrelation function $G(\tau)$. Prove that the intensity I is the integral of the power spectral density $S(\nu)$.

12.1-6 **Mutual Intensity.** The mutual intensity of an optical wave at points on the x axis is given by

$$G(x_1, x_2) = I_0 \exp\left[-\frac{(x_1^2 + x_2^2)}{W_0^2}\right] \exp\left[\frac{-(x_1 - x_2)^2}{\rho_c^2}\right],$$

where I_0, W_0, and ρ_c are constants. Sketch the intensity distribution as a function of x. Derive an expression for the normalized mutual intensity $g(x_1, x_2)$ and sketch it as a function of $x_1 - x_2$. What is the physical meaning of the parameters I_0, W_0, and ρ_c?

12.1-7 **Mutual Coherence Function.** An optical wave has a mutual coherence function at points on the x axis,

$$G(x_1, x_2, \tau) = \exp\left(-\frac{\pi \tau^2}{2\tau_c^2}\right) \exp[j 2\pi u(x_1, x_2) \tau] \exp\left[-\frac{(x_1 - x_2)^2}{\rho_c^2}\right],$$

where $u(x_1, x_2) = 5 \times 10^{14} \text{s}^{-1}$ for $x_1 + x_2 > 0$, and $6 \times 10^{14} \text{s}^{-1}$ for $x_1 + x_2 < 0$, $\rho_c = 1$ mm, and $\tau_c = 1$ μs. Determine the intensity, the power spectral density, the coherence length, and the coherence distance in the transverse plane. Which of these quantities is position dependent? If this wave were recorded on color film, what would the recorded image look like?

12.1-8 **Coherence Length.** Show that light of narrow spectral width has a coherence length $l_c \approx \lambda^2 / \Delta\lambda$, where $\Delta\lambda$ is the linewidth in wavelength units. Show that for light of broad uniform spectrum extending between the wavelengths $\lambda_{\min}$ and $\lambda_{\max} = 2\lambda_{\min}$, the coherence length $l_c = \lambda_{\max}$.

12.1-9 **Effect of Spectral Width on Spatial Coherence.** A point source at the origin $(0, 0, 0)$ of a Cartesian coordinate system emits light with a Lorentzian spectrum and coherence time $\tau_c = 10$ ps. Determine an expression for the normalized mutual intensity of the light at the points $(0, 0, d)$ and $(x, 0, d)$, where $d = 10$ cm. Sketch the magnitude of the normalized mutual intensity as a function of x.

12.1-10 **Gaussian Mutual Intensity.** An optical wave in free space has a mutual coherence function $G(\mathbf{r}_1, \mathbf{r}_2, \tau) = J(\mathbf{r}_1 - \mathbf{r}_2) \exp(j 2\pi \nu_0 \tau)$.

(a) Show that the function $J(\mathbf{r})$ must satisfy the Helmholtz equation $\nabla^2 J + k_o^2 J = 0$, where $k_o = 2\pi \nu_0 / c$.

(b) An approximate solution of the Helmholtz equation is the Gaussian-beam solution

$$J(\mathbf{r}) = \frac{1}{q(z)} \exp\left[-\frac{j k_o (x^2 + y^2)}{2 q(z)}\right] \exp(-j k_o z),$$

where $q(z) = z + j z_0$ and z_0 is a constant. This solution has been studied extensively in Chapter 3 in connection with Gaussian beams. Determine an expression for the coherence area near the z axis and show that it increases with $|z|$, so that the wave gains coherence with propagation away from the origin.

12.2-1 **Effect of Spectral Width on Fringe Visibility.** Light from a sodium lamp of Lorentzian spectral linewidth $\Delta\nu = 5 \times 10^{11}$ Hz is used in a Michelson interferometer. Determine the maximum pathlength difference for which the visibility of the interferogram $\mathcal{V} > \frac{1}{2}$.

12.2-2 **Number of Observable Fringes in Young's Interferometer.** Determine the number of observable fringes in Young's interferometer if each of the sources in Table 12.1-2 is used. Assume full spatial coherence in all cases.

12.2-3 **Spectrum of a Superposition of Two Waves.** An optical wave is a superposition of two waves $U_1(t)$ and $U_2(t)$ with identical spectra $S_1(\nu) = S_2(\nu)$, which are Gaussian with spectral width $\Delta\nu$ and central frequency ν_0. The waves are not necessarily uncorrelated. Determine an expression for the power spectral density $S(\nu)$ of the superposition $U(t) = U_1(t) + U_2(t)$. Explore the possibility that $S(\nu)$ is also Gaussian, with a shifted central frequency $\nu_1 \neq \nu_0$. If this were possible, our faith in using the Doppler shift as a method to determine the velocity of stars would be shaken, since frequency shifts could originate from something other than the Doppler effect.

*12.3-1 **Partially Coherent Gaussian Beam.** A quasi-monochromatic light wave of wavelength λ travels in free space in the z direction. Its intensity in the $z = 0$ plane is a Gaussian function $I(x) = I_0 \exp(-2x^2/W_0^2)$ and its normalized mutual intensity is also a Gaussian function $g(x_1, x_2) = \exp[-(x_1 - x_2)^2/\rho_c^2]$. Show that the intensity at a distance z satisfying conditions of the Fraunhofer approximation is also a Gaussian function $I_z(x) \propto \exp[-2x^2/W^2(z)]$ and derive an expression for the beam width $W(z)$ as a function of z and the parameters W_0, ρ_c, and λ. Discuss the effect of spatial coherence on beam divergence.

*12.3-2 **Fourier-Transform Lens.** Quasi-monochromatic spatially incoherent light of uniform intensity illuminates a transparency of intensity transmittance $f(x, y)$ and the emerging light is transmitted between the front and back focal planes of a lens. Determine an expression for the intensity of the observed light. Compare your results with the case of coherent light in which the lens performs the Fourier transform (see Sec. 4.2).

*12.3-3 **Light from Two-Point Incoherent Source.** A spatially incoherent quasi-monochromatic source of light emits only at two points separated by a distance $2a$. Determine an expression for the normalized mutual intensity at a distance d from the source (use the Fraunhofer approximation).

*12.3-4 **Coherence of Light Transmitted Through a Fourier-Transform Optical System.** Light from a quasi-monochromatic spatially incoherent source with uniform intensity is transmitted through a thin slit of width $2a$ and travels between the front and back focal planes of a lens. Determine an expression for the normalized mutual intensity in the back focal plane.

12.4-2 **Partially Polarized Light.** The intensities of the two components of a partially polarized wave are $I_x = I_y = \frac{1}{2}$, and the argument of the cross-correlation coefficient g_{xy} is $\pi/2$.

(a) Plot the degree of polarization $\mathbb{P}$ versus the magnitude of the cross-correlation coefficient $|g_{xy}|$.

(b) Determine the coherency matrix if $\mathbb{P} = 0$, 0.5, and 1, and describe the nature of the light in each case.

(c) If the light is transmitted through a polarizer with its axis in the x direction, what is the intensity of the light transmitted?

CHAPTER

13

PHOTON OPTICS

13.1 THE PHOTON   516
    A. Photon Energy
    B. Photon Polarization
    C. Photon Position
    D. Photon Momentum
    E. Photon Interference
    F. Photon Time

13.2 PHOTON STREAMS   529
    A. Photon Flow
    B. Randomness of Photon Flow
    C. Photon-Number Statistics
    D. Random Partitioning of Photon Streams

*13.3 QUANTUM STATES OF LIGHT   541
    A. Coherent States
    B. Quadrature-Squeezed States
    C. Photon-Number-Squeezed States
    D. Two-Photon Light

Max Planck (1858–1947) suggested that the emission and absorption of light by matter takes the form of quanta of energy.

Albert Einstein (1879–1955) advanced the hypothesis that light itself comprises quanta of energy.

Fundamentals of Photonics, Third Edition. Bahaa E. A. Saleh and Malvin Carl Teich.
©2019 John Wiley & Sons, Inc. Published 2019 by John Wiley & Sons, Inc.

Electromagnetic optics, introduced in Chapter 5, provides the most complete treatment of light within the confines of **classical optics**. It encompasses wave optics, which in turn encompasses ray optics (Fig. 13.0-1). Though classical electromagnetic theory is capable of providing explanations for the preponderance of effects in optics, as attested to by the earlier chapters of this book, it nevertheless fails to account for certain optical phenomena. This failure, which became evident in about 1900, ultimately led to the formulation of a quantum electromagnetic theory known as **quantum electrodynamics** (QED). When applied to optical phenomena, QED is usually called **quantum optics**. Quantum optics properly describes almost all known optical phenomena.

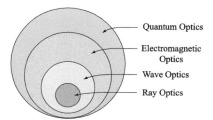

Figure 13.0-1 The theory of quantum optics explains virtually all optical phenomena. It is more general than electromagnetic optics, which was shown earlier to encompass wave optics and ray optics.

In the mathematical framework of quantum electrodynamics, the vectors **E** and **H** that are used to describe the electric and magnetic fields of classical electromagnetic optics, respectively, are promoted to operators in a Hilbert space. These operators are assumed to satisfy certain operator equations and commutation relations that govern their time dynamics and interdependence. Though the equations of QED describe the interactions of electromagnetic fields with matter in much the same way as Maxwell's equations, QED leads to results that are uniquely quantum in nature. In spite of its vast successes, quantum optics is nevertheless not the final arbiter of *all* optical effects. That distinction currently belongs to the **electroweak theory**, which combines quantum electrodynamics with the theory of weak interactions.[†] Continuing efforts are underway to combine electroweak theory with the theories of strong and gravitational interactions in an attempt to forge a general unified theory that accommodates all four fundamental forces of nature, as they are currently understood.

This Chapter

A formal treatment of quantum optics is beyond the scope of this book. Nevertheless, it is possible to describe many of the quantum properties of light, and its interaction with matter, by supplementing electromagnetic optics with several simple relationships drawn from quantum optics; these embody the corpuscularity, localization, and fluctuations of quantum fields and energy. This set of rules, which we call **photon optics**, permits us to deal with optical phenomena that lie beyond the reach of classical theory, while retaining classical optics as a limiting case. However, photon optics is not capable of accommodating all of the optical effects that can be explained by quantum optics.

In Sec. 13.1 we introduce the concept of the photon and examine its properties. Using electromagnetic optics as a point of departure, we impose a number of rules that

---

[†] For example, parity-nonconserving small rotations of the plane of polarization of light upon passage through certain materials cannot be accommodated by quantum electrodynamics but are successfully explained by electroweak theory; see, e.g., P. A. Vetter, D. M. Meekhof, P. K. Majumder, S. K. Lamoreaux, and E. N. Fortson, Precise Test of Electroweak Theory from a New Measurement of Parity Nonconservation in Atomic Thallium, *Physical Review Letters*, vol. 74, pp. 2658–2661, 1995.

govern the behavior of photon energy, polarization, position, momentum, interference, and time. These rules, which are deceptively simple, form the basis of photon optics and have far-reaching implications. This is followed, in Sec. 13.2, by a discussion of the properties of collections of photons and photon streams. The number of photons emitted by a fixed-intensity light source in a sequence of fixed time intervals is almost always random, with statistical properties that depend on the nature of the source. The photon-number statistics for commonly encountered optical sources, such as lasers and thermal radiators, are set forth. The effect of simple optical components, such as beamsplitters and filters, on the randomness of photon streams is also examined. In Sec. 13.3, we study the random fluctuations of the magnitude, phase, and photon number associated with the electromagnetic field from the perspective of quantum optics. We provide a brief introduction to coherent, quadrature-squeezed, photon-number-squeezed, and entangled-photon states of light, and indicate several generation mechanisms and applications. The interactions of photons with atoms and semiconductors are described in Secs. 14.3 and 17.2, respectively.

13.1 THE PHOTON

From a quantum perspective, light consists of particles called **photons**. A photon carries electromagnetic energy and momentum, as well as intrinsic angular momentum (or spin) associated with its polarization properties. It can also carry orbital angular momentum. The photon has zero rest mass and travels at c_o, the speed of light in vacuum; its speed in dielectric materials is reduced to $c < c_o$. A photon concomitantly has a wavelike character that determines its localization properties in space and time, and governs how it interferes and diffracts.

The notion of the photon initially grew out of an attempt by Max Planck in 1900 to resolve a long-standing conundrum concerning the spectrum of blackbody radiation emanating from a cavity held at a fixed temperature T (this topic is discussed in Sec. 14.4B). Planck ultimately resolved the problem by assuming that the allowed energies of the atoms in the walls of the cavity were quantized to discrete values. In 1905, Albert Einstein proposed that the quantization be imposed directly on the energy of the electromagnetic radiation, rather than on the atoms, which led to the concept of the photon. This enabled Einstein to successfully explain the photoelectric effect (this topic is discussed in Sec. 19.1A). The term "photon" was introduced by Gilbert Lewis in 1926.

The concept of the photon and the rules of photon optics are introduced by considering light inside an optical resonator (cavity). This is a convenient choice because it restricts the space under consideration to a simple geometry. However, the presence of the resonator turns out not to be an important feature of the argument; the results can be shown to be independent of the form of the resonator, and even of its presence.

Electromagnetic-Optics Theory of Light in a Resonator

In accordance with electromagnetic optics, light inside a lossless resonator of volume V is completely characterized by an electromagnetic field that takes the form of a superposition of discrete orthogonal modes of different spatial distributions, different frequencies, and different polarizations. The electric-field vector, $\mathcal{E}(\mathbf{r}, t) = \text{Re}\{\mathbf{E}(\mathbf{r}, t)\}$, can therefore be expressed in terms of the complex electric field $\mathbf{E}(\mathbf{r}, t)$ via

$$\mathbf{E}(\mathbf{r}, t) = \sum_{\mathbf{q}} A_{\mathbf{q}} U_{\mathbf{q}}(\mathbf{r}) \exp(j2\pi \nu_{\mathbf{q}} t) \, \widehat{\mathbf{e}}_{\mathbf{q}}. \qquad (13.1\text{-}1)$$

The qth mode has complex envelope $A_{\mathbf{q}}$, frequency $\nu_{\mathbf{q}}$, polarization along the direction of the unit vector $\widehat{\mathbf{e}}_{\mathbf{q}}$, and a spatial distribution characterized by the complex function

$U_\mathbf{q}(\mathbf{r})$, which is normalized such that $\int_V |U_\mathbf{q}(\mathbf{r})|^2\, d\mathbf{r} = 1$. The expansion functions $U_\mathbf{q}(\mathbf{r})$, $\exp(j2\pi\nu_\mathbf{q} t)$, and $\hat{\mathbf{e}}_\mathbf{q}$ are not unique; other choices are available, including those comprising polychromatic modes.

In a cubic resonator of dimension d, a convenient choice for the spatial expansion functions is the set of standing waves

$$U_\mathbf{q}(\mathbf{r}) = \left(\frac{2}{d}\right)^{3/2} \sin\left(q_x \frac{\pi}{d} x\right) \sin\left(q_y \frac{\pi}{d} y\right) \sin\left(q_z \frac{\pi}{d} z\right), \qquad (13.1\text{-}2)$$

where the integers q_x, q_y, and q_z are usually specified in the form (q_x, q_y, q_z) [Sec. 11.3C and Fig. 13.1-1(a)]. In accordance with (5.4-9), the energy density associated with mode $\mathbf{q}$ is $\tfrac{1}{2}\epsilon |A_\mathbf{q}|^2 |U_\mathbf{q}(\mathbf{r})|^2$, so that the energy contained in mode $\mathbf{q}$ is

$$E_\mathbf{q} = \tfrac{1}{2}\epsilon \int_V |A_\mathbf{q}|^2 |U_\mathbf{q}(\mathbf{r})|^2\, d\mathbf{r} = \tfrac{1}{2}\epsilon |A_\mathbf{q}|^2, \qquad (13.1\text{-}3)$$

where V is the modal volume. In classical electromagnetic theory, the energy $E_\mathbf{q}$ can assume any nonnegative value, no matter how small, and the total energy is the sum of the energies in all modes.

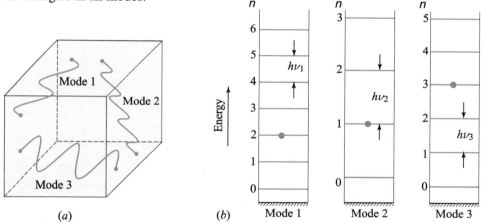

Figure 13.1-1 (a) Schematic of three electromagnetic modes of different frequencies and directions in a cubic resonator. (b) Allowed energy levels of three modes in the context of photon optics. Modes 1, 2, and 3 have frequencies ν_1, ν_2, and ν_3, respectively. In the example presented in the figure, modes 1, 2, and 3 contain $n = 2$, 1, and 3 photons, respectively, as represented by the filled circles.

Photon-Optics Theory of Light in a Resonator

The electromagnetic-optics theory described above is maintained in photon optics, but a restriction is placed on the energy that each mode is permitted to carry. Rather than assuming a continuous range, with no minimum allowed value, the modal energy is restricted to discrete values separated by a fixed energy $h\nu$, where ν is the frequency of the mode [Fig. 13.1-1(b)]. The energy of a mode is thus quantized, with only integral units of this fixed energy permitted. Each unit of energy is carried by a single photon and the mode may carry an arbitrary number of photons.

> Light in a resonator comprises a set of modes, each of which contains an integral number of identical photons. Characteristics of the mode, such as its frequency, spatial distribution, direction of propagation, and polarization, are assigned to the photons.

A. Photon Energy

Photon optics provides that the energy of an electromagnetic mode is quantized to discrete levels separated by the energy of a photon [Fig. 13.1-1(b)]. The energy of a photon in a mode of frequency ν is

$$E = h\nu = \hbar\omega, \qquad (13.1\text{-}4)$$
Photon Energy

where $h = 6.6261 \times 10^{-34}$ J·s is **Planck's constant** and $\hbar \equiv h/2\pi$. Energy may be added to, or taken from this mode only in units of $h\nu$.

A mode containing zero photons nevertheless carries energy $E_0 = \frac{1}{2}h\nu$, which is called the **zero-point energy** and is associated with the fluctuations of the **vacuum state** (Fig. 13.3-3). When it carries n photons, a mode therefore has total energy

$$E_n = (n + \tfrac{1}{2})h\nu, \qquad n = 0, 1, 2, \ldots . \qquad (13.1\text{-}5)$$

This expression is identical to that for the energy levels of a quantum-mechanical harmonic oscillator, as provided in (13.3-4); the connection will be established in Sec. 13.3. In most experiments, the zero-point energy is not directly observable because only energy differences [e.g., $E_2 - E_1$ in (13.1-5)] are measured. However, the zero-point energy is not innocuous since it constitutes a source of noise ("shot noise") that limits the sensitivity of certain precision measurements. This will become evident in Sec. 13.3B, where we will demonstrate how the vacuum state can be manipulated ("squeezed") by configuring an experiment in a particular way so as to reduce the deleterious effects of vacuum-state fluctuations. Zero-point fluctuations are also responsible for the process of spontaneous emission from an atom, as discussed in Sec. 14.3. Moreover, are the origin of the **Casimir effect**, a small attractive force that acts between two parallel uncharged conducting plates located in close proximity.

The order of magnitude of the photon energy is readily estimated. An infrared photon of wavelength $\lambda_o = 1$ μm in free space has a frequency $\nu \approx 3 \times 10^{14}$ Hz, by virtue of the relation $\lambda_o \nu = c_o$, and a period $T = 1/\nu$. Its energy is thus $h\nu \approx 1.99 \times 10^{-19}$ J. In units of electron volts, the photon energy becomes $h\nu/e = (1.99 \times 10^{-19})/(1.6 \times 10^{-19}) = 1.24$ eV; this is equivalent to the kinetic energy imparted to an electron when it is accelerated through a potential difference of 1.24 V. Another example is provided by a microwave photon with a wavelength of 1 cm; the photon energy is then 10^4 times smaller, namely $h\nu = 1.24 \times 10^{-4}$ eV. A convenient conversion formula between wavelength (μm) and photon energy (eV) is therefore expressible as

$$E\,(\text{eV}) \approx \frac{1.24}{\lambda_o\,(\mu\text{m})}. \qquad (13.1\text{-}6)$$

The reciprocal wavelength is also frequently used as a unit of energy, often in chemistry. It is specified in cm^{-1} and is determined by expressing the wavelength in cm and simply taking the inverse. Thus, 1 cm^{-1} corresponds to $1.24/10\,000$ eV and 1 eV corresponds to 8065 cm^{-1}. Conversions among photon wavelength, frequency, period, and energy are illustrated in Fig. 13.1-2.

Because the photon energy increases with frequency, the particle nature of light becomes increasingly prevalent as the frequency of the radiation increases. X-rays and gamma-rays almost always behave like particles, and wavelike effects such as diffraction and interference are difficult to discern. In contrast, radio waves almost always behave like waves. The frequency of light in the optical region is such that both particle-like and wavelike behavior are readily observed, thus spurring the need for photon optics.

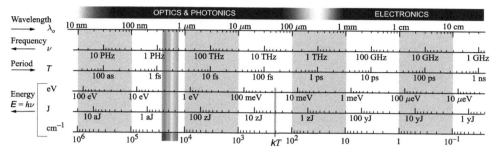

Figure 13.1-2 Relationships among photon wavelength λ_o, frequency ν, period T, and energy E (specified in units of eV, J, and reciprocal wavelength $1/\lambda_o$ in cm^{-1}). A photon of free-space wavelength $\lambda_o = 1$ μm has frequency $\nu = 300$ THz, period $T = 3.33$ fs, and energy $E = 1.24$ eV = 199 zJ = 10^4 cm^{-1}. At room temperature ($T = 300°$ K), the thermal energy $kT = 26$ meV = 4.17 zJ = 210 cm^{-1}. Two spectral domains are indicated: 1) optics & photonics, and 2) electronics.

B. Photon Polarization

As indicated earlier, light is characterized by a set of modes of different frequencies, directions, and polarizations, each occupied by an integral number of photons. For each monochromatic plane wave traveling in a particular direction, there are two polarization modes. The polarization of a photon is that of the mode it occupies. For example, the photon may be linearly polarized in the x direction, or right circularly polarized. Since the polarization modes of free space are degenerate, they are not unique. One may use modes with linear polarization in the x and y directions, linear polarization in two other orthogonal directions, say x' and y', or right- and left-circular polarizations. The choice of a particular set is a matter of convenience. A problem arises when a photon occupying a given mode (say linear polarization in the x direction) is to be observed in a different set of modes (say linear polarization in the x' and y' directions). Since the photon energy cannot be split between the two modes, a probabilistic interpretation is called for.

In classical electromagnetic optics, the state of polarization of a plane wave is described by a Jones vector, whose components (A_x, A_y) are the components of the complex envelope in the x and y directions, respectively (Sec. 6.1A). The same wave may also be represented in a different coordinate system (x', y'), e.g., one that makes a 45° angle with the initial coordinate system, by a Jones vector with components

$$A_{x'} = \tfrac{1}{\sqrt{2}}(A_x - A_y), \qquad A_{y'} = \tfrac{1}{\sqrt{2}}(A_x + A_y), \qquad (13.1\text{-}7)$$

as described in Sec. 6.1B. Therefore, a wave that is linearly polarized in the x direction is described by a Jones vector with components $(A_0, 0)$ in the x–y coordinate system, where A_0 is the complex envelope. In the (x', y') coordinate system, the Jones vector has components $(\tfrac{1}{\sqrt{2}}A_0, \tfrac{1}{\sqrt{2}}A_0)$.

> The state of polarization of a single photon is described by a Jones vector with complex components (A_x, A_y), normalized such that $|A_x|^2 + |A_y|^2 = 1$. The coefficients A_x and A_y are interpreted as complex probability amplitudes, and their squared magnitudes, $|A_x|^2$ and $|A_y|^2$, represent the probabilities that the photon is observed in the x and y linear polarization modes, respectively.

The components (A_x, A_y) are transformed from one coordinate system to another in the same manner as ordinary Jones vectors, and the new components represent complex probability amplitudes in the new modes. Thus, a single photon may exist, probabilistically, in more than one mode. This concept is illustrated by the following examples.

Linearly Polarized Photon

A photon is linearly polarized in the x direction. In terms of the x–y linearly polarized modes, the photon is described by a Jones vector with components $(1, 0)$. In a set of linearly polarized modes in the x' and y' directions at $45°$, these components are $(\frac{1}{\sqrt{2}}, \frac{1}{\sqrt{2}})$ so that the probabilities of observing the photon in a linear polarization mode along the x' or y' directions are both $1/2$. This is illustrated schematically in Fig. 13.1-3.

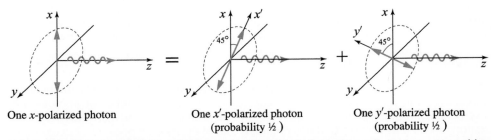

One x-polarized photon   One x'-polarized photon (probability ½)   One y'-polarized photon (probability ½)

Figure 13.1-3 A photon in the x linear polarization mode is the same as a photon in a superposition of the x' and y' linear polarization modes, each with probability $1/2$.

EXAMPLE 13.1-1. *Transmission of a Linearly Polarized Photon Through a Polarizer.*
Consider the transmission of a photon that is linearly polarized in the x direction through a linear polarizer whose transmission axis is along the x' direction at an angle θ, as illustrated in Fig. 13.1-4. The polarizer transmits light that is linearly polarized in the x' direction but blocks light in the orthogonal y' direction. The probability that the photon is transmitted through the polarizer is determined by writing the Jones vector of the photon polarization state in the x'–y' coordinate system as $(\cos\theta, -\sin\theta)$ [see (6.1-21) and (6.1-22)]. The probability of observing the photon in the mode with x' linear polarization is therefore $\cos^2\theta$, which represents the probability of passage of the photon through the polarizer: $p(\theta) = \cos^2\theta$. The probability that the photon is blocked is therefore $1 - p(\theta) = \sin^2\theta$. Classical polarization optics reveals that the intensity transmittance of a polarizer in this same configuration is $\cos^2\theta$ (Prob. 6.1-7). We conclude that the probability of transmission of a single photon is identical to the classical transmittance, i.e., $p(\theta) = \mathcal{T}(\theta)$.

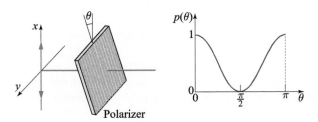

Figure 13.1-4 Probability of a linearly polarized photon passing through a polarizer. The axis of the polarizer is at an angle θ with respect to the photon polarization.

Circularly Polarized Photon

A circularly polarized photon is described by a Jones vector that has components $\frac{1}{\sqrt{2}}(1, \pm j)$, where the + and − signs correspond to right- and left-handed polarization, respectively. This description is based on an x–y coordinate system, i.e., linearly polarized modes. Therefore, the probability of the photon passing through a linear polarizer pointing in either the x or y direction is $1/2$. It can also be shown that this result prevails whatever the direction of the linear polarizer. The circularly polarized photon may be regarded as equivalent to the probabilistic superposition of one photon with linear polarization in the x direction and another in the y direction, each with probability $1/2$.

Right- and left-circular polarizations may also be used as modes (as a coordinate system). In that description, a linearly polarized photon may be regarded as a probabilistic superposition of right- and left-circularly polarized photons, each with probability $1/2$, as illustrated in Fig. 13.1-5.

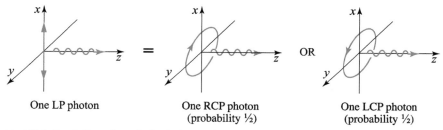

One LP photon   One RCP photon (probability ½)   One LCP photon (probability ½)

Figure 13.1-5 A linearly polarized photon is equivalent to the superposition of a right- and a left-circularly polarized photon, each with probability $1/2$.

C. Photon Position

Associated with each photon of frequency ν is a wave described by the complex wavefunction $U(\mathbf{r})\exp(j2\pi\nu t)$ of the mode. However, when a photon impinges on a detector of small area dA located normal to the direction of propagation, at the position $\mathbf{r}$, its indivisibility causes it to be either wholly detected or not detected at all. The location at which the photon is registered is not precisely determined. Rather, it is governed by the optical intensity $I(\mathbf{r}) \propto |U(\mathbf{r})|^2$, in accordance with the following probabilistic law:

> The probability $p(\mathbf{r})\,dA$ of observing a photon at the position $\mathbf{r}$ within an incremental area dA, at any time, is proportional to the local optical intensity $I(\mathbf{r}) \propto |U(\mathbf{r})|^2$, so that
> $$p(\mathbf{r})\,dA \propto I(\mathbf{r})\,dA. \qquad (13.1\text{-}8)$$
> **Photon Position**

The photon is therefore more likely to be found at those locations where the intensity is high. A photon in a mode described by a standing wave with the intensity distribution $I(x,y,z) \propto \sin^2(\pi z/d)$, where $0 \le z \le d$, for example, is most likely to be detected at $z = d/2$, but will never be detected at $z = 0$ or $z = d$. In contrast to waves, which are extended in space, and particles, which are localized in space, optical photons behave as extended *and* localized entities. This behavior is called **wave–particle duality**. The localized nature of photons becomes evident when they are detected.

EXERCISE 13.1-1

Photon in a Gaussian Beam.

(a) Consider a single photon described by a Gaussian beam (the $\text{TEM}_{0,0}$ mode of a spherical-mirror resonator; see Secs. 3.1B, 5.4A, and 11.2B). What is the probability of detecting the photon at a point within a circle whose radius is the waist radius of the beam W_0? Recall from (3.1-12) that at the waist, $I(\rho, z = 0) \propto \exp(-2\rho^2/W_0^2)$, where ρ is the radial coordinate.

(b) If the beam carries a large number n of independent photons, estimate the average number of photons that lie within this circle.

Transmission of a Single Photon Through a Beamsplitter

An ideal beamsplitter is an optical device that losslessly splits a beam of light into two beams that emerge at right angles. It is characterized by an intensity transmittance $\mathcal{T}$ and an intensity reflectance $\mathcal{R} = 1 - \mathcal{T}$. The intensity of the transmitted wave I_t and the intensity of the reflected wave I_r can be calculated from the intensity of the incident wave I using the electromagnetic relations $I_t = \mathcal{T}I$ and $I_r = (1 - \mathcal{T})I$.

Because a photon is indivisible, it must choose between the two possible directions permitted by the beamsplitter. A single photon incident on the device will follow these directions in accordance with the probabilistic photon-position rule (13.1-8). The probability that the photon is transmitted is proportional to I_t and is therefore equal to the transmittance $\mathcal{T} = I_t/I$. The probability that it is reflected is $1 - \mathcal{T} = I_r/I$. From the point of view of probability, the problem is identical to that of flipping a biased coin. Figure 13.1-6 illustrates the process.

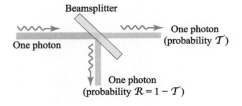

Figure 13.1-6 Probabilistic reflection or transmission of a photon at a lossless beamsplitter.

Single-Photon Imaging

As described in Sec. 4.4 and in (A.3-3) of Appendix A, a coherent imaging system is characterized by an impulse response function $h(x, y; x', y')$ that links its output and input fields, $U_o(x, y)$ and $U_i(x, y)$, respectively, via the two-dimensional convolution

$$U_o(x, y) = \iint_{-\infty}^{\infty} U_i(x', y') \, h(x, y; x', y') \, dx' \, dy'. \tag{13.1-9}$$

The very same relationship characterizes the single-photon wavefunctions at the output and input of a single-photon imaging system, where $|U_o(x)|^2$ represents the probability density function of the photon position in the image plane.

D. Photon Momentum

In classical electromagnetic optics, as discussed in Sec. 5.4A, an electromagnetic plane wave carries a linear momentum density $(W/c)\hat{\mathbf{k}}$, where W is the energy density (per unit volume) and $\hat{\mathbf{k}}$ is a unit vector in the direction of the wavevector $\mathbf{k}$.

In photon optics, the linear momentum of a photon is $\mathbf{p} = (E/c)\hat{\mathbf{k}}$ where $E = \hbar\omega = \hbar c k$ is the photon energy, so that:

> The linear momentum associated with a photon in a plane-wave mode of wavevector $\mathbf{k}$ is
> $$\mathbf{p} = \hbar\mathbf{k}. \qquad (13.1\text{-}10)$$
> Photon Momentum
>
> The magnitude of the momentum is $p = \hbar k = \hbar\omega/c = \hbar 2\pi/\lambda$, so that
> $$p = E/c = h/\lambda. \qquad (13.1\text{-}11)$$

*Momentum of a Localized Wave

A wave more general than a plane wave, with a complex wavefunction of the form $U(\mathbf{r})\exp(j2\pi\nu t)$, can be expanded as a sum of plane waves of different wavevectors by using the techniques of Fourier optics (Chapter 4). The component with wavevector $\mathbf{k}$ may be written in the form $A(\mathbf{k})\exp(-j\mathbf{k}\cdot\mathbf{r})\exp(j2\pi\nu t)$, where $A(\mathbf{k})$ is its amplitude.

> The momentum of a photon described by an arbitrary complex wavefunction $U(\mathbf{r})\exp(j2\pi\nu t)$ is uncertain. It assumes the value
> $$\mathbf{p} = \hbar\mathbf{k}, \qquad (13.1\text{-}12)$$
> with probability proportional to $|A(\mathbf{k})|^2$, where $A(\mathbf{k})$ is the amplitude of the plane-wave Fourier component of $U(\mathbf{r})$ with wavevector $\mathbf{k}$.

If $f(x,y) = U(x,y,0)$ is the complex amplitude at the $z = 0$ plane, the plane-wave Fourier component with wavevector $\mathbf{k} = (k_x, k_y, k_z)$ has an amplitude $A(\mathbf{k}) = F(k_x/2\pi, k_y/2\pi)$, where $F(\nu_x, \nu_y)$ is the two-dimensional Fourier transform of $f(x,y)$, as described in Chapter 4. Because the functions $f(x,y)$ and $F(\nu_x, \nu_y)$ form a Fourier transform pair, their widths are inversely related and satisfy the position–direction relation provided in (A.2-6) of Appendix A. The uncertainty relation between the position of the photon and the direction of its momentum is established because the position of the photon at the $z = 0$ plane is probabilistically determined by $|U(\mathbf{r})|^2 = |f(x,y)|^2$, and the direction of its momentum is probabilistically determined by $|A(\mathbf{k})|^2 = |F(k_x/2\pi, k_y/2\pi)|^2$. Thus if, at the plane $z = 0$, σ_x is the positional uncertainty in the x direction, and $\sigma_\theta = \sin^{-1}(\sigma_{k_x}/k) \approx (\lambda/2\pi)\sigma_{k_x}$ is the angular uncertainty about the z axis (which is assumed to be $\ll 1$), then the uncertainty relation $\sigma_x \sigma_{k_x} \geq 1/2$ is equivalent to $\sigma_x \sigma_\theta \geq \lambda/4\pi$.

A plane-wave photon has a known momentum (fixed direction and magnitude), so that $\sigma_\theta = 0$, but its position is totally uncertain ($\sigma_x = \infty$); it is equally likely to be detected anywhere in the $z = 0$ plane. When a plane-wave photon passes through an aperture, its position becomes localized at the expense of a spread in the direction of its momentum. The position–direction uncertainty of the photon therefore parallels the theory of diffraction described in Sec. 4.3. At the other extreme from the plane wave is the spherical-wave photon. It is well localized in position (at the center of the wave), but the direction of its momentum is totally uncertain.

Radiation Pressure

Because a photon carries momentum, and momentum is conserved, the atom emitting the photon experiences a recoil of magnitude $h\nu/c$. The momentum associated with a photon can also be transferred to an object, giving rise to a force that results in mechanical motion. As an example, light beams can be used to deflect atomic beams or to contain collections of atoms or small dielectric particles (Sec. 14.3F). The term **radiation pressure** is often used to describe this phenomenon (pressure = force/area).

EXERCISE 13.1-2

Photon-Momentum Recoil. Determine the recoil velocity imparted to a ^{198}Hg atom that emits a photon of energy 4.88 eV. Compare this with the root-mean-square thermal velocity v of the atom at a temperature $T = 300°$ K (this is obtained by setting the average kinetic energy equal to the average thermal energy, $\frac{1}{2}mv^2 = \frac{3}{2}kT$, where $k = 1.38 \times 10^{-23}$ J/K is Boltzmann's constant).

Photon Spin Angular Momentum

The intrinsic **spin angular momentum** of a photon associated with circular polarization is characterized by an electric-field vector that rotates in a circle normal to the direction of propagation. Because the photon travels at the speed of light, the projection of its spin vector lies either parallel or antiparallel to the wavevector, so that its **helicity** is quantized to two values,

$$\mathbb{S} = \pm\hbar, \qquad (13.1\text{-}13)$$
Photon Spin

where the plus (minus) signs are associated with right-handed (left-handed) circular polarization, respectively. Linearly polarized photons have equal probability of exhibiting parallel and antiparallel spin. Just as photons can transfer linear momentum to an object, so too can circularly polarized photons exert a torque on an object. A circularly polarized photon will, for example, exert a torque on a half-wave plate.

Bosons and fermions. Fundamental particles are divided into two broad classes: **Bosons**, such as photons and other force-carrier particles, have a spin that is an integer multiple of $\hbar$, as do quasiparticles such as plasmons, polaritons, and phonons. In contrast **fermions**, such as electrons, protons, neutrons, and other material particles, have a spin that is a half-integer multiple of $\hbar$.

Photon Orbital Angular Momentum

Aside from the spin angular momentum associated with polarization, an electromagnetic wave may carry angular momentum by virtue of the twisting of its wavefront about the axis of propagation. For example, the Laguerre–Gaussian beam described by the complex wavefunction $U_{l,m}(\rho, \phi, z)$ in (3.4-1), which has an azimuthal phase dependence $\exp(-jl\phi)$ and therefore a helical wavefront, has an angular momentum (for $l \neq 0$) that is independent of its state of polarization. To distinguish this from spin angular momentum, it is referred to as **orbital angular momentum**. A photon in such a spatial mode possesses an orbital angular momentum $L = l\hbar$.

Another example is provided by a photon in a whispering-gallery mode (WGM) of a cylindrical resonator (Sec. 11.3B). In the context of ray optics, the mode is described by a ray tracing the cross-sectional circular boundary of the resonator. In the context of wave optics, the wavelength satisfies the resonance condition $2\pi a = q\lambda$, where a is the radius of the circle and $q = 1, 2, \ldots$. The linear momentum of the photon is $p = \hbar k = \hbar 2\pi/\lambda = q\hbar/a$, and its angular momentum is therefore $ap = q\hbar$. Similarly, a photon in a WGM mode of a microsphere resonator (Sec. 11.4C) of radius a has an angular momentum $L = \ell\hbar$, where the integer ℓ is associated with the resonance wavelength for an optical path that traces a great circle. The quantity ℓ may be regarded as an angular-momentum quantum number similar to that used to describe the hydrogen atom (Sec. 14.1A).

E. Photon Interference

Young's double-pinhole interference experiment is generally invoked to demonstrate the wave nature of light (Exercise 2.5-2). In fact, Young's experiment can be carried out even when there is only a single photon in the apparatus at a given time. The outcome of this experiment can be understood in the context of photon optics by using the photon-position rule. The intensity at the observation plane is calculated using wave optics and the result is converted to a probability density function that specifies the random position of the detected photon. The interference arises from phase differences associated with the two possible paths.

Consider a plane wave illuminating a screen with two pinholes separated by a distance $2a$, as portrayed in Fig. 13.1-7 (see also Fig. 2.5-6). The line joining the holes defines the x axis. Two spherical waves are generated that interfere at the observation plane. As is understood from Exercise 2.5-2, in the paraboloidal-wave approximation, these give rise to a sinusoidal intensity that behaves in accordance with

$$I(x) \approx 2I_0 \left(1 + \cos \frac{2\pi x \theta}{\lambda}\right). \tag{13.1-14}$$

Here I_0 is the intensity of each of the waves individually at the observation plane, λ is the wavelength, and θ is the angle subtended by the two pinholes at the observation plane (Fig. 13.1-7). The result provided in (13.1-14) describes the intensity pattern that is experimentally observed for classical light.

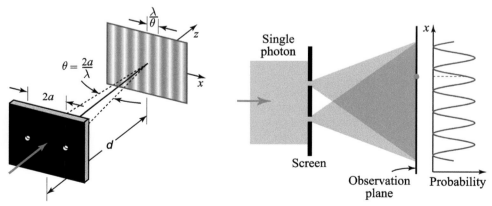

Figure 13.1-7 Young's double-pinhole experiment with a single photon. The interference pattern $I(x)$ determines the probability density of detecting the photon at position x.

Now, if only a single photon is present in the apparatus, in accordance with (13.1-8) the probability of detecting it at position x is proportional to $I(x)$. It is most likely to be detected at those values of x for which $I(x)$ is a maximum and it will never be detected at values of x for which $I(x) = 0$. If a histogram of the locations of the detected photon is constructed by repeating the experiment many times, as Taylor did in 1909, the classical interference pattern obtained by carrying out the experiment only once with a strong beam of light emerges. The classical interference pattern does indeed represent the probability density of the position at which a single photon is observed.

The occurrence of the interference is a result of the extended nature of the photon, which permits it to pass through *both* holes of the apparatus. This endows the photon with knowledge of the geometry of the entire experiment when it reaches the observation plane, where it is detected as a single entity. If one of the holes were to be covered, the interference pattern would disappear.

EXERCISE 13.1-3

Single Photon in a Mach–Zehnder Interferometer. Consider a plane wave of light of wavelength λ that is split into two parts at a beamsplitter (Sec. 13.1C) and recombined in a Mach–Zehnder interferometer, as shown in Fig. 13.1-8 [see also Fig. 2.5-3(a)]. If the wave contains only a single photon, plot the probability of finding it at the detector as a function of d/λ (for $0 \leq d/\lambda \leq 1$), where d is the difference between the two optical paths of the light. Assume that the mirrors and beamsplitters are perfectly flat and lossless, and that the beamsplitters have $\mathcal{T} = \mathcal{R} = 1/2$. Where might the photon be located when the probability of finding it at the detector is not unity?

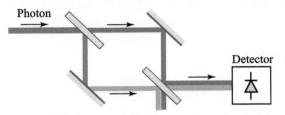

Figure 13.1-8 Single photon in a Mach–Zehnder interferometer.

F. Photon Time

The modal expansion provided in (13.1-1) comprises monochromatic modes that are "eternal" harmonic functions of time; a photon in a monochromatic mode is equally likely to be detected at any time. As indicated earlier, however, a modal expansion of the radiation inside (or outside) a resonator is not unique. A more general expansion comprises polychromatic modes such as time-localized wavepackets (Sec. 2.6A). The probability of detecting a photon described by the complex wavefunction $U(\mathbf{r}, t)$, at any position and in the incremental time interval between t and $t + dt$, is proportional to $I(\mathbf{r}, t) \, dt \propto |U(\mathbf{r}, t)|^2 \, dt$. The photon-position rule of photon optics presented in (13.1-8) may therefore be generalized to include photon time localization:

The probability $p(\mathbf{r}, t) \, dA \, dt$ of observing a photon at position $\mathbf{r}$ within an incremental area dA, and during an incremental time interval dt following time t, is proportional to the optical intensity of the mode at $\mathbf{r}$ and t, i.e.,

$$p(\mathbf{r}, t) \, dA \, dt \propto I(\mathbf{r}, t) \, dA \, dt \propto |U(\mathbf{r}, t)|^2 \, dA \, dt. \qquad (13.1\text{-}15)$$
Photon Position and Time

Time–Energy Uncertainty

The time during which a photon in a monochromatic mode of frequency ν may be detected is totally uncertain, whereas the value of its frequency ν (and its energy $h\nu$) is absolutely certain. In contrast, a photon in a wavepacket mode with an intensity function $I(t)$ of duration σ_t must be localized within this time. Bounding the photon time in this way engenders an uncertainty in its frequency (and energy) by virtue of the properties of the Fourier transform, and the result is a polychromatic photon. Suppressing the **r** dependence for simplicity, the frequency uncertainty is readily determined by carrying out a Fourier-transform expansion of $U(t)$ in terms of its harmonic components,

$$U(t) = \int_{-\infty}^{\infty} V(\nu) \exp(j2\pi\nu t) \, d\nu, \qquad (13.1\text{-}16)$$

where $V(\nu)$ is the Fourier transform of $U(t)$ (Sec. A.1 of Appendix A). The width σ_ν of $|V(\nu)|^2$ represents the spectral width. If σ_t is the power-RMS width of the function $|U(t)|^2$, then σ_t and σ_ν must satisfy the duration–bandwidth reciprocity relation $\sigma_\nu \sigma_t \geq 1/4\pi$ or, equivalently, $\sigma_\omega \sigma_t \geq 1/2$ (the definitions of σ_t and σ_ν that lead to this uncertainty relation are provided in Sec. A.2 of Appendix A).

The energy of the photon $\hbar\omega$ cannot then be specified to an accuracy better than $\sigma_E = \hbar\sigma_\omega$. It follows that the energy uncertainty of a photon, and the time during which it may be detected, must satisfy

$$\boxed{\sigma_E \sigma_t \geq \frac{\hbar}{2},} \qquad (13.1\text{-}17)$$
Time–Energy Uncertainty

which is known as the **time–energy uncertainty relation**. This relation is analogous to that between position and wavenumber (momentum), which sets a limit on the precision with which the position and momentum of a photon can be simultaneously specified. The average energy $\overline{E}$ of this polychromatic photon is $\overline{E} = h\overline{\nu} = \hbar\overline{\omega}$.

To summarize: a monochromatic photon ($\sigma_\nu \to 0$) has an eternal duration within which it can be observed ($\sigma_t \to \infty$). In contrast, a photon associated with an optical wavepacket is localized in time and is thus polychromatic with a corresponding energy uncertainty. Thus, a *wavepacket photon* can be viewed as a confined traveling packet of energy.

EXERCISE 13.1-4

Single Photon in a Gaussian Wavepacket. Consider a plane-wave wavepacket (Sec. 2.6A) containing a single photon traveling in the z direction, with complex wavefunction

$$U(\mathbf{r}, t) = \mathfrak{a}\left(t - \frac{z}{c}\right) \qquad (13.1\text{-}18)$$

where

$$\mathfrak{a}(t) = \exp\left(-\frac{t^2}{4\tau^2}\right) \exp(j2\pi\nu_0 t). \qquad (13.1\text{-}19)$$

(a) Show that the uncertainties in its time and z position are $\sigma_t = \tau$ and $\sigma_z = c\sigma_t$, respectively.

(b) Show that the uncertainties in its energy and momentum satisfy the minimum uncertainty relations:

$$\sigma_E \sigma_t = \hbar/2 \tag{13.1-20}$$

$$\sigma_z \sigma_p = \hbar/2. \tag{13.1-21}$$

Equation (13.1-21) is the minimum-uncertainty limit of the **Heisenberg position–momentum uncertainty relation** provided in (A.2-7) of Appendix A.

Summary

Electromagnetic radiation may be described in terms of a sum of modes, e.g., monochromatic uniform plane waves of the form

$$\mathbf{E}(\mathbf{r},t) = \sum_\mathbf{q} A_\mathbf{q} \exp(-j\mathbf{k_q} \cdot \mathbf{r}) \exp(j2\pi\nu_\mathbf{q} t)\,\widehat{\mathbf{e}}_\mathbf{q}. \tag{13.1-22}$$

Each plane wave has two orthogonal polarization states (e.g., vertical/horizontal linearly polarized, right/left circularly polarized) represented by the vectors $\widehat{\mathbf{e}}_\mathbf{q}$. When the energy of a mode is measured, the result is an integer (in general, random) number of energy quanta (photons). Each of the photons associated with the mode **q** has the following properties:

- Energy $E = h\nu_\mathbf{q}$.
- Momentum $\mathbf{p} = \hbar \mathbf{k_q}$.
- Helicity (spin angular momentum) $\mathbb{S} = \pm\hbar$, if it is circularly polarized.
- The photon is equally likely to be found anywhere in space, and at any time, since the wavefunction of the mode is a monochromatic plane wave.

The choice of modes is not unique. A modal expansion in terms of non-monochromatic (quasi-monochromatic), non-plane waves, is also possible:

$$\mathbf{E}(\mathbf{r},t) = \sum_\mathbf{q} A_\mathbf{q}\, U_\mathbf{q}(\mathbf{r},t)\,\widehat{\mathbf{e}}_\mathbf{q}. \tag{13.1-23}$$

Each of the photons associated with the mode **q** then has the following properties:

- The photon position and time are governed by the complex wavefunction $U_\mathbf{q}(\mathbf{r},t)$. The probability of observing the photon at position **r** within an incremental area dA, and during an incremental time interval dt following time t, is proportional to $|U_\mathbf{q}(\mathbf{r},t)|^2\, dA\, dt$.
- If $U_\mathbf{q}(\mathbf{r},t)$ has a finite time duration σ_t, i.e., if the photon is localized in time, then the photon energy $h\nu_\mathbf{q}$ has an uncertainty $h\sigma_\nu \geq h/4\pi\sigma_t$.
- If $U_\mathbf{q}(\mathbf{r},t)$ has a finite spatial extent in the transverse ($z = 0$) plane, i.e., if the photon is localized in the x direction, for example, then the direction of photon momentum is uncertain. The spread in photon momentum can be determined by analyzing $U_\mathbf{q}(\mathbf{r},t)$ as a sum of plane waves, the wave with wavevector **k** corresponding to photon momentum $\hbar \mathbf{k}$. Spatial localization of the photon in the transverse plane results in an increase in the uncertainty of the photon-momentum direction.

13.2 PHOTON STREAMS

In Sec. 13.1 we concentrated on the properties and behavior of single photons. We now consider the properties of collections of photons. As a result of the processes by which photons are created (e.g., atomic emissions, as discussed in Chapter 14), the number of photons occupying any mode is generally random. Photon streams often contain numerous propagating modes, each carrying a random number of photons. If an experiment is carried out in which a weak stream of photons falls on a photosensitive surface, the individual photons are registered (detected) at random localized instants of time and at random points in space, in accordance with (13.1-15). (This space–time process can be discerned by viewing a barely illuminated object with the naked, dark-adapted eye.)

The temporal and spatial behavior of the photon registrations can be examined individually. The *temporal pattern* can be highlighted by making use of a detector that integrates light over a finite area A, but has good temporal resolution (e.g., a photodiode), as illustrated in Fig. 13.2-1. According to (13.1-15), the probability of detecting a photon in the incremental time interval between t and $t + dt$ is proportional to $P(t) = \int_A I(\mathbf{r}, t)\, dA$, which is the optical power at time t. The photons are registered at random times.

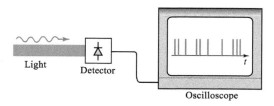

Figure 13.2-1 Photon registrations at random localized instants of time for a detector that integrates light over an area A.

The *spatial pattern* of photon registrations, on the other hand, is readily manifested by making use of a detector that integrates light over a fixed exposure time T but has good spatial resolution (e.g., photographic film). In accordance with (13.1-15), the probability of detecting a photon in an incremental area dA surrounding the point $\mathbf{r}$ is then proportional to $\int_0^T I(\mathbf{r}, t)\, dt$, which is the integrated local intensity. The photons are registered at random locations, as illustrated by the grainy image of Max Planck provided in Fig. 13.2-2. This image was obtained by rephotographing, under very low light conditions, the image of Max Planck presented on page 514. Each white dot in the photograph represents a random photon registration; the density of these registrations follows the local intensity.

Figure 13.2-2 Random photon registrations exhibit a spatial density that follows the local optical intensity. This image of Max Planck, illuminated by a sparse stream of photons, should be compared with that on page 514, illuminated with high-intensity light.

A. Photon Flow

We begin by introducing a number of definitions that relate the mean flow of photons to classical electromagnetic intensity, power, and energy. These definitions are inspired by (13.1-15), which dictates the position and time at which a single-photon detection occurs. We then turn to randomness in the photon flow and consider the photon-number statistics for different sources of light. Finally, we consider the random partitioning of a photon stream by a beamsplitter or photodetector.

Mean Photon-Flux Density

Monochromatic light of frequency ν and constant classical intensity $I(\mathbf{r})$ (watts/cm^2) carries a mean **photon-flux density**

$$\phi(\mathbf{r}) = \frac{I(\mathbf{r})}{h\nu}. \qquad (13.2\text{-}1)$$
Mean Photon-Flux Density

Since each photon carries energy $h\nu$, this equation provides a straightforward conversion from a classical measure (units of energy/s-cm^2) into a quantum measure (units of photons/s-cm^2). For quasi-monochromatic light of central frequency $\bar{\nu}$, all photons have approximately the same energy $h\bar{\nu}$, so that the mean photon-flux density can be approximately expressed as

$$\phi(\mathbf{r}) \approx \frac{I(\mathbf{r})}{h\bar{\nu}}. \qquad (13.2\text{-}2)$$

Typical values of $\phi(\mathbf{r})$ for some commonly encountered sources of light are provided in Table 13.2-1. It is clear from these numbers that trillions of photons rain down on each square centimeter of us each second.

Table 13.2-1 Mean photon-flux density for various sources of light.

Source	Mean Photon-Flux Density (photons/s-cm^2)
Starlight	10^6
Moonlight	10^8
Twilight	10^{10}
Indoor light	10^{12}
Sunlight	10^{14}
Laser light[a]	10^{22}

[a] A 10-mW He–Ne laser beam at $\lambda_o = 633$ nm focused to a 20-μm-diameter spot.

Mean Photon Flux

The mean photon flux Φ (units of photons/s) is obtained by integrating the mean photon-flux density over a specified area,

$$\Phi = \int_A \phi(\mathbf{r})\, dA = \frac{P}{h\bar{\nu}}, \qquad (13.2\text{-}3)$$
Mean Photon Flux

where the optical power (watts) is defined as

$$P = \int_A I(\mathbf{r})\, dA \qquad (13.2\text{-}4)$$

and $h\bar{\nu}$ is again the average energy of a photon. As an example, 1 nW of optical power at a wavelength $\lambda_o = 0.2$ μm delivers to an object an average photon flux $\Phi \approx 10^9$ photons/s. Roughly speaking, one photon thus strikes the object every nanosecond, i.e.,

$$\boxed{1 \text{ nW at } \lambda_o = 0.2 \text{ μm} \Rightarrow 1 \text{ photon/ns.}} \quad (13.2\text{-}5)$$

A photon of wavelength $\lambda_o = 1$ μm carries one-fifth as much energy, in which case 1 nW corresponds to an average of 5 photons/ns.

Mean Number of Photons

The mean number of photons $\bar{n}$ detected in the area A and in the time interval T is obtained by multiplying the mean photon flux Φ in (13.2-3) by the time duration, whereupon

$$\bar{n} = \Phi T = \frac{E}{h\bar{\nu}}, \quad (13.2\text{-}6)$$
Mean Photon Number

where $E = PT$ is the optical energy (joules).

To summarize, the relations between the classical and quantum measures of photon flow are:

Classical		Quantum	
Optical intensity	$I(\mathbf{r})$	Photon-flux density	$\phi(\mathbf{r}) = I(\mathbf{r})/h\bar{\nu}$
Optical power	P	Photon flux	$\Phi = P/h\bar{\nu}$
Optical energy	E	Photon number	$\bar{n} = E/h\bar{\nu}$

Spectral Densities

For polychromatic light of nonnegligible bandwidth, it is useful to define spectral densities of the classical intensity, power, and energy, and their respective quantum counterparts: spectral photon-flux density, spectral photon flux, and spectral photon number:

Classical		Quantum	
I_ν	(W/cm²-Hz)	$\phi_\nu = I_\nu/h\nu$	(photons/s-cm²-Hz)
P_ν	(W/Hz)	$\Phi_\nu = P_\nu/h\nu$	(photons/s-Hz)
E_ν	(J/Hz)	$\bar{n}_\nu = E_\nu/h\nu$	(photons/Hz)

As an example, $P_\nu \, d\nu$ represents the optical power in the frequency range between ν and $\nu + d\nu$ while $\Phi_\nu \, d\nu$ represents the flux of photons in that frequency range.

Time-Varying Light

If the light intensity is time varying, it follows that the mean photon-flux density given in (13.2-1) is a function of time, i.e.,

$$\phi(\mathbf{r}, t) = \frac{I(\mathbf{r}, t)}{h\overline{\nu}}. \qquad (13.2\text{-}7)$$
Mean Photon-Flux Density

The mean photon flux and optical power are then also functions of time,

$$\Phi(t) = \int_A \phi(\mathbf{r}, t)\, dA = \frac{P(t)}{h\overline{\nu}}, \qquad (13.2\text{-}8)$$
Mean Photon Flux

where

$$P(t) = \int_A I(\mathbf{r}, t)\, dA. \qquad (13.2\text{-}9)$$

Consequently, the mean number of photons registered in a time interval between $t = 0$ and T, which is obtained by integrating the photon flux, also varies with time,

$$\overline{n} = \int_0^T \Phi(t)\, dt = \frac{E}{h\overline{\nu}}, \qquad (13.2\text{-}10)$$
Mean Photon Number

where the mean optical energy (intensity integrated over time and area) is given by

$$E = \int_0^T P(t)\, dt = \int_0^T \int_A I(\mathbf{r}, t)\, dA\, dt. \qquad (13.2\text{-}11)$$

B. Randomness of Photon Flow

When the classical intensity $I(\mathbf{r}, t)$ is constant, the time and position at which a single photon is detected are governed by (13.1-15), which provides that the probability density of detecting that photon at the space–time point $(\mathbf{r}, t)$ is proportional to $I(\mathbf{r}, t)$. The classical electromagnetic intensity $I(\mathbf{r}, t)$ governs the behavior of photon streams as well as single photons, but the interpretation ascribed to $I(\mathbf{r}, t)$ differs:

> For photon streams, the classical intensity $I(\mathbf{r}, t)$ determines the mean photon-flux density $\phi(\mathbf{r}, t)$. The fluctuations of $\phi(\mathbf{r}, t)$ are determined by the properties of the light source that emits the photons.

Consider a detector that integrates over space, such as that illustrated in Fig. 13.2-1. If the intensity I is constant in time, then so too is the power P. The mean photon-flux density is then $\phi = I/h\overline{\nu}$ and the mean photon flux is $\Phi = P/h\overline{\nu}$. Nevertheless, the times at which the photons arrive are random, as illustrated schematically in Fig. 13.2-3(a); the statistical properties are determined by the nature of the source emitting the photons, as set forth in Sec. 13.2C. An example of how random photon arrivals might arise may be understood from the following example. Consider a source of optical power $P = 1$ nW that emits at a wavelength $\lambda_o = 1$ μm so it delivers an *average* photon flux of $\Phi = 5$ photons/ns or 0.005 photons/ps. Since only integral numbers of photons may be detected, this signifies that if 10^5 time intervals are examined, each of duration $T = 1$ ps, then most intervals will be empty (zero photons will be registered), about 500 intervals will contain one photon, and very few intervals will contain two or more photons.

If the optical power $P(t)$ does vary with time, the mean density of photon detections will follow the function $P(t)$, as schematically illustrated in Fig. 13.2-3(b). The mean photon flux $\Phi(t) = P(t)/h\bar{\nu}$ accommodates the fact that there are more photon arrivals when the power is large than when it is small. This variation is in addition to the fluctuations in photon occurrence times associated with the character of the source.

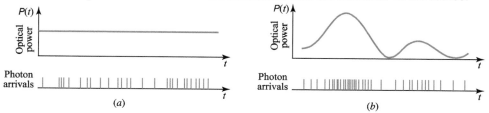

Figure 13.2-3 (a) Constant optical power and a sample function of the randomly arriving photons, whose statistical properties are determined by the nature of the source. (b) Time-varying optical power and a sample function of the randomly arriving photons, whose statistical properties are determined both by the fluctuations of the optical power and by the nature of the source, as considered in Sec. 13.2C.

The image of Max Planck in Fig. 13.2-2 illustrates analogous behavior in the spatial domain. The locations of the detected photons generally follow the classical intensity distribution, with a high photon density where the intensity is large and a low photon density where the intensity is small. But there is considerable graininess (noise) in the image corresponding to the fluctuations in photon occurrence positions associated with the source of illumination. These fluctuations are most discernible when the mean photon-flux density is small, as is the case in Fig. 13.2-2. When the mean photon-flux density becomes large everywhere, as it is in the image of Max Planck on page 514, the graininess disappears and the classical intensity distribution is recovered.

C. Photon-Number Statistics

An understanding of photon-number statistics is important for applications such as the reduction of noise in weak images and the optimization of optical information transmission. In an optical fiber communications system, for example, information is carried in the form of pulses of light (Chapter 25). Only the mean number of photons per pulse is controllable at the source; the actual number of photons emitted is unpredictable and varies from pulse to pulse, resulting in errors in the transmission of information.

The statistical distribution of the number of photons depends on the nature of the light source and must generally be treated in the context of quantum optics, as described briefly in Sec. 13.3. Under certain conditions, however, the arrival of photons may be regarded as the independent occurrences of a sequence of random events at a rate equal to the photon flux, which is proportional to the optical power. The optical power may be deterministic (as with coherent light) or a time-varying random process (as with partially coherent light). For partially coherent light (Chapter 12), the power fluctuations are correlated, so that the photon arrivals will no longer form a sequence of independent events and the photon statistics are significantly altered.

Coherent Light

Coherent light has constant optical power P. The corresponding mean photon flux $\Phi = P/h\bar{\nu}$ (photons/s) is also constant, but the actual photon registration times are random, as portrayed schematically in Fig. 13.2-4. Given a time interval of duration T, called the **counting time**, let the n signify the number of detected photons, called the **photon number**. We already know from (13.2-6) that the **mean photon number** is $\bar{n} = \Phi T = PT/h\bar{\nu}$. We now seek to establish the **photon-number distribution** $p(n)$, i.e., the probability $p(0)$ of detecting zero photons, the probability $p(1)$ of detecting one photon, etc., in the counting time T.

Figure 13.2-4 Random arrival of photons for a coherent light source of power P. Consecutive counting times of duration T are indicated. Though the optical power is constant, the photon number n observed in each counting time is random.

An expression for the photon-number distribution, $p(n)$ vs. n, suitable for coherent light can be derived under the assumption that the photon registrations are statistically independent events, as derived below. The result, known as the **Poisson distribution**, takes the form

$$p(n) = \frac{\bar{n}^n \exp(-\bar{n})}{n!}, \qquad n = 0, 1, 2, \ldots . \qquad (13.2\text{-}12)$$
Poisson Distribution

Equation (13.2-12) is displayed on a semilogarithmic plot in Fig. 13.2-5 for several values of the mean photon number $\bar{n}$.

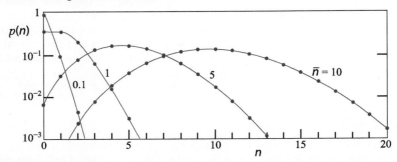

Figure 13.2-5 Semilogarithmic plot of the Poisson photon-number distribution, $p(n)$ vs. n, for several values of the mean photon number $\bar{n}$. The curves become progressively broader as $\bar{n}$ increases.

The Poisson photon-number distribution is suitable for describing the photon statistics of the coherent light emitted by an ideal, amplitude-stabilized, single-mode laser operated well above its threshold of oscillation (Chapter 16). This same result emerges in the context of quantum optics (Sec. 13.3A). Moreover, the Poisson distribution provides a good approximation for describing the photon statistics of a number of other sources of light, such as multimode thermal light.

□ **Derivation of the Poisson Distribution.** Divide the time interval T shown in Fig. 13.2-4 into a large number N of subintervals, each of sufficiently short duration T/N such that each subinterval carries one photon with probability $p = \bar{n}/N$ and zero photons with probability $1-p$. The probability of finding n independent photons in the N subintervals, like the flips of a biased coin, then follows the binomial distribution:

$$p(n) = \frac{N!}{n!(N-n)!} p^n (1-p)^{N-n}$$

$$= \frac{N!}{n!(N-n)!} \left(\frac{\bar{n}}{N}\right)^n \left(1 - \frac{\bar{n}}{N}\right)^{N-n}.$$

In the limit as the number of subintervals $N \to \infty$, we have $N!/(N-n)! \, N^n \to 1$, $(1-\bar{n}/N)^{-n} \to 1$, and $(1-\bar{n}/N)^N \to \exp(-\bar{n})$, which yields (13.2-12). ∎

Photon-Number Mean and Variance

The mean and variance are typically used to characterize a random variable. The mean photon number is expressed as

$$\bar{n} = \sum_{n=0}^{\infty} n\, p(n), \qquad (13.2\text{-}13)$$

while the variance, which is the average of the squared deviation from the mean, is given by

$$\sigma_n^2 = \sum_{n=0}^{\infty} (n - \bar{n})^2 p(n). \qquad (13.2\text{-}14)$$

The standard deviation σ_n, which is the square root of the variance, is a measure of the width of the distribution. The quantities $p(n)$, $\bar{n}$, and σ_n are collectively called **photon-number statistics**. Though the distribution $p(n)$ contains information beyond the mean and variance, these two parameters provide a rough outline of its nature.

It is not difficult to show, by inserting (13.2-12) into (13.2-13) and (13.2-14), that the mean of the Poisson distribution is indeed $\bar{n}$ and that its variance is equal to its mean:

$$\boxed{\sigma_n^2 = \bar{n}\,.} \qquad (13.2\text{-}15)$$
Variance
Poisson Distribution

Taking $\bar{n} = 100$, for example, the standard deviation is $\sigma_n = 10$, which signifies that the observation of 100 photons on average is accompanied by an uncertainty of roughly ± 10 photons.

Signal-to-Noise Ratio

Photon-number randomness constitutes a fundamental source of noise that must be contended with when using light to transmit a signal. A useful measure of the performance of an information transmission system is the photon-number-based signal-to-noise ratio (SNR). Representing the signal by its mean $\bar{n}$, and the noise by its standard deviation σ_n, the SNR is defined as

$$\text{SNR} = \frac{(\text{mean})^2}{\text{variance}} = \frac{\bar{n}^2}{\sigma_n^2}. \qquad (13.2\text{-}16)$$

If the light obeys Poisson photon-number statistics, then $\sigma_n^2 = \bar{n}$ from (13.2-15), so that

$$\boxed{\text{SNR} = \bar{n}\,.} \qquad (13.2\text{-}17)$$
Signal-to-Noise Ratio
Poisson Distribution

The Poisson signal-to-noise ratio increases linearly with the mean photon number.

Though the SNR is a useful measure of the randomness of a signal, in some applications it is necessary to make use of the full probability distribution. For example, an important measure of the performance of a digital fiber optic communications system is the probability of error. If such a system is used to transmit information using bits with a mean photon number of, say $\bar{n} = 20$, (13.2-12) dictates that the probability of no photons being received is $p(0) \approx 2 \times 10^{-9}$. Since the receipt of zero photons represents an error, as discussed in Sec. 25.2B, the full probability distribution is required to calculate system performance.

Thermal Light

When the photon arrival times are not independent, as is the case for thermal light, the photon-number statistics differ from Poisson. Thermal light is generated in an optical resonator whose walls are maintained at a fixed temperature T and whose atoms emit photons into the modes of the resonator. In accordance with the laws of statistical mechanics under conditions of thermal equilibrium, the probability of occupancy of energy level E_n in a system satisfies the **Boltzmann probability distribution**

$$P(E_n) \propto \exp\left(-\frac{E_n}{kT}\right). \quad (13.2\text{-}18)$$

Boltzmann Distribution

This exponential distribution is sketched in Fig. 13.2-6 with $P(E_n)$ plotted along the abscissa. The occupancy of each energy level is random and higher energies are relatively less probable than lower energies. The distribution is parameterized by kT, where k is **Boltzmann's constant** ($k = 1.38 \times 10^{-23}$ J/°K). At $T = 300°$ K (room temperature), $kT = 26$ meV $= 4.14$ zJ $= 209$ cm^{-1}, as illustrated in Fig. 13.1-2. The origin of the Boltzmann distribution is discussed in Sec. 14.2.

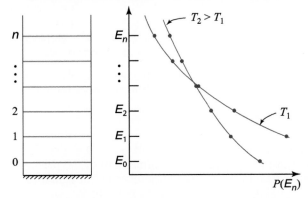

Figure 13.2-6 Boltzmann probability distribution $P(E_n)$ (plotted along the abscissa) versus energy E_n (plotted along the ordinate) for two values of the temperature T. The lower the temperature, the less likely that higher energy levels are occupied. The allowed energy levels of a collection of photons in a mode of frequency ν are illustrated at left.

We now assume that the collection of photons in a mode of frequency ν inside the resonator behaves as a gas in thermal equilibrium at temperature T that obeys the Boltzmann distribution, and that the mode has allowed energy levels given by $E_n = (n + \frac{1}{2})h\nu$, as provided in (13.1-5). It then follows that the probability of finding n photons in the mode is given by

$$p(n) \propto \exp\left(-\frac{nh\nu}{kT}\right) = \left[\exp\left(-\frac{h\nu}{kT}\right)\right]^n, \qquad n = 0, 1, 2, \ldots . \quad (13.2\text{-}19)$$

Normalizing (13.2-19) so that it sums to unity, i.e., imposing the condition $\sum_{n=0}^{\infty} p(n) = 1$, provides the normalization constant $[1 - \exp(-h\nu/kT)]$. The zero-point energy $E_0 = \frac{1}{2}h\nu$ disappears into the normalization and does not affect the results.

The probability distribution for the number of photons n in a resonator mode of frequency ν given in (13.2-19) is most simply written in terms of the mean photon number $\bar{n}$ as

$$p(n) = \frac{1}{\bar{n}+1}\left(\frac{\bar{n}}{\bar{n}+1}\right)^n, \quad (13.2\text{-}20)$$

Bose–Einstein Distribution

where

$$\bar{n} = \frac{1}{\exp(h\nu/kT) - 1},\qquad(13.2\text{-}21)$$

which has been determined from (13.2-13). It is reassuring that (13.2-21) accords with the mean photon number calculated in (14.4-7) for a collection of photons interacting with atoms in thermal equilibrium, as will be seen in Sec. 14.4A. In the parlance of probability theory, the distribution displayed in (13.2-20) is known as the **geometric distribution** since $p(n)$ is a geometrically decreasing function of n. In the physics literature, it is generally referred to as the **Bose–Einstein distribution** since it was first set forth by Bose based on a statistical argument for counting the states of indistinguishable particles such as photons. Einstein recognized that (13.2-20) was also applicable for describing bosons whose numbers are conserved, and he predicted the possibility of a condensation to the lowest energy state in a bosonic atomic gas cooled below a critical temperature (Sec. 14.3F).

The Bose–Einstein distribution is displayed on a semilogarithmic plot in Fig. 13.2-7 for several values of the mean photon number $\bar{n}$ [or, equivalently, several values of the temperature T via (13.2-21)]. Its exponential character is apparent from the straight-line behavior on this semilogarithmic plot. Comparing Fig. 13.2-7 with Fig. 13.2-5 for the Poisson distribution demonstrates that the photon-number distributions for thermal light decrease monotonically from $n = 0$ and are far broader than those for coherent light.

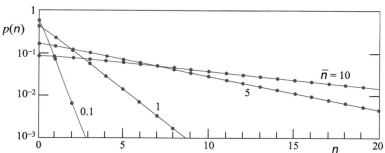

Figure 13.2-7 Semilogarithmic plot of the Bose–Einstein photon-number distribution, $p(n)$ vs. n, for several values of the mean photon number $\bar{n}$. The curves broaden substantially as $\bar{n}$ increases.

The photon-number variance of the Bose–Einstein distribution, which is readily calculated via (13.2-14), turns out to be

$$\boxed{\sigma_n^2 = \bar{n} + \bar{n}^2,}\qquad(13.2\text{-}22)$$
Variance
Bose–Einstein Distribution

where $\bar{n}$ is the photon-number mean. Comparing the Bose–Einstein and Poisson variances given in (13.2-22) and (13.2-15), respectively, reveals that, for $\bar{n} > 1$, the former grows quadratically with $\bar{n}$ while the latter grows linearly. The photon-number fluctuations of the Bose–Einstein distribution are clearly far greater than those of the Poisson distribution, as is apparent in the comparison of Figs. 13.2-7 and 13.2-5. This large variability is consistent with the random nature of thermal light, as described in Sec. 12.1. The noisiness of the Bose–Einstein distribution is crisply highlighted in its signal-to-noise ratio, which, in accordance with (13.2-16), is given by

$$\text{SNR} = \frac{\bar{n}}{\bar{n}+1}.\qquad(13.2\text{-}23)$$

The Bose–Einstein SNR always remains smaller than unity no matter how large $\bar{n}$, confirming that thermal light is generally too noisy to be used for the transmission of information.

EXERCISE 13.2-1

Average Energy of a Resonator Mode in Thermal Equilibrium. Show that the average energy of a resonator mode of frequency ν, under conditions of thermal equilibrium at temperature T, is given by

$$\bar{E} = kT \frac{h\nu/kT}{\exp(h\nu/kT) - 1}. \tag{13.2-24}$$

Sketch the dependence of $\bar{E}$ on $h\nu$ for several values of kT. Use a Taylor-series expansion of the denominator to obtain an expression for $\bar{E}$ in the limit $h\nu/kT \ll 1$. Explain the result on a physical basis.

---

*Doubly Stochastic Photon-Number Statistics

As indicated earlier, coherent light has constant intensity $I(\mathbf{r}, t)$, constant optical power P, and constant photon flux $\Phi = P/h\bar{\nu}$. The arriving photons behave as independent events with a Poisson photon-number distribution $p(n) = \bar{n}^n e^{-\bar{n}}/n!$, where the mean photon number $\bar{n} = \Phi T = PT/h\bar{\nu}$ is constant. However, if the light is partially coherent and its intensity varies in time, then so too does the optical power [as portrayed in Fig. 13.2-3(b)], the photon flux, and the mean photon number $\bar{n}$. In accordance with (13.2-10) and (13.2-11), in that case the mean photon number, which we denote as w rather than $\bar{n}$ for notational convenience, can be expressed as

$$w \equiv \bar{n} = \frac{1}{h\bar{\nu}} \int_0^T P(t)\, dt = \frac{1}{h\bar{\nu}} \int_0^T \!\!\int_A I(\mathbf{r}, t)\, dA\, dt. \tag{13.2-25}$$

The integrated intensity w, which has units of mean photon number and is thus dimensionless, varies in time for partially coherent light.

Variations in the mean photon number arising from intensity fluctuations cause the photon-number distribution to depart from Poisson behavior, as we now show. If the fluctuations of w are described by a probability density function $p(w)$, the new photon-number probability distribution is obtained by averaging the Poisson distribution conditioned on w being constant, $p(n|w) = w^n e^{-w}/n!$, over the permitted values of w dictated by $p(w)$. The alert reader will notice that we have co-opted the symbol n for the new photon number. The resultant photon-number distribution then takes the form

$$\boxed{p(n) = \int_0^\infty \frac{w^n e^{-w}}{n!} p(w)\, dw,} \tag{13.2-26}$$
Mandel's Formula

which is known as **Mandel's formula**. Equation (13.2-26) is a **doubly stochastic photon-number distribution** by virtue of its two contributing sources of randomness: 1) the random arrivals of the photons themselves, which locally behave in Poisson fashion; and 2) the integrated-intensity fluctuations arising from the partially coherent nature of the illumination. An underlying sequence of random photon arrivals that leads to doubly stochastic photon-number statistics is known as a **doubly stochastic Poisson process (DSPP)**.

The photon-number mean and variance for the doubly stochastic photon-number distribution are obtained by using (13.2-13) and (13.2-14) in conjunction with (13.2-26); the results are

$$\overline{n} = \overline{w} \tag{13.2-27}$$

and

$$\sigma_n^2 = \overline{n} + \sigma_w^2, \tag{13.2-28}$$

respectively. Here σ_w^2 signifies the variance of w. The photon-number variance thus comprises two contributions: 1) the basic Poisson contribution $\overline{n}$; and 2) a (positive) contribution arising from the intensity fluctuations. It is worth noting that the theory of photon statistics presented here is applicable only for **classical light**; a more general theory encompassing **nonclassical light** requires a quantum approach (Sec. 13.3).

An instructive example makes use of an integrated intensity that obeys the exponential probability density function

$$p(w) = \begin{cases} \dfrac{1}{\overline{w}} \exp\left(-\dfrac{w}{\overline{w}}\right), & w \geq 0 \\ 0, & w < 0. \end{cases} \tag{13.2-29}$$

Equation (13.2-29) is appropriate for quasi-monochromatic light whose independent and identically distributed real and imaginary complex-field-amplitude components are Gaussian. It is applicable for a source of light whose spectral width is sufficiently small such that its coherence time τ_c is much greater than the counting time T, and whose coherence area A_c is much greater than the detector integration area A (see Chapter 12). The photon-number distribution $p(n)$ that corresponds to (13.2-29) is determined by substituting this result into (13.2-26) and evaluating the integral. The outcome turns out to be the Bose–Einstein distribution given in (13.2-20). Hence, the photon statistics of the Gaussian-distributed optical field described above are identical to those of single-mode thermal light. Multimode thermal light emerges when the counting time T and photodetector integration area A are not small (Probs. 13.2-6–13.2-8).

D. Random Partitioning of Photon Streams

A photon stream is said to be partitioned when it is subjected to the removal of some of its photons. The process is called **random partitioning** when the removed photons are diverted and **random selection** when they are annihilated. There are numerous ways in which this can occur. Perhaps the simplest example of random partitioning is provided by an ideal lossless beamsplitter. Photons are randomly selected to exit either of the two output ports (Fig. 13.2-8). An example of random selection is provided by the action of a photodetector. Photons incident on the photosensitive material are selected either to be absorbed and create photoelectrons or to pass through it and be lost.

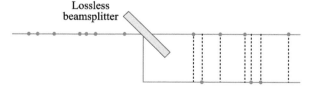

Figure 13.2-8 Random partitioning of a stream of photons by a beamsplitter.

We restrict the treatment presented here to situations in which the probability of each photon being partitioned behaves in accordance with an independent **Bernoulli trial** (coin toss). In terms of the beamsplitter, this is satisfied if a photon stream impinges on only one of its input ports (Fig. 13.2-8). This eliminates the possibility of interference, which in general invalidates the independent-trial assumption (see, e.g., Example 13.3-3). The results derived below apply to both random partitioning and random selection.

Consider a lossless beamsplitter of transmittance $\mathcal{T}$ and reflectance $\mathcal{R} = 1 - \mathcal{T}$. The result of a single photon impinging on this device was examined in Sec. 13.1C, where it was shown that the probability of a photon being transmitted is equal to the transmittance of the beamsplitter $\mathcal{T}$ and the probability of the photon being reflected is equal to $1 - \mathcal{T}$ (Fig. 13.1-6). We now consider a photon stream of mean flux Φ is incident on the beamsplitter, so that a mean number of photons $\bar{n} = \Phi T$ impinges on it in the time interval T. The mean number of transmitted and reflected photons is then $\mathcal{T}\bar{n}$ and $(1 - \mathcal{T})\bar{n}$, respectively.

We proceed to determine how the photon-number statistics of the incident photon stream are modified upon partitioning. If the incident stream consists of precisely n photons, the probability $p(m)$ that m photons are transmitted is the same as that of flipping a biased coin n times and obtaining m heads, when the probability of obtaining a head is $\mathcal{T}$. From elementary probability theory we know that $p(m)$ is characterized by the binomial distribution

$$p(m) = \binom{n}{m} \mathcal{T}^m (1-\mathcal{T})^{n-m}, \qquad m = 0, 1, \ldots, n, \qquad (13.2\text{-}30)$$

where $\binom{n}{m} = n!/m!\,(n-m)!$. By symmetry, the results for the reflected photons are identical, with $1 - \mathcal{T}$ replacing $\mathcal{T}$. The statistics of the binomial distribution yield the mean number of transmitted photons

$$\bar{m} = \mathcal{T}n, \qquad (13.2\text{-}31)$$

and its variance

$$\sigma_m^2 = \mathcal{T}(1-\mathcal{T})n = (1-\mathcal{T})\bar{m}. \qquad (13.2\text{-}32)$$

The signal-to-noise ratio specified in (13.2-16) is thus SNR $= \bar{m}^2/\sigma_m^2 = \bar{m}/(1-\mathcal{T})$, which increases linearly with the mean number of transmitted photons $\bar{m}$. When the incident wave is strong, the photons will therefore be partitioned between the transmitted and reflected beams in good agreement with $\mathcal{T}$ and $(1-\mathcal{T})$, respectively, confirming that the classical-optics result is recovered.

The expressions provided above for a fixed number of incident photons permit the photon-number statistics of the partitioned stream to be determined. The calculation proceeds by recognizing that in the general case the number of photons n at the input to the beamsplitter is random rather than fixed. Let the probability be $p_0(n)$ that in a specified time interval there are n photons present at the beamsplitter input. The photon-number probability distribution for the transmitted stream will then be a weighted sum of binomial distributions, with the weighting established by the probability of n photons being present. The photon-number distribution $p(m)$ at the output of the beamsplitter, for an input photon-number distribution $p_0(n)$, is therefore $p(m) = \sum_n p(m|n)\, p_0(n)$, where the observation of m photons conditioned on n being fixed is the binomial distribution $p(m|n) = \binom{n}{m} \mathcal{T}^m (1-\mathcal{T})^{n-m}$. Finally, then, we obtain a formula that yields $p(m)$ in terms of $p_0(n)$ and $\mathcal{T}$:

$$\boxed{p(m) = \sum_{n=m}^{\infty} \binom{n}{m} \mathcal{T}^m (1-\mathcal{T})^{n-m}\, p_0(n).} \qquad (13.2\text{-}33)$$

Photon-Number Distribution Under Random Partitioning

This formula is also applicable for the *detection* of photons, as considered in Sec. 19.6A.

When the input photon-number distribution $p_0(n)$ is Poisson (for coherent light) or Bose–Einstein (for single-mode thermal light), the results turn out to be quite simple: the form of the partitioned photon-number distribution $p(m)$ exactly matches that of the incident photon-number distribution $p_0(n)$. Thus, single-mode laser light transmitted through a beamsplitter remains Poisson and thermal light remains Bose–Einstein, although of course the photon-number mean is reduced by the factor $\mathcal{T}$. In contrast, photon-number-squeezed light (Sec. 13.3C) does not retain its form under random partitioning, a property that is responsible for its lack of robustness. In particular, number-state light, which comprises a deterministic photon number, obeys the binomial photon-number distribution after partitioning.

The signal-to-noise ratio for m is readily determined for partitioned photon streams. For coherent light and single-mode thermal light, the results are, respectively,

$$\text{SNR} = \begin{cases} \mathcal{T}\bar{n} & \text{coherent light} \qquad (13.2\text{-}34) \\ \dfrac{\mathcal{T}\bar{n}}{\mathcal{T}\bar{n}+1} & \text{thermal light.} \qquad (13.2\text{-}35) \end{cases}$$

Since $\mathcal{T} \leq 1$ it is clear that random partitioning decreases the signal-to-noise ratio. Another way of stating this is to say that random partitioning introduces noise.

*13.3 QUANTUM STATES OF LIGHT

The number of photons in an electromagnetic mode is generally a random quantity. In this section it will be shown that in the context of quantum optics the electric field itself is also generally random. Consider a monochromatic plane-wave electromagnetic mode in a volume V, described by the electric field $\mathcal{E}(\mathbf{r},t) = \text{Re}\{\mathbf{E}(\mathbf{r},t)\}$, where

$$\mathbf{E}(\mathbf{r},t) = A \exp(-j\mathbf{k}\cdot\mathbf{r})\exp(j2\pi\nu t)\,\widehat{\mathbf{e}}. \qquad (13.3\text{-}1)$$

According to classical electromagnetic optics, as provided in (13.1-3), the energy of the mode is fixed at $\tfrac{1}{2}\epsilon|A|^2 V$. We define a complex variable a, such that $\tfrac{1}{2}\epsilon|A|^2 V = h\nu|a|^2$, thereby allowing $|a|^2$ to be interpreted as the energy of the mode in units of photon number. The electric field may then be written as

$$\mathbf{E}(\mathbf{r},t) = \sqrt{\frac{2h\nu}{\epsilon V}}\, a \exp(-j\mathbf{k}\cdot\mathbf{r})\exp(j2\pi\nu t)\,\widehat{\mathbf{e}}, \qquad (13.3\text{-}2)$$

where the complex variable a determines the complex amplitude of the field.

In classical electromagnetic optics, $a\exp(j2\pi\nu t)$ is a rotating phasor whose projection on the real axis determines the sinusoidal electric field, as portrayed in Fig. 13.3-1. The real and imaginary parts of $a = x + jp$, which are $x = \text{Re}\{a\}$ and $p = \text{Im}\{a\}$, respectively, are termed the quadrature components of the phasor a because they are a quarter cycle (90°) out of phase with each other. They determine the amplitude and phase of the sine wave that represents the temporal variation of the electric field. The rotating phasor $a\exp(j2\pi\nu t)$ also describes the motion of a harmonic oscillator; the real component x is proportional to position and the imaginary component p to momentum. From a mathematical point of view, then, a *classical* monochromatic mode of the electromagnetic field and a classical harmonic oscillator behave identically. A parallel argument can be constructed to show that a *quantum* monochromatic electromagnetic mode and a one-dimensional quantum-mechanical harmonic oscillator also have identical behavior. To facilitate the comparison, we review the quantum theory of a simple harmonic oscillator before proceeding.

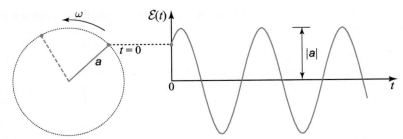

Figure 13.3-1 The real and imaginary parts of the variable $a\exp(j2\pi\nu t)$, which govern the complex amplitude of a classical electromagnetic mode of frequency ν. The time dynamics are identical to those of a classical harmonic oscillator with angular frequency $\omega = 2\pi\nu$.

Quantum Theory of the Harmonic Oscillator

A particle of mass m, position x, momentum p, and potential energy $V(x) = \frac{1}{2}\kappa x^2$, where κ is the elastic constant, represents a one-dimensional harmonic oscillator of total energy $\frac{1}{2}p^2/m + \frac{1}{2}\kappa x^2$ and oscillation frequency $\omega = \sqrt{\kappa/m}$. Without loss of generality, we take $m = 1$ so that the energy is $\frac{1}{2}(p^2 + \omega^2 x^2)$.

In accordance with the laws of quantum mechanics, the behavior of this quantum system in a stationary state is described by a complex wavefunction $\psi(x)$ that satisfies the time-independent Schrödinger equation [see (14.1-3)],

$$-\frac{\hbar^2}{2m}\frac{d^2\psi(x)}{dx^2} + V(x)\psi(x) = E\psi(x), \qquad (13.3\text{-}3)$$

where E is the energy of the particle. The solutions of the Schrödinger equation for the harmonic oscillator, for which $V(x) = \frac{1}{2}\omega^2 x^2$, give rise to discrete energy values given by

$$E_n = \left(n + \tfrac{1}{2}\right)h\nu, \qquad n = 0, 1, 2, \ldots . \qquad (13.3\text{-}4)$$

Adjacent energy levels are seen to be separated by a quantum of energy $h\nu = \hbar\omega$. The corresponding wavefunctions $\psi_n(x)$ are normalized Hermite–Gaussian functions,

$$\psi_n(x) = \frac{1}{\sqrt{2^n\, n!}}\left(\frac{\omega}{\pi\hbar}\right)^{1/4}\mathbb{H}_n\!\left[\sqrt{\frac{\omega}{\hbar}}\,x\right]\exp\!\left(-\frac{\omega x^2}{2\hbar}\right), \qquad (13.3\text{-}5)$$

where $\mathbb{H}_n(x)$ is the Hermite polynomial of order n [see (3.3-6)–(3.3-8) and (3.3-11)].

An arbitrary wavefunction $\psi(x)$ may be expanded in terms of the set of orthonormal eigenfunctions $\{\psi_n(x)\}$ in terms of the superposition $\psi(x) = \sum_n c_n \psi_n(x)$. Given the wavefunction $\psi(x)$, which governs the state of the system, the behavior of the particle is determined as follows:

- The probability $p(n)$ that the harmonic oscillator carries n quanta of energy is given by the coefficient $|c_n|^2$.
- The probability density of finding the particle at the position x is given by $|\psi(x)|^2$.
- The probability density that the momentum of the particle is p is given by $|\phi(p)|^2$, where $\phi(p)$ is proportional to the inverse Fourier transform of $\psi(x)$ evaluated at the (spatial) frequency p/h,

$$\phi(p) = \frac{1}{\sqrt{h}}\int_{-\infty}^{\infty}\psi(x)\exp\!\left(j2\pi\frac{p}{h}x\right)dx. \qquad (13.3\text{-}6)$$

As shown in Sec. A.2 of Appendix A, the Fourier-transform relation between

$\psi(x)$ and $\phi(p)$ indicates that there is an uncertainty relation between the power-RMS widths of x and p/h given by

$$\frac{\sigma_x \sigma_p}{h} \geq \frac{1}{4\pi} \quad \text{or} \quad \sigma_x \sigma_p \geq \frac{\hbar}{2}. \tag{13.3-7}$$

This is the well-known Heisenberg position–momentum uncertainty relation provided in (A.2-7) of Appendix A.

Analogy Between an Optical Mode and a Harmonic Oscillator

The energy of an electromagnetic mode is $h\nu|a|^2 = h\nu(x^2 + p^2)$. The analogy with a harmonic oscillator of energy $\frac{1}{2}(\omega^2 x^2 + p^2)$ is established by effecting the connections

$$x = \left(1/\sqrt{2\hbar\omega}\right)\omega x \quad \text{and} \quad p = \left(1/\sqrt{2\hbar\omega}\right)p. \tag{13.3-8}$$

The modal energy then becomes $h\nu(x^2 + p^2) = \frac{1}{2}(\omega^2 x^2 + p^2)$, which leads to the harmonic-oscillator energy levels provided in (13.3-4). Because the analogy is complete, we conclude that the energy of a quantum, monochromatic electromagnetic mode, like that of a one-dimensional, quantum-mechanical harmonic oscillator, is quantized to the values $E_n = (n + \frac{1}{2})h\nu$, as initially suggested in (13.1-5). With proper scaling normalization factors, the behavior of the position and momentum of the harmonic oscillator, x and p, also describe the quadrature components of the electromagnetic field, x and p, respectively.

Properties of a Quantum Electromagnetic Mode

An quantum electromagnetic mode of frequency ν is described by a complex wavefunction $\psi(x)$ that governs the uncertainties of the quadrature components x and p of the electromagnetic field, as well as the statistics of the number of photons in the mode.

- The probability $p(n)$ that the mode contains n photons is given by $|c_n|^2$, where the c_n are coefficients of the expansion of $\psi(x)$ in terms of the eigenfunctions $\psi_n(x)$, i.e., $\psi(x) = \sum_n c_n \psi_n(x)$.
- The probability density functions of the quadrature components x and p are given by $|\psi(x)|^2$ and $|\phi(p)|^2$, respectively, where $\psi(x)$ and $\phi(p)$ are related by

$$\phi(p) = \frac{1}{\sqrt{\pi\omega}} \int_{-\infty}^{\infty} \psi(x) \exp(j2px)\, dx. \tag{13.3-9}$$

This equation is derived from (13.3-6) by use of the transformation (13.3-8) and by noting that $|\psi(x)|^2$ and $|\phi(p)|^2$ integrate to unity.

- The uncertainty relation between the power-RMS widths of the quadrature components is given by

$$\boxed{\sigma_x \sigma_p \geq \tfrac{1}{4},} \tag{13.3-10}$$

Quadrature Uncertainty

so that these components cannot be simultaneously determined with arbitrary precision.

A. Coherent States

The quadrature uncertainty product $\sigma_x\sigma_p$ attains its minimum value of $1/4$ when the function $\psi(x)$ is Gaussian (Sec. A.2 of Appendix A). In that case

$$\psi(x) \propto \exp[-(x-\alpha_x)^2], \qquad (13.3\text{-}11)$$

so that its Fourier transform is also Gaussian,

$$\phi(p) \propto \exp[-(p-\alpha_p)^2]. \qquad (13.3\text{-}12)$$

Here, α_x and α_p are arbitrary values that represent the means of x and p, respectively. The quadrature uncertainties, determined from $|\psi(x)|^2$ and $|\phi(p)|^2$, are then given by

$$\sigma_x = \sigma_p = 1/2. \qquad (13.3\text{-}13)$$

Under these conditions the electromagnetic field is said to be in a **coherent state**. The one-standard-deviation range of uncertainty in the quadrature components x and p, as well as in the complex amplitude a and in the electric field $\mathcal{E}(t)$, are illustrated in Fig. 13.3-2 for coherent-state light. The uncertainties are most pronounced when α_x and α_p are small. The squared-magnitude coefficients $|c_n|^2$ for the expansion of $\psi(x)$ in the Hermite–Gaussian basis are $\overline{n}^n \exp(-\overline{n})/n!$, where $\overline{n} = \alpha_x^2 + \alpha_p^2$, so the photon-number distribution $p(n)$ is Poisson, as suggested in the discussion surrounding (13.2-12). Unlike its status in electromagnetic optics, in the context of quantum optics coherent-state light is *not* deterministic. The coherent state is generated by an ideal, amplitude-stabilized, single-mode laser operated well above its threshold of oscillation.

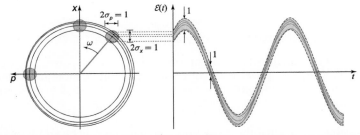

Figure 13.3-2 Quadrature and electric-field uncertainties for the coherent state. Representative values of $\mathcal{E}(t) \propto a\exp(j2\pi\nu t)$ are traced for several arbitrary points within the uncertainty circle; the coefficient of proportionality is chosen to be unity for convenience.

Figure 13.3-3 displays the quadrature uncertainties and time behavior of the electric field for the coherent state when $\alpha_x = \alpha_p = 0$, which is called the **vacuum state**.

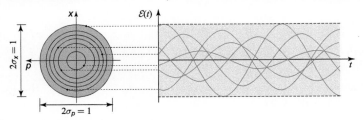

Figure 13.3-3 Quadrature and electric-field uncertainties for the vacuum state. This state is a limiting case for both the coherent state ($\alpha_x = \alpha_p = 0$) and the number state ($n = 0$). The mode carries zero photons and has only the residual zero-point energy $\frac{1}{2}h\nu$. The circle of uncertainty is squeezed into an ellipse for the squeezed vacuum state; however, squeezing the vacuum endows it with a finite mean photon number.

B. Quadrature-Squeezed States

Though the uncertainty product $\sigma_x \sigma_p$ cannot be reduced below its minimum value of $1/4$, the uncertainty of one of the quadrature components can be reduced ("squeezed") below $1/2$ but this entails increased uncertainty in the other component. This form of light, which is distinctly nonclassical, is said to be **quadrature-squeezed**. For example, a state for which $\psi(x)$ is a Gaussian function with a squeezed width $\sigma_x = 1/2s$, where $s > 1$, corresponds to a Gaussian function $\phi(p)$ with a stretched width $\sigma_p = s/2$. The product $\sigma_x \sigma_p$ maintains its minimum value of $1/4$, but the uncertainty circle of the phasor a for the coherent state (Fig. 13.3-2) is squeezed into an ellipse, as displayed in Fig. 13.3-4.

The particular quadrature-squeezed state illustrated in Fig. 13.3-4 is known as an amplitude-squeezed state since the phasor amplitude uncertainty is reduced at the expense of its phase uncertainty. The asymmetry in the uncertainties of the two quadratures is manifested in the time course of the electric field by periodic occurrences of decreased uncertainty, followed each quarter cycle later by occurrences of increased uncertainty. If the field is measured only at those times when its uncertainty is minimal, its noise will be reduced below that of the coherent state. The selection of those times may be achieved by heterodyning the squeezed field with a coherent optical field of appropriate phase (heterodyne receivers are discussed in Sec. 25.4).

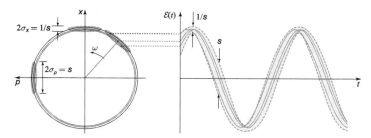

Figure 13.3-4 Quadrature and electric-field uncertainties for a quadrature-squeezed state (specifically, an *amplitude-squeezed state*). The uncertainty circle associated with the coherent state (Fig. 13.3-2) is squeezed into an ellipse of the same area. If the uncertainty ellipse were to be rotated by 90°, so that its long axis lay along the phasor rather than perpendicular to it, the result would be a *phase-squeezed state* since the phase uncertainty would then be reduced at the expense of the amplitude uncertainty. The squeezed vacuum state takes the form of an ellipse at the origin.

Generation and applications of quadrature-squeezed light. Though not robust in the face of optical losses, quadrature-squeezed light has garnered an important niche in precision measurements because of the reduced noisiness of one of its quadratures. The textbook example is the merit of using quadrature-squeezed light in the LIGO gravitational-wave interferometer. In the interferometer configuration considered in Example 2.5-1, for example, all beamsplitter ports are covered by impinging light beams except for the output port, which is exposed to the normal vacuum. Improved interferometer sensitivity is obtained by injecting squeezed vacuum (of appropriate phase) into the output port.[†] Quadrature-squeezed light may be generated in a number of ways, including via optical parametric downconversion in a cavity (Sec. 22.2C).

[†] See, e.g., M. C. Teich and B. E. A. Saleh, Squeezed and Antibunched Light, *Physics Today*, vol. 43, no. 6, pp. 26–34, 1990.

C. Photon-Number-Squeezed States

Quadrature-squeezed light exhibits reduced uncertainty in one of its quadrature field components relative to that of the coherent state. Another form of nonclassical light is **photon-number-squeezed** light, also referred to as **intensity-squeezed** or **sub-Poisson** light. This form of light has a photon-number variance that is squeezed below the coherent-state (Poisson) value, i.e., $\sigma_n^2 < \bar{n}$. Photon-number fluctuations obeying this relation are nonclassical since (13.2-28) cannot be satisfied.

An electromagnetic mode described by the harmonic oscillator eigenstate $\psi(x) = \psi_{n_0}(x)$ provides an example of photon-number-squeezed light. This state is referred to as the **number state** because $p(n) = |c_n|^2 = 1$ for a fixed number of photons $n = n_0$; the number state is also called the **Fock state**. Since the number of photons carried by the mode is deterministic, we have $\bar{n} = n_0$ and $\sigma_n^2 = 0$. The case $n_0 = 1$ corresponds to a single photon in the mode. The uncertainties associated with number-state light are illustrated in Fig. 13.3-5. Though the quadrature components as well as the phasor magnitude and phase are all uncertain, the photon number is fixed. For $n_0 = 0$, the number state reduces to the vacuum state displayed in Fig. 13.3-3. Photon-number-squeezing can also be generated using other states, e.g., the binomial state (Prob. 13.3-1).

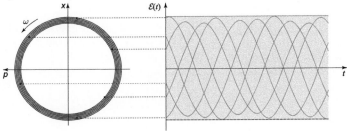

Figure 13.3-5 Representative uncertainties for the number state. The mode contains a fixed number of photons, $n = n_0$. This state is photon-number-squeezed but not quadrature-squeezed.

Generation and applications of photon-number-squeezed light. Just as with quadrature-squeezed light, losses adversely affect photon-number-squeezed light (Sec. 13.2D). Nevertheless, this form of nonclassical light enjoys a range of applications in quantum information processing, communications, computing, and cryptography. Photon-number-squeezed light may be generated via mechanisms that include: 1) the use of feedback to impart anticorrelations to an otherwise Poisson sequence of photons; 2) sub-Poisson excitation (e.g., via a stream of electrons subject to space charge) of an atom or other entity that emits a single photon in response to the excitation; 3) conversion from two-photon light whereby the arrival of one photon of a pair serves to herald the presence of its simultaneously arriving companion, as described in Sec. 13.3D; and 4) emission from a quantum dot, defect, or similar entity embedded in a photonic structure, as considered in Sec. 17.2E.

D. Two-Photon Light

Two-photon light, yet another form of nonclassical light, is described by a joint wavefunction that represents its joint probability amplitude. For example, its spatial properties are described by the wavefunction $U(x_1, x_2)$, where $|U(x_1, x_2)|^2$ is the joint probability density of finding the photons at positions x_1 and x_2 in the transverse plane. The corresponding directional (or spatial-frequency) properties are described by the two-dimensional Fourier transform of $U(x_1, x_2)$, which is denoted $V(\nu_{x1}, \nu_{x2})$. The joint probability density of finding photons with wavevectors $\mathbf{k}_1$ and $\mathbf{k}_2$ (momenta

$\hbar\mathbf{k}_1$ and $\hbar\mathbf{k}_2$) is given by $|V(\nu_{x1}, \nu_{x2})|^2$, where $k_{x1} = 2\pi\nu_{x1}$ and $k_{x2} = 2\pi\nu_{x2}$ are the transverse components of the wavevectors. The temporal and spectral properties are similarly described by the joint wavefunctions $U(t_1, t_2)$ and its two-dimensional Fourier transform $V(\nu_1, \nu_2)$.

While the polarization state of classical or single-photon light is described by the Jones vector $\mathbf{J} = \begin{bmatrix} A_x \\ A_y \end{bmatrix}$, that of two-photon light is described by a 2×2 joint matrix $\mathbf{J} = \begin{bmatrix} A_{xx} & A_{yx} \\ A_{xy} & A_{yy} \end{bmatrix}$, where the first and second subscripts denote the polarizations of the first and second photons, respectively. For example, $|A_{yx}|^2$ represents the probability of detecting the first and second photons in the y (vertical or V) and x (horizontal or H) polarizations, respectively.

Entangled Photons

If the two photons are independent, their joint wavefunction factors into a product of the individual wavefunctions. In that case

$$U(x_1, x_2) = U_1(x_1)U_2(x_2), \tag{13.3-14}$$

so that the probability of detecting the two photons also factors, indicating that the positions of the two photons are statistically independent and uncorrelated.

If the joint wavefunction is not factorable into a product, however, the photons are said to be **entangled**. An extreme case is the wavefunction

$$U(x_1, x_2) = U_s(x_1)\delta(x_1 - x_2), \tag{13.3-15}$$

which describes two photons that are always detected at the same position ($x_1 = x_2$), although that position is random with probability density $|U_s(x_1)|^2$. In that case, the photons are said to be **maximally entangled**. The Fourier transform of (13.3-15) is

$$V(\nu_{x1}, \nu_{x2}) = V_s(\nu_{x1})\delta(\nu_{x1} + \nu_{x2}), \tag{13.3-16}$$

where $V_s(\nu_x)$ is the one-dimensional Fourier transform of $U_s(x)$. The transverse components of the wavevector are then anticorrelated ($k_{x1} = -k_{x2}$), indicating that the photon directions are also anticorrelated.

An example of two photons that are maximally entangled in polarization is provided by the polarization state

$$\mathbf{J} = \frac{1}{\sqrt{2}}\begin{bmatrix} 0 & 1 \\ 1 & 0 \end{bmatrix}. \tag{13.3-17}$$

There are then two polarization possibilities and they have equal probabilities: photon 1 is y polarized and photon 2 is x polarized, or *vice-versa*. Each photon can be either vertically or horizontally polarized, but if one is vertically polarized, the other *must* be horizontally polarized. In the quantum-optics literature, the photon pair is said to be in a VH+HV superposition state, where V and H represent the vertical (y) and horizontal (x) polarizations, respectively. An entangled photon pair is also called a **biphoton** and two-photon light is sometimes called **twin-beam light**.

Generation and applications of two-photon light. Entangled two-photon light may be generated by means of **spontaneous parametric downconversion (SPDC)**, a nonlinear optical process whereby some fraction of a beam of photons incident on a nonlinear optical crystal are split into pairs of photons, while conserving energy and momentum (Sec. 22.2C). Given that the energy of a photon is $E = \hbar\omega$, conservation of energy dictates that the frequency of the incident photon (called the *pump*) must equal the sum of the energies of the two downconverted photons (called the *signal* and *idler*), i.e., $\omega_p = \omega_1 + \omega_2$ or $(\omega_1 - \frac{1}{2}\omega_p) = -(\omega_2 - \frac{1}{2}\omega_p)$. If the pump is monochromatic, ω_p is fixed and ω_1 and ω_2 are anticorrelated variables. By virtue of the Fourier-transform relation between the spectral and temporal wavefunctions, this implies that the emission times are fully correlated, i.e., $t_1 = t_2$.

Similarly, given that the momentum of a photon is $\mathbf{p} = \hbar\mathbf{k}$, conservation of momentum dictates that the pump wavevector must equal the sum of the wavevectors of the two downconverted photons, i.e., $\mathbf{k}_p = \mathbf{k}_1 + \mathbf{k}_2$. If the pump is a plane-wave traveling along the axial (z) direction, the transverse components k_{1x} and k_{2x} of the wavevectors of the generated photons must then sum to zero. Thus, $0 = k_{1x} + k_{2x}$ or $k_{2x} = -k_{1x}$, revealing that these components are anticorrelated; the positions from which the two photons are emitted are thus correlated, i.e., $x_1 = x_2$. This two-photon wavefunction thus assumes the form specified in (13.3-15), where $U_s(x)$ is determined by the dimensions and optical properties of the nonlinear crystal in which the entangled photons are generated. Downconverted photons can also exhibit polarization entanglement if the nonlinear process requires that they have orthogonal polarizations. SPDC is useful for generating entangled two-photon light in crystal-based bulk optics (Example 22.2-3) as well as in monolithic semiconductor chips. Hollow-core photonic-crystal fibers (Sec. 10.4) filled with noble gases such as argon can also be configured to generate two-photon light via a modulation instability.

Two-photon light can be used to generate single-photon light by using the arrival of one member of the photon pair to herald the presence of the other. Entangled photons find use in applications such as secure quantum communications, cryptography, sensing, and imaging. Entanglement can be distributed over long distances — satellite-based entanglement distribution has been achieved for locations on earth separated by more than 1200 km.

Two-Photon Optics

The transmission of two-photon light through a linear optical system obeys equations based on classical optics, as we proceed to describe.

Polarization optics. When a classical plane wave whose polarization described by the Jones vector $\mathbf{J}_i$ is transmitted through a polarization element with Jones matrix $\mathbf{T}$, the Jones vector of the outgoing wave is $\mathbf{J}_o = \mathbf{T}\mathbf{J}_i$ (Sec. 6.1B). This relation, which is also applicable for single-photon light, is readily generalized to two-photon light by applying the matrix $\mathbf{T}$ twice, once for each photon, which gives rise to

$$\mathbf{J}_o = \mathbf{T}\mathbf{J}_i\mathbf{T}^\mathsf{T}, \qquad (13.3\text{-}18)$$

where the superscript T signifies the matrix transpose.

EXAMPLE 13.3-1. *Polarization Rotator.* The Jones matrix of a 45° polarization rotator is $\mathbf{T} = \frac{1}{\sqrt{2}} \begin{bmatrix} 1 & -1 \\ 1 & 1 \end{bmatrix}$ [see (6.1-20)]. Upon transmission through such a device, polarization-entangled photons described by the VH+HV Jones vector $\mathbf{J}_i = \frac{1}{\sqrt{2}} \begin{bmatrix} 0 & 1 \\ 1 & 0 \end{bmatrix}$ [see (13.3-17)] are characterized by $\mathbf{J}_o = \frac{1}{\sqrt{2}} \begin{bmatrix} -1 & 0 \\ 0 & 1 \end{bmatrix}$. This Jones vector represents entangled photons in the VV−HH state, for which the outgoing photons must be of the same polarization; the probability that they are of orthogonal polarization is zero. It can be demonstrated that maximally entangled light remains maximally entangled when subjected to polarization rotation for an arbitrary angle.

Spatial optics. When a classical wave with wavefunction $U_i(x,y)$ is transmitted through an optical system with impulse response function $h(x,y;x',y')$, the outgoing wave $U_o(x,y)$ is described by the integral (13.1-9), which also applies for single-photon light. This result may be generalized to two-photon light by applying the impulse response function twice, once for each photon. Ignoring the y dependence for simplicity, we then have

$$U_o(x_1, x_2) = \int\!\!\!\int_{-\infty}^{\infty} h(x_1; x_1')\, h(x_2; x_2')\, U_i(x_1', x_2')\, dx_1'\, dx_2', \qquad (13.3\text{-}19)$$

where the subscripts 1 and 2 denote photons 1 and 2, respectively. Note that (13.3-19) is analogous to (12.3-5) in Sec. 12.3A, which describes the propagation of partially coherent light with mutual intensity $G(x_1, x_2)$ through a linear optical system. But note also that the conjugation in (12.3-5) is absent in (13.3-19).

EXAMPLE 13.3-2. *Fourier-Transform System.* For a 2-f optical system that implements a Fourier-transform operation (Sec. 4.2B), i.e., $h(x;x') \propto \exp(-jxx'/\lambda f)$, where λ is the wavelength and f is the focal length of the lens, we have $U_o(x_1, x_2) \propto V_i(x_1/\lambda f, x_2/\lambda f)$, where $V_i(\nu_{x1}, \nu_{x2})$ is the two-dimensional Fourier transform of $U_i(x_1, x_2)$. The 2-f system manifests the directional, or momentum, characteristics of two-photon light. Correlated photons are converted by the Fourier-transforming lens into anticorrelated photons.

Two-beam optics. Two-photon light in the form of two beams, labeled a and b, is described by a 2×2 joint matrix $\mathbf{J}_i = \frac{1}{\sqrt{2}} \begin{bmatrix} A_{\mathrm{aa}} & A_{\mathrm{ab}} \\ A_{\mathrm{ba}} & A_{\mathrm{bb}} \end{bmatrix}$, where, for example, $|A_{\mathrm{ba}}|^2$ is the probability that the first and second photons are found in beams b and a, respectively. If the two beams are mixed at a lossless device characterized by a scattering matrix $\mathbf{S}$ (see Sec. 7.1A), the outgoing pair of beams of two-photon light is described by the matrix $\mathbf{J}_o = \mathbf{S}\mathbf{J}_i\mathbf{S}^\mathsf{T}$, a relation that assumes the same form as (13.3-18).

EXAMPLE 13.3-3. *Hong–Ou–Mandel (HOM) Interferometer.* A lossless symmetric beam-splitter is characterized by the scattering matrix $\mathbf{S} = \frac{1}{\sqrt{2}} \begin{bmatrix} 1 & j \\ j & 1 \end{bmatrix}$, as provided in (7.1-18). For incoming light described by $\mathbf{J}_i = \frac{1}{\sqrt{2}} \begin{bmatrix} 0 & 1 \\ 1 & 0 \end{bmatrix}$, indicating that there is a single photon in each beam, we obtain $\mathbf{J}_o = \mathbf{S}\mathbf{J}_i\mathbf{S}^\mathsf{T} = \frac{j}{\sqrt{2}} \begin{bmatrix} 1 & 0 \\ 0 & 1 \end{bmatrix}$. This reveals that the outgoing photons always emerge together as a pair, out of one or the other port, and there is zero probability that each photon emerges from a different port.[†] Hence, the two

---

[†] See C. K. Hong, Z. Y. Ou, and L. Mandel, Measurement of Subpicosecond Time Intervals Between Two Photons by Interference, *Physical Review Letters*, vol. 59, pp. 2044–2046, 1987.

indistinguishable photons that were initially separate, each entering a different input port of the beamsplitter, exit the beamsplitter "stuck together" from one of the (randomly chosen) output ports. This outcome may be understood by observing that there are two paths the input photons could follow at the beamsplitter for them to emerge from *different* output ports — both could undergo reflection or both could undergo transmission. However, the probability amplitudes for these two possibilities are equal since the photons are indistinguishable, and they cancel because of the phase shift associated with reflection at the beamsplitter. Hence, the only remaining possibilities are that the photons emerge together from the same output port. The beamsplitter thus serves to convert two indistinguishable photons at its input ports into two entangled photons at its output ports, since the two photons always emerge from the same output port, although which port is random. This form of **two-photon interference**, known as **Hong–Ou–Mandel interference**, is widely used in quantum optics, quantum information, and quantum imaging.

Applications of two-photon optics. Optical systems employing two-photon light find application in a number of areas. An example in the domain of imaging is provided by quantum optical coherence tomography[†] (QOCT), a two-photon interferometric technique that allows a multilayered medium to be axially sectioned. This imaging modality makes use of two indistinguishable photons; a Hong–Ou–Mandel interferometer in which one of the photons is reflected from a movable mirror while the other is reflected from the sample before being presented to the two input ports of the beamsplitter; and a pair of photodetectors connected to register the photon-coincidence rate at the output ports of the beamsplitter. The depths of the reflective layers in the sample are revealed by determining the path delays of the movable mirror that lead to dips in the coincidence rate, thereby indicating the presence of HOM interference. QOCT is the two-photon analog of optical coherence tomography (OCT), discussed in Sec. 12.2B, which makes use of partially coherent light of short coherence length; a Michelson interferometer in which one the mirrors is replaced by the sample; and a photodetector responsive to intensity (Fig. 12.2-3). For OCT, the depths of the reflective layers in the sample are revealed by determining the path delays of the movable mirror that lead to interference fringes in the intensity. Though QOCT is more complex to implement than OCT, it has the merit that it is immune to even-order group velocity dispersion (GVD) in the sample, which serves to increase the attainable resolution and sectioning depth; the technique simultaneously permits the GVD coefficients of the media that comprise the sample to be determined.

The implementation of two-photon optical systems has been substantially advanced by the development of **integrated quantum photonics**, a platform in which **on-chip quantum circuits** generate and manipulate quantum states of light such as single photons, entangled photons, and complex states in which entanglement is shared among multiple modes.

READING LIST

Quantum Optics

Z.-Y. J. Ou, *Quantum Optics for Experimentalists*, World Scientific, 2017.

M. Orszag, *Quantum Optics: Including Noise Reduction, Trapped Ions, Quantum Trajectories, and Decoherence*, Springer-Verlag, 3rd ed. 2016.

C. W. Gardiner and P. Zoller, *The Quantum World of Ultra-Cold Atoms and Light. Book I: Foundations of Quantum Optics*, Imperial College Press, 2014.

[†] See M. B. Nasr, B. E. A. Saleh, A. V. Sergienko, and M. C. Teich, Dispersion-Cancelled and Dispersion-Sensitive Quantum Optical Coherence Tomography, *Optics Express*, vol. 12, pp. 1353–1362, 2004.

G. S. Agarwal, *Quantum Optics*, Cambridge University Press, 2012.

G. Grynberg, A. Aspect, and C. Fabre, *Introduction to Quantum Optics: From the Semi-Classical Approach to Quantized Light*, Cambridge University Press, 2010.

U. Leonhardt, *Essential Quantum Optics: From Quantum Measurements to Black Holes*, Cambridge University Press, 2010.

D. F. Walls and G. J. Milburn, *Quantum Optics*, Springer-Verlag, 2nd ed. 2008.

J. C. Garrison and R. Y. Chiao, *Quantum Optics*, Oxford University Press, 2008.

P. Meystre and M. Sargent III, *Elements of Quantum Optics*, Springer-Verlag, 4th ed. 2007.

M. Fox, *Quantum Optics: An Introduction*, Oxford University Press, 2006.

W. Vogel and D.-G. Welsch, *Quantum Optics*, Wiley–VCH, 3rd ed. 2006.

C. C. Gerry and P. L. Knight, *Introductory Quantum Optics*, Cambridge University Press, 2005.

H.-A. Bachor and T. C. Ralph, *A Guide to Experiments in Quantum Optics*, Wiley–VCH, 2nd ed. 2004.

W. P. Schleich, *Quantum Optics in Phase Space*, Wiley–VCH, 2001.

R. Loudon, *The Quantum Theory of Light*, Oxford University Press, 3rd ed. 2000.

V. Peřinová, A. Lukš, and J. Peřina, *Phase in Optics*, World Scientific, 1998.

M. O. Scully and M. S. Zubairy, *Quantum Optics*, Cambridge University Press, paperback ed. 1997.

G. S. Agarwal, ed., *Selected Papers on Fundamentals of Quantum Optics*, SPIE Optical Engineering Press (Milestone Series Volume 103), 1995.

R. P. Feynman (with an introduction by A. Zee), *QED: The Strange Theory of Light and Matter*, Princeton University Press, 1985, reissued 2014.

S. Weinberg, Light as a Fundamental Particle, *Physics Today*, vol. 28, no. 6, pp. 32–37, 1975.

J. R. Klauder and E. C. G. Sudarshan, *Fundamentals of Quantum Optics*, Benjamin, 1968; Dover, reissued 2006.

Coherence, Quantum Fluctuations, Antibunching, and Photon Statistics

B. R. Masters, Satyendra Nath Bose and Bose–Einstein Statistics, *Optics & Photonics News*, vol. 24, no. 4, pp. 40–47, 2013.

H. J. Carmichael, *Statistical Methods in Quantum Optics 2: Non-Classical Fields*, Springer-Verlag, 2010.

R. J. Glauber, *Quantum Theory of Optical Coherence: Selected Papers and Lectures*, Wiley–VCH, 2007.

R. J. Glauber, Nobel Lecture: One Hundred Years of Light Quanta, *Reviews of Modern Physics*, vol. 78, pp. 1267–1278, 2006.

Z. Ficek and S. Swain, *Quantum Interference and Coherence: Theory and Experiments*, Springer-Verlag, 2005.

J. Peřina, ed., *Coherence and Statistics of Photons and Atoms*, Wiley, 2001.

H. J. Carmichael, *Statistical Methods in Quantum Optics 1: Master Equations and Fokker–Planck Equations*, Springer-Verlag, 1999.

L. Mandel and E. Wolf, *Optical Coherence and Quantum Optics*, Cambridge University Press, 1995.

L. Mandel and E. Wolf, eds., *Selected Papers on Coherence and Fluctuations of Light (1850–1966)*, SPIE Optical Engineering Press (Milestone Series Volume 19), 1990.

R. A. Campos, B. E. A. Saleh, and M. C. Teich, Quantum-Mechanical Lossless Beam Splitter: SU(2) Symmetry and Photon Statistics, *Physical Review A*, vol. 40, pp. 1371–1384, 1989.

E. R. Pike and H. Walther, eds., *Photons and Quantum Fluctuations*, CRC Press, 1988.

M. C. Teich and B. E. A. Saleh, Photon Bunching and Antibunching, in E. Wolf, ed., *Progress in Optics*, North-Holland, 1988, vol. 26, pp. 1–104.

J. Peřina, *Coherence of Light*, Reidel, 2nd ed. 1985.

E. Wolf, Einstein's Researches on the Nature of Light, *Optics News*, vol. 5, no. 1, pp. 24–39, 1979.

B. E. A. Saleh, *Photoelectron Statistics*, Springer-Verlag, 1978.

H. J. Kimble, M. Dagenais, and L. Mandel, Photon Antibunching in Resonance Fluorescence, *Physical Review Letters*, vol. 39, pp. 691–695, 1977.

W. H. Louisell, *Quantum Statistical Properties of Radiation*, Wiley, 1973; Wiley–VCH, reprinted 1990.

L. Mandel and E. Wolf, eds., *Selected Papers on Coherence and Fluctuations of Light*, Volumes 1 and 2, Dover, 1970.

L. Mandel, Fluctuations of Light Beams, in E. Wolf, ed., *Progress in Optics*, North-Holland, 1963, vol. 2, pp. 181–248.

S. N. Bose, Plancks Gesetz und Lichtquantenhypothese (Planck's Law and the Light-Quantum Hypothesis), *Zeitschrift für Physik*, vol. 26, pp. 178-181, 1924.

Quadrature-Squeezed and Photon-Number-Squeezed Light

X. He, N. F. Hartmann, X. Ma, Y. Kim, R. Ihly, J. L. Blackburn, W. Gao, J. Kono, Y. Yomogida, A. Hirano, T. Tanaka, H. Kataura, H. Htoon, and S. K. Doorn, Tunable Room-Temperature Single-Photon Emission at Telecom Wavelengths from sp^3 Defects in Carbon Nanotubes, *Nature Photonics*, vol. 11, pp. 577–582, 2017.

H. Vahlbruch, M. Mehmet, K. Danzmann, and R. Schnabel, Detection of 15 dB Squeezed States of Light and their Application for the Absolute Calibration of Photoelectric Quantum Efficiency, *Physical Review Letters*, vol. 117, 110801, 2016.

J. Aasi *et al.* (LIGO Scientific Collaboration), Enhanced Sensitivity of the LIGO Gravitational Wave Detector by Using Squeezed States of Light, *Nature Photonics*, vol. 7, pp. 613–619, 2013.

H. J. Carmichael, *Statistical Methods in Quantum Optics 2: Non-Classical Fields*, Springer-Verlag, 2010.

M. C. Teich and B. E. A. Saleh, Squeezed and Antibunched Light, *Physics Today*, vol. 43, no. 6, pp. 26–34, 1990 (Erratum: vol. 43, no. 11, pp. 123–124, 1990).

M. C. Teich and B. E. A. Saleh, Squeezed States of Light, *Quantum Optics: Journal of the European Physical Society B*, vol. 1, pp. 153–191, 1989.

R. E. Slusher, L. W. Hollberg, B. Yurke, J. C. Mertz, and J. F. Valley, Observation of Squeezed States Generated by Four-Wave Mixing in an Optical Cavity, *Physical Review Letters*, vol. 55, pp. 2409–2412, 1985.

M. C. Teich and B. E. A. Saleh, Observation of Sub-Poisson Franck–Hertz Light at 253.7 nm, *Journal of the Optical Society of America B*, vol. 2, pp. 275–282, 1985.

D. Stoler, B. E. A. Saleh, and M. C. Teich, Binomial States of the Quantized Radiation Field, *Optica Acta (Journal of Modern Optics)*, vol. 32, pp. 345–355, 1985.

R. Short and L. Mandel, Observation of Sub-Poissonian Photon Statistics, *Physical Review Letters*, vol. 51, pp. 384–387, 1983.

C. M. Caves, Quantum-Mechanical Noise in an Interferometer, *Physical Review D*, vol. 23, pp. 1693–1708, 1981.

Entangled-Photon and Two-Photon Light

J. Wang, S. Paesani, Y. Ding, R. Santagati, P. Skrzypczyk, A. Salavrakos, J. Tura, R. Augusiak, L. Mančinska, D. Bacco, D. Bonneau, J. W. Silverstone, Q. Gong, A. Acín, K. Rottwitt, L. K. Oxenløwe, J. L. O'Brien, A. Laing, and M. G. Thompson, Multidimensional Quantum Entanglement With Large-Scale Integrated Optics, *Science*, vol. 360, pp. 285–291, 2018.

M. A. Finger, T. Sh. Iskhakov, N. Y. Joly, M. V. Chekhova, and P. St. J. Russell, Raman-Free, Noble-Gas-Filled Photonic-Crystal Fiber Source for Ultrafast, Very Bright Twin-Beam Squeezed Vacuum, *Physical Review Letters*, vol. 115, 143602, 2015.

D. Kang, M. Kim, H. He, and A. S. Helmy, Two Polarization-Entangled Sources from the Same Semiconductor Chip, *Physical Review A*, vol. 92, 013821, 2015.

F. Boitier, A. Orieux, C. Autebert, A. Lemaître, E. Galopin, C. Manquest, C. Sirtori, I. Favero, G. Leo, and S. Ducci, Electrically Injected Photon-Pair Source at Room Temperature, *Physical Review Letters*, vol. 112, 183901, 2014.

M. F. Saleh, G. Di Giuseppe, B. E. A. Saleh, and M. C. Teich, Photonic Circuits for Generating Modal, Spectral, and Polarization Entanglement, *IEEE Photonics Journal*, vol. 2, pp. 736–752, 2010.

M. B. Nasr, S. Carrasco, B. E. A. Saleh, A. V. Sergienko, M. C. Teich, J. P. Torres, L. Torner, D. S. Hum, and M. M. Fejer, Ultrabroadband Biphotons Generated via Chirped Quasi-Phase-Matched Optical Parametric Down-Conversion, *Physical Review Letters*, vol. 100, 183601, 2008.

J. Audretsch, *Entangled Systems*, Wiley–VCH, 2007.

J. S. Bell (with an introduction by A. Aspect), *Speakable and Unspeakable in Quantum Mechanics*, Cambridge University Press, 2nd ed. 2004.

B. E. A. Saleh, A. F. Abouraddy, A. V. Sergienko, and M. C. Teich, Duality Between Partial Coherence and Partial Entanglement, *Physical Review A*, vol. 62, 043816, 2000.

D. N. Klyshko, *Photons and Nonlinear Optics*, Nauka (Moscow), 1980 [Translation: Gordon and Breach, New York, 1988).

A. Aspect, P. Grangier, and G. Roger, Experimental Tests of Realistic Local Theories via Bell's Theorem, *Physical Review Letters*, vol. 47, pp. 460-463, 1981.

D. C. Burnham and D. L. Weinberg, Observation of Simultaneity in Parametric Production of Optical Photon Pairs, *Physical Review Letters*, vol. 25, pp. 84–87, 1970.

D. Magde and H. Mahr, Study in Ammonium Dihydrogen Phosphate of Spontaneous Parametric Interaction Tunable from 4400 to 16 000 Å, *Physical Review Letters*, vol. 18, pp. 905–907, 1967.

S. E. Harris, M. K. Oshman, and R. L. Byer, Observation of Tunable Optical Parametric Fluorescence, *Physical Review Letters*, vol. 18, pp. 732–734, 1967.

A. Einstein, B. Podolsky, and N. Rosen, Can Quantum-Mechanical Description of Physical Reality Be Considered Complete?, *Physical Review*, vol. 47, pp. 777–780, 1935.

E. Schrödinger, Die gegenwärtige Situation in der Quantenmechanik, *Die Naturwissenschaften*, vol. 23, pp. 807–812, 823–828, 844–849; 1935 [Translation: The Present Situation in Quantum Mechanics: A Translation of Schrödinger's 'Cat Paradox' Paper, *Proceedings of the American Philosophical Society*, vol. 124, pp. 323–338, 1980].

Quantum Imaging, Metrology, Lithography, and Spectroscopy

D. S. Simon, G. Jaeger, A. V. Sergienko, *Quantum Metrology, Imaging, and Communication*, Springer-Verlag, 2017.

M. A. Taylor, J. Janousek, V. Daria, J. Knittel, B. Hage, H.-A. Bachor, and W. P. Bowen, Subdiffraction-Limited Quantum Imaging within a Living Cell, *Physical Review X*, vol. 4, 011017, 2014.

T. Ono, R. Okamoto, and S. Takeuchi, An Entanglement-Enhanced Microscope, *Nature Communications* 4, 2426 doi: 10.1038/ncomms3426, 2013.

M. C. Teich, B. E. A. Saleh, F. N. C. Wong, and J. H. Shapiro, Variations on the Theme of Quantum Optical Coherence Tomography: A Review, *Quantum Information Processing*, vol. 11, pp. 903–923, 2012.

M. B. Nasr, D. P. Goode, N. Nguyen, G. Rong, L. Yang, B. M. Reinhard, B. E. A. Saleh, and M. C. Teich, Quantum Optical Coherence Tomography of a Biological Sample, *Optics Communications*, vol. 282, pp. 1154–1159, 2009.

M. B. Nasr, B. E. A. Saleh, A. V. Sergienko, and M. C. Teich, Demonstration of Dispersion-Cancelled Quantum-Optical Coherence Tomography, *Physical Review Letters*, vol. 91, 083601, 2003.

A. F. Abouraddy, B. E. A. Saleh, A. V. Sergienko, and M. C. Teich, Entangled-Photon Fourier Optics, *Journal of the Optical Society of America B*, vol. 19, pp. 1174–1184, 2002.

A. F. Abouraddy, B. E. A. Saleh, A. V. Sergienko, and M. C. Teich, Role of Entanglement in Two-Photon Imaging, *Physical Review Letters*, vol. 87, 123602, 2001.

A. F. Abouraddy, B. E. A. Saleh, A. V. Sergienko, and M. C. Teich, Quantum Holography, *Optics Express*, vol. 9, pp. 498–505, 2001.

A. N. Boto, P. Kok, D. S. Abrams, S. L. Braunstein, C. P. Williams, and J. P. Dowling, Quantum Interferometric Optical Lithography: Exploiting Entanglement to Beat the Diffraction Limit, *Physical Review Letters*, vol. 85, pp. 2733-2736, 2000.

B. E. A. Saleh, B. M. Jost, H.-B. Fei, and M. C. Teich, Entangled-Photon Virtual-State Spectroscopy, *Physical Review Letters*, vol. 80, pp. 3483–3486, 1998.

M. C. Teich and B. E. A. Saleh, Entangled-Photon Microscopy, Spectroscopy, and Display, *U.S. Patent 5,796,477*, Patented August 18, 1998.

Quantum Information, Communications, Computing, and Cryptography

I. Khan, B. Heim, A. Neuzner, and C. Marquardt, Satellite-Based QKD, *Optics & Photonics News*, vol. 29, no. 2, pp. 26–33, 2018.

M. M. Wilde, *Quantum Information Theory*, Cambridge University Press, 2nd ed. 2017.

J. Yin *et al.*, Satellite-Based Entanglement Distribution Over 1200 Kilometers, *Science*, vol. 356, pp. 1140–1144, 2017.

M. Hayashi, S. Ishizaka, A. Kawachi, G. Kimura, and T. Ogawa, *Introduction to Quantum Information Science*, Springer-Verlag, 2015.

J. Carolan, C. Harrold, C. Sparrow, E. Martín-López, N. J. Russell, J. W. Silverstone, P. J. Shadbolt, N. Matsuda, M. Oguma, M. Itoh, G. D. Marshall, M. G. Thompson, J. C. F. Matthews, T. Hashimoto, J. L. O'Brien, and A. Laing, Universal Linear Optics, *Science*, vol. 349, pp. 711–716, 2015.

I. A. Walmsley, Quantum Optics: Science and Technology in a New Light, *Science*, vol. 348, pp. 525–530, 2015.

J. A. Bergou and M. Hillery, *Introduction to the Theory of Quantum Information Processing*, Springer-Verlag, 2013.

J. A. Jones and D. Jaksch, *Quantum Information, Computation and Communication*, Cambridge University Press, 2012.

R. T. Perry, *Quantum Computing from the Ground Up*, World Scientific, 2012.

A. Zeilinger, *Dance of the Photons: From Einstein to Quantum Teleportation*, Farrar, Straus and Giroux, 2010.

S. M. Barnett, *Quantum Information*, Oxford University Press, 2009.

G. Jaeger, *Entanglement, Information, and the Interpretation of Quantum Mechanics*, Springer-Verlag, 2009.

J. L. O'Brien, A. Furusawa, and J. Vučković, Photonic Quantum Technologies, *Nature Photonics*, vol. 3, pp. 687–695, 2009.

H. Paul, *Introduction to Quantum Optics: From Light Quanta to Quantum Teleportation*, Cambridge University Press, 2004.

M. A. Nielsen and I. L. Chuang, *Quantum Computation and Quantum Information*, Cambridge University Press, 2002.

PROBLEMS

13.1-5 **Photon Energy.**
   (a) What voltage should be applied to an electrode to accelerate an electron from zero velocity such that it acquires the same energy as a photon of wavelength $\lambda_o = 0.87$ μm?
   (b) A photon of wavelength 1.06 μm is combined with a photon of wavelength 10.6 μm to create a photon whose energy is the sum of the energies of the two photons. What is the wavelength of the resultant photon? This process, known as sum-frequency generation, is illustrated in Fig. 22.2-6.

13.1-6 **Position of a Single Photon at a Screen.** Consider monochromatic light of wavelength λ_o incident on an infinite screen in the plane $z = 0$, with an intensity $I(\rho) = I_0 \exp(-\rho/\rho_0)$, where $\rho = \sqrt{x^2 + y^2}$. Assume that the intensity of the source is reduced to a level at which only a single photon strikes the screen.
   (a) Find the probability that the photon strikes the screen within a radius ρ_0 of the origin.
   (b) If the incident light contains exactly 10^6 photons, how many photons strike within a circle of radius ρ_0 on average?

13.1-7 **Photon Momentum.** Compare the magnitude of the total momentum in all of the photons in a 10-J laser pulse with that of:
   (a) a 1-g mass moving at a velocity of 1 cm/s.
   (b) an electron moving at a velocity $c_o/10$.

*13.1-8 **Momentum of a Photon in a Gaussian Beam.**
   (a) What is the probability that the momentum vector of a photon associated with a Gaussian beam of waist radius W_0 lies within the beam divergence angle θ_0? Refer to Sec. 3.1 for definitions.
   (b) Does the relation $p = E/c_o$ hold in this case? Explain.

13.1-9 **Levitation by Light Pressure.** An isolated hydrogen atom has mass 1.67×10^{-27} kg.
  (a) Find the gravitational force on this hydrogen atom near the surface of the earth (assume that at sea level the gravitational acceleration constant $g = 9.8$ m/s^2).
  (b) Let an upwardly directed laser beam emitting 1-eV photons be focused in such a way that the full momentum of each of its photons is transferred to the atom. Find the average upward force on the atom provided by one photon striking each second.
  (c) Find the number of photons that must strike the atom per second, and the corresponding optical power, for it not to fall under the effect of gravity, assuming ideal conditions in vacuum.
  (d) How many photons per second would be required to keep the atom from falling if it were perfectly reflecting?

*13.1-10 **Single Photon in a Fabry–Perot Resonator.** Consider a Fabry–Perot resonator of length $d = 1$ cm that contains nonabsorbing material of refractive index $n = 1.5$ and perfectly reflecting mirrors. Assume that there is exactly one photon in the mode described by the standing wave $\sin(10^5 \pi x/d)$.
  (a) Determine the photon wavelength and energy (in eV).
  (b) Estimate the uncertainty in the photon's position and momentum (magnitude and direction). Compare with the value obtained from the relation $\sigma_x \sigma_p = \hbar/2$.

13.1-11 **Single-Photon Beating (Time Interference).** Consider a detector illuminated by a polychromatic plane wave consisting of two superposed monochromatic waves traveling in the same direction. The constituent waves have complex wavefunctions given by

$$U_1(t) = \sqrt{I_1} \exp(j2\pi\nu_1 t) \quad \text{and} \quad U_2(t) = \sqrt{I_2} \exp(j2\pi\nu_2 t),$$

with frequencies ν_1 and ν_2 and intensities I_1 and I_2, respectively. In accordance with Sec. 2.6B, the intensity of this wave is given by $I(t) = I_1 + I_2 + 2\sqrt{I_1 I_2} \cos[2\pi(\nu_2 - \nu_1)t]$. Assume that the two constituent plane waves have equal intensities ($I_1 = I_2$) and that the wave is sufficiently weak so that only a single polychromatic photon reaches the detector during the time interval $T = 1/|\nu_2 - \nu_1|$.
  (a) Sketch the probability density $p(t)$ for detecting the photon in the interval $0 \le t \le 1/|\nu_2 - \nu_1|$. At what time instant during T is the probability density zero that the photon will be detected?
  (b) An attempt to discover from which of the two constituent waves the photon arrives entails an energy measurement with a precision better than

$$\sigma_E < h|\nu_2 - \nu_1|.$$

  Use the time–energy uncertainty relation to show that the time required for such a measurement is of the order of the beat-frequency period. This signifies that the very process of measurement washes out the interference and thereby precludes the measurement from being made.

13.1-12 **Photon Momentum Exchange at a Beamsplitter.** Consider a single photon, in a mode described by a plane wave, impinging on a lossless beamsplitter. What is the momentum vector of the photon before it impinges on the mirror? What are the possible values of the momentum vector of the photon, and the probabilities of observing these values, after passage through the beamsplitter?

13.2-2 **Photon Flux.** Demonstrate that the power of a monochromatic optical beam that carries an average of one photon per optical cycle is inversely proportional to the square of the wavelength.

13.2-3 **The Poisson Distribution.** Verify that the Poisson probability distribution provided in (13.2-12) is normalized to unity and has mean $\bar{n}$ and variance $\sigma_n^2 = \bar{n}$.

13.2-4 **Photon Statistics of a Coherent Gaussian Beam.** Assume that a 100-pW single-mode He–Ne laser emits light at 633 nm in a TEM$_{00}$ Gaussian beam (Sec. 3.1).
  (a) What is the mean number of photons crossing a circle of radius equal to the waist radius of the beam W_0 in a time $T = 100$ ns?
  (b) What is the root-mean-square value of the number of photons in (a)?
  (c) What is the probability that the number of photons is zero in (a)?

13.2-5 **The Bose–Einstein Distribution.**
(a) Verify that the Bose–Einstein probability distribution provided in (13.2-20) is normalized and has mean $\bar{n}$ and variance $\sigma_n^2 = \bar{n} + \bar{n}^2$.
(b) If a stream of photons obeying Bose–Einstein statistics contains an average of $\Phi = 1$ photon per nanosecond, what is the probability that zero photons will be detected in a 20-ns time interval?

*13.2-6 **The Negative–Binomial Distribution.** It is well known in the literature of probability theory that the sum of $\mathcal{M}$ identically distributed random variables, each with a geometric (Bose–Einstein) distribution, obeys the negative binomial distribution with overall mean $\bar{n}$,

$$p(n) = \binom{n + \mathcal{M} - 1}{n} \frac{(\bar{n}/\mathcal{M})^n}{(1 + \bar{n}/\mathcal{M})^{n+\mathcal{M}}}.$$

Verify that the negative-binomial distribution reduces to the Bose–Einstein distribution for $\mathcal{M} = 1$ and to the Poisson distribution as $\mathcal{M} \to \infty$.

*13.2-7 **Photon-Number Statistics for Multimode Thermal Light in a Cavity.** Consider $\mathcal{M}$ modes of thermal radiation sufficiently close to each other in frequency that each can be considered to be occupied in accordance with a Bose–Einstein distribution of the same mean photon number $1/[\exp(h\nu/kT) - 1]$. Show that the variance of the *total* number of photons n is related to its mean by

$$\sigma_n^2 = \bar{n} + \frac{\bar{n}^2}{\mathcal{M}},$$

which indicates that multimode thermal light has less variance than does single-mode thermal light. The presence of the multiple modes provides averaging, thereby reducing the noisiness of the light.

*13.2-8 **Photon-Number Statistics for a Beam of Multimode Thermal Light.** A multimode thermal light source that carries $\mathcal{M}$ identical modes, each with an exponentially distributed (random) integrated rate, has an overall probability density $p(w)$ describable by the gamma density

$$p(w) = \frac{1}{(\mathcal{M} - 1)!} \left(\frac{\langle w \rangle}{\mathcal{M}}\right)^{-\mathcal{M}} w^{\mathcal{M}-1} \exp\left(-\frac{w}{\langle w \rangle / \mathcal{M}}\right), \qquad w \geq 0.$$

Use Mandel's formula (13.2-26) to show that the resulting photon-number distribution assumes the form of the negative-binomial distribution defined in Prob. 13.2-6.

*13.2-9 **Mean and Variance of the Doubly Stochastic Poisson Distribution.** Prove (13.2-27) and (13.2-28).

*13.2-10 **Random Partitioning of Coherent Light.**
(a) Use (13.2-33) to show that the photon-number distribution of randomly partitioned coherent light retains its Poisson form.
(b) Show explicitly that the mean photon number for light reflected from a lossless beamsplitter is $(1 - \mathcal{T})\bar{n}$.
(c) Prove (13.2-34) for coherent light.

13.2-11 **Random Partitioning of Single-Mode Thermal Light.**
(a) Use (13.2-33) to show that the photon-number distribution of randomly partitioned single-mode thermal light retains its Bose–Einstein form.
(b) Show explicitly that the mean photon number for light reflected from a lossless beamsplitter is $(1 - \mathcal{T})\bar{n}$.
(c) Prove (13.2-35) for single-mode thermal light.

*13.2-12 **Exponential Decay of Mean Photon Number in an Absorber.**
(a) Consider an absorptive material of thickness d and absorption coefficient α (cm^{-1}). If the average number of photons entering the material is $\bar{n}_0$, write a differential equation to find the average number of photons $\bar{n}(x)$ at position x, where x is the depth into the material ($0 \leq x \leq d$).
(b) Solve the differential equation. State why your result turns out to be the exponential decay law obtained in electromagnetic optics (Sec. 5.5A).

(c) If the incident light is coherent, write an expression for the photon-number distribution $p(n)$ at an arbitrary position x into the absorber.

(d) What is the probability that a single photon incident on the absorber survives passage through it?

*13.3-1 **Statistics of the Binomial Photon-Number Distribution.** The binomial probability distribution, which is written as

$$p(n) = \frac{M!}{(M-n)!\,n!}\, p^n (1-p)^{M-n},$$

describes the photon-number statistics for certain sources of photon-number-squeezed light.[†]

(a) Indicate a plausible mechanism whereby number-state light is converted into light described by binomial photon statistics.

(b) Prove that the binomial probability distribution is normalized to unity.

(c) Find the photon-number mean $\bar{n}$ and the photon-number variance σ_n^2 of the binomial probability distribution in terms of its two parameters, p and M.

(d) Find an expression for the SNR in terms of $\bar{n}$ and p. Evaluate the SNR for the limiting cases $p \to 0$ and $p \to 1$. What is the nature of the light corresponding to those two limits?

*13.3-2 **Noisiness of the Uniform Photon-Number Distribution.** Consider a light source that generates a photon stream with the discrete-uniform photon-number distribution

$$p(n) = \begin{cases} \dfrac{1}{2\bar{n}+1}, & 0 \leq n \leq 2\bar{n} \\ 0, & \text{otherwise.} \end{cases}$$

(a) Verify that the distribution is normalized to unity and has mean $\bar{n}$. Calculate the photon-number variance σ_n^2 and the signal-to-noise ratio (SNR) and compare these quantities with those for the Bose–Einstein and Poisson distributions of the same mean.

(b) In terms of SNR, would this source be quieter or noisier than an ideal single-mode laser when $\bar{n} < 2$? When $\bar{n} = 2$? When $\bar{n} > 2$?

(c) By what factor is the SNR for this light larger than that for single-mode thermal light?

(d) Suggest a mechanism for generating light with this photon-number distribution.

Useful formulas:

$$1+2+3+\cdots+j = \frac{j(j+1)}{2}, \qquad 1^2+2^2+3^2+\cdots+j^2 = \frac{j(j+1)(2j+1)}{6}.$$

[†] See D. Stoler, B. E. A. Saleh, and M. C. Teich, Binomial States of the Quantized Radiation Field, *Optica Acta (Journal of Modern Optics)*, vol. 32, pp. 345–355, 1985.

APPENDIX

A

FOURIER TRANSFORM

This appendix provides a brief review of the Fourier transform, and its properties, for functions of one and two variables.

A.1 ONE-DIMENSIONAL FOURIER TRANSFORM

The harmonic function $F \exp(j2\pi\nu t)$ plays an important role in science and engineering. It has frequency ν and complex amplitude F. Its real part $|F|\cos(2\pi\nu t + \arg\{F\})$ is a cosine function with amplitude $|F|$ and phase $\arg\{F\}$. The variable t usually represents time; the frequency ν has units of cycles/s or Hz. The harmonic function is regarded as a building block from which other functions may be obtained by a simple superposition.

In accordance with the Fourier theorem, a complex-valued function $f(t)$, satisfying some rather unrestrictive conditions, may be decomposed as a superposition integral of harmonic functions of different frequencies and complex amplitudes,

$$f(t) = \int_{-\infty}^{\infty} F(\nu) \exp(j2\pi\nu t)\, d\nu. \qquad \text{(A.1-1)}$$
Inverse Fourier Transform

The component with frequency ν has a complex amplitude $F(\nu)$ given by

$$F(\nu) = \int_{-\infty}^{\infty} f(t) \exp(-j2\pi\nu t)\, dt. \qquad \text{(A.1-2)}$$
Fourier Transform

$F(\nu)$ is termed the **Fourier transform** of $f(t)$, and $f(t)$ is the **inverse Fourier transform** of $F(\nu)$. The functions $f(t)$ and $F(\nu)$ form a **Fourier transform pair**; if one is known, the other may be determined.

In this book we adopt the convention that $\exp(j2\pi\nu t)$ is a harmonic function with positive frequency, whereas $\exp(-j2\pi\nu t)$ represents negative frequency. The opposite convention is used by some authors, who define the Fourier transform in (A.1-2) with a positive sign in the exponent, and use a negative sign in the exponent of the inverse Fourier transform (A.1-1).

In communication theory, the functions $f(t)$ and $F(\nu)$ represent a signal, with $f(t)$ its time-domain representation and $F(\nu)$ its frequency-domain representation. The absolute-squared value $|f(t)|^2$ is called the **signal power**, and $|F(\nu)|^2$ is the energy spectral density. If $|F(\nu)|^2$ extends over a wide frequency range, the signal is said to have a wide bandwidth.

Fundamentals of Photonics, Third Edition. Bahaa E. A. Saleh and Malvin Carl Teich.
©2019 John Wiley & Sons, Inc. Published 2019 by John Wiley & Sons, Inc.

Properties of the Fourier Transform

Some important properties of the Fourier transform are provided below. These properties can be proved by direct application of the definitions (A.1-1) and (A.1-2) (see any of the books in the Reading List).

- *Linearity.* The Fourier transform of the sum of two functions is the sum of their Fourier transforms.
- *Scaling.* If $f(t)$ has a Fourier transform $F(\nu)$, and τ is a real scaling factor, then $f(t/\tau)$ has a Fourier transform $|\tau|F(\tau\nu)$. This means that if $f(t)$ is scaled by a factor τ, its Fourier transform is scaled by a factor $1/\tau$. Thus, if $\tau > 1$, then $f(t/\tau)$ is a stretched version of $f(t)$, whereas $F(\tau\nu)$ is a compressed version of $F(\nu)$. The Fourier transform of $f(-t)$ is $F(-\nu)$.
- *Time Translation.* If $f(t)$ has a Fourier transform $F(\nu)$, the Fourier transform of $f(t-\tau)$ is $\exp(-j2\pi\nu\tau)F(\nu)$. Thus, delay by time τ is equivalent to multiplication of the Fourier transform by a phase factor $\exp(-j2\pi\nu\tau)$.
- *Frequency Translation.* If $F(\nu)$ is the Fourier transform of $f(t)$, the Fourier transform of $f(t)\exp(j2\pi\nu_0 t)$ is $F(\nu - \nu_0)$. Thus, multiplication by a harmonic function of frequency ν_0 is equivalent to shifting the Fourier transform to a higher frequency ν_0.
- *Symmetry.* If $f(t)$ is real, then $F(\nu)$ has Hermitian symmetry, i.e., $F(-\nu) = F^*(\nu)$. If $f(t)$ is real and symmetric, then $F(\nu)$ is also real and symmetric.
- *Convolution Theorem.* If the Fourier transforms of $f_1(t)$ and $f_2(t)$ are $F_1(\nu)$ and $F_2(\nu)$, respectively, the inverse Fourier transform of the product

$$F(\nu) = F_1(\nu)F_2(\nu) \qquad (A.1\text{-}3)$$

is

$$f(t) = \int_{-\infty}^{\infty} f_1(\tau)f_2(t-\tau)\,d\tau. \qquad (A.1\text{-}4)$$
Convolution

The operation defined in (A.1-4) is known as the **convolution** of $f_1(t)$ with $f_2(t)$. Convolution in the time domain is therefore equivalent to multiplication in the Fourier domain.

- *Correlation Theorem.* The **correlation** between two complex functions is defined as

$$f(t) = \int_{-\infty}^{\infty} f_1^*(\tau)f_2(t+\tau)\,d\tau. \qquad (A.1\text{-}5)$$
Correlation

The Fourier transforms of $f_1(t)$, $f_2(t)$, and $f(t)$ are related by

$$F(\nu) = F_1^*(\nu)F_2(\nu). \qquad (A.1\text{-}6)$$

If $f_2(t) = f_1(t)$, (A.1-5) is called the **autocorrelation**.

- *Parseval's Theorem.* The signal energy, which is the integral of the signal power $|f(t)|^2$, equals the integral of the energy spectral density $|F(\nu)|^2$, so that

$$\int_{-\infty}^{\infty} |f(t)|^2\,dt = \int_{-\infty}^{\infty} |F(\nu)|^2\,d\nu. \qquad (A.1\text{-}7)$$
Parseval's Theorem

A.1 ONE-DIMENSIONAL FOURIER TRANSFORM

Examples

The Fourier transforms of some important functions are provided in Table A.1-1. The Fourier transforms of many other functions are readily obtained by making use of the properties of linearity, scaling, delay, and frequency translation. The functions used in Table A.1-1 have the following definitions:

- $\text{rect}(t) \equiv 1$ for $|t| \leq \frac{1}{2}$, and $= 0$ elsewhere, i.e., it is a pulse of unit height and unit width centered about $t = 0$.
- $\delta(t)$ is the impulse function (also called the Dirac delta function), which is defined as $\delta(t) \equiv \lim_{\alpha \to \infty} \alpha \, \text{rect}(\alpha t)$. It is the limit of a rectangular pulse of unit area as its width approaches zero so that its height approaches infinity.
- $\text{sinc}(t) \equiv \sin(\pi t)/(\pi t)$ is a symmetric function with a peak value of unity at $t = 0$ and with zeros at $t = \pm 1, \pm 2, \ldots$.

Table A.1-1 Selected functions and their Fourier transforms.

Function	$f(t)$	$F(\nu)$
Uniform	1	$\delta(\nu)$
Impulse	$\delta(t)$	1
Rectangular	$\text{rect}(t)$	$\text{sinc}(\nu)$
Exponential [a]	$\exp(-|t|)$	$\dfrac{2}{1+(2\pi\nu)^2}$
Gaussian	$\exp(-\pi t^2)$	$\exp(-\pi\nu^2)$
Hyperbolic secant	$\text{sech}(\pi t)$	$\text{sech}(\pi\nu)$
Chirp [b]	$\exp(j\pi t^2)$	$e^{j\pi/4} \exp(-j\pi\nu^2)$
$M = 2S+1$ Impulses	$\sum\limits_{m=-S}^{S} \delta(t-m)$	$\dfrac{\sin(M\pi\nu)}{\sin(\pi\nu)}$
Comb	$\sum\limits_{m=-\infty}^{\infty} \delta(t-m)$	$\sum\limits_{m=-\infty}^{\infty} \delta(\nu-m)$

[a] The double-sided exponential function is shown. The Fourier transform of the single-sided exponential, $f(t) = \exp(-t)$ with $t \geq 0$, is $F(\nu) = 1/[1 + j2\pi\nu]$. Its magnitude is $1/\sqrt{1+(2\pi\nu)^2}$.
[b] The functions $\cos(\pi t^2)$ and $\cos(\pi\nu^2)$ are shown. The function $\sin(\pi t^2)$ is shown in Fig. 4.3-6.

A.2 TIME DURATION AND SPECTRAL WIDTH

It is often useful to have a measure of the width of a function. The width of a function of time $f(t)$ is its time duration and the width of its Fourier transform $F(\nu)$ is its spectral width (or bandwidth). Since there is no unique definition for the width, a plethora of definitions are in use. *All definitions, however, share the property that the spectral width is inversely proportional to the temporal width, in accordance with the scaling property of the Fourier transform.* The following definitions are used at different places in this book.

The Root-Mean-Square Width

The *root-mean-square (RMS) width* σ_t of a nonnegative real function $f(t)$ is defined by

$$\sigma_t^2 = \frac{\int_{-\infty}^{\infty}(t-\bar{t})^2 f(t)\,dt}{\int_{-\infty}^{\infty} f(t)\,dt}, \quad \text{where} \quad \bar{t} = \frac{\int_{-\infty}^{\infty} t f(t)\,dt}{\int_{-\infty}^{\infty} f(t)\,dt}. \tag{A.2-1}$$

If $f(t)$ represents a mass distribution (t representing position), then $\bar{t}$ represents the centroid and σ_t the radius of gyration. If $f(t)$ is a probability density function, these quantities represent the mean and standard deviation, respectively. As an example, the *Gaussian function* $f(t) = \exp(-t^2/2\sigma_t^2)$ has an RMS width σ_t. Its Fourier transform is given by $F(\nu) = (1/\sqrt{2\pi}\,\sigma_\nu)\exp(-\nu^2/2\sigma_\nu^2)$, where

$$\sigma_\nu = \frac{1}{2\pi\sigma_t} \tag{A.2-2}$$

is the RMS spectral width.

This definition is not appropriate for functions with negative or complex values. For such functions the RMS width of the absolute-squared value $|f(t)|^2$ is used,

$$\sigma_t^2 = \frac{\int_{-\infty}^{\infty}(t-\bar{t})^2 |f(t)|^2\,dt}{\int_{-\infty}^{\infty} |f(t)|^2\,dt}, \quad \text{where} \quad \bar{t} = \frac{\int_{-\infty}^{\infty} t|f(t)|^2\,dt}{\int_{-\infty}^{\infty} |f(t)|^2\,dt}.$$

We call this version of σ_t the *power-RMS width*.

With the help of the Schwarz inequality, it can be shown that the product of the power RMS widths of an arbitrary function $f(t)$ and its Fourier transform $F(\nu)$ must be equal to or greater than $1/4\pi$,

$$\boxed{\sigma_t \sigma_\nu \geq \frac{1}{4\pi},} \tag{A.2-3}$$

Duration–Bandwidth
Reciprocity Relation

where the spectral width σ_ν is defined by

$$\sigma_\nu^2 = \frac{\int_{-\infty}^{\infty}(\nu-\bar{\nu})^2 |F(\nu)|^2\,d\nu}{\int_{-\infty}^{\infty} |F(\nu)|^2\,d\nu}, \quad \text{where} \quad \bar{\nu} = \frac{\int_{-\infty}^{\infty} \nu|F(\nu)|^2\,d\nu}{\int_{-\infty}^{\infty} |F(\nu)|^2\,d\nu}.$$

Thus the time duration and the spectral width cannot simultaneously be made arbitrarily small. The *Gaussian function* $f(t) = \exp(-t^2/4\sigma_t^2)$, for example, has a power-RMS width σ_t. Its Fourier transform is also a Gaussian function, $F(\nu) = (1/2\sqrt{\pi}\,\sigma_\nu)\exp(-\nu^2/4\sigma_\nu^2)$, with power-RMS width

$$\sigma_\nu = \frac{1}{4\pi\sigma_t}. \tag{A.2-4}$$

Since $\sigma_t\sigma_\nu = 1/4\pi$, the Gaussian function has the minimum permissible value of the duration–bandwidth product. In terms of the angular frequency $\omega = 2\pi\nu$,

$$\sigma_t\sigma_\omega \geq \tfrac{1}{2}. \tag{A.2-5}$$

If the variables t and ω, which usually describe time and angular frequency (rad/s), are replaced with the position variable x and the spatial angular frequency k (rad/m), respectively, then (A.2-5) becomes

$$\sigma_x\sigma_k \geq \tfrac{1}{2}. \tag{A.2-6}$$

In quantum mechanics, the position x of a particle is described by the wavefunction $\psi(x)$, and the wavenumber k is described by a function $\phi(k)$, which is the Fourier transform of $\psi(x)$. The uncertainties of x and k are the RMS widths of the probability densities $|\psi(x)|^2$ and $|\phi(k)|^2$, respectively, so that σ_x and σ_k are interpreted as the uncertainties of position and wavenumber. Since the particle momentum is $p = \hbar k$ (where $\hbar = h/2\pi$ and h is Planck's constant), the position–momentum uncertainty product satisfies the inequality

$$\boxed{\sigma_x\sigma_p \geq \frac{\hbar}{2},} \tag{A.2-7}$$
Heisenberg Uncertainty Relation

which is known as the **Heisenberg position–momentum uncertainty relation**.

The Power-Equivalent Width

The power-equivalent width of a signal $f(t)$ is the signal energy divided by the peak signal power. If $f(t)$ has its peak value at $t = 0$, for example, then the power-equivalent width is

$$\tau = \int_{-\infty}^{\infty} \frac{|f(t)|^2}{|f(0)|^2}\,dt. \tag{A.2-8}$$

The *double-sided exponential function* $f(t) = \exp(-|t|/\tau)$, for example, has a power-equivalent width τ, as does the Gaussian function $f(t) = \exp(-\pi t^2/2\tau^2)$. This definition is used in Sec. 12.1, where the coherence time of light is defined as the power-equivalent width of the complex degree of temporal coherence.

The power-equivalent spectral width is similarly defined by

$$\mathcal{B} = \int_{-\infty}^{\infty} \frac{|F(\nu)|^2}{|F(0)|^2}\,d\nu. \tag{A.2-9}$$

If $f(t)$ is real, so that $|F(\nu)|^2$ is symmetric, and if it has its peak value at $\nu = 0$, the power-equivalent spectral width is usually defined as the positive-frequency width,

$$B = \int_{0}^{\infty} \frac{|F(\nu)|^2}{|F(0)|^2}\,d\nu. \tag{A.2-10}$$

In the case when $F(\nu) = \tau/(1 + j2\pi\nu\tau)$, for example, we have

$$B = \frac{1}{4\tau}. \qquad (A.2\text{-}11)$$

This definition is used in Sec. 19.6A to describe the bandwidth of photodetector circuits susceptible to photon and circuit noise (see also Prob. 19.6-5).

Using Parseval's theorem (A.1-7), together with the relation $F(0) = \int_{-\infty}^{\infty} f(t)\, dt$, (A.2-10) may be written in the form

$$B = \frac{1}{2T}, \qquad (A.2\text{-}12)$$

where

$$T = \frac{\left[\int_{-\infty}^{\infty} f(t)\, dt\right]^2}{\int_{-\infty}^{\infty} f^2(t)\, dt} \qquad (A.2\text{-}13)$$

is yet another definition of the time duration [the square of the area under $f(t)$ divided by the area under $f^2(t)$]. In this case, the duration–bandwidth product $BT = 1/2$.

The 1/e-, Half-Maximum, and 3-dB Widths

Another type of measure of the width of a function is its duration at a prescribed fraction of its maximum value ($1/\sqrt{2}$, $1/2$, $1/e$, or $1/e^2$, are examples). Either the half-width or the full width on both sides of the peak may be used. Two commonly encountered measures are the full-width at half-maximum (FWHM) and the half-width at $1/\sqrt{2}$-maximum, called the 3-dB width. The following are three important examples:

- The *exponential function* $f(t) = \exp(-t/\tau)$ for $t \geq 0$ and $f(t) = 0$ for $t < 0$, which describes the response of a number of electrical and optical systems, has a $1/e$-maximum width $\Delta t_{1/e} = \tau$. The magnitude of its Fourier transform $F(\nu) = \tau/(1 + j2\pi\nu\tau)$ has a 3-dB width (half-width at $1/\sqrt{2}$-maximum)

$$\Delta\nu_{3\text{-dB}} = \frac{1}{2\pi\tau}. \qquad (A.2\text{-}14)$$

- The *double-sided exponential function* $f(t) = \exp(-|t|/\tau)$ has a half-width at $1/e$-maximum $\Delta t_{1/e} = \tau$. Its Fourier transform $F(\nu) = 2\tau/[1 + (2\pi\nu\tau)^2]$, known as the *Lorentzian distribution*, has a full-width at half-maximum

$$\Delta\nu_{\text{FWHM}} = \frac{1}{\pi\tau}, \qquad (A.2\text{-}15)$$

and is usually written in the form $F(\nu) = (\Delta\nu/2\pi)/[\nu^2 + (\Delta\nu/2)^2]$, where $\Delta\nu = \Delta\nu_{\text{FWHM}}$. The Lorentzian distribution describes the spectrum of certain light emissions (see Sec. 14.3D).
- The *Gaussian function* $f(t) = \exp(-t^2/2\tau^2)$ has a full-width at $1/e$-maximum $\Delta t_{1/e} = 2\sqrt{2}\,\tau$. Its Fourier transform $F(\nu) = \sqrt{2\pi}\,\tau \exp(-2\pi^2\tau^2\nu^2)$ has a full-width at $1/e$-maximum

$$\Delta\nu_{1/e} = \frac{\sqrt{2}}{\pi\tau} \qquad (A.2\text{-}16)$$

and a full-width at half-maximum

$$\Delta\nu_{\text{FWHM}} = \frac{\sqrt{2\ln 2}}{\pi\tau}, \quad \text{(A.2-17)}$$

so that

$$\Delta\nu_{\text{FWHM}} = \sqrt{\ln 2}\,\Delta\nu_{1/e} = 0.833\,\Delta\nu_{1/e}\,. \quad \text{(A.2-18)}$$

The Gaussian function is also used to describe the spectrum of certain light emissions (see Sec. 14.3D), as well as to describe the spatial distribution of light beams (see Sec. 3.1).

A.3 TWO-DIMENSIONAL FOURIER TRANSFORM

We now consider a function of two variables $f(x,y)$. If x and y represent the coordinates of a point in a two-dimensional space, then $f(x,y)$ represents a spatial pattern (e.g., the optical field in a given plane). The harmonic function $F\exp[-j2\pi(\nu_x x + \nu_y y)]$ is regarded as a building block from which other functions may be composed by superposition. The variables ν_x and ν_y represent spatial frequencies in the x and y directions, respectively. Since x and y have units of length (mm), ν_x and ν_y have units of cycles/mm, or lines/mm. Examples of two-dimensional harmonic functions are illustrated in Fig. A.3-1.

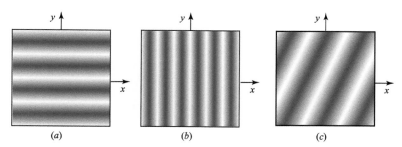

Figure A.3-1 The real part, $|F|\cos[2\pi\nu_x x + 2\pi\nu_y y + \arg\{F\}]$, of a two-dimensional harmonic function: (a) $\nu_x = 0$; (b) $\nu_y = 0$; (c) arbitrary ν_x and ν_y. For this illustration we have assumed that $\arg\{F\} = 0$ so that the white and dark regions represent positive and negative values of the function, respectively.

The Fourier theorem may be generalized to functions of two variables. A function $f(x,y)$ may be decomposed as a superposition integral of harmonic functions of x and y,

$$f(x,y) = \iint_{-\infty}^{\infty} F(\nu_x,\nu_y)\exp\left[-j2\pi(\nu_x x + \nu_y y)\right]d\nu_x\,d\nu_y\,, \quad \text{(A.3-1)}$$

Inverse Fourier Transform

where the coefficients $F(\nu_x, \nu_y)$ are determined by use of the two-dimensional Fourier transform

$$F(\nu_x, \nu_y) = \iint_{-\infty}^{\infty} f(x, y) \exp\left[j 2\pi(\nu_x x + \nu_y y)\right] dx\, dy. \qquad \text{(A.3-2)}$$

Fourier Transform

Our definitions of the two- and one-dimensional Fourier transforms, (A.3-2) and (A.1-2), respectively, differ in the signs of their exponents. The choice of these signs are, of course, arbitrary, as long as opposite signs are used in the Fourier and inverse Fourier transforms. In this book we have adopted the convention that $\exp(j 2\pi \nu t)$ has positive temporal frequency ν, while $\exp[-j 2\pi(\nu_x x + \nu_y y)]$ has positive spatial frequencies ν_x and ν_y. We have elected to use different signs in the spatial (two-dimensional) and temporal (one-dimensional) cases to simplify the notation used in Chapter 4 (Fourier optics), in which the traveling wave $\exp(+j 2\pi \nu t) \exp[-j(k_x x + k_y y + k_z z)]$ has temporal and spatial dependences of opposite sign.

Properties

Many properties of the two-dimensional Fourier transform are clear generalizations of those of the one-dimensional Fourier transform, but others are unique to the two-dimensional case:

- *Convolution Theorem.* The two-dimensional convolution of two functions, $f_1(x, y)$ and $f_2(x, y)$, with Fourier transforms $F_1(\nu_x, \nu_y)$ and $F_2(\nu_x, \nu_y)$, respectively, is written as

$$f(x, y) = \iint_{-\infty}^{\infty} f_1(x', y') f_2(x - x', y - y')\, dx'\, dy'. \qquad \text{(A.3-3)}$$

The Fourier transform of the convolution $f(x, y)$ is

$$F(\nu_x, \nu_y) = F_1(\nu_x, \nu_y) F_2(\nu_x, \nu_y), \qquad \text{(A.3-4)}$$

so that convolution in the spatial domain is equivalent to multiplication in the Fourier domain, as in the one-dimensional case.

- *Separable Functions.* If $f(x, y) = f_x(x) f_y(y)$ is the product of one function of x and another of y, then its two-dimensional Fourier transform is a product of one function of ν_x and another of ν_y. The two-dimensional Fourier transform of $f(x, y)$ is then related to the product of the one-dimensional Fourier transforms of $f_x(x)$ and $f_y(y)$ by $F(\nu_x, \nu_y) = F_x(-\nu_x) F_y(-\nu_y)$. We provide two examples: (1) the Fourier transform of $\delta(x - x_0)\, \delta(y - y_0)$, which represents an impulse located at (x_0, y_0), is the harmonic function $\exp[j 2\pi(\nu_x x_0 + \nu_y y_0)]$; (2) the Fourier transform of the Gaussian function $\exp[-\pi(x^2 + y^2)]$ is the Gaussian function $\exp[-\pi(\nu_x^2 + \nu_y^2)]$.

- *Circularly Symmetric Functions.* The Fourier transform of a circularly symmetric function is also circularly symmetric. Consider, for example, the *circ function*, which is denoted by the symbol $\mathrm{circ}(x, y)$, and is given by

$$f(x, y) = \begin{cases} 1, & \sqrt{x^2 + y^2} \le 1 \\ 0, & \text{otherwise.} \end{cases} \qquad \text{(A.3-5)}$$

Its Fourier transform is given by

$$F(\nu_x, \nu_y) = \frac{J_1(2\pi\nu_\rho)}{\nu_\rho}, \quad \nu_\rho = \sqrt{\nu_x^2 + \nu_y^2}, \qquad (A.3\text{-}6)$$

where J_1 is the Bessel function of order 1. Both of these circularly symmetric functions are illustrated in Fig. A.3-2.

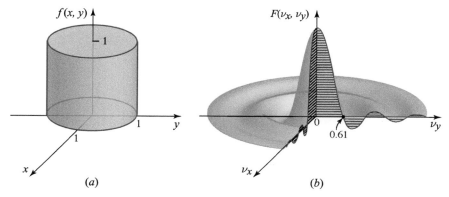

Figure A.3-2 (a) The circ function and (b) its two-dimensional Fourier transform.

READING LIST

L. Chaparro, *Signals and Systems using MATLAB*, Academic Press, 2nd ed. 2015.
G. R. Cooper and C. D. McGillem, *Probabilistic Methods of Signal and System Analysis*, Oxford University Press, 3rd ed. 2007.
B. P. Lathi, *Linear Systems and Signals*, Oxford University Press, 2nd ed. 2004.
S. Haykin and B. Van Veen, *Signals and Systems*, Wiley, 2nd ed. 2003.
A. V. Oppenheim, A. S. Willsky, and S. H. Nawab, *Signals and Systems*, Prentice Hall, 1983, 2nd ed. 1997.
R. N. Bracewell, *The Fourier Transform and Its Applications*, McGraw–Hill, 3rd ed. 2000.
J. D. Gaskill, *Linear Systems, Fourier Transforms, and Optics*, Wiley, 1978.
L. E. Franks, *Signal Theory*, Prentice Hall, 1969, revised ed. 1981.
  1986.

APPENDIX B

LINEAR SYSTEMS

This appendix provides a review of the essential characteristics of one- and two-dimensional linear systems.

B.1 ONE-DIMENSIONAL LINEAR SYSTEMS

Consider a system whose input and output are the functions $f_1(t)$ and $f_2(t)$, respectively. The system is characterized by a rule that relates the output to the input. In general, the rule may take the form of a simple mathematical operation such as $f_2(t) = \log[f_1(t)]$, an integral transform, or a differential equation. An example is a harmonic oscillator that undergoes a displacement $f_2(t)$ in response to a time-varying force $f_1(t)$.

Linear Systems

A system is said to be *linear* if it satisfies the principle of superposition, i.e., if its response to the sum of any two inputs is the sum of its responses to each of the inputs separately. The output at time t is, in general, a weighted superposition of the input contributions at different times τ,

$$f_2(t) = \int_{-\infty}^{\infty} h(t;\tau) f_1(\tau) \, d\tau, \qquad (\text{B.1-1})$$

where $h(t;\tau)$ is a weighting function representing the contribution of the input at time τ to the output at time t. If the input is an impulse at time τ, so that $f_1(t) = \delta(t - \tau)$, then (B.1-1) yields $f_2(t) = h(t;\tau)$. Thus $h(t;\tau)$ is the **impulse response function** of the system (also known as the **Green's Function**).

Linear Shift-Invariant Systems

A linear system is said to be **time-invariant** or **shift-invariant** if, when its input is shifted in time, its output shifts by an equal time, but otherwise remains the same. The impulse response function is then a function of the time difference $h(t;\tau) = h(t - \tau)$. Under these conditions (B.1-1) becomes

$$f_2(t) = \int_{-\infty}^{\infty} h(t-\tau) f_1(\tau) \, d\tau. \qquad (\text{B.1-2})$$

The output $f_2(t)$ is then the convolution of the input $f_1(t)$ with the impulse response function $h(t)$ [see (A.1-4)]. If $f_1(t) = \delta(t)$, then $f_2(t) = h(t)$; if $f_1(t) = \delta(t-\tau)$, then $f_2(t) = h(t-\tau)$, as illustrated in Fig. B.1-1.

Fundamentals of Photonics, Third Edition. Bahaa E. A. Saleh and Malvin Carl Teich.
©2019 John Wiley & Sons, Inc. Published 2019 by John Wiley & Sons, Inc.

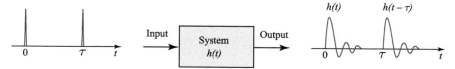

Figure B.1-1 Response of a linear shift-invariant system to impulses.

The Transfer Function

In accordance with the convolution theorem discussed in Appendix A [see (A.1-3)], the Fourier transforms $F_1(\nu)$, $F_2(\nu)$, and $H(\nu)$ of $f_1(t)$, $f_2(t)$, and $h(t)$, respectively, are related by

$$F_2(\nu) = H(\nu)F_1(\nu). \qquad \text{(B.1-3)}$$

If the input $f_1(t)$ is a harmonic function $F_1(\nu)\exp(j2\pi\nu t)$, the output $f_2(t) = H(\nu)F_1(\nu)\exp(j2\pi\nu t)$ is also a harmonic function of the same frequency but with a modified complex amplitude $F_2(\nu) = F_1(\nu)H(\nu)$, as illustrated in Fig. B.1-2. The multiplicative factor $H(\nu)$ is known as the system's **transfer function**; it is the Fourier transform of the impulse response function. Equation (B.1-3) embodies the essence of the usefulness of Fourier methods in the analysis of linear shift-invariant systems. To determine the output of a system for an arbitrary input, we simply decompose the input into its harmonic components, multiply the complex amplitude of each harmonic function by the transfer function at the appropriate frequency, and superpose the resultant harmonic functions.

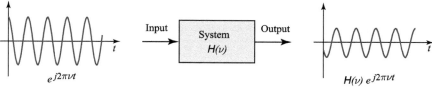

Figure B.1-2 Response of a linear shift-invariant system to a harmonic function.

Examples

- *Ideal system:* $H(\nu) = 1$ and $h(t) = \delta(t)$; the output is a replica of the input.
- *Ideal system with delay:* $H(\nu) = \exp(-j2\pi\nu\tau)$ and $h(t) = \delta(t - \tau)$; the output is a replica of the input delayed by time τ.
- *System with exponential response:* $H(\nu) = \tau/(1 + j2\pi\nu\tau)$ and $h(t) = e^{-t/\tau}$ for $t \geq 0$, and $h(t) = 0$ otherwise; this represents the response of a system described by a first-order linear differential equation, e.g., that representing an RC circuit with time constant τ. An impulse at the input results in an exponentially decaying response.
- *Chirped system:* $H(\nu) = \exp(-j\pi\nu^2)$ and $h(t) = e^{-j\pi/4}\exp(j\pi t^2)$; the system distorts the input by imparting to it a phase shift proportional to ν^2. An impulse at the input generates an output in the form of a chirped signal, i.e., a harmonic function whose instantaneous frequency (the derivative of the phase) increases linearly with time. This system describes the propagation of optical pulses through media with a frequency-dependent phase velocity; it also describes changes in the spatial distribution of light waves as they propagate through free space (see Secs. 23.3A and 4.1C, respectively).

Linear Shift-Invariant Causal Systems

The impulse response function $h(t)$ of a linear shift-invariant *causal* system must vanish for $t < 0$ since the system's response cannot begin before the application of the input. The function $h(t)$ is therefore not symmetric and its Fourier transform, the transfer function $H(\nu)$, must be complex. It can be shown[†] that if $h(t) = 0$ for $t < 0$, then the real and imaginary parts of $H(\nu)$, denoted $H'(\nu)$ and $H''(\nu)$, respectively, are related by

$$H'(\nu) = \frac{1}{\pi} \int_{-\infty}^{\infty} \frac{H''(s)}{s - \nu} ds \qquad \text{(B.1-4)}$$

$$H''(\nu) = \frac{1}{\pi} \int_{-\infty}^{\infty} \frac{H'(s)}{\nu - s} ds, \qquad \text{(B.1-5)}$$
Hilbert Transform

where the Cauchy principal values of the integrals are to be evaluated, i.e.,

$$\int_{-\infty}^{\infty} \equiv \lim_{\Delta \to 0} \left(\int_{-\infty}^{\nu-\Delta} + \int_{\nu+\Delta}^{\infty} \right), \quad \Delta > 0.$$

Functions that satisfy (B.1-4) and (B.1-5) are said to form a **Hilbert transform pair**, $H''(\nu)$ being the **Hilbert transform** of $H'(\nu)$.

If the impulse response function $h(t)$ is also real, its Fourier transform must be symmetric, $H(-\nu) = H^*(\nu)$ (see Appendix A, Sec. A.1). As a result, the real part $H'(\nu)$ then has even symmetry, and the imaginary part $H''(\nu)$ has odd symmetry. The integrals in (B.1-4) and (B.1-5) may then be rewritten as integrals over the interval $(0, \infty)$, and the resultant equations are known as the **Kramers–Kronig relations**:

$$H'(\nu) = \frac{2}{\pi} \int_{0}^{\infty} \frac{s H''(s)}{s^2 - \nu^2} ds \qquad \text{(B.1-6)}$$

$$H''(\nu) = \frac{2}{\pi} \int_{0}^{\infty} \frac{\nu H'(s)}{\nu^2 - s^2} ds. \qquad \text{(B.1-7)}$$
Kramers–Kronig Relations

In summary, the Hilbert-transform relations, or the Kramers–Kronig relations, relate the real and imaginary parts of the transfer function of a linear shift-invariant causal system, so that if one part is known at all frequencies, the other part may be determined.

Example: The Harmonic Oscillator

The linear system described by the differential equation

$$\left(\frac{d^2}{dt^2} + \zeta \frac{d}{dt} + \omega_0^2 \right) f_2(t) = f_1(t) \qquad \text{(B.1-8)}$$

describes a harmonic oscillator with displacement $f_2(t)$ under an applied force $f_1(t)$, where ω_0 is the resonance angular frequency and ζ is a coefficient representing damping effects. The transfer function $H(\nu)$ of this system may be obtained by substituting

---

[†] See, e.g., L. E. Franks, *Signal Theory*, Prentice Hall, 1969, revised ed. 1981.

$f_1(t) = \exp(j2\pi\nu t)$ and $f_2(t) = H(\nu)\exp(j2\pi\nu t)$ in (B.1-8), which yields

$$H(\nu) = \frac{1}{(2\pi)^2} \frac{1}{\nu_0^2 - \nu^2 + j\nu\Delta\nu}, \tag{B.1-9}$$

where $\nu_0 = \omega_0/2\pi$ is the resonance frequency, and $\Delta\nu = \zeta/2\pi$. The real and imaginary parts of $H(\nu)$ are therefore, respectively,

$$H'(\nu) = \frac{1}{(2\pi)^2} \frac{\nu_0^2 - \nu^2}{(\nu_0^2 - \nu^2)^2 + (\nu\Delta\nu)^2} \tag{B.1-10}$$

$$H''(\nu) = -\frac{1}{(2\pi)^2} \frac{\nu\Delta\nu}{(\nu_0^2 - \nu^2)^2 + (\nu\Delta\nu)^2}. \tag{B.1-11}$$

Since the system is causal, $H'(\nu)$ and $H''(\nu)$ satisfy the Kramers–Kronig relations. When $\nu_0 \gg \Delta\nu$, $H'(\nu)$ and $H''(\nu)$ are narrow functions centered about ν_0. For $\nu \approx \nu_0$, $(\nu_0^2 - \nu^2) \approx 2\nu_0(\nu_0 - \nu)$, whereupon (B.1-10) and (B.1-11) may be approximated by

$$H''(\nu) = -\frac{1}{(2\pi)^2} \frac{\Delta\nu/4\nu_0}{(\nu_0 - \nu)^2 + (\Delta\nu/2)^2} \tag{B.1-12}$$

$$H'(\nu) = 2\frac{\nu - \nu_0}{\Delta\nu} H''(\nu). \tag{B.1-13}$$

Equation (B.1-12) has a Lorentzian form. The transfer function of the harmonic-oscillator system is used in Secs. 5.5 and 15.1 to describe dielectric and atomic systems.

B.2 TWO-DIMENSIONAL LINEAR SYSTEMS

A two-dimensional system relates a pair of two-dimensional functions, $f_1(x,y)$ and $f_2(x,y)$, called the input and output functions. These functions may, for example, represent optical fields at two parallel planes, with (x,y) representing position variables; the system comprises the free space and optical components that lie between the two planes.

The concepts of linearity and shift invariance defined in the one-dimensional case are easily generalized to the two-dimensional case. The output $f_2(x,y)$ of a *linear* system is related to its input $f_1(x,y)$ by a superposition integral

$$f_2(x,y) = \iint_{-\infty}^{\infty} h(x,y;x',y') f_1(x',y') \, dx' \, dy', \tag{B.2-1}$$

where $h(x,y;x',y')$ is a weighting function that represents the effect of the input at the point (x',y') on the output at the point (x,y). The function $h(x,y;x',y')$ is the **impulse response function** of the system (also known as the **point-spread function**).

The system is said to be **shift-invariant** (or **isoplanatic**) if shifting its input in some direction shifts the output by the same distance and in the same direction without otherwise altering it (see Fig. B.2-1). The impulse response function then depends on differences of position, $h(x,y;x',y') = h(x-x', y-y')$, whereupon (B.2-1) becomes the two-dimensional convolution of $h(x,y)$ with $f_1(x,y)$:

$$f_2(x,y) = \iint_{-\infty}^{\infty} f_1(x',y')\, h(x-x', y-y')\, dx'\, dy'. \qquad \text{(B.2-2)}$$

Applying the two-dimensional convolution result provided in (A.3-4) of Appendix A yields

$$F_2(\nu_x, \nu_y) = H(\nu_x, \nu_y)\, F_1(\nu_x, \nu_y), \qquad \text{(B.2-3)}$$

where $F_2(\nu_x, \nu_y)$, $H(\nu_x, \nu_y)$, and $F_1(\nu_x, \nu_y)$ are the Fourier transforms of $f_2(x,y)$, $h(x,y)$, and $f_1(x,y)$, respectively.

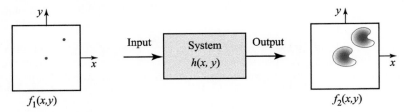

Figure B.2-1 Response of a two-dimensional linear shift-invariant system.

A harmonic input of complex amplitude $F_1(\nu_x, \nu_y)$ therefore produces a harmonic output of the same spatial frequency but with complex amplitude $F_2(\nu_x, \nu_y) = H(\nu_x, \nu_y)\, F_1(\nu_x, \nu_y)$, as illustrated in Fig. B.2-2. The multiplicative factor $H(\nu_x, \nu_y)$ is the system **transfer function**, which is the Fourier transform of its impulse response function. Either of these functions allows us to characterize the system completely and enables us to determine the output for an arbitrary input.

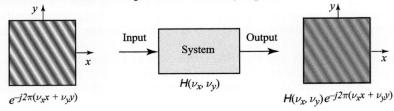

Figure B.2-2 Response of a two-dimensional linear shift-invariant system to harmonic functions.

In summary, a two-dimensional linear shift-invariant system is characterized by its impulse response function $h(x,y)$ or its transfer function $H(\nu_x, \nu_y)$. For example, a system with $h(x,y) = \mathrm{circ}(x/\rho_s, y/\rho_s)$ smears each point of the input into a patch in the form of a circle of radius ρ_s. It has a transfer function $H(\nu_x, \nu_y) = \rho_s J_1(2\pi\rho_s \nu_\rho)/\nu_\rho$, where $\nu_\rho = \sqrt{\nu_x^2 + \nu_y^2}$, which has the shape illustrated in Fig. A.3-2(b). The system severely attenuates spatial frequencies higher than $0.61/\rho_s$ lines/mm.

READING LIST

See the reading list in Appendix A.

APPENDIX C

MODES OF LINEAR SYSTEMS

This Appendix provides a brief overview of modes of linear systems that are described explicitly by input–output relations that take the form of a matrix or integral operation, or implicitly by a linear ordinary or linear partial differential equation.

Consider first a linear system described by an explicit input–output relation characterized by a linear operator $\mathcal{L}$ that operates on an input vector $\mathbf{X}$ to generate a corresponding output vector $\mathbf{Y}$:

$$\mathbf{Y} = \mathcal{L}\mathbf{X}. \tag{C.1-1}$$

The vector $\mathbf{X}$ may be an array of complex numbers, represented by a column matrix, or a complex function of one or more variables. The **modes** of such a system are those special inputs that remain unaltered (except for a multiplicative constant) upon passage through the system. They thus obey

$$\mathcal{L}\mathbf{X}_q = \lambda_q \mathbf{X}_q, \tag{C.1-2}$$
Eigenvalue Problem

where q is an index that labels the mode. The vector $\mathbf{X}_q$ is known as the **eigenvector**, and the associated multiplicative constant λ_q, which is generally a complex number, is called the **eigenvalue**. The condition set forth in (C.1-2) is known as an **eigenvalue problem**.

Consider next a linear dynamical system whose state is described by N continuous variables constituting a vector $\mathbf{X}(t)$. The evolution of *any* of the N variables of this N-dimensional vector is, in general, dependent on all N variables. However, the same system may be described in a new coordinate system whereupon the N new variables evolve independently, so that the description of the system decomposes into N independent one-dimensional systems. These decoupled variables are the modes of the system.

Consider finally a linear system characterized implicitly by a linear partial differential equation that may be cast in the form of (C.1-2), where $\mathcal{L}$ is a differential operator and $\mathbf{X}$ is a complex function of one or several variables. In this case, the modes are simply solutions of the differential equation and the eigenvectors are called **eigenfunctions**. The notion of an input and an output is not meaningful in these circumstances.

We proceed to describe a number of applications of modal analysis in photonics. Before commencing, however, we briefly review a number of geometrical concepts from linear algebra. Associated with each pair of vectors $\mathbf{X}$ and $\mathbf{Y}$ is a complex scalar quantity $(\mathbf{X}, \mathbf{Y})$ called the **inner product**. The square root of the inner product of a vector $\mathbf{X}$ with itself, $(\mathbf{X}, \mathbf{X})$, is known as the **norm** of $\mathbf{X}$ and is a measure of its "length." The inner product of two vectors of unit norm can be thought of as the cosine of the "angle" between them. Two vectors are said to be **orthogonal** if their inner product is zero. If the vectors comprise arrays of complex numbers, $\{X_i\}$ and $\{Y_i\}$,

$i = 1, 2, \ldots, N$, then $(\mathbf{X}, \mathbf{Y}) = \sum_{i=1}^{N} X_i^* Y_i$. If, on the other hand, the vectors are complex functions $X(t)$ and $Y(t)$, then $(\mathbf{X}, \mathbf{Y}) = \int_{-\infty}^{\infty} X^*(t) Y(t) dt$.

Two classes of operators $\mathcal{L}$ that lead to solutions of the eigenvalue problem with special properties are considered in turn:

Hermitian operators. Hermitian operators are defined by the property $(\mathbf{X}, \mathcal{L}\mathbf{Y}) = (\mathcal{L}\mathbf{X}, \mathbf{Y})$, i.e., the inner product is the same no matter to which of the two vectors the operator is applied. The eigenvalues of a Hermitian operator are real and the eigenvectors are orthogonal. Further, the eigenvectors of a Hermitian operator obey the **variational principle**, which is based on a scalar $E_{\text{var}} = \frac{1}{2}(\mathbf{X}, \mathcal{L}\mathbf{X})/(\mathbf{X}, \mathbf{X})$, called the variational energy. This principle states that the eigenvector $\mathbf{X}_1$ with the lowest eigenvalue minimizes E_{var}; the eigenvector $\mathbf{X}_2$ with the next lowest eigenvalue minimizes E_{var}, subject to the condition that it is orthogonal to $\mathbf{X}_1$, and so on.

Unitary operators. Passive, lossless physical systems are described by unitary operators, which are defined by the norm-preserving property $(\mathcal{L}\mathbf{X}, \mathcal{L}\mathbf{X}) = (\mathbf{X}, \mathbf{X})$. An example is the "rotation" operation. The eigenvalues of unitary operators are unimodular, i.e., $|\lambda_q| = 1$, and therefore represent pure phase.

1) Modes of a Discrete Linear System

A discrete linear system is described by a matrix relation $\mathbf{Y} = \mathbf{M}\mathbf{X}$, where the input vector $\mathbf{X}$ is a set of N complex numbers $(X_1, X_2, \ldots, X_N)$ arranged in a column matrix, $\mathbf{M}$ is an $N \times N$ matrix that represents the linear system, and the output vector $\mathbf{Y}$ is also a column matrix of dimension N. The modes are those input vectors that remain parallel to themselves upon transmission through the system, so that the matrix equation

$$\mathbf{M}\mathbf{X}_q = \lambda_q \mathbf{X}_q \tag{C.1-3}$$

is obeyed. Thus, the modes of the system are the eigenvectors $\mathbf{X}_q$ of the matrix $\mathbf{M}$, and the scalars λ_q are the corresponding eigenvalues, which are determined by solving the algebraic equation $\det(\mathbf{M} - \lambda \mathbf{I}) = 0$, where $\mathbf{I}$ is the identity matrix. There are N such modes, labeled by the index $q = 1, 2, \ldots, N$.

The special case of binary systems ($N = 2$) is particularly important in optics. In a binary system, each vector is a pair of complex numbers (X_1, X_2) arranged in a column matrix $\mathbf{X}$. The system is characterized by a 2×2 square matrix $\mathbf{M}$ whose elements are denoted A, B, C, and D. The relation $\mathbf{Y} = \mathbf{M}\mathbf{X}$ signifies

$$\begin{bmatrix} Y_1 \\ Y_2 \end{bmatrix} = \begin{bmatrix} A & B \\ C & D \end{bmatrix} \begin{bmatrix} X_1 \\ X_2 \end{bmatrix}.$$

The eigenvalues are determined by solving the algebraic equation $(A - \lambda)(D - \lambda) - BC = 0$ for the two eigenvalues λ_1 and λ_2.

The following are examples of optical systems described by binary linear systems:

Application: Polarization matrix optics. In polarization matrix optics (Sec. 6.1B), the vector (X_1, X_2) represents the components of the input electric field in two orthogonal directions (the Jones vector), and (Y_1, Y_2) similarly represents the output electric field. The matrix $\mathbf{M}$ is the Jones matrix of the system. In this case, the modes are the polarization states that are maintained as light is transmitted through the system.

Application: Ray matrix optics. In geometrical paraxial optics (Sec. 1.4), the position and angle of an optical ray are described by a vector (X_1, X_2), and the effect of optical components, such as lenses and mirrors, is described by a matrix $\mathbf{M}$, called the ray-transfer matrix or the *ABCD* matrix. For a closed optical system, such as a resonator, the modes are ray positions and angles that self-reproduce after a round trip, so that they are confined within the resonator.

Application: Multilayer matrix optics. In multilayer matrix optics (Sec. 7.1A) light is reflected and refracted at each boundary, so that there are forward- and backward-traveling waves at each plane, with amplitudes described by a vector $\mathbf{X} = (X_1, X_2)$. A system containing a set of boundaries between an input and an output plane is described by a wave-transfer matrix $\mathbf{M}$. The modes of such a system are the vectors that self reproduce upon transmission through the system, so that if the system is replicated periodically, as in a 1D photonic crystal (Sec. 7.2), the propagation modes are the modes of the system $\mathbf{M}$.

2) Modes of a Continuous System Described by an Integral Operator

Linear systems represented by integral operators are discussed in Appendix B. Consider, for example, a function of time $f(t)$, such as an optical pulse or a broadband optical field, transmitted through a linear time-invariant system such as an optical filter. The system is described by the convolution operation (A.1-4):

$$g(t) = \int_{-\infty}^{\infty} h(t-\tau) f(\tau)\, d\tau. \tag{C.1-4}$$

In this system, the vectors $\mathbf{X}$ and $\mathbf{Y}$ are the functions $f(t)$ and $g(t)$, respectively, and the operator $\mathcal{L}$ is an integral operator. The modes of this system are the harmonic functions $\exp(j2\pi\nu t)$. This is evident since the input function $\exp(j2\pi\nu t)$ generates another harmonic output function $H(\nu)\exp(j2\pi\nu t)$, where $H(\nu)$ is the Fourier transform of $h(t)$. In this case, there is a continuum of modes with continuous eigenvalues $H(\nu)$. Here, the index q is the frequency ν, which takes continuous values.

Another example is a linear shift-invariant system that operates on a two-dimensional (2D) function $f(x, y)$ of the position (x, y), as described in (B.2-2):

$$g(x,y) = \iint_{-\infty}^{\infty} h(x-x', y-y') f(x', y')\, dx'\, dy'. \tag{C.1-5}$$

The eigenfunctions are 2D harmonic functions $\exp[j2\pi(\nu_x x + \nu_y y)]$, and the eigenvalues are $H(\nu_x, \nu_y)$, the 2D Fourier transform of $h(x, y)$. Again, there is a continuum of eigenfunctions, labeled by the spatial frequencies (ν_x, ν_y).

Translational symmetry and harmonic modes. It is not surprising that harmonic functions are the modes of a shift-invariant system. Because the harmonic function is invariant to time shift, i.e., it remains a harmonic function if translated in time, it is the eigenfunction of the time-invariant (stationary) linear system. Similarly, because 2D harmonic functions are invariant to translation in the plane, they are the eigenfunctions of the space-invariant (homogeneous) linear system.

If the linear system is not space-invariant, i.e., does not enjoy translational symmetry, then in the 2D case it is represented by the more general linear operation described in (B.2-1):

$$g(x,y) = \iint_{-\infty}^{\infty} h(x,y; x', y') f(x', y')\, dx'\, dy'. \tag{C.1-6}$$

The eigenfunctions, which are now not necessarily harmonic functions, are determined by solving the eigenvalue problem posed in (C.1-2), which in this case takes the form of an integral equation

$$\iint_{-\infty}^{\infty} h(x,y; x', y')\, f_q(x', y')\, dx'\, dy' = \lambda_q\, f_q(x, y), \quad q = 1, 2, \ldots . \tag{C.1-7}$$

The functions $f_q(x, y)$ and the constants λ_q are the eigenfunctions and eigenvalues of the system, respectively, and the index q labels a discrete set of modes.

Application: Optical resonator modes. An example is provided by light traveling between the two parallel mirrors of a laser resonator (Sec. 11.2E). The distributions of the optical field in the transverse plane at the beginning and at the end of a single round trip are the input and output of the system. The modes of the resonator are those field distributions that maintain their form after one round trip. The kernel $h(x, y; x', y')$ in (C.1-7) represents propagation in free space and reflection from one of the mirrors, followed by backward free-space propagation and reflection from the other mirror. Clearly, the presence of curved mirrors, or mirrors of finite extent, makes this system shift-variant. If the mirrors are spherical and are assumed to modulate the incoming light by a phase factor that is a quadratic function of the radial distance, then the resonator modes are Hermite–Gaussian functions of x and y (see Sec. 3.3). In the presence of apertures, (C.1-7) can only be solved numerically, as considered in Sec. 11.2E.

3) Modes of a System Described by an Ordinary Differential Equation

The dynamics of certain physical systems are characterized by a set of coupled ordinary differential equations. For example, the dynamics of N coupled oscillators are described by N differential equations that are conveniently written in matrix form as

$$\ddot{\mathbf{X}} = -\mathbf{MX}, \tag{C.1-8}$$

where $\mathbf{X}$ is a column matrix with components $(X_1, X_2, \ldots, X_N)$, $\ddot{\mathbf{X}} \equiv d^2\mathbf{X}/dt^2$, and $\mathbf{M}$ is an $N \times N$ matrix with time-independent coefficients, so that the system is time invariant.

Time invariance requires that the modes be harmonic functions that take the form $\exp(j\omega t)$, i.e., the vector $\mathbf{X}(t) = \mathbf{X}(0)\exp(j\omega t)$. Hence, substitution in (C.1-8) yields

$$\mathbf{MX} = \omega^2 \mathbf{X}. \tag{C.1-9}$$

This equation represents a discrete-system eigenvalue problem. Its eigenvalues provide the resonance frequencies $\omega_1, \omega_2, \ldots, \omega_N$ of the modes, and its eigenvectors are called the **normal modes**. All components of the eigenvector $\mathbf{X}_q$ of mode q oscillate at the same resonance frequency ω_q, without alteration of their relative amplitudes or phases. In this sense, the modes are stationary solutions that are decoupled from one another.

4) Modes of a System Described by a Partial Differential Equation

Fields and waves are described by partial differential equations such as Maxwell's equations, which characterize the dynamics of the electric and magnetic fields in a dielectric medium. Similarly, the Schrödinger equation expresses the dynamics of the wavefunction of a particle subject to a specified potential. If these physical systems are stationary, i.e., if the dielectric medium and the potential are time independent, then each mode must be a harmonic function of time that takes the form $\exp(j\omega t)$ with some frequency ω. The wave equation is then converted into the generalized Helmholtz equation

$$\nabla \times [\eta(\mathbf{r})\, \nabla \times \mathbf{H}] = \frac{\omega^2}{c_o^2}\mathbf{H}, \tag{C.1-10}$$

where $\eta(\mathbf{r}) = \epsilon_o/\epsilon(\mathbf{r})$ is the electric impermeability of the dielectric medium [see (7.0-2)]. Analogously, the Schrödinger equation (14.1-1) yields the time-independent Schrödinger equation (14.1-3),

$$\left[-\frac{\hbar^2}{2m}\nabla^2 + V(\mathbf{r})\right]\psi(\mathbf{r}) = E\psi(\mathbf{r}), \tag{C.1-11}$$

where $V(\mathbf{r})$ is the potential and $E = \hbar\omega$.

Both of these equations take the form of an eigenvalue problem (C.1-2), where $\mathcal{L}$ is a Hermitian differential operator characterized by the function $\eta(\mathbf{r})$ or $V(\mathbf{r})$. The eigenvalues, which are real, provide the frequencies ω_q of the modes (and hence the corresponding energies E_q in the case of the Schrödinger equation). The eigenfunctions are the spatial distributions of the electromagnetic field (or the wavefunction) for each mode. Note that the field (or the wavefunction) of the qth mode evolves with time as $\exp(j\omega_q t)$ at all positions, so that each mode is stationary, as required.

Modes of fields/waves in a homogeneous medium with boundary conditions. If the dielectric medium is homogeneous, i.e., the impermeability $\eta(\mathbf{r})$ is constant, then the system is shift-invariant. To be consistent with this translational symmetry, the modes of the electromagnetic system must be harmonic functions of position, i.e., plane waves. Similarly, if the potential $V(\mathbf{r})$ is constant, then the modes are plane-wave wavefunctions, so that the particle is equally likely to be found anywhere.

In other situations, $\eta(\mathbf{r})$ and $V(\mathbf{r})$ are constant within a finite region bounded by a surface that imposes certain boundary conditions. For example, the electromagnetic modes of a cavity resonator with perfectly conducting surfaces can be determined by requiring that the parallel components of the electric field vanish at the surface. For a rectangular resonator, the modes are harmonic functions of position — standing waves oscillating in unison (see Sec. 11.3C). Similarly, the modes of a particle in a quantum box (dot) are obtained by requiring that the wavefunction vanishes at the boundaries (see Sec. 17.1G).

In yet another geometry, a homogeneous dielectric medium may be bounded in one direction, e.g., by two parallel planar mirrors. Here, the boundary conditions correspond to a discrete set of standing waves in the direction orthogonal to the mirrors (transverse direction), with traveling waves in the parallel (axial) direction, so that the modes travel in this optical waveguide as harmonic functions in the axial direction, without altering their transverse distributions (see Sec. 9.1). If β_q is the propagation constant of mode q, then the eigenvalue is the phase factor $\exp(-j\beta_q z)$.

Modes of fields/waves in a periodic medium. As is evident from the previous examples, the modes of a system described by a partial differential equation are dictated by the spatial distribution of the medium, e.g., the function $\eta(\mathbf{r})$ or $V(\mathbf{r})$. If this function is constant, the modes must be invariant to arbitrary translation. If it is periodic, then the modes must be invariant to translation by a period. This type of translational symmetry requires that the modes be Bloch waves (see Sec. 7.2A). For example, if the medium is homogeneous in the x and y directions but periodic in the z direction, a Bloch mode takes the form of a harmonic function $\exp(-jKz)$, modulated by a periodic standing wave $p_K(z)$ with period equal to that of the medium; the dependence on x and y is, of course, harmonic. For a given value of K, the frequencies of the modes and the shapes of the corresponding standing waves $p_K(z)$ depend on the shape of the periodic function $\eta(\mathbf{r})$ or $V(\mathbf{r})$. This type of translational symmetry results in a spectrum of eigenvalues (and hence frequencies ω or energies $E = \hbar\omega$) in the form of bands that are separated by bandgaps within which no modes are allowed. Thus, an electron in a periodic potential distribution exhibits the well-known band structure of solids (see Secs. 14.1D and 17.1A). Likewise, an optical field in a periodic dielectric medium, i.e., a photonic crystal, exhibits a band structure with photonic bandgaps (see Secs. 7.2 and 7.3).

READING LIST

D. C. Lay, S. R. Lay, and J. J. McDonald, *Linear Algebra and its Applications*, Pearson, 5th ed. 2015.

S. Axler, *Linear Algebra Done Right*, Springer-Verlag, 3rd ed. 2015.

S. J. Leon, *Linear Algebra with Applications*, Pearson, 9th ed. 2014.

G. Strang, *Introduction to Linear Algebra*, Wellesley Cambridge Press, 4th ed. 2009.

SYMBOLS AND UNITS

Roman Symbols and Acronyms

- a = Radius of an aperture or fiber [m]; also, Radius of a spherical scattering particle [m]; also, Radius of a circle [m]; also, Distance between locations [m]; also, Lattice constant [m]; also, Length of a thin metallic rod [m]; also, Chirp parameter for an optical pulse
- a_0 = Bohr radius (radius of ground state of Bohr hydrogen atom; $a_0 \approx 0.53$ Å) [m]
- a = Complex amplitude or magnitude of an optical wave; also, Normalized complex amplitude of an optical field ($|a|^2$ = photon-flux density)
- a = Normalized field amplitude in a cavity ($|\mathsf{a}|^2$ = field energy in units of photon number)
- a = Acceleration of a carrier [m·s^{-2}]
- $\mathbf{a}$ = Primitive vector defining a lattice unit cell [m]
- $\mathbf{a}$ = Complex-amplitude vector
- A = Complex envelope of a monochromatic plane wave; also, Pulse amplitude
- $A(\mathbf{r})$ = Complex envelope of a monochromatic wave
- $A(\nu)$ = Fourier transform of the complex envelope of an optical pulse
- $\mathbf{A}$ = Complex vector envelope of a monochromatic plane wave; also, Vector potential [V·s·m^{-1}]
- $\mathcal{A}$ = Absorbance
- $\mathcal{A}(\mathbf{r}, t)$ = Complex envelope of a polychromatic (e.g., pulsed) wave
- $\mathcal{A}(t)$ = Complex envelope of an optical pulse
- A = Area [m^2]; also, Element of the $ABCD$ ray-transfer and wave-transfer matrices $\mathbf{M}$
- A_c = Coherence area [m^2]
- A_{ij} = Element of Jacobian transformation matrix
- A_r = Relative atomic mass
- $\mathrm{Ai}(\cdot)$ = Airy function
- $\mathbf{A}$ = Jacobian transformation matrix
- A = Einstein A coefficient [s^{-1}]
- AC = Alternating current
- ACS = American Chemical Society
- ADC = Analog-to-digital converter
- ADM = Add–drop multiplexer
- ADP = Ammonium dihydrogen phosphate
- AGIL = All gas-phase iodine laser
- AM = Amplitude modulation
- AMLCD = Active-matrix liquid-crystal display
- AMOLED = Active-matrix organic light-emitting display
- AND = AND logic gate

Fundamentals of Photonics, Third Edition. Bahaa E. A. Saleh and Malvin Carl Teich.
©2019 John Wiley & Sons, Inc. Published 2019 by John Wiley & Sons, Inc.

AOC = Active optical cables
AOM = Acousto-optic modulator
APD = Avalanche photodiode
APS = American Physical Society
ASE = Amplified spontaneous emission
ASK = Amplitude shift keying
AWG = Arrayed waveguides

b = Radius of a circle [m]; also, Chirp coefficient [s^2]
b = Bowing parameter
B = Magnetic flux-density complex amplitude [Wb · m^{-2} or T]; also, Bandwidth [Hz]; also, Bandwidth of an electrical circuit [Hz]; also, Spatial bandwidth [m^{-1}]; also, Spectral width supporting net gain in a laser medium [Hz]
B = Magnetic flux-density complex amplitude vector [Wb · m^{-2} or T]
B_0 = Bit rate [b · s^{-1}]
$\mathcal{B}$ = Power-equivalent spectral width [Hz]
$\mathcal{B}$ = Magnetic flux density vector [Wb · m^{-2} or T]
B = Element of the $ABCD$ ray-transfer and wave-transfer matrices **M**
$\mathbb{B}$ = Einstein B coefficient [m^3 · J^{-1} · s^{-2}]
BBO = Beta barium borate
BER = Bit error rate
BGR = Bragg grating reflector
BPF = Bandpass filter
BPP = Bulk plasmon polariton
BPSK = Binary phase shift keying
BRF = Birefringent filter
BSO = Bismuth silicon oxide

c = Speed of light [m · s^{-1}]; also, Phase velocity [m · s^{-1}]
c_o = Speed of light in free space [m · s^{-1}]
C = Electrical capacitance [F]
$C(\cdot)$ = Fresnel integral
$\mathcal{C}$ = Coupling coefficient in a directional coupler [m^{-1}]
C = Element of the $ABCD$ ray-transfer and wave-transfer matrices **M**
C-band = Conventional optical fiber telecommunications band (1530–1565 nm)
CAD = Computer-aided design
CAPD = Conventional avalanche photodiode
CARS = Coherent anti-Stokes Raman scattering
CATV = Cable television
CCD = Charge-coupled device
CCT = Correlated color temperature
CCW = Counterclockwise
CD = Compact-disc
CDM = Code-division multiplexing
CDMA = Code-division multiple access
CFL = Compact fluorescent lamp
CLSM = Confocal laser-scanning microscopy
CMOS = Complementary metal-oxide-semiconductor
COB = Chip-on-board light-emitting diode
COIL = Chemical oxygen–iodine laser

CPA = Chirped-pulse amplification
CRI = Color rendering index
CVD = Chemical vapor deposition
CW = Continuous-wave; also, Clockwise
CWDM = Coarse wavelength-division multiplexing

$d\mathbf{r}$ = Incremental volume [m^3]
ds = Incremental length [m]
d = Coefficient of second-order optical nonlinearity [C · V^{-2}]
d$_{\text{eff}}$ = Effective coefficient of second-order optical nonlinearity [C · V^{-2}]
d$_{ijk}$ = Component of second-order optical nonlinearity tensor [C · V^{-2}]
d$_{iJ}$ = Component of second-order optical nonlinearity tensor (contracted indices) [C · V^{-2}]
d($\omega_3; \omega_1, \omega_2$) = Coefficient of second-order optical nonlinearity (dispersive medium) [C · V^{-2}]
d = Distance, Length, Thickness [m]
d_{b} = Propagation length of a plasmon wave along its boundary [m]
d_{d} = Thickness of dielectric layer in a layered metamaterial [m]
d_{ex} = Mean distance traveled by a photon in a random laser before exiting [m]
d_{m} = Thickness of metallic layer in a layered metamaterial [m]
$d_{\min}$ = Minimum distance [m]
d_p = Penetration depth [m]
d_{p} = Distance of coupling prism from a waveguide [m]
d_{pulse} = Length of a mode-locked optical pulse [m]
d_{S} = Length along a small dimension [m]
d_{st} = Mean distance traveled by a photon in a random laser before stimulating a clone photon [m]
D = Diameter [m]; also, Electric flux-density complex amplitude [C · m^{-2}]; also, Width of an optical beam [m]
D_s = Width of an acoustic beam [m]
D^* = Specific detectivity of a photodetector [cm · $\sqrt{\text{Hz}}$ · W^{-1}]
D = Electric flux-density complex amplitude vector [C · m^{-2}]
D_w = Waveguide dispersion coefficient [s · m^{-2}]
D_x, D_y = Lateral widths [m]
D_λ = Material dispersion coefficient [s · m^{-2}]
D_ν = Material dispersion coefficient [s^2 · m^{-1}]
$\mathfrak{D}$ = Electric flux density vector [C · m^{-2}]
D = Element of the $ABCD$ ray-transfer and wave-transfer matrices **M**
DBR = Distributed Bragg reflector
DC = Direct current
DCF = Dispersion-compensating fiber
DD = Direct detection
DESY = Deutsches Elektronen-Synchrotron
DEW = Directed-energy weapon
DFB = Distributed-feedback
DFF = Dispersion-flattened fiber
DFG = Difference-frequency generation
DGD = Differential group delay
DH = Double-heterostructure
DIP = Dual-inline package
DKDP = Deuterated potassium dihydrogen phosphate
DMD = Digital micromirror device
DMUX = Demultiplexer (also abbreviated as DEMUX)

DNG = Double-negative medium
DPC = Digital photon-counting device
DPS = Double-positive medium
DPSS = Diode-pumped solid-state
DQPSK = Differential quaternary phase shift keying
DRO = Doubly resonant oscillator
DSF = Dispersion-shifted fiber
DSP = Digital signal processing
DSPP = Doubly stochastic Poisson process
DUV = Deep ultraviolet, stretching from 200 to 300 nm
DVD = Digital-video-disc
DWDM = Dense wavelength-division multiplexing
DWELL = Quantum-dot-in-well
DWELL-QDIP = Quantum-dot-in-well quantum-dot infrared photodetector

e = Magnitude of electron charge [C]
$\hat{e}_x$ = Unit vector in the x direction
E = Electric-field complex amplitude [V · m^{-1}]; also, Steady or slowly varying field [V · m^{-1}]
E_L = Local-oscillator electric-field complex amplitude [V · m^{-1}]
E_s = Signal electric-field complex amplitude [V · m^{-1}]
$\mathbf{E}$ = Electric-field complex amplitude vector [V · m^{-1}]
$\mathbf{E}_i$ = Electric-field complex amplitude vector within a scattering sphere [V · m^{-1}]
$\mathbf{E}_s$ = Scattered electric-field complex amplitude vector [V · m^{-1}]
$\mathbf{E}_0$ = Incident electric-field complex amplitude vector [V · m^{-1}]
$\mathcal{E}$ = Electric field vector [V · m^{-1}]
E = Energy [J]; also, Radiant energy [J]
E_A = Acceptor energy level [J]; also, Activation energy [J]
E_{beam} = Electron-beam energy in a free-electron laser [GeV]
E_c = Energy at the bottom of the conduction band [J]
E_D = Donor energy level [J]
E_f = Fermi energy [J]
E_{fc} = Quasi-Fermi energy for the conduction band [J]
E_{fv} = Quasi-Fermi energy for the valence band [J]
E_g = Bandgap energy [J]
E_k = Kinetic energy [J]
$E_{\max}$ = Maximum kinetic energy [J]
E_r = Rotational energy [J]
E_v = Energy at the top of the valence band [J]
E_v = Luminous energy [lm · s]
E_{var} = Variational energy
E_ν = Energy spectral density [J · Hz^{-1}]
E-band = Extended optical fiber telecommunications band (1360–1460 nm)
EAM = Electroabsorption modulator
ECLD = External-cavity laser diode
EDFA = Erbium-doped fiber amplifier
e-e-o = Extraordinary-extraordinary-ordinary designations for waves 1, 2, and 3
EIT = Electromagnetically induced transparency
ELI = Extreme Light Infrastructure
E/O = Electronic-to-optical
e-o-e = Extraordinary-ordinary-extraordinary designations for waves 1, 2, and 3

e-o-o = Extraordinary-ordinary-ordinary designations for waves 1, 2, and 3
EQE = External quantum efficiency
EUV = Extreme-ultraviolet, stretching from 10 to 100 nm

f = Focal length of a lens [m]; also, Frequency [Hz]; also, Frequency of sound [Hz]
$f(E)$ = Fermi function
f_a = Probability that absorption condition is satisfied
$f_c(E)$ = Fermi function for the conduction band
f_{col} = Collision rate [s^{-1}]
f_e = Probability that emission condition is satisfied
f_g = Fermi inversion factor
$f_v(E)$ = Fermi function for the valence band
f = Frequency of sound [Hz]; also, Modulation frequency [Hz]
f = Volume fraction of a homogeneous medium occupied by scatterers (filling ratio)
F = Focal point of an optical system; also, Excess noise factor of a photodetector
F_P = Purcell factor
$F_\#$ = F-number of a lens
$\mathcal{F}$ = Finesse of a resonator; also, Force [kg · m · s^{-2}]
$\mathcal{F}_p$ = Ponderomotive force [kg · m · s^{-2}]
$F_{m\ell}$ = Elements of the matrix **F**
F = Hermitian matrix for generalized Helmholtz equation posed as an eigenvalue problem
FB = Fiber bundle
FBG = Fiber Bragg grating
FDM = Frequency-division multiplexing
FDMA = Frequency-division multiple access
FEL = Free-electron laser
FET = Field-effect transistor
FFT = Fast Fourier transform
FIR = Far infrared, stretching from 20 to 300 μm
FM = Frequency-modulated
FON = Fiber-optic network
FPA = Focal-plane array
FPI = Fabry–Perot interferometer
FROG = Frequency-resolved optical gating
FSK = Frequency shift keying
FTIR = Frustrated total internal reflection
FUV = Far ultraviolet, stretching from 100 to 200 nm
FWHM = Full-width at half-maximum
FWM = Four-wave mixing

g = Resonator g-parameter; also, Gravitational acceleration constant at earth's surface [m · s^{-2}]
$g(\mathbf{r}_1, \mathbf{r}_2)$ = Normalized mutual intensity
$g(\mathbf{r}_1, \mathbf{r}_2, \tau)$ = Complex degree of coherence
$g(\nu)$ = Lineshape function of a transition [Hz^{-1}]
$g(\tau)$ = Complex degree of temporal coherence
g_0 = Gain factor
$g_{\nu 0}(\nu)$ = Electron–photon collisionally broadened lineshape function in a semiconductor [Hz^{-1}]
g = Coupling coefficient in a parametric interaction [m^{-3}]
g = Fundamental spatial frequency of a periodic structure; also, Degeneracy parameter
g = Primitive vector defining a reciprocal-lattice unit cell [m^{-1}]

G = Gain of an amplifier; also, Gain of a photodetector; also, Conductance $[\Omega^{-1}]$
$G(\mathbf{r}_1, \mathbf{r}_2)$ = Mutual intensity $[\text{W} \cdot \text{m}^{-2}]$
$G(\mathbf{r}_1, \mathbf{r}_2, \tau)$ = Mutual coherence function $[\text{W} \cdot \text{m}^{-2}]$
$G(\nu)$ = Gain of an optical amplifier
$G(\tau)$ = Temporal coherence function $[\text{W} \cdot \text{m}^{-2}]$
$G_\mathcal{A}(\tau)$ = Pulse-envelope autocorrelation function
$G_I(\tau)$ = Intensity autocorrelation function $[\text{J}^2 \cdot \text{m}^{-4} \cdot \text{s}^{-1}]$
G_R = Gain of a Raman amplifier
$\mathbf{G}$ = Coherency matrix $[\text{W} \cdot \text{m}^{-2}]$; also, Gyration vector of an optically active medium; also, Wavevector of a phase grating $[\text{m}^{-1}]$
G = Photoionization rate in a photorefractive material
G_0 = Rate of thermal electron–hole generation in a semiconductor $[\text{m}^{-3} \cdot \text{s}^{-1}]$
$\mathbf{G}$ = Reciprocal-lattice vector $[\text{m}^{-1}]$
$\mathbb{G}_n(\cdot)$ = Hermite–Gaussian function of order n
GB = Gain–bandwidth product [Hz]
GFP = Group-IV photonics; also, Green fluorescent protein
GR = Generation–recombination
GRIN = Graded-index
GVD = Group velocity dispersion

h = Complex round-trip amplitude attenuation factor in a resonator; also, Planck's constant $[\text{J} \cdot \text{s}]$
$h(t)$ = Impulse response function of a linear system
$h(x, y)$ = Impulse response function of a two-dimensional linear system
$h_D(t)$ = Photodetector impulse response function
$\hbar = h/2\pi$ $[\text{J} \cdot \text{s}]$
H = Principal point of an optical system; also, Magnetic-field complex amplitude $[\text{A} \cdot \text{m}^{-1}]$
$\mathbf{H}$ = Magnetic-field complex amplitude vector $[\text{A} \cdot \text{m}^{-1}]$
$\mathcal{H}$ = Magnetic field vector $[\text{A} \cdot \text{m}^{-1}]$
$H(\nu)$ = Transfer function of a linear system [$H(f)$ for low-frequency signals]
$H'(\nu)$ = Real part of the transfer function of a linear system
$H''(\nu)$ = Imaginary part of the transfer function of a linear system
$H(\nu_x, \nu_y)$ = Transfer function of a two-dimensional linear system
$H_e(f)$ = Envelope transfer function of a linear system
H_0 = Transfer-function magnitude
$\text{H}_\ell^{(1)}(\cdot)$ = Hankel function of the first kind of order ℓ
$\mathbb{H}_n(\cdot)$ = Hermite polynomial of order n
H = Horizontal polarization
HAPLS = High-repetition-rate Advanced Petawatt Laser System
HD = High definition
HG = Hermite–Gaussian
HHG = High-harmonic generation
HNLF = Highly nonlinear fiber
HOE = Holographic optical element
HOM = Hong–Ou–Mandel
HOMO = Highest occupied molecular orbital
HVPE = Hydride vapor-phase epitaxy
HXR = Hard-X-ray

i = Electric current [A]; also, Integer; also, $\sqrt{-1}$
i_d = Dark current [A]

i_e = Electron current [A]
i_h = Hole current [A]
i_p = Photoelectric current (photocurrent) [A]
i_s = Reverse current in a semiconductor p–n diode [A]
i_t = Threshold current of a laser diode [A]
i_T = Transparency current for a laser-diode amplifier [A]
I = Optical intensity (also called Irradiance) [W · m^{-2}]
$I(t)$ = Intensity of an optical pulse [W · m^{-2}]
I_L = Local-oscillator intensity [W · m^{-2}]
I_s = Saturation optical intensity of an amplifier or absorber [W · m^{-2}]; also, Acoustic intensity [W · m^{-2}]; also, Signal intensity [W · m^{-2}]
I_s = Optical intensity of a scattered wave [W · m^{-2}]
I_t = Threshold intensity of a laser [W · m^{-2}]
I_ν = Spectral intensity [W · m^{-2} · Hz^{-1}]
I_0 = Optical intensity of an incident wave [W · m^{-2}]
$I_0(\cdot)$ = Modified Bessel function of order zero
$\mathcal{I}$ = Fourier transform of intensity profile; also, Moment of inertia [kg · m^2]
I = Identity matrix
I = In-phase component of the field
IC = Integrated circuit
ICL = Interband cascade laser
IEEE = Institute of Electrical and Electronics Engineers
IF = Intermediate frequency
IG = Ince–Gaussian
IM = Intensity modulation
IP = Internet protocol
IR = Infrared
IRE = Institute of Radio Engineers
ISI = Intersymbol interference
ITO = Indium tin oxide
ITU = International Telecommunications Union

$j = \sqrt{-1}$; also, Integer
J = Electric current density [A · m^{-2}]
J_e = Electron current density [A · m^{-2}]
J_h = Hole current density [A · m^{-2}]
$J_\ell(\cdot)$ = Bessel function of the first kind of order ℓ
J_p = Photoelectric current density [A · m^{-2}]
J_t = Threshold current density of a laser diode [A · m^{-2}]
J_T = Transparency current density of a laser-diode amplifier [A · m^{-2}]
J = Jones vector
$\mathcal{J}$ = Total angular-momentum quantum number
$\boldsymbol{\mathcal{J}}$ = Electric current density vector [A · m^{-2}]

k = Wavenumber [m^{-1}]; also, Integer; also, Spatial angular frequency [rad · m^{-1}]
k_e = Complex effective wavenumber of a host medium with embedded scatterers [m^{-1}]
k_o = Free-space wavenumber [m^{-1}]
k_r = Reflected wavenumber [m^{-1}]; also, Upshifted wavenumber of a Bragg-reflected wave [m^{-1}]
k_s = Downshifted wavenumber of a Bragg-reflected wave [m^{-1}]

k_s = Wavenumber inside a scattering medium [m^{-1}]
$k_T = \sqrt{k_x^2 + k_y^2}$ = Transverse component of the wavevector [m^{-1}]
k_x, k_y = Wavevector components in x and y directions [m^{-1}]; also, Spatial angular frequencies in x and y directions [rad · m^{-1}]
k_0 = Central wavenumber [m^{-1}]
$\mathbf{k}$ = Wavevector [m^{-1}]
$\mathbf{k}_g$ = Grating wavevector [m^{-1}]
$\mathbf{k}_r$ = Reflected wavevector [m^{-1}]; also, Upshifted wavevector of a Bragg-reflected wave [m^{-1}]
$\mathbf{k}_s$ = Downshifted wavevector of a Bragg-reflected wave [m^{-1}]
k = Ionization ratio for an avalanche photodiode
k = Boltzmann's constant [J · K^{-1}]
K = Undulator (magnetic-deflection) parameter in a free-electron laser
$\mathrm{K}_m(\cdot)$ = Modified Bessel function of the second kind of order m
Kα = Designation of X-ray line arising from transition from n = 2 to n = 1 atomic shells
K = Bloch wavenumber [m^{-1}]
$\mathbf{K}$ = Bloch wavevector [m^{-1}]
KDP = Potassium dihydrogen phosphate
KGW = Potassium gadolinium tungstate
KTP = Potassium titanyl phosphate
KYW = Potassium yttrium tungstate

l = Length [m]; also, Integer
l_c = Coherence length [m]
ℓ = Azimuthal quantum number
ℓ_0 = Optical pathlength of the central frequency component of a pulse
L = Length [m]; also, Distance [m]; also, Electrical inductance [H]; also, Loss factor; also, Number of incoming optical beams to an acousto-optic switch; also, Integer
L_c = Coherence length in a parametric interaction [m]
L_p = Soliton–soliton interaction period [m]
L_v = Luminance [cd · m^{-2}]
$L_0 = \pi/2\mathcal{C}$ = Coupling length (transfer distance) in a directional coupler [m]
$\mathcal{L}$ = Linear operator
L = Orbital angular momentum quantum number
$^{2S+1}\mathrm{L}_J$ = Term symbol for angular-momentum quantum numbers with LS coupling
L = Angular momentum [J · s]
$\mathbb{L}_m(\cdot)$ = Laguerre polynomial of degree m
$\mathbb{L}_m^l(\cdot)$ = Generalized Laguerre polynomial of degree m, order l, and index (m, l)
L-band = Long optical fiber telecommunications band (1565–1625 nm)
LB_0 = Bit-rate–distance product [km · Gb · s^{-1}]
LAN = Local-area network
LANL = Los Alamos National Laboratory
LASER = Light amplification by stimulated emission of radiation
LaWS = Laser Weapon System (U.S. Navy)
LBO = Lithium triborate
LC = Liquid crystal
LCD = Liquid-crystal display
LCLS = Linac Coherent Light Source at SLAC National Accelerator Laboratory operated by Stanford University
LCP = Left-circularly polarized
LD = Laser diode

LED = Light-emitting diode
LEP = Light-emitting polymer material
LG = Laguerre–Gaussian
LHS = Left-hand side
LIGO = Laser Interferometer Gravitational-wave Observatory
LINAC = Linear accelerator
LLNL = Lawrence Livermore National Laboratory
LMA = Large mode-area
LO = Local oscillator
LP = Linearly polarized
LPE = Liquid-phase epitaxy
LSP = Localized surface plasmon (localized surface plasmon polariton)
LuAG = Lutetium aluminum garnet
LUMO = Lowest unoccupied molecular orbital
LWFA = Laser wakefield acceleration
LWI = Lasing without inversion
LWIR = Long-wavelength infrared, stretching from 8 to 14 μm

m = Mass of a particle [kg]; also, Free electron mass [kg]; also, Integer; also, Contrast or modulation depth
m_c = Effective mass of a conduction-band electron [kg]
m_p = Proton mass [kg]
m_r = Reduced mass of an electron–hole pair in a semiconductor [kg]
m_v = Effective mass of a valence-band hole [kg]
m_0 = Free electron mass [kg]
$\mathfrak{m}$ = Magnetic dipole moment [A · m^2]
m = Photon number; also, Photoelectron number
m_0 = Photoelectron-number sensitivity of an optical receiver
m = Magnetic quantum number
M = Magnification in an image system; also, Number of modes; also, Magnetization density complex amplitude [A · m^{-1}]; also, Number of harmonics; also, Integer
M_v = Illuminance [lx]
M = Magnetization density complex amplitude vector [A · m^{-1}]
$\mathfrak{M}$ = Figure of merit indicating strength of acousto-optic effect in a material [m^2 · W^{-1}]
$\boldsymbol{\mathfrak{M}}$ = Magnetization density vector [A · m^{-1}]
M = Mass of an atom or molecule [kg]
$M(\nu)$ = Density of modes in a resonator [m^{-3}·Hz^{-1} for 3D resonator; m^{-1}·Hz^{-1} for 1D resonator]
M_r = Reduced mass of an atom or molecule [kg]
M = Ray-transfer matrix; also, Wave-transfer matrix
$\mathbb{M}^2$ = Factor representing deviation of optical-beam profile from Gaussian form
MAN = Metropolitan-area network
MBE = Molecular-beam epitaxy
MC = Mode converter
MCF = Multicore fiber
MCP = Microchannel plate
MEMS = Microelectromechanical system
MI = Michelson interferometer
MIM = Metal–insulator–metal
MIMO = Multiple-input multiple-output
MIR = Mid infrared, stretching from 2 to 20 μm

MIRACL = Mid-infrared advanced chemical laser
MIS = Metal–insulator–semiconductor
MKS = Meter/kilogram/second unit system
MMF = Multimode fiber
MOCVD = Metalorganic chemical vapor deposition (same as MOVPE)
MOFA = Master-oscillator fiber-amplifier
MOPA = Master-oscillator power-amplifier
MOSFET = Metal-oxide-semiconductor field-effect transistor
MOT = Magneto-optical trap
MOVPE = Metalorganic vapor phase epitaxy (same as MOCVD)
MPM = Multiphoton microscopy
MQD = Multiquantum dot
MQW = Multiquantum well
M-ROADM = Multi-degree reconfigurable optical add–drop multiplexer
MUV = Mid ultraviolet, stretching from 200 to 300 nm
MUX = Multiplexer
MWIR = Medium-wavelength infrared, stretching from 3 to 5 μm
MZI = Mach–Zehnder interferometer
MZM = Mach–Zehnder modulator

n = Refractive index; also, Integer
$n(\mathbf{r})$ = Refractive index of an inhomogeneous medium
$n(\theta)$ = Refractive index of extraordinary wave in a uniaxial crystal
n_b = Effective refractive index associated with SPP at metal–dielectric boundary
n_e = Extraordinary refractive index
n_o = Ordinary refractive index
n_p = Refractive index of a prism
n_s = Refractive index of a scattering volume
n_2 = Optical Kerr coefficient (nonlinear refractive index) [$\mathrm{m}^2 \cdot \mathrm{W}^{-1}$]
$\mathfrak{n}$ = Photon-number density [m^{-3}]
$\mathfrak{n}_s$ = Saturation photon-number density [m^{-3}]
n = Photon number
$\bar{n}$ = Mean photon number
$\bar{n}_\nu$ = Spectral photon number [Hz^{-1}]
n_0 = Number-state photon number; also, Photon-number sensitivity of an optical receiver
n = Principal quantum number
n = Concentration of electrons in a semiconductor [m^{-3}]
n_i = Concentration of electrons/holes in an intrinsic semiconductor [m^{-3}]
n_0 = Equilibrium concentration of electrons in a semiconductor [m^{-3}]
N = Group index; also, Integer; also, Number of atoms; also, Number of stages; also, Number of optical cycles; also, Number of resolvable spots of a scanner; also, Order of a higher-order soliton
N_F = Fresnel number
N = Number density [m^{-3}]; also, $N = N_2 - N_1$ = Population density difference [m^{-3}]
N_a = Atomic number density [m^{-3}]
N_A = Number density of ionized acceptor atoms in a semiconductor [m^{-3}]
N_D = Number density of ionized donor atoms in a semiconductor [m^{-3}]; also, Number density of donor atoms in a photorefractive material [m^{-3}]
N_D^+ = Number density of ionized donor atoms in a photorefractive material [m^{-3}]
N_s = Number density of scatterers [m^{-3}]

N_t = Laser threshold population difference [m^{-3}]
N_0 = Steady-state population difference in the absence of amplifier radiation [m^{-3}]
NA = Numerical aperture
NEA = Negative-electron-affinity
NEP = Noise-equivalent power
NIF = National Ignition Facility
NIM = Negative-index material
NIR = Near infrared, stretching from 0.760 to 2 μm
NL = Nonlinear
NLDC = Nonlinear directional coupler
NOLM = Nonlinear optical loop mirror
NRI = Negative refractive index
NRZ = Non-return-to-zero
NUV = Near ultraviolet, stretching from 300 to 390 nm
NZ-DSF = Non-zero dispersion shifted fiber

O-band = Original optical fiber telecommunications band (1260–1360 nm)
OA = Optical amplifier
OADM = Optical add–drop multiplexer
OAM = Orbital angular momentum
OC = Optical carrier
OCT = Optical coherence tomography
ODMUX = Optical demultiplexer (also abbreviated as ODEMUX)
O/E = Optical-to-electronic
o-e-e = Ordinary-extraordinary-extraordinary designations for waves 1, 2, and 3
OEIC = Optoelectronic integrated circuit
o-e-o = Ordinary-extraordinary-ordinary designations for waves 1, 2, and 3
OEO = Optical-electrical-optical
OFA = Optical fiber amplifier
OFC = Optical frequency comb; also, Optical frequency conversion
OH = Hydroxyl radical
OLED = Organic light-emitting diode
OM = Optical mode
OMUX = Optical multiplexer
o-o-e = Ordinary-ordinary-extraordinary designations for waves 1, 2, and 3
OOK = ON–OFF keying
OPA = Optical parametric amplifier
OPC = Optical phase conjugation
OPCPA = Optical parametric chirped-pulse amplification
OPD = Organic photodetector
OPO = Optical parametric oscillator
OR = OR logic gate
OSA = Optical Society of America
OTDM = Optical time-division multiplexer
OXC = Optical cross-connect

p = Probability; also, Momentum [kg $\cdot$ m $\cdot$ s^{-1}]; also, Graded-index fiber profile parameter
$p(n)$ = Probability of n events
$p(n)$ = Photon-number distribution
$p(x, y)$ = Aperture function or pupil function

p_{ab} = Probability density for absorption (mode containing one photon) [s^{-1}]
p_{sp} = Probability density for spontaneous emission (into one mode) [s^{-1}]
p_{st} = Probability density for stimulated emission (mode containing one photon) [s^{-1}]
p = Electric dipole moment [C · m]
p = Normalized electric-field quadrature component
p = Elasto-optic (strain-optic) coefficient
p$_{ijkl}$ = Component of the photoelasticity tensor
p$_{IK}$ = Component of the photoelasticity tensor (contracted indices)
p = Concentration of holes in a semiconductor [m^{-3}]
p$_0$ = Equilibrium concentration of holes in a semiconductor [m^{-3}]
P = Electric polarization-density complex amplitude [C · m^{-2}]; also, Probability of impact ionization
$P(\nu_x, \nu_y)$ = Fourier transform of the aperture function $p(x, y)$
P_{ab} = Probability density for absorption (mode containing many photons) [s^{-1}]
$P_{\text{m}}^{\ell}(\cdot)$ = Microsphere-resonator adjoint Legendre function
P_{NL} = Complex amplitude of the nonlinear component of the polarization density [C · m^{-2}]
P_{sp} = Probability density for spontaneous emission (into any mode) [s^{-1}]
P_{st} = Probability density for stimulated emission (mode containing many photons) [s^{-1}]
P = Electric polarization-density complex amplitude vector [C · m^{-2}]
$\mathcal{P}$ = Electric polarization density vector [C · m^{-2}]
$\mathcal{P}_{\text{L}}$ = Linear component of the polarization density [C · m^{-2}]
$\mathcal{P}_{\text{NL}}$ = Nonlinear component of the polarization density [C · m^{-2}]
P = Optical power (also called Radiant power or Radiant flux) [W]
P_e = Electrical power [W]
P_i = Incident optical power [W]
P_L = Local-oscillator power [W]
P_o = Output optical power [W]; also, Average output optical power [W]
P_p = Peak pulse power [W]
P_p = Optical pump power [W]
P_r = Received optical power [W]
P_s = Signal (optical) power [W]
P_s = Optical power of a scattered wave [W]
P_t = Threshold optical pump power [W]
P_v = Luminous flux [lm]
P_ν = Power spectral density [W · Hz^{-1}]
P_π = Half-wave optical power in a Kerr medium [W]
$\mathbb{P}$ = Degree of polarization
P = Optical power [dBm]
P$_c$ = Optical power loss associated with splicing and coupling [dBm]
P$_m$ = Optical power allotted for safety margin [dBm]
P$_r$ = Receiver optical power sensitivity [dBm]
P$_s$ = Source optical power [dBm]
PAL-SLM = Parallel aligned spatial light modulator
PBG = Photonic bandgap
PBS = Polarizing beamsplitter
PCB = Printed circuit board
PCF = Photonic-crystal fiber
PC-LED = Phosphor-conversion LED
PCM = Pulse code modulation
PD = Photodetector

PDE = Photon detection efficiency
PET = Positron-emission tomography
PG-FROG = Polarization-gated frequency-resolved optical gating
PIC = Photonic integrated circuit
PIN = p-type–i-type–n-type photodiode
PLASER = Powder laser
PLC = Planar lightwave circuit
PLED = Polymer organic light-emitting diode
PM = Phase modulation
PMD = Polarization mode dispersion
PMT = Photomultiplier tube
P-OLED = Polymer light-emitting diode
PPLN = Periodically poled lithium niobate
PPV = Poly(p-phenylene vinylene)
PROM = Pockels readout optical modulator
PRR = Pulse repetition rate
PSK = Phase shift keying
PWFA = Plasma wakefield acceleration
PWM = Pulse-width modulation

q = Electric charge [C]; also, Wavenumber of an acoustic wave [m^{-1}]; also, Integer (mode index, diffraction order, quantum number); also, Spatial angular frequency [rad · m^{-1}]
$q(z)$ = Complex Gaussian-beam parameter [m]
q = Quantum defect
$\mathbf{q}$ = Wavevector of an acoustic wave [m^{-1}]
Q = Electric charge [C]; also, Quality factor of an optical resonator or a resonant circuit
Q_a = Absorption efficiency
Q_s = Scattering efficiency
$\mathbf{Q}$ = Quadrature component of the field
QAM = Quadrature amplitude modulation
QCL = Quantum cascade laser
QCSE = Quantum-confined Stark effect
QD = Quantum dot
QDIP = Quantum-dot infrared photodetector
QED = Quantum electrodynamics
QOCT = Quantum optical coherence tomography
QPM = Quasi-phase matching; also, Quadratic phase modulator
QPSK = Quaternary phase shift keying
QWIP = Quantum-well infrared photodetector

r = Radial distance in spherical and cylindrical coordinates [m]
r_n = Radii of allowed electron orbits in Bohr atom [m]
$\mathbf{r}$ = Position vector [m]
$\hat{\mathbf{r}}$ = Unit vector in radial direction in spherical coordinates
r = Complex amplitude reflectance; also, Complex round-trip amplitude attenuation factor in a resonator
$|\mathrm{r}|$ = Magnitude of round-trip amplitude attenuation factor in a resonator
r_+ = Frequency-upshifted Bragg amplitude reflectance
r_- = Frequency-downshifted Bragg amplitude reflectance
$r(\nu)$ = Rate of photon emission/absorption from a semiconductor [s^{-1} · m^{-3} · Hz^{-1}]

r = Linear electro-optic (Pockels) coefficient [m · V^{-1}]; also, Rotational quantum number
r_{ijk} = Component of the linear electro-optic (Pockels) tensor [m · V^{-1}]
r_{Ik} = Component of the linear electro-optic (Pockels) tensor (contracted indices) [m · V^{-1}]
r = Electron–hole recombination coefficient [m^3 · s^{-1}]
r_{nr} = Nonradiative electron–hole recombination coefficient [m^3 · s^{-1}]
r_r = Radiative electron–hole recombination coefficient [m^3 · s^{-1}]
rect(·) = Pulse of unit height and unit width centered about 0
R = Radius of curvature [m]; also, Electrical resistance [Ω]
$R(z)$ = Radius of curvature of a Gaussian beam [m]
$R(\tau)$ = Field autocorrelation function
R_L = Load resistance [Ω]
R_m = Distance from the focal point to the mth ring of a Fresnel zone plate [m]
R_0 = Radius of cylinder in which a meridional ray is confined [m]
$\mathbf{R}(\theta)$ = Jones matrix for coordinate rotation by an angle θ
$\mathcal{R}$ = Intensity or power reflectance; also, approximate reflectance of a Bragg reflector
$\mathcal{R}_e$ = Exact intensity or power reflectance of a Bragg reflector
R = Pumping rate [s^{-1} · m^{-3}]; also, Recombination rate in a semiconductor [s^{-1} · m^{-3}]; also, Electron–hole injection rate in a semiconductor [s^{-1} · m^{-3}]
R_t = Laser threshold pumping rate [s^{-1} · m^{-3}]
$\mathbf{R}$ = Lattice vector [m]
R = Responsivity of a photon source [W·A^{-1}]; also, Responsivity of a photon detector [A·W^{-1}]
R_d = Differential responsivity of a laser diode [W · A^{-1}]
$\mathbb{R}_{n\ell}(r)$ = Hydrogen-atom associated Laguerre function of order ℓ and index n
R = Ratio of complex-envelope polarization-component magnitudes
RC = Resistor–capacitor combination
RC = Resonant-cavity
RCLED = Resonant-cavity light-emitting diode
RCP = Right-circularly polarized
REFA = Rare-earth-doped fiber amplifier
RF = Radio-frequency
RFA = Raman fiber amplifier
RFID = Radio-frequency identification
RFL = Raman fiber laser
RHS = Right-hand side
RMS = Root-mean square
ROADM = Reconfigurable optical add–drop multiplexer
RW = Ridge waveguide
Rx = Receiver
RZ = Return-to-zero

s = Length or distance [m]; also, Scale factor
$s(x, t)$ = Strain wavefunction
s_{ij} = Component of the strain tensor
s = Photorefractivity proportionality constant for photoionization cross section
$s(\mathbf{r}_1, \mathbf{r}_2, \nu)$ = Normalized cross-spectral density
s = Quadratic electro-optic (Kerr) coefficient [m^2 · V^{-2}]; also, Spin quantum number
s_{ijkl} = Component of the quadratic electro-optic (Kerr) tensor [m^2 · V^{-2}]
s_{IK} = Component of the quadratic electro-optic (Kerr) tensor (contracted indices) [m^2 · V^{-2}]
sinc(·) = Symmetric function with peak value of unity at 0 [sinc$(t) \equiv \sin(\pi t)/(\pi t)$]
S = Poynting-vector magnitude [W·m^{-2}]; also, Transition strength (oscillator strength) [m^2·Hz]

$S(\mathbf{r})$ = Complex amplitude for a radiation source [V · m^{-3}]
$S(\cdot)$ = Fresnel integral
S_0 = Strain amplitude
$\mathbf{S}$ = Complex Poynting vector [W · m^{-2}]
$\mathcal{S}(t)$ = Source of optical radiation created by an incident field [V · m^{-3}]; also, Spin angular-momentum quantum number
$\boldsymbol{\mathcal{S}}$ = Poynting vector [W · m^{-2}]
$S(\mathbf{r})$ = Eikonal [m]
$S(\mathbf{r}_1, \mathbf{r}_2, \nu)$ = Cross-spectral density [W · m^{-2} · Hz^{-1}]
$S(\lambda_o)$ = Wavelength power spectral density [W · m^{-1}];
$S(\nu)$ = Spectral intensity of an optical wave or pulse [W · m^{-2} · Hz^{-1}]; also, Power spectral density [W · m^{-2} · Hz^{-1}]
$S(\nu, t)$ = Spectrogram of an optical pulse $[S(\nu, t) = |\Phi(\nu, t)|^2]$
$\mathbf{S}$ = Scattering matrix
$\mathbb{S}$ = Projection of photon-spin angular momentum along the wavevector (helicity) [J · s]
$S_{[\cdot]}$ = Stokes parameters
S-band = Short optical fiber telecommunications band (1460–1530 nm)
SACM = Separate absorption, charge, and multiplication
SAM = Separate absorption and multiplication
SAPD = Superlattice avalanche photodiode; also, Staircase avalanche photodiode
SASE = Self-amplified spontaneous emission
SBN = Strontium barium niobate
SBS = Stimulated Brillouin scattering
SCF = Single-core fiber
SCG = Supercontinuum generation
SCIDCM = Single-carrier-injection double-carrier multiplication
SCISCM = Single-carrier-injection single-carrier multiplication
SDH = Synchronous digital hierarchy
SDL = Semiconductor disk laser
SDM = Space-division multiplexing
SEED = Self-electro-optic-effect device
SESAM = Semiconductor saturable-absorber mirror
SFG = Sum-frequency generation
SGDFB = Sampled-grating distributed-feedback
SH = Second harmonic
SHG = Second-harmonic generation
SHG-FROG = Second-harmonic generation frequency-resolved optical gating
SI = International system of units; also, Step-index
SLA = Semiconductor laser amplifier
SLAC = National Accelerator Laboratory at Stanford University
SLED = Superluminescent diode
SLM = Spatial light modulator
SMD = Surface-mounted device
SMF = Single-mode fiber
SMOLED = Small-molecule organic light-emitting diode
SNG = Single-negative medium
SNOM = Scanning near-field optical microscopy
SNR = Signal-to-noise ratio
SNSPD = Superconducting nanowire single-photon detector
SOA = Semiconductor optical amplifier

SOI = Silicon-on-insulator
SONET = Synchronous optical network
SOS = Silica-on-silicon
SPAD = Single-photon avalanche diode
SPASER = Surface-plasmon amplification by stimulated emission of radiation
SPDC = Spontaneous parametric downconversion
SPE = Single-photon emitter
SPIE = The International Society for Optical Engineering
SPM = Self-phase modulation
SPP = Surface plasmon polariton
SPR = Surface plasmon resonance
SQUID = Superconducting quantum-interference device
SQW = Single quantum well
SRO = Singly resonant oscillator
SRS = Stimulated Raman scattering
SSFS = Soliton self-frequency shift
SSPD = Superconducting single-photon detector
STS = Synchronous transport signal
SVE = Slowly varying envelope
SXR = Soft-X-ray

t = Time [s]
t_i = Ionization time [s]
t_r = Recombination time [s]
$t_{\rm sp}$ = Spontaneous lifetime [s]; also, Effective spontaneous lifetime [s]
t = Complex amplitude transmittance; also, Normalized time for an optical pulse
T = Temperature [°K]
$\mathbf{T}$ = Jones matrix
$\mathcal{T}$ = Intensity or power transmittance; also, Power-transfer or power-transmission ratio
T = Transit time [s]; also, Counting time [s]; also, Switching time [s]; also, Bit time interval [s]; also, Resolution time ($T = 1/2B$ where $B = $ Bandwidth) [s]; also, Period of a wave ($T = 1/\nu$ where $\nu = $ frequency) [s]; also, Transit time of sound across an optical beam
$T_F = 1/\nu_F$ = Inverse of Fabry–Perot resonator-mode frequency spacing ($T_F = 2d/c$) [s]; also, Period of a mode-locked laser pulse train [s]
$T_{\ell m}$ = Element of transmission or interconnection matrix
T_0 = Ground-state orbital period in Bohr hydrogen atom ($T_0 \approx 150$ as) [s]
T_2 = Electron–phonon collision time [s]
$\mathbf{T}$ = Transmission matrix; also, Interconnection matrix
TADF = Thermally activated delayed fluorescence
TDM = Time-division multiplexing
TDMA = Time-division multiple access
TE = Transverse electric
TEM = Transverse electromagnetic
TES = Transition-edge sensor
TFF = Thin-film filter
TFT = Thin-film transistor
TGG = Terbium gallium garnet
THG = Third-harmonic generation
TIR = Total internal reflection
TM = Transverse magnetic

1322   SYMBOLS AND UNITS

TMD = Transition-metal dichalcogenide
TOAD = Terahertz optical asymmetric demultiplexer
TPD = Triphenyl diamine derivative
TPLSM = Two-photon laser scanning fluorescence microscopy
TSI = Time-slot interchange
TST = Time–space–time
TV = Television receiver
Tx = Transmitter
2PM = Two-photon microscopy
3PM = Three-photon microscopy

u = Displacement [m]
$u(\mathbf{r}, t)$ = Wavefunction of an optical wave
$\hat{\mathbf{u}}$ = Unit vector
u = Number of electrons in a subshell
$U(\mathbf{r})$ = Complex amplitude of a monochromatic optical wave
$U(\mathbf{r}, t)$ = Complex wavefunction of an optical wave
$U(t)$ = Complex wavefunction of an optical pulse
$U(t_1, t_2)$ = Joint temporal wavefunction for two-photon light
$U(x_1, x_2)$ = Joint spatial wavefunction for two-photon light
$U_s(x)$ = Probability amplitude of location of maximally entangled photons
$U_s(\mathbf{r})$ = Complex amplitude of a scattered monochromatic optical wave
$U_0(\mathbf{r})$ = Complex amplitude of an incident monochromatic optical wave
$\mathcal{U}(x)$ = Unit step function [$\mathcal{U}(x) = 1$ if $x > 0$ and $\mathcal{U}(x) = 0$ if $x < 0$]
U-band = Ultra-long optical fiber telecommunications band (1625–1675 nm)
ULSI = Ultra-large-scale integration
UV = Ultraviolet
UVA = Ultraviolet-A band, stretching from 315 to 400 nm
UVB = Ultraviolet-B band, stretching from 280 to 315 nm
UVC = Ultraviolet-C band, stretching from 100 to 280 nm

v = Group velocity of a wave [m · s^{-1}]
$v(\mathbf{r}, \nu)$ = Fourier transform of the wavefunction of an optical wave
v_s = Velocity of sound [m · s^{-1}]
v = Velocity of an atom or object [m · s^{-1}]; also, Drift velocity of a carrier [m · s^{-1}]
v_e = Velocity of an electron [m · s^{-1}]
v_h = Velocity of a hole [m · s^{-1}]
$\mathbf{v}$ = Velocity vector of a charge carrier [m · s^{-1}]
v = Vibrational quantum number
V = Vertex; also, Volume [m^3]; also, Modal volume [m^3]; also, Voltage [V]
$V(\mathbf{r}, \nu)$ = Fourier transform of the complex wavefunction of an optical wave
$V(\lambda_o)$ = Photopic luminosity function (photopic luminous efficiency function)
$V(\nu)$ = Fourier transform of the complex wavefunction of an optical pulse
$V(\nu_1, \nu_2)$ = Joint spectral wavefunction for two-photon light
$V(\nu_{x1}, \nu_{x2})$ = Joint wavevector wavefunction for two-photon light [$k_{x1} = 2\pi\nu_{x1}$ and $k_{x2} = 2\pi\nu_{x2}$]
V_B = Battery voltage [V]
V_c = Critical voltage for a liquid-crystal cell [V]
$V_s(\nu_x)$ = Fourier transform of $U_s(x)$
V_π = Half-wave voltage of an electro-optic retarder or modulator [V]

V_0 = Built-in potential difference in a p–n junction [V]; also, Switching voltage of a directional coupler [V]
$\mathcal{V}$ = Visibility
V = Verdet constant [rad · m^{-1} · T^{-1}]
V = Fiber V parameter
$V(\mathbf{r})$ = Potential energy [J]
$\mathbb{V}$ = Abbe number of a dispersive medium
V = Vertical polarization
VCSEL = Vertical-cavity surface-emitting laser
VECSEL = Vertical external-cavity surface-emitting laser
VLSI = Very-large-scale integration
VOx = Vanadium oxide
VPE = Vapor-phase epitaxy
VUV = Vacuum ultraviolet, stretching from 10 to 200 nm

w = Width [m]; also, Radius of a thin metallic rod [m]; also, Length of an acousto-optic cell [m]
w = Integrated photon flux (integrated optical power in units of photon number)
w_d = Width of the absorption region in an avalanche photodiode [m]
w_m = Width of the multiplication region in an avalanche photodiode [m]
W = Time-averaged electromagnetic energy density [J · m^{-3}]
$W(t)$ = Window function; also, Window function for short-time Fourier transform
$W(z)$ = Width (radius) of a Gaussian beam at an axial distance z from the beam center [m]
W_0 = Waist radius of a Gaussian beam [m]
$\mathcal{W}$ = Electromagnetic energy density [J · m^{-3}]
W = Probability density for absorption of pump light [s^{-1}]
W_i = Probability density for absorption and stimulated emission [s^{-1}]
$\mathbb{W}$ = Ionization energy of an atom [J]; also, Photoelectric work function [J]
WAN = Wide-area network
WC = Wavelength converter
WCI = Wavelength-channel interchange
WDM = Wavelength-division multiplexing
WDMA = Wavelength-division multiple access
WG = Waveguide
WGM = Whispering-gallery mode
WGR = Waveguide grating router
WKB = Wentzel–Kramers–Brillouin
WLAN = Wireless local-area network
WOLED = White organic light-emitting diode
WPE = Wall-plug efficiency
WSS = Wavelength-selective switch

x = Position coordinate [m]; also, Displacement [m]
$\hat{\mathbf{x}}$ = Unit vector in the x direction in Cartesian coordinates
$\mathsf{x}(t)$ = Inverse Fourier transform of the susceptibility of a dispersive medium $\chi(\nu)$
x = Normalized electric-field quadrature component
X = Normalized photon-flux density at the input to an optical amplifier
$\mathbf{X}$ = Input vector to a linear system
$\mathbf{X}_q$ = Eigenvectors associated with an eigenvalue problem
$\mathcal{X}(u)$ = Real function associated with the Hermite–Gaussian beam
$\mathcal{X}^{(2)}(\omega_1, \omega_2)$ = Second-order nonlinear susceptibility

$X(\cdot)$ = Normalized rate of change of the radial distribution in the core of a step-index fiber
XC = Cross-connect
XGM = Cross-gain modulation
XOR = Exclusive OR gate
XPM = Cross-phase modulation
XUV = Extreme ultraviolet

y = Position coordinate [m]
$\hat{y}$ = Unit vector in the y direction in Cartesian coordinates
Y = Normalized photon-flux density at the output of an optical amplifier
$\mathbf{Y}$ = Output vector from a linear system
$\mathcal{Y}(v)$ = Real function associated with the Hermite–Gaussian beam
$Y(\cdot)$ = Normalized rate of change of the radial distribution in the cladding of a step-index fiber
YAG = Yttrium aluminum garnet
YIG = Yttrium iron garnet
YLF = Yttrium lithium fluoride

z = Position coordinate (Cartesian or cylindrical coordinates) [m]
$z_{\min}$ = Location of the minimum width for a chirped Gaussian pulse [m]
z_{NL} = Nonlinear characteristic length of a Kerr medium [m]
z_p = Soliton period [m]
z_{T} = Talbot distance [m]
z_0 = Rayleigh range of a Gaussian beam [m]
$|z_0|$ = Dispersion length of a Gaussian pulse traveling through a dispersive medium [m]
$\hat{z}$ = Unit vector in the z direction in Cartesian coordinates
z = Normalized distance for an optical pulse
Z = Atomic number; also, Electrical-circuit impedance [Ω]
$\mathcal{Z}(\cdot)$ = Real function associated with the Hermite–Gaussian beam

Greek Symbols

α = Apex angle of a prism; also, Twist coefficient of a twisted nematic liquid crystal [m^{-1}]; also, Attenuation or absorption coefficient [m^{-1}]; also, Intensity extinction coefficient: $\alpha = \alpha_a + \alpha_s$ [m^{-1}]; also, Linewidth-enhancement factor for a laser diode
α_a = Absorption coefficient of a scattering material [m^{-1}]
α_e = Electron ionization coefficient in a semiconductor [m^{-1}]
α_h = Hole ionization coefficient in a semiconductor [m^{-1}]
α_m = Loss coefficient of a resonator attributed to a mirror [m^{-1}]
α_p = Mean value of p for a coherent state
α_r = Effective overall distributed loss coefficient [m^{-1}]
α_s = Loss coefficient of a laser medium [m^{-1}]
α_s = Scattering coefficient [m^{-1}]
α_x = Mean value of x for a coherent state
α_ν = Angular dispersion coefficient [Hz^{-1}]
α = Attenuation coefficient of an optical fiber [dB/km]

$\beta = k_z$ = Propagation constant [m^{-1}]; also, Phase-retardation coefficient of a twisted nematic liquid crystal [m^{-1}]
β' = First derivative of β with respect to ω [m$^{-1} \cdot$ s]
β'' = Second derivative of β with respect to ω [m$^{-1} \cdot$ s^2]
$\beta(\nu)$ = Propagation constant in a dispersive medium [m^{-1}]

$\beta_0 = \beta(\nu_0)$ = Propagation constant at the central frequency ν_0 [m^{-1}]
β = Spontaneous-emission coupling coefficient

γ = Field attenuation coefficient [m^{-1}]; also, Field extinction coefficient [m^{-1}]; also, Lateral decay coefficient in a waveguide [m^{-1}]; also, Lorentz factor in special relativity ($\gamma = [1-(v/c)^2]^{-1/2}$); also, Coupling coefficient in a parametric device [m^{-1}]; also, Nonlinear coefficient in soliton theory
γ_b = Amplitude attenuation coefficient of a traveling surface plasmon polariton [m^{-1}]
γ_B = Magnetogyration coefficient [m$^2 \cdot$ Wb^{-1}]
γ_m = Field extinction coefficient of an evanescent wave [m^{-1}]
γ_R = Photorefractivity recombination-rate coefficient
γ = Gain coefficient of an optical amplifier [m^{-1}]
$\gamma(\nu)$ = Gain coefficient of an optical amplifier [m^{-1}]
γ_m = Maximum gain coefficient of a quantum-well laser-diode amplifier [m^{-1}]
γ_p = Peak gain coefficient of a laser-diode amplifier [m^{-1}]
γ_R = Raman gain coefficient [m^{-1}]
$\gamma_\beta(\nu)$ = Gain coefficient of a subset of atoms [m^{-1}]
$\gamma_0(\nu)$ = Small-signal gain coefficient of an optical amplifier [m^{-1}]
Γ = Phase retardation; also, Power confinement factor in a laser diode or waveguide photodiode; also, Crystallographic symbol for irreducible Brillouin zone
Γ_m = Power confinement factor in a waveguide

δ = Secondary-emission gain random variable
$\delta(\cdot)$ = Delta function or impulse function
δx = Increment of x
$\delta\theta$ = Angular divergence of an optical beam
$\delta\theta_s$ = Angular divergence of an acoustic beam
$\delta\nu$ = Spectral width of a resonator mode [Hz]
Δ = Thickness of a thin optical component [m]; also, Fractional refractive-index change in an optical fiber or waveguide ($\Delta \approx (n_1 - n_2)/n_1$)
Δn = Concentration of excess electron–hole pairs [m^{-3}]
Δn$_T$ = Concentration of injected carriers for a semiconductor optical amplifier at transparency [m^{-3}]
Δx = Increment of x
$\Delta\nu$ = Spectral width or linewidth [Hz]; also, Atomic linewidth or transition linewidth [Hz]
$\Delta\nu_c = 1/\tau_c$ = Spectral width [Hz]
$\Delta\nu_D$ = Doppler linewidth [Hz]
$\Delta\nu_{\text{FWHM}}$ = Full-width-at-half-maximum spectral width [Hz]
$\Delta\nu_L$ = Laser linewidth [Hz]
$\Delta\nu_s$ = Linewidth of a saturated amplifier [Hz]
$\Delta\nu_{\text{ST}}$ = Schawlow–Townes minimum laser linewidth [Hz]

ϵ = Electric permittivity of a medium [F $\cdot$ m^{-1}]; also, Focusing error [m^{-1}]
ϵ' = Real part of the electric permittivity of a medium [F $\cdot$ m^{-1}]
ϵ'' = Imaginary part of the electric permittivity of a medium [F $\cdot$ m^{-1}]
ϵ_b = Effective permittivity associated with SPP at metal–dielectric boundary [F $\cdot$ m^{-1}]
ϵ_c = Effective permittivity of a conductive medium [F $\cdot$ m^{-1}]
ϵ_d = Electric permittivity of the dielectric medium in a layered metamaterial [F $\cdot$ m^{-1}]
ϵ_e = Effective permittivity of a host medium with embedded objects [F $\cdot$ m^{-1}]
ϵ_{ij} = Component of the electric permittivity tensor [F $\cdot$ m^{-1}]
ϵ_m = Electric permittivity of the metallic medium in a layered metamaterial [F $\cdot$ m^{-1}]

ϵ_o = Electric permittivity of free space [F·m^{-1}]
ϵ_r = Relative permittivity or dielectric constant
ϵ_s = Electric permittivity of a scattering volume [F·m^{-1}]
$\boldsymbol{\epsilon}$ = Electric permittivity tensor [F·m^{-1}]

$\zeta(z)$ = Excess axial phase of a Gaussian beam; also, Fraction of electron–hole pairs that successfully contribute to detector photocurrent
ζ = Damping coefficient of a harmonic oscillator [s^{-1}]; also, Scattering rate or collision frequency ($\zeta = 1/\tau$ where τ = scattering time or collision time) [s^{-1}]

η = Impedance of a dielectric medium [Ω]; also, Photodetector quantum efficiency; also, Photon detection efficiency
η_c = Power-conversion efficiency (also called Overall efficiency and Wall-plug efficiency)
η_d = External differential quantum efficiency
η_e = Extraction efficiency; also, Transmission efficiency
η_{ex} = External efficiency
η_i = Internal quantum efficiency
η_o = Impedance of free space [Ω]
η_o = Optical-to-optical efficiency
η_{OFC} = Optical frequency conversion efficiency
η_s = Differential power-conversion efficiency (also called Slope efficiency)
η_s = Optical-to-optical slope efficiency
η_{SHG} = Second-harmonic generation efficiency
η = Electric impermeability
η_{ij} = Component of the electric impermeability tensor
$\boldsymbol{\eta}$ = Electric impermeability tensor

θ = Angle; also, Twist angle in a liquid crystal; also, Deflection angle of a prism
$\bar{\theta} = 90° - \theta$ = Complement of angle θ
θ_a = Acceptance angle
θ_B = Brewster angle
$\theta_{\mathcal{B}}$ = Bragg angle
θ_c = Critical angle
$\bar{\theta}_c$ = Complementary critical angle
θ_d = Deflection angle of a prism
$\theta_{\max}$ = Maximum angle
θ_p = Angle of incidence of a ray at the surface of a prism from its interior
θ_r = Angle of reflection
θ_s = Deviation angle of maximum spatial frequency; also, Angle subtended by a source
θ_0 = Divergence angle of a Gaussian beam
$\hat{\boldsymbol{\theta}}$ = Unit vector in polar direction in spherical coordinates
ϑ = Threshold
$\Theta_{\ell m}(\theta)$ = Hydrogen-atom associated Legendre function

κ = Elastic constant of a harmonic oscillator [J·m^{-2}]

λ = Wavelength [m]
λ_A = Long-wavelength limit [m]; also, Activation wavelength [m]
λ_c = Cutoff wavelength [m]
λ_F = Wavelength spacing of adjacent Fabry–Perot resonator modes [m]

λ_{FEL} = Wavelength of light emitted by a free-electron laser [m]
λ_g = Bandgap wavelength (long-wavelength limit) of a semiconductor [m]
λ_o = Free-space wavelength [m]
λ_p = Plasma wavelength of a metal [m]
λ_{p} = Wavelength of maximum blackbody energy density [m]; also, Peak wavelength [m]
λ_q = Eigenvalues associated with an eigenvalue problem
λ_{ZD} = Zero-dispersion wavelength of a medium [m]
λ_0 = Central wavelength [m]
λ_{dB} = de Broglie wavelength [m]
Λ = Spatial period of a grating or periodic structure [m]; also, Wavelength of acoustic wave [m]
Λ_{u} = Spatial period of undulator in a free-electron laser [m]

μ = Magnetic permeability of a medium [H · m^{-1}]; also, Carrier mobility in a semiconductor [m^2 · s^{-1} · V^{-1}]; also, Mean of a random variable
μ' = Real part of the magnetic permeability of a medium [H · m^{-1}]
μ'' = Imaginary part of the magnetic permeability of a medium [H · m^{-1}]
μ_e = Effective magnetic permeability of a host medium with embedded objects [H · m^{-1}]
μ_e = Electron mobility [m^2 · s^{-1} · V^{-1}]
μ_h = Hole mobility [m^2 · s^{-1} · V^{-1}]
μ_{ij} = Component of the magnetic permeability tensor [H · m^{-1}]
μ_o = Magnetic permeability of free space [H · m^{-1}]
$\boldsymbol{\mu}$ = Magnetic permeability tensor [H · m^{-1}]

ν = Frequency [Hz]; also, Spatial frequency [m^{-1}]
ν_A = Anti-Stokes-shifted frequency [Hz]
ν_B = Brillouin frequency [Hz]
$\nu_{\mathcal{B}}$ = Bragg frequency [Hz]
ν_c = Cutoff frequency [Hz]
ν_F = Frequency spacing between adjacent Fabry–Perot modes (free spectral range) [Hz]
ν_i = Offset frequency for an optical frequency comb [Hz]
ν_{i} = Instantaneous frequency [Hz]
ν_p = Pump frequency [Hz]
ν_{p} = Frequency of maximum blackbody energy density [Hz]; also, Peak frequency of electroluminescence spectrum [Hz]
ν_q = Frequency of mode q [Hz]
ν_R = Raman frequency [Hz]
ν_s = Spatial bandwidth of an imaging system [m^{-1}]; also, Signal frequency [Hz]; also, Laser frequency [Hz]
ν_S = Stokes-shifted frequency [Hz]
ν_x, ν_y = Spatial frequencies in the x and y directions [m^{-1}]
ν_0 = Central frequency [Hz]
ν_ρ = Radial component of the spatial frequency: $\nu_\rho = \sqrt{\nu_x^2 + \nu_y^2}$ [m^{-1}]

ξ = Coupling coefficient in four-wave mixing
$\xi_{\text{sp}}(\nu)$ = Amplifier noise photon-flux density per unit length [m^{-3} · s^{-1}]

ρ = Rotatory power of an optically active medium [m^{-1}]; also, Resistivity [Ω · m]; also, $\rho = \sqrt{x^2 + y^2}$ = radial distance in a cylindrical coordinate system [m]
ρ_c = Coherence distance [m]
ρ_m = Radius of the center of the mth ring of a Fresnel zone plate [m]

ρ_s = Radius of the Airy disk [m]; also, Radius of the blur spot of an imaging system [m]
ϱ = Mass density of a medium [kg · m^{-3}]; also, Charge density [C · m^{-3}]; also, Retardance of a birefringent medium [m]
$\varrho(k)$ = Wavenumber density of states [m^{-2}]
$\varrho(\nu)$ = Spectral energy density [J · m^{-3} · Hz^{-1}]; also, Optical joint density of states [m^{-3} · Hz^{-1}]
$\varrho_c(E)$ = Density of states near the conduction band edge [m^{-3} · J^{-1} in a bulk semiconductor]
$\varrho_v(E)$ = Density of states near the valence band edge [m^{-3} · J^{-1} in a bulk semiconductor]
$\rho(\nu)$ = Normalized Lorentzian cavity mode [Hz^{-1}]

σ = Conductivity [Ω^{-1} · m^{-1}]
$\sigma(\nu)$ = Transition cross section [m^2]
$\overline{\sigma}(\nu)$ = Average transition cross section [m^2]
σ_a = Absorption cross section [m^2]
σ_{ab} = Effective transition cross section for absorption [m^2]
σ_{em} = Effective transition cross section for emission [m^2]
σ_f = Transfer-function bandwidth [Hz]
σ_{max} = Maximum transition cross section [m^2]
σ_q = Circuit-noise parameter
σ_r = Circuit-noise current RMS value [A]
σ_s = Scattering cross section [m^2]
σ_x = Standard deviation of a random variable x; RMS width of a function of x
σ_x^2 = Variance of a random variable x
$\sigma_0 = \sigma(\nu_0)$ = Transition cross section at the central frequency ν_0 [m^2]
σ_0 = Conductivity at low frequencies [Ω^{-1} · m^{-1}]
σ_λ = Spectral width [m]
σ_τ = Temporal width [s]
$\boldsymbol{\sigma}$ = Conductivity tensor [Ω^{-1} · m^{-1}]

τ = Lifetime [s]; also, Decay time [s]; also, Pulse width [s]; also, Relaxation time [s]; also, Scattering time [s]; also, Collision time [s]; also, Width of a function of time [s]; also, Excess-carrier electron–hole recombination lifetime in a semiconductor [s]
τ_c = Coherence time [s]
τ_{col} = Mean time between collisions [s]
τ_d = Delay time [s]
τ_e = Electron transit time [s]
τ_h = Hole transit time [s]
τ_m = Multiplication time in an avalanche photodiode [s]
τ_{nr} = Nonradiative electron–hole recombination lifetime [s]
τ_p = Resonator photon lifetime [s]
τ_{pulse} = Duration of a mode-locked optical pulse [s]
τ_r = Radiative electron–hole recombination lifetime [s]
τ_R = Receiver-circuit time constant [s]
τ_{RC} = RC time constant [s]
τ_s = Saturation time constant of a laser transition [s]
τ_{21} = Lifetime of a transition between energy levels 2 and 1 [s]

ϕ = Angle in a cylindrical or spherical coordinate system; also, Photon-flux density [m^{-2} · s^{-1}]
$\phi(p)$ = Particle momentum wavefunction [s$^{1/2}$ · kg$^{-1/2}$ · m$^{-1/2}$]
$\phi(p)$ = Wavefunction for p quadrature component of the electric field
$\phi_s(\nu)$ = Saturation photon-flux density [m^{-2} · s^{-1}]

ϕ_ν = Spectral photon-flux density [m^{-2} · s^{-1} · Hz^{-1}]
$\widehat{\boldsymbol{\phi}}$ = Unit vector in azimuthal direction in spherical coordinates
φ = Phase or phase difference or phase shift
$\varphi(t)$ = Phase of the complex envelope of an optical pulse
$\varphi(\nu)$ = Phase-shift coefficient of an optical amplifier [m^{-1}]
φ_L = Local-oscillator phase
φ_0 = Phase shift from reflection at a resonator mirror
Φ = Photon flux [s^{-1}]
$\Phi(\nu, \tau)$ = Short-time Fourier transform; also, Wigner distribution function
$\Phi_\mathrm{m}(\phi)$ = Hydrogen-atom harmonic function
Φ_ν = Spectral photon flux [s^{-1} · Hz^{-1}]

χ = Electric susceptibility; also, Electron affinity [J]
χ' = Real part of the electric susceptibility χ
χ'' = Imaginary part of the electric susceptibility χ
$\chi(\nu)$ = Electric susceptibility of a dispersive medium
χ_e = Effective electric susceptibility
χ_{ij} = Component of the electric susceptibility tensor
χ_m = Electric susceptibility of a metal at high frequencies ($\omega \gg \omega_p$)
$\chi^{(3)}$ = Coefficient of third-order optical nonlinearity [C · m · V^{-3}]
$\chi^{(3)}_{ijkl}$ = Component of the third-order optical nonlinearity tensor [C · m · V^{-3}]
$\chi^{(3)}_\mathrm{I}$ = Imaginary part of the nonlinear third-order susceptibility $\chi^{(3)}$
$\chi^{(3)}_{IK}$ = Component of the third-order optical nonlinearity tensor (contracted indices) [C · m · V^{-3}]
$\chi^{(3)}_\mathrm{R}$ = Real part of the nonlinear third-order susceptibility $\chi^{(3)}$
X = Polarization-ellipse angle of ellipticity
$\boldsymbol{\chi}$ = Electric susceptibility tensor

ψ = Normalized amplitude of an optical pulse
$\psi(\mathbf{r}, t)$ = Particle wavefunction [m$^{-3/2}$ · s$^{-1/2}$]
$\psi(x)$ = Particle position wavefunction [m$^{-1/2}$]
$\psi(x)$ = Wavefunction for x quadrature component of the electric field
$\psi(\nu)$ = Spectral phase of an optical pulse
$\boldsymbol{\psi}$ = Polarization-ellipse orientation of major axis
$\Psi(\mathcal{E})$ = Nonlinear polarization density [C · m^{-2}]
$\Psi_e(f)$ = Envelope transfer function phase

ω = Angular frequency [rad · s^{-1}]
ω_B = Bragg angular frequency [rad · s^{-1}]
ω_i = Instantaneous angular frequency [rad · s^{-1}]
ω_I = Heterodyne intermediate (angular) frequency [rad · s^{-1}]
ω_L = Local-oscillator frequency [rad · s^{-1}]
ω_p = Plasma frequency of a metal [rad · s^{-1}]; also, Pump angular frequency [rad · s^{-1}]
ω_r = Upshifted angular frequency of a Bragg-reflected wave [rad · s^{-1}]
ω_s = Downshifted angular frequency of a Bragg-reflected wave [rad · s^{-1}]; also, Angular frequency below which a surface plasmon polariton can exist [rad · s^{-1}]; also, Signal frequency [rad · s^{-1}]
ω_0 = Central angular frequency [rad · s^{-1}]; also, Localized surface plasmon resonance frequency [rad · s^{-1}]
Ω = Angular frequency of an acoustic wave [rad · s^{-1}]; also, Angular frequency of a harmonic electric signal [rad · s^{-1}]; also, Solid angle [sr]

Mathematical Symbols

$\doteqdot$ = Result decreased to the nearest integer
$\doteq$ = Result increased to the nearest integer
$\det\{\cdot\}$ = Determinant of a matrix
$\text{Tr}\{\cdot\}$ = Trace of a matrix
$\{\cdot\}^{\mathsf{T}}$ = Transpose of a matrix
$\{\cdot\}^{-1}$ = Inverse of a matrix
$\bar{x}$ = Mean of the quantity x
$\langle x \rangle$ = Ensemble average over x
d = Differential
∂ = Partial differential
∇ = Gradient operator
$\nabla \cdot$ = Divergence operator
$\nabla \times$ = Curl operator
∇^2 = Laplacian operator ($\nabla^2 = \partial^2/\partial x^2 + \partial^2/\partial y^2 + \partial^2/\partial z^2$ in Cartesian coordinates)
∇_T^2 = Transverse Laplacian operator ($\nabla_T^2 = \partial^2/\partial x^2 + \partial^2/\partial y^2$ in Cartesian coordinates)

AUTHORS

Bahaa E. A. Saleh has been Distinguished Professor and Dean of CREOL, The College of Optics and Photonics at the University of Central Florida, since 2009. He was at Boston University in 1994–2008, serving as Chair of the Department of Electrical and Computer Engineering (ECE) in 1994–2007, and becoming Professor Emeritus in 2008. He received the Ph.D. degree from the Johns Hopkins University in 1971 and was a faculty member at the University of Wisconsin-Madison from 1977 to 1994, serving as Chair of the ECE Department from 1990 to 1994. He held faculty and research positions at the University of Santa Catarina in Brazil, Kuwait University, the Max Planck Institute in Germany, the University of California-Berkeley, the European Molecular Biology Laboratory, Columbia University, and the University of Vienna.

His research contributions cover a broad spectrum of topics in optics and photonics including statistical optics, nonlinear optics, quantum optics, and image science. He is the author of *Photoelectron Statistics* (Springer-Verlag, 1978) and the co-author of *Fundamentals of Photonics* (Wiley, *First Edition* 1991, *Second Edition* 2007, *Third Edition* 2019) and *Introduction to Subsurface Imaging* (Cambridge University Press, 2011). He has published more than 600 papers in technical journals and conference proceedings. He holds nine patents.

Saleh served as the founding editor of the OSA journal *Advances in Optics and Photonics* (2008–2013), editor-in-chief of the *Journal of the Optical Society of America A* (1991–1997), and Chairman of the Board of Editors of OSA (1997–2001). He served as Vice President of the International Commission of Optics (ICO) (2000–2002), and as a member of the Board of Directors of the Laser Institute of America (2010–2011). He is a Fellow of the Institute of Electrical and Electronics Engineers (IEEE), the Optical Society of America (OSA), the International Society for Optics and Photonics (SPIE), the American Physical Society (APS), and the Guggenheim Foundation. He received the 1999 OSA Beller Medal for outstanding contributions to optical science and engineering education, the 2004 SPIE BACUS award for his contribution to photomask technology, the 2006 Kuwait Prize, the 2008 OSA Distinguished Service Award, and the 2013 OSA Mees Medal. He is a member of Phi Beta Kappa, Sigma Xi, and Tau Beta Pi.

Malvin Carl Teich received the S.B. degree in physics from the Massachusetts Institute of Technology in 1961, the M.S. degree in electrical engineering from Stanford University in 1962, and the Ph.D. degree from Cornell University in 1966. His first professional affiliation, in 1966, was with MIT Lincoln Laboratory. He joined the faculty at Columbia University in 1967, where he served as a member of the Electrical Engineering Department (as Chairman from 1978 to 1980), the Applied Physics and Applied Mathematics Department, the Columbia Radiation Laboratory in the Department of Physics, and the Fowler Memorial Laboratory at the Columbia College of Physicians & Surgeons. During his tenure at Columbia, he carried out research in the areas of photon statistics and point processes; quantum heterodyne detection; the generation of nonclassical light; noise in avalanche photodiodes and fiber-optic amplifiers; and information transmission in biological sensory systems. In 1996 he became Professor Emeritus at Columbia.

From 1995 to 2011, he served as a faculty member at Boston University in the Departments of Electrical & Computer Engineering, Biomedical Engineering, and Physics. He was the Director of the Quantum Photonics Laboratory and a Member of the Photonics Center, the Hearing Research Center, and the Graduate Program for Neuroscience. In 2011, he was appointed Professor Emeritus at Boston University.

Since 2011, Dr. Teich has been pursuing his research interests as Professor Emeritus at Columbia University and Boston University, and as a member of the Boston University Photonics Center. He is also a consultant to government and private industry and has served as an expert in numerous patent conflict cases. He is most widely known for his work in photonics and for his studies of fractal stochastic processes and information transmission in biological systems. His current efforts in photonics are directed toward the characterization of noise in photon streams. His work in fractals focuses on elucidating the information-carrying properties of sensory-system action-potential patterns and the nature of heart-rate variability in patients with coronary disorders. His efforts in neuroscience are directed toward auditory and visual perception, neural information transmission, and sensory detection. During periods of sabbatical leave, he served as a visiting faculty member at the University of Colorado at Boulder, the University of California at San Diego, and the University of Central Florida at Orlando.

Dr. Teich is a Life Fellow of the Institute of Electrical and Electronics Engineers (IEEE), and a Fellow of the Optical Society of America (OSA), the International Society for Optics and Photonics (SPIE), the American Physical Society (APS), the American Association for the Advancement of Science (AAAS), and the Acoustical Society of America (ASA). He is a member of Sigma Xi and Tau Beta Pi. In 1969 he received the IEEE Browder J. Thompson Memorial Prize for his paper "Infrared Heterodyne Detection." He was awarded a Guggenheim Fellowship in 1973. In 1992 he was honored with the Memorial Gold Medal of Palacký University in the Czech Republic, and in 1997 he received the IEEE Morris E. Leeds Award. In 2009, he was honored with the Distinguished Scholar Award of Boston University. He has authored or coauthored some 350 refereed journal articles/book chapters and some 550 conference presentations/lectures; he holds six patents. He is the co-author of *Fundamentals of Photonics* (Wiley, *First Edition* 1991, *Second Edition* 2007, *Third Edition* 2019) and of *Fractal-Based Point Processes* (Wiley, 2005, with S. B. Lowen).

INDEX

k surface
    anisotropic medium, 234–237
ABCD
    inverse ray-transfer matrix, 31
    law, 97–99
    ray-transfer matrix, 28, 1302
    transmission matrix, 380
    wave-transfer matrix, 259, 1303
g-parameters, 448
q-parameter, 81, 97

Absorption
    and scattering, 198–199
    atoms, 584, 589–593
    coefficient, 182, 198, 777
    coefficient in resonant medium, 189
    cross section, 199
    dielectric media, 181–184
    efficiency, 199
    indirect-bandgap semiconductor, 771
    left-handed media, 309
    occupancy probability, 773
    semiconductors, 769
    strong, 184
    transition rate, 592, 775
    weak, 183
Acceptance angle, 19, 25, 395
Acousto-optic devices, 958–967
    acoustic spectrum analyzer, 963
    figure of merit, 973
    filter, 966
    frequency shifter, 967
    intensity modulator, 959, 973
    intensity reflectance, 973
    interconnection capacity, 966
    optical isolator, 967
    reflector, 973
    scanner, 961
    spatial light modulator, 965
    switch, photonic, 959, 964, 1194
Acousto-optics, 943–974
    anisotropic media, 967–971
    Born approximation, 953, 973

    Bragg angle, 948
    Bragg condition, 948
    Bragg condition tolerance, 949
    Bragg diffraction, 945–958, 972, 973
    Bragg diffraction by beam, 956, 961
    Bragg diffraction by thin beam, 957
    Bragg diffraction, downshifted, 952
    Bragg diffraction, upshifted, 949
    coupled-wave equations, 953, 954
    Debye–Sears scattering, 957, 973
    Doppler shift, 950
    elasto-optic coefficient, 946
    figure of merit, 946
    index-ellipsoid modification, 970
    introduction to, 944–945
    Raman–Nath scattering, 973
    reflectance, amplitude, 948
    reflectance, intensity, 950
    scattering theory, 953, 973
    strain-optic coefficient, 946
Airy
    beam, 106–107
    beam generation, 107, 158
    disk, 131
    formulas, 261
    function, 107
    pattern, 131
Amplified spontaneous emission (ASE), 651–653, 699, 821, 830
Amplifier
    Brillouin fiber, 644
    chirped pulse, 1095
    electronic, 620, 621
    laser, 619–656, 659
    phase, 70
    phase-sensitive, 1062
    Raman fiber, 642–644
    reflection, 1065
    semiconductor, 817–831
    transmission, 1065
Anharmonic oscillator, 1071–1073
Anisotropic media
    k surface, 234–237

acousto-optics of, 967–971
arbitrary propagation, 232–233
biaxial crystal, 229
conductive, 882
dispersion relation, 234–237
double refraction, 238–239, 254
electro-optics of, 989–996
ellipsoid of revolution, 318
energy transport, 234–237
extraordinary refractive index, 229
four-wave mixing, 1074
gyration vector, 241
hyperbolic, 339–340, 352
hyperboloid of revolution, 318
impermeability tensor, 230
index ellipsoid, 230, 232–233
liquid crystals, 244–246, 996–1005
magneto-optics, 242–243
magnetogyration coefficient, 243
mixed-sign permittivities, 317, 352
negative uniaxial crystal, 229
negative-index, 338–339
nonlinear and dispersive, 1073
nonlinear optics of, 1066–1069
normal modes, 230, 232–233, 241
normal surface, 234–237
optic axis, 229
optical activity, 240–242
optics of, 227–239
ordinary refractive index, 229
permittivity tensor, 228
photoelastic effect, 968
positive uniaxial crystal, 229
principal axes, 229
principal refractive indices, 229
principal-axis propagation, 230–231
quadric representation, 229, 230, 317
rays, 234–237
refraction of rays, 239, 254
refractive indices of, 228–230
three-wave mixing, 1066–1069, 1073
uniaxial crystal, 229, 233, 236, 317
Verdet constant, 243
wavefronts, 234–237
Antenna
optical, 332–334
plasmonic, 332–334
Antibunching, photon, 780
Antireflection coating, 264, 301
Aperture
function, 129, 138
ring, 67, 118, 159
Approximation
Born, 192, 953, 973, 1020, 1076

Fraunhofer, 124, 125, 130
Fresnel, 49, 121, 125, 130
paraboloidal, 49
paraxial, 9, 50
quasi-static, 196, 197
quasi-stationary, 944
SVE, 51, 563, 954, 1050, 1064, 1086, 1111, 1127, 1133
Taylor-series, 10, 203, 1127
Volterra-series, 1070
WKB, 407
Array detectors, *see* Photodetectors
Atoms
absorption, 584, 589–593
atom amplifier, 602
atom chip, 602
atom interferometry, 601
atom optics, 601
atomic number, 564, 568
Bose–Einstein condensate, 601
bosonic, 568, 602
Doppler cooling, 599
electron configuration, 565, 566, 569
energy levels, 562
evaporative cooling, 600
fermionic, 568
Hartree method, 565
hydrogen, 564
interaction with broadband light, 590
interaction with light, 583–602
ionization energy, 568
isotopes, 568
laser cooling, 599
laser trapping, 600
lineshape function, 586
magnetic trap, 602
magneto-optical trap, 601
matter waves, 601
mononuclidic, 568
multielectron, 565–568
optical lattices, 601
optical molasses, 599
optical tweezers, 600
Pauli exclusion principle, 565
periodic table, 566, 567
photon recoil, 524, 555, 600
polarization gradient cooling, 600
relative atomic mass, 568
shells and subshells, 565
Sisyphus cooling, 600
spontaneous emission, 583, 587–588
spontaneous lifetime, 587, 588
spontaneous lifetime, effective, 589
stimulated emission, 585, 589–593

term symbol, 566, 569
transition strength, 586
Attenuation
  and scattering, 198, 416, 432
  coefficient, 182, 415–417
  optical fiber, 415–417
Attosecond optics, 1141–1145
Auger recombination, 752, 821
Autocorrelation, 1288
Avalanche photodiodes, 895–907
  avalanche buildup time, 899
  breakdown voltage, 901
  comparison with photodiodes, 927
  conventional (CAPD), 895
  dead space, 903
  excess noise factor, 917–921
  gain, 896, 898, 940
  gain noise, 917–921
  Geiger-mode, 904
  group-IV, 902
  initial-energy effects, 903
  ionization coefficients, 895
  ionization ratio, 896
  materials, 901
  position-dependent gradients, 904
  position-dependent ionization, 904
  punchthrough voltage, 901
  reach-through device, 899
  response time, 899
  responsivity, 898
  SACM device, 899, 902
  SAM device, 899, 901, 902
  SCIDCM devices, 897, 917
  SCISCM devices, 896
  sick space, 903
  single-photon (SPAD), 904
Axicon, 14, 57, 106, 137

Band-structure engineering, 851
Bandgap
  direct, 737
  energy, 576, 733, 738
  indirect, 737
  wavelength, 738, 768
Bardeen, John, 731
Basov, Nikolai G., 619
Beam, optical, 79–109
  accelerating, 107
  Airy, 106–107, 158
  Bessel, 57, 105, 136, 137
  Bessel–Gaussian, 106, 181
  Bessel/Gaussian comparison, 106
  divergence–size tradeoff, 85, 106
  donut, 103, 109
  elliptic Gaussian, 102, 109
  elliptic Hermite–Gaussian, 102
  Gaussian, 51, 80–99
  Hermite–Gaussian, 99–102, 105
  holographic generation, 155
  Ince–Gaussian, 105
  introduction to, 80
  Laguerre–Gaussian, 102–105, 155
  nondiffracting, 57, 105, 136, 137
  quality, 90
  self-healing, 107
  toroidal, 103, 109
  vector, 179, 180
  vortex, 104
Beamsplitter, 13, 63, 301, 549
  dielectric-slab, 268
  phase shift, 64, 269
  polarizing, 239, 247
  random partitioning, 539, 555
  scattering matrix, 263
  single-photon transmittance, 522
  two-photon transmittance, 550
  wave-transfer matrix, 263
Bell Laboratories, 1163
Bessel
  -Gaussian beam, 106
  beam, 57, 105, 136, 137
  beam vs. Gaussian beam, 106
  function, 105, 131, 398, 467, 1295
Bessel, Friedrich Wilhelm, 79
Bioluminescence, 608
Biprism, 14, 57
  Fresnel, 14, 57
Bistability, optical
  amplifying NL element, 1218, 1219
  coupled microring lasers, 1219
  devices, 1216–1220
  dispersive NL element, 1216, 1220
  dissipative NL element, 1217
  embedded system, 1215
  Fabry–Perot, nonlinear, 1217, 1218
  hybrid, 1220
  hysteresis, 1211
  intrinsic system, 1215
  MZI, nonlinear, 1216, 1218
  photonic logic, 1223
  principles of, 1213–1215
  quantum-confined Stark effect, 1220
  saturable absorber, 1217
  self-electro-optic-effect device, 1220
  spatial light modulator, 1220
  systems, 1211
Blackbody radiation, 602–607
  1D, 618, 923

2D, 617
    spectrum, 604
    correlated color temperature, 812
    emission comparison, 617
    Rayleigh–Jeans formula, 463, 606
    Stefan–Boltzmann law, 606, 618
    thermal equilibrium, 602
    thermal light, 602
    thermography, 606
    Wien's law, 617
Bloch
    modes, 257, 277–282, 295, 734, 1305
    phase, 281
    wavenumber, 278
Bloch, Felix, 255
Bloembergen, Nicolaas, 1015
Bohr
    atom, 564
    period, 564, 705
    radius, 564, 784
Bohr, Niels, 561
Boltzmann distribution, 536, 581
Bonding
    covalent, 573, 576, 743
    ionic, 573, 576
    metallic, 576
    van der Waals, 573, 576, 743
Born approximation
    for acousto-optics, 953, 973
    for nonlinear optics, 1020
    for scattering, 192–193
Born postulate, 563
Born, Max, 473
Bose–Einstein
    condensate, 601, 602
    distribution, 536
Boson, 524, 537, 568
Boundary
    conditions, 164, 221
    planar, 11
    spherical, 14
Bragg
    angle, 67, 270, 948
    condition, 270, 948
    condition, tolerance, 949
    diffraction, 945–958, 972, 973
    diffraction by acoustic beam, 956, 961
    diffraction of optical beam, 955
    diffraction, quantum view, 952
    diffraction, Raman–Nath, 957
    frequency, 270
    reflection, 67
Bragg grating, 269–276, 698
    chirp filter, 1099, 1161
    distributed Bragg reflector, 269, 434, 466, 471, 798
    fiber, 269, 687, 692
    total reflection, 273
    waveguide, 385
Bragg, William Henry, 943
Bragg, William Lawrence, 943
Brattain, Walter H., 731
Bremsstrahlung, 562
Brewster
    angle, 224, 225, 247, 253
    window, 225
Brillouin
    fiber amplifier, 644
    scattering, 613
    zone, 278, 279, 283, 735, 738
    zone, irreducible, 295
Bulk optics, 305, 354

Catadioptric system, 15
Cathodoluminescence, 607, 608, 618
Caustic
    curve, 9, 17
    surface, 397
Characteristic equation
    optical fiber, 400
Chemiluminescence, 608
Cherenkov radiation, 562, 875
Chirp
    coefficient, 1090
    coefficient, angular-dispersion, 1097
    coefficient, Bragg-grating, 1099
    function, 1289
    parameter, 1082
    pulse amplifier, 1095
Chirp filter, 1088–1099
    angular-dispersion, 1096
    arbitrary phase filter, 1090
    Bragg-grating, 1099, 1161
    cascaded, 1090
    chirp coefficient, 1090
    diffraction-grating, 1098
    envelope transfer function, 1088
    for compression, 1094
    for expansion, 1094
    ideal filter, 1089
    implementations of, 1095–1099
    impulse response function, 1090
    optical-fiber, 1103–1111, 1161
    phase filter, 1089
    prism, 1097, 1105, 1161
    vs. quadratic-phase modulator, 1100
Circular

dichroism, 254
  polarization, 213, 253, 524
Circulator, optical, 251, 1168, 1173
CMOS technology, 375, 907, 909
  scientific CMOS (sCMOS), 909
Coherence
  area, 485, 495
  coherent light, 62, 152, 474, 544
  cross-spectral density, 486
  cross-spectral purity, 486
  degree of coherence, 483
  degree of temporal coherence, 478
  distance, 495, 505
  enhancement via filtering, 481, 489
  extended radiator, 485
  gain via propagation, 502
  incoherent light, 485
  introduction to, 474–475
  length, 479
  length, wave-mixing, 1032
  longitudinal, 487
  mutual coherence function, 483
  mutual intensity, 485, 498, 502, 549
  partial polarization, 506–510
  power spectral density, 479
  quantum, 483
  quasi-monochromatic light, 485
  random wavepacket sequence, 482
  role in interference, 489–496
  role in interferometry, 490, 493
  spatial, 483–487
  speckle, 406
  spectral density, 479
  spectral width, 481
  temporal, 476–483
  temporal coherence function, 477
  time, 478
  transmission, 497
  van Cittert–Zernike theorem, 503
  Wiener–Khinchin theorem, 480
  Wolf equations, 484
Coherent
  anti-Stokes Raman scattering, 614
  detection, 74, 1266
  lidar, 75
  optical amplifier, 620
Colladon, Jean-Daniel, 353
Collimator, 9, 15
Color-rendering index, 812, 816
Comb, optical frequency, 721, 1141
  applications, 722
Complex
  amplitude, 46, 80
  analytic signal, 71

  degree of coherence, 483
  degree of temporal coherence, 478
  envelope, 47, 73, 80, 175, 211, 1080
  representation, 45
  wavefunction, 45, 72
Conductive medium, 320–326
  power reflectance, 321
Conductivity, 320, 882
  tensor, 882
Constitutive relation, 166
Converter
  evanescent-to-propagating, 317
  frequency, 548, 610, 611, 1027, 1054
  incoherent-to-coherent, 989, 1005
  indistinguishable-to-entangled, 550
  mode, 789, 1265
  space-to-time, 1119
  wavelength, 1205
Convolution, 1288, 1294, 1296, 1300
Corkum, Paul B., 1078
Correlation, 1288
Coupled-wave theory
  acousto-optics, 953, 954
  four-wave mixing, 1059
  three-wave mixing, 1047, 1127
  waveguide, 378
Coupler
  prism, 329
Critical angle, 12, 223, 225, 310, 327
  complementary, 363
Cross section
  absorption, 199
  effective absorption, 585, 634
  effective emission, 585, 634
  Füchtbauer–Ladenburg equation, 588
  lineshape function, 586
  scattering, 195, 196
  transition, 584, 588, 634, 644
  transition strength, 586
Cross-phase modulation, 1041, 1197
Crystal
  biaxial, 229
  negative uniaxial, 229
  positive uniaxial, 229
  structure, 294
  symmetry, 991
  uniaxial, 229
Cutoff frequency
  waveguide, 359, 366

Debye–Sears scattering, 973
Degeneracy parameter, 582
Delay, group, 200
Detectors, *see* Photodetectors

Dichroism, 247
  circular, 254
Dielectric
  boundary, reflection from, 53
  boundary, refraction at, 53
  constant, 167
Diffraction, 129–136
  aperture function, 129
  Bragg, 945–958
  circular aperture, 131
  dispersion analogy, 1112
  focused optical beam, 132
  Fraunhofer, 130–132, 158
  Fresnel, 132–137, 158
  Gaussian aperture, 134
  grating, 59, 78, 158
  nondiffracting beams, 105–107
  nondiffracting waves, 136
  periodic aperture, 135
  rectangular aperture, 130
  single-photon, 523
  slit, 133
  Talbot effect, 135
  two pinholes, 158
Diffusion equation, 1111, 1112
Diode lasers, see Laser diodes
Dipole
  electric, 177
  magnetic, 178
  moment, 177, 178
  wave, 177, 194
Dirac equation, 565
Directional coupler, 381, 984
  integrated-photonic, 986
  nonlinear, 1186
  soliton router, 1187
Dispersion, 184–191
  -compensating fiber, 423
  -flattened fiber, 423
  -shifted fiber, 422
  angular, 1096
  anomalous, 202
  chromatic, 422
  coefficient, 201, 202, 419
  compensation, 685, 1110, 1111, 1254
  diffraction analogy, 1112
  diffractive, 1096
  group velocity, 201, 207, 370, 425, 550, 1104
  in multi-resonance medium, 204
  interferometric, 1096
  management, 1255
  material, 419, 1096
  material and modal, 420
  measures, 185
  modal, 370, 389, 418, 1096
  multipath, 1096
  nonlinear, 425, 1096
  normal, 202
  optical-fiber, 418–426
  polarization mode, 423, 1096
  pulse propagation with, 199–203
  spatial, 1096
  waveguide, 421, 1096
Dispersion relation
  anisotropic medium, 234–237
  photonic crystal, 282, 285, 296
  waveguide, 359, 370
Dispersive medium
  anisotropic and nonlinear, 1073
  four-wave mixing, 1074
  nonlinear optics of, 1069–1074
  three-wave mixing, 1073, 1076, 1162
Distributed Bragg reflector, 269, 434, 466, 471, 798
  fiber, 687, 692, 720
  laser, 843
Doped dielectric media, 568–572
  actinide metals, 572
  lanthanide metals, 571, 572
  transition metals, 569
Doppler
  effect, 75
  radar, optical, 75
  shift, 950
Double refraction, 238–239, 254
Double-slit experiment, 65, 493, 525
  Michelson stellar interferometer, 505
  role of source size, 495
  role of spectral width, 495
Downconversion, 1077
Downconversion, parametric, 1027
Drude model, 322–326, 388
  group velocity, 352
  simplified, 324–326
Drude, Paul Karl Ludwig, 303
Drude–Lorentz model, see Drude model

Efficiency
  absorption, 199
  differential power-conversion, 670
  external, 800
  external differential, 836, 838
  extraction, 670, 838
  internal, 754, 796, 838
  optical-to-optical, 671
  optical-to-optical slope, 671
  overall, 670

photon detection, 876
power-conversion, 670, 837, 838
scattering, 196
slope, 670, 837
wall-plug, 670, 838
Eigenvalue problem, 220, 279, 563, 1301
Eikonal equation, 26, 52, 343, 407
Einstein
    $\mathbb{A}$ and $\mathbb{B}$ coefficients, 591
    postulates, 591
Einstein, Albert, 515, 561, 871
Electric
    dipole, 177
    dipole moment, 177
    field, 162
    field, internal, 196
    flux density, 163
Electro-optic devices
    directional coupler, 381, 984
    dynamic wave retarder, 980
    intensity modulator, 981, 995, 1014
    interferometric photonic switch, 1192
    liquid-crystal SLM, 1002–1005
    phase modulator, 973, 979, 995, 1013, 1014
    self-electro-optic-effect device, 1220
    spatial light modulator, 987–989
    strain sensor, 1013
    switch, 985
Electro-optics, 975–1014
    anisotropic media, 989–996
    double refraction, 1014
    index-ellipsoid modification, 989
    introduction to, 976–977
    Kerr effect, 978, 994, 1037, 1077
    Pockels effect, 978, 991–994, 1024, 1077
    principles of, 977–989
Electroabsorption, 1010–1012
    Franz–Keldysh effect, 1010
    modulator, 1010, 1011
    switch, 1010
Electrochromism, 1009
Electroluminescence, 608, 789
    injection, 609, 790
Electromagnetic optics, 160–208
    constitutive relation, 166
    introduction to, 161–162
    material equation, 173
    relation to wave optics, 180
Electromagnetic wave
    amplitude-modulated, 207
    dipole wave, 177, 194
    energy, 165

    energy density, 176
    Gaussian beam, 179
    in a medium, 163
    in absorptive medium, 181, 309–310
    in anisotropic medium, 170
    in conductive medium, 320–326
    in dielectric medium, 166–180
    in dispersive medium, 170, 184, 199
    in double-negative medium, 308
    in double-positive medium, 308
    in free space, 162
    in inhomogeneous medium, 168, 174
    in negative-index material, 314
    in nonlinear medium, 171
    in resonant medium, 186
    in single-negative medium, 308
    intensity, 165, 172, 176
    momentum, 165
    monochromatic, 172–180
    plane wave, 175, 307
    power, 165, 172
    Poynting vector, 165
    scattered, 194, 196
    spherical wave, 177
    TEM wave, 175, 307
    traveling standing, 207
    vector beam, 180
Electron configuration, 566
    actinide metals, 569
    He, 566
    lanthanide metals, 569
    Ne, 566, 701
    Ne^+, 701
    Ne-like, 699
    Ni-like, 699
    rare-earth elements, 569
    transition metals, 569
Electroweak theory, 515
Elements
    actinide metals, 569, 572
    lanthanide metals, 569, 571
    periodic table, 566, 567
    rare-earth, 571, 682, 687
    semiconductor, 686
    transition metals, 569, 745
Energy
    activation, 885
    anisotropic transport, 234–237
    density, electromagnetic, 176
    electromagnetic, 165, 172
    optical, 44
    photon, 518
Energy levels
    alexandrite, 570

azimuthal quantum number, 564
bandgap energy, 576, 578, 733, 738
Bohr atom, 564
Boltzmann distribution, 581
C^{5+}, 564
CO_2 molecule, 574
conduction band, 576
Cr^{3+}:crysoberyl, 570
Cr^{3+}:sapphire, 570
crystal-field theory, 570
degeneracy, 572, 582
diatomic molecules, rotating, 573
diatomic molecules, vibrating, 573
dye molecule, 575
Fermi–Dirac distribution, 582
fine structure, 565
forbidden band, 576
H, 564
He, 566
hyperfine structure, 565
ionization energy, 568
ions, 568–572
lanthanide-ion manifolds, 572
ligand field theory, 570
magnetic quantum number, 564
manifold, 566, 572, 634, 683
molecular, 573
multielectron atoms, 565–568
Nd^{3+}:glass, 571, 595
Nd^{3+}:YAG, 571, 595
Ne, 566
occupation, 581
principal quantum number, 564
quantum dots, 580
relativistic effects, 565
rotational quantum number, 573
ruby, 570
spin–orbit coupling, 565
Stark splitting, 568, 601, 1010
triatomic molecule, vibrating, 574
valence band, 576
vibrational quantum number, 573
vibrational–rotational, 574
Zeeman splitting, 568, 599, 601
Etalon
  Fabry–Perot, 69, 265–269, 679, 728
Evanescent wave
  amplified, 316
  at DPS-DPS boundary, 224
  at DPS-SNG boundary, 310
  decay, 120, 146
  nanolaser, 862
  near-field imaging, 146
  waveguide, 368

Excess noise factor
  CAPD, 917, 920, 940
  dark-noise, 921
  dead-space APD, 918
  definition, 911, 916
  modified, 920
  photoconductive detector, 940
  PMT, 916, 920, 940
  position-dependent APD, 919
  relation to gain variance, 916
  SACM APD, 918
  SAM APD, 918
  SCIDCM CAPD, 940
  SCISCM CAPD, 940
  staircase APD, 919–921
  superlattice APD, 921
Excitons
  bulk semiconductors, 767
  electroabsorption, 1010
  organic semiconductors, 784, 810
  quantum dots, 580, 850
  quantum-confined, 779
Extinction
  coefficient, field, 198, 311, 312, 368
  coefficient, intensity, 198
  coefficient, waveguide, 368
  ratio, 983

Fabry, Charles, 433
Fabry–Perot
  etalon, 69, 265–269, 471, 679, 728
  finesse, 69, 266, 441, 443
  free spectral range, 267, 438, 446
  internal intensity, 68, 441
  loss coefficient, 443
  loss factor, 443, 444
  losses, 440, 442–446
  modal spectral width, 442
  off-axis modes, 446
  photon lifetime, 444, 471
  quality factor, 445
  resonance frequencies, 437, 446
  resonator, 69, 265, 436–446, 661
  spectral width, 440
  transmittance, 266–268
Faraday
  effect, 242–243
  rotator, 250, 1173
Fast light, 204
Fermat's principle
  maximum time, 38
  minimum time, 6
Fermat, Pierre de, 3
Fermi

-Dirac distribution, 582, 747
energy, 582
function, 582, 746
inversion factor, 777
level, 747, 748, 785
tail, 748
velocity, 744
Fermion, 524, 565, 568
Dirac-, 744
Fiber optics, 391–432
infrared, 429
introduction to, 392–393
Fiber, optical, 18, 19
V parameter, 399
absorption bands, 416
acceptance angle, 19, 25, 395
attenuation, 415
Bragg grating, 269, 687, 692, 720
characteristic equation, 400
chromatic dispersion, 422
cladding, 392
communications, 1224–1286
core, 392
coupler, 414, 432
differential group delay, 423, 425
dispersion, 418–426
dispersion relation, 400
dispersion-compensating, 423, 1110
dispersion-flattened, 423
dispersion-shifted, 422
extrinsic absorption bands, 417
fiber optics, 391–432
fiber parameter, 399
grade-profile parameter, 396, 412
graded-index, 392, 396, 409, 432
group velocity, 404, 412, 432
guided rays, 393–397
guided waves, 397–415
holey, 256, 426–428
hybrid, 429
infrared, 429
introduction to fiber optics, 392–393
material and modal dispersion, 420
material dispersion, 419, 432
materials, 429–430
meridional ray, 394
metamaterial, 428
modal dispersion, 392, 418, 432
mode cutoff, 402
modes, 400, 409, 432
multicore, 413, 1263
multicore couplers, 413
multimode, 392
nonlinear dispersion, 425

number of modes, 403, 410
numerical aperture, 19, 395
optimal index profile, 412
photonic lantern, 414
photonic-crystal, 426–428
polarization mode dispersion, 423
polarization-maintaining, 406
power transmission ratio, 416
propagation constant, 404, 411, 432
quasi-plane wave solutions, 407, 432
second-harmonic generation, 1023
silica-glass, 392
single-mode, 392, 405
skewed ray, 394
specialty, 429
SPM, 1130
step-index, 392–396, 398–405, 408
waveguide, 392
waveguide dispersion, 421
Finesse, 69, 266, 441
First-order optics, see Paraxial optics
Fluorescence, 609, 811
fluorophore, 610
three-photon, 610
two-photon, 610
up-conversion, 610, 611
Focal length
arbitrary paraxial system, 31
cylindrical lens, 17
Fresnel zone plate, 67, 118
optical Kerr lens, 1039
paraboloidal mirror, 9
pulse focusing system, 1110
spherical lens, 16, 17
spherical mirror, 11
time lens, 1113
Focal point
SELFOC lens, 24
arbitrary paraxial system, 31
cylindrical lens, 346
paraboloidal mirror, 8
Four-wave mixing, 1042, 1059, 1074
degenerate, 1044, 1045
in supercontinuum generation, 1140
switch, photonic, 1199
Fourier optics, 110–159
amplitude modulation, 116
far-field Fourier transform, 124–126
Fourier transform via lens, 158, 549
free-space impulse response, 122
free-space propagation, 113–123
free-space transfer function, 119, 316
frequency modulation, 116
Fresnel approximation, 120–121

Fresnel zone plate, 118
Huygens–Fresnel principle, 123
imaging, 117
impulse response function, 112
introduction to, 111–112
optical Fourier transform, 124–128
periodic media, 286–289
plane wave, 113
pulsed waves, 1117
scanning, 117
spatial frequency, 111
spatial frequency and angle, 113
spatial harmonic function, 113
spatial spectral analysis, 114
transfer function, 112
Fourier transform
autocorrelation, 1288
circularly symmetric function, 1294
convolution, 1288, 1294
correlation, 1288
far-field, 124–126
one-dimensional, 1287–1293
optical, 124–128, 1118
pairs, 1289
Parseval's theorem, 914, 1288
properties, 1288, 1294
separable functions, 1294
short-time, 1083
spectroscopy, 491
table, 1289
two-dimensional, 1293–1295
using lens, 126–128, 158
window function, 1083
Fourier, Jean-Baptiste Joseph, 110
Franken, Peter, 1015
Franz–Keldysh effect, 1010
Fraunhofer
approximation, 124
diffraction, 130–132, 158, 1090
Fraunhofer, Josef von, 110
Frequency
-resolved optical gating, 1156
beat, 74, 967, 1153, 1267
conversion, 1027, 1054
instantaneous, 1081
modulation, spatial, 116
of light, 42
shifter, acousto-optic, 967
Fresnel
approximation, 49, 120–121, 1118
biprism, 14, 57
diffraction, 132–137, 158, 1090
equations, 223, 253, 258
integrals, 134

lens, 18
number, 50, 121, 125, 458
reflection, 639, 797
zone plate, 67, 118
Fresnel, Augustin-Jean, 210

Gabor, Dennis, 110
Gain
amplifier, parametric, 1056
amplifier, reflection, 1065
CAPD, 896
conversion, 1271
laser amplifier, 623
laser diode, 834
photoconductive detector, 884
Raman, 642, 1041
saturated, 647, 649
SCIDCM CAPD, 898, 940
SCISCM staircase APD, 940
secondary-emission, 916
semiconductor optical amplifier, 818
Gain coefficient
amplifier, parametric, 1057
peak, 821, 832, 869
quantum-well SOA, 828
Raman, 1041
saturated, 646, 649, 660
small-signal, 622, 660
SOA, 820
Gate, photonic logic, 1165, 1211
Gauss, Carl Friedrich, 79
Gaussian
chirped pulse, 1084–1085
pulse, 1084, 1161
pulse, Gaussian-beam analogy, 1113
Gaussian beam, 51, 80–99
M^2 factor, 90
$ABCD$ law, 97–99
q-parameter, 81, 97
Bessel-beam comparison, 106
characterization, 88
collimation, 95
complex amplitude, 80–82, 123
complex envelope, 81
confocal parameter, 85
depth of focus, 85
divergence angle, 85, 109
elliptic, 102, 109
expansion, 95
focusing, 93, 109, 1122
Gaussian-pulse analogy, 1113
Gaussian-pulsed, 1120
Gouy effect, 86
intensity, 82

parameters, 89, 90, 108
paraxial approximation, 89
phase, 86
power, 83
properties, 82–91, 451
pulsed, 1087
quality, 90
radius of curvature, 86
Rayleigh range, 81
reflection from spherical mirror, 96
relaying, 94
shaping, 93
single-photon, 522
spherical-mirror resonator, 451, 453
spot size, 84, 109
standing wave, 471
through arbitrary system, 97
through components, 91–99
through free space, 98
through graded-index slab, 109
through thin lens, 91–93
through transparent plate, 99
vector, 179
waist radius, 84
wavefronts, 86
width, 84
Gaussian optics, *see* Paraxial optics
General Electric Corporation, 787
Geometrical optics, *see* Ray optics
Goos–Hänchen effect
shift, 253
waveguide, 371
Gordon, James P., 1078
Gouy effect, 86
Graded-index
GRIN material, 20, 343
SELFOC slab, 23
fiber, 24, 40, 396–397, 409
lens, 24
optics, 20–25, 343
ray equation, 20–21
ray equation, paraxial, 21
slab, 22
Graphene photonics, 719, 744
Grating
coupler, 377
diffraction, 59, 78
Grazing incidence, 223, 311, 434, 698
Group
delay, 200
index, 201
velocity, 200, 352
velocity dispersion, 201, 207, 370, 425, 550, 1104

velocity, fiber, 404, 412
velocity, photonic crystal, 284
velocity, waveguide, 360, 370
Group-IV photonics, 737, 743
2D materials, 745, 780
allotropes, 743
array detector, 909
avalanche photodiode, 902
GeSn-on-Si laser, 860
graphene photonics, 719, 744
microcavity laser, 859
photodiode, 892
Schottky-barrier photodiode, 894
SiC Schottky diode, 809
silicon photonics, 780, 808
transition-metal dichalcogenides, 745
Guided-wave optics, 353–390
Gyration vector, 241

Harmonic oscillator, 170, 323, 1072, 1298
analogy with optical mode, 543
energy, 542
quantum theory, 542, 543, 573
Heaviside, Oliver, 164
Helmholtz equation, 46, 81, 173, 192
coupled, 954, 1049, 1060
coupled paraxial, 1050
generalized, 257, 343, 1304
nonlinear, 1134
optical fiber, 397
paraxial, 51, 78, 81, 102
two-dimensional, 105
Hermite polynomials, 100, 542
Hermite–Gaussian beam, 99–102, 105
axial phase, 109
complex amplitude, 101
elliptic, 102
excess phase, 101
Hermite polynomials, 100
Hermite–Gaussian functions, 101
intensity, 102
power confinement, 109
superposition, 103, 109
Hermite–Gaussian functions, 542
Hermitian operator, 1302
Hero's principle, 6
Hertz, Heinrich, 871
Heterodyne
optical, 74, 545, 967, 1153, 1267
Heterostructures
organic semiconductors, 810
photoconductor, 886
photodiode, 891
SOAs, 825

High-harmonic generation, 1141–1145
    in Ar, 1144
        optical frequency comb, 723
        recollisional model, 1141, 1145
Hilbert transform, 625, 1298
Hole burning
    spatial, 673
    spectral, 651, 674
Hologram, 148
    computer-generated, 155
    grating vector, 153
    holographic optical element, 155
    interconnection, 1222
    interconnection capacity, 1170
    Laguerre–Gaussian beam, 104, 155
    metasurface, 342
    oblique plane wave, 149, 159
    point source, 149, 155, 159
    rainbow, 154
    volume, 153
Holography, 147–155
    ambiguity term, 150
    apparatus, 152
    computer-generated, 155, 1171
    dynamic, 1009
    Fourier-transform, 150
    hologram, 148
    holographic code, 148
    object wave, 148
    off-axis, 150
    optical correlation, 159
    real-time, 1045
    reconstructed wave, 149
    reference wave, 148
    spatial filters, 151
    surface-relief, 1172
    volume, 153
Huygens, Christiaan, 41
Huygens–Fresnel principle, 123, 193
Hyperbolic
    media: Type-I & Type-II, 352
    medium, 317–320
    metamaterial, 339–340, 352

IBM Corporation, 787
Imaging, 137–147, 499
    2-f, 127, 549
    4-f, 139
    4-f impulse response function, 140
    4-f system transfer function, 139
    diffraction limit, 146
    equation, 11, 17, 32, 58, 138, 315
    equation, incoherent light, 499
    focal-plane array (FPA), 907
    functional, 610
    hyperbolic medium, 318–319
    image intensifier, 875
    incoherent light, 499–502
    incoherent vs. coherent, 500–502
    multiphoton microscopy, 611
    near-field, 146, 334
    negative-index slab, 315–317
    optical correlation, 159
    paraxial system, 31
    perfect, 315, 318
    phase object, 158
    point-spread function, 499
    scanning near-field microscopy, 147
    single-lens, 58, 137, 142, 500
    single-lens impulse response function, 142, 500
    single-lens transfer function, 144
    single-photon, 522
    spatial filtering, 141, 159
    spherical mirror, 11
    structural, 610
    subwavelength, 146, 315, 334
    thick lens, 32
    thin lens, 31
    three-photon microscopy, 611
    two-photon, 549, 550
    two-photon microscopy, 610
    two-point resolution, 159
    X-ray, 705
Impedance, 176
    complex, 183
    imaginary, 308, 310
Impermeability, electric
    effect of electric field, 990
    in magneto-optic material, 243
    tensor, 230, 989
Impulse response function, 1296, 1299
    4-f imaging, 140
    free space, 122
    incoherent light, 499
    single-lens imaging, 137, 142, 500
Ince–Gaussian beam, 105
Incoherent light, *see* Coherence
Index ellipsoid, 230, 232–233, 989
    acousto-optic modification, 968–970
    electro-optic modification, 989–990
Index of refraction, *see* Refractive index
Infrared
    frequencies, 42, 853
    LWIR band, 42, 853
    molecular-fingerprint region, 853
    MWIR band, 42, 853
    optical fiber, 429

sensor card, 612
wavelengths, 42, 853
Injection lasers, *see* Laser diodes
Instantaneous frequency, 1081
Insulators
  band structure, 577
Integrated
  optics, 354, 375, 808
  photonics, 354, 375, 808
Intensity
  autocorrelation function, 1150
  average, 475
  Bessel beam, 106
  Bessel-like beam, 107
  electromagnetic, 165, 172, 176
  elliptic Gaussian beam, 102
  Gaussian beam, 82, 106
  Hermite–Gaussian beam, 102
  instantaneous, 476
  irradiance, 44, 812
  Laguerre–Gaussian beam, 103
  measurement for pulse, 1146
  optical, 44, 46
  partially coherent light, 475
  polychromatic light, 72
  random, 476
  scattered, 193, 194
  spectral density, 480
Interconnect, optical, 1166–1178
  circulator, 1168, 1173
  computer-com, 1174–1177
  diffractive, 1168
  free-space, 1168–1172
  guided-wave, 1172
  holographic, 1177
  inter-board, 1174
  inter-chip, 1174
  interconnection matrix, 1166
  intrachip, 1176
  introduction to, 1164–1165
  isolator, 1167
  nonreciprocal, 1167, 1173
  nonreciprocal, multiport, 1167
  optochip, 1174
  rationale, 1177
  refractive, 1168
Interference, 61–71, 489–496
  Bragg reflection, 67
  double-slit experiment, 65, 493, 525
  equation, 61, 489
  finite number of waves, 66, 1183
  Fourier-transform spectroscopy, 491
  Fresnel zone plate, 67
  infinite number of waves, 68

light from extended source, 495
multiple waves, 65–71, 75
oblique-plane-wave, 64
OCT, 492
partially coherent light, 489–496
plane-wave and spherical-wave, 64
QOCT, 550
single-photon, 525, 555
spherical-wave, 65, 493
two-photon, 550
two-wave, 61, 74, 489
visibility, 78, 490
Interferometer, 62–71
  double-slit, 65, 493, 495, 505, 512, 525
  Fabry–Perot, 69, 70
  Fabry–Perot, nonlinear, 1217
  finesse, 69, 266, 441
  gravitational-wave, 70–71, 545
  gyroscope, 63
  Hong–Ou–Mandel, 550
  interferogram, 491
  LIGO, 70, 545
  Mach–Zehnder, 62, 1181
  Michelson, 62, 70, 75, 78, 491
  Michelson stellar, 505
  multipath, 66, 1183
  MZI, NL, 1185, 1216, 1218, 1223
  nonlinear, 1154, 1156, 1162
  role of spatial coherence, 493
  role of temporal coherence, 490
  Sagnac, 62
  Sagnac, nonlinear, 1185, 1201
  self-referenced spectral, 1154
  single-photon, 526
  spectral, 1153
  stellar, 505
  temporal, 74, 1153
Invisibility cloak, 347–348
  metamaterials, 348
Ionization
  coefficients, 895
  coefficients, history-dependent, 903
  coefficients, position-dependent, 903
  energy of a donor electron, 742
  energy of an atom, 568
  energy of Ar, 1144
  energy of H, 564, 568, 742
  ratio, 896
Ions
  actinide metals, 572
  electron configuration, 569
  lanthanide metals, 568, 571, 572
  noble-gas lasers, 568

term symbol, 569
transition metals, 569
Irradiance, 44, 812
Isolator, optical, 250, 967, 1167

John, Sajeev, 255
Jones matrix, 217–221
    cascaded devices, 219, 253
    coordinate transformation, 219
    diagonal, 222
    field version, 509
    half-wave retarder, 219, 220
    linear polarizers, cascaded, 254
    normal modes, 220
    polarization rotator, 219, 549
    polarizer, 217, 253
    quarter-wave retarder, 218
    wave retarder, 218, 253
    wave-retarder cascade, 219, 253
Jones vector, 215–221
    coordinate transformation, 219
    field version, 507
    normal modes, 220
    orthogonal expansion, 215, 216
    two-photon, 547, 548

Kaminow, Ivan Paul, 1224
Kao, Sir Charles Kuen, 1224
Keck, Donald B., 391
Kerr
    coefficient, 978, 1037, 1077
    effect, 978, 994, 1037
    lens, 1039
    medium, 1036, 1129, 1149
    medium characteristic length, 1130
    switch, 1148
Kerr, John, 975
Kramers–Kronig relations, 186, 1298

Laguerre
    -Gaussian beam, 102–105, 155
    generalized polynomial, 103
    polynomial, 103
Laguerre, Edmond Nicolas, 79
Laser, 657–730
    Q-switched, 708, 712, 716, 730
    Ag^{19+}, 700, 706
    alexandrite, 570, 645, 706
    Ar^+-ion, 568, 645, 678, 706, 720
    ArF exciplex, 645, 706
    Brillouin fiber, 692
    broadening, homogeneous, 672
    broadening, inhomogeneous, 673
    C^{5+}, 645, 699, 706
    cascaded Raman fiber, 692
    cascaded silicon Raman, 692
    cavity dumping, 708, 730
    ceramic hosts, 681
    chemical, 696
    CO, 854
    CO_2, 645, 706, 720, 854
    coherent-state generation, 544
    cooling, 599
    Cr^{2+}:ZnS, 645, 686, 706, 854
    Cr^{2+}:ZnSe, 686, 854
    Cr^{3+}:colquiriite, 682
    Cr^{3+}:crysoberyl, 570, 645, 706
    Cr^{3+}:sapphire, 570, 645, 657, 664, 681, 706, 716, 730
    Cr^{4+}:forsterite, 645, 686, 706, 720
    crystalline hosts, 681
    Cu Kα, 645, 701
    dopant ions, 681
    DPSS, 682, 690, 789
    dye, 696, 697
    efficiency, 669
    Er^{3+}:silica fiber, 645, 689, 706, 720
    excimer, 695
    exciplex, 695
    extreme-ultraviolet, 697–702
    Fe^{2+}:ZnS, 686
    Fe^{2+}:ZnSe, 686
    fiber, 687–691
    four-level pumping, 630, 633
    free-electron, 702–705
    frequency pulling, 664
    gain clamping, 666
    gain switching, 707, 711, 730
    gas, 695–696
    glass hosts, 681
    H_2O, 706
    HAPLS, 640, 849, 1095
    HCN, 706
    He–Ne, 645, 706, 720
    in-band pumping, 634, 686, 689, 697
    incoherent-feedback, 693
    InGaAsP, 645
    inner-shell photopumped, 700
    intracavity tilted etalon, 679
    ion, 695
    ionized-atom plasma, 699
    Kr^+-ion, 568, 706
    KrF exciplex, 706
    lasing without inversion, 663
    linewidth, 680
    loss coefficient, 661
    metal-nanocavity, 863
    methanol, 706

INDEX 1347

microcavity, 854–862
microdisk, 859
microring, 859
microring, coupled, 1219
mode locking, active, 719
mode locking, passive, 719
mode-locked, 685, 709, 716–723
molecular, 695
MOPA, 688
multiple-mirror resonator, 679
multiquantum-dot, 850
multiquantum-well, 845
multiquantum-wire, 849
nanocavity, 862–864
nanoring, 863
nanosphere, 864
Nd^{3+}:CaF_2, 572
Nd^{3+}:glass, 571, 595, 645, 706, 720
Nd^{3+}:YAG, 571, 645, 683, 706, 720
Nd^{3+}:YVO_4, 645, 682, 706
Ne Kα, 645, 701, 706
non-plasmonic nanocavity, 863
number of modes, 672
optical vortex, 860
oscillation conditions, 662
oscillation frequencies, 664, 665
output characteristics, 666–680
overall efficiency, 706
petawatt, 640, 849, 1095
phase noise, 680
phonon-terminated, 686
photon lifetime, 662
photonic-bandgap fiber, 688
photonic-crystal, 860
photonic-crystal array, 861
plaser, 693
plasmonic nanocavity, 863
polarization, 677, 678
powder, 693
power-conversion efficiency, 706
pulsed, 707–721
pumping, 630, 635
quantum cascade, 851–854
quantum-confined, 844–854
quantum-dot, 850
quantum-well, 845
quantum-wire, 849
quasi-three-level pumping, 633
quasi-two-level pumping, 634
radar, 75
Raman fiber, 691–692
random, 693–695
rate equations, 709
rhodamine-6G dye, 645, 706, 720

ribbon fiber, 688
ruby, 570, 645, 657, 664, 681, 706, 716, 730
SASE, 703
Schawlow–Townes linewidth, 680
Se^{24+}, 699
seed, 688, 700, 701
SGDFB, 853
silicon Raman, 692, 809
slab-waveguide fiber, 688
solid-state, 681–686
spatial distribution, 675
spectral distribution, 671
spontaneous lifetime, 644
strained-layer QW, 846
theory of oscillation, 659–665
thin-disk, 683, 684, 706
three-level pumping, 632, 633
threshold, 662, 663, 698
thresholdless, 861
Ti^{3+}:sapphire, 645, 685, 706, 720
Tm^{3+}:silica-fiber, 690, 706
transient effects, 709–721
transition parameters, 644
trapping, 600
two-level pumping, 655
U^{3+}:CaF_2, 572
unipolar, 851
unstable-resonator, 677
vibronic, 570, 685, 686
W^{46+}, 699
wall-plug efficiency, 706
wavelengths, 644, 706
X-ray, 697–702
X-ray free-electron, 704–706
Yb^{3+}:silica fiber, 639, 688, 706, 720
Yb^{3+}:YAG, 645, 729
Yb^{3+}:YAG thin-disk, 684, 706
ZnO, 693, 694
Laser amplifier, 619–656, 659
    ASE, 651–653, 699, 821, 830
    bandwidth, 624
    broadband, 655
    coherent, 620
    Doppler-broadened medium, 650
    Er^{3+}:silica fiber, 641–642
    four-level pumping, 630, 633, 729
    gain, 623
    gain coefficient, 622, 660
    gain coefficient, saturated, 646, 649
    gain, saturated, 647, 649
    homogeneously broadened, 645–649
    in-band pumping, 634, 640, 642
    in-line, 636

incoherent optical, 620
inhomogeneously broadened, 649
line, 636
MOFA, 636
MOPA, 636, 688
National Ignition Facility, 639, 688
Nd^{3+}:glass, 638–640, 1095
noise, 651–653
nonlinearity, 645–651
optical fiber, 640–642
phase-shift coefficient, 625
photon statistics, 653, 656
population inversion, 582, 622–629
postamplifier, 636
power amplifier, 636
preamplifier, 636
pumping, 626–635
quasi-three-level pumping, 633
quasi-two-level pumping, 634
rare-earth-doped fiber, 640
rate equations, 626–630
rates and decay times, 626
ruby, 636–637
saturation, 645–651
saturation time constant, 629
spontaneous lifetime, 644
steady-state, 626
theory, 622–625
three-level pumping, 632, 633
transition characteristics, 644
two-level pumping, 655
wavelengths, 644
Laser diodes (LDs), 831–844
bipolar, 851
broad-area, 834, 848, 849
buried-heterostructure, 848
communications component, 1232
compare with LEDs, 839, 841
compare with SLEDs, 839, 841
confinement factor, 833
differential responsivity, 837
distributed Bragg reflector, 842
distributed-feedback, 842, 848
double-heterostructure, 825, 845
efficiency, 836
external differential efficiency, 836
external-cavity, 843
extraction efficiency, 838
far-field radiation pattern, 842
gain condition, 834
gain-guided, 834
Ge, 772
GeSn-on-Si, 860
group-IV, 859
III–antimonide, 854
III–V quantum-dot-on-Si, 859
in-band pumping, 817
index-guided, 834
interband, 851
interband cascade, 854
IV–VI, 854
lead-salt, 854
light–current curve, 837, 839
linewidth, 843
linewidth-enhancement factor, 843
mode-locked, 844
multimode MQW, 848
PbSnSe, 854
PbSnTe, 854
phase noise, 843
power output, 836
power-conversion efficiency, 838
ridge-waveguide, 847
Schawlow–Townes linewidth, 843
single-mode, 842
single-mode MQW, 847
slope efficiency, 837
spatial characteristics, 841
spectral characteristics, 839
threshold, 834
VCSEL, 856–858
VECSEL, 721, 859
wall-plug efficiency, 838
wavelength-tunable, 843
Layered media, 258–276
off-axis wave, 264
Lens, 16–18
aspheric, 17
biconcave, 17
biconvex, 17
collimating, 95
compound, 17
converging, 17
cylindrical, 17, 24, 40, 117, 118
diverging, 17
dome, 799, 814, 815
double-convex, 58
electro-optic, 976
expanding, 95
focal length, 16
focal point, 33
focusing, 58, 93, 109
Fourier transform, 126, 128, 158, 513, 549
Fresnel, 18
Fresnel zone plate, 67, 118
graded-index, 24, 60
hyperlens, 319

imaging, 58, 137, 139, 142, 158, 500
Kerr, 719, 1039
LED, 799
meniscus, 17
perfect, 315–317
plano-concave, 17
plano-convex, 17, 58
principal point, 33
relaying, 94
sequence, 36, 37
shaping, 93
spherical, 16
superlens, 316
thick, 33
thin, 57, 91
time, 1113, 1162
vertex point, 33
Lidar, 789
   coherent, 75
   PIC, 1238
Light
   classical, 4, 539
   guide, 18
   interaction with atoms, 583–602
   interaction with semiconductors, 766
   line, 285
   nonclassical, 4, 539, 545, 546
Light-emitting diodes (LEDs), 789–817
   additive color mixing, 814–815
   AMOLED, 1004
   arrays, 815
   bioinspired, 798
   Ce^{3+}:YAG phosphor, 814
   characteristics, 794–803
   chip-on-board (COB), 815
   color rendering index (CRI), 812
   communications component, 1233
   compare with incandescent, 812, 816
   compare with LDs, 839, 841
   compare with SLEDs, 839, 841
   complementary colors, 813
   correlated color temperature, 812
   device structures, 803–817
   die geometries, 798
   discrete, 813
   edge-emitting, 804
   electronic circuitry, 803, 816
   external efficiency, 800
   extraction efficiency, 796, 869
   illumination applications, 788, 811
   indication applications, 788
   infrared applications, 806
   internal efficiency, 795, 800
   light–current curve, 801, 839
   lighting, 811–817
   materials, 803–817
   optics for, 15, 799
   organic, 810, 816
   output photon flux, 799
   overall efficiency, 800
   phosphor-conversion, 814
   photonic-crystal, 798
   plasmonic, 796
   power-conversion efficiency, 800
   quantum-dot, 807, 814
   resonant-cavity, 800
   response time, 802
   responsivity, 801
   retrofit lamps, 816
   roughened-surface, 798
   solid-state lighting, 811–817
   spatial pattern, 799
   spectral distribution, 802, 868
   surface-emitting, 804
   surface-mounted device, 814
   trapping of light, 19, 39
   ultraviolet applications, 808
   visible applications, 807
   wall-plug efficiency, 800
   white, 813–815
   WOLED, 810
Line broadening, 593–597
   collision, 595
   Doppler, 597
   homogeneous, 572, 595
   inhomogeneous, 572, 595
   lifetime, 593
Linear system
   causal, 1298
   Hilbert transform, 1298
   impulse response function, 1299
   isoplanatic, 1299
   Kramers–Kronig relations, 1298
   modes, 1301–1305
   one-dimensional, 1296–1299
   point-spread function, 1299
   shift-invariant, 1296, 1299
   supermodes, 383
   time-invariant, 1296
   transfer function, 1297, 1300
   two-dimensional, 1299–1300
Lineshape function, 586
Linewidth
   -enhancement factor, 843
   laser, 680
   laser diode, 843
   Schawlow–Townes, 680, 843
   transition, 586, 644

Liquid crystal
   cholesteric, 244
   electro-optics of, 996–1005
   ferroelectric, 1001
   modulator, 997–1001
   nematic, 244, 996
   optics of, 244–246
   parameters, 999
   smectic, 244, 1001
   switch, photonic, 1194
   twisted nematic, 244–246, 999
   wave retarder, 997–1001
Liquid-crystal display (LCD), 1002
   active matrix, 1003
   passive-matrix, 1002
   segmented, 1002
Lithography
   electron-beam, 348, 468, 696
   EUV, 696, 702
   focused ion-beam, 348, 696
   holographic, 298
   micro-, 298
   multiphoton, 299, 611
   X-ray, 696
LLNL, 639, 688, 698, 699, 849, 1095
Localized surface plasmon, 330–332
   LED, 796
   nanolaser, 862
   resonance, 330
Logic
   photonic logic gates, 1211
Lorentz
   oscillator, 186, 1018, 1072
   relativistic factor, 703
Lorentzian, 189, 483, 512, 593–595, 624, 625, 647, 730, 1080, 1292
Luminescence, 607–612
   betaluminescence, 607
   bioluminescence, 608
   cathodoluminescence, 607, 618
   chemiluminescence, 608
   electroluminescence, 608
   fluorescence, 609, 811
   multiphoton fluorescence, 610
   phosphorescence, 609, 811
   photoluminescence, 609–612, 813
   radioluminescence, 609
   sonoluminescence, 607
   up-conversion fluorescence, 611
Luminous
   efficacy, 812, 817
   flux, 812, 817

Magnetic
   dipole moment, 178
   field, 162
   flux density, 163
Magnetization density, 164
Magneto-optics, 242–243
   Faraday effect, 242, 1013
   magnetogyration coefficient, 243
   switch, photonic, 1195
   Verdet constant, 243
Magnification
   Gaussian beam, 92
   shift-variant, 144
   spherical boundary, 15
   spherical lens, 17
   spherical mirror, 11
   time lens, 1162
Maiman, Theodore H., 657
Manley–Rowe relations, 1029, 1051, 1063
Maser, 659, 680
   astrophysical, 694
Master-oscillator fiber-amplifier, 688
   examples, 689
Master-oscillator power-amplifier
   examples, 636, 639, 688, 689
Material
   double-negative (DNG), 306
   double-positive (DPS), 306
   equation, 173, 228, 241, 243
   hyperbolic, 317–320, 352
   left-handed, 306, 309
   negative-index (NIM), 309, 314–317
   single-negative (SNG), 306
Matrix optics, 27–37
   $ABCD$ matrix, 28, 1302
   arbitrary paraxial system, 31
   Bragg grating, 271
   layered media, 258–276, 1303
   periodic media, 280–286
   periodic systems, 33–37, 447–450
   ray-transfer matrix, 27, 1302
   scattering matrix, 259–265
   scattering vs. wave-transfer, 260
   simple components, 29
   thick lens, 32
   transmission matrix, 380, 383
   wave-transfer matrix, 258–265, 1303
Maurer, Robert D., 391
Maxwell's equations
   boundary conditions, 164, 221
   complex permeability, 309–310
   complex permittivity, 309–310
   dielectric constant, 167
   electric field, 162
   electric flux density, 163

impermeability tensor, 230
in a medium, 163, 172
in conductive medium, 320–326
in free space, 162
magnetic field, 162
magnetic flux density, 163
magnetization density, 164
permeability, 163, 167, 306, 343
permeability tensor, 343
permittivity, 163, 167, 306
permittivity tensor, 170, 228, 343
polarization density, 164, 1017
relative permittivity, 167
speed of light, 163, 167
vector potential, 177
wave equation, 163, 167, 169, 171
Maxwell, James Clerk, 160
Media, *see* Materials
Metals
    band structure, 577
    bound-electron absorption, 324
    conductive media, 320–326
    conductivity, 577
    Drude model, 322–326
    group velocity, 352
    loss, 470
    optics of, 304, 320–326
    photoemission, 873
    plasma frequency, 698
    plasmonics, 320–334
    reflectance, 324
    work function, 873
Metamaterials
    holey metallic film, 342
    hyperbolic, 339–340, 352
    invisibility cloak, 348
    metasurfaces, 340–343
    negative-index, 338–339
    negative-permeability, 337–338
    negative-permittivity, 336–337
    optical fiber, 428
    optics of, 304, 334–343
    photonic-crystal, 256, 334, 428
    point-dipole approximation, 335
Metameric white light, 813
Metasurfaces, 340–343
    complementary, 340
    holey metallic film, 342
    hologram, 342
    phase modulator, 341
    reflection, 342
    refraction, 342
    Snell's law, modified, 342
Michelson stellar interferometer, 505

Micro-optics, 305, 1168
Microcavity, *see* Microresonator
Microcavity lasers, 854–862
Microresonator
    microcavity, 463
    microdisk, 465
    micropillar, 465
    microsphere, 466–468
    microtoroid, 465
    modal density, 465
    modal volume, 464
    photonic-crystal, 468
    quality factor, 464
    rectangular, 464–465
Microscopy
    multiphoton, 610, 611
    near-field, 147
    three-photon, 611
    two-photon, 610
Mie scattering, 197
Miniband, 579, 764, 779, 852
    QCL, 852
Mirror
    collimator, 9
    elliptical, 9
    paraboloidal, 8, 10
    planar, 8, 53, 78
    spherical, 9, 30, 78, 96
    variable-reflectance spherical, 97
MIT Lincoln Laboratory, 787
Mixing, optical, 74, 967, 1153, 1267
Mode locking, 76, 709, 716–723
    applications, 721
    examples, 685–690, 696, 720
    external-cavity LDs, 844
    fiber lasers, 844
    harmonic, 721
    Kerr-lens, 719, 1039
    methods, 719
    optical frequency comb, 721
    parameters, 718
    properties, 716, 718, 730
    QCLs, 721, 854
    QD lasers, 851
    saturable absorber, 719
    SESAM, 719
    VECSELs, 721, 859
Modes
    discrete linear system, 1302
    eigenfunction, 1301
    eigenvalue, 1301
    eigenvector, 1301
    homogeneous medium, 1305
    integral operator, 1303

linear system, 1301–1305
normal, 1304
ordinary differential equation, 1304
partial differential equation, 1304
periodic medium, 1305
resonator, 1304
supermodes, 383
zero-point energy, 518, 573, 586
Modulation
amplitude shift keying, 1259
binary phase shift keying, 1259, 1273
constellation, 1259
differential phase shift keying, 1259
digital, 1258
field, 1257, 1266
frequency shift keying, 1259
intensity, 1258
multilevel coding, 1259
on–off keying, 931, 1259, 1272
phase shift keying, 1259
pulse code, 1258
quadrature amplitude, 1259
quaternary phase shift keying, 1259
spectral efficiency, 1259
Modulator
acousto-optic, 959
electroabsorption, 1010, 1011
intensity, 998
intensity, acousto-optic, 973
intensity, electro-optic, 981–983, 995
intensity, magneto-optic, 1013
intensity, push–pull, 1014
interferometric, 981
liquid-crystal, 997–1001
Mach–Zehnder, 981, 1014
multiquantum-well, 1010
optically addressed SLM, 1004
parallel-aligned SLM, 1005
phase, 998
phase, acousto-optic, 973
phase, cascaded, 1014
phase, electro-optic, 973, 979, 995
phase, opto-optic, 1076
quadratic phase, 1100
spatial light, 987–989
spatial light, acousto-optic, 965
spatial light, liquid-crystal, 1002
Molecules, 572–575
covalent bonding, 572
dye, 575
ionic bonding, 572
rotating diatomic, 573
van der Waals bonding, 572
vibrating diatomic, 573
vibrating triatomic, 574
Momentum
electromagnetic, 165
localized photon, 523
localized wave, 523
photon, 523–524, 554, 555
radiation pressure, 524, 555
Momentum, angular
photon orbital, 524
photon spin, 524
Mourou, Gérard, 1078
Multiphoton
absorption, 1017
detection, 939, 1156, 1162
fluorescence, 610
lithography, 298, 299, 611
microscopy, 610, 611
photoluminescence, 610
Multiple access
code-division, 1276
frequency-division, 1276
time-division, 1276
Multiplexing
code-division, 1261
CWDM, 1263
DWDM, 1263
electronic, 1261
frequency-division, 1260
optical, 1261
space-division, 1263–1266
time-division, 1208, 1260
wavelength-division, 1179, 1262
Multiquantum
-dot lasers, 850
-well lasers, 845
-wire lasers, 849
well, 764, 1010

Nano-optics, *see* Nanophotonics
Nanocavity lasers, 862–864
Nanophotonics, 193–199, 305, 326–343, 386–388, 428, 469–470
nanolasers, 862
subwavelength imaging, 147, 315
Nanoresonator, 469–470
metallic nanodisk, 469
metallic nanosphere, 470
Nanosphere
dielectric, 196–197
metallic, 330–332
scattering from, 196–197, 330–332
Near-field imaging, 146, 334
Negative-index
materials, 314

metamaterials, 338–339
Network, fiber-optic, 1274–1281
    broadcast-and-select, 1277
    bus, 1275
    interface, 1276
    local-area (LAN), 1274
    mesh, 1275
    multi-hop broadcast-and-select, 1278
    ring, 1275
    star, 1275
    topologies, 1275
    wavelength-routed, 1279
    WDM, 1277
Newton, Sir Isaac, 3
Nobel laureates, 110, 160, 619, 657, 731, 867, 943, 1015, 1224
    Nobel lectures, 157, 431, 551, 615, 616, 726, 727, 782, 783, 867, 868, 938, 972, 1075
Noise
    $1/f$, 923
    ASE, 651, 830
    background, 911
    circuit, 910, 922–930
    dark-current, 911
    gain, 910, 915
    generation–recombination (GR), 940
    laser phase, 843
    optical amplifier, 651
    photoconductor, 940
    photocurrent, 912
    photodetector, 909–923
    photoelectron, 539, 910, 912
    photon, 533–539, 910, 911
    pink, 923
    semiconductor optical amplifier, 821
    shot, 518, 703, 912, 913
    superluminescent diode, 830
    thermal, 922
Nondiffracting beams, 105–107
Nondiffracting waves, 136
Nonlinear optical coefficients, 1018, 1019, 1025, 1037, 1066, 1068
Nonlinear optics, 171, 1015–1077
    anharmonic oscillator, 1071–1073
    anisotropic dispersive medium, 1073
    anisotropic medium, 1066–1069, 1073
    Born approximation, 1020, 1076
    chi-two medium, 1021
    coherence length, wave-mixing, 1032
    coupled waves, 1047
    coupled-waves, 1059, 1127
    cross-phase modulation, 1041
    DFG, 854, 1027
    differential-equation description, 1071
    dispersive medium, 1069–1074
    downconversion, 1027, 1077
    electro-optic effect, 1024
    extreme, 1141
    five-wave mixing, 1077
    four-wave mixing, 1042, 1044, 1059, 1074, 1077, 1140
    frequency conversion, 1027, 1054
    HHG, 723, 1141
    holography, real-time, 1045
    idler, 1027
    integral-transform description, 1070
    introduction to, 1016–1017
    Kerr effect, optical, 1037, 1042
    Kerr medium, 1036
    Manley–Rowe, 1029, 1051, 1076
    Miller's rule, 1073
    nonlinear coefficients, 1018, 1019, 1025, 1037, 1066, 1068
    nonparametric, 1017, 1028
    parametric, 1017
    parametric interactions, 1026–1029
    periodic poling, 1035, 1036
    phase conjugation, 1044, 1063
    phase matching, 1026, 1029–1033
    phase-mismatching tolerance, 1076
    photonic-crystal soliton, 1141
    plane-wave conjugation, 1045
    polarization density, 1017–1019
    poling, 1036
    pump, 1027
    quasi-phase matching, 1034, 1076
    Raman gain, 1041
    rectification, optical, 1023
    rectification, pulsed optical, 1128
    refraction, nonlinear, 1038
    scattering theory, 1020, 1076
    Schrödinger equation, nonlinear, 1040, 1135, 1138
    second-order, 1021–1036, 1047–1059
    self-focusing, 1039
    SFG, 1027
    SFG and SHG combined, 1077
    SHG, 1021, 1027, 1049, 1051
    SHG efficiency, 1022, 1052
    SHG phase mismatch, 1053
    signal, 1027
    solitary wave, 1131
    soliton self-frequency shift, 1140
    soliton, spatial, 1039
    soliton, spatiotemporal, 1139
    soliton, temporal, 425, 1130–1139
    SPDC, 548, 1027

spherical-wave conjugation, 1045
SPM, 425, 1038, 1129, 1130, 1140
supercontinuum generation, 1139
THG, 1037, 1063
third-order, 1036–1047, 1059–1066
three-wave mixing, 545, 1025, 1026, 1043, 1050, 1061, 1066, 1073, 1076, 1077, 1127, 1162
THz pulse generation, 1128
tuning curves, 1029
two-wave mixing, 1009, 1026, 1042
ultrafast, 1126–1145
up-conversion, 1054, 1055
Volterra-series expansion, 1070
walk-off effect, 1126
wave equation, 1019, 1047, 1133
wave restoration, 1046

Nonlinear-optic devices
amplifier, parametric, 1027, 1056
amplifier, phase-sensitive, 1062
amplifier, reflection, 1065
amplifier, transmission, 1065
DFG, 1027
downconverter, 1027, 1077
FPI, 1217, 1218
intensity autocorrelator, 1150, 1162
lens, optical Kerr, 1039
loop mirror, 1186
modulator, opto-optic phase, 1076
MZI, 1216, 1218
oscillator, doubly resonant, 1058
oscillator, parametric, 1027, 1057
oscillator, phase-conjugation, 1066
oscillator, singly resonant, 1058
router, directional-coupler, 1186
router, nonlinear MZI, 1185
router, nonlinear Sagnac, 1185
router, soliton, 1187
SFG, 1027
streak camera, 1149
switch, FWM, 1199
switch, optical Kerr, 1148, 1198
switch, photonic, 1196–1211
switch, Sagnac, 1201
switch, SFG, 1197
switch, SHG, 1147
switch, XPM, 1197–1199
up-converter, 1027

Nonparametric processes, 1028
Normal modes
anisotropic medium, 230, 232–233
optically active medium, 241
polarization system, 220, 221

Numerical aperture, 19, 25, 39, 181, 366, 395, 432

Ohm's law, 320, 882
Optical
frequency comb, 721, 1141
Kerr effect, 1037, 1042
lattices, 601
molasses, 599
OADM, 1179, 1180, 1279
pathlength, 5, 6, 53, 78
phase conjugation, 1044, 1063
sectioning, 492, 550
tweezers, 600
Optical activity, 240–242
gyration vector, 241
normal modes, 241
rotatory power, 242
Optical coherence, see Coherence
Optical coherence tomography, 492
frequency-domain, 493
quantum, 550
time-domain, 492
Optical components, 8–20, 53–61
active-matrix LCD, 1003
active-matrix OLED, 1004
antireflection coatings, 264, 301
axicons, 14, 57, 106, 137
backlight, 1002
beam combiners, 13
beam directors, 14
beamsplitters, 13, 63, 247, 263, 268, 301, 522, 539, 549
catadioptric, 15
circulator, 1173
collimators, 9, 15, 799
diffraction gratings, 59, 78
electro-optic prism, 983
fiber couplers, 413
fibers, 392, 422–423, 426, 429
filter, acousto-optic, 966
frequency shifter, acousto-optic, 967
graded-index, 22–24, 40, 60, 396
integrated, 354, 808
isolator, 250, 967, 1167
LED optics, 15, 799
lenses, 16–18, 57, 58, 319
mirrors, 9, 10, 53
modulators, acousto-optic, 959
passive-matrix LCD, 1002
photonic lantern, 414
plates, 39, 55
polarizers, 217, 253, 520
prisms, 13, 14, 56, 57

resonators, 434, 447, 459, 463, 516
scanners, acousto-optic, 961
segmented LCD, 1002
space switches, acousto-optic, 964
spatial light modulator, 965, 1002
spectrum analyzer, AO, 963
spiral phase plate, 104
superprism, 1180
thin films, 264, 301
waveguide couplers, 376–383
waveguides, 355, 363, 372, 386
Optical fiber amplifier, 640–642
communications component, 1234
compare with SOA, 830
Optical fiber communications, 1224–1286
analog, 1253
analog coherent, 1271
attenuation, 1246, 1247, 1252
attenuation compensation, 1253
balanced homodyne receiver, 1268
balanced mixer, 1268
bit error rate, 1245
coherent, 1266–1274
coherent receiver advantages, 1269
components, 1226–1238
direct vs. heterodyne, 1274
direct vs. homodyne, 1274
dispersion, 1246, 1249, 1252
dispersion compensation, 1254, 1255
dispersion management, 1255
evolution, 1240–1243
eye diagram, 1245
fibers, 1226–1231
heterodyne receiver, 1267
heterodyne vs. direct, 1274
homodyne BPSK, 1273
homodyne OOK, 1272
homodyne QPSK, 1273
homodyne vs. direct, 1274
homodyne vs. heterodyne, 1274
introduction to, 1225–1226
local oscillator, 1267
modulation, 1257–1260
multiplexing, 1260–1263
networks, 1274–1281
optical amplifier, 1234
performance, 1243–1246
photodetector, 1235–1237
photonic integrated circuit, 1238
power budget, 1247
receiver sensitivity, 1246
soliton, 1256
SONET standard, 1239, 1276
sources, 1232–1234
systems, 1238–1257
time budget, 1249
Optical indicatrix, *see* Index ellipsoid
Optical materials
2D, 745, 780
fused silica, 19, 186, 191, 204, 205, 207, 394, 416, 417, 419, 641, 961, 1110
GaAs, 207
glass, BK7, 1098, 1105, 1107, 1116
glass, phosphate, 639, 643
glass, phosphosilicate, 643, 692
host glass, 640
$LiNbO_3$, 254, 375, 973, 980, 992, 995, 1006, 1036
$LiTaO_3$, 992, 995
periodically poled, 1035, 1036
quartz, 240, 242, 254, 1015
soft glasses, 429
TeO_2, 958
TMDs, 745, 780
Optical receiver
analog receiver sensitivity, 929–930
bipolar-transistor amplifier, 925
bit error rate, 911, 932
circuit-noise parameter, 923–925
digital receiver sensitivity, 931–934
FET amplifier, 925
on–off keying (OOK), 931
resistance-limited, 924
sensitivity, 911
signal-to-noise ratio, 925–929
SNR dependence on APD gain, 928
SNR dependence on bandwidth, 928
SNR dependence on photon flux, 926
Optical system, periodic, 33–37, 447–450
GRIN plate, 40
harmonic trajectory, 35
lens sequence, 36
lens-pair sequence, 37
periodic trajectory, 35
ray position, 33
resonator, 37, 40
Organic semiconductors, 742, 876
Oscillator
harmonic, 170, 323, 542, 543, 573, 1072, 1298
Lorentz, 186, 1018, 1072
optical parametric, 1027, 1057
phase-conjugation, 1066

Paraboloidal
approximation, 49
mirror, 8

surface, 86
wave, 49, 51, 81
Parametric
    amplifier, 1027, 1056
    downconverter, 1027
    oscillator, 1027, 1057
    oscillator, doubly resonant, 1058
    oscillator, singly resonant, 1058
    processes, 1017
    switches, 1197
Paraxial
    approximation, 9, 27, 50
    Helmholtz equation, 51, 78
    imaging system, 31
    optics, 5, 9, 27–37
    ray equation, 21, 22, 40
    rays, 9
    system, focal length, 31
    system, focal point, 31
    system, principal point, 31
    system, vertex point, 31
    wave, 50, 1086
    wave equation, generalized, 1124
Parseval's theorem, 914, 1288
Partially coherent light, see Coherence
Pauli exclusion principle, 565, 582, 734
Pendry, Sir John, 303
Penetration depth, 308, 312
Periodic media
    2D periodic structure, 292
    3D periodic structure, 294
    Fourier optics, 286–289
    matrix optics, 280–281
Periodic table
    elements, 566, 567
    semiconductors, 737
Permeability, magnetic, 163, 167, 343
    complex, 306, 309–310
    negative, 306
    tensor, 343
Permittivity, electric, 163, 167, 306
    complex, 183, 198, 306, 309–310
    effect of electric field, 990
    effective, 198, 320
    frequency-dependent, 184
    in magneto-optic material, 243
    negative, 306
    relative, 167
    tensor, 170, 228, 343
Perot, Alfred, 433
Phase
    -mismatching tolerance, 1076
    -sensitive amplifier, 1062
    -shift coefficient, 660

amplifier, 70
matching, 288, 1026, 1029–1145
noise, 680, 843
spiral, 103, 155
velocity, 48, 200, 284, 352
Phosphorescence, 609, 811
Photoconductors, 875, 883–886
    detection circuit, 939
    doped extrinsic, 886
    extrinsic, 885
    gain, 884
    intrinsic, 883
    noise, 940
    response time, 885
    spectral response, 885
Photodetectors, 871–942
    array detectors, 907, 1149
    array readout circuitry, 908
    avalanche photodiodes, 895–907
    bolometer, 872
    CCD readout circuitry, 908
    charge-coupled device (CCD), 908
    circuit noise, 922–930
    CMOS readout circuitry, 909
    communications component, 1235
    digital photon-counting device, 906
    electron-multiplying CCD, 909
    external photoeffect, 873
    extrinsic photoconductive, 886
    focal-plane array (FPA), 907
    FROG, 1156
    gain, 879
    gain noise, 915–921
    general properties, 876–883
    Golay cell, 872
    intensified CCD (ICCD), 909
    intensity, pulse, 1146–1151
    intensity-autocorrelation, 1150
    internal photoeffect, 873
    introduction to, 872
    microbolometer, 872, 907
    microchannel plate, 875
    minimum-detectable signal, 911
    negative-electron-affinity, 874
    noise, 909–923
    noise-equivalent power, 911
    optical-pulse, 1146–1158
    optical-pulse phase, 1152–1156
    organic, 876
    performance measures, 910
    photoconductors, 875, 883–886
    photodiodes, 887–894
    photoelectric, 872
    photoelectric emission, 873–875

photoemission equation, 873
photomultiplier, 874, 904
photon detection efficiency, 876
photon-number-resolving, 904, 905
phototube, 874
plasmonic, 878
pyroelectric, 872
QDIP, 886
quantum efficiency, 876
QWIP, 886
Ramo's theorem, 880
RC time constant, 883
reach-through APD, 899
resonant-cavity, 878
response time, 880
responsivity, 878
SACM APD, 899, 902
SAM APD, 899, 901, 902
secondary emission, 874
signal-to-noise ratio, 910, 913, 916
silicon photomultiplier (SiPM), 905
single-photon, 904
specific detectivity, 911
spectral intensity, pulse, 1151–1152
spectrogram, pulse, 1156–1158
streak camera, 1149
superconducting, 907
thermal, 872
thermocouple, 872
thermopile, 872
transit-time spread, 880, 939
transition-edge sensor (TES), 907
two-photon, 939, 1156, 1162
vacuum photodiode, 874
Photodiodes, 887–894
    p–i–n junction, 889
    p–n junction, 887
    avalanche, 895–907
    comparison with APDs, 927
    depletion layer, 887
    edge-illuminated, 890, 892
    evanescently coupled, 890
    group-IV, 892, 894
    heterostructure, 891
    metal–semiconductor, 893
    modes of operation, 888
    open-circuit operation, 888
    photon-trapping microstructures, 891
    photovoltaic operation, 888
    response time, 888
    reverse-biased operation, 889
    Schottky-barrier, 893, 894
    short-circuit operation, 888
    solar cell, multi-junction, 892

    traveling-wave configuration, 890
Photoelastic
    effect, 968
    tensor, 969
Photoluminescence, 608–612, 813
    applications, 610
    multiphoton, 610
    quantum dots, 580
Photometry
    illuminance, 812
    luminous efficacy, 812
    luminous flux, 812
    photopic luminosity function, 812
Photon, 516–529
    antibunching, 780
    at beamsplitter, 522
    boson, 524, 537
    energy, 518
    Gaussian-wavepacket, 527
    helicity, 524
    in Fabry–Perot resonator, 555
    in Gaussian beam, 522
    in Mach–Zehnder, 526
    in Young interferometer, 525
    interference, 525
    lifetime, 662
    momentum, 523–524, 554, 555
    monochromatic, 527
    optics, 514–541
    orbital angular momentum, 104, 524
    polarization, 519
    polychromatic, 527, 555
    position, 521
    position and time, 527
    spin angular momentum, 524
    time, 526
    transmission through polarizer, 520
    wavepacket, 527
Photon detectors, *see* Photodetectors
Photon stream, 529–541
    number of photons, 531, 532
    partitioned, 539–541
    photon flux, 530, 532
    photon-flux density, 530, 532
    photon-number statistics, 533–541
    randomness, 532
    spectral density, 531
Photon-number statistics, 533–541
    Bernoulli, 539
    Bernoulli trial, 912
    binomial, 540, 541, 546, 557
    Bose–Einstein, 536, 537, 539, 556
    coherent light, 533
    counting time, 533

doubly stochastic, 538, 556
exponential density function, 539
geometric, 537
Mandel's formula, 538
mean, 533, 535
mean under absorption, 556
negative-binomial, 556
noncentral-chi-square density, 656
noncentral-negative-binomial, 653
partitioned distribution, 540, 912
partitioned SNR, 541, 912
photon number, 533
photon-number distribution, 533
Poisson, 534, 544, 555
random partitioning, 539, 556
random selection, 539, 912
signal-to-noise ratio, 535, 911
sub-Poisson, 546
thermal light, 536
uniform, 557
variance, 535
Photonic
   bandgap, 282, 285, 296
   integrated circuits, 354, 809, 1176, 1238
   lantern, 414
Photonic crystal, 277–302
   2D, 292–294
   3D, 294–299
   band structure, 282, 285, 296
   bandgap, 282, 285, 296
   Bloch modes, 277, 1305
   dispersion relation, 285
   fabrication, 297, 299
   fibers, 426–428
   group velocity, 284
   holes and poles, 298
   holes on diamond lattice, 297
   inverse-opal, 297, 298
   laser, 860
   laser array, 861
   lattice defects, 298
   metal–dielectric array, 388
   metamaterial, 256, 334
   omnidirectional reflection, 290
   one-dimensional, 277–291
   optics, 255–302
   optics, introduction to, 256–258
   phase velocity, 284
   point defects, 298
   projected dispersion diagram, 285
   silicon, 297
   soliton generation, 1141
   switch, photonic, 1200

   thresholdless laser, 861
   waveguide, 385–386
   woodpile, 297
   Yablonovite, 297
Photorefractivity, 1005–1009
   applications, 1009
   real-time holography, 1009
   simplified theory, 1006–1009
Planck
   Planck's constant, 518
   spectrum, 605
Planck, Max, 515, 516, 529
Plane wave, 47, 111, 113
   acoustic, 946
   conjugate, 1045
   electromagnetic TEM, 175, 307
   partially coherent, 487
   pulsed, 1086
   quasi-plane wave, 407
   reflection, 221
   refraction, 221, 238
   wavefronts, 48
   wavefunction, 48
Plasma
   frequency, 322
   wavelength, 322
Plasmonics, 320–334, 387
   LEDs, 796
   nanolasers, 862
   photodetectors, 878
   resonator, 469–470
   switch, photonic, 1201
   waveguide, 386–388
Pockels
   coefficient, 978, 1025, 1077
   effect, 978, 991–994, 1024
   readout modulator (PROM), 988
Pockels, Friedrich, 975
Poincaré sphere, 213, 508
Point-spread function, 1299
Poisson, Siméon Denis, 871
Polarization, 211–221
   p, 222
   s, 222
   -maintaining fiber, 406
   autocorrelation, 506
   circular, 213, 253, 509, 524
   coherency matrix, 507
   complex envelope, 211, 214
   complex polarization ratio, 213
   coordinate transformation, 219
   cross-correlation, 506
   degree, 509
   electric-field vector, 211

INDEX 1359

ellipse, 211–212
Jones matrix, 217–221, 509
Jones vector, 215–221, 507
linear, 212, 508, 524
matrix representation, 215–221, 1302
mode dispersion, 423
normal modes, 220, 221
optics, 209–254
optics, introduction to, 210–211
orthogonal, 222, 252
orthogonal expansion, 216
orthogonal Jones vectors, 215
parallel, 222
partial, 506–510
photon, 519
Poincaré sphere, 213, 508
rotator, 219, 245, 249, 252, 253, 549
Stokes parameters, 213, 507
transverse-electric, 222, 223
transverse-magnetic, 222, 224
two-photon, 547
unpolarized light, 508
Polarization devices
rotator, 252
Polarization density, 164, 1017
Polarization devices, 247–251
anti-glare screen, 253
cascade, 219, 253, 254
circulator, 251
coordinate transformation, 219
dichroic, 247
Faraday rotator, 250, 1173
fast and slow axes, 218, 248, 253
half-wave retarder, 219, 220, 253
intensity control, 249
isolator, 250
linear-polarizer cascade, 254
nonlinear Kerr switch, 1148, 1198
nonreciprocal, 250–251
normal modes, 220
polarizer, 217, 247–248, 253
polarizing beamsplitter, 248
quarter-wave retarder, 218
rotator, 219, 249, 549
router, 1184
wave retarder, 218, 248, 253
wave-retarder cascade, 219, 253
Polychromatic light, 71–76
intensity, 72
Ponderomotive force, 703, 1142
Postulates
Born postulate, 563
Einstein postulates, 591
ray optics, 5–8

wave optics, 43–44
Power
electromagnetic, 165, 172
Gaussian beam, 83
optical, 44
scattered, 194, 195
spectral density, 479
Poynting
theorem, 168
vector, 165, 173, 312
Principal
axes, 229
refractive indices, 229
Principal point
SELFOC lens, 24
arbitrary paraxial system, 31
Prism, 13, 14, 56
chirp filter, 1097, 1105, 1161
coupler, 329, 352, 377
dispersion compensation, 685
electro-optic, 983
laser-line selector, 677
spatial, 116
superprism, 1180
Prokhorov, Aleksandr M., 619
Propagation
along principal axis, 230–231
anisotropic medium, 232–233
free space, 113–123
gain of spatial coherence, 502
homogeneous medium, 6
partially coherent light, 497–498
van Cittert–Zernike theorem, 503
Pulse, optical
attosecond, 1141
characteristics, 1079–1083
chirp filtering of, 1088–1099
chirp parameter, 1082
chirped, 1082
chirped-Gaussian, 1084–1085
complex envelope, 73, 1080
compression, 1099, 1109, 1130, 1144
detection, 1146–1158
dispersion length, 1106
down-chirped, 1082
Fourier-transform-limited, 1084
frequency-to-space mapping, 1101
FROG, 1156
Gaussian, 1080, 1084–1085, 1161
Gaussian beam, 1087, 1120
Gaussian, bandwidth-limited, 1084
in dispersive media, 199–203
instantaneous frequency, 1081
intensity autocorrelation, 1150, 1162

intensity detection of, 1146–1151
light bullet, 1139
linear filtering, 1088
nonseparability, 1117, 1125
phase detection of, 1152–1156
plane wave, 73, 1086
propagation in fiber, 1102–1115
rectification, 1128
sech, 1040, 1080, 1133, 1136, 1161
shaping, 1101–1102
slowly varying, 1086
soliton, 1133
soliton self-frequency shift, 1140
soliton, spatiotemporal, 1139
soliton, temporal, 1130–1139
spatial characteristics, 1086–1088
spectral intensity, 1080
spectral intensity detection, 1151
spectral phase, 1080
spectral shift, 1117
spectral width, 1080
spectrogram, 1083, 1156–1158
spherical wave, 78, 1086
SPM, 1129, 1130, 1140
spread, 201
streak-camera detection, 1149
sub-femtosecond, 1141
supercontinuum light, 1139
temporal broadening, 1116, 1117
temporal width, 1080
three-wave mixing, 1127, 1162
THz pulse generation, 1128
time-to-space mapping, 1102
time-varying spectrum, 1083
transform-limited Gaussian, 1084
up-chirped, 1082
wavepacket, 73
Pupil function, 138, 142, 500
generalized, 143
Purcell factor, 598, 796, 800, 855, 861

Quadric representation, 229, 230
Quantum
circuits, 550
cutting, 610
defect, 634, 690
electrodynamics, 4, 515
entanglement, 547
mechanics, 562
optics, 4, 515, 541–550
Quantum cascade laser (QCL), 851–854
heterogeneous, 852
single-frequency, 853
strained-layer, 853
tunable, 853
Quantum dot
infrared photodetector (QDIP), 886
Quantum dots, 579–580, 607, 766
applications, 580
artificial atoms, 580
core–shell, 580
excitons, 580, 850
fabrication, 579
lasers, 850–851
LEDs, 807, 814
mode-locked lasers, 851
photoluminescence, 580
self-assembly, 580, 807, 850
silicon photonics, 780, 808
single-photon emitter, 546, 780
SOAs, 829
synthesis, 579
Quantum number
azimuthal, 564
magnetic, 564
orbital angular momentum, 566
overall angular momentum, 566
principal, 564
rotational, 573
spin, 564, 566
vibrational, 573
Quantum state, 541–550
amplitude-squeezed, 545
binomial, 546, 557
biphoton, 548
coherent, 544
entangled, 547
Fock, 546
intensity-squeezed, 546
number, 546
phase-squeezed, 545
photon-number-squeezed, 546
quadrature-squeezed, 545
sub-Poisson, 546
thermal, 536
twin-beam, 548
two-photon, 546
Quantum well, 578, 761–764
infrared photodetector (QWIP), 886
lasers, 845
SOAs, 826–830
Quantum wire, 579, 765
lasers, 849
Quantum-confined
excitons, 779
lasers, 844–854
structures, 760–766
Quasi-phase matching, 1034, 1076

Radar
    Doppler, 75
    laser, 789
Radiometry
    irradiance, 812
    radiant flux, 812
Raman
    -Nath scattering, 957, 973
    cascaded fiber laser, 692
    cascaded silicon laser, 692
    distributed fiber amplifier, 643
    fiber amplifier, 642
    fiber laser, 691
    gain, 1041
    lumped fiber amplifier, 643
    scattering, 613
    silicon laser, 692, 809
    stimulated scattering, 613, 642, 691, 1140
    Stokes shift, 643, 691
Ramo's theorem, 880, 939
Random light, *see* Statistical optics
Rate equations
    amplifier radiation absent, 626
    amplifier radiation present, 627
    broadband radiation, 617
    photon-number, 709
    population-difference, 710
    thermal light, 604
Ray equation, 20–21
    paraxial, 21
Ray optics, 3–40
    introduction to, 4–5
    postulates, 5–8
    relation to electromagnetic optics, 4
    relation to quantum optics, 4
    relation to wave optics, 4, 26, 42, 52
Ray-transfer matrix, 27–33
    4×4, 40
    SELFOC plate, 40
    air followed by lens, 31
    arbitrary paraxial system, 31
    cascade of components, 30
    determinant, 31
    free-space, 29
    inverse, 31
    lens system, 40
    parallel transparent plates, 30
    planar boundary, 29
    planar mirror, 29
    skewed rays, 40
    special forms, 28
    spherical boundary, 29
    spherical mirror, 30
    spherical-mirror resonator, 447
    thick lens, 32
    thin lens, 29, 31
Rayleigh
    -Jeans formula, 463, 606
    inverse fourth-power law, 194, 417
    range, 81
    scattering, 193–197, 416, 612
Rayleigh, Lord (John William Strutt), 160
Reciprocal
    lattice, 292
Rectification, optical, 1023
    pulsed, 1128
Reflectance
    at boundary, 222–225
    between GaAs and air, 226
    between glass and air, 226, 253
    from a Bragg grating, 950
    from a plate, 227
    of a conductive medium, 321
    power, 226
    quarter-wave film, 264, 301
Reflection, 53, 221–227
    at a dielectric boundary, 7
    at a DPS-SNG boundary, 310
    at a metasurface, 342
    at a mirror, 7
    at an absorbing boundary, 253
    Bragg grating, 273
    Brewster angle, 224, 225, 247
    circularly polarized light, 253
    critical angle, 12, 223, 225, 310, 327
    external, 223, 224
    frustrated total internal, 329, 378
    internal, 223, 225
    omnidirectional, 290, 302
    phase shift, 224
    total internal, 12, 13, 15, 18, 223, 253, 353, 392, 698
    transverse-electric polarization, 223
    transverse-magnetic polarization, 224
Refraction, 53, 221–227
    all-angle, 320
    at a dielectric boundary, 7, 15
    at a hyperbolic medium, 319
    at a metasurface, 342
    at a spherical boundary, 14
    at normal incidence, 346
    Brewster angle, 253
    conical, 254
    double, 238–239, 254
    external, 11
    internal, 11
    negative, 314–317

nonlinear, 1038
rays in anisotropic media, 239, 254
transverse-electric polarization, 223
transverse-magnetic polarization, 224
without reflection, 345
Refractive index, 5, 168, 183
air, 208
anisotropic media, 228–230
extraordinary, 229, 233, 236
frequency-dependent, 184
group, 201
negative, 308
optical materials, 184, 191
ordinary, 229, 233, 236
resonant medium, 189
Sellmeier equation, 191
Resolution
acousto-optic filter, 967
acousto-optic scanner, 961
electro-optic scanner, 983
liquid-crystal display, 1003
multiphoton lithography, 611
multiphoton microscopy, 611
optical-pulse detection, 1148
routing device, 1182
wavelength demultiplexer, 1184
Resonant medium, 186
anharmonic oscillator, 1071–1073
Lorentz oscillator, 186, 1018, 1072
Resonator
g-parameters, 448
axial modes, 456
bow-tie, 439–440
circular (2D), 460–461, 469
cold, 662
concentric, 449
conditionally stable, 448
confinement condition, 448–450, 453
confocal, 449
diffraction loss, 457, 458, 1304
energy per mode, 538
Fabry–Perot, 69, 265, 436, 661
fiber-ring, 434
finesse, 69, 266, 441
finite apertures, 457, 1304
free spectral range, 267, 438, 446
frequencies, 437, 439, 446, 460
guided-wave, 434, 471
integrated-optic-ring, 434, 692
losses, 440
microdisk, 464
micropillar, 464
microresonator, 463–468
microring, 1180

microsphere, 464
microtoroid, 464
modal density (1D), 440
modal density (2D), 460
modal density (3D), 462–463
modal volume, 435
modes, 1304
modes, standing wave, 437, 471
modes, traveling wave, 438
multiple microring, 1181
nanodisk, 469
nanoresonator, 469–470
nanosphere, 330–332, 470
number of modes, 472
optics, 433–472, 661
periodic optical system, 37, 40, 448
photonic modes, 469
photonic-crystal, 464, 468
planar-mirror, 436–446
plano–concave, 453
plasmonic, 330–334, 469
quality factor, 435, 445, 708
rectangular (2D), 459–460
rectangular (3D), 461–465, 517, 1305
ring, 434, 439–440
size vs. resonance wavelength, 435
spectral width, 440
spherical-mirror, 447–458
stability diagram, 449, 471
stable, 448
symmetric, 449
transmittance, 729
transverse modes, 456
traveling wave, 439
unstable, 448, 471
whispering-gallery modes, 461
Resonator, spherical-mirror, 447–458
axial modes, 456
free spectral range, 455, 456
Gaussian modes, 451
Hermite–Gaussian modes, 455
modes, 450–458
ray-transfer matrix, 37, 447, 448
resonance frequencies, 455, 456
symmetric, 449, 453, 471, 472
transverse modes, 456
Responsivity
avalanche photodiode, 898
differential, 837
LD, 839
LED, 801, 839
photodetector, 878, 1146
SLED, 839
Retarder, 218, 248

half-wave, 219, 220, 253
quarter-wave, 218, 254
quartz, 254
Ring
    aperture, 67, 118, 159
    benzene, 742
    network, 1275
    resonator, 434, 439–440
Rotator
    Faraday, 250, 1173
    nonreciprocal polarization, 251, 1173
    polarization, 245, 249, 252
Router, passive optical, 1178–1187
    add–drop multiplexer, 1179, 1279
    arrayed waveguides, 1182, 1223
    broadcast-and-select, 1179
    demultiplexer, 1179
    intensity-based, 1185
    introduction to, 1164–1165
    Mach–Zehnder interferometer, 1181
    multipath interferometer, 1183
    multiplexer, 1179
    phase-based, 1185
    polarization-based, 1184
    waveguide grating, 1182
    wavelength-based, 1179–1184
    wavelength-division multiplexer, 1179
Russell, Philip St John, 1224

Saturable absorber, 649
    bistable device, 1217
Scalar wave optics, *see* Wave optics
Scanner
    acousto-optic, 961
    electro-optic, 983
    holographic, 117
Scattering, 612–614
    and absorption, 198–199
    and attenuation, 198, 416, 432
    anti-Stokes, 613
    Bragg diffraction, 953, 973
    Brillouin, 613
    CARS, 614
    coefficient, 198
    Debye–Sears, 957, 973
    dielectric nanosphere, 196–197
    efficiency, 196
    elastic, 192–197
    Huygens–Fresnel principle, 193
    Maxwell-Garnett mixing rule, 199
    metallic nanosphere, 330–332
    Mie, 197, 613
    quasi-static approximation, 196, 197
    Raman, 613
    Raman–Nath, 957, 973
    Rayleigh, 193–197, 416, 432, 612
    small scatterers, 193
    stimulated Brillouin, 614, 644, 692
    stimulated Raman, 613, 642, 691, 1140
    Stokes, 613
    strong, 196
    volume fraction, 198
    weak, 192–194
Scattering matrix, 259–265, 549
    beamsplitter, 263
    dielectric boundary, 262, 264
    dielectric slab, 263, 265
    homogeneous medium, 260
    lossless medium, 261
    lossless symmetric system, 262
    relation to wave-transfer matrix, 260
Schawlow, Arthur, 657, 1015
Schrödinger equation
    nonlinear, 1040, 1135, 1138
    time-dependent, 562, 1304
    time-independent, 542, 563, 1304
Schultz, Peter C., 391
Second-harmonic generation, 730, 1021, 1049, 1051
    efficiency, 1022, 1052
    phase mismatch, 1053
    switch, 1147
Self-focusing, 1039
Self-phase modulation, 425, 1038
    in supercontinuum generation, 1140
    pulse, optical, 1129, 1130
Sellmeier equation, 191, 781
Semiconductor optical amplifiers, 817
    bandwidth, 819, 869
    compare with OFAs, 830
    double-heterostructure, 825
    gain, 818
    gain coefficient, 820
    heterostructures, 825
    peak gain coefficient, 821, 869
    pumping, 822
    quantum-dot, 829
    quantum-well, 826–830
    SLEDs, 830
    switch, photonic, 1193, 1196
    waveguide, 829
Semiconductors, 577–580
    k-selection rule, 770
    p–i–n junction, 759, 889
    p–n junction, 756, 887
    p–n junction, biased, 757
    absorption, 768, 777

AlGaAs, 806
AlGaN, 807
AlInGaN, 808
AlInGaP, 806
allotropes, 743
alloy broadening, 794
Auger recombination, 752
bandgap energy, 576, 733, 738
bandgap wavelength, 738, 768
bowing parameter, 785
Brillouin zone, 735
bulk, 578, 766–778
carrier concentrations, 748, 751
carrier generation, 752
carrier injection, 753
carrier recombination, 752
carriers, 734
degenerate, 750
density of states, 745
density of states, joint, 770
depletion layer, 756, 887
direct-bandgap, 737
dopants, 741
doped extrinsic, 886
drift velocity, 880
effective mass, 736
electroluminescence, 789–794
electron affinity, 873
elemental, 737
energy bands, 577, 733, 1305
energy–momentum relations, 735
excitons, 767, 784, 810, 811, 1010
extrinsic, 741
Fermi function, 746
Fermi inversion factor, 777
fundamentals, 731–766
GaAs, 805
GaAsP, 805
gain coefficient, 776
GaN, 807
heterojunction, 759
II–VI materials, 740
III–nitride materials, 738, 807–808
III–V materials, 738–739, 803–808
impact ionization, 876, 895
indirect-bandgap, 737
InGaAs, 805
InGaAsP, 805
InGaAsSb, 806
InGaN, 807
interaction with light, 766–781
internal efficiency, 754
intrinsic, 741
IV–VI materials, 741
Kronig–Penney model, 733
law of mass action, 750
minibands, 579, 764, 779, 852
mobility, 880
multiquantum-well, 764
nanocrystals, 579
nonradiative recombination, 752
occupancy probabilities, 746, 773
optics, 766–786
organic, 742
periodic table, 737
photoconductors, 875, 883–886
photoemission, 873
quantum dots, 579, 766
quantum wells, 578, 761–764
quantum wires, 579, 765
quantum-confined, 760, 779
quasi-equilibrium, 751
recombination coefficient, 752
recombination lifetime, 753
refractive index, 781
SESAM, 719
Shockley equation, 758
Si photonics, 375
SiC, 737
silicon photonics, 692, 780, 808, 1238
superlattice, 579, 764, 779, 852
transition probabilities, 774
Vegard's law, 739, 785
Semimetals
band structure, 577, 740, 741
graphene, 744
massless Dirac fermions, 744
Shockley, William B., 731
Shot noise, 518, 703, 912, 913
Silicon photonics, 375, 692, 808, 1176
direct-mounting integration, 809
flip-chip integration, 809
heteroepitaxy, 809, 859
heterogeneous integration, 809
hybrid approach, 809
microring laser, 859
OEIC, 1238
PIC, 809, 1238
PLC, 1238
quantum dots, 780
Single-mode
fiber, 405
waveguide, 359, 366
Skin depth, *see* Penetration depth
Slow light, 204
Snell's law, 54, 184, 221
at a boundary, 7

at a metamaterial boundary, 346
at a metasurface, 342
modified, 238, 342
negative refractive index, 307, 314
proof, 8
Solids, 575–580
covalent, 575
doped dielectric media, 568
ionic, 575
metallic, 576
molecular, 576
van der Waals, 573, 576, 745
Soliton
N-soliton wave, 1136
collision, 1199
condition, 1132
dark, 1138
directional-coupler router, 1187
envelope, 1133
fundamental, 1135, 1162
generation, 1137
higher-order, 1136
interaction, 1137
laser, 1138
optical fiber communications, 1256
period, 1136
photonic-crystal, 1141
sech pulse, 1133, 1136
self-frequency shift, 1140
solitary wave, 1131
spatial, 1039
spatial–temporal analogy, 1138
spatiotemporal, 1139
switch, photonic, 1199
temporal, 1130–1139
vector, 1200
SONET, 1239, 1276
Sonoluminescence, 607, 608
Spatial
coherence, 483
dispersiveness, 241
filter, 141, 151
frequency, 111, 113
harmonic function, 113
hole burning, 673
Lambertian pattern, 799
laser emission pattern, 675
LD emission pattern, 841
LED emission pattern, 799
solitons, 1039
spectral analysis, 114
Spatial light modulator
acousto-optic, 965
bistable, 1220

digital micromirror device, 1190
electro-optic, 987–989
liquid-crystal, 1002–1005
optically addressed, 987, 1004
parallel-aligned (PAL-SLM), 1005
Pockels readout optical (PROM), 988
Speckle, 406
Spectral
density, 479
hole burning, 651
packet, 596
width, 481
Spectrogram, 1083, 1156–1158
Wigner distribution function, 1157
Spectrum analyzer
acoustic, 963
interferometric, 1151
optical, 1151
Speed of light, 5, 42, 43, 163, 167, 184, 200, 204, 516, 1016
Spherical wave, 48, 177
complex amplitude, 77
conjugate, 1045
intensity, 77
paraboloidal approximation, 49
partially coherent, 488
pulsed, 1086
reflection, 78
wavefronts, 48
wavefunction, 48
Spin
-allowed transitions, 609, 811
-forbidden transitions, 609, 811
-orbit coupling, 565, 566, 811
-spin coupling, 565
angular momentum, 524
electron, 564, 746
multiplicity, 566
photon, 516, 524, 746
singlet state, 566, 575, 609, 811
triplet state, 566, 575, 609, 811
Spontaneous emission
atoms, 583, 587–588
enhanced, 464, 597, 796, 800
inhibited, 464, 617
occupancy probability, 773
peak rate, 786
Purcell factor, 598, 796, 855, 861
semiconductors, 769
spectral intensity, 775
transition rate, 592, 775
Stark effect, 568, 601
light shift, 600
quantum-confined, 1010

Statistical optics, 473–513
    gain of spatial coherence, 502
    imaging with incoherent light, 499
    interference, 489
    interferometry, 490, 493, 550
    introduction to, 474–475
    longitudinal coherence, 487
    optical intensity, 475
    partial polarization, 506
    quantum interferometry, 550
    spatial coherence, 483
    spectrum, 476
    temporal coherence, 476
    transmission of random light, 497
Step-index fiber, 393–396, 398–405, 408
Stimulated Brillouin scattering, 614, 644
Stimulated emission
    atoms, 585, 589–593
    occupancy probability, 773
    semiconductors, 769
    semiconductors, indirect-gap, 772
    transition rate, 592, 775
Stimulated Raman scattering, 613, 642, 691, 1140
Stokes
    parameters, 214, 216, 507
    vector, 214
Stokes, George Gabriel, 210
Strain
    sensor, 1013
    tensor, 968
Sum-frequency generation, 1027, 1197
Supercontinuum generation, 1139
Superlattice, 579, 764, 779, 852
    quantum cascade laser, 852
Superluminescent diodes (SLEDs), 830
    compare with LDs, 839, 841
    compare with LEDs, 839, 841
    light–current curve, 839
Surface plasmon
    resonance spectroscopy, 329
    resonance, 330
Surface plasmon polariton, 326
    at DPS-SNG boundary, 311, 351
    at metal–dielectric boundary, 326
    nanolaser, 862
    surface-charge wave, 311
Susceptibility, electric, 166
    complex, 181
    frequency-dependent, 184
    resonant medium, 188
    tensor, 170
Switch
    acousto-optic, 959

electro-optic, 985
electroabsorption, 1010
ferroelectric liquid crystal, 1001
waveguide, 381
Switch, photonic, 1187–1211
    acousto-optic, 964, 1194
    all-optical, 1196–1211
    architectures, 1187
    Banyan switch, 1210
    buffer, 1209
    characteristics, 1188, 1196
    circuit switching, 1209
    configurations, 1189
    electro-optic, 1192
    energy required, 1202
    Franz–Keldysh effect, 1193
    FWM, 1199
    heat dissipation, 1203
    implementations, 1188
    introduction to, 1164–1165
    liquid-crystal, 1194
    magneto-optic, 1195
    mechano-optic, 1190
    MEMS, 1190
    nonlinear Kerr, 1148, 1198
    nonlinear optical retardation, 1198
    nonlinear Sagnac interferometer, 1201
    opto-optic, 1196–1211
    optoelectronic, 1188
    optomechanical, 1190
    packet, 1209
    parametric, 1197
    photonic-crystal, 1200
    plasmonic, 1201
    programmable delay, 1209
    QCSE, 1193
    quantum limit, 1202
    quantum-confined Stark effect, 1010
    ROADM, 1204, 1280
    second-harmonic generation, 1147
    semiconductor, 1193
    SFG, 1197
    SOA, 1193, 1196
    soliton, vector, 1200
    soliton-collision, 1199
    space, 964, 1187–1203
    space–wavelength, 1204
    switching time, 1190, 1201
    thermo-optic, 1195
    time–space–time, 1207
    time-division demultiplexer, 1207
    time-division multiplexer, 1207
    time-domain, 1206–1209
    time-slot interchange, 1208, 1223

TOAD, 1201
wavelength converter, 1205
wavelength selector, 1203
wavelength-channel interchange, 1204
wavelength-selective, 1203–1206
XGM, 1196
XPM, 1197–1199

Tail
   band, 750, 821, 870, 1010
   Fermi, 748
   Urbach, 778, 1010
Talbot effect, 135
TEM wave, 175, 307, 355, 506
Temperature
   BEC formation, 602
   blackbody spectrum, 605
   correlated color, 812, 816
   Doppler cooling limit, 600
   earth, 606
   single-photon recoil limit, 600
   sub-recoil cooling, 600
   sun, 606
   thermographic images, 606
Tensor
   conductivity, 882
   constraints, 969, 1066, 1074
   elasto-optic, 969
   electric permittivity, 170, 228, 343
   electric susceptibility, 170
   first-rank, 228
   fourth-rank, 969, 990, 1066
   geometrical representation, 228
   impermeability, 230, 989
   index ellipsoid, 230
   linear electro-optic, 990
   magnetic permeability, 343
   photoelasticity, 969
   quadratic electro-optic, 990
   quadric representation, 229, 230
   second-order nonlinearity, 1066
   second-rank, 170, 228–230, 343, 882
   strain, 968
   strain-optic, 969
   third-order nonlinearity, 1066
   third-rank, 990, 1066
   zeroth-rank, 228
Terahertz
   frequencies, 42, 853, 854, 1128
Term symbol
   actinide metals, 569
   atoms, 566
   He, 566
   lanthanide metals, 569
   occupied subshells, 566
   rare-earth elements, 569
   transition metals, 569
Thermal light, 602–607
   blackbody radiation, 602
   rate equation, 604
   Rayleigh–Jeans formula, 606
   spectrum, 604
   Stefan–Boltzmann law, 606, 618
   thermography, 606
   Wien's law, 617
Thermo-optic effect, 1195, 1198
Third-harmonic generation, 1037, 1063
Three-wave mixing, 545, 1025, 1043, 1050, 1061, 1066, 1073, 1076
   pulsed, 1127, 1162
Time
   -varying spectrum, 1083
   lens, 1113, 1162
Townes, Charles H., 619
Transfer function, 1297, 1300
   4-f imaging system, 139
   free space, 119–122, 316
   single-lens imaging, 144–145
Transformation optics, 343–348
   cylindrical focusing, 346
   refraction at normal incidence, 346
   refraction without reflection, 345
   transformation principle, 344–347
Transition
   absorption, 584, 589
   excitonic, 767, 779
   free-carrier, 767
   impurity-to-band, 767
   interband, 766, 779
   intersubband, 779
   intraband, 767
   miniband, 779
   phonon, 767
   spontaneous, 583, 587
   stimulated, 585, 589
Transmittance
   biprism, 57
   complex-amplitude, 54, 222–225
   diffraction grating, 59
   Fabry–Perot, 266–268
   Fresnel biprism, 57
   graded-index plate, 60
   optical components, 54, 497
   plate of varying thickness, 55
   power, 226
   prism, 56
   thin lens, 57
   transparent plate, 55

Tyndall, John, 353

Ultrafast optics, 1078–1162
    introduction to, 1079
    linear, 1115–1125
    nonlinear, 1126–1145
    pulse characteristics, 1079–1088
    pulse compression, 1088–1102
    pulse detection, 1146–1158
    pulse propagation in fibers, 1102
    pulse shaping, 1088–1102
Ultraviolet
    EUV band, 697
    frequencies, 42
    UVA band, 42, 808
    UVB band, 42
    UVC band, 42, 808
    VUV band, 42
    wavelengths, 42
Uncertainty relation
    duration–bandwidth, 1290
    field quadratures, 543
    Heisenberg, 528, 543, 1291
    position–momentum, 543
    time–energy, 527, 555
Undulator, 702
Uniaxial crystal, 229, 233, 236, 317–320
Units, radiometric and photometric, 812
    illuminance, 812
    irradiance, 812
    luminous efficacy, 812
    luminous flux, 812
    radiant flux, 812
Up-conversion, 1027, 1054, 1055
    fluorescence, 611

Vacuum state, 518, 544
van Cittert–Zernike theorem, 503–505
VCSELs, 856–858
VECSELs, 721, 859
Vector
    beam, 179, 180
    potential, 177
Velocity
    group, 200, 204, 284, 352, 360, 370, 404, 412
    information, 204
    phase, 48, 200, 204, 284, 352
Verdet constant, 243
Veselago, Victor Georgievich, 303
Visibility, 78, 490
Visible
    frequencies, 42
    wavelengths, 42

Vortex, optical, 104
    laser, 860
    phase singularity, 104
    topological charge, 104, 155

Wave
    -particle duality, 521
    acoustic plane, 946
    beating, 74, 967, 1153, 1267
    complex amplitude, 46
    complex analytic signal, 71
    complex envelope, 47, 80, 175
    complex representation, 45
    complex wavefunction, 45, 72
    conjugate, 78, 1045
    cylindrical, 78
    evanescent, 120, 146, 224, 316, 368
    in a GRIN slab, 78
    localized, 523
    monochromatic, 44–52
    nondiffracting, 136
    paraboloidal, 49, 51, 81
    paraxial, 50, 1086
    partially coherent, 487, 488
    plane, 47, 111, 113
    polychromatic, 71–76
    pulsed, 73, 78, 1086
    quasi-monochromatic, 72
    quasi-plane, 407
    restoration, 1046
    retarder, 218, 248
    retarder, dynamic, 980
    retarder, liquid-crystal, 997–1001
    retarder, voltage-controlled, 998
    spherical, 48, 77, 78
    standing, 78
    stationary, 136
    wavefronts, 47
    wavefunction, 43, 45, 71
    wavelength, 47
Wave equation, 43, 45, 72
    diffusion equation, 1111
    generalized paraxial, 1124
    in free space, 163
    in homogeneous medium, 171
    in inhomogeneous medium, 169, 953
    in medium, 167
    in nonlinear dispersive medium, 1133
    in nonlinear medium, 1019, 1047
    SVE, 1111
    SVE, nonlinear, 1133
Wave optics, 41–78
    introduction to, 42–43
    postulates, 43–44

vs. electromagnetic optics, 180
vs. ray optics, 52
Wave-transfer matrix, 258–265, 1303
   antireflection film, 264
   beamsplitter, 263
   cascade of elements, 259
   dielectric boundary, 262, 264
   dielectric slab, 263, 265
   homogeneous medium, 260
   lossless medium, 261
   lossless system, 262
   relation to scattering matrix, 260
Wavefronts
   anisotropic medium, 234–237
   helical, 103, 155, 524
   plane-wave, 48
   spherical-wave, 48
Wavefunction
   complex, 45, 716, 1080, 1158
   electron, 562, 582, 734, 1291, 1304
   entangled-photon, 547
   harmonic-oscillator, 542
   plane-wave, 48
   quantum mode, 543
   single-photon, 521
   spherical-wave, 48
   two-photon, 546
Waveguide, optical, 354
   arrays, 383
   asymmetric planar, 372
   bounce angles, 357, 365, 390
   Bragg-grating, 385
   channel, 374
   cladding, 363
   confinement factor, 369, 390
   core, 363
   coupled-mode theory, 378–384
   couplers, 378–383
   coupling, 376–384, 390, 985
   cutoff, 359, 366, 390
   cylindrical, 392
   dispersion relation, 359, 370
   evanescent wave, 368
   extinction coefficient, 368
   fiber, 392
   field distributions, 358, 367, 389, 390
   GaAs/AlGaAs, 375
   glass, 375
   Goos–Hänchen effect, 371
   group velocity, 360, 370
   InGaAsP, 375
   input coupling, 376
   $LiNbO_3$, 375
   materials, 375
   metal–insulator–metal, 387
   metal–slab, 388, 390
   modes, 355–357, 363–366
   multimode fields, 362
   number of modes, 359, 366, 367, 390
   numerical aperture, 366
   optical-power flow, 359, 363
   periodic, 384
   photonic-crystal, 385–386
   planar dielectric, 363–372, 389, 390
   planar-mirror, 355–363, 1305
   plasmonic, 386–388
   power-transfer ratio, 381
   propagation constants, 357, 365
   rectangular dielectric, 373, 390
   rectangular mirror, 372
   side coupling, 377
   silica-on-silicon, 375
   silicon-on-insulator, 375
   single-mode, 359, 366, 390
   switch, 381
   transfer distance, 378, 381, 985
   two-dimensional, 372–375
Wavelength, 47
   γ-ray, 697
   -division multiplexer, 1179
   activation, 885
   bandgap, 738, 768
   converter, 1205
   de Broglie, 601, 761
   infrared, 42, 853
   laser, 644, 706
   laser amplifier, 644
   plasmon, 312
   ultraviolet, 42, 697
   visible, 42
   X-ray, 697
   zero-dispersion, 1140
Wavenumber, 48, 173
   complex, 182
Wavepacket, 73
   mode, 527
   random sequence, 482
   single-photon, 527, 594
   velocity, 360
Wavevector, 47
Width of a function
   $1/e$-, 1292
   3-dB, 1292
   FWHM, 1292
   Gaussian, 1290–1292
   measures of, 1290–1293
   power-equivalent, 1291
   power-RMS, 1290

root-mean-square (RMS), 1290
Wiener–Khinchin theorem, 480, 512
Wigner distribution function, 1157
WOLED, 810, 816, 1002
Wolf, Emil, 473

X-ray
    energies, 697
    hard-X-ray (HXR) band, 697
    imaging, 705
    optical components, 698
    soft-X-ray (SXR) band, 697
    wavelengths, 697

Yablonovitch, Eli, 255
Yablonovite, 297
Young, Thomas, 41

Zeeman effect, 568, 599, 601
Zero-point energy, 518, 536, 573, 586
Zone plate, Fresnel, 67, 118